Szegő's Theorem and Its Descendants

M. B. PORTER LECTURES

RICE UNIVERSITY, DEPARTMENT OF MATHEMATICS

SALOMON BÔCHNER, FOUNDING EDITOR

RICE

H. Furstenberg, *Recurrence in Ergodic Theory and Combinatorial Number Theory* (1981)

M. Atiyah and N. Hitchin, *The Geometry and Dynamics of Magnetic Monopoles* (1988)

Yuri I. Manin, *Topics in Noncommutative Geometry* (1991)

János Kollár, *Shafarevich Maps and Automorphic Forms* (1995)

Shmuel Weinberger, *Computers, Rigidity, and Moduli: The Large-Scale Fractal Geometry of Riemannian Moduli Space* (2005)

Barry Simon, *Szegő's Theorem and Its Descendants: Spectral Theory for L^2 Perturbations of Orthogonal Polynomials* (2011)

Szegő's Theorem and Its Descendants

Spectral Theory for L^2 Perturbations of Orthogonal Polynomials

Barry Simon

PRINCETON UNIVERSITY PRESS
PRINCETON AND OXFORD

Published by Princeton University Press, 41 William Street,
Princeton, New Jersey 08540

In the United Kingdom: Princeton University Press, 6 Oxford Street,
Woodstock, Oxfordshire OX20 1TW

press.princeton.edu

Library of Congress Cataloging-in-Publication Data

Simon, Barry, 1946–
Szegő's theorem and its descendants : spectral theory for L^2 perturbations of orthogonal polynomials / Barry Simon.
p. cm.
Includes bibliographical references and index.
ISBN 978-0-691-14704-8 (cloth : alk. paper)
1. Spectral theory (Mathematics). 2. Orthogonal polynomials.
I. Title.
QC20.7.S64S56 2011
515′.55—dc22 2010023223

British Library Cataloging-in-Publication Data is available

This book has been composed in Times

Printed on acid-free paper. ∞

Typeset by S R Nova Pvt Ltd, Bangalore, India

Printed in the United States of America

10 9 8 7 6 5 4 3 2 1

Contents

Preface

In September 2006, I gave the Milton Brockett Porter Lectures at Rice University. Broadly speaking, these are the notes of those lectures. Of course, five hours of lectures did not cover 600 pages of material. Roughly, Chapters 1, 2, 3, 5, and 4 + 8 covered the topics of the five lectures, and I planned all along to include the material in Chapters 6 and 10, but Chapters 7 and 9 cover developments subsequent to Fall 2006.

The core motivation for the notes was to expose the developments of sum rule techniques that grew out of my paper [225] with Rowan Killip. Another motivation was a lure David Damanik used to convince me to produce notes. He pointed out the irony that while the theory of periodic orthogonal polynomials was developed in the real line case (OPRL), there was a comprehensive presentation [400] only for the unit circle case (OPUC), and suggested that I could include the OPRL case in these notes. It was in this regard that I "threw in" the chapter on Toda lattices (Chapter 6).

As mentioned, the inclusion of material that did not even exist in September 2006 added roughly 150 pages to the total. The biggest part concerns material on finite gap OPRL (Chapter 9). I intended the capstone of the lectures to be the work that Damanik, Killip, and I were then writing up [97] on perturbations of periodic OPUC using sum rules. This describes perturbations of Jacobi matrices whose essential spectrum is a finite union $\mathfrak{e} = [\alpha_1, \beta_1] \cup \cdots \cup [\alpha_{\ell+1}, \beta_{\ell+1}]$ of $\ell + 1$ disjoint subsets of $\mathbb{R}$.

But not all such finite gap sets arise in this way. Given $\mathfrak{e}$, one defines harmonic measure $\rho_{\mathfrak{e}}$ on $\mathfrak{e}$, the equilibrium measure for logarithmic potential theory. For there to be an underlying periodic problem, each $\rho_{\mathfrak{e}}([\alpha_j, \beta_j])$ must be rational.

One of the issues [97] wrestled with involved existence of a limit when Szegő conditions hold. In the summer before the Rice lectures, I talked at a conference in Lund on this material, and Peter Yuditskii pointed out to me that he and Peherstorfer had, in fact, solved this problem in a more general context. This informed me of his wonderful, but too much ignored, work with Sodin and Peherstorfer [413, 343] on general finite gap problems. I returned from Rice intending to work on understanding and extending this work with two Caltech postdocs, Jacob Christiansen and Maxim Zinchenko, and eventually three papers [86, 87, 88, 89] and Chapter 9 resulted.

One subject I expected to reluctantly leave out of these lectures was the Denisov–Rakhmanov theorem, but two developments changed that. First, in the spring of 2007, Remling [366] found his spectacular approach to a.c. spectrum that extended the Denisov–Rakhmanov theory to the general finite gap case. Then Christiansen, Zinchenko, and I found that Remling's result could be used as part of a new approach to Szegő asymptotics in the finite gap case. This belonged in Chapter 9, but required the inclusion of Remling's theorem, and so Chapter 7.

A final addition was the beautiful developments of Lubinsky [287, 288]. While these were somewhat peripheral to the main thrust of these lectures, I succumbed to the temptation to include them in Sections 2.14–2.17, 3.11–3.12, and 5.11–5.12.

Another factor in the length of these notes was the decision to expose some of the mathematical background not so readily available in monographs/textbook literature. This is especially true of Sections 9.2–9.4.

The reader may wonder why, given my earlier two-volume opus [399, 400], there is a need for the current book and if this book is not merely a subset of the earlier one. There is indeed some overlap, especially heavy in Chapters 2 and 3 of this book, but the overwhelming bulk of the material here is not covered in those earlier books. This current book has a rather different focus, concentrating on topics connected with sum rules and also on the real line case rather than the unit circle. But there is another factor. It has been five years since I finished the other books and there have been some remarkable developments during the period. I have in mind not only some of my own work on finite gap systems (which are not OPUC in any event) but especially the beautiful work of Lubinsky on (Christoffel–Darboux) CD kernels and Remling on a.c. spectrum, both of which are presented here. It is also true that in that period I have learned some things about OPUC that existed when [399, 400] were written but were not included—notable here are the Helson–Lowdenslager approach to the singular component of the measure in the classical Szegő setup and the Máté–Nevai–Totik work on asymptotics of the diagonal CD kernel.

I would emphasize that while I have endeavored to be accurate in the historical notes, I am not a serious historian of science, and that these notes are affected by where I learned of some results and the generally accepted beliefs. When I was a graduate student, I took a course in the quantum theory of solids with Eugene Wigner. When he got to the discussion of the Brillouin zone, he remarked they actually appeared first in the work of Bloch. When a student asked why it was called the Brillouin zone and not the Bloch zone, Wigner's face turned red and he stammered out: "Well, I learned of them from Brillouin's paper and named them and the name stuck." The notes are here because they make the material more interesting and provide further references, rather than because they are intended as definitive history.

For these notes I owe many debts of gratitude: David Damanik and Mike Wolf for their hospitality at Rice; Jacob Christiansen, Mamikon Mnatsakanian, and Mihai Stoiciu for help with figures; Cherie Galvez for manuscript preparation and correcting; and especially Jacob Christiansen, David Damanik, Sasha Pushnitski, Rowan Killip, and Maxim Zinchenko for the joy of collaboration central to this work. I also thank Danny Calegari, Rostyslav Kozhan, Doron Lubinsky, Milivoje Lukic, Nick Makarov, Franz Peherstorfer, and Peter Yuditskii for useful comments. Finally, Martha, whose love is so important and not only to this book.

Barry Simon
Los Angeles
September 2009

Szegő's Theorem and Its Descendants

Chapter One

Gems of Spectral Theory

The central theme of this monograph is the view of a remarkable 1915 theorem of Szegő as a result in spectral theory. We use this theme to present major aspects of the modern analytic theory of orthogonal polynomials. In this chapter, we bring together the major results that will flow from this theme.

1.1 WHAT IS SPECTRAL THEORY?

Broadly defined, spectral theory is the study of the relation of things to their spectral characteristics. By "things" here we mean mathematical objects, especially ones that model physical situations. Think of the brain modeled by a density function, or a piece of ocean with possible submarines again modeled by a density function. Other examples are the surface of a drum with some odd shape, a quantum particle interacting with some potential, or a vibrating string with a density function. To pass to more abstract mathematical objects, consider a differentiable manifold with Riemannian metric. To get into number theory, this manifold might have arithmetic significance, say, the upper half-plane with the Poincaré metric quotiented by a group of fractional linear transformations induced by some set of matrices with integral coefficients.

By spectral characteristics, mathematicians and physicists originally meant characteristic frequencies of the object—modes of vibration of the drum or, to state the example that gives the subject its name, the light spectrum produced by a chemical like Helium inside the sun.

Eventually, it was realized that besides the discrete set of frequencies associated with a drum, vibrating string, or compact Riemannian manifold, there were objects with continuous spectrum where the spectral characteristics become scattering or related data. For example, in the case of a brain, the spectral data is the raw output of a computer tomography machine. For quantum scattering on the line, it might be the reflection coefficient.

The process of going from the object to the spectral data or of going from some property of the object to some property of the data is called the *direct spectral problem* (or *direct problem*). The process of going from the spectral data to the object or from some aspect of the spectral data to some aspect of the object is the *inverse spectral problem* (or *inverse problem*).

The general wisdom is that direct problems are easier than inverse problems, and this is true on two levels: first, on the level of mere existence and/or even specifying the domain of definition; and second, in proving theorems that say if some property holds on one side, then some other property holds on the other.

Almost all these models (tomography is an exception) are described by a differential equation—ordinary or partial—or by a difference equation. In most cases, the object is a selfadjoint operator on some Hilbert space. In that case, the direct problem is usually solved via some variant of the spectral theorem, which says:

Theorem 1.1.1. *If A is a selfadjoint operator on a Hilbert space, $\mathcal{H}$, and $\varphi \in \mathcal{H}$, there is a measure $d\mu$ on $\mathbb{R}$ so that*

$$\langle \varphi, e^{-itA}\varphi\rangle = \int e^{-ixt}\, d\mu(x) \tag{1.1.1}$$

for all $t \in \mathbb{R}$.

Remarks. 1. All our Hilbert spaces are complex and $\langle \cdot\,, \cdot\rangle$ is linear in the second factor and antilinear in the first.

2. For a proof, see [14, 361, 369]. Also see Section 1.3 later for the case of bounded A.

3. I have ignored subtle points here when A is an unbounded operator (as happens for differential operators) concerning what it means to be selfadjoint, how e^{-itA} is defined, and so on. Because we look at difference equations in most of these notes, our A is bounded, and then for $n = 0, 1, 2, \dots,$ (1.1.1) is equivalent to

$$\langle \varphi, A^n\varphi\rangle = \int x^n\, d\mu(x) \tag{1.1.2}$$

4. We will also consider unitary operators, U, where $d\mu$ is now on $\partial\mathbb{D} = \{z \mid |z| = 1\}$ and

$$\langle \varphi, U^n\varphi\rangle = \int z^n\, d\mu(z) \tag{1.1.3}$$

for $n \in \mathbb{Z}$.

Notice that a spectral measure requires both an operator and a vector, φ. Sometimes there is a natural φ, sometimes not. Sometimes the full spectral measure is overkill—for example, the problem made famous by Mark Kac [212]: "Can you hear the shape of a drum?" asks about whether the eigenvalues of the Laplace–Beltrami operator of a (two-dimensional) compact surface determine the metric up to isometry. The spectral measure typically has point masses at the eigenvalues but also has weights for those masses so has more data than the eigenvalues alone.

It is worth noting that it is arguable whether the shape of a drum problem is a direct or an inverse problem. It asks if the direct map from isometry classes of manifolds to their eigenvalue spectrum is one-one. But on a different level, it asks if an inverse map exists!

By the way, the answer to Kac's question is no (see [181]). For a review of more on this question and its higher-dimensional analogs, see [40, 64, 65, 180, 427].

Here is an example that shows we often do not understand the range of the direct map, and hence also the domain of the inverse map. Let $H_0 = -d^2/dx^2$ on $L^2(-\infty, \infty)$ and consider a function $V(x) \in L^1_{\text{loc}}(\mathbb{R})$ so that $(H_0 + 1)^{-1}(V + i)^{-1} \times (H_0+1)^{-1}$ is compact (e.g., this holds if $V(x) \to \infty$ as $|x| \to \infty$ but it also holds for $V = W^2 + W'$ with $W = x^2(2 + \sin(e^x))$ where V is unbounded below). Then

$$H = H_0 + V \tag{1.1.4}$$

has spectrum a set of eigenvalues $\{E_n\}_{n=1}^\infty$ where $E_n \to \infty$. It is well known that this is not sufficient spectral data to determine V.

Here is some additional data that is sufficient. Let H_D be H with a Dirichlet boundary condition at $x = 0$, that is,

$$H_D = H_D^+ \oplus H_D^- \tag{1.1.5}$$

where H_D^+ acts on $L^2(0, \infty)$ and H_D^- acts on $L^2(-\infty, 0)$, and selfadjointness is guaranteed by demanding $u(0) = 0$ boundary conditions.

Let E_n^D be the eigenvalues of H_D. It is not hard to prove the following:

(i) $E_n \le E_n^D \le E_{n+1}$

(ii) $E_n^D = E_n \Leftrightarrow u_n(0) = 0 \Leftrightarrow E_n^D = E_{n-1}^D$

Here u_n is the eigenfunction for H with eigenvalue E_n. Notice that (i) says each (E_n, E_{n+1}) contains at most one eigenvalue, and if there, it is simple. On the other hand, if $E_n^D \in \{E_j\}_{j=1}^\infty$, then it is a doubly degenerate eigenvalue.

If $E_n^D \in (E_n, E_{n+1})$, as noted E_n^D is simple, so we have a sign $\sigma_n^D \in \{\pm 1\}$, so E_n^D is an eigenvector of $H_D^{\sigma_n^D}$. If $E_n^D \in \{E_n, E_{n+1}\}$, σ_n^D is undefined. We will see shortly that $\{E_n\}_{n=1}^\infty \cup \{E_n^D, \sigma_n^D\}_{n=1}^\infty$ is a complete set of spectral data and that $\{V \mid E_n(V) = E_n(V_0)\}$ is an infinite-dimensional set of potentials. In a situation like this, where some set of the "spectral data" is distinguished but not determining, the set of objects whose spectral data in this subset is the same as for object$_0$ is called the *isospectral set* of object$_0$. It is usually a manifold, so we will often call it the *isospectral manifold* even if we have not proven it is a manifold!

Here is the theorem that describes what I have just indicated:

Theorem 1.1.2 ([165, 166]). *If $V, W \in L^1_{\text{loc}}$ and $E_n(V) = E_n(W)$, $E_n^D(V) = E_n^D(W)$, $\sigma_n^D(V) = \sigma_n^D(W)$, then $V = W$ (i.e., $V \mapsto \{E_n(V), E_n^D(V), \sigma_n^D(V)\}_{n=1}^\infty$ is one-one). Moreover, if $V \in L^1_{\text{loc}}$ and $N < \infty$ are given and $\tilde{E}_n$, $\tilde{E}_n^D$, σ_n^D are such that*

$$\begin{aligned} \tilde{E}_n &= E_n(V) && \text{all } n \\ \tilde{E}_n^D &= E_n^D(V) && \text{all } n > N \\ \tilde{\sigma}_n^D &= \sigma_n^D(V) && \text{all } n > N \end{aligned}$$

$\{E_n, E_n^D\}$ obey (i) *and* (ii) *above, then there is a W with*

$$E_n(W) = \tilde{E}_n \qquad E_n^D(W) = \tilde{E}_n^D \qquad \sigma_n^D(W) = \sigma_n^D$$

for all n.

It is an interesting exercise to fix N and picture the topology of the allowed $\tilde{E}_n^D, \tilde{\sigma}_n^D$. Alas, it is not known precisely what direct data $\{\tilde{E}_n^D, \sigma_n^D\}$ can occur for a given V. It is definitely not all $\{\tilde{E}_n, \sigma_n^D\}$ obeying (i), (ii). For example, it cannot happen that

$$E_n^D = \tfrac{1}{4} E_n + \tfrac{3}{4} E_{n+1} \tag{1.1.6}$$

for all n.

Open Question 1. What is the range of the map $V \mapsto \{E_n(V), E_n^D(V), \sigma_n^D(V)\}$ as V runs through all L^1_{loc} functions with $(H_0+1)^{-1/2}(V+i)^{-1}(H_0+1)^{-1/2}$ compact, or through all continuous functions obeying $V(x) \to \infty$ as $|x| \to \infty$.

Even the most basic isospectral manifolds such as $V(x) = x^2$ where $E_n(V) = 2n+1$ are not understood.

Open Question 2. Prove that the isospectral manifold of continuous V's with $V(x) \to \infty$ as $x \to \infty$ and $E_n(V) = 2n+1$ is connected.

I have described this example in detail to emphasize how little we understand even some basic spectral problems.

Having set the stage with a very general overview, we are now going to focus in these notes on two classes of spectral problems: those associated with orthogonal polynomials on the real line (OPRL) and orthogonal polynomials on the unit circle (OPUC). These are the most simple and most basic of spectral setups for several reasons:

(a) As we will see, the construction of the inverse is not only simple and basic, but historically these problems appeared initially as what we will end up thinking of as an inverse problem.

(b) The objects are connected with difference—not differential—operators, so various technicalities that might cause difficulty concerning differentiability, unbounded operators, and so on are absent.

(c) They are, in essence, half-line problems; the parameters in the difference equation are indexed by $n = 1, 2, \dots$ or $n = 0, 1, 2, \dots$.

(c) is a virtue and a flaw. It is a virtue in that, as is typical for half-line problems, one can precisely describe the range of the direct map. It is a flaw in that the methods one develops are often not relevant to go to higher dimensions or, sometimes, even to whole-line problems.

OPRL appear initially in Section 1.2 and OPUC in Section 1.7.

Remarks and Historical Notes. The centrality of spectral theory to modern science can be seen by contemplating the variety of Nobel prizes that are related to the theory—from the 1915 physics prize awarded to the Braggs to the 1979 medicine prize for computer tomography.

1.2 OPRL AS A SOLUTION OF AN INVERSE PROBLEM

Let $d\rho$ be a measure on $\mathbb{R}$. All our measures will be positive with finite total weight. Normally, we will demand that ρ is a probability measure, that is, $\rho(\mathbb{R}) = 1$. But for now we only suppose $\rho(\mathbb{R}) < \infty$. ρ is called *trivial* if $L^2(\mathbb{R}, d\rho)$ is finite-dimensional; equivalently, if $\text{supp}(d\rho)$ is a finite set. Otherwise we call ρ *nontrivial*.

If

$$\int |x^n|\, d\rho(x) < \infty \tag{1.2.1}$$

for all n, we say $d\rho$ has *finite moments*. We will always suppose this. Indeed, we will soon mainly restrict ourselves to the case where ρ has bounded support.

If ρ is nontrivial with finite moments, every polynomial, P, obeys

$$0 < \int |P(x)|^2 \, d\rho(x) < \infty \tag{1.2.2}$$

since the integral can only be zero if ρ is supported on the finite set of zeros of P.

Thus, $\{x^n\}_{n=0}^\infty$ are independent in $L^2(\mathbb{R}, d\rho)$. They may or may not span L^2. If the support is bounded, they are spanning by the Weierstrass approximation theorem. In the case where the support is unbounded, there is a beautiful theory of when the polynomials span—it is presented in Section 3.8. One of the simplest examples of a case where they are not spanning is $\exp(-\sqrt{|x|})\,dx$ (see Example 3.8.1 in Sections 3.8 and 3.9 for a discussion).

Thus, we can define monic orthogonal polynomials $\{P_n(x)\}_{n=0}^\infty$ of degree n by

$$P_n = \pi_n^\perp[x^n] \tag{1.2.3}$$

where π_n is the projection onto the n-dimensional space of polynomials of degree at most $n-1$ and

$$\pi_n^\perp = \mathbf{1} - \pi_n \tag{1.2.4}$$

So P_n is determined by

$$\begin{aligned} P_n(x) &= x^n + \text{lower order} \\ P_n &\perp x^j \qquad j = 0, \dots, n-1 \end{aligned} \tag{1.2.5}$$

By an obvious induction, we have

Proposition 1.2.1. $\{P_j\}_{j=0}^n$ *span* $\operatorname{Ran}(\pi_{n+1})$. *In particular,* $P_j/\|P_j\|$ *are an orthonormal basis of this* $n+1$*-dimensional space. So if* $Q \in \operatorname{Ran}(\pi_{n+1})$,

$$Q = \sum_{j=0}^n \langle P_j, Q\rangle \|P_j\|^{-2} P_j \tag{1.2.6}$$

One gets (1.2.6) by noting $Q-$rhs of (1.2.6) $\perp$ P_k for $k = 0, \dots, n$ since $\langle P_j, P_k\rangle = \|P_j\|^2 \delta_{jk}$. Here is a key fact for OPRL:

Proposition 1.2.2.

$$\langle P_j, xP_n\rangle = 0 \qquad \text{if } j < n-1 \tag{1.2.7}$$

Proof. $\langle P_j, xP_n\rangle = \langle xP_j, P_n\rangle = 0$ since xP_j has degree $j+1 < n$. □

This leads to the recursion relation obeyed by OPRL:

Theorem 1.2.3. *For any nontrivial measure with finite moments, there exist* $\{b_j\}_{j=1}^\infty$ *in* $\mathbb{R}^\infty$ *and* $\{a_j\}_{j=1}^\infty$ *in* $(0,\infty)^\infty$ *so that for* $n \ge 0$,

$$xP_n(x) = P_{n+1}(x) + b_{n+1}P_n(x) + a_n^2 P_{n-1}(x) \tag{1.2.8}$$

where $P_{-1}(x) \equiv 0$ *(so we do not need* $a_n = 0$*).*

Proof. $xP_n(x) - P_{n+1}(x)$ is a polynomial of degree n (since the x^{n+1} terms cancel) and so orthogonal to P_{n+1}, that is,

$$\langle P_{n+1}, xP_n\rangle = \|P_{n+1}\|^2 \tag{1.2.9}$$

which means the coefficient of P_{n+1} in (1.2.6) with $Q = xP_n$ is 1. Moreover, the coefficient of P_{n-1} is

$$\langle P_{n-1}, xP_n\rangle \|P_{n-1}\|^{-2} = \langle P_n, xP_{n-1}\rangle \|P_{n-1}\|^{-2} \tag{1.2.10}$$

$$= \left(\frac{\|P_n\|}{\|P_{n-1}\|}\right)^2 \tag{1.2.11}$$

where (1.2.10) follows from the reality of P_j and x, and (1.2.11) uses (1.2.9) for n replaced by $n - 1$.

So we set

$$a_{n+1} = \frac{\|P_n\|}{\|P_{n+1}\|} \qquad b_{n+1} = \langle P_n, xP_n\rangle \|P_n\|^{-2} \tag{1.2.12}$$

and (1.2.6) becomes (1.2.8) on account of (1.2.7). □

The a_n's and b_n's are called *Jacobi parameters*. We start labeling with $n = 1$, but some authors start with $n = 0$ or even label b from $n = 0$ but a from $n = 1$. Also, some reverse the a's and b's or use other letters.

The formula (1.2.12) for a_n implies

Theorem 1.2.4. *We have that*

$$\|P_n\| = a_n \dots a_1 \rho(\mathbb{R}) \tag{1.2.13}$$

The orthonormal polynomials

$$p_n(x) = \frac{P_n(x)}{\|P_n\|} \tag{1.2.14}$$

obey

$$xp_n(x) = a_{n+1}p_{n+1}(x) + b_{n+1}p_n(x) + a_n p_{n-1}(x) \tag{1.2.15}$$

and multiplication by x in the orthonormal set $\{p_j\}_{j=0}^\infty$ has the matrix

$$J = \begin{pmatrix} b_1 & a_1 & 0 & & \\ a_1 & b_2 & a_2 & \ddots & \\ 0 & a_2 & b_3 & \ddots & \\ & \ddots & \ddots & \ddots & \ddots \end{pmatrix} \tag{1.2.16}$$

Remarks. 1. Matrices of the form (1.2.16) are called *Jacobi matrices*.

2. When supp$(d\rho)$ is bounded, $\{p_n\}_{n=0}^\infty$ is a basis, as we have seen. Shortly we will restrict to this case.

We now have our direct equation: $\{a_n, b_n\}_{n=1}^\infty$ defines a second-order difference equation for $n = 1, 2, 3, \dots$,

$$u_{n+1} = a_n^{-1}((\lambda - b_n)u_n - a_{n-1}u_{n-1}) \tag{1.2.17}$$

where a_0 is picked in a convenient way and λ is a parameter. The solution with

$$u_0 = 0 \qquad u_1 = 1 \tag{1.2.18}$$

is

$$u_n = p_{n-1}(\lambda) \tag{1.2.19}$$

In Section 1.3, we will turn to the direct problem of going from $\{a_n, b_n\}_{n=1}^{\infty}$ to $d\rho$, but we see that at the heart of OPRL is an inverse spectral problem! Central to this language is the idea that going from a difference/differential equation is a direct question.

We will eventually see (Section 3.2) that the inverse problem has a second method of solution. We note that the $P_n(x)$ for $d\rho$ and $c_0\, d\rho$ for any c_0 are the same and so also for Jacobi parameters. Thus, we will eventually mainly restrict to $\rho(\mathbb{R}) = 1$.

Before leaving this introduction, we want to discuss two other ways of understanding OPRL that actually work for positive measures on $\mathbb{C}$, so we pause to define OPs in that case. Let $d\zeta(z)$ be a positive measure on $\mathbb{C}$ so that

$$\int |z|^n \, d\zeta(z) < \infty \tag{1.2.20}$$

which is nontrivial (i.e., $\operatorname{supp}(d\zeta)$ is not a finite set of points). Thus, we can form monic orthogonal polynomials $\Xi_n(z)$.

Unlike OPRL, $\Xi_n(z)$ do not obey a three-term recurrence relation because Proposition 1.2.2 uses reality (in general, $\langle \Xi_j, z\Xi_n\rangle = \langle \bar{z}\Xi_j, \Xi_n\rangle$). Indeed, only OPRL and OPUC (and polynomials for sets affinely related to $\mathbb{D}$ and $\partial\mathbb{D}$) are known to obey finite-order recursion relations, and so fit into the scheme of "spectral theory."

We note that $\{z^n\}_{n=0}^{\infty}$ may not span $L^2(\mathbb{C}, d\zeta)$ even if $\operatorname{supp}(d\zeta)$ is bounded. For example, if there is an open set $U \subset \mathbb{C}$ and c so that

$$d\zeta \geq c\chi_U \, d^2z \tag{1.2.21}$$

then they are not dense since the closure of the set of polynomials is analytic on U (see the Notes). And, as we will see (Section 2.11, especially Theorem 2.11.5), for measures on $\partial\mathbb{D}$, the issue of density is subtle. But we can define $\{\Xi_n(z)\}_{n=0}^{\infty}$ in any event.

Theorem 1.2.5 (Christoffel Variational Principle). *Let $\mathcal{M}_n$ be the monic polynomials of degree n, that is, $Q \in \mathcal{M}_n$ means*

$$Q(z) = z^n + \textit{lower order}$$

Then

$$\|\Xi_n\|^2 = \min_{Q \in \mathcal{M}_n} \|Q\|^2 \tag{1.2.22}$$

that is, for all $Q \in \mathcal{M}_n$,

$$\int |\Xi_n(z)|^2 \, d\zeta(z) \leq \int |Q(z)|^2 \, d\zeta(z) \tag{1.2.23}$$

with equality if and only if $Q = \Xi_n$.

Proof. This follows from the minimization property of orthogonalization, that is, if π is any orthogonal projection in a Hilbert space,

$$\|(1-\pi)u\|^2 = \min_{v\in\mathrm{Ran}(\pi)} \|u-v\|^2 \tag{1.2.24}$$

□

It is remarkable how powerful this principle is, given its simplicity.

The other general theorem concerns zeros.

Theorem 1.2.6. *Let $d\zeta$ be a measure obeying* (1.2.20) *and M_z multiplication by z on polynomials. Let π_n be the orthogonal projection in $L^2(\mathbb{C}, d\zeta)$ onto the n-dimensional space of polynomials of degree at most $n-1$. Let*

$$A = \pi_n M_z \pi_n \tag{1.2.25}$$

on $\mathrm{Ran}(\pi_n)$. *Then*

(i) *The eigenvalues of A are precisely the zeros of $\Xi_n(z)$.*

(ii) *Each eigenvalue of A has geometric multiplicity* 1.

(iii) *Each eigenvalue z_0 of A has algebraic multiplicity equal to the order of z_0 as a zero of $\Xi_n(z)$.*

(iv) *We have that*

$$\det(z-A) = \Xi_n(z) \tag{1.2.26}$$

Remark. Recall the geometric multiplicity of z_0 is the dimension of $\{v \mid (A - z_0)v = 0\}$. The algebraic multiplicity is the dimension of $\{v \mid (A-z_0)^\ell v = 0 \text{ for some } \ell\}$. It is the order of the zero in $\det(z-A)$.

Proof. Let $v \in \mathrm{Ran}(\pi_{n+1})$. Then $\pi_n v = 0$ if and only if $v = c\,\Xi_n$. Thus, if $w \in \mathrm{Ran}(\pi_n)$, $w \neq 0$, then $(A-z_0)w = 0 \Leftrightarrow \pi_n(z-z_0)w = 0 \Leftrightarrow (z-z_0)w = c\,\Xi_n$. Moreover, $w \neq 0$ implies $(z-z_0)w \neq 0$, so $c \neq 0$.

$$\Xi_n(z) = c^{-1}(z-z_0)w \tag{1.2.27}$$

implies $\Xi_n(z_0) = 0$, so (i) is half proven. Conversely, if $\Xi(z_0) = 0$, (1.2.27) is solved precisely by

$$w(z) = \frac{c\,\Xi_n(z)}{z-z_0} \tag{1.2.28}$$

which lies in $\mathrm{Ran}(\pi_n)$. Thus, (i) is proven and so is (ii).

The same analysis shows $(A-z_0)^\ell w = 0$ with $(A-z_0)^{\ell-1} w \neq 0$ if and only if z_0 is a zero of $\Xi_n(z)$ of order at least ℓ, and this proves (iii).

(iv) holds since both sides are monic polynomials of degree n with the same zeros counting orders. □

Corollary 1.2.7 (Fejér's Theorem). *Zeros of $\Xi_n(z)$ lie in the convex hull of* $\mathrm{supp}(d\zeta)$.

Proof. If $\Xi_n(z_0) = 0$, there is $w \in \mathrm{Ran}(\pi_n)$, with $\|w\|_{L^2} = 1$, so $\pi_n(z - z_0)w = 0$. Thus, $\langle w, (z - z_0)w\rangle = 0$, so

$$z_0 = \langle w, zw\rangle = \int z|w(z)|^2 \, d\zeta(z) \tag{1.2.29}$$

Since $\|w\| = 1$, $|w|^2 \, d\zeta$ is a probability measure, so the integral lies in the convex hull of $\mathrm{supp}(w^2 d\zeta)$, which lies in the convex hull of $\mathrm{supp}(d\zeta)$. □

Corollary 1.2.8. *Suppose that $d\rho$ is a measure on $\mathbb{R}$, with $a = \min \mathrm{supp}(d\rho)$, $b = \max \mathrm{supp}(d\rho)$. Then all the zeros of $P_n(x; d\rho)$ lie in $[a, b]$.*

Corollary 1.2.9. *Let $d\mu$ be a measure on $\partial\mathbb{D}$ and $\Phi_n(z; d\mu)$ the monic orthogonal polynomials. Then the zeros of Φ_n lie in $\overline{\mathbb{D}}$.*

Remark. One can show that if the convex hull of the support of $d\zeta$ does not lie in a straight line, then zeros lie in the interior of the convex hull of the support of the measure. In particular, in the case of Corollary 1.2.9, the zeros lie in $\mathbb{D}$, not merely $\overline{\mathbb{D}}$. We will prove this explicitly in Theorem 1.8.4.

Often, one has an explicit matrix representation of the operator A of (1.2.27), and so an explicit version of (1.2.24). For OPRL, one can take the basis $\{p_j\}_{j=0}^{n-1}$ and so get

Theorem 1.2.10. *Let $J_{n;F}$ be the $n \times n$ cutoff Jacobi matrix*

$$J_{n;F} = \begin{pmatrix} b_1 & a_1 & 0 & & & \\ a_1 & b_2 & a_2 & \ddots & & \\ 0 & a_2 & b_3 & \ddots & \ddots & \\ & & & \ddots & \ddots & \ddots \\ & & & & b_{n-1} & a_{n-1} \\ & & & & a_{n-1} & b_n \end{pmatrix} \tag{1.2.30}$$

Then

$$P_n(x) = \det(x - J_{n,F}) \tag{1.2.31}$$

Since $\det(x - A) = x^n - \mathrm{Tr}(A)x^{n-1} + O(x^{n-2})$ for $n \times n$ matrices, we see that

$$P_n(x) = x^n - \left(\sum_{j=1}^n b_j\right) x^{n-1} + O(x^{n-2}) \tag{1.2.32}$$

and, by (1.2.13)/(1.2.14),

$$p_n(x) = (a_1 \dots a_n)^{-1} \left[x^n - \left(\sum_{j=1}^n b_j\right) x^{n-1} \right] + O(x^{n-2}) \tag{1.2.33}$$

This provides another way of understanding the recursion (1.2.8). Expand $\det(x - J_{n+1,F})$ in minors in the last row. The minor of $x - b_{n+1}$ is $P_n(x)$ and the minor of $-a_n$ is $a_n P_{n-1}(x)$.

Remarks and Historical Notes. I would be remiss if I did not mention the "classical" OPRL: Jacobi, Laguerre, and Hermite associated, respectively, to the measures $(1+x)^\alpha(1-x)^\beta\,dx$ on $[-1,1]$ with $\alpha > -1$, $\beta > -1$, $x^\alpha e^{-x}$ on $[0,\infty)$ with $\alpha > -1$, and Hermite with $e^{-x^2}\,dx$. Jacobi polynomials with $\alpha=\beta=0$ are Legendre, and with $|\alpha| = |\beta| = \frac{1}{2}$ are Chebyshev (of four kinds depending on the signs of α and β). Chebyshev with $\alpha=\beta=-\frac{1}{2}$ and $\alpha=\beta=\frac{1}{2}$ (of the first and second kind) will occur repeatedly later in these notes. They obey (up to normalization; U_n is normalized but not monic, while T_n is neither the normalized nor monic OP), respectively,

$$T_n(\cos\theta) = \cos(n\theta) \tag{1.2.34}$$

$$U_n(\cos\theta) = \frac{\sin((n+1)\theta)}{\sin\theta} \tag{1.2.35}$$

These and other specific examples are discussed in detail in Szegő [434] and Ismail [204].

The classical polynomials obey many other relations like the Rodriguez formula and second-order (in x) differential equations. This is specific to them; indeed, there is a theorem of Bochner (see [48, 188, 371] and [204, Section 20.1]) that says any set of orthogonal polynomials that obeys a second-order differential equation of the proper form is one of the classical ones!

The question of when $\{x^n\}_{n=0}^\infty$ are dense in $L^2(\mathbb{R}, d\rho)$ is intimately connected to the issue of determinacy of the moment problem discussed in the Notes to the next section. We will return to this issue in Sections 3.8 and 3.9.

Analyticity often places restrictions on the density of polynomials. If $U \subset \mathbb{C}$ is open and $d\zeta \geq c\chi_U\, d^2z$ for some measure on $\mathbb{C}$ for which (1.2.20) holds, then by the Cauchy integral formula, for any compact $K \subset U$, we have

$$\sup_{z\in K} |f(z)| \leq C_K \|f\|_{L^2(\mathbb{C},d\zeta)}$$

for any function analytic in U and in L^2. It follows that any f in the L^2-closure of the polynomials is analytic on U since the locally uniform limit of analytic functions is analytic. Thus, when (1.2.21) holds, the polynomials do not span L^2. A celebrated theorem of Mergelyan (for a proof, see Greene–Krantz [183, Ch. 12]) says that if K is compact, with $\mathbb{C}\setminus K$ connected, then the L^∞-closure of the polynomials is the functions continuous on K and analytic on K^{int}.

OPRL have their roots in work of Legendre, Gauss, and Jacobi. As a general abstract theory, the key figures were Chebyshev, Markov, Christoffel, and especially, Stieltjes. You can find more history in the books of Szegő [434], Chihara [82], Freud [141], Nevai [320], and Ismail [204].

Closely entwined to the history is the idea of continued fraction expansions of resolvents, an issue we return to in Sections 2.5 and 3.2 and which was pioneered by Jacobi for finite matrices (hence the name Jacobi matrix for (1.2.30)) and Stieltjes.

Variational principles like (1.2.22) for OPRL go back to Christoffel. Their use in OPUC with a twist (see Section 2.12 later) is due to Szegő [434]. As a spectral theory tool, they have been especially advocated and exploited by Freud [141] and Nevai [321].

That zeros of OPRL are eigenvalues of truncated Jacobi matrices is well known in the Schrödinger operator community. I am unsure who noted it first. The extension to measures on $\mathbb{C}$ where there is the complication of nontrivial algebraic multiplicity was arrived at in discussions I had with E. Brian Davies.

1.3 FAVARD'S THEOREM, THE SPECTRAL THEOREM, AND THE DIRECT PROBLEM FOR OPRL

What the orthogonal polynomial community calls Favard's theorem is the assertion that the map from measures on $\mathbb{R}$ (with finite moments) to Jacobi parameters is onto $\{a_n, b_n\}_{n=1}^{\infty}$ with $a_n > 0$ and $b_n \in \mathbb{R}$. It is intimately connected to the spectral theorem; indeed, we will prove the spectral theorem for bounded selfadjoint operators in this section (modulo some remarks in the Notes that go from Jacobi matrices to general operators). In the bounded case, we will see the map is also one-one if we restrict to probability measures.

Our discussion will be in three stages: first, finite Jacobi matrices, then bounded, and finally, unbounded (where we will assume, rather than prove, the spectral theorem).

Consider a trivial probability measure, that is,

$$d\rho = \sum_{j=1}^{N} \rho_j \delta_{x_j} \tag{1.3.1}$$

for

$$x_1 > x_2 > \cdots > x_N \tag{1.3.2}$$

and

$$\sum_{j=1}^{N} \rho_j = 1 \tag{1.3.3}$$

As usual, we can use Gram–Schmidt to define monic polynomials $P_0, \dots, P_{N-1}$ since our proof of independence of $\{x^j\}_{j=0}^{\infty}$ in the nontrivial case shows that $\{x^j\}_{j=0}^{N-1}$ are independent in this case. We can also use (1.2.3) to define $P_N(x)$ as the zero vector in $L^2(\mathbb{R}, d\rho)$, which, among monic Nth degree polynomials, is unique, namely,

$$P_N(x) = \prod_{j=1}^{N} (x - x_j) \tag{1.3.4}$$

The P's obey a recursion relation of the form (1.2.8) for $n = 0, 1, 2, \dots, N-1$ and so define $b_1, \dots, b_N, a_1, \dots, a_{N-1}$ and an $N \times N$ finite Jacobi matrix.

To go backward, we start with an $N \times N$ finite Jacobi matrix, that is, $\{b_j\}_{j=1}^{N}$ and $\{a_j\}_{j=1}^{N-1}$ are given with $a_j > 0$ and $b_j \in \mathbb{R}$, and we do not (yet) know they come from a measure.

We do not have a measure yet, so we cannot define P_j by orthogonality, but we do have recursion coefficients, so we define $\{P_j\}_{j=0}^N$ inductively by (1.2.8) with $P_0(x) \equiv 1$, $P_{-1}(x) \equiv 0$ (they could also be defined directly by (1.2.31)!), then p_j for $j = 0, 1, 2, \dots, N-1$ by $p_0(x) = 1$, and for $1 \le j \le N-1$,

$$p_j(x) = \frac{P_j(x)}{a_1 \dots a_j} \tag{1.3.5}$$

Then p_n obey (1.2.15) for $n = 0, 1, 2, \dots, N-2$ and

$$(b_N - x)p_{N-1}(x) + a_{N-1}p_{N-2}(x) = -(a_1 \dots a_{N-1})^{-1} P_N(x) \tag{1.3.6}$$

Proposition 1.3.1. *Let* $J \equiv J_{N;F}$ *be a finite Jacobi matrix given by* (1.2.30).

(a) *Define the vector* $\vec{v}(x) \in \mathbb{C}^N$ *by*

$$v_j(x) = p_{j-1}(x) \qquad j = 1, 2, \dots, N \tag{1.3.7}$$

Then

$$(J - x)\vec{v}(x) = -(a_1 \dots a_{N-1})^{-1}\delta_{jN} P_N(x) \tag{1.3.8}$$

(b) *If* $\vec{w} \in \mathbb{C}^N$ *obeys*

$$[(J - x)\vec{w}]_j = 0 \qquad j = 1, \dots, N-1 \tag{1.3.9}$$

then

$$w_j = w_1 p_{j-1}(x) \tag{1.3.10}$$

(c) *The eigenvalues of* J *are exactly the set of zeros of* $P_N(x)$ *and each zero has geometric multiplicity* 1.

(d) *The zeros of* P_N *are simple and real.*

(e) *If the zeros of* P_N *are labeled by* (1.3.2) *and*

$$(\varphi_\ell)_j = \frac{p_{j-1}(x_\ell)}{(\sum_{j=1}^N |p_{j-1}(x_\ell)|^2)^{1/2}} \tag{1.3.11}$$

then the φ_ℓ *are an orthonormal basis of eigenvectors.*

(f) *If*

$$\rho_\ell = |(\varphi_\ell)_1|^2 = \left(\sum_{j=1}^N |p_{j-1}(x_\ell)|^2 \right)^{-1} \tag{1.3.12}$$

then (1.3.3) *holds and* $\{P_j(x)\}_{j=0}^N$ *are the OPRL for the measure* (1.3.1).

Proof. (a) (1.3.8) is just (1.2.15) for $j = 1, \dots, N-1$ and (1.3.6) for $j = N$.

(b) (1.3.10) holds trivially for $j = 1$ and then inductively by subtracting (1.3.8) from (1.3.10), and noting this implies

$$(w_{j+1} - w_1 p_j(x)) = (a_j)^{-1}[(x - b_j)(w_j - w_1 p_{j-1}(x)) - a_{j-1}(w_{j-1} - w_1 p_{j-2}(x))] \tag{1.3.13}$$

for $j = 1, 2, \dots, N-1$ (with $a_{-1} \equiv 0$).

(c) Any eigenvector obeys (1.3.9) and so must be a multiple of $\vec{v}$. It obeys $[(J - x)\vec{v}(x)]_N = 0$ if and only if $P_N(x) = 0$ by (1.3.8). This argument shows any eigenvector is a multiple of φ_j given by (1.3.9), and so the geometric multiplicity is 1.

(d) Define φ_j by (1.3.9). Then $\langle \varphi_k, J\varphi_\ell\rangle = \langle J\varphi_k, \varphi_\ell\rangle$ implies, using $J\varphi_\ell = x_\ell \varphi_\ell$, that

$$(\bar{x}_k - x_\ell)\langle \varphi_k, \varphi_\ell\rangle = 0 \tag{1.3.14}$$

Taking $k = \ell$, we see x_k is real since $(\varphi_\ell)_1 \neq 0$ implies $\langle \varphi_\ell, \varphi_\ell\rangle \neq 0$.

To see that zeros are simple, suppose $P_N'(x_j) = 0$. Let

$$\vec{w} = \frac{\partial \vec{v}}{\partial x}\bigg|_{x=x_j} \tag{1.3.15}$$

(the components of v are polynomials, hence differentiable). Since $P_N'(x_1) = 0$, (1.3.8) implies

$$(J - x_j)w = v(x_j) \tag{1.3.16}$$

That cannot be since it implies

$$\begin{aligned}\langle v(x_j), v(x_j)\rangle &= \langle v(x_j), (J - x_j)w\rangle \\ &= \langle (J - x_j)v(x_j), w\rangle \\ &= 0\end{aligned}$$

and $v_1(x_j) = 1$, so $\langle v(x_j), v(x_j)\rangle \neq 0$.

(e) $\|\varphi_\ell\|^2 = 1$ is immediate and $\langle \varphi_j, \varphi_\ell\rangle = 0$ for $j \neq \ell$ by (1.3.14). Since $P_N(x)$ has N zeros, the φ_ℓ's must span the space.

(f) Since $\{\varphi_\ell\}_{\ell=1}^N$ are an orthonormal basis, $U_{k\ell} = (\varphi_\ell)_k$ obeys

$$\sum_k \bar{U}_{k\ell} U_{kj} = \delta_{\ell j}$$

that is, $U^*U = \mathbf{1}$, so since it is finite-dimensional, $UU^* = 1$, that is (using $(\varphi_\ell)_j$ real to drop bars),

$$\sum_\ell (\varphi_\ell)_j (\varphi_\ell)_k = \delta_{jk} \tag{1.3.17}$$

This says, by the definitions (1.3.11) and (1.3.12),

$$\sum_\ell \rho_\ell\, p_{j-1}(x_\ell)\, p_{k-1}(x_\ell) = \delta_{jk} \tag{1.3.18}$$

Taking $j = k = 1$ and using $p_0(x) = 1$, we see that (1.3.3) holds and (1.3.18) implies that the $\{p_j\}_{j=0}^{N-1}$ are orthonormal polynomials for the measure (1.3.1), so $\{P_j\}_{j=0}^{N-1}$ are the monic OPRL. Since $P_N(x_j) = 0$, P_N is the monic OPRL for $d\rho$. $\square$

Remarks. 1. To be self-contained, we have given the standard argument that symmetric matrices have real eigenvalues and have algebraic multiplicities equal to geometric ones.

2. Notice that we have, in essence, just proven the spectral theorem for finite Jacobi matrices.

3. For a more conventional proof that the zeros of OPRL are all real and simple, see Subsection 5 of Section 1.2 of [399].

We have thus proven

Theorem 1.3.2 (Favard's Theorem for Trivial Measures). *Every finite $N \times N$ Jacobi matrix is the Jacobi matrix of some measure supported on N points.*

Proof. (f) of the last theorem says the $\{P_j\}_{j=0}^N$ are the OPRL for $d\rho$ defined by (1.3.12) and $P_N(x_j) = 0$. The Jacobi parameters of P_j are the given Jacobi matrix since the polynomials alone obeying (1.2.8) determine a and b inductively by looking at the x^N and x^{N-1} terms on both sides of (1.2.8). For example, if $x_\ell^{(n)}$ are the roots of $P_n(x)$,

$$b_n = \left[\sum_{\ell=1}^{n} x_\ell^{(n)}\right] - \left[\sum_{\ell=1}^{n-1} x_\ell^{(n)}\right]$$

as will occur prominently in Section 8.5. □

Theorem 1.3.3. *The map from $d\rho$ of the form* (1.3.1)/(1.3.3) *to* $\{a_j\}_{j=0}^{N-1} \cup \{b_j\}_{j=0}^N$ *is one-one* (*and onto by Theorem 1.3.2*).

First Proof. Given the Jacobi matrix, J_N, of $d\rho$, following the construction of Theorem 1.3.2, construct a measure $d\rho' = \sum_{j=1}^N \rho'_j \delta_{x'_j}$. By construction, x'_j are the zeros of $P_N(x; d\rho)$, which are exactly the x_j's, that is, after renumbering $x'_j = x_j$. Moreover, the construction shows the normalized eigenvectors with positive first component are (1.3.11), so since φ_ℓ in $L^2(\mathbb{R}, d\rho)$ or $L^2(\mathbb{R}, d\rho')$ is the function $f(x) = \delta_{x_\ell x}$, we have

$$\begin{aligned} \rho_\ell &= \langle \varphi_\ell, 1\rangle_{L^2(\mathbb{R},d\rho)} \\ &= \langle \varphi_\ell, p_0\rangle_{L^2(\mathbb{R},d\rho)} \\ &= (\varphi_\ell)_1 \\ &= \text{given by (1.3.12)} \end{aligned}$$

showing $\rho_\ell = \rho'_\ell$. □

We want to give a second proof, not because this result is so important or so difficult, but because a slightly more involved proof will yield tools that are useful in the $N = \infty$ case.

Proposition 1.3.4. (a) *Two* (*probability*) *measures $d\rho, d\rho'$* (*supports can be infinite*) *have the same Jacobi parameters up to n, $\{a_j\}_{j=1}^{n-1} \cup \{b_j\}_{j=0}^n$, if and only if*

$$\int x^k \, d\rho = \int x^k \, d\rho' \tag{1.3.19}$$

$k = 0, 1, \ldots, 2n - 1$.

(b) *Two measures, $d\rho, d\rho'$, each supported at n* (*possibly different*) *points are equal if and only if* (1.3.19) *holds for $k = 0, \ldots, 2n - 1$.*

Proof. (a) By (1.2.8), we see that if Jacobi parameters are equal, then

$$P_j(x; d\rho) = P_j(x; d\rho') \tag{1.3.20}$$

Multiplying by x^ℓ, $\ell = 0, \dots, j-1$ and integrating, we see

$$\int x^{\ell+j} d\rho = \text{function of} \left\{ \int x^{\ell+k}\, d\rho \right\}_{k=0}^{j-1} \tag{1.3.21}$$

$$= \int x^{\ell+j}\, d\rho' \tag{1.3.22}$$

where the function is the same by (1.3.20), and (1.3.19) then follows by induction. As j runs from 0 to n and ℓ from 0 to $j-1$, $\ell + j$ goes from 0 to $2n-1$.

Conversely, if (1.3.19) for $k = 0, \dots, 2n-1$, the Gram matrices $\{\langle x^j, x^\ell \rangle\}_{0 \le j, \ell \le n-1}$ are equal, which, by the Gram–Schmidt process, implies $p_j(x; d\rho) = p_j(x; d\rho')$ for $0 \le j \le n-1$, and so

$$P_j(x; d\rho) = P_j(x; d\rho') \tag{1.3.23}$$

Since

$$P_n(x) = x^n - \sum_{j=0}^{n-1} \langle p_j, x^n \rangle p_j(x)$$

the moments $\int x^{n+\ell}\, d\rho$, $\ell = 0, \dots, n$, then also determine P_n so (1.3.23) also holds for $j = n$. As noted above, the polynomials determine the a's and b's in the recursion relation.

(b) As noted in (a), the stated moments determine $P_N(x)$ and so its zeros, and so $\{x_j\}_{j=1}^N$ and $\{x'_j\}_{j=1}^N$ are identical sets. Then the ρ's are determined by the equations

$$\sum_{j=1}^N \rho_j x_j^{\ell-1} = \int x^{\ell-1}\, d\rho \tag{1.3.24}$$

for $\ell = 1, 2, \dots, N$ since the Vandermonde determinant

$$\det(x_j^{\ell-1}) = \prod_{i<j} (x_i - x_j) \tag{1.3.25}$$

is nonzero. □

Second Proof of Theorem 1.3.3. By (a) of Proposition 1.3.4, the Jacobi parameters determine the first $2n$ moments and then, knowing the support is n points, the measures by (b) of the proposition. □

One can combine Theorems 1.3.2 and 1.3.3 and more in

Theorem 1.3.5. *Fix N. There is a one-one correspondence among each of*

(i) *Jacobi parameters $\{a_j\}_{j=1}^{N-1} \cup \{b_j\}_{j=1}^N$ with $b_j \in \mathbb{R}$ and $a_j > 0$.*

(ii) *Trivial measures of the form* (1.3.1) *where* (1.3.3) *holds and $\rho_j > 0$.*

(iii) *Unitary equivalence classes of symmetric $N \times N$ matrices A with a distinguished cyclic vector, φ.*

Remarks. 1. φ is called cyclic if $\{A^j\varphi\}_{j=0}^\infty$ span the space. For $N \times N$ matrices, we can instead take $\{A^j\varphi\}_{j=0}^{N-1}$ since if $P(A)$ is the (monic) characteristic polynomial of A, $P(A)A^\ell\varphi = 0$ shows inductively that $\{A^j\varphi\}_{j=N}^\infty$ are functions of $\{A^j\varphi\}_{j=0}^{N-1}$.

2. (A, φ) and (A', φ') are unitarily equivalent if and only if there is a unitary $U: \mathbb{C}^N \to \mathbb{C}^N$ so $UAU^{-1} = A'$ and $U\varphi = \varphi'$.

Proof. (i) $\Leftrightarrow$ (ii) is precisely the construction of Section 1.2 combined with Theorems 1.3.2 and 1.3.3.

It is easy to see that $\delta_1 = (1, 0, \dots, 0)^t$ is cyclic for a finite Jacobi matrix J. Indeed, if $\{p_\ell\}_{\ell=0}^{n-1}$ are the orthonormal polynomials, then $\delta_\ell = p_{\ell-1}(J)\delta_1$, so each Jacobi matrix with distinguished δ_1 is in an equivalence class.

Conversely, if φ is cyclic for A, $\{A^j\varphi\}_{j=0}^{N-1}$ must be independent (since they span $\mathbb{C}^N$). Thus, by Gram–Schmidt, we can find polynomials $\{p_j(A)\}_{j=0}^{N-1}$ with $p_0(A) = \mathbf{1}$ so $\varphi_j = p_{j-1}(A)\varphi$, $j = 0, \dots, N-1$, is an orthonormal basis. By the Gram–Schmidt construction, $\langle A^k\varphi, p_j(A)\varphi\rangle = 0$ if $k < j$. So by the same argument as in Section 1.2, there are constant $\{b_j\}_{j=1}^N$, $\{a_j\}_{j=1}^{N-1}$, so

$$A\varphi_j = a_{j+1}\varphi_{j+1} + b_{j+1}\varphi_j + a_j\varphi_{j-1} \tag{1.3.26}$$

for $j = 0, \dots, N-1$ where we interpret a_N and a_0 as 0. Thus, $\langle\varphi_j, A\varphi_k\rangle$ is a Jacobi matrix! The construction is unitarily invariant so the map is from equivalence classes to Jacobi matrices.

The two constructions are inverses showing the one-one correspondence. □

Now we turn to the case of bounded semi-infinite Jacobi matrices.

Proposition 1.3.6. *A Jacobi matrix* (1.2.16) *is bounded on* ℓ^2 *if and only if*

$$\sup_n |a_n| + \sup_n |b_n| < \infty \tag{1.3.27}$$

Proof. Let δ_n be the vector with components δ_{nj}. $b_n = \langle\delta_n, J\delta_n\rangle$ while $a_n = \langle\delta_{n+1}, J\delta_n\rangle$ so $|b_n| \le \|J\|$ and $|a_n| \le \|J\|$. Thus, J bounded implies (1.3.27). A diagonal matrix $D = \{d_n\delta_{nm}\}$ has $\|D\| = \sup_n |d_n|$, and if A, B are the diagonal matrices with elements a and b, and if $S\delta_n = \delta_{n+1}$, then

$$J = AS^* + B + SA \tag{1.3.28}$$

so

$$\|J\| \le 2\sup_n |a_n| + \sup_n |b_n| \tag{1.3.29}$$

We have thus proven

$$\sup_n |a_n| + \sup_n |b_n| \le 2\|J\| \le 4\big(\sup_n |a_n| + \sup_n |b_n|\big) \tag{1.3.30}$$

□

We can now turn to the main theorem of this section (given our interest in the bounded support regime):

Theorem 1.3.7 (Favard's Theorem for Bounded Jacobi Matrices). *Let $\{a_n\}_{n=1}^\infty$, $\{b_n\}_{n=1}^\infty$ be a set of Jacobi parameters obeying (1.3.27). Then there is a nontrivial measure, $d\rho$, of bounded support so that its Jacobi parameters are the given ones.*

Proof. Let J be a Jacobi matrix and $J_{n;F}$ its finite truncations. By Theorem 1.3.2, there are trivial n-point measures, $d\rho_n$, whose Jacobi parameters are $\{a_j\}_{j=0}^{n-1} \cup \{b_j\}_{j=0}^{n}$. By Proposition 1.3.4,

$$\int x^\ell \, d\rho_n = \int x^\ell \, d\rho_{n'} \tag{1.3.31}$$

for $\ell = 0, 1, \dots, 2\min(n, n') - 1$. In particular, for each ℓ, $\int x^\ell \, d\rho_n$ is constant for n large, so

$$\lim_{n\to\infty} \int x^\ell \, d\rho_n$$

exists for each ℓ.

By construction, $d\rho_n$ is supported on the eigenvalues of $J_{n;F}$ and so on $[-\|J_{n;F}\|, \|J_{n;F}\|]$, and so on $[-\|J\|, \|J\|]$. Thus, the $d\rho_n$'s are supported in a fixed compact set. Since the polynomials are dense in $C([-\|J\|, \|J\|])$, the probability measures, $d\rho_n$, have a weak limit $d\rho$. This weak limit, by (1.3.31), obeys

$$\int x^\ell \, d\rho_n = \int x^\ell \, d\rho \qquad \ell = 0, \dots, 2n-1 \tag{1.3.32}$$

By Proposition 1.3.4, the Jacobi parameters of $d\rho$ are J. □

Remark. Modulo discussion in the Notes, we have just proven the spectral theorem for bounded operators!

In the following, we could also discuss cyclic vectors, but we will not (see the Notes):

Theorem 1.3.8. *There is a one-one correspondence between bounded Jacobi matrices and nontrivial probability measures of bounded support under the map of measures to Jacobi parameters.*

Proof. Clearly, if $d\rho$ has support $[-C, C]$, then

$$|b_n| \le \int |x| \, |p_n(x)|^2 \, d\rho \le C$$

$$|a_n| \le \int |x| \, |p_n(x)| \, |p_{n-1}(x)| \, d\rho \le C$$

so J is bounded. By Favard's theorem, the map from measures of bounded support to bounded Jacobi parameters is onto. By Proposition 1.3.4, it is one-one. □

In this monograph, we are mainly interested in the bounded support case, so we will state Favard's theorem in the unbounded case without giving the proof for now. We will essentially prove it in Section 3.8; see Theorem 3.8.4.

Theorem 1.3.9 (Favard's Theorem). *For any set of Jacobi parameters, there is a measure, $d\rho$, on $\mathbb{R}$ with $\int |x|^n \, d\rho(x) < \infty$ for all n, which has those Jacobi parameters.*

The measure may not be unique. This is discussed in Sections 3.8 and 3.9.

Remarks and Historical Notes. Favard's theorem is named after Favard [128] but goes back to Stieltjes [422]. The close connection to the spectral theorem also predates Favard in work of Stone [423] and Wintner [461]; see also Natanson [313], Perron [345], Sherman [382], and the discussion in Marcellán and Álvarez-Nodarse [295]. I am not aware of the approach here appearing elsewhere, but it will not surprise experts and I suspect is known to some.

Given any bounded selfadjoint operator, A, on a separable Hilbert space, $\mathcal{H}$, it is not hard to see that one can find $\{\varphi_j\}_{j=1}^N$ (N finite or infinite) so that for any $\ell, m, j \neq k$, $\langle A^\ell \varphi_j, A^m \varphi_k \rangle = 0$ and so that $\{A^\ell \varphi_j\}_{j,\ell}$ span $\mathcal{H}$. Thus, Theorem 1.3.7 and Gram–Schmidt imply there is a unitary U from $\mathcal{H}$ onto $\oplus_{j=1}^N L^2(\mathbb{R}, d\mu_j)$ so that $(UAU^{-1}f)_m(x) = xf_m(x)$. This is the spectral theorem for bounded operators.

The same idea shows that if A has a cyclic vector, φ, then applying Gram–Schmidt to $\{A^j\varphi\}_{j=0}^\infty$ yields an orthonormal basis in which J is a cyclic vector, allowing the two-part equivalence of Theorem 1.3.8 to extend to the three-part equivalence of Theorem 1.3.5.

1.4 GEMS OF SPECTRAL THEORY

In order to explain what I will mean by a gem of spectral theory, I begin by describing a pair of beautiful theorems in the spectral theory of OPRL:

Theorem 1.4.1 (Blumenthal–Weyl). *Let J be a Jacobi matrix with Jacobi parameters $\{a_n, b_n\}_{n=1}^\infty$. If*

$$a_n \to 1 \qquad \text{and} \qquad b_n \to 0 \tag{1.4.1}$$

then

$$\sigma_{\text{ess}}(J) = [-2, 2] \tag{1.4.2}$$

Remarks. 1. Recall (see Reed–Simon [364, Section XIII.4]) that σ_{ess} is defined by $\sigma_{\text{ess}}(J) = \sigma(J) \setminus \sigma_d(J)$, where $\sigma(J)$, the spectrum of J, is $\{\lambda \mid (J - \lambda)$ does not have a bounded inverse$\}$, and $\sigma_d(J)$ are isolated points λ_0 of $\sigma(J)$, where $\oint_{|z-\lambda_0|=\varepsilon} (z - J)^{-1} \, dz$ is finite rank. For J's with cyclic vector (like Jacobi matrices) and spectral measure $d\rho$, $\sigma_{\text{ess}}(J)$ is the set of nonisolated points of $\text{supp}(d\rho)$.

2. See the Notes for a discussion of proof and history.

3. For any $a, b \in \mathbb{R}$ with $a > 0$, $N(a, b)$, the *Nevai class*, is the set of measures where $a_n \to a$, $b_n \to b$. By scaling, $\sigma_{\text{ess}}(J) = [b - 2a, b + 2a]$ if $J \in N(a, b)$.

Theorem 1.4.2 (Denisov–Rakhmanov). *Let J be a Jacobi matrix with measure $d\rho$ and Jacobi parameters $\{a_n, b_n\}_{n=1}^{\infty}$. Suppose* (1.4.2) *holds and*

$$d\rho(x) = f(x)\,dx + d\rho_{\mathrm{s}}(x) \tag{1.4.3}$$

where $d\rho_{\mathrm{s}}$ is singular and (*modulo sets of measure* 0)

$$\{x \mid f(x) > 0\} = [-2, 2]$$

Then (1.4.1) *holds.*

Remark. See the Notes for a discussion of proof and history. We will return to this theorem and prove it in Section 7.6.

These theorems are illuminated by the following:

Example 1.4.3. Let $a_n \equiv \frac{1}{2}$ and b_n be the sequence $(1, -1, 1, 1, -1, -1, 1, 1, 1, -1, -1, -1, \dots)$, that is, 1 k times followed by -1 k times for $k = 1, 2, \dots$. It is not hard to show $\sigma(J) = \sigma_{\mathrm{ess}}(J) = [-2, 2]$, so (1.4.2) is not sufficient for (1.4.1) to hold. □

Thus, we have a pair of deep theorems that go in opposite directions, but they do not set up equivalences. This leads us to:

Definition. By a *gem of spectral theory*, I mean a theorem that describes a class of spectral data and a class of objects so that an object is in the second class if and only if its spectral data lie in the first class.

This idea will be illuminated as we describe gems for OPUC and for OPRL in Sections 1.8 and 1.10 and a nongem in Section 1.9. In a sense, the overriding purpose of this book is to explore gems of OPRL/OPUC that depend on sum rules with positive coefficients. As we will see, the focus is somewhat narrower than that! And we will discuss some descendants of Szegő's theorem that are not gems (yet).

Remarks and Historical Notes. I find that some listeners object strongly to my use of the term "gem." I respond that it is a definition and I add that for a mathematician, a definition is not something that can be "wrong." But if I called them the "Jims of Spectral Theory," I wouldn't get the same reaction. And, of course, I used gems because of its connotation. Gems of spectral theory are typically beautiful and hard—but there can be beautiful and hard results that are not necessary and sufficient: Theorem 1.4.2 comes to mind.

The Blumenthal–Weyl theorem is named after contributions of Blumenthal [46] and Weyl [457]; Denisov–Rakhmanov after results of Rakhmanov [358, 359] and Denisov [107]; see Sections 9.1 and 9.2 of [400] for further history.

Theorem 1.4.1 is a consequence of Weyl's theorem (see Reed–Simon [364, Section XIII.4]) that if C is compact and selfadjoint and A bounded and selfadjoint, then $\sigma_{\mathrm{ess}}(A + C) = \sigma_{\mathrm{ess}}(A)$. In Theorem 1.4.1, $A = J_0$, the Jacobi matrix with $a_n \equiv 1$, $b \equiv 0$, and $C = J - J_0$ is compact when (1.4.1) holds.

Rakhmanov's theorem for OPUC is proven in Chapter 9 of [400]. Theorem 1.4.2 is proven in Section 13.4 of that book. As mentioned, we will provide a proof of a more general result in Chapter 7 of the present monograph.

1.5 SUM RULES AND THE PLANCHEREL THEOREM

The basic tool we will use is to establish sum rules with positive terms. In this section, we illustrate this with the granddaddy of all spectral sum formulae: the fact that if $A = \{a_{ij}\}_{1 \le i,j \le N}$ is a finite matrix and $\{\lambda_j\}_{j=1}^N$ are its eigenvalues, then

$$\sum_{j=1}^{N} \lambda_j = \mathrm{Tr}(A) \equiv \sum_{j=1}^{N} a_{jj} \tag{1.5.1}$$

The left side is spectral theoretic and the right side involves the coefficients of the object.

One standard proof of (1.5.1) is to prove invariance of trace under similarity and the fact that there is a similarity taking A to upper triangular (even Jordan) form. But for us, the "right" proof is to note that the λ_j are the roots of the characteristic polynomials, so

$$\det(\lambda \mathbf{1} - A) = \prod_{j=1}^{N} (\lambda - \lambda_j) \tag{1.5.2}$$

Since, by expanding the determinant

$$\det(\lambda \mathbf{1} - A) = \lambda^n - \mathrm{Tr}(A)\lambda^{n-1} + \cdots \tag{1.5.3}$$

we get (1.5.1). The idea that sum rules occur as Taylor coefficients of suitable analytic functions recurs throughout this book.

In the infinite-dimensional case, there are convergence and other issues. Let X be a Banach space. A bounded linear map $A \colon X \to X$ is called *finite rank* if $\mathrm{Ran}(A)$ is finite-dimensional. Every such map has the form

$$Ax = \sum_{j=1}^{N} \ell_j(x) x_j \tag{1.5.4}$$

For some $\{\ell_j\}_{j=1}^N \subset X^*$ and $\{x_j\}_{j=1}^N \subset X$. It is not hard to show that

$$\mathrm{Tr}(A) = \sum_{j=1}^{N} \ell_j(x_j) \tag{1.5.5}$$

is independent of the ℓ's and x's used in the representation (1.5.4) (essentially by the invariance of trace in the finite-dimensional case). One defines the trace norm of a finite-rank operator by

$$\|A\|_1 = \inf \left\{ \sum_{j=1}^{n} \|\ell_j\|_{X^*} \|x_j\|_X \;\middle|\; A = \sum \ell_j(\cdot) x_j \right\} \tag{1.5.6}$$

The nuclear operators, $N(X)$, are the completion of the finite-rank operators in $\|\cdot\|_1$. It is not hard to see that every such object is associated to an operator and that one can define $\mathrm{Tr}(\cdot)$ on $N(X)$ since

$$|\mathrm{Tr}(A)| \le \|A\|_1 \tag{1.5.7}$$

If X is a Hilbert space, then $N(X)$ is called the *trace class operators*. A celebrated theorem of Lidskii says that

Theorem 1.5.1 (Lidskii's Theorem). *If A is a trace class operator on a Hilbert space, $\mathcal{H}$, then $\sigma_{\text{ess}}(A) = \{0\}$ and A has nonzero eigenvalues $\{\lambda_j\}_{j=1}^N$ (counting algebraic multiplicity) so that*

$$\sum_{j=1}^{N} \lambda_j = \operatorname{Tr}(A) \tag{1.5.8}$$

There are two limitations to note. First, on general Banach spaces, this result is false. Indeed, there is a Banach space, X, with a nuclear operator A so that $A^2 = 0$ (so any eigenvalue is 0) but $\operatorname{Tr}(A) = 1$! (See the Notes.)

Second, consider the operator, C, on ℓ^2, which is a direct sum $C_1 \oplus C_2 \oplus \dots$ of 2×2 matrices

$$C_j = \begin{pmatrix} \alpha_j & \alpha_j \\ -\alpha_j & -\alpha_j \end{pmatrix} \tag{1.5.9}$$

$C_j^2 = 0$, so C has only eigenvalue zero. Indeed, it is easy to see that $\sigma(C) = \{0\}$. If $\sum_{j=1}^{\infty} |\alpha_j| = \infty$, but $\alpha_j \to 0$, then C is compact but not trace class. The sum of the eigenvalues is 0. As for the "trace," the sum of the diagonal matrix elements of C is conditionally convergent to zero, so it looks like a success. But conditionally convergent sums can be rearranged to any value! And rearranged sums are just rearranged bases. The moral is that, due to cancellations, (1.5.8) is subtle as soon as one leaves trace class, and it is unlikely that there is any kind of necessary and sufficient condition directly related to (1.5.8).

However, positivity can rescue something. It is not hard to prove

Theorem 1.5.2. *Let A be a bounded selfadjoint operator on a Hilbert space. Then A^2 is trace class if and only if A has a pure point spectrum with eigenvalues $\{\lambda_j(A)\}_{j=1}^{\infty}$ obeying*

$$\sum_{j=1}^{\infty} \lambda_j(A)^2 < \infty \tag{1.5.10}$$

In fact, if one writes $\operatorname{Tr}(A^2) = \infty$ if A^2 is not trace class and $\sum \lambda_j(A)^2 = \infty$ if A has any nonpoint spectrum, Theorem 1.5.2 comes from a sum rule

$$\operatorname{Tr}(A^2) = \sum_j \lambda_j(A)^2 \tag{1.5.11}$$

There are no cancellations because of positivity.

On $\ell^2(\partial\mathbb{D}, \frac{d\theta}{2\pi})$, one can specialize to operators of the form

$$(A_f g)(\theta) = \int f(\theta - \psi) g(\psi) \, \frac{d\psi}{2\pi} \tag{1.5.12}$$

where $\theta - \psi$ is computed mod 2π. Then $\lambda_j(A_f)$ are the Fourier coefficients, Theorem 1.5.2 is the Plancherel theorem, and the sum rule (1.5.11) is Parseval's equality. As we will see in Section 2.11, Szegő's theorem can be viewed as a kind of nonlinear Plancherel theorem.

Remarks and Historical Notes. The view of Theorem 1.5.2 as a sum rule with positivity, and so a model of Szegő's theorem as a sum rule, has been pushed especially by Killip [222].

For a proof of Lidskii's theorem, see, for example, [389], which obtains it from an equality for trace class operators

$$\det(1+zA) = \prod_{j=1}^{\infty}(1+z\lambda_j(A)) \tag{1.5.13}$$

An analog of (1.5.14) for Hilbert–Schmidt integral operators, namely,

$$\det[(1+zA)e^{-zA}] = \prod_{j=1}^{\infty}[(1+z\lambda_j(A))e^{-z\lambda_j(A)}] \tag{1.5.14}$$

goes back to Carleman [74] in 1921. One can regard him as the father of Theorem 1.5.2. Lidskii's theorem is named after [281], although the theorem was found somewhat earlier by Grothendieck [187]. Unaware of Grothendieck's work, Simon [389] rediscovered his approach to the problem.

For an introduction to nuclear operators on a general Banach space, see Chapter 10 of Simon [390]. (This book also discusses trace class, Lidskii's theorem, and proves (1.5.13) and (1.5.14); another reference on those subjects is Gohberg–Krein [170].) In particular, the example mentioned of a nuclear operator with $A^2 = 0$, but $\mathrm{Tr}(A) = 1$ is from Grothendieck [186].

1.6 PÓLYA'S CONJECTURE AND SZEGŐ'S THEOREM

Pólya and Szegő have linked names much like Hardy and Littlewood or Laurel and Hardy. This is most of all because of their great two-volume encyclopedia of analysis [353] and because, as part of Szegő's establishing of a great school of mathematics at Stanford, he brought Pólya to Palo Alto. But they are also linked in the initial history of the main theme of this monograph.

As we will see in Section 3.8, Hankel matrices, that is, finite matrices of the form $\{c_{j+k}\}_{jk=1}^n$ are fundamental to the theory of the moment problem on $\mathbb{R}$ (since they arise as Gram matrices for $\{x^j\}_{j=0}^{n-1}$). A *Toeplitz matrix*, T, is one of the form

$$t_{jk} = c_{j-k} \qquad 1 \le j, k \le n \tag{1.6.1}$$

Just as in the Hankel case, a situation of special interest is when c are the moments of a measure but now on $\partial\mathbb{D}$:

$$c_k = \int_0^{2\pi} e^{-ik\theta}\, d\mu(\theta) \tag{1.6.2}$$

We will, for now, restrict to the case $d\mu_s = 0$ where

$$d\mu(\theta) = \frac{w(\theta)}{2\pi}\, d\theta + d\mu_s \tag{1.6.3}$$

that is, to the case

$$c_k = \int e^{-ik\theta}\, w(\theta)\, \frac{d\theta}{2\pi} \tag{1.6.4}$$

Define $D_n(w)$ (more generally, $D_n(d\mu)$) to be the determinant of the $(n+1) \times (n+1)$ Toeplitz matrix

$$D_n(w) = \det \begin{vmatrix} c_0 & c_1 & \dots & c_n \\ c_{-1} & c_0 & \dots & c_{n-1} \\ \vdots & \vdots & & \vdots \\ c_{-n} & \dots & \dots & c_0 \end{vmatrix} \tag{1.6.5}$$

Because of a flurry of activity about moment problems on $\partial\mathbb{D}$ unleashed by Carathéodory in 1907 (see the Notes to Section 1.3 of [399]), Toeplitz matrices were all the rage from 1910–1915, and Pólya, a young postdoc, conjectured in [352] that if $w > 0$ and in L^1, then

$$\lim_{n\to\infty} D_n(w)^{1/n} = \exp\biggl(\int \log(w(\theta))\, \frac{d\theta}{2\pi}\biggr) \tag{1.6.6}$$

In a visit back to his native Budapest, Pólya mentioned this conjecture to Szegő, then an undergraduate, and he proved the theorem below, published in 1915 [428]. At the time, Szegő was nineteen, and when the paper was published, he was serving in the Austrian Army in World War I! Here is the first version of Szegő's theorem:

Theorem 1.6.1 (Szegő's Theorem). *If* $w(\theta) \geq 0$ *and*

$$\int w(\theta)\, \frac{d\theta}{2\pi} < \infty \tag{1.6.7}$$

then (1.6.6) *holds.*

Remarks. 1. Since $\log_+(x) \equiv \max(0, \log(x)) \leq x$ (i.e., $x \leq e^x$ for $x \geq 1$), (1.6.7) implies $\int \log_+(w(\theta))\frac{d\theta}{2\pi} < \infty$, so $\int \log(w(\theta))\frac{d\theta}{2\pi}$ is either convergent or $-\infty$. In the latter case, we interpret the right side of (1.6.6) as 0.

2. Szegő (following a suggestion of Fekete) actually proved a stronger result, namely, that

$$\frac{D_{n+1}}{D_n} \to \text{RHS of (1.6.6)} \tag{1.6.8}$$

Since $D_n^{1/n} = D_0^{1/n}(\prod_{j=0}^{n-1} \frac{D_{j+1}}{D_j})^{1/n}$, (1.6.8) implies (1.6.6).

This theorem (in an extended form) is the subject of Chapter 2 where it is proven. For now, it does not appear to have a spectral content—its transformation to that form is the subject of the next two sections. But we note (1.6.6) is an equality (sum rule) with something involving a measure on one side and something rather different on the other, so my assertion that there is a gem lurking nearby should not be too surprising.

Remarks and Historical Notes. Not only did Szegő find the leading term in the asymptotics of $\log(D_n)$ in 1915, but he found the second term [433] thirty-seven years later! The strongest form of this second-term asymptotics is the following:

Theorem 1.6.2 (Sharp Form of the Strong Szegő Theorem). *If c_k is given by* (1.6.4), *if $w(\theta) \geq 0$, if* (1.6.7) *holds, and if*

$$\widehat{L}_k = \int_0^{2\pi} e^{-ik\theta} \log(w(\theta)) \frac{d\theta}{2\pi} \tag{1.6.9}$$

then

$$\lim_{n\to\infty} \frac{D_n}{e^{(n+1)\widehat{L}_0}} = \exp\left(\sum_{k=1}^{\infty} k|\widehat{L}_k|^2 \right) \tag{1.6.10}$$

Chapter 6 of [399] has six different proofs of this theorem, due in this strong form to Ibragimov [203]. There is a seventh proof in Section 9.10 of [400]; see also [401]. In Section 1.12, we show this implies a gem. We note that there are no general terms in an asymptotic series—for nonvanishing analytic w's, after the first two terms, the error is $O(e^{-cn})$.

Since $\log(\det(A)) = \mathrm{Tr}(\log(A))$ for positive matrices, A, if $T_{n+1}(w)$ is the matrix whose determinant is in (1.1.5), then (1.6.6) can be rewritten:

$$\lim_{n\to\infty} \frac{1}{n} \mathrm{Tr}(\log(T_n(w))) = \int \log(w(\theta)) \frac{d\theta}{2\pi} \tag{1.6.11}$$

More generally, one can prove that (see Theorem 2.7.13 of [399])

Theorem 1.6.3. *If f is a continuous function on* $[0, \infty)$ *with* $\lim_{x\to\infty} f(x)/x = 0$, *then*

$$\lim_{n\to\infty} \frac{1}{n} \mathrm{Tr}(f(T_n(w))) = \int f(w(\theta)) \frac{d\theta}{2\pi} \tag{1.6.12}$$

We will focus on Szegő's theorem and its descendants within spectral theory, but it has given birth to many other children. In the period 1950–1970, it was a major theme in a program called Function Algebras looking at fairly abstract Banach algebras. This work is discussed in the Notes to Section 2.6 of [399].

1.7 OPUC AND SZEGŐ'S RESTATEMENT

In 1920, Szegő [430] revisited his theorem realizing, in part, that it could be restated in terms of orthogonal polynomials on the unit circle (OPUC). We will discuss that here. Another critical realization in that paper—rephrasing the theorem as a variational principle—will be discussed in Section 2.12. A third—asymptotics of OPUC—will appear in Section 2.9.

Let $\{f_j\}_{j=1}^N$ be a set of independent vectors in a Hilbert space. Let $\{g_j\}_{j=1}^N$ be the set obtained by unnormalized Gram–Schmidt, that is,

$$g_j = f_j + \sum_{k=1}^{j-1} h_{jk} f_k \tag{1.7.1}$$

and

$$\langle g_j, g_k \rangle = 0 \qquad \text{if } j \neq k \tag{1.7.2}$$

Let H be the $N \times N$ matrix

$$h_{jk} = \begin{cases} 1 & k = j \\ h_{jk} & k < j \\ 0 & k > j \end{cases} \tag{1.7.3}$$

Then the Gram matrices are clearly related by

$$G(f)_{jk} \equiv \langle f_j, f_k \rangle \qquad G(g)_{jk} \equiv \langle g_j, g_k \rangle \tag{1.7.4}$$

$$G(g) = H^* G(f) H \tag{1.7.5}$$

We thus conclude:

Theorem 1.7.1.

$$\det(G(f)) = \prod_{j=1}^{N} \|g_j\|^2 \tag{1.7.6}$$

Proof. By (1.7.3),

$$\det(H) = \det(H^*) = 1$$

and by (1.7.5),

$$\begin{aligned} \det(G(g)) &= |\det(H)|^2 \det(G(f)) \\ &= \det(G(f)) \end{aligned}$$

Since $G(g)_{jk} = \|g_j\|^2 \delta_{jk}$, (1.7.6) is immediate. □

Given a nontrivial measure, $d\mu$, on $\partial\mathbb{D}$, define the monic OPUC by

$$\Phi_n(z) = z^n + \text{ lower order} \tag{1.7.7}$$

$$\int \bar{z}^j \Phi_n(z)\, d\mu(z) = 0 \qquad j = 0, 1, \dots, n-1 \tag{1.7.8}$$

We will use $\Phi_n(z; d\mu)$ if we want the $d\mu$ dependence to be explicit.

Thus, if $f_j = z^{j-1}$, $j = 1, \dots, N$, and $g_j = \Phi_{j-1}$, $j = 1, \dots, N$, we see that f and g are related by (1.7.1)/(1.7.2). Recognizing $G(f)$ as the Toeplitz matrix with

$$c_{k-j} = \langle z^k, z^j \rangle = \int e^{-i(k-j)\theta}\, d\mu(\theta) \tag{1.7.9}$$

we obtain from (1.7.6) that

Corollary 1.7.2. *The* $(N+1) \times (N+1)$ *Toeplitz determinant,* $D_N(d\mu)$, *obeys*

$$D_N(d\mu) = \prod_{j=0}^{N} \|\Phi_j(z; d\mu)\|^2 \tag{1.7.10}$$

Szegő also realized a special feature of OPUC that makes Fekete's remark that D_{N+1}/D_N has the same limit as $D_N^{1/N}$ transparent, namely,

Proposition 1.7.3. *For each n,*

$$\|\Phi_n\| \le \|\Phi_{n-1}\| \tag{1.7.11}$$

Thus, $\lim_{n\to\infty} \|\Phi_n\|^2$ *exists and equals*

$$\lim_{N\to\infty} D_N(d\mu)^{1/N} \tag{1.7.12}$$

Proof. Since Φ_n is orthogonal to any polynomial of degree $n-1$, it minimizes $\{\|\Phi_n + g\| \mid \deg(g) \le n-1\}$. As a consequence,

$$\|\Phi_n\| = \min\{\|P_n\| \mid P_n(z) = z^n + \text{lower order}\}$$

Thus, since $z\Phi_{n-1}$ is monic and $|z| = 1$ on $\text{supp}(d\mu)$,

$$\|\Phi_n\| \le \|z\Phi_{n-1}\| = \|\Phi_{n-1}\|$$

proving (1.7.11). Since $\|\Phi_n\|$ is decreasing and positive, it has a limit and, of course, $(\|\Phi_0\|^2 \dots \|\Phi_n\|^2)^{1/n}$ then converges to $\lim \|\Phi_n\|^2$. $\square$

We thus see an equivalent form of Szegő's theorem, Theorem 1.6.1:

Theorem 1.7.4. (1.6.6) *is equivalent to*

$$\lim_{n\to\infty} \left\| \Phi_n\left(z, \frac{w(\theta)}{2\pi}\, d\theta\right)\right\|^2 = \exp\left(\int \log(w(\theta))\, \frac{d\theta}{2\pi}\right) \tag{1.7.13}$$

Remarks and Historical Notes. Szegő's great 1920–1921 paper [430] was the first systematic exploration of OPUC, although he had earlier discussed OPs on curves [429].

1.8 VERBLUNSKY'S FORM OF SZEGŐ'S THEOREM

In this section, we give the final reformulation of Szegő's theorem as a sum rule and see that it implies a gem of spectral theory. The first element we need is the recursion relation obeyed by the monic OPUC, $\Phi_n(z)$, that will give us the parameters of the direct problem.

We first define natural maps $\delta_n : L^2(\partial\mathbb{D}, d\mu)$ to itself by

$$(\delta_n f)(e^{i\theta}) = e^{in\theta}\, \overline{f(e^{i\theta})} \tag{1.8.1}$$

Proposition 1.8.1. (i) δ_n *is an anti-unitary map of* L^2 *to* L^2.

(ii) *If* π_n *is the orthogonal projection onto the span of* $\{z^j\}_{j=0}^{n-1}$, *then* δ_n *maps* $\text{Ran}(\pi_{n+1})$ *to itself. Indeed, if* $P \in \text{Ran}(\pi_{n+1})$, *then*

$$(\delta_n P)(z) = z^n\, \overline{P(1/\bar z)} \tag{1.8.2}$$

Equivalently, if

$$P(z) = \sum_{j=0}^{n} c_j z^j \tag{1.8.3}$$

then

$$(\delta_n P)(z) = \sum_{j=0}^{n} \bar{c}_{n-j}\, z^j \tag{1.8.4}$$

(iii) *If* $f \in \operatorname{Ran}(\pi_{n+1})$ *and* $f \perp \{z, z^2, \dots, z^n\}$, *then* f *is a multiple of* $\delta_n(\Phi_n)$.

Proof. (i) Multiplication by $e^{in\theta}$ is unitary and $f \to \bar{f}$ is anti-unitary.

(ii) Immediate from $\delta_n(z^j) = z^n \overline{z^j} = z^{n-j}$.

(iii) Since δ_n is anti-unitary and $\delta_n(z^j) = z^{n-j}$, $\delta_n f \perp \{1, z, \dots, z^{n-1}\}$, so $\delta_n f = c\Phi_n$, and thus $f = \delta_n^2(f) = \bar{c}\delta_n(\Phi_n)$. □

We now shift to standard, albeit unfortunate, notation and use Φ_n^* for $\delta_n(\Phi_n)$ and, more generally, P^* for $\delta_n(P)$. It is hoped that in context, the value of n is clear. But for $d\mu = \frac{d\theta}{2\pi}$, $\Phi_n(z) = z^n$, $\Phi_n^* = 1$, and $(\Phi_n^*)^* = z^n$ becomes $1^* = z^n$. The notation is awful but, as I said, standard.

Theorem 1.8.2 (Szegő Recursion Relations). *For any nontrivial measure* $d\mu$ *on* $\partial\mathbb{D}$, *there exist constants* $\alpha_n \in \mathbb{C}$ *so that*

$$\Phi_{n+1} = z\Phi_n - \bar{\alpha}_n \Phi_n^* \tag{1.8.5}$$

Proof. Since Φ_n and Φ_{n+1} are monic, $\Phi_{n+1} - z\Phi_n$ is a polynomial of degree n. Moreover, if $j = 1, \dots, n$,

$$\langle z^j, \Phi_{n+1} - z\Phi_n \rangle = \langle z^j, \Phi_{n+1} \rangle - \langle z^{j-1}, \Phi_n \rangle = 0$$

so, by (iii) of Proposition 1.8.1, there is α_n, so (1.8.5) holds. □

Applying * on $n+1$ degree polynomials, we obtain

$$\Phi_{n+1}^* = \Phi_n^* - \alpha_n z \Phi_n \tag{1.8.6}$$

The $\{\alpha_n\}_{n=0}^{\infty}$ are called *Verblunsky coefficients*. They are the analog of the Jacobi parameters for OPRL. The reason for the minus sign and complex conjugate will become clear later (see Theorem 2.5.2). Setting $z = 0$ in (1.8.5) and using the fact that Φ_n monic implies

$$\Phi_n^*(0) = 1 \tag{1.8.7}$$

we have

$$\alpha_n = -\overline{\Phi_{n+1}(0)} \tag{1.8.8}$$

The following is critical:

Theorem 1.8.3. *We have that*

$$\|\Phi_{n+1}\|^2 = (1 - |\alpha_n|^2)\|\Phi_n\|^2 \tag{1.8.9}$$

so that

$$|\alpha_n| < 1 \tag{1.8.10}$$

and if $\mu(\partial\mathbb{D}) = 1$, *then*

$$\|\Phi_n\|^2 = \prod_{j=0}^{n-1}(1 - |\alpha_j|^2) \tag{1.8.11}$$

Remark. Of course, more generally,

$$\|\Phi_n\|^2 = \left[\prod_{j=0}^{n-1}(1 - |\alpha_j|^2)\right]\mu(\partial\mathbb{D}) \tag{1.8.12}$$

Proof.

$$\begin{aligned}\|\Phi_n\|^2 &= \|z\Phi_n\|^2 = \|\Phi_{n+1} + \bar{\alpha}_n\Phi_n^*\|^2 \\ &= \|\Phi_{n+1}\|^2 + |\alpha_n|^2\|\Phi_n\|^2\end{aligned} \tag{1.8.13}$$

since $\|\Phi_n^*\| = \|\Phi_n\|$ and $\Phi_n^* \perp \Phi_{n+1}$. (1.8.13) implies (1.8.9). $\|\Phi_n\|^2 > 0$ implies (1.8.10) and (1.8.11) follows by induction. □

From (1.8.11) and (1.8.5)/(1.8.6), we obtain

$$\varphi_{n+1} = \rho_n^{-1}(z\varphi_n - \bar{\alpha}_n\varphi_n^*) \tag{1.8.14}$$

$$\varphi_{n+1}^* = \rho_n^{-1}(\varphi_n^* - \alpha_n z\varphi_n) \tag{1.8.15}$$

where

$$\rho_n = (1 - |\alpha_n|^2)^{1/2} \tag{1.8.16}$$

The same calculation that led to (1.8.13) implies

Theorem 1.8.4. *If* $\Phi_n(z_0) = 0$, *then* $|z_0| < 1$. *If* $\Phi_n^*(z_0) = 0$, *then* $|z_0| > 1$.

Proof. Since $|z_0| < 1 \Leftrightarrow |1/z_0| > 1$, the first sentence implies the second. If $\Phi_n(z_0) = 0$, let $P(z) = \Phi_n(z)/(z - z_0)$, which is a polynomial of degree $n - 1$, so orthogonal to Φ_n. Then

$$\begin{aligned}\|P\|^2 &= \|zP\|^2 = \|(z - z_0)P + z_0P\|^2 \\ &= \|\Phi_n + z_0P\|^2 \\ &= \|\Phi_n\|^2 + |z_0|^2\|P\|^2\end{aligned} \tag{1.8.17}$$

Since $\|\Phi_n\|^2 > 0$, $|z_0| < 1$. □

By Theorem 1.8.3, $d\mu \mapsto \{\alpha_n(d\mu)\}_{n=0}^\infty$ maps the nontrivial measure to $\mathbb{D}^\infty$. The following is fundamental to thinking of OPUC as a spectral problem:

Theorem 1.8.5 (Verblunsky's Theorem). *The map of* $d\mu \mapsto \{\alpha_n(d\mu)\}_{n=0}^\infty$ *is a one-one map of nontrivial probability measures onto* $\mathbb{D}^\infty$.

We will prove this in Section 2.5 (see Theorem 2.5.3); see also the Notes to this section. We can now state Verblunsky's form of Szegő's theorem; by (1.8.11), the limit on the left of (1.7.13) is just an infinite product:

Theorem 1.8.6 (Verblunsky's Form of Szegő's Theorem). *For any nontrivial probability measure $d\mu$ on $\partial\mathbb{D}$ with w given by* (1.6.3), *we have*

$$\prod_{n=0}^{\infty}(1-|\alpha_n|^2)=\exp\biggl(\int \log(w(\theta))\,\frac{d\theta}{2\pi}\biggr) \tag{1.8.18}$$

This is the version we will prove in Chapter 2; see Section 2.7. We note that it has two differences from Szegő's theorem, even the variant in Theorem 1.7.4. First, we have written it in terms of Verblunsky coefficients, and second, unlike Szegő's original version, this allows $d\mu_s \neq 0$. One has the remarkable fact that the left side of (1.8.18) is independent of $d\mu_s$!

(1.8.18) *always* holds, although both sides can be zero connected with a "divergent product" on the left and a diverging integral on the right. The two sides are nonzero at the same time, so we get the following gem:

Corollary 1.8.7. *For nontrivial probability measures $d\mu$ on $\partial\mathbb{D}$ obeying* (1.6.3),

$$\sum_{n=0}^{\infty}|\alpha_n|^2<\infty \Leftrightarrow \int \log(w(\theta))\,\frac{d\theta}{2\pi}>-\infty \tag{1.8.19}$$

Remarks and Historical Notes. The Szegő recursion, (1.8.5)/(1.8.6), appeared first in 1939 in his famous book on orthogonal polynomials [434]. But at roughly the same time, they appeared in work of Geronimus [156, 157]. The history is murky, but especially as their proofs and presentations are different, it seems like Geronimus' work was independent but several months later. Interestingly enough, an equivalent form was rediscovered by Levinson [277] about ten years later, and the engineering literature sometimes calls it the Levinson or Levinson–Szegő algorithm.

Five years before Szegő, the α_n appeared in work of Verblunsky in two remarkable papers [452, 453] that were mainly ignored for almost seventy years! Verblunsky did not define the α_n via a recursion relation, but in [452], he proved there were rational functions $\zeta_n(c_0, c_1, \dots, c_{n-1}; \bar{c}_0, \dots, \bar{c}_{n-1}) \in \mathbb{C}$ and $R_n(c_0, c_1, \dots, c_{n-1}; \bar{c}_0, \dots, \bar{c}_{n-1}) \in (0,\infty)$ so that if $\{c_j\}_{j=0}^{n-1}$ were moments of some nontrivial measure on $\partial\mathbb{D}$, then the allowed values of c_n for nontrivial measures were all the possible values in the open disk of radius R_n in $\mathbb{C}$ centered at ζ_n. He then defined α_{n-1} by

$$c_n=\zeta_n+\alpha_{n-1}R_n \tag{1.8.20}$$

This is discussed in Section 3.1 of [399]. Interestingly enough, the analog of this approach for OPRL was rediscovered by Krein [252], Karlin–Studden [213], and Krein–Nudel'man [253], and codified in a book by Dette–Studden [112] who included the analysis of OPUC, thus reinventing [452]!

Theorem 1.8.4 goes back to Szegő [430]. The proof we give is due to Landau [263]. [399] has six proofs of the theorem.

In [452], Verblunsky also proved Theorem 1.8.5 using his definition of $\{\alpha_n\}_{n=0}^{\infty}$. Other proofs of this theorem are presented in [399] and [398]. In particular, we mention the spectral theory proof, the analog of the proof of Favard's theorem that we gave in Section 1.3. Of course, for that we need an analog of Jacobi matrices. The proper analog, the CMV matrix, will be discussed in Section 2.11. It is due to Cantero, Moral, and Velázquez [70] but essentially was discovered earlier by Amar, Gragg, Reichel, and Watson (see [403]) as a tool in numerical matrix analysis. See Chapter 4 of [399] and [403] for further discussions. Before [399, 400] introduced "Verblunsky coefficient," the α_n's had a wide variety of names: reflection coefficient, Schur parameter, Szegő parameter, and Geronimus coefficient.

In [453], Verblunsky proved Theorem 1.8.6. In particular, he had the sum rule (1.8.18) and he had a proof that allowed a singular part of the measure. Much of the literature since has attributed this singular-part-allowed result to Kolmogorov and Krein, whose work was later and which only proved $\sum|\alpha_n|^2=\infty \Leftrightarrow \int \log(w(\theta))\frac{d\theta}{2\pi} = -\infty$ with a singular part allowed. Others attributed the general result to Geronimus or Szegő—again based on later work.

It is also true that KdV sum rules should be viewed as analogs of Verblunsky's sum rule, but the connection was not realized until many years later. Indeed, the Killip–Simon sum rules discussed in Section 1.10 were discovered in a chain going back to KdV sum rules without knowing of Verblunsky's work. It was in tracking down the history of (1.8.18) that we uncovered [452, 453].

One of the consequences of Corollary 1.8.7 is the existence of mixed spectrum consistent with ℓ^2 decay: Given any measure $d\rho_s$ with $\int d\rho_s < 1$, there is a measure with a.c. support all of $\partial\mathbb{D}$, that $d\rho_s$, and with $\sum_{j=0}^{\infty}|\alpha_j|^2 < \infty$. Not knowing of this, the existence of analogous mixed spectral results for Schrödinger operators was regarded as a significant problem around 2000.

1.9 BACK TO OPRL: SZEGŐ MAPPING AND THE SHOHAT–NEVAI THEOREM

We can translate the gem for OPUC to a result for OPRL using an interesting connection that Szegő found in 1922 [431, 434]. It is connected to the natural conformal bijection of $\mathbb{D} \to \mathbb{C} \cup \{\infty\} \setminus [-2, 2]$ by

$$z \to E = z + z^{-1} \tag{1.9.1}$$

This maps $\partial\mathbb{D}$ two-one to $[-2, 2]$ by

$$e^{i\theta} \xrightarrow{Q} 2\cos\theta \tag{1.9.2}$$

We can use this to map $C([-2, 2])$, the continuous functions on $[-2, 2]$, to $C(\partial\mathbb{D})$:

$$(Q_\sharp f)(e^{i\theta}) = f(Q(e^{i\theta})) = f(2\cos\theta) \tag{1.9.3}$$

Notice $\operatorname{Ran}(Q_\sharp)$ is exactly the set of all functions invariant under $e^{i\theta} \to e^{-i\theta}$. Duality then induces a map $Q_\sharp^*\colon \mathcal{M}_{+1,1}(\partial\mathbb{D}) \to \mathcal{M}_{+,1}([-2, 2])$ between the probability measures by

$$\int f(x)[Q_\sharp^*(d\mu)](x) = \int (Q_\sharp f)(e^{i\theta})\, d\mu(\theta) \tag{1.9.4}$$

$Q_\sharp^*$ is onto $\mathcal{M}_{+,1}([-2,2])$, but it is not one-one. For example, if f is any nonnegative L^1 function with $f(\theta) + f(2\pi - \theta) = 1$ and $\int f(\theta)\frac{d\theta}{2\pi} = 1$, then $Q_\sharp(f\frac{d\theta}{2\pi}) = Q_\sharp(\frac{d\theta}{2\pi}) = \pi^{-1}(4-x^2)^{-1/2}\,dx$. However, restricted to measures invariant under $\theta \to -\theta$, $Q_\sharp$ is one-one, and we denote its restriction to even measures by Sz for Szegő mapping. Thus, $d\rho = \mathrm{Sz}(d\mu)$ if and only if $d\mu(\theta) = d\mu(-\theta)$ and

$$\int f(\theta)\, d\mu(\theta) = \int f\left(\arccos\left(\frac{x}{2}\right)\right) d\rho(x) \tag{1.9.5}$$

for any f obeying $f(-\theta) = f(\theta)$. Sz is a bijection between nontrivial even probability measures on $\partial\mathbb{D}$ and nontrivial probability measures on $[-2,2]$.

Because of the impact of symmetry on Szegő recursion, we see

$$d\mu \text{ even} \Leftrightarrow \overline{\Phi_n(z)} = \Phi_n(\bar z) \Leftrightarrow \alpha_n \in \mathbb{R} \text{ for all } n \tag{1.9.6}$$

Szegő [431, 434] proved the following:

Theorem 1.9.1. *Let $d\rho = \mathrm{Sz}(d\mu)$ for nontrivial probability measures on $[-2,2]$ and $\partial\mathbb{D}$. Let P_n, p_n be the monic and orthonormal OPRL for $d\rho$ and Φ_n, φ_n the monic and orthonormal OPUC for $d\mu$. Then*

$$P_n\left(z + \frac{1}{z}\right) = [1 - \alpha_{2n-1}(d\mu)]^{-1} z^{-n} [\Phi_{2n}(z) + \Phi_{2n}^*(z)] \tag{1.9.7}$$

$$\|P_n\|_{L^2(d\rho)}^2 = 2(1 - \alpha_{2n-1})^{-1} \|\Phi_{2n}\|_{L^2(d\mu)}^* \tag{1.9.8}$$

$$p_n\left(z + \frac{1}{z}\right) = [2(1 - \alpha_{2n-1})]^{-1/2} z^{-n} (\varphi_{2n}(z) + \varphi_{2n}^*(z)) \tag{1.9.9}$$

Sketch. (For details, see Theorem 13.1.5 of [400].) The right side of (1.9.7) is a Laurent polynomial of the form $\sum_{j=-n}^{n} c_j z^j$ invariant under $z \to \frac{1}{z}$ on account of (1.9.6). Every such Laurent polynomial has the form $Q_n(z + \frac{1}{z})$ for $Q_n(\cdot)$ of degree n.

Since $\Phi_{2n}(0) = -\bar\alpha_{2n-1}$, $\Phi_{2n}^*(z) = -\alpha_{2n-1} z^{2n} + \cdots$, so Q_n is monic. Moreover, by (1.9.5) for $\ell < n$,

$$\int Q_n(x) Q_\ell(x)\, d\rho(x) = \int \overline{\Phi_{2n} + \Phi_{2n}^*}\,(z)\, z^{n-\ell}(\Phi_{2\ell} + \Phi_{2\ell}^*)\, d\mu(z) \tag{1.9.10}$$

$$= 0$$

since $\Phi_{2n} \perp \{z, \dots, z^{2n-1}\}$ and $\Phi_{2n}^* \perp \{z, \dots, z^{2n-1}\}$. Thus, the Q_n's are the monic OPRL for $d\rho$, that is, we have proven (1.9.7).

(1.9.8) follows from (1.9.7) and

$$\begin{aligned}\langle \Phi_{2n}, \Phi_{2n}^* \rangle &= \langle \Phi_{2n}, \Phi_{2n-1}^* - \alpha_{2n-1} z \Phi_{2n-1} \rangle \\ &= -\alpha_{2n-1} \langle \Phi_{2n}, \Phi_{2n} + \bar\alpha_{2n-1} \Phi_{2n-1}^* \rangle \\ &= -\alpha_{2n-1} \|\Phi_{2n}\|^2 \end{aligned} \tag{1.9.11}$$

by using Szegő recursion and orthogonality. (1.9.9) is immediate from (1.9.7) and (1.9.8). □

There are several other relations we want to note because we will need them in Section 3.11. First, (1.9.9) can be written

$$p_n\left(z+\frac{1}{z}\right) = [2(1-\alpha_{2n-1})]^{-1/2}\left(z^{-n}\varphi_{2n}(z) + z^n\varphi_{2n}\left(\frac{1}{z}\right)\right) \tag{1.9.12}$$

By the same method, one can see

$$p_n\left(z+\frac{1}{z}\right) = [2(1+\alpha_{2n-1})]^{-1/2}\left(z^{-(n-1)}\varphi_{2n-1}(z) + z^{(n-1)}\varphi_{2n-1}\left(\frac{1}{z}\right)\right) \tag{1.9.13}$$

Besides $d\rho = \mathrm{Sz}(d\mu)$, there is a second (nonprobability) measure one can associate to $d\mu$, namely,

$$d\rho_1(x) \equiv \mathrm{Sz}_1(d\mu)(x) = \tfrac{1}{4}(4-x^2)\,d\rho(x) \tag{1.9.14}$$

Its orthonormal polynomials are denoted by $q_n(x)$. As with the derivation of (1.9.9), one finds

$$\tfrac{1}{2}q_{n-1}\left(z+\frac{1}{z}\right) = [2(1+\alpha_{2n-1})]^{-1/2}\left(\frac{z^{-n}\varphi_{2n}(z) - z_n\varphi_{2n}(\frac{1}{z})}{z-z^{-1}}\right) \tag{1.9.15}$$

$$= [2(1-\alpha_{2n-1})]^{-1/2}\left(\frac{z^{-(n-1)}\varphi_{2n-1}(z) - z^{(n-1)}\varphi_{2n-1}(\frac{1}{z})}{z-z^{-1}}\right) \tag{1.9.16}$$

This leads to

$$z^{-n}\varphi_{2n}(z) = \left[\tfrac{1}{2}(1-\alpha_{2n-1})\right]^{1/2} p_n\left(z+\frac{1}{z}\right) + \left[\tfrac{1}{2}(1+\alpha_{2n-1})\right]^{1/2}\left(\frac{z-z^{-1}}{2}\right)q_{n-1}\left(z+\frac{1}{z}\right) \tag{1.9.17}$$

$$z^{-(n-1)}\varphi_{2n-1}(z) = \left[\tfrac{1}{2}(1+\alpha_{2n-1})\right]^{1/2} p_n\left(z+\frac{1}{z}\right) + \left[\tfrac{1}{2}(1-\alpha_{2n-1})\right]^{1/2}\left(\frac{z-z^{-1}}{2}\right)q_{n-1}\left(z+\frac{1}{z}\right) \tag{1.9.18}$$

When $z = e^{i\theta}$, $p_n(2\cos\theta)$ and $q_{n-1}(2\cos\theta)$ are real, but $(z-z^{-1})/2 = i\sin\theta$ is pure imaginary, so the absolute value square has no cross-term. Thus, we find the formula we will need in Section 3.11

$$|\varphi_{2n}(e^{i\theta})|^2 + |\varphi_{2n-1}(e^{i\theta})|^2 = |p_n(2\cos\theta)|^2 + \sin^2\theta|q_{n-1}(2\cos\theta)|^2 \tag{1.9.19}$$

where we used $([\frac{1}{2}(1+\alpha_{2n-1})]^{1/2})^2 + ([\frac{1}{2}(1-\alpha_{2n-1})]^{1/2})^2 = 1$ to miraculously have α_{2n-1} drop out!

From Theorem 1.9.1, we get the formula relating a_n, b_n and α_n:

Theorem 1.9.2 (Direct Geronimus Relations). *Let $d\rho = \text{Sz}(d\mu)$ for nontrivial probability measures on $[-2,2]$ and $\partial\mathbb{D}$. Let $\{a_n, b_n\}_{n=1}^\infty$ be the Jacobi parameters for $d\rho$ and $\{\alpha_n\}_{n=0}^\infty$ the Verblunsky coefficients for $d\mu$. Then*

$$\text{(i)} \qquad (a_1 \dots a_n)^2 = 2(1+\alpha_{2n-1}) \prod_{j=0}^{2n-2} (1-\alpha_j^2) \tag{1.9.20}$$

$$\text{(ii)} \qquad a_{n+1}^2 = (1+\alpha_{2n+1})(1-\alpha_{2n}^2)(1-\alpha_{2n-1}) \tag{1.9.21}$$

$$\text{(iii)} \qquad b_{n+1} = (1-\alpha_{2n-1})\alpha_{2n} - (1+\alpha_{2n-1})\alpha_{2n-2} \tag{1.9.22}$$

Remark. (i) holds for $n \geq 1$ and (ii)/(iii) for $n \geq 0$. For $n = 1$, (1.9.20) says $a_1^2 = 2(1+\alpha_1)(1-\alpha_0^2)$, so (1.9.21) holds for $n = 1$ if we define

$$\alpha_{-1} = -1 \tag{1.9.23}$$

While α_{-2} enters in (1.9.22) for $n = 0$, it is multiplied by $(1+\alpha_{-1}) = 0$, so only the "boundary condition" (1.9.23) is needed.

Sketch. (For details, see Theorems 13.1.7 and 13.1.12 of [400].)

(i) Since

$$\frac{1-\alpha_{2n-1}^2}{1-\alpha_{2n-1}} = 1 + \alpha_{2n-1} \tag{1.9.24}$$

this is a rewriting of (1.9.8) using (1.8.11) and (1.2.13).

(ii) This follows from dividing (i) for $n+1$ by (i) for n using (1.9.24).

(iii) This comes from (1.9.7) looking at the $O(z^{n-1})$ terms. By a simple induction from (1.2.8),

$$P_n(x) = x^n - \left(\sum_{j=1}^n b_j\right) x^{n-1} + O(x^{n-2}) \tag{1.9.25}$$

From (1.8.5) and (1.8.6), we get that if

$$\Phi_n(z) = z^n + C_n z^{n-1} + O(z^{n-2}) \tag{1.9.26}$$

$$\Phi_n^*(z) = -\alpha_{n-1} z^n + D_n z^{n-1} + O(z^{n-2}) \tag{1.9.27}$$

then, by induction,

$$C_n = \sum_{j=0}^{n-1} \bar{\alpha}_j \alpha_{j-1} \tag{1.9.28}$$

(where, as usual, $\alpha_{-1} = -1$) and

$$D_n = -\alpha_{n-2} - \alpha_{n-1} C_{n-1} \tag{1.9.29}$$

These formulae and (1.9.7) imply that

$$-\sum_{j=1}^{n} b_j = C_{2n-1} - \alpha_{n-2} \tag{1.9.30}$$

and this yields (1.9.22). □

This lets us "translate" Corollary 1.8.7 to OPRL:

Theorem 1.9.3 (Shohat–Nevai Theorem). *Let*

$$d\rho(x) = f(x)\,dx + d\rho_{\mathrm{s}}(x)$$

be supported on $[-2, 2]$. *Then*

$$\int_{-2}^{2} (4 - x^2)^{-1/2} \log(f(x))\,dx > -\infty \tag{1.9.31}$$

if and only if

$$\limsup a_1 \dots a_n > 0 \tag{1.9.32}$$

If these conditions hold, then

$$\lim a_1 \dots a_n \tag{1.9.33}$$

exists in $(0, \infty)$ *and*

$$\sum_{n=1}^{\infty} (a_n - 1)^2 + b_n^2 < \infty \tag{1.9.34}$$

and

$$\sum_{n=1}^{N} (a_n - 1) \qquad \textit{and} \qquad \sum_{n=1}^{N} b_n \tag{1.9.35}$$

have limits in $(-\infty, \infty)$.

Remarks. 1. We emphasize (1.9.32) is lim sup, that is, it allows lim inf to be 0 so long as some subsequence stays away from 0.

2. This can be rephrased as saying $a_1 \dots a_n$ always has a limit when $\mathrm{supp}(d\rho) \subset [-2, 2]$ since the negation of (1.9.32) is $\lim a_1 \dots a_n = 0$. This is discussed further in Section 3.6.

Proof. Let μ be defined by $\mathrm{Sz}(d\mu) = d\rho$. By (1.9.30),

$$(a_1 \dots a_n)^2 \leq 4 \prod_{j=0}^{2n-2} (1 - \alpha_j^2)$$

so (1.9.32) implies $\lim \prod_{j=0}^{\infty}(1 - \alpha_j^2)$ (the limit always exists) is strictly positive and thus, $\sum_{j=0}^{\infty} \alpha_j^2 < \infty$. Conversely, if $\sum_j \alpha_j^2 < \infty$, then $\alpha_j \to 0$ and so, by (1.9.20), $\lim a_1 \dots a_n$ exists in $(0, \infty)$. We have thus proven that

$$(1.9.32) \Rightarrow \sum_{j=0}^{\infty} \alpha_j^2 < \infty \Rightarrow \lim a_1 \dots a_n \text{ exists in } (0, \infty) \tag{1.9.36}$$

On the other hand, if

$$d\mu = w(\theta)\,\frac{d\theta}{2\pi} + d\mu_s \tag{1.9.37}$$

then, by (1.9.5),

$$w(\theta) = 2\pi\,|\sin\theta|\,f(2\cos\theta) \tag{1.9.38}$$

It follows that (changing variables, using $x = 2\cos\theta \Rightarrow dx = 2\sin\theta\, d\theta$ or $d\theta = (4-x^2)^{-1/2}\,dx$)

$$\int \log(w(\theta))\,\frac{d\theta}{2\pi} > -\infty \Leftrightarrow \int \log(f(x))(4-x^2)^{-1/2}\,dx > -\infty \tag{1.9.39}$$

Thus, (1.8.19), (1.9.36), and (1.9.39) imply

$$\limsup(a_1 \dots a_n) > 0 \Leftrightarrow (1.9.31)$$

and if this holds, then (1.9.33) has a limit.

Since b_{n+1} and $a_{n+1}^2 - 1$ are functions of α_{2n+j} $(j = -2, -1, 0, 1)$, we see that if $\sum_{j=0}^\infty \alpha_j^2 < \infty$, then $\sum b_n^2 < \infty$ and $\sum (a_n^2 - 1)^2 < \infty$. Since $(a_n + 1) \geq 1$, $(a_n - 1)^2 = (a_n^2 - 1)^2/(a_n + 1)^2 \leq (a_n^2 - 1)^2$, so (1.9.34) holds.

Finally, when $\sum_{j=0}^\infty \alpha_j^2 < \infty$, $a_{n+1}^2 - 1$ and b_{n+1} are the sum of an L^1 sequence and a telescoping sequence, so $a_{n+1}^2 - 1$ and b_{n+1} are summable. Since $(a_j^2 - 1) - 2(a_j - 1) = (a_j - 1)^2$ is summable, we see that so is $a_{n+1} - 1$. □

We want to emphasize that while Corollary 1.8.7, on which Theorem 1.9.3 is based, is a gem (equivalence of purely spectral condition to purely sufficient condition), Theorem 1.9.3 is not. For it makes the a priori condition that $\operatorname{supp}(d\rho) \subset [-2, 2]$, that is, it is the equivalence of

$$(1.9.31) + \operatorname{supp}(d\rho) \subset [-2, 2] \tag{1.9.40}$$

to

$$(1.9.32) + \operatorname{supp}(d\rho) \subset [-2, 2] \tag{1.9.41}$$

(1.9.40) is purely spectral, but (1.9.41) is not a condition only about the Jacobi parameters. Indeed, $\operatorname{supp}(d\rho) \subset [-2, 2]$ is a very strong restriction if $\limsup(a_1 \dots a_n) > 0$. Indeed, it implies strong conditions on the b_n's ($\sum_{n=1}^\infty b_n^2 < \infty$ and $\sum_{n=1}^N b_n$ conditionally convergent).

Remarks and Historical Notes. The Szegő mapping was introduced by Szegő in [431] and further discussed by him in [434]. Its purpose was to carry over asymptotics of OPUC when the Szegő condition holds to asymptotics of OPRL when the OPRL Szegő condition holds (see Section 3.7).

$d\mu$ and $d\rho = \operatorname{Sz}(d\mu)$ can be related via their natural transforms

$$F(z) = \int \frac{e^{i\theta} + z}{e^{i\theta} - z}\,d\mu(\theta) \qquad m(z) = \int \frac{d\rho(x)}{x - z} \tag{1.9.42}$$

namely,

$$F(z) = 2(z - z^{-1})m(z + z^{-1}) \tag{1.9.43}$$

This formula is from Geronimus [159]; see also the proof of Theorem 13.1.2 in [400].

The map $z \to E = z + z^{-1}$ may seem miraculous, but it is canonical and uniquely determined. By the Riemann mapping theorem, there is an analytic bijection, g, of $\mathbb{D}$ to $\mathbb{C} \cup \{\infty\} \setminus [-2, 2]$ and it is uniquely determined by $g(0) = \infty$ and $\lim_{z\to 0} zg(z) > 0$. This unique map, abstractly guaranteed, is $g(z) = z + z^{-1}$. This will become a major theme in Chapter 9.

Geronimus [159, 160] found the relations (1.9.21)/(1.9.22). Other proofs can be found in Damanik–Killip [96], Killip–Nenciu [223], and Faybusovich–Gekhtman [129]. The latter two proofs are discussed in Section 13.2 of [400] and in Section 13.3 of the expected second edition of [400], which is posted online at `http://www.math.caltech.edu/opuc/newsection13-3.pdf`.

Szegő found a second natural map on nontrivial symmetric probability measures on $\partial\mathbb{D}$ to a large subset of measures on $[-2, 2]$, the map we called Sz_1 in (1.9.14). There are, in fact, four natural maps discussed in Section 13.2 of [400] and references therein. We note that all the original papers prior to 2000 use $[-1, 1]$ not $[-2, 2]$, and $z \to \frac{1}{2}(z + z^{-1})$. [400] discusses normalized measures (one needs to multiply $d\rho_1$ by $2[(1 - |\alpha_0|^2)(1 - \alpha_1)]^{-1}$ to normalize). For our purposes in Section 3.11, the unnormalized measure that leads to (1.9.19) is more convenient.

Szegő's book [434] includes (1.9.12)–(1.9.15) (in Section 11.5) and he noted their inverses (in Section 6 of his appendix). The compact consequence in (1.9.19) is from Máté–Nevai–Totik [302].

It is interesting to check these formulae in case $d\mu = \frac{d\theta}{2\pi}$. Then

$$\mathrm{Sz}(d\mu)(x) = \frac{1}{\pi}\frac{1}{\sqrt{4 - x^2}}\,dx \tag{1.9.44}$$

$$= \frac{1}{\pi}\,d\left(\arccos\left(\frac{x}{2}\right)\right) \tag{1.9.45}$$

and (Chebyshev polynomials of the first and second kinds)

$$p_n(2\cos\theta) = \sqrt{2}\,\cos(n\theta) \tag{1.9.46}$$

$$q_n(2\cos\theta) = \sqrt{2}\,\frac{\sin((n+1)\theta)}{\sin\theta} \tag{1.9.47}$$

$\alpha_{2n-1} = 0$ and, for example, (1.9.18) says

$$e^{-in\theta}e^{2ni\theta} = \frac{1}{\sqrt{2}}\sqrt{2}\,\cos(n\theta) + \frac{1}{\sqrt{2}}\,i\sin\theta\sqrt{2}\,\frac{\sin(n\theta)}{\sin\theta} \tag{1.9.48}$$

Theorem 1.9.3 first appeared in Nevai [320] using in part ideas in Shohat [384].

We will eventually see (Theorem 3.6.1) that Theorem 1.9.3 can be extended to situations where there is some point spectrum outside $[-2, 2]$, namely, we will need $\sigma_{\mathrm{ess}}(d\mu) = [-2, 2]$ and

$$\sum_{\substack{E\in\mathrm{supp}(d\mu)\\ E\notin[-2,2]}} \mathrm{dist}(E, \sigma_{\mathrm{ess}}(d\mu))^{1/2} < \infty \tag{1.9.49}$$

1.10 THE KILLIP–SIMON THEOREM

As we noted, Theorem 1.9.3 is a spectral result about OPRL related to Szegő's theorem, but not a gem as we defined it. Here is an OPRL gem that is related to Szegő's theorem.

It will involve the free Jacobi matrix, J_0, whose Jacobi parameters are

$$a_n \equiv 1 \qquad b_n \equiv 0 \tag{1.10.1}$$

The OPs for this case are (as is easy to check obey the recursion relations on account of trigonometric addition formulae; these are essentially the Chebyshev polynomials of the second kind; see (1.2.35))

$$P_n(2\cos\theta) = \frac{\sin(n+1)\theta}{\sin\theta} \tag{1.10.2}$$

The spectral measure is

$$d\rho_0(x) = \frac{1}{2\pi}(4-x^2)^{1/2}\,dx \tag{1.10.3}$$

so that

$$\sigma(J_0) = \sigma_{\text{ess}}(J_0) = \sigma_{\text{ac}}(J_0) = [-2,2] \tag{1.10.4}$$

Theorem 1.10.1 (Killip–Simon Theorem). *Let $\{a_n, b_n\}_{n=1}^\infty$ be the Jacobi parameters of a Jacobi matrix, J. Then*

$$\sum_{n=1}^\infty (a_n-1)^2 + b_n^2 < \infty \tag{1.10.5}$$

if and only if

(a)

$$\sigma_{\text{ess}}(J) = \sigma_{\text{ess}}(J_0) \qquad \textit{(Blumenthal–Weyl)} \tag{1.10.6}$$

(b) *The eigenvalues $E_n \notin \sigma_{\text{ess}}(J_0)$ obey*

$$\sum_{n=1}^\infty \text{dist}(E_n, \sigma_{\text{ess}}(J_0))^{3/2} < \infty \qquad \textit{(Lieb–Thirring)} \tag{1.10.7}$$

(c) *The function f of* (1.4.3) *obeys*

$$\int_{\sigma(J_0)} \text{dist}(x, \mathbb{R}\setminus\sigma(J_0))^{1/2}\log(f(x))\,dx > -\infty \qquad \textit{(Quasi-Szegő)} \tag{1.10.8}$$

Remarks. 1. (1.10.5) is equivalent to $J - J_0$ being a Hilbert–Schmidt operator (see [170, 381]).

2. (1.10.8) is called "quasi-Szegő" because it looks like the Szegő condition (1.9.30) except $-\frac{1}{2}$ has become $\frac{1}{2}$, allowing a larger class of f's. Similarly, (1.10.7) looks like (1.9.49) except that $\frac{1}{2}$ has become $\frac{3}{2}$.

The proof of Theorem 1.10.1 will be the main topic of Chapter 3, but to set the stage we want to say something about it. As with Szegő's theorem, the key is a sum rule. It will involve two somewhat complicated-looking functions, F defined on $\mathbb{R}\setminus[-2,2]$ and G on $(0,\infty)$:

$$F(\beta+\beta^{-1}) = \tfrac{1}{4}[\beta^2-\beta^{-2}-\log(\beta^4)] \qquad \beta\in\mathbb{R}\setminus[-1,1] \tag{1.10.9}$$

$$G(a) = a^2-1-\log(a^2) \tag{1.10.10}$$

Notice that $\beta\mapsto\beta+\beta^{-1}$ is a bijection of $\mathbb{R}\setminus[-1,1]$ to $\mathbb{R}\setminus[-2,2]$ so (1.10.9) defines F.

We will eventually show that (Lemma 3.5.3)

$$F(E) = \tfrac{1}{2}\int_2^{|E|}(E^2-4)^{1/2}\,dE \tag{1.10.11}$$

which implies

$$F(E)>0 \qquad \text{on } \mathbb{R}\setminus[-2,2] \tag{1.10.12}$$

and

$$F(E) = \tfrac{2}{3}\,(|E|-2)^{3/2} + O((|E|-2)^{5/2}) \tag{1.10.13}$$

We also see that (Lemma 3.5.2)

$$G(a)>0 \qquad \text{on } (0,\infty)\setminus\{1\} \tag{1.10.14}$$

$$G(a) = 2(a-1)^2 + O((a-1)^3) \tag{1.10.15}$$

We also need to define

$$Q(\rho) = \frac{1}{4\pi}\int_{-2}^{2}\log\biggl(\frac{\sqrt{4-x^2}}{2\pi f(x)}\biggr)\sqrt{4-x^2}\,dx \tag{1.10.16}$$

which, given (1.10.3), can be rewritten

$$Q(\rho) = -\tfrac{1}{2}\int\log\biggl[\biggl(\frac{d\rho}{d\rho_0}\biggr)^{-1}\biggr]\,d\rho_0 \tag{1.10.17}$$

whose integral is a relative entropy (see (2.2.1)). As we will show (Theorem 2.2.3), using Jensen's inequality, $Q(\rho)\geq 0$. The sum rule is

Theorem 1.10.2. *Let $d\rho$ be a nontrivial probability measure with associated Jacobi parameters $\{a_n,b_n\}_{n=1}^\infty$ and $\sigma_{\text{ess}}(d\rho)=[-2,2]$. Then*

$$Q(\rho)+\sum F(E_n) = \sum_{n=1}^{\infty}[\tfrac{1}{4}\,b_n^2+\tfrac{1}{2}\,G(a_n)] \tag{1.10.18}$$

This is called the P_2 sum rule. Notice that all terms on both sides are positive so the sums always make sense, but they may be infinite. Moreover, $\sigma_{\text{ess}}(d\rho)=[-2,2]$ and the left-hand side of (1.10.18) $<\infty$ if and only if (a)–(c) of Theorem 1.10.1 holds, on account of (1.10.13) and (1.10.16). On the other hand, using Theorem 1.4.1 and (1.10.15), $\sigma_{\text{ess}}(d\rho)=[-2,2]$ and the right-hand side of (1.10.18) $< \infty$ if and only if (1.10.5) holds. Thus, Theorem 1.10.2 implies Theorem 1.10.1.

Where will complicated objects like F and G come from? The sum rule of Verblunsky (1.8.18) is a form of Jensen's equality for analytic functions, hence the logs. In this case, the function is nonvanishing. The sum rule (1.10.18) will come from a Jensen–Poisson equality and involves two Taylor coefficients: the zeroth, which has logs, and the second without logs. There are terms from the zeros in this case, hence the logs in the sum involving F. These details will unfold in Chapter 3.

Remarks and Historical Notes. Theorems 1.10.1 and 1.10.2 are from Killip–Simon [225]. For historical context and the name "P_2," see the Notes to Sections 3.1 and 3.4.

1.11 PERTURBATIONS OF THE PERIODIC CASE

The material in Chapters 5, 6, and 8 is all connected with analyzing Szegő-like theorems for OPRL (and some related OPUC) where the $[-2, 2]$ of Theorem 1.10.1 is replaced by a union of a finite number of closed bounded intervals, especially the case of perturbations of periodic OPRL. Chapters 5 and 6 discuss periodic OPRL themselves, that is, Jacobi matrices, J_0, where

$$a_{n+p}^{(0)} = a_n^{(0)} \qquad b_{n+p}^{(0)} = b_n^{(0)} \tag{1.11.1}$$

for some $p \geq 2$ and all $n = 1, 2, \ldots$. (In Section 5.14, we also discuss OPUC when $\alpha_{n+p}^{(0)} = \alpha_n^{(0)}$, mainly with p even.) Rather than studying a_n, b_n, which approach $a_n \equiv 1, b_n \equiv 0$ in some sense, we want to discuss approach to J_0. J_0 is obviously parametrized by $\mathbb{R}^{2p} = \{(a_n^{(0)}, b_n^{(0)})_{n=1}^p\}$.

We begin the discussion by describing $\sigma(J_0)$, the spectrum of J_0 (see Sections 5.2, 5.3, and 5.4):

Theorem 1.11.1. *$\sigma_{\text{ess}}(J_0)$ is the disjoint union of $k + 1 \leq p$ distinct bounded intervals*

$$\sigma_{\text{ess}}(J_0) = \bigcup_{j=1}^{k+1} [c_j, d_j] \tag{1.11.2}$$

where

$$c_1 < d_1 < c_2 < \cdots < c_{k+1} < d_{k+1}$$

Each of the k gaps (d_j, c_{j+1}), $j = 1, \ldots, k$, has zero or one point mass.

Generically, $k = p - 1$. Indeed, $\{(a_n^{(0)}, b_n^{(0)}) \mid k < p - 1\}$ is a variety of codimension 2 in $\mathbb{R}^{2p}$. If $k = p - 1$, we say "all gaps are open."

While we will not say a lot about the proof now, we do want to mention one of the key tools. There is a natural polynomial in x, $\Delta(x; \{a_n^{(0)}, b_n^{(0)}\}_{n=1}^p) = \Delta(x; J_0)$ of exact degree p, so

$$\sigma_{\text{ess}}(J_0) = \Delta^{-1}([-2, 2]) \tag{1.11.3}$$

We are interested in the analog Theorem 1.10.1 when J_0 is a periodic Jacobi matrix. The conjectured analog of the spectral side is obvious: (1.10.6)–(1.10.8) were carefully stated in terms of $\sigma_{\rm ess}(J_0)$ rather than $[-2, 2]$ precisely because they will be one side of the proper periodic theorem.

There is an obvious guess for an analog of (1.10.5), namely,

$$\sum_{n=1}^{\infty}(a_n - a_n^{(0)})^2 + (b_n - b_n^{(0)})^2 < \infty \tag{1.11.4}$$

This cannot be right for the following reason. The map

$$J_1 = \{(a_n^{(1)}, b_n^{(1)}) \mid a_{n+p}^{(1)} = a_n^{(1)},\ b_{n+p}^{(1)} = b_n^{(1)}\} \to \Delta(x, J_1) \tag{1.11.5}$$

is a map of $\mathbb{R}^{2p}$ to $\mathbb{R}^{p+1}$, since Δ has $p+1$ coefficients. As one would expect, generic inverse images of a fixed Δ are of dimension $2p - (p+1) = p - 1$. In fact, we will show (see Section 5.13):

Theorem 1.11.2. *For fixed periodic J_0, $\{J_1 \mid \Delta(x, J_1) = \Delta(x, J_0)\}$ is a torus of dimension k where*

$$k + 1 = \#\ \textit{of components of}\ \sigma_{\rm ess}(J_0) \tag{1.11.6}$$

This set is called the *isospectral torus* of J_0, which we denote $\mathcal{T}_{J_0}$. By (1.11.3), if $J_1 \in \mathcal{T}_{J_0}$, $\sigma_{\rm ess}(J_1) = \sigma_{\rm ess}(J_0)$, and so J_1 also obeys (1.10.6)–(1.10.8), but J_1 does not obey (1.11.4). What we need is not ℓ^2 approach to a fixed J_0 but rather to $\mathcal{T}_{J_0}$. We define

$$d_m((a_n, b_n)_{n=1}^{\infty}, (a_n', b_n')_{n=1}^{\infty}) = \sum_{j=m}^{\infty} e^{-|j-m|}[|a_j - a_j'| + |b_j - b_j'|] \tag{1.11.7}$$

which measures the distances of the tails from each other. We also define

$$d_m((a_n, b_n)_{n=1}^{\infty}, \mathcal{T}_{J_0}) = \min_{(a_n', b_n') \in \mathcal{T}_{J_0}} d_m((a, b), (a', b')) \tag{1.11.8}$$

It can happen that the minimizing (a', b') is m-dependent and that $d_m((a, b), \mathcal{T}_{J_0}) \to 0$ as $m \to \infty$ without $d_m((a, b), J_1) \to 0$ for any J_1 (although, by compactness of $\mathcal{T}_{J_0}$, there will be J_1 and a subsequence for which $d_{m_\ell}((a, b), J_1) \to 0$ as $\ell \to \infty$).

Damanik–Killip–Simon [97] have proven:

Theorem 1.11.3 (DKS [97]). *Let J_0 be a fixed periodic Jacobi matrix of period p with all gaps open (i.e., $k = p - 1$). Let J be another bounded Jacobi matrix with Jacobi parameters $(a_n, b_n)_{n=1}^{\infty}$. Then the following are equivalent:*

(a) (1.10.6), (1.10.7), *and* (1.10.8) *hold.*

(b)

$$\sum_{m=1}^{\infty} d_m((a, b), \mathcal{T}_{J_0})^2 < \infty \tag{1.11.9}$$

The proof of this theorem is the main goal of Chapter 8. A key tool will be the study of the matrix $\Delta(J; J_0)$, that is, the matrix obtained by placing J for x in the polynomial $\Delta(x; J_0)$. Since Δ has degree p, $\Delta(J)$ will be a matrix of band width $2p+1$, that is, p diagonals strictly above, p strictly below, and on the main diagonal. Such a matrix can be thought of as "tridiagonal" if we replace a's and b's by $p \times p$ blocks. We will prove a Killip–Simon theorem for such block Jacobi matrices in Chapter 4, and that will be a main tool in proving Theorem 1.11.3.

In the periodic case, $\sigma_{\text{ess}}(J_0)$ is a disjoint union, (1.11.2). But not every such union is $\sigma_{\text{ess}}(J_0)$ for some periodic J_0. Basically, there is a natural map (harmonic measure),

$$\mathcal{M}\colon \{c_1 < d_1 < c_2 < \cdots < d_{k+1}\} \to \left\{ (\theta_j)_{j=1}^{k+1} \,\middle|\, \theta_j > 0;\ \sum_{j=1}^{k+1} \theta_j = 1 \right\}$$

which is continuous and onto. The allowed $\sigma_{\text{ess}}(J_0)$ for periodic J_0's with all gaps open is $\mathcal{M}((c,d)) = (\frac{1}{p}, \dots, \frac{1}{p})$, and if we drop the demand that all gaps are open, then the range is the set of rational θ's.

For other finite band sets, $\sigma_{\text{ess}}(J_0)$ can be that set if we allow certain almost periodic J_0's. There is no Killip–Simon-type theorem known in this case, but one-half of a Shohat–Nevai-type theorem is known due to work of Akhiezer, Widom, Aptekarev, and Peherstorfer–Yuditskii. It will be the subject of Chapter 9. Chapter 10 will discuss Killip–Simon-like theorems for perturbations of the graph Laplacian on a Bethe–Cayley tree.

Remarks and Historical Notes. As noted, Theorem 1.11.3 is from Damanik–Killip–Simon [97]. Prior results and historical context are discussed in the Notes to Section 8.1. The history of results mentioned in the last paragraph are in the Notes to Section 9.13.

1.12 OTHER GEMS IN THE SPECTRAL THEORY OF OPUC

While gems are the leitmotif of this chapter, our choice of topics is motivated by looking at relatives of Szegő's theorem. We will see that in this section by mentioning some other gems for OPUC (the Notes discuss OPRL) that will not be discussed further. Here are three theorems in particular:

Theorem 1.12.1 (Baxter's Theorem). *Let μ be a probability measure on $\partial\mathbb{D}$ of the form* (1.6.3) *and let $\{\alpha_n\}_{n=1}^\infty$ be its Verblunsky coefficients. Then the following are equivalent:*

(i)

$$\sum_{n=0}^\infty |\alpha_n| < \infty \tag{1.12.1}$$

(ii) $d\mu_s = 0$,

$$\inf w(\theta) > 0 \tag{1.12.2}$$

$$\sum_{n=-\infty}^\infty |\widehat{w}_n| < \infty \tag{1.12.3}$$

where

$$\widehat{w}_n = \int e^{-in\theta} w(\theta) \frac{d\theta}{2\pi} \tag{1.12.4}$$

Remark. (1.12.3) implies w is continuous, so the inf in (1.12.2) is a min.

Theorem 1.12.2 (Ibragimov's Form of the Strong Szegő Theorem). *Let μ be a probability measure on $\partial\mathbb{D}$ of the form* (1.6.3) *and let $\{\alpha_n\}_{n=1}^\infty$ be its Verblunsky coefficients. Then the following are equivalent:*

(i)

$$\sum_{n=0}^{\infty} n|\alpha_n|^2 < \infty \tag{1.12.5}$$

(ii) $d\mu_s = 0$*, the Szegő condition* (1.8.19) *holds, and*

$$\sum_{n=1}^{\infty} n|\widehat{L}_n|^2 < \infty \tag{1.12.6}$$

where

$$\widehat{L}_n = \int e^{-in\theta} \log(w(\theta)) \frac{d\theta}{2\pi} \tag{1.12.7}$$

Theorem 1.12.3 (Nevai–Totik Theorem). *Let μ be a probability measure on $\partial\mathbb{D}$ of the form* (1.6.3) *and let $\{\alpha_n\}_{n=1}^\infty$ be its Verblunsky coefficients. Let $R > 1$. Then the following are equivalent:*

(i) $\limsup |\alpha_n|^{1/n} \leq R^{-1}$

(ii) $\mu_s = 0$ *and the Szegő function D, defined by* (2.9.14), *has $D^{-1}(z)$ analytic in* $\{z \mid |z| < R\}$.

There are two distinctions between these results and Szegő's theorem. These only involve μ's with $\mu_s = 0$ and with more rapid decay than just ℓ^2. If $\alpha_n \sim Cn^{-s}$; Szegő requires $s > \frac{1}{2}$, but these require $s > 1$ (and exponential decay in the case of the Nevai–Totik theorem).

Remarks and Historical Notes. Baxter's theorem is from Baxter [32] and is discussed in [399, Chapter 5]. Ibragimov's form is from Ibragimov [203] and related to Szegő's work on the second term in Toeplitz determinant asymptotics discussed in the Notes to Section 1.6 where references appear. The Nevai–Totik theorem is from Nevai–Totik [323] and discussed in [399, Chapter 7].

For analogs of Theorems 1.12.1 and 1.12.2 for OPRL, see Ryckman [375, 376]. For an OPRL analog of Theorem 1.12.3, see Damanik–Simon [100].

Chapter Two

Szegő's Theorem

> In algebra, when one says $a = b$, it is a tautology and so uninteresting; while in analysis, when one says $a = b$, it is two deep inequalities.
>
> —attributed to *S. Bochner*

> If one only proves $a = b$ by showing $a \leq b$ and $b \leq a$, one has not understood the true reason that $a = b$.
>
> —attributed to *E. Noether*

In this chapter we will prove Szegő's theorem in Verblunsky's form (Theorem 1.8.6). Our main thrust will be a proof that extends to the other situations we wish to discuss in later chapters. The Szegő case is simpler than these later ones because the underlying analytic functions have neither zeros nor poles in $\mathbb{D}$, so we will only need that if f is nonvanishing and analytic in $\mathbb{D}$ and $\log(f(z))$ is in some Hardy class H^p ($p \geq 1$), then $f(0) = \exp(\int \log(f(e^{i\theta}))\frac{d\theta}{2\pi})$. In later chapters, we have to use Blaschke products to accommodate poles and zeros that can occur.

Section 2.1 lays out the strategy of this approach. The last steps establish the sum rule by proving complementary inequalities. One inequality will depend on the realization of integrals involving logs as a relative entropy and semicontinuity properties of entropy—the subject of Section 2.2. Section 2.3 is a minicourse on functions on $\mathbb{D}$ and on $\mathbb{C}_+ = \{z \mid \operatorname{Im} f > 0\}$ relevant to spectral theory.

In Sections 2.4 and 2.5, we turn from generalities back to the specifics of OPUC. By discussing second kind polynomials and Weyl solutions, we can prove the basics, especially coefficient stripping, the relation between $d\mu$ and $d\mu^{(1)}$ defined by $\alpha_n(d\mu^{(1)}) = \alpha_{n+1}(d\mu)$. With those basics, in Section 2.6 we construct the function needed for Step 1 in our strategy, and then we implement this strategy in Section 2.7.

The next six sections are extensions and alternate approaches. Section 2.8 discusses higher-order Szegő theorems, Section 2.12 presents Szegő's variational approach to his theorem, three sections (2.9, 2.10, and 2.13) discuss asymptotics of OPUC and of Weyl solutions, and Section 2.11 has several additional topics.

In the last four sections, we study asymptotics of the CD kernel, a subject we return to in Sections 3.11, 3.12, and 5.11.

2.1 STATEMENT AND STRATEGY

Given a nontrivial probability measure on $\partial\mathbb{D}$,

$$d\mu(\theta) = w(\theta)\,\frac{d\theta}{2\pi} + d\mu_s(\theta) \tag{2.1.1}$$

with $d\mu_s$ singular, recall that we define monic OPUC, $\Phi_n(z)$, and orthonormal $\varphi_n(z) = \Phi_n(z)/\|\Phi_n\|$. Recall that the Verblunsky coefficients $\{\alpha_n(d\mu)\}_{n=0}^\infty$ are given by

$$\alpha_n = -\overline{\Phi_{n+1}(0)} \tag{2.1.2}$$

The Szegő dual $\Phi_n^*(z)$ is given by

$$\Phi_n^*(z) = z^n\,\overline{\Phi_n(1/\bar z)} \tag{2.1.3}$$

and the Szegő recursion relations by

$$\Phi_{n+1}(z) = z\Phi_n(z) - \bar\alpha_n\Phi_n^*(z) \tag{2.1.4}$$

$$\Phi_{n+1}^*(z) = \Phi_n^*(z) - \alpha_n z\Phi_n(z) \tag{2.1.5}$$

$$z\varphi_n(z) = \rho_n\varphi_{n+1}(z) + \bar\alpha_n\varphi_n^*(z) \tag{2.1.6}$$

$$\varphi_n^*(z) = \rho_n\varphi_{n+1}^*(z) + \alpha_n z\varphi_n(z) \tag{2.1.7}$$

where

$$\rho_n = (1-|\alpha_n|^2)^{1/2} \tag{2.1.8}$$

Moreover, if

$$\kappa_n = \prod_{j=0}^{n-1}\rho_j^{-1} \tag{2.1.9}$$

then

$$\|\Phi_n\| = \kappa_n^{-1} \qquad \varphi_n(z) = \kappa_n z^n + \text{lower order} \tag{2.1.10}$$

We discussed several variants of Szegő's theorem in the last chapter. In this chapter, our goal is to prove the following (which implies the others and has the gem, Corollary 1.8.7, as a consequence):

Theorem 2.1.1 (Verblunsky's Form of Szegő's Theorem). *For any nontrivial probability measure on $\partial\mathbb{D}$, we have that*

$$\prod_{n=0}^\infty(1-|\alpha_n|^2) = \exp\biggl(\int \log(w(\theta))\,\frac{d\theta}{2\pi}\biggr) \tag{2.1.11}$$

Recall $\int \log(w(\theta))\frac{d\theta}{2\pi}$ can only diverge to $-\infty$, in which case we interpret the right side as $e^{-\infty}=0$. The product $\prod_{n=1}^N(1-|\alpha_n|^2)$ is monotone decreasing in N, so the limit exists although it may be zero.

In this section, we describe the overall strategy that we will use. The first problem with (2.1.11) is how one can hope to prove it when $\int\log(w(\theta))\frac{d\theta}{2\pi} = -\infty$ where both sides are singular. Our strategy will be to find a result that is always finite and always holds.

Let $d\mu_1$ be the measure defined by dropping α_0 and shifting the other α's down, that is,

$$\alpha_j(d\mu_1) = \alpha_{j+1}(d\mu) \tag{2.1.12}$$

We call the process "coefficient stripping." Write

$$d\mu_1(\theta) = w_1(\theta)\,\frac{d\theta}{2\pi} + d\mu_{s,1}(\theta) \tag{2.1.13}$$

More generally, let $d\mu_N$ be given by

$$\alpha_j(d\mu_N) = \alpha_{j+N}(d\mu) \tag{2.1.14}$$

and

$$d\mu_N = w_N(\theta)\,\frac{d\theta}{2\pi} + d\mu_{s,N} \tag{2.1.15}$$

Formally, if (2.1.11) holds for $d\mu$ and $d\mu_1$ and we divide, we get what we will call the step-by-step sum rule

$$(1 - |\alpha_0|^2) = \exp\biggl(\int \biggl[\log \frac{w(\theta)}{w_1(\theta)}\biggr]\frac{d\theta}{2\pi}\biggr) \tag{2.1.16}$$

The key to our proof of (2.1.11) will be to prove that (2.1.16) is always true if suitably interpreted.

The phrase "if suitably interpreted" is needed because $w(\theta)$ and/or $w_1(\theta)$ may vanish on a set of positive measure. What we will prove is that there is a nonnegative function $g(\theta)$ so $\log(g(\theta)) \in \cap_{p<\infty} L^p(\partial\mathbb{D}, \frac{d\theta}{2\pi})$ so that

$$(1 - |\alpha_0|^2) = \exp\biggl(\int \log(g(\theta))\,\frac{d\theta}{2\pi}\biggr) \tag{2.1.17}$$

and for a.e. θ where $w(\theta) \neq 0$, we have $w_1(\theta)$ is also nonzero and $g(\theta) = w(\theta)/w_1(\theta)$. We will eventually keep track of this subtlety associated with zeros of $w(\theta)$, but for the rest of this section we will ignore it.

Suppose $d\mu$ is such that for some N,

$$\alpha_n(d\mu) = 0 \qquad \text{for } n \geq N \tag{2.1.18}$$

Then w_N, given by (2.1.14) and (2.1.15), is 1. Thus, iterating (2.1.16), we find

$$\prod_{j=0}^{N-1}(1 - |\alpha_j|^2) = \exp\biggl(\int \log(w(\theta))\frac{d\theta}{2\pi}\biggr) \tag{2.1.19}$$

This is (2.1.11) for μ's obeying (2.1.18). We note that while this special case of (2.1.11) follows easily from (2.1.16), we will also see (see the remark after Theorem 2.7.2) that it has a simple direct proof.

The next step is to approximate any $d\mu$ given by $d\mu^{(N)}$ defined by

$$\alpha_j(d\mu^{(N)}) = \begin{cases} 0 & j \geq N \\ \alpha_j(d\mu) & j < N \end{cases} \tag{2.1.20}$$

Be careful to distinguish $d\mu^{(N)}$ from $d\mu_N$, which we have defined by (2.1.14). $d\mu_N$ strips N α's off the "bottom" while $d\mu^{(N)}$ leaves the bottom N α's and sets the others to zero.

By a simple induction using $\int \Phi_j(e^{i\theta})\, d\mu(\theta) = \delta_{j0}$, $\{\Phi_j(z)\}_{j=0}^N$ determine the moments $\{c_j(d\mu)\}_{j=0}^N$ of (1.6.2). Thus,

$$c_j(d\mu^{(N)}) = c_j(d\mu) \qquad \text{if } j \leq N \tag{2.1.21}$$

and so

$$d\mu^{(N)} \xrightarrow{w} d\mu \tag{2.1.22}$$

By (2.1.19) for $d\mu^{(N)}$,

$$\prod_{j=0}^{N-1} (1 - |\alpha_j(d\mu)|^2) = \exp\biggl(\int \log(w^{(N)}(\theta))\, \frac{d\theta}{2\pi}\biggr) \tag{2.1.23}$$

Clearly, as $N \to \infty$, the left-hand side of (2.1.25) converges monotonically to the LHS of (2.1.11). To complete the proof of (2.1.11), we need only prove that

$$\lim_{N\to\infty} \int \log(w^{(N)}(\theta))\, \frac{d\theta}{2\pi} = \int \log(w(\theta))\, \frac{d\theta}{2\pi} \tag{2.1.24}$$

This is certainly true since it is equivalent to (2.1.11), but I know no direct proof. All that one gets from general principles is a semicontinuity. To state it, we shift to positive quantities:

$$\eta_j(d\mu) = -\log \rho_j(d\mu)^2 \tag{2.1.25}$$

$$N(d\mu) = -\int \log(w(\theta))\, \frac{d\theta}{2\pi} \tag{2.1.26}$$

Then (2.1.23) says

$$\sum_{j=0}^{N-1} \eta_j(d\mu) = N(d\mu^{(N)}) \tag{2.1.27}$$

The semicontinuity we will prove in the next section is

Theorem 2.1.2 (see Theorem 2.2.3). *Let $d\mu_\ell, d\mu$ be nontrivial probability measures on $\partial\mathbb{D}$ so that $d\mu_\ell \to d\mu$ weakly (in the dual topology defined by $C(\partial\mathbb{D})$). Then*

$$\liminf N(d\mu_\ell) \geq N(d\mu) \tag{2.1.28}$$

Thus, (2.1.27) implies

$$\sum_{j=0}^{\infty} \eta_j(d\mu) \geq N(d\mu) \tag{2.1.29}$$

Clearly, to prove (2.1.11), we need that

$$\sum_{j=0}^{\infty} \eta_j(d\mu) \leq N(d\mu) \tag{2.1.30}$$

If $\log(w) \notin L^1$, then $N(d\mu) = \infty$ and (2.1.30) is trivial.

If $N(d\mu) < \infty$, we can separate $\log(w)$ and $\log(w_1)$ in (2.1.16), which becomes

$$\eta_0(d\mu) + N(d\mu_1) \leq N(d\mu) \tag{2.1.31}$$

Iterating,

$$\sum_{j=0}^{N-1} \eta_j(d\mu) + N(d\mu_N) \leq N(d\mu) \tag{2.1.32}$$

Here positivity saves us! $N(d\mu_N) \geq 0$, so we have

$$\sum_{j=0}^{N} \eta_j(d\mu) \leq N(d\mu) \tag{2.1.33}$$

Taking $N \to \infty$, we get (2.1.30).

This completes our sketch of the proof of Theorem 2.1.1. To summarize, the steps involved (which will reappear in Chapters 3, 4, and 9) are:

(1) Prove a step-by-step sum rule with positive terms from some kind of Jensen equality.
(2) Get a sum rule for cases of "compact support spectral data" by iterating the step-by-step sum rule N times.
(3) Get an inequality by going to infinity using semicontinuity of an entropy.
(4) Get the opposite inequality using positivity and the step-by-step sum rule.

Remarks and Historical Notes. The basic strategy here was invented to prove the Killip–Simon theorem by them [225] and honed by Simon–Zlatoš [410] and Simon [396]. Parts of it appear applied to the Szegő theorem in Chapter 2 of [399]. There is some overlap with ideas in Verblunsky's proof [453].

Theorem 2.1.1 implies that

$$\sum_{n=0}^{\infty} |\alpha_n|^2 < \infty \Rightarrow |\mathbb{D} \setminus \{\theta \mid w(\theta) > 0\}| = 0 \tag{2.1.34}$$

where $|\cdot|$ is Lebesgue measure. This is an optimal result in the sense that for any $p > 2$, there are measures μ, which are purely singular but

$$\sum_{n=0}^{\infty} |\alpha_n|^p < \infty \tag{2.1.35}$$

All constructions of such measures have some subtlety but there are many such constructions at this point such as:

(i) A method, dubbed Totik's workshop in [399, Section 2.10], due to Totik [442] that shows for any measure, γ, with $\operatorname{supp}(\gamma) = \mathbb{D}$, there is μ mutually equivalent to γ so (2.1.35) holds for all $p > 2$.
(ii) Using Riesz products, Khrushchev [220] constructed singular continuous measures with (2.1.35) for all $p > 2$; see [399, Section 2.11].
(iii) As discussed in [400, Section 12.7], if $\{\alpha_j(\omega)\}_{j=0}^{\infty}$ are independent random variables with $\mathbb{E}(\alpha_j(\omega)) = \mathbb{E}(\alpha_j(\omega)^2) = 0$, $\sup_{\omega,j}|\alpha_j(\omega)| < 1$, $\sup_{\omega}|\alpha_j(\omega)| \to 0$, and for $\Gamma > 0$,

$$\mathbb{E}(|\alpha_j(\omega)|^2)^{1/2} = \Gamma j^{-1/2}$$

for j large (e.g., if $\beta_j(\omega)$ are independent, identically distributed random variables, uniformly distributed on $\{z \mid |z| = \frac{1}{2}\}$, one can take $\alpha_j(\omega) = \min(\frac{1}{2}, 2\Gamma j^{-1/2})\beta_j(\omega))$, then for a.e. ω, the corresponding measure has no a.c. spectrum. If $\Gamma^2 > 1$, μ is pure point, and if $\Gamma^2 \leq 1$, the spectrum is purely singular continuous of Hausdorff dimension $1 - \Gamma^2$. While the OPUC case is from [400], it is motivated by an OPRL paper of Kiselev, Last, and Simon [227]; see [400] for earlier papers on OPRL with decaying random potentials.

(iv) It is known that generically slow decay yields purely singular continuous spectrum; see [400, Section 12.4]. Explicitly, for any $p > 2$ and $C < 1$, a dense G_δ in $\{\{\alpha_j\}_{j=0}^\infty \mid \sup_j|\alpha_j| \leq C, \sum_{j=0}^\infty|\alpha_j|^p < \infty\}$ in the ℓ^p metric has an associated measure with purely singular continuous spectrum. Also, for any $k < \frac{1}{2}$ and $C > 0$, a dense G_δ in $\{\{\alpha_j\}_{j=0}^\infty \mid \|\alpha\|_{C,k} = \sup_j(j+1+C)^k|\alpha_j| \leq 1$ and $j^k|\alpha_j| \to 0\}$ in $\|\cdot\|_{C,k}$ norm has an associated measure with purely singular continuous spectrum. This relies on the Wonderland theorem of Simon [393].

(v) One can construct sparse (i.e., α_j mainly zeros with the nonzero values very far apart) $\{\alpha_j\}_{j=0}^\infty$ in ℓ^p for all $p > 2$ so that the associated measures are purely singular continuous; see Golinskii [175] and [400, Section 12.5] and see the notes for the motivating Schrödinger operator papers.

Lest one thinks decay slower than $n^{-1/2}$ always means no a.c. spectrum, we note (see [400, Section 12.1] and the reference to Golinskii–Nevai [177] and earlier works there) that if $\sum_{n=0}^\infty|\alpha_{n+1} - \alpha_n| < \infty$, then there is pure a.c. spectrum on $\partial\mathbb{D} \setminus \{1\}$.

2.2 THE SZEGŐ INTEGRAL AS AN ENTROPY

In this section, we will prove Theorem 2.1.2 as a special case of a more general result concerning *relative entropy*. This object is defined by

Definition. Let μ, ν be two (positive) measures on a compact metric space. Define their relative entropy by

$$S(\mu \mid \nu) = \begin{cases} -\infty & \text{if } \mu \text{ is not } \nu\text{-a.c.} \\ -\int \log\left(\frac{d\mu}{d\nu}\right) d\mu & \text{if } \mu \text{ is } \nu\text{-a.c.} \end{cases} \tag{2.2.1}$$

Notice that if $d\nu$ is fixed and $d\mu = g\,d\nu$, then

$$S(g\,d\nu \mid d\nu) = \int -g\log(g)\,d\nu \tag{2.2.2}$$

$x \mapsto -x\log(x)$ is concave (its second derivative is $-1/x$) and is sometimes called the entropy function. If $d\nu$ is a counting measure on a finite set and $d\mu$ a probability measure on the same set with $\mu(\{j\}) = g_j$, then the right-hand side of (2.2.2) is $\sum_j -g_j\log(g_j)$, the familiar entropy of statistical mechanics courses.

Example 2.2.1. Let $d\mu_0 = \frac{d\theta}{2\pi}$ and $d\mu$ given by (1.6.3). Then, if $w > 0$ for a.e. θ, $d\mu_0$ is $d\mu$-a.c., and $d\mu_0/d\mu = w^{-1}$ so $\log(d\mu_0/d\mu) = -\log(w)$ and

$$S\biggl(\frac{d\theta}{2\pi} \biggm| w\,\frac{d\theta}{2\pi} + d\mu_s\biggr) = \int \log(w(\theta))\,\frac{d\theta}{2\pi} \tag{2.2.3}$$

If $w > 0$ for a.e. θ is false, then $d\mu_0$ is not $d\mu$-a.c., and both sides of (2.2.3) are $-\infty$ and (2.2.3) still holds. We thus see that the Szegő integral is a relative entropy. That will also be the case for other objects in sum rules, for example, two times the negative of the function Q in (1.10.16) (see (1.10.17)).

There is nothing special about the μ_0 here. If μ, ν are arbitrary measures on a compact metric space, one can write

$$d\nu = w\,d\mu + d\nu_s$$

(Lebesgue decomposition) and so that

$$S(\mu \mid \nu) = \int \log(w(x))\,d\mu(x) \tag{2.2.4}$$

□

The key to controlling S is

Proposition 2.2.2 (Linear Variational Principle for the Entropy). *Let $\mathcal{E}(X)$ be the family of strictly positive continuous functions on X. Then*

$$S(\mu \mid \nu) = \inf_{f\in\mathcal{E}(X)} \mathcal{S}(f;\mu,\nu) \tag{2.2.5}$$

where

$$\mathcal{S}(f;\mu,\nu) = \int f(x)\,d\nu(x) - \int (1+\log(f(x))\,d\mu(x) \tag{2.2.6}$$

Sketch. (For details, see Lemma 2.3.3 of [399].) Define for $b > 0$, $x > 0$,

$$Q_b(x) = xb^{-1} - 1 - \log(x) \tag{2.2.7}$$

Then $Q_b'(x) = b^{-1} - x^{-1}$ and $Q_b''(x) = x^{-2}$. Thus, Q_b is convex in x and its derivative vanishes at $x = b$ where $Q_b(x) = -\log(b)$. Since a smooth convex function with a zero derivative at some point takes its minimum at the point where the derivative vanishes, we have

$$Q_b(x) \geq -\log(b) \tag{2.2.8}$$

Suppose $d\mu$ is $d\nu$-a.c. Let $g = d\mu/d\nu$ and $A = \{x \mid g(x) \neq 0\}$. Then

$$d\nu = \chi_{X\setminus A}d\nu + g^{-1}\,d\mu \tag{2.2.9}$$

and, for $f \in \mathcal{E}$,

$$\mathcal{S}(f;\mu,\nu) = \int_{X\setminus A} f(x)\,d\nu(x) + \int_A Q_{g(x)}(f(x))\,d\mu(x) \tag{2.2.10}$$

$$\geq -\int \log(g(x))\,d\mu(x) \tag{2.2.11}$$

$$= S(\mu \mid \nu) \tag{2.2.12}$$

where (2.2.11) follows from (2.2.8).

If g is continuous and strictly positive, choose $f = g$. Then

$$\mathcal{S}(g; g\,d\nu, \nu) = S(\mu \mid \nu) \tag{2.2.13}$$

which proves (2.2.5) in case $d\mu = g\,d\nu$ with g continuous and nonvanishing.

The proof can be completed using two approximation arguments. One approximates any g by strictly positive continuous g's to prove (2.2.5) in the general case where μ is ν-a.c. The other uses very large g's approximately supported on a set, A, where $\nu(A) = 0$, $\mu(A) > 0$ to show the right-hand side of (2.2.5) is $-\infty$ if μ is not ν-a.c. $\square$

As an immediate corollary, we have

Theorem 2.2.3. *$S(\mu \mid \nu)$ is jointly concave and jointly weakly upper semicontinuous in μ, ν. Moreover, if*

$$\mu(X) = \nu(X) = 1 \tag{2.2.14}$$

then

$$S(\mu \mid \nu) \le 0 \tag{2.2.15}$$

Remarks. 1. Joint concavity means for $0 \le \theta \le 1$,

$$S(\theta\mu_1 + (1-\theta)\mu_0 \mid \theta\nu_1 + (1-\theta)\nu_0) \ge \theta S(\mu_1 \mid \nu_1) + (1-\theta) S(\mu_0 \mid \nu_0) \tag{2.2.16}$$

2. Upper semicontinuity means $\mu_n \xrightarrow{w} \mu$, $\nu_n \xrightarrow{w} \nu$ implies

$$\limsup S(\mu_n, \nu_n) \le S(\mu, \nu) \tag{2.2.17}$$

Proof. $\mathcal{S}(f; \mu, \nu)$ is linear and weakly continuous jointly in μ, ν for any $f \in \mathcal{E}(X)$. Thus, by (2.2.5), $S(\mu \mid \nu)$ is concave and upper semicontinuous. Noticing that if (2.2.14) holds, then $\mathcal{S}(f \equiv 1; \mu, \nu) = 0$, we obtain (2.2.15) from (2.2.5). $\square$

Corollary 2.2.4 (≡ Theorem 2.1.2). *If N is given by* (2.1.26), *then* (2.1.28) *holds.*

Proof. Follows from (2.2.17) since

$$N(d\mu) = -S\biggl(\frac{d\theta}{2\pi} \biggm| \mu\biggr) \tag{2.2.18}$$

by (2.2.3). $\square$

Example 2.2.5. Here are some examples that show S is only upper semicontinuous and not continuous. Let

$$d\mu_\infty = \frac{d\theta}{2\pi} \qquad d\mu_N = \frac{1}{N}\sum_{j=0}^{N-1} \delta_{2\pi j/N}$$

Then $d\mu_N \xrightarrow{w} d\mu_\infty$ but

$$S(d\mu_\infty \mid d\mu_N) = -\infty$$
$$S(d\mu_\infty \mid d\mu_\infty) = 0$$

(2.2.17) holds, but clearly, there is no equality.

Another example where measures are mutually a.c. is

$$d\mu_\infty = \frac{d\theta}{2\pi} \qquad d\mu_N = \left(1 + \frac{1}{2}\cos(N\theta)\right)\frac{d\theta}{2\pi}$$

Then $d\mu_N \xrightarrow{w} d\mu_\infty$, and by scaling,

$$S(d\mu_\infty \mid d\mu_N) = S(d\mu_\infty \mid d\mu_1) < 0 = S(d\mu_\infty \mid d\mu_\infty)$$

Finally, we note the more usual proof of (2.2.15). It depends on

Theorem 2.2.6 (Jensen's Inequality). *If F is convex on $\mathbb{R}^n$, then for any probability measure $d\mu$ on $\mathbb{R}^n$,*

$$F\left(\int \vec{x}\, d\mu(\vec{x})\right) \le \int F(\vec{x})\, d\mu(\vec{x}) \tag{2.2.19}$$

Remark. As our proof shows, this result holds if F is defined on a convex set, A, in $\mathbb{R}^n$ so long as $d\mu$ is supported there.

Proof. Convexity implies for each j, $(D_j^+ F)(x_0) = \lim_{y\downarrow 0}[F(x_0 + y\delta_j) - F(x_0)]/y$ exists for each $x_0 \in \mathbb{R}^n$, and for all x,

$$F(x) - F(x_0) \ge (x - x_0)\cdot (D^+F)(x_0) \tag{2.2.20}$$

Pick $x_0 = \int \vec{x}\, d\mu(x)$ and integrate (2.2.20) $d\mu_0$ to get (2.2.19). □

Alternate Proof of (2.2.15). Since $-\log(\cdot)$ is convex on $(0,\infty)$, Jensen's inequality implies that if $d\mu = g\,d\nu$ and $A = \{x \mid g(x) \ne 0\}$, then

$$\begin{aligned} S(\mu \mid \nu) &= \int_A \log(g^{-1})\, d\mu \\ &\le \log\left(\int_A g^{-1}\, d\mu\right) \\ &= \log(\nu(A)) \le 0 \end{aligned}$$

□

Remarks and Historical Notes. Entropy was discovered in thermodynamics and understood in statistical mechanics. That entropy has the form of $-\sum p_j \log(p_j)$ is a discovery of Boltzmann. Variational principles go back to Gibbs. His variational principle in this context says:

$$S(\mu \mid \nu) = \inf_{g\in C(X)}\left[\log \int e^g\, d\nu - \int g\, d\mu\right] \tag{2.2.21}$$

It is not hard to prove his relation from (2.2.5); see Section 10.6.

For discussion of entropy in statistical mechanics, see Israel [205], Ruelle [373, 374], or Simon [392]. For a mathematical discussion of entropy, see Carl–Stephani [73], Ellis [122], Gray [182], Ohya–Petz [327], or Parry [331].

While he did not know it was entropy he was using, Verblunsky [453] proves the Gibbs variational principle for the Szegő integral, namely,

$$\inf\left[\frac{\int e^g\, d\mu}{\exp(\int g\,\frac{d\theta}{2\pi})}\right] = \exp\left[\int \log(w(\theta))\,\frac{d\theta}{2\pi}\right]$$

and used it to prove a semicontinuity result. The use of entropy in proving sum rules was then rediscovered by Killip–Simon [225].

2.3 CARATHÉODORY, HERGLOTZ, AND SCHUR FUNCTIONS

One of the surprises (but which I have already strongly hinted at) is that complex analysis is a central tool in the spectral analysis of orthogonal polynomials. We will eventually see that techniques from Riemann surface theory, namely, Abelian integrals (see Sections 5.12, 5.13, and 9.11) and covering spaces (see Sections 9.2–9.5) will enter. In this section, we discuss more conventional boundary value theory.

Definition. A *Carathéodory function* is an analytic function, $F(z)$, on $\mathbb{D}$ so

$$\operatorname{Re} F(z) > 0 \qquad F(0) = 1 \tag{2.3.1}$$

A *Herglotz function* is an analytic function, $G(z)$, on

$$\mathbb{C}_+ = \{z \mid \operatorname{Im} z > 0\} \tag{2.3.2}$$

so that on $\mathbb{C}_+$,

$$\operatorname{Im} G(z) > 0 \tag{2.3.3}$$

A *Schur function* is an analytic function, f, on $\mathbb{D}$ so that

$$|f(z)| \leq 1 \tag{2.3.4}$$

Remarks. 1. Herglotz functions are also called Pick functions or Nevanlinna functions.

2. By the maximum principle, either (2.3.4) can be strengthened to

$$|f(z)| < 1$$

that is, $f\colon \mathbb{D} \to \mathbb{D}$, or else f is constant

$$f(z) = w_0 \in \partial\mathbb{D} \tag{2.3.5}$$

Example 2.3.1. The following shows the close connection between Herglotz functions and OPRL. Let $d\rho$ be a measure on $\mathbb{R}$ with

$$\int (1 + |x|)^{-1}\, d\rho(x) < \infty \tag{2.3.6}$$

Let

$$m(z) = \int \frac{d\rho(x)}{x - z} \tag{2.3.7}$$

Then

$$\operatorname{Im} m(z) = \operatorname{Im} z \int \frac{d\rho(x)}{|x - z|^2} \tag{2.3.8}$$

so m is Herglotz.

Suppose now that $d\rho$ has compact support and $\int d\rho(x) = 1$. Then writing

$$(x - z)^{-1} = -z^{-1} + x(x - z)^{-1} z^{-1} \tag{2.3.9}$$

we see that

$$m(z) = -z^{-1} + O(z^{-2}) \tag{2.3.10}$$

□

This motivates a definition:

Definition. A *discrete m-function* is a Herglotz function, $m(z)$, so that for some bounded interval $I \subset \mathbb{R}$, we have that $m(z)$ has an analytic continuation from $\mathbb{C}_+$ to $\mathbb{C} \setminus I$ with

$$z \in \mathbb{R} \setminus I \Rightarrow \operatorname{Im} m(z) = 0 \tag{2.3.11}$$

and (2.3.10) holds.

It is easy to see that, given the analyticity assumption, (2.3.11) is equivalent to

$$m(\bar{z}) = \overline{m(z)} \tag{2.3.12}$$

We will shortly prove (see Theorem 2.3.6) that every discrete m-function has the form (2.3.7) for a probability measure $d\rho$ on I. For now, we note

Proposition 2.3.2. *Suppose $m(z)$ has the form* (2.3.7) *where*

$$\operatorname{supp}(d\rho) \subset [-R, R] \tag{2.3.13}$$

for some R. Let c_n be the moments of $d\rho$:

$$c_n = \int x^n \, d\rho(x) \tag{2.3.14}$$

Then for $|z| > R$, we have an absolutely convergent series

$$m(z) = -\sum_{n=0}^{\infty} c_n z^{-(n+1)} \tag{2.3.15}$$

Proof. Immediate from the geometric series expansion, uniformly and absolutely convergent on $|z| > R + \varepsilon$ for each $\varepsilon > 0$,

$$(x - z)^{-1} = -\sum_{n=0}^{\infty} x^n z^{-(n+1)} \tag{2.3.16}$$

□

If R is the minimum value for which (2.3.13) holds, it is easy to see the Taylor series at infinity (2.3.15) diverges if $|z| < R$. We will eventually find Padé approximants (see the remark after Proposition 3.2.8) that converge on all of $\mathbb{C} \setminus I$. Indeed, the numerator and denominator will be orthogonal polynomials! Equivalently, we will find continued fraction expansions in terms of the Jacobi parameters. We will see all this in Section 3.2 and its OPUC analog in Section 2.5.

To figure out the OPUC analog of (2.3.7), we need the complex Poisson representation:

Proposition 2.3.3. *Let f be analytic in a neighborhood of $\overline{\mathbb{D}}$. Then for $z \in \mathbb{D}$,*

$$f(z) = i \operatorname{Im} f(0) + \int \frac{e^{i\theta}+z}{e^{i\theta}-z} \operatorname{Re} f(e^{i\theta}) \frac{d\theta}{2\pi} \tag{2.3.17}$$

Proof. f has a Taylor series converging for $|z| < 1+\varepsilon$

$$f(z) = \sum_{n=0}^{\infty} a_n z^n \tag{2.3.18}$$

so

$$\operatorname{Re} f(e^{i\theta}) = \operatorname{Re} a_0 + \tfrac{1}{2} \sum_{n=1}^{\infty} (a_n e^{in\theta} + \bar{a}_n e^{-in\theta})$$

Thus,

$$\int e^{-in\theta} \operatorname{Re} f(e^{i\theta}) \frac{d\theta}{2\pi} = \begin{cases} \operatorname{Re} a_0 & \text{if } n=0 \\ \frac{1}{2} a_n & \text{if } n>0 \end{cases} \tag{2.3.19}$$

On the other hand, for $|w| < 1$,

$$\frac{1+w}{1-w} = (1+w)\biggl(\sum_{n=0}^{\infty} w^n\biggr) = 1 + 2\sum_{n=1}^{\infty} w^n \tag{2.3.20}$$

so that for $|z| < 1$,

$$\frac{e^{i\theta}+z}{e^{i\theta}-z} = \frac{1+ze^{-i\theta}}{1-ze^{-i\theta}} = 1 + 2\sum_{n=1}^{\infty} z^n e^{-in\theta} \tag{2.3.21}$$

Therefore, by (2.3.19),

$$\begin{aligned} \text{RHS of (2.3.17)} &= i \operatorname{Im} a_0 + \biggl(\operatorname{Re} a_0 + 2\sum_{n=1}^{\infty} z^n (\tfrac{1}{2} a_n)\biggr) \\ &= f(z) \end{aligned}$$

□

It is useful to note that

$$\operatorname{Re}\biggl(\frac{e^{i\theta}+z}{e^{i\theta}-z}\biggr) = \frac{1-|z|^2}{|e^{i\theta}-z|^2} > 0 \tag{2.3.22}$$

In particular,

$$\operatorname{Re}\biggl(\frac{e^{i\theta}+re^{i\varphi}}{e^{i\theta}-re^{i\varphi}}\biggr) = \frac{1-r^2}{1+r^2-2r\cos(\theta-\varphi)} \tag{2.3.23}$$

the celebrated *Poisson kernel*.

Proposition 2.3.3 motivates:

Definition. The *Carathéodory function* of a probability measure $d\mu$ on $\partial\mathbb{D}$ is given by

$$F(z) = \int \frac{e^{i\theta} + z}{e^{i\theta} - z}\, d\mu(\theta) \tag{2.3.24}$$

This is a Carathéodory function since (2.3.22) implies $\operatorname{Re} F(z) > 0$ and $\int d\mu(\theta) = 1$ implies $F(0) = 1$.

Our three classes of functions are clearly related. The map

$$z \to i\left[\frac{1+z}{1-z}\right] \tag{2.3.25}$$

maps $\mathbb{D}$ bijectively and biholomorphically onto $\mathbb{C}_+$. If G is a function on $\mathbb{C}_+$ with values in $\mathbb{C}$, then G is Herglotz if and only if $\operatorname{Im} G(i) > 0$ and

$$F(z) = \frac{-i\left[G(i\left[\frac{1+z}{1-z}\right]) - \operatorname{Re} G(i)\right]}{\operatorname{Im} G(i)} \tag{2.3.26}$$

is a Carathéodory function. And the association

$$F(z) = \frac{1 + zf(z)}{1 - zf(z)} \tag{2.3.27}$$

sets up a one-one correspondence between Carathéodory functions and Schur functions. To see this, one needs the Schwarz lemma:

Proposition 2.3.4 (Schwarz Lemma). *If f is a Schur function with $f(0) = 0$, then $f(z)/z$ is also a Schur function.*

Proof. Let $g(z) = f(z)/z$. Then g is analytic and for $0 < r < 1$,

$$\max_{|z|\le r} |g(z)| = \max_{|z|=r} |g(z)| \le r^{-1} \max_{|z|=r} |f(z)| \le \frac{1}{r}$$

Taking $r \uparrow 1$, we see

$$\max_{|z|<1} |g(z)| \le 1 \tag{2.3.28}$$

□

We note the inverse formula to (2.3.27),

$$f(z) = z^{-1}\left(\frac{F(z) - 1}{F(z) + 1}\right) \tag{2.3.29}$$

If $d\mu$ is a probability measure on $\partial\mathbb{D}$ and F is its Carathéodory function, the function f given by (2.3.27)/(2.3.28) is called its *Schur function*. A main result for Carathéodory functions is:

Theorem 2.3.5 (Herglotz Representation for Carathéodory Functions). *$F(z)$ is a Carathéodory function if and only if there is a probability measure μ obeying (2.3.24).*

Proof. Write

$$F(z) = 1 + 2\sum_{n=1}^{\infty} c_n z^n \tag{2.3.30}$$

Since $\operatorname{Re} F(z) > 0$, for $0 < r < 1$,

$$d\mu_r(\theta) = \operatorname{Re} F(re^{i\theta}) \frac{d\theta}{2\pi} \tag{2.3.31}$$

defines a measure, with $\int d\mu_r(\theta) = 1$ since $F(0) = 1$. Moreover, since $F(rz)$ is analytic in a neighborhood of $\overline{\mathbb{D}}$, (2.3.17) implies

$$F(rz) = \int \frac{e^{i\theta} + z}{e^{i\theta} - z}\, d\mu_r(\theta) \tag{2.3.32}$$

which, by (2.3.21), implies

$$\int e^{in\theta}\, d\mu_r(\theta) = r^n c_n \qquad (n > 0)$$

It follows that $d\mu_r(\theta)$ is a family of measures where $\lim_{r\to 1} \int e^{in\theta}\, d\mu_r(\theta) = c_n$ for $n > 0$, and by reality, the limit is $\bar{c}_{-n}$ for $n < 0$. Thus, by the fact that $|\int f(e^{i\theta})\, d\mu_r(\theta)| \le \|f\|_\infty$ and the density of Laurent polynomials in $\mathbb{C}(\partial\mathbb{D})$, $d\mu_r$ has a weak limit $d\mu$. Taking $r \to 1$ in (2.3.32), we obtain

$$F(z) = \int \frac{e^{i\theta} + z}{e^{i\theta} - z}\, d\mu(\theta) \tag{2.3.33}$$

$\square$

By (2.3.26), this translates to a Herglotz representation for Herglotz functions (see the Notes). One could use that to analyze discrete m-functions, but we will instead use a direct argument that mimics the above proof.

Theorem 2.3.6 (Herglotz Representation for Discrete m-functions). *A function $m(z)$ on $\mathbb{C}_+$ is a discrete m-function if and only if m has the form* (2.3.7) *for some probability measure $d\rho$ supported on a bounded interval in $\mathbb{R}$.*

Proof. Suppose (2.3.13) holds and pick $\delta > 0$ and $M > R + \delta + 1$. Let Γ_1 be the contour going clockwise around the rectangle centered at 0 with width $2(R+1)$ and height 2δ and Γ_2 be the circle of radius M centered at zero going counterclockwise. If $y \in \mathbb{R}$ and $R + \delta + 1 < |y| < M$, we have

$$m(y) = \frac{1}{2\pi i} \int_{\Gamma_1 \cup \Gamma_2} \frac{m(z)}{z - y}\, dz \tag{2.3.34}$$

The contribution of Γ_2 is dominated in absolute value by

$$\frac{1}{2\pi} (2\pi M) \frac{1}{M - |y|} \sup_{|z|=M} |m(z)|$$

which goes to zero as $M \to \infty$. Thus, for $|y| > R + \delta + 1$,

$$m(y) = \frac{1}{2\pi i} \int_{\Gamma_1} \frac{m(z)}{z - y}\, dz \tag{2.3.35}$$

A similar analysis just closing the contour in the upper half-plane shows for $y' \neq \delta > 0$ that

$$\frac{1}{2\pi i}\int_{-\infty}^{\infty}\frac{m(x+i\delta)}{x+i\delta-iy'}\,dx = \begin{cases} m(iy') & y' > \delta \\ 0 & 0 < y' < \delta \end{cases}$$

Taking this for $y' > \delta$ and $y'' = \delta-(y'-\delta)$ and subtracting, we get for $\delta < y' < 2\delta$ that

$$\int_{-\infty}^{\infty} m(x+i\delta)\,\frac{dx(y'-\delta)}{x^2+(y'-\delta)^2} = \pi m(iy')$$

By analyticity, we deduce this for all y with $\operatorname{Re} y > \delta$ and, in particular, for $y \in (\delta,\infty)$. Taking the imaginary part, multiply by y', and taking $y' \to \infty$ (using $\operatorname{Im} m(x+iy) = O(1/x^2)$), we get

$$\int_{-\infty}^{\infty}\frac{1}{\pi}\operatorname{Im} m(x+i\delta)\,dx = 1 \tag{2.3.36}$$

Now let

$$d\rho_\delta(x) = \chi_{[-R,R]}(x)\,\frac{1}{\pi}\operatorname{Im} m(x+i\delta)\,dx \tag{2.3.37}$$

and see (2.3.36) implies, as $\delta \downarrow 0$,

$$\int d\rho_\delta(x) \to 1 \tag{2.3.38}$$

and that (2.3.35) implies for y real with $|y| > R+1$,

$$\int \frac{d\rho_\delta(x)}{x-y} \to m(y) \tag{2.3.39}$$

Since $\{(x-y)^{-1} \mid |y| > R+1\}$ is total in $\mathbb{C}([-R,R])$, we have that $d\rho_\delta$ has a limit $d\rho$ and (2.3.7) holds first for $z \in \mathbb{R}\setminus[-R-1,R+1]$ and then by analytic continuation for $z \in \mathbb{C}_+$. □

Our proofs showed

Theorem 2.3.7. *If F is a Carathéodory function with associated measure $d\mu$ in* (2.3.26), *then*

$$\operatorname*{w-lim}_{r\uparrow 1} \operatorname{Re} F(re^{i\theta})\,\frac{d\theta}{2\pi} = d\mu(\theta) \tag{2.3.40}$$

If m is a discrete m-function and $d\rho$ the associated measure in (2.3.7), *then*

$$\operatorname*{w-lim}_{\delta\downarrow 0} \frac{1}{\pi}\operatorname{Im} m(x+i\delta)\,dx = d\rho(x) \tag{2.3.41}$$

Define $H^p(\mathbb{D})$ by

Definition. Let $0 < p < \infty$. An analytic function, f, on $\mathbb{D}$ is said to lie in $H^p(\mathbb{D})$ if and only if

$$\|f\|_p \equiv \sup_{r<1}\left(\int |f(re^{i\theta})|^p\,\frac{d\theta}{2\pi}\right)^{1/p} < \infty \tag{2.3.42}$$

and in $H^\infty(\mathbb{D})$ if

$$\|f\|_\infty = \sup_{|z|<1} |f(z)| < \infty \tag{2.3.43}$$

If $0 < p < q \le \infty$, $H^q \subset H^p$ and $\|f\|_p \le \|f\|_q$. $\|\cdot\|_p$ is a norm if $p \ge 1$ (but not if $p < 1$). I will quote results about H^p functions largely without proof, leaving that for the books, for example, Duren [118] and Rudin [372].

We will require the following version of M. Riesz's celebrated theorem on conjugate harmonic functions:

Proposition 2.3.8 (M. Riesz's Theorem). *Let $1 < p < \infty$. If f is analytic on $\mathbb{D}$ and*

$$\sup_{0<r<1} \int |\text{Re}\, f(re^{i\theta})|^p \,\frac{d\theta}{2\pi} < \infty \tag{2.3.44}$$

then $f \in H^p(\mathbb{D})$.

Indeed, one has for some constant C_p that

$$\|f\|^p_{H^p(\mathbb{D})} \le C_p[\text{LHS of (2.3.44)} + \text{Im}|f(0)|^p] \tag{2.3.45}$$

In this monograph, we could prove our main results only knowing Riesz's theorem for $p = 2$ where it is trivial since

$$\int f(re^{i\theta})^2 \,\frac{d\theta}{2\pi} = f(0)^2 \tag{2.3.46}$$

so

$$\int |f(re^{i\theta})|^2 \,\frac{d\theta}{2\pi} \le 2\int (\text{Re}\, f(re^{i\theta}))^2 \,\frac{d\theta}{2\pi} + (\text{Im}\, f(0))^2 \tag{2.3.47}$$

Proposition 2.3.9. *If $f \in H^p$ for any $p > 0$, then*

$$f(e^{i\theta}) \equiv \lim_{r\uparrow 1} f(re^{i\theta}) \tag{2.3.48}$$

exists for a.e. θ. Moreover,

$$\left(\int |f(e^{i\theta})|^p \,\frac{d\theta}{2\pi}\right)^{1/p} \le \|f\|_p \tag{2.3.49}$$

and for $1 \le p < \infty$,

$$\lim_{r\uparrow 1} \int |f(re^{i\theta}) - f(e^{i\theta})|^p \,\frac{d\theta}{2\pi} = 0 \tag{2.3.50}$$

Note that (2.3.50) for $p = 1$ implies that (by the Cauchy formula):

$$\int f(e^{i\theta}) \,\frac{d\theta}{2\pi} = \lim_{r\uparrow 1} \int f(re^{i\theta}) \,\frac{d\theta}{2\pi} = f(0) \tag{2.3.51}$$

and more generally,

$$\int e^{-in\theta} f(e^{i\theta}) \,\frac{d\theta}{2\pi} = \begin{cases} f^{(n)}(0)/n! & n > 0 \\ 0 & n < 0 \end{cases} \tag{2.3.52}$$

Indeed, the Poisson representation (2.3.17) holds for any $f \in H^1$ by taking limits for $f(rz)$ as $r \uparrow 1$ using (2.3.50).

In particular, if f is a Schur function, it lies in H^∞ and so has boundary values in the sense of (2.3.48). Thus, by (2.3.27) and (2.3.26), we have:

Proposition 2.3.10. *For any Carathéodory function, F, one has*

$$F(e^{i\theta}) = \lim_{r\uparrow 1} F(re^{i\theta}) \tag{2.3.53}$$

exists for $\frac{d\theta}{2\pi}$-a.e. θ. For any Herglotz function G and dx-a.e. $x \in \mathbb{R}$,

$$\lim_{\varepsilon\downarrow 0} G(x+i\varepsilon) \equiv G(x+i0) \tag{2.3.54}$$

exists.

(2.3.40)/(2.3.41) are complemented by the following:

Proposition 2.3.11. *Let $d\mu$ on $\partial\mathbb{D}$ have the form* (1.6.3) *and let F be its Carathéodory function. Then*

$$w(\theta) = \operatorname{Re} F(e^{i\theta}) \tag{2.3.55}$$

Let $d\rho$ on $[a,b]$ have the form (1.4.3) *and let m be its m-function. Then*

$$f(x) = \pi^{-1} \operatorname{Im} m(x+i0) \tag{2.3.56}$$

One can also relate the singular part of the measure to the boundary values of F and m:

Proposition 2.3.12. *Let $d\mu$ on $\partial\mathbb{D}$ have the form* (1.6.3) *and let F be its Carathéodory function. Then $d\mu_{\rm s}$ is supported on $\{e^{i\theta} \mid \lim_{r\uparrow 1} \operatorname{Re} F(re^{i\theta}) = \infty\}$. Moreover, for any θ_0,*

$$\mu(\{e^{i\theta_0}\}) = \lim_{r\uparrow 1} \tfrac{1}{2}(1-r)F(re^{i\theta_0}) \tag{2.3.57}$$

Let $d\rho$ on $[a,b]$ have the form (1.4.3) *and let m be its m-function. Then $d\rho_{\rm s}$ is supported on $\{x \in [a,b] \mid \lim_{\varepsilon\downarrow 0} \operatorname{Im} m(x+i\varepsilon) = \infty\}$, and for any $x_0 \in [a,b]$,*

$$\rho(\{x_0\}) = \lim_{\varepsilon\downarrow 0} (-i\varepsilon)m(x_0+i\varepsilon) \tag{2.3.58}$$

Proposition 2.3.11 has the following consequence:

Corollary 2.3.13. *If $d\mu$ is a purely singular (positive not necessarily normalized) measure on $\partial\mathbb{D}$ and F is its Carathéodory function, then*

$$g(z) = e^{-F(z)} \tag{2.3.59}$$

obeys g analytic in $\mathbb{D}$ and

$$|g(z)| \le 1 \qquad \lim_{r\uparrow 1} |g(re^{i\theta})| = 1 \quad \text{for a.e. } \theta \tag{2.3.60}$$

Conversely, any everywhere nonvanishing analytic function on $\mathbb{D}$ obeying (2.3.60) *with $g(0)$ positive has the form* (2.3.59) *for F, the Carathéodory function of a purely singular measure.*

Remark. Any Schur function obeying (2.3.60) is called an *inner function*. Ones of the form (2.3.59) are called *singular inner functions*.

Proof. By (2.3.55), if $d\rho$ is purely singular, then $\text{Re}\, F(e^{i\theta}) = 0$ for a.e. θ. Since $\text{Re}\, F(z) > 0$ for $z \in \mathbb{D}$, (2.3.60) is immediate.

Conversely, if g is everywhere nonvanishing and $|g(z)| \leq 1$, $-\log(g)$ defines a function F with $\text{Re}\, F > 0$ and $g(0)$ positive implies $F(0)$ is real. F is thus a positive multiple of a Carathéodory function. By (2.3.60) and (2.3.55), the corresponding measure is purely singular. □

Other inner functions have to vanish somewhere in $\mathbb{D}$. For each $z_0 \in \mathbb{D}$, define

$$M_{z_0}(z) = \frac{z - z_0}{1 - z\bar{z}_0} \tag{2.3.61}$$

It will enter as a building block for inner functions but also in the discussion of the Schur algorithm. It has the following critical properties:

Proposition 2.3.14. *M_{z_0} maps $\mathbb{D}$ bijectively onto $\mathbb{D}$. M_{z_0} is analytic in a neighborhood of $\mathbb{D}$ and obeys*

$$|M_{z_0}(e^{i\theta})| = 1 \tag{2.3.62}$$

for all $\theta \in [0, 2\pi)$. Any analytic bijection, M, of $\mathbb{D}$ onto $\mathbb{D}$ has the form

$$M(z) = e^{i\theta} M_{z_0}(z) \tag{2.3.63}$$

for $z_0 \in \mathbb{D}$ and $e^{i\theta} \in \partial\mathbb{D}$.

Proof. A simple computation shows

$$M_{z_0} \circ M_{-z_0} = \mathbf{1} \tag{2.3.64}$$

and that

$$M_{z_0}(e^{i\theta}) = e^{i\theta}\left[\frac{1 - e^{-i\theta}z_0}{1 - e^{i\theta}\bar{z}_0}\right] \tag{2.3.65}$$

so (2.3.62) holds. By the maximum principle, M_{z_0} maps $\mathbb{D}$ to $\mathbb{D}$ so, by (2.3.64), it is a bijection.

For the converse, suppose first $M(0) = 0$. Then the Schwarz lemma (Proposition 2.3.4) implies

$$\left|\frac{M(z)}{z}\right| \leq 1 \tag{2.3.66}$$

Since M^{-1} is also a bijection taking 0 to 0,

$$\left|\frac{M^{-1}(z)}{z}\right| \leq 1$$

so letting $w = M(z)$,

$$\left|\frac{w}{M(w)}\right| \leq 1$$

that is, $|M(z)/z| = 1$. By the maximum principle, $M(z) = e^{i\theta}z$ for some $e^{i\theta} \in \partial\mathbb{D}$.

For any bijection, pick z_0 so $M(z_0) = 0$. Let

$$M_1 = M \circ M_{z_0}^{-1}$$

so $M_1(0) = 0$ and thus, $M_1(z) = e^{i\theta}z$, which implies (2.3.63). □

It will be convenient for some purposes to pick $e^{i\theta}$ in (2.3.63) of a special form:

Definition. A *Blaschke factor*, defined for each $z_0 \in \mathbb{D} \setminus \{0\}$, is defined by

$$b_{z_0}(z) = -\frac{|z_0|}{z_0} \frac{z - z_0}{1 - \bar{z}_0 z} \tag{2.3.67}$$

The phase is picked so $b_{z_0}(0) > 0$ (minimizing $|1 - b_{z_0}(0)|$). A *Blaschke product* is a product of Blaschke factors. There are three main results about Blaschke products:

Proposition 2.3.15. *An inner function, which has an analytic continuation to a neighborhood of* $\overline{\mathbb{D}}$*, has a finite number of zeros in* $\mathbb{D}$ *and is a phase factor* $e^{i\theta}$ *times the Blaschke product of those zeros.*

Proposition 2.3.16. *Let* $\{z_j\}_{j=1}^\infty$ *be a sequence of points in* $\mathbb{D} \setminus \{0\}$*. Then either*

(i)

$$\sum_{j=1}^\infty (1 - |z_j|) = \infty \tag{2.3.68}$$

in which case $\prod_{j=1}^N b_{z_j}(z)$ *converges uniformly as* $N \to \infty$ *on compact subsets of* $\mathbb{D}$ *to* 0*, or*

(ii)

$$\sum_{j=1}^\infty (1 - |z_j|) < \infty \tag{2.3.69}$$

in which case $\prod_{j=1}^N b_{z_j}(z)$ *converges uniformly on compact subsets of* $\mathbb{D}$ *as* $N \to \infty$ *to a limit* $B_\infty(z)$*, which is an inner function vanishing on* $\mathbb{D}$ *exactly at the points* $\{z_j\}_{j=1}^\infty$*.*

Proposition 2.3.17. *If* $f \in H^p(\mathbb{D})$ *for some* $p > 0$ *and* $\{z_j\}_{j=1}^\infty$ *are its zeros in* $\mathbb{D}$*, then* (2.3.69) *holds.*

From these, one gets two important results about H^p functions:

Theorem 2.3.18. *Every inner function is uniquely a product of a Blaschke product obeying* (2.3.69)*,* z^ℓ *for some* $\ell \geq 0$*, a constant phase factor and a singular inner function.*

Definition. An *outer function*, $f(z)$, is a function on $\mathbb{D}$ of the form

$$f(z) = \exp\biggl(\int \frac{e^{i\theta} + z}{e^{i\theta} - z}\, g(e^{i\theta})\, \frac{d\theta}{2\pi}\biggr) \tag{2.3.70}$$

where g is real and in $L^1(\partial\mathbb{D}, \frac{d\theta}{2\pi})$.

Theorem 2.3.19. *Any* H^p *function,* f*,* $p > 0$*, can be written uniquely as a product of an inner and an outer function, that is, if* $\{z_j\}_{j=1}^N$ *are its zeros in* $\mathbb{D} \setminus \{0\}$ *(*N *finite* *or infinite), then for some* $e^{i\theta}$, $\ell \geq 0$, *and nonnegative singular measure,* $d\nu$, *on* $\partial\mathbb{D}$,

$$f(z) = e^{i\theta} z^{\ell} \left[\prod_{j=1}^{N} b_{z_j}(z) \right] \exp\left(\int \frac{e^{i\varphi} + z}{e^{i\varphi} - z} \left[\log|f(e^{i\varphi})| \, \frac{d\varphi}{2\pi} - d\nu(\varphi) \right] \right) \tag{2.3.71}$$

If f is analytic in a neighborhood of $\overline{\mathbb{D}}$, then $d\nu = 0$ and $N < \infty$. In that case when $\ell = 0$, (2.3.71) at $z = 0$ is a celebrated formula of Jensen. Thus, (2.3.71) is sometimes called the *Poisson–Jensen formula*. Its importance is due to Nevanlinna.

The next fact we need concerns the issue of whether boundary values determine an H^p function. The key fact is:

Theorem 2.3.20. *Let* $f \in H^p(\mathbb{D})$, $0 < p < \infty$, *be not identically zero, and let* $f(e^{i\theta})$ *be its boundary values. Then* $\{e^{i\theta} \mid f(e^{i\theta}) = 0\}$ *has* $\frac{d\theta}{2\pi}$ *measure zero. If* $f, g \in H^p$ *and* $f(e^{i\theta}) = g(e^{i\theta})$ *for* $e^{i\theta} \in \Sigma$ *with* $|\Sigma| > 0$, *then* $f = g$.

Sketch. Using Jensen's formula, one proves that $\int \log_-|f(e^{i\theta})| \frac{d\theta}{2\pi} < \infty$ (where, for $x > 0$, $\log_-(x) = \max(0, -\log(x))$ and this implies $|\{e^{i\theta} \mid f(e^{i\theta}) = 0\}| = 0$. The second statement follows from the first since $f - g \in H^p$. □

Theorem 2.3.21. (a) *If* f *and* g *are distinct Carathéodory functions, then* $\Sigma = \{e^{i\theta} \mid f(e^{i\theta}) = g(e^{i\theta})\}$ *has* $|\Sigma| = 0$. *Similarly,* $|\{e^{i\theta} \mid f(e^{i\theta}) = c\}| = 0$ *for each* $c \in \mathbb{C}$ *(except for the case* $c = 1$, $f \equiv 1$*).*

(b) *If* m, n *are distinct discrete m-functions,*

$$|\{x \in \mathbb{R} \mid m(x + i0) = n(x + i0)\}| = 0 \tag{2.3.72}$$

Similarly, $|\{x \in \mathbb{R} \mid m(x + i0) = c\}| = 0$ *for any* $c \in \mathbb{C}$.

Proof. (a) $e^{-f}, c^{-g} \in H^\infty(\mathbb{D})$, so this follows from the previous theorem.

(b) e^{im}, e^{in} are bounded and, by mapping $\mathbb{C}_+$ to $\mathbb{D}$, we can apply the previous theorem. □

We have just seen that if $\Sigma \subset \mathbb{R}$ has positive measure, then the map $m \to m(\cdot + i0) \restriction \Sigma$ is one-one. Later (see Theorem 7.4.4), we will need to know this map has a continuous inverse. Since $\operatorname{Im} m(x + i0) \geq 0$, for $t \geq 0$, $|e^{itm(x+i0)}| \leq 1$, so we will topologize the functions by saying $m_n \to m$ *weakly on* Σ if and only if for each $g \in L^2(\Sigma, dx)$ and t positive and rational, we have

$$\int_\Sigma g(x) e^{itm_n(x+i0)} \, dx \to \int_\Sigma g(x) e^{itm(x+i0)} \, dx \tag{2.3.73}$$

Theorem 2.3.22. *Let* $\mathfrak{e} \subset \mathbb{R}$ *be compact and let* $\Sigma \subset \mathfrak{e}$ *be a Borel set of strictly positive Lebesgue measure. Let* $\mathcal{M}(\mathfrak{e})$ *be the set of functions,* $m(z)$, *which are discrete m-functions of the form* (2.3.7) *where* $\operatorname{supp}(d\rho) \subset \mathfrak{e}$. *Topologize* $\mathcal{M}(\mathfrak{e})$ *with the topology of uniform convergence on compact subsets of* $\mathbb{C}_+$. *Topologize functions on* Σ *with positive imaginary part by* (2.3.73). *Then*

$$R_\Sigma : m \mapsto m(\cdot + i0) \restriction \Sigma$$

is a continuous map with a compact range and a continuous inverse on this range.

Proof. $\mathcal{M}(\mathfrak{e})$ is compact and we have proven above that R_Σ is one-one, so continuity automatically implies a compact range and a continuous inverse.

In $H^2(\mathbb{D})$, uniform convergence on compacts is equivalent to weak convergence of boundary values in $L^2(\partial\mathbb{D})$ (since $\{e^{in\theta}\}$ are total and given by Taylor coefficients). By mapping $\mathbb{C}_+$ to $\mathbb{D}$ and noting that e^{itm} is bounded, we get functions in H^∞, so in H^2. □

There is one final topic concerning Schur functions that we want to discuss: the Schur algorithm and Schur parameters. Given a Schur function, f, either $f(z) \equiv e^{i\theta} = \gamma_0$ or else

$$f(0) \equiv \gamma_0 \in \mathbb{D} \tag{2.3.74}$$

In the latter case, we can look at $M_{\gamma_0}(f(z))$, which is a Schur function vanishing at 0, so, by the Schwarz lemma,

$$f_1(z) = \frac{1}{z}\,\frac{f(z) - \gamma_0}{1 - \bar{\gamma}_0 f(z)} \tag{2.3.75}$$

is also a Schur function and

$$f(z) = \frac{\gamma_0 + z f_1(z)}{1 + z\bar{\gamma}_0 f_1(z)} \tag{2.3.76}$$

We can iterate this process. If $f_1(z) \equiv e^{i\theta} = \gamma_1$, we stop. Otherwise we set $f_1(0) = \gamma_1$ and define f_2 via the analog of (2.3.75). In this way we associate with a Schur function a sequence of numbers $\{\gamma_j\}_{j=0}^N$ with either $N < \infty$ and then

$$|\gamma_j| < 1 \quad j = 0, \dots, N-1 \qquad |\gamma_N| = 1 \tag{2.3.77}$$

or else $N = \infty$ and then for all j,

$$|\gamma_j| < 1 \qquad j = 0, 1, \dots \tag{2.3.78}$$

We also get a sequence $\{f_j\}_{j=0}^\infty$ (with $f = f_0$) of Schur functions called the *Schur iterates*.

Define $\mathcal{S}$ to be the set of potential Schur parameters, that is, finite sequences obeying (2.3.77) or infinite sequences obeying (2.3.78). $\mathcal{S}$ has a natural topology of convergence of each γ_j (with the rule that if the limit $\gamma^{(\infty)}$ has $N < \infty$, we only require $\gamma_j^{(n)} \to \gamma_j^{(\infty)}$ for $j \le N$) in which $\mathcal{S}$ is compact. We use $\gamma(f) \in \mathcal{S}$ for the Schur parameters of f.

The main theorem about the Schur algorithm is:

Theorem 2.3.23. *The map from Schur functions to $\mathcal{S}$ is a bijection and for a sequence, $\{g_n\}_{n=1}^\infty$, of Schur functions and another Schur function, g_∞, convergence of $\gamma(g_n)$ to $\gamma(g_\infty)$ is equivalent to convergence of $g_n(z)$ to $g(z)$ uniformly on compact subsets of $\mathbb{D}$. Moreover, $\gamma(f)$ has $N < \infty$ if and only if f is a phase factor times a finite Blaschke product.*

We want to sketch the proof of this theorem.

Proposition 2.3.24. *A Schur function f has $N < \infty$ if and only if it is a Blaschke product of order N times a phase factor.*

Proof. We will use induction in N. $N = 0$ is obvious.

f has $N = n > 0$ if and only if there is $\gamma_0 \in \mathbb{D}$ and f_1 with $N = n - 1$ so (2.3.76) holds. If f_1 is a Blaschke product of order $n - 1$, then zf_1 has order n and so is analytic in a neighborhood of $\overline{\mathbb{D}}$ with winding number n as a map of $\partial\mathbb{D}$ to $\partial\mathbb{D}$. Since $e^{i\theta} \to (\gamma_0 + e^{i\theta})/(1 + \bar{\gamma}_0 e^{i\theta})$ is a bijection of $\partial\mathbb{D}$ with positive derivative, it preserves winding number, so f also has winding number n, and so, by the argument principle, it has n zeros. By Proposition 2.3.15, it is a Blaschke product of order n.

Conversely, if f is a Blaschke product of order n, the above winding number argument shows zf_1 has n zeros, and so f_1 has $n-1$ zeros. So, by Proposition 2.3.11 again, f_1 is a Blaschke product of order $n - 1$.

Thus, f is a Blaschke product of order n (times a phase factor) if and only if f_1 is a Blaschke product of order $n - 1$. This plus induction completes the proof. □

Lemma 2.3.25. *For any finite sequence* $(\gamma_0, \dots, \gamma_{n-1}) \in \mathbb{D}^n$, *there is a function* f *whose Schur parameters are* $\gamma_0, \gamma_1, \dots, \gamma_{n-1}, 0, 0, \dots, 0, \dots$.

Proof. If $n = 0$, take $f(z) = 0$, which has $\gamma_0 = 0$, $f_1(z) = 0$, and so Schur parameters $(0, 0, \dots)$. Now given any finite sequence, we can suppose inductively we have f_1 with Schur parameters $(\gamma_1, \gamma_2, \dots, \gamma_{n-1}, 0, \dots)$ and define f by (2.3.76). □

Lemma 2.3.26. *Let* f *and* g *be two Schur functions with Schur parameters* $\gamma_j(f)$ *and* $\gamma_j(g)$. *Then*

$$f^{(j)}(0) = g^{(j)}(0) \qquad j = 0, 1, \dots, n \tag{2.3.79}$$

if and only if

$$\gamma_j(f) = \gamma_j(g) \qquad j = 0, 1, \dots, n \tag{2.3.80}$$

Proof. Since $w \to M_{-\gamma_0}(w)$ is a smooth bijection near $w = 0$ and $M'_{-\gamma_0}(0) = 1 - |\gamma_0|^2$, we have for w small,

$$M_{-\gamma_0}(w) = (1 - |\gamma_0|^2)w + O(w^2) \tag{2.3.81}$$

Thus, if $\gamma_0(f) = \gamma_0(g)$,

$$f(z) - g(z) = (1 - |\gamma_0|^2)z(f_1(z) - g_1(z)) + O(z^2(f_1(z) - g_1(z)) \tag{2.3.82}$$

which proves that if $\gamma_0(f) = \gamma_0(g)$, then

$$f(z) - g(z) = O(z^n) \Leftrightarrow f_1(z_0) - g_1(z) = O(z^{n-1})$$

By induction, one obtains the result. □

Theorem 2.3.27. *If* f *and* g *are Schur functions and* (2.3.79) *holds, then*

$$|f(z) - g(z)| \le 2|z|^{n+1} \tag{2.3.83}$$

for all $z \in \mathbb{D}$.

Proof. Let $h(z) = \frac{1}{2}(f(z) - g(z))$, which is a Schur function that, by Lemma 2.3.26, obeys $h(z) = O(z^{n+1})$. By repeated use of the Schwarz lemma, $h(z)/z^{n+1}$ is a Schur function so $|h(z)| \le |z|^{n+1}$, which is (2.3.83). □

Proof of Theorem 2.3.23. If $\gamma_j(f) = \gamma_j(g)$ for all j and $\gamma_j(f) \in \mathbb{D}^\infty$, then $|f(z) - g(z)| \leq 2|z|^n$ for all n, so taking $n \to \infty$, $f = g$ on $\mathbb{D}$. If $\gamma_N(f) \in \partial\mathbb{D}$ for some N, f and g can both be obtained via a finite Schur algorithm, and so are equal. Thus, $f \to \gamma(f)$ is one-one.

If $\gamma \in \mathbb{D}^\infty$, let $f^{[N]}$ be the Schur function with Schur parameters $(\gamma_0, \dots, \gamma_N, 0, 0, \dots)$ guaranteed by Lemma 2.3.25. Thus, by Theorem 2.3.27, if $M > N$,

$$|f^{[N]}(z) - f^{[M]}(z)| \leq 2|z|^N \tag{2.3.84}$$

so $f^{[N]}(z)$ is Cauchy and converges uniformly on compact subsets of $\mathbb{D}$ to $f(z)$ obeying

$$|f(z) - f^{[N]}(z)| \leq 2|z|^N \tag{2.3.85}$$

By Lemma 2.3.26, $\gamma_j(f) = \gamma_j(f^{[N]}) = \gamma_j$ for $j < N$. Thus, we have shown $f \to \gamma.(f)$ is onto. □

Remarks and Historical Notes. Most of the material is standard analysis textbook fare; see Duren [118] or Rudin [372]. The exception is the material on the Schur algorithm due to Schur [380]. The argument we use for that follows Schur, except that he has a weaker bound than (2.3.83). That bound is from [119].

Riesz's theorem (Proposition 2.3.8) fails for $p = 1$ or $p = \infty$. However, it is a theorem of Kolmogorov (see [118]) that if (2.3.44) holds for $p = 1$, then $f \in \bigcap_{p<1} H^p(\mathbb{D})$. In particular, since Carathéodory functions have

$$\int_0^{2\pi} |\text{Re}\, F(re^{i\theta})| \,\frac{d\theta}{2\pi} = \text{Re}\, F(0) = 1$$

we see that any Carathéodory function lies on $\bigcap_{p<1} H^p(\mathbb{D})$.

The standard references ([118, 372]) provide details of the fact that restriction of boundary values to sets of positive measures is injective (Theorem 2.3.20). For further discussion of the continuity result (Theorem 2.3.22), see [400, Section 10.11].

The general form of the representation theorem for Herglotz functions is that (see [14, Theorem 2 of Section 59] or [240, Chapter 6]) there exists a measure $d\rho$ on $\mathbb{R}$ with

$$\int \frac{d\rho(x)}{1+x^2} < \infty \tag{2.3.86}$$

and $A \geq 0$ so that

$$G(z) = \text{Re}\, G(i) + Az + \int \left(\frac{1}{x-z} - \frac{x}{1+x^2} \right) d\rho(x) \tag{2.3.87}$$

The fact that the Poisson representation, (2.3.17), holds for H^p functions, $p \geq 1$, will be critical later on (e.g., in Section 3.3) in showing a certain function has no singular inner part. For us, $p = 2$ suffices, so let us mention the easy proof in this case (and the full result is proven from the case $p = 2$ and Blaschke product technology). If $f(z) = \sum_{n=0}^\infty a_n z^n$ and $f_r(e^{i\theta}) = f(re^{i\theta})$, then $\|f_r\|^2_{L^2(d\theta/2\pi)} = \sum_{n=0}^\infty |a_n|^2 r^{2n}$, so $f \in H^2 \Leftrightarrow \sum_{n=0}^\infty |a_n|^2 < \infty$. In that case, one can define

$f(e^{i\theta})$ as the L^2 function with $\{a_n\}$ as Fourier coefficients. Since $\|f_1 - f_r\|_2^2 = \sum_{n=0}^{\infty}|a_n|^2(1 - r^{2n}) \to 0$ (by dominated convergence), we get L^2, so L^1 convergence. Since (2.3.17) holds for $f(rz)$, it holds for $f(z)$.

2.4 WEYL SOLUTIONS

One of the key tools of spectral analysis as originally developed in the study of Sturm–Liouville operators is the connection of spectral measures with solutions L^2 at infinity. For ODEs, such solutions were first emphasized by Weyl, so we will call them Weyl solutions. In this section, we will discuss Weyl solutions for OPUC associated with the work of Geronimus, Geronimo, Golinskii, Nevai, and Peherstorfer.

Of course, we have to ask: Solutions of what? Define for $\alpha \in \mathbb{D}$ and $z \in \mathbb{C}$,

$$\Lambda(\alpha, z) = \begin{pmatrix} z & -\bar{\alpha} \\ -\alpha z & 1 \end{pmatrix} \tag{2.4.1}$$

Then the Szegő recursion (2.1.4) and its dual (2.1.5) can be written

$$\begin{pmatrix} \Phi_{n+1}(z) \\ \Phi_{n+1}^*(z) \end{pmatrix} = \Lambda(\alpha_n, z) \begin{pmatrix} \Phi_n(z) \\ \Phi_n^*(z) \end{pmatrix} \tag{2.4.2}$$

Using (1.8.9), we can also write the recursion for the orthonormal polynomials (this is (2.1.6) and (2.1.7))

$$\begin{pmatrix} \varphi_{n+1}(z) \\ \varphi_{n+1}^*(z) \end{pmatrix} = A(\alpha_n, z) \begin{pmatrix} \varphi_n(z) \\ \varphi_n^*(z) \end{pmatrix} \tag{2.4.3}$$

where

$$A(\alpha, z) = \rho^{-1}\Lambda(\alpha, z) \qquad \rho = (1 - |\alpha|^2)^{1/2} \tag{2.4.4}$$

It is easy to see that

$$A(\alpha, z)^{-1} = \frac{1}{\rho z} \begin{pmatrix} 1 & \bar{\alpha} \\ \alpha z & z \end{pmatrix} \tag{2.4.5}$$

so (2.4.3) implies the inverse recursion relations:

$$z\varphi_n(z) = \rho_n^{-1}(\varphi_{n+1}(z) + \bar{\alpha}_n \varphi_{n+1}^*(z)) \tag{2.4.6}$$

$$\varphi_n^*(z) = \rho_n^{-1}(\varphi_{n+1}^*(z) + \alpha_n \varphi_{n+1}(z)) \tag{2.4.7}$$

We define the *transfer matrix* by

$$T_n(\{\alpha_j\}_{j=0}^{n-1}, z) = A(\alpha_{n-1}, z)A(\alpha_{n-2}, z) \dots A(\alpha_0, z) \tag{2.4.8}$$

so

$$\begin{pmatrix} \varphi_n(z) \\ \varphi_n^*(z) \end{pmatrix} = T_n(z) \begin{pmatrix} 1 \\ 1 \end{pmatrix} \tag{2.4.9}$$

By a *solution of the Szegő recursion* at a point $z \in \mathbb{C}$, we mean $v_n \in \mathbb{C}^2$ obeying

$$v_{n+1} = A(\alpha_n, z)v_n \quad \Leftrightarrow \quad v_n = T_n(z)v_0 \tag{2.4.10}$$

Notice this is for fixed z and there is no dual relation between the two components of v_n. (Indeed, since we do not have a solution at $1/z$, there could not be a dual relation!)

One of the main results in this section is

Theorem 2.4.1. *Let $z \in \mathbb{D}$. The solution of* (2.4.10) *with*

$$v_0 = \begin{pmatrix} 1 \\ zf(z) \end{pmatrix} \tag{2.4.11}$$

is in ℓ^2 (i.e., $\sum_{n=0}^{\infty} \|v_n\|^2 < \infty$) and it is the only such solution up to a constant multiple.

We will actually prove that v_n with v_0 given by (2.4.11) obeys more than an ℓ^2 condition. In fact, $\|v_n\| \le Cz^n$ (see Theorem 2.4.5) and not only is this the only solution in ℓ^2 but no other solution has $\|v_n\| \to 0$. By giving us a direct link between the spectral measures and the recursion relation, Theorem 2.4.1 will be the key to coefficient stripping and a key step in the strategy outlined in Section 2.1. We will prove Theorem 2.4.1 by using a certain set of polynomials:

Definition. The *monic second kind polynomials*, $\Psi_n(z)$, are defined by

$$\Psi_0(z) = 1 \tag{2.4.12}$$

and for $n \ge 1$,

$$\Psi_n(z) = \int \frac{e^{i\theta} + z}{e^{i\theta} - z} \left(\Phi_n(e^{i\theta}) - \Phi_n(z)\right) d\mu(\theta) \tag{2.4.13}$$

Theorem 2.4.2. *Ψ_n is a monic polynomial of degree n and obeys for $n =$ $1, 2, 3, \ldots,$*

$$\Psi_n^*(z) = -z^n \int \frac{e^{i\theta} + z}{e^{i\theta} - z} \left(\overline{\Phi_n(e^{i\theta})} - \overline{\Phi_n(1/\bar{z})}\right) \tag{2.4.14}$$

and for $n = 0, 1, 2, \ldots,$

$$\Psi_{n+1}(z) = z\Psi_n(z) + \overline{\alpha_n(d\mu)}\, \Psi_n^*(z) \tag{2.4.15}$$

Proof. If $Q(e^{i\theta}, z) = (e^{i\theta} + z)/(e^{i\theta} - z)$, then

$$\overline{Q(e^{i\theta}, 1/\bar{z})} = \frac{e^{-i\theta} + \frac{1}{z}}{e^{-i\theta} - \frac{1}{z}} = -Q(e^{i\theta}, z)$$

which implies (2.4.14).

We first check (2.4.15) for $n = 0$. Since $\int \Phi_1(e^{i\theta})\, d\mu(\theta) = 0$, we have, since $\Phi_1(z) = z - \bar{\alpha}_0$, that

$$\bar{\alpha}_0 = \int e^{i\theta}\, d\mu(\theta) \tag{2.4.16}$$

By (2.4.13),

$$\begin{aligned}\Psi_1(z) &= \int \frac{e^{i\theta}+z}{e^{i\theta}-z}\,[(e^{i\theta}-\bar\alpha_0)-(z-\bar\alpha_0)]\,d\mu(\theta)\\ &= \int (e^{i\theta}+z)\,d\mu(\theta)\\ &= z+\bar\alpha_0\end{aligned}$$

by (2.4.16). This is (2.4.15) for $n=0$ since $\Psi_0=\Psi_0^*=1$.

Now suppose $n\geq 1$. By Szegő recursion and (2.4.14),

$$\text{LHS of (2.4.15)} - \text{RHS of (2.4.15)} = \boxed{1}+\boxed{2} \tag{2.4.17}$$

where

$$\boxed{1} = \int \frac{e^{i\theta}+z}{e^{i\theta}-z}\,(e^{i\theta}-z)\Phi_n(e^{i\theta})\,d\mu(\theta) \tag{2.4.18}$$

$$\boxed{2} = -\bar\alpha_n \int \frac{e^{i\theta}+z}{e^{i\theta}-z}\,(e^{in\theta}-z^n)\,\overline{\Phi_n(e^{i\theta})}\,d\mu(\theta) \tag{2.4.19}$$

Since $n\geq 1$, $\int \Phi_n(e^{i\theta})\,d\mu(\theta)=\langle\Phi_n,1\rangle=0$ while

$$\begin{aligned}\int e^{i\theta}\Phi_n(e^{i\theta})\,d\mu(\theta) &= \int (\Phi_{n+1}(e^{i\theta})+\bar\alpha_n\Phi_n^*(e^{i\theta}))\,d\mu(\theta)\\ &= \bar\alpha_n\langle 1,\Phi_n^*\rangle\\ &= \bar\alpha_n\langle\Phi_n,z^n\rangle\end{aligned}$$

so

$$\boxed{1} = \int (e^{i\theta}+z)\Phi_n(e^{i\theta})\,d\mu(\theta) = \bar\alpha_n\langle\Phi_n,z^n\rangle$$

In computing $\boxed{2}$, we note

$$\frac{e^{i\theta}+z}{e^{i\theta}-z}\,(e^{in\theta}-z^n) = (e^{i\theta}+z)\sum_{j=0}^{n-1} e^{ij\theta}z^{(n-1-j)}$$

and that

$$\int e^{ij\theta}\,\overline{\Phi_n(e^{i\theta})}\,d\mu(\theta)=0 \qquad j=0,1,\dots,n-1$$

so

$$\begin{aligned}\boxed{2} &= -\bar\alpha_n\int e^{in\theta}\,\overline{\Phi_n(e^{i\theta})}\,d\mu(\theta)\\ &= -\bar\alpha_n\langle\Phi_n,z^n\rangle\end{aligned}$$

Therefore, $\boxed{1}+\boxed{2}=0$ and (2.4.17) implies (2.4.15). □

This theorem says that Ψ_n are the monic OPUC for the measure $d\mu_{-1}$ with

$$\alpha_j(d\mu_{-1}) = -\alpha_j(d\mu) \tag{2.4.20}$$

Of course, we have not yet proven Verblunsky's theorem (Theorem 1.8.5), so we do not know such a $d\mu_{-1}$ exists, and we are currently in the middle of a sequence of arguments that will lead to Verblunsky's theorem. So we do not want to use $d\mu_{-1}$. Fortunately, we will not need to. We may not have $d\mu_{-1}$ but we have

$$\rho_j = (1 - |-\alpha_j|^2)^{1/2} = (1 - |\alpha_j|^2)^{1/2}$$

so we still define

$$\psi_n(z) = \left(\prod_{j=0}^{n-1} \rho_j^{-1}\right) \Psi_n(z) \tag{2.4.21}$$

Since

$$\begin{pmatrix} \psi_{n+1} \\ \psi_{n+1}^* \end{pmatrix} = A(-\alpha_n, z) \begin{pmatrix} \psi_n \\ \psi_n^* \end{pmatrix} \tag{2.4.22}$$

and

$$A(-\alpha_n, z) = \begin{pmatrix} 1 & 0 \\ 0 & -1 \end{pmatrix} A(\alpha, z) \begin{pmatrix} 1 & 0 \\ 0 & -1 \end{pmatrix}^{-1} \tag{2.4.23}$$

we have

$$\begin{pmatrix} \psi_{n+1} \\ -\psi_{n+1}^* \end{pmatrix} = A(\alpha_n, z) \begin{pmatrix} \psi_n \\ -\psi_n^* \end{pmatrix} \tag{2.4.24}$$

so if $T_n(z)$ is the transfer matrix for $\{\alpha_j\}_{j=0}^\infty$, we have

$$\begin{pmatrix} \psi_n \\ -\psi_n^* \end{pmatrix} = T_n(z) \begin{pmatrix} 1 \\ -1 \end{pmatrix} \tag{2.4.25}$$

(2.4.25) and (2.4.9) imply that

$$T_n(z) = \begin{pmatrix} \frac{1}{2}(\varphi_n(z) + \psi_n(z)) & \frac{1}{2}(\varphi_n(z) - \psi_n(z)) \\ \frac{1}{2}(\varphi_n^*(z) - \psi_n^*(z)) & \frac{1}{2}(\varphi_n^*(z) + \psi_n^*(z)) \end{pmatrix} \tag{2.4.26}$$

Next, we define

$$g_n(z) = \psi_n(z) + F(z)\varphi_n(z) \tag{2.4.27}$$

$$g_n^*(z) = -\psi_n^*(z) + F(z)\varphi_n^*(z) \tag{2.4.28}$$

which is also a solution of (2.4.10). This solution is natural since $F(z)\varphi_n(z)$ cancels a part of the integral that defines $\psi_n(z)$.

As an aside, we note the * in g_n^* is *not* supposed to denote $z^n\, \overline{g_n(1/\bar{z})}$. Indeed,

Proposition 2.4.3. *If* $\operatorname{supp}(d\mu) \neq \partial\mathbb{D}$, *then* $F(z)$ *has an analytic continuation to* $\mathbb{C} \setminus \operatorname{supp}(d\mu)$ *which obeys*

$$F(z) = -\overline{F(1/\bar{z})} \tag{2.4.29}$$

and in that case, $g_n(z)$ *has an analytic continuation and*

$$g_n^*(z) = -z^n\, \overline{g_n(1/\bar{z})} \tag{2.4.30}$$

Proof. The definition of F, (2.3.24), shows F is analytic for $z \in \mathbb{C} \setminus \operatorname{supp}(d\mu)$. Moreover, by (2.3.55), $\operatorname{Re} F(e^{i\theta}) = 0$ for $e^{i\theta} \in \partial\mathbb{D} \setminus \operatorname{supp}(d\mu)$, so (2.4.29) holds at such $e^{i\theta}$, and so by analytic continuation for all z. (2.4.30) follows immediately from (2.4.27)–(2.4.29). □

The first key fact about g_n, g_n^* is

Proposition 2.4.4. *We have that*

$$\begin{pmatrix} g_n(z) \\ g_n^*(z) \end{pmatrix} = T_n(z) \begin{pmatrix} F(z)+1 \\ F(z)-1 \end{pmatrix} \tag{2.4.31}$$

$$= (F(z)+1) T_n(z) \begin{pmatrix} 1 \\ zf(z) \end{pmatrix} \tag{2.4.32}$$

Proof. (2.4.31) is immediate from (2.4.9) and (2.4.25)–(2.4.28). (2.4.32) then follows from (2.3.29). □

Here is a strong form of half of Theorem 2.4.1:

Theorem 2.4.5. *We have that*

$$g_n(z) = \int \frac{e^{i\theta}+z}{e^{i\theta}-z}\, \varphi_n(e^{i\theta})\, d\mu(\theta) \tag{2.4.33}$$

$$g_n^*(z) = z^n \int \frac{e^{i\theta}+z}{e^{i\theta}-z}\, \overline{\varphi_n(e^{i\theta})}\, d\mu(\theta) \tag{2.4.34}$$

Moreover,

$$|g_n(z)| \le 2|z|^n (1-|z|)^{-1} \tag{2.4.35}$$

$$|g_n^*(z)| \le 2|z|^n (1-|z|)^{-1} \tag{2.4.36}$$

$$|z^{-n} g_n^*(z)| \to 0 \qquad \text{as } n \to \infty \tag{2.4.37}$$

Remark. If $d\mu(\theta) = \frac{d\theta}{2\pi}$, then $g_n(z) = 2z^n$, so (2.4.35) cannot be much improved and, in particular, (2.4.37) may not hold for g_n.

Proof. (2.4.33) is immediate from (2.4.13) and the definition of F, as is (2.4.34) from (2.4.14). To get (2.4.35), we use

$$\frac{e^{i\theta}+z}{e^{i\theta}-z} = 1 + 2\sum_{j=1}^{\infty} e^{-ij\theta} z^j \tag{2.4.38}$$

and $\int e^{-ij\theta} \varphi_n(e^{i\theta})\, d\mu(\theta) = 0$ for $j = 0, 1, \dots, n-1$ plus

$$\left| \int e^{ij\theta} \varphi_n(e^{i\theta})\, d\mu(\theta) \right|^2 \le \int |\varphi_n(e^{i\theta})|^2\, d\mu(\theta) = 1 \tag{2.4.39}$$

to see that

$$|g_n(z)| \le 2 \sum_{j=n}^{\infty} |z|^j = \text{RHS of (2.4.35)}$$

On the other hand, for all θ,

$$\left| \frac{e^{i\theta} + z}{e^{i\theta} - z} \right| \le \frac{2}{1 - |z|} \tag{2.4.40}$$

so

$$h_z(\theta) = \frac{e^{i\theta} + z}{e^{i\theta} - z}$$

lies in $L^2(d\mu)$ and obeys

$$\|h_z\|_{L^2(d\mu)} \le 2(1 - |z|)^{-1}$$

Thus, (2.4.36) holds by the Schwarz inequality in $L^2(d\mu)$. Moreover, by the orthonormality of φ_n and Bessel's inequality,

$$\sum_{n=0}^{\infty} |z^{-n} g_n^*(z)|^2 \le 4(1 - |z|)^{-2} \tag{2.4.41}$$

which implies (2.4.37). □

The following completes the proof of Theorem 2.4.1:

Proposition 2.4.6. *Let $z \in \mathbb{D}$ be fixed and let*

$$\begin{pmatrix} a_n \\ b_n \end{pmatrix} = T_n(z) \begin{pmatrix} a_0 \\ b_0 \end{pmatrix} \tag{2.4.42}$$

Then

$$a_n g_n^*(z) - b_n g_n(z) = z^n (a_0 g_0^*(z) - b_0 g_0(z)) \tag{2.4.43}$$

In particular, if $|a_n| + |b_n| \to 0$, then for some c,

$$a_0 = c g_0(z) \qquad b_0 = c g_0^*(z) \tag{2.4.44}$$

Proof. Let N_n be the matrix $\begin{pmatrix} a_n & g_n(z) \\ b_n & g_n^*(z) \end{pmatrix}$. Then

$$N_n = T_n(z) N_0$$

(2.4.43) is just $\det(N_n) = \det(T_n(z)) \det N_0(0)$ given that

$$\det(A(\alpha, z)) = z \tag{2.4.45}$$

If $|a_n| + |b_n| \to 0$ then, by (2.4.35)/(2.4.36), $|z|^{-n} |a_n g_n^*(z) - b_n g_n(z)| \to 0$, so $a_0 g_0^*(z) - b_0 g_0(z) = 0$, which implies (2.4.43). □

Remark. (2.4.37) shows we only need $|b_n| \to 0$ and $\sup_n |a_n| < \infty$ to conclude (2.4.43).

The above arguments show that:

Theorem 2.4.7.

$$|g_m^*(z)| \le |z|\, |g_m(z)| \qquad \textit{for } m \textit{ and all } z \in \mathbb{D} \tag{2.4.46}$$

Proof. Fix m. Consider the measure $d\mu_m$ of (2.1.14). $\binom{a_n}{b_n} = T_{m+n}(z)\binom{1}{zf(z)}$ is ℓ^2 and of the form $T_n(z; \{\alpha_j(d\mu_m)\}_{j=0}^{\infty})\binom{g_m(z)}{g_m^*(z)}$. It follows that

$$g_m^*(z) = zf(z; d\mu_m)g_m(z) \tag{2.4.47}$$

which implies (2.4.46) since $|f(z)| \leq 1$. □

One interesting application of Weyl solutions is to an explicit formula for $F(z)$ when $\alpha_j(d\mu) = 0$ for $j \geq N$:

Theorem 2.4.8. *If*

$$\alpha_j(d\mu) = 0 \qquad j \geq N \tag{2.4.48}$$

then for $|z| < 1$,

$$F(z; d\mu) = \frac{\Psi_N^*(z)}{\Phi_N^*(z)} = \frac{\psi_N^*(z)}{\varphi_N^*(z)} \tag{2.4.49}$$

Remark. By Theorem 1.8.4, $\Phi_N^*(z)$ is nonvanishing on $\mathbb{D}$.

Proof. (2.4.48) implies (by (2.1.7)) that

$$\psi_{N+j}^*(z) = \psi_N^*(z) \qquad \varphi_{N+j}^*(z) = \varphi_N^*(z) \tag{2.4.50}$$

so

$$g_{N+j}^*(z) = g_N^*(z) \tag{2.4.51}$$

which implies

$$g_N^*(z) = 0 \tag{2.4.52}$$

since $g_{N+j}^*(z) \to 0$ as $j \to \infty$. (2.4.52) is equivalent to $F(z) = \psi_N^*(z)/\varphi_N^*(z)$ and the equality of normalization factors implies all of (2.4.49). □

Corollary 2.4.9. *For any measure, $d\mu$, uniformly on compact subsets of $\mathbb{D}$,*

$$F(z; d\mu) = \lim_{N\to\infty} \frac{\psi_N^*(z)}{\varphi_N^*(z)} \tag{2.4.53}$$

Proof. Immediate from (2.1.22) and (2.4.49). □

Corollary 2.4.10. *If* (2.4.48) *holds, then*

$$d\mu(\theta) = \frac{1}{|\varphi_N(e^{i\theta})|^2}\frac{d\theta}{2\pi} \tag{2.4.54}$$

In particular, for any measure $d\mu$,

$$\int \frac{1}{|\varphi_k(e^{i\theta})|^2}\frac{d\theta}{2\pi} = 1 \tag{2.4.55}$$

for all k.

Proof. By the same determinantal argument that led to (2.4.43), we have for any measure that

$$\varphi_k^*(z)\psi_k(z) + \varphi_k(z)\psi_k^*(z) = 2z^k \tag{2.4.56}$$

(since $\varphi_0^*\psi_0 + \varphi_0\psi_0^* = 2$). This implies that at $z = e^{i\theta}$,

$$\mathrm{Re}(\bar{\varphi}_k(e^{i\theta})\psi_k(e^{i\theta})) = 1 \tag{2.4.57}$$

Now suppose (2.4.48) holds. By Theorem 1.8.4 and (2.4.49), $F(z)$ is bounded as $|z| \uparrow 1$ so, by Proposition 2.3.12, $d\mu_s = 0$. By (2.4.48) using $F(e^{i\theta})$ for the limit of $F(re^{i\theta})$ as $r \uparrow 1$,

$$\begin{aligned} \mathrm{Re}\, F(e^{i\theta}) &= \mathrm{Re}\left(\frac{\psi_N^*(e^{i\theta})}{\varphi_N^*(e^{i\theta})}\right) \\ &= \mathrm{Re}\left(\frac{\psi_N(e^{i\theta})}{\varphi_N(e^{i\theta})}\right) \\ &= \frac{1}{|\varphi_N(e^{i\theta})^2|}\, \mathrm{Re}(\,\overline{\varphi_N(e^{i\theta})}\, \psi_N(e^{i\theta})) \\ &= \frac{1}{|\varphi_N(e^{i\theta})|^2} \end{aligned}$$

by (2.4.57).

The fact that $d\mu$ is normalized implies (2.4.55) for $d\mu$ obeying (2.4.48) and $k = N$. Since

$$\varphi_k(z; d\mu) = \varphi_k(z; d\mu^{(k)}) \tag{2.4.58}$$

this implies (2.4.55) in general. □

We will see later (see the remark after Theorem 2.7.2) that (2.4.54) implies Szegő's theorem for μ's obeying (2.4.48).

Corollary 2.4.11. *For any measure $d\mu$,*

$$d\mu = \text{w-lim}\ \frac{1}{|\varphi_n(e^{i\theta})|^2}\, \frac{d\theta}{2\pi} \tag{2.4.59}$$

Proof. Immediate from (2.1.22), (2.4.54), and (2.4.58). □

(2.4.59) is called the *Bernstein–Szegő approximation.*

Remarks and Historical Notes. It was Weyl [458] who first studied solutions of $-u'' + Vu = \lambda u$ L^2 at infinity and Titchmarsh [440] who emphasized the spectral nature of these solutions (Weyl–Titchmarsh m-function; see Bennewitz–Everitt [39, 124] for further history).

For OPRL, Nevanlinna [325], codified by Akhiezer [13], emphasized the study of ℓ^2 solutions in the moment problem; see also Simon [395].

For OPUC, second kind polynomials were introduced by Geronimus [158, 160, 161]. Weyl solutions in that language were introduced by Geronimo [153], Peherstorfer [336], and Golinskii–Nevai [177], although Golinskii earlier considered the unnormalized analog of g_n, g_n^*. Our approach here owes a lot to

Geronimo [153] and Golinskii [173]. The Golinskii–Nevai [177] approach (see also [399]) uses Weyl circles instead.

Theorem 2.4.8 is due to Verblunsky [453]. He did not define the $\Phi_N(z)$ but defined polynomials via explicit determinants that are essentially the $\Phi_N^*(z)$ and $\Psi_N^*(z)$ and proved (2.4.49). He also proved Corollary 2.4.11.

These results were rediscovered by Geronimus and were central to his work in [158, 160, 161].

Bernstein and Szegő did not write down the approximation in (2.4.59), but they studied measures of the form $\frac{d\theta}{2\pi}|P_n(e^{i\theta})|^{-2}$ for some polynomial, P, and even used them to prove some pointwise limit theorems for OPUC; see [43, 434].

2.5 COEFFICIENT STRIPPING, GERONIMUS' AND VERBLUNSKY'S THEOREMS, AND CONTINUED FRACTIONS

As we saw in the proof of Theorem 2.4.7,

$$T_n(z, \{\alpha_j\}_{j=0}^{n-1})\begin{pmatrix} 1 \\ zf(z; d\mu) \end{pmatrix} = c\begin{pmatrix} 1 \\ zf(z; d\mu_n) \end{pmatrix} \tag{2.5.1}$$

In particular,

$$A(\alpha_0, z)\begin{pmatrix} 1 \\ zf(z; d\mu) \end{pmatrix} = c\begin{pmatrix} 1 \\ zf(z; d\mu_1) \end{pmatrix} \tag{2.5.2}$$

so

$$zf(z; d\mu_1) = \frac{zf(z) - z\alpha_0}{z - \bar{\alpha}_0 zf(z)} \tag{2.5.3}$$

since c drops out. For $z \neq 0$, we can cancel z in the right side of (2.5.3) and then take limits as $z \to 0$. We thus have:

Theorem 2.5.1 (Coefficient Stripping). *$f(z; d\mu)$ and $f(z; d\mu_1)$ are related by*

$$zf(z; d\mu_1) = \frac{f(z; d\mu) - \alpha_0}{1 - \bar{\alpha}_0 f(z; d\mu)} \tag{2.5.4}$$

$$f(z; d\mu) = \frac{\alpha_0 + zf(z; d\mu_1)}{1 + \bar{\alpha}_0 zf(z; d\mu_1)} \tag{2.5.5}$$

Proof. (2.5.4) follows from (2.5.3) as noted. (2.5.5) then follows from (2.3.64) □

Recognizing that (2.5.5) is the Schur algorithm (2.3.76) since it implies $f(0; d\mu) = \alpha_0$, we see that

Theorem 2.5.2 (Geronimus' Theorem). *Let $d\mu$ be a nontrivial probability measure on $\partial\mathbb{D}$. Let $f(z; d\mu)$ be the Schur function of $d\mu$ (given by (2.3.26)/(2.3.29)) and $\{\gamma_n\}_{n=0}^\infty$ its Schur parameters. Let $\alpha_n(d\mu)$ be the Verblunsky coefficients of $d\mu$. Then*

$$\alpha_n(d\mu) = \gamma_n \tag{2.5.6}$$

If $f_n(z; d\mu)$ are the Schur iterates of $f(z; d\mu)$, then

$$f_n(z; d\mu) = f(z; d\mu_n) \tag{2.5.7}$$

Remarks. 1. Trivial measures lead to f's, which are Blaschke products, hence, by Proposition 2.3.24, to finite sequence $\{\gamma_n\}_{n=0}^N$ with $\gamma_N \in \partial\mathbb{D}$. Here the number of pure points in $d\mu$ is $N+1$. Thus, one can form $\Phi_0, \Phi_1, \dots, \Phi_{N+1}$ where Φ_{N+1} is the monic polynomial of degree $N+1$ that vanishes exactly at the $N+1$ points. (2.5.6) still holds, that is, $\gamma_j = -\overline{\Phi_j(0)}$ for $j = 0, \dots, N$ and γ_N is in $\partial\mathbb{D}$.

2. The choice of minus sign and the bar in (1.8.5) is picked so that (2.5.6) holds.

Proof. As noted, (2.5.6) for $n = 0$ and (2.5.7) for $n = 1$ are (2.5.5). Now just iterate. □

Given theorem 2.3.23, we see that $d\mu \to \gamma_n(f(z; d\mu))$ is a bijection of nontrivial measures and $\mathbb{D}^\infty$, so (2.5.6) immediately implies

Theorem 2.5.3 (Verblunsky's Theorem ≡ Theorem 1.8.5)**.** *The map $d\mu \mapsto \{\alpha_j(d\mu)\}_{j=0}^\infty$ of nontrivial probability measures on $\partial\mathbb{D}$ to $\mathbb{D}^\infty$ is a bijection.*

Now that we have wedded the Schur algorithm to OPUC, it is a good time (but definitely an aside) to discuss continued fractions, a language that is a recurring theme in understanding F and m-functions. We begin with the simple continued fraction expansion for a real number $x > 1$. Let $\xi_0 \equiv [x]$ be the greatest integer less than x and $(x) = x - [x]$ its fractional part. Then $0 \le (x) < 1$, so $y_1(x) = \frac{1}{(x)} \in (1, \infty]$ with ∞ allowed and

$$x = \xi_0 + \frac{1}{y_1(x)} \tag{2.5.8}$$

so long as $y_1 \neq \infty$, that is, $x \notin \mathbb{Z}$, we can iterate and so define

$$y_n(x) = \xi_n + \frac{1}{y_{n+1}(x)} \tag{2.5.9}$$

If ever $y_n(x) = \infty$, we stop. In that case, it is easy to see x is rational and the converse is also true (see the Notes). Thus, any irrational $x \in (1, \infty)$ is described by a sequence $(\xi_0, \xi_1, \xi_2, \dots)$ of integers in $\{1, 2, 3, \dots\}$. In a moment we will see that the map is onto all such sequences.

The iteration of (2.5.8) leads to

$$x = \xi_0 + \cfrac{1}{\xi_1 + \cfrac{1}{\xi_2 + \cdots + \cfrac{1}{\xi_n + \cfrac{1}{y_{n+1}(x)}}}} \tag{2.5.10}$$

If y_{n+1} is replaced by ∞, we get the nth *continued fraction approximate*, $x_n(x)$. (2.5.9) is connected with the fractional linear transformation (FLT)

$$f_n(y) = \xi_n + \frac{1}{y} \tag{2.5.11}$$

$$= \frac{\xi_n y + 1}{y} \tag{2.5.12}$$

(2.5.10) is an iterated FLT. Such iterations, discussed extensively in Section 9.2, are most easily studied by using the equivalence on the vectors $\binom{v_1}{w_1}$, $\binom{v_2}{w_2}$ in $\mathbb{C}^2\setminus\{0\}$,

$$\begin{pmatrix} v_1 \\ w_1 \end{pmatrix} \doteq \begin{pmatrix} v_2 \\ w_2 \end{pmatrix} \Leftrightarrow \exists c \in \mathbb{C} \text{ so that } \begin{pmatrix} v_1 \\ w_1 \end{pmatrix} = c \begin{pmatrix} v_2 \\ w_2 \end{pmatrix} \tag{2.5.13}$$

Equivalence classes are lines through 0 (with 0 removed). The set of equivalence classes is the Riemann sphere viewed as the complex projective line.

The general FLT

$$f(y) = \frac{\alpha y + \beta}{\gamma y + \delta} \tag{2.5.14}$$

with $\alpha\delta - \beta\gamma \neq 0$ is equivalent to

$$\begin{pmatrix} f(y) \\ 1 \end{pmatrix} \doteq \begin{pmatrix} \alpha & \beta \\ \gamma & \delta \end{pmatrix} \begin{pmatrix} y \\ 1 \end{pmatrix} \tag{2.5.15}$$

and composition of FLT is just matrix multiplication. The particular $y \mapsto f_n(y)$ corresponds to $\left(\begin{smallmatrix} \xi_n & 1 \\ 1 & 0 \end{smallmatrix}\right)$ and if

$$T_n = \begin{pmatrix} \xi_0 & 1 \\ 1 & 0 \end{pmatrix} \begin{pmatrix} \xi_1 & 1 \\ 1 & 0 \end{pmatrix} \cdots \begin{pmatrix} \xi_n & 1 \\ 1 & 0 \end{pmatrix} \tag{2.5.16}$$

we have

$$\begin{pmatrix} x_n \\ 1 \end{pmatrix} \doteq T_n \begin{pmatrix} 1 \\ 0 \end{pmatrix} \tag{2.5.17}$$

for $\binom{y_{n+1}}{1} \doteq \binom{1}{1/y_{n+1}}$ goes to $\binom{1}{0}$ as $y_{n+1} \to \infty$.

T_n has the form, by induction,

$$T_n = \begin{pmatrix} q_n & q_{n-1} \\ p_n & p_{n-1} \end{pmatrix} \tag{2.5.18}$$

where for $n \geq 0$, we have the *Euler–Wallis equations*:

$$q_n = \xi_n q_{n-1} + q_{n-2} \qquad p_n = \xi_n p_{n-1} + p_{n-2} \tag{2.5.19}$$

with boundary conditions

$$q_{-1} = 1 \qquad q_{-2} = 0 \qquad p_{-1} = 0 \qquad p_{-2} = 1$$

By (2.5.17),

$$x_n = \frac{q_n}{p_n} \tag{2.5.20}$$

It is not hard to see that (2.5.19) implies p_n, q_n are relatively prime. The following shows the power of the matrix formalism:

Theorem 2.5.4. *For each n,*

$$\frac{q_{2n}}{p_{2n}} < \frac{q_{2n+2}}{p_{2n+2}} < x < \frac{q_{2n+1}}{p_{2n+1}} < \frac{q_{2n-1}}{p_{2n-1}} \tag{2.5.21}$$

$$\frac{q_n}{p_n} - \frac{q_{n+1}}{p_{n+1}} = \frac{(-1)^{n+1}}{p_n p_{n+1}} \tag{2.5.22}$$

$$\lim_{n\to\infty} \frac{q_n}{p_n} = x \tag{2.5.23}$$

Remark. The same proof shows that if x is not given, but $\xi_0, \xi_1, \ldots \geq 1$ are, then q_n/p_n has a limit, call it x, and then $\xi_0, \xi_1, \ldots$ are the sequence associated to x.

Proof. We start by noting that f_n given by (2.5.11) is order reversing. Hence,

$$y_n(x) = f_n(y_{n+1}(x)) > f_n(\infty)$$

Each time we apply f_j, the order reverses so

$$(-1)^n \frac{q_n}{p_n} = (-1)^n f_0 \circ \cdots \circ f_n(\infty) < (-1)^n f_0 \circ \cdots \circ f_n(y_{n+1}(x)) = (-1)^n x \tag{2.5.24}$$

Thus,

$$\frac{q_{2n}}{p_{2n}} < x < \frac{q_{2n+1}}{p_{2n+1}} \tag{2.5.25}$$

Next, note that since $\det \left(\begin{smallmatrix} \xi & 1 \\ 1 & 0 \end{smallmatrix}\right) = -1$, we have

$$\det T_{n+1} = (-1)^n \tag{2.5.26}$$

so, by (2.5.12),

$$q_{n+1}p_n - p_{n+1}q_n = (-1)^n \tag{2.5.27}$$

Dividing by $-p_n p_{n+1}$ gives (2.5.22). Since $p_{n+1} > p_{n-1}$ (on account of (2.5.19)), we see

$$\frac{1}{p_{n+1}p_n} < \frac{1}{p_n p_{n-1}}$$

and thus, by (2.5.22),

$$\left| \frac{q_{n+1}}{p_{n+1}} - \frac{q_n}{p_n} \right| < \left| \frac{q_{n-1}}{p_{n-1}} - \frac{q_n}{p_n} \right| \tag{2.5.28}$$

which, with (2.5.25), proves (2.5.21).

Finally, by $\xi_n \geq 1$, we have $p_n \geq p_{n-1} \geq p_{n-2}$ and so $p_n \geq 2p_{n-2}$. Thus, for $n \geq 0$,

$$p_{2n+1} \geq p_{2n} \geq 2^n$$

Thus, by (2.5.21)/(2.5.22),

$$\left| x - \frac{q_n}{p_n} \right| \leq \frac{1}{p_n p_{n+1}} \leq \frac{1}{2^{2n-1}} \tag{2.5.29}$$

which implies (2.5.23). □

There are two extensions of this that are often also called continued fractions. First, f_n is often replaced by

$$f_n(y) = \xi_n + \frac{\zeta_n}{y} \tag{2.5.30}$$

These are sometimes called "extended continued fractions." Second, ξ_n (and ζ_n) may become functions of an external variable, x or z. This happens most especially when one or both are linear functions of x or z. In the context of (2.5.30), we find (2.5.19) becomes

$$p_n = \xi_n p_{n-1} + \zeta_n p_{n-2} \tag{2.5.31}$$

If $\zeta_n = -a_{n-1}^2$ and $\xi_n = x - b_n$, (2.5.31) has the form of the OPRL recursion relation (1.2.8), which suggests that continued fractions have a close connection to OPRL. We will see this in Section 3.2—indeed, $m(z)$ has a continued fraction expansion, (3.2.31), whose approximants, (3.2.43), have the OPRL P_n in the denominator.

Because

$$\frac{\alpha + w}{1 - \bar\alpha w} = \alpha + \frac{1 - |\alpha|^2}{-\bar\alpha + \frac{1}{w}} \tag{2.5.32}$$

the Schur algorithm has the flavor of a continued fraction also (see the Notes). In any event, it suggests that we use a matrix approach to iterate the Schur algorithm and, in essence, the resulting matrix is the inverse of (2.5.1).

We have the following:

Theorem 2.5.5. *Given $\{\alpha_j\}_{j=0}^\infty$, for $j = 0, 1, 2, \dots,$ there exist polynomials A_j and B_j of degree at most j so that*

$$T_n(z) = \prod_{j=0}^{n-1} (1 - |\alpha_j|^2)^{-1/2} \begin{pmatrix} zB_{n-1}^*(z) & -A_{n-1}^*(z) \\ -zA_{n-1}(z) & B_{n-1}(z) \end{pmatrix} \tag{2.5.33}$$

Moreover,

$$\begin{pmatrix} 1 \\ zf(z) \end{pmatrix} \doteq \begin{pmatrix} B_{n-1}(z) & A_{n-1}^*(z) \\ zA_{n-1}(z) & zB_{n-1}^*(z) \end{pmatrix} \begin{pmatrix} 1 \\ zf_n(z) \end{pmatrix} \tag{2.5.34}$$

In particular,

$$f^{[n]}(z) = \frac{A_n(z)}{B_n(z)} \tag{2.5.35}$$

Remarks. 1. As in the proof of Theorem 2.3.23, $f^{[n]}(z)$ is the Schur function with Schur parameters $(\gamma_0, \gamma_1, \dots, \gamma_n, 0, 0 \dots)$.

2. A_n and B_n are called the *Wall polynomials.*

3. A_n, B_n are not monic; indeed, they may not be of degree n. They are normalized by $B_n(0) = 1$, so $B_n^*(z)$ is monic.

4. The * in (2.5.33) is the one suitable for degree $n-1$ polynomials.

Proof. Since $A(\alpha, z)$ is affine in z, $T_n(z)$ has components that are polynomials of degree at most n. For the * relations note that if $J = J^{-1} = \left(\begin{smallmatrix} 0 & 1 \\ 1 & 0 \end{smallmatrix}\right)$, then $JA(\alpha, z)J^{-1} = z\,\overline{A(\alpha, 1/\bar z)}$, so

$$JT_n(z)J^{-1} = z^n\, \overline{T_n(1/\bar z)} \tag{2.5.36}$$

which leads to the *-relation.

(2.5.34) follows from (2.5.1) and the lemma below. (2.5.35) comes from setting $f_{n+1} = 0$ in the relation (2.5.34) with n replaced by $n+1$. □

Lemma 2.5.6. *Let* $ad - bc \neq 0$ *and*

$$v \doteq \begin{pmatrix} a & b \\ c & d \end{pmatrix} w \tag{2.5.37}$$

then

$$w \doteq \begin{pmatrix} d & -b \\ -c & a \end{pmatrix} v \tag{2.5.38}$$

Proof. Follows immediately from

$$\begin{pmatrix} a & -b \\ -c & d \end{pmatrix} \begin{pmatrix} a & b \\ c & d \end{pmatrix} = (ad - bc)\mathbf{1} \tag{2.5.39}$$

$\square$

The recursion $T_n(z) = A(\alpha_{n-1}, z)T_{n-1}(z)$ immediately implies recursion relations for the Wall polynomials

$$A_n(z) = A_{n-1}(z) + \gamma_n z B_{n-1}^*(z) \tag{2.5.40}$$

$$B_n(z) = B_{n-1}(z) + \gamma_n z A_{n-1}^*(z) \tag{2.5.41}$$

$$A_n^*(z) = z A_{n-1}^*(z) + \bar{\gamma}_n B_{n-1}(z) \tag{2.5.42}$$

$$B_n^*(z) = z B_{n-1}^*(z) + \bar{\gamma}_n A_{n-1}(z) \tag{2.5.43}$$

$$A_0(z) = \gamma_0 \tag{2.5.44}$$

$$B_0(z) = 1 \tag{2.5.45}$$

Moreover, (2.4.26) and (2.5.33) lead to the Pintér–Nevai formulae

$$\Phi_n(z) = z B_{n-1}^*(z) - A_{n-1}^*(z) \tag{2.5.46}$$

$$\Psi_n(z) = z B_{n-1}^*(z) + A_{n-1}^*(z) \tag{2.5.47}$$

Remarks and Historical Notes. Coefficient stripping formulae for OPUC go back to Peherstorfer [338]. The name is from [399].

Geronimus' theorem was found by him in 1944 [158]. There are four proofs of it in [399] and one in [398]. See the notes in [399], especially Section 3.2, for a discussion of the history of these proofs.

Verblunsky's theorem was the main result in his paper [452], although his definition of α_n was different (see the Notes to Section 1.8). [399] has four proofs of it. In the later literature, it is often called "Favard's Theorem for the Unit Circle."

Continued fractions for reals were formally defined by Wallis in the 1600s with important contributions by Euler, Lagrange, and Galois. Knopp [231] and Ryll-Nardzewski [377] found a Lebesgue a.c. measure under which $x \to y_1(x)$ is ergodic and so they determined the a.e. typical distributions of the ξ's.

That a rational, x, has a finite string of ξ_n's is a consequence of the Euclidean algorithm. Write $x = p/q$ and then

$$\begin{aligned} p &= \xi_0 q + r_1 \\ q_1 &= \xi_1 r_1 + r_2 \\ r_1 &= \xi_2 r_2 + r_3 \\ &\vdots \end{aligned}$$

until $r_k = 0$.

Theorem 2.5.4 is largely classical. Khinchin's book [218] is the standard reference.

When the ζ's become functions, we have the "analytic theory of continued fractions" whose founders are Jacobi, Chebyshev, Markov, and Stieltjes. Wall's book [455] is the classic exposition.

The name "Wall polynomials" is due to Khrushchev [219], although the polynomials appear already in Schur's paper [380]. The Pintér–Nevai formula appeared in their paper [346]. As discussed by Geronimus [158] and Khrushchev [219], there is a continued fraction expansion associated to f for which the $f^{[N]}$ are only half the approximants.

2.6 THE RELATIVE SZEGŐ FUNCTION AND THE STEP-BY-STEP SUM RULE

Our goal in this section is to prove (2.1.16). Suppose we can find a nonvanishing analytic function $(\delta D)(z)$, which is in $H^p(\mathbb{D})$ so that

$$|(\delta D)(e^{i\theta})|^2 = \frac{w(\theta)}{w_1(\theta)} \tag{2.6.1}$$

Then Jensen's formula, essentially the fact that $\log|\delta D(z)|^2$ is harmonic, will lead to

$$|\delta D(0)|^2 = \exp\biggl(\frac{1}{2\pi}\int \log\biggl(\frac{w(\theta)}{w_1(\theta)}\biggr) d\theta\biggr) \tag{2.6.2}$$

This is a simplification. The existence of singular inner functions shows that $f \in H^p(\mathbb{D})$ and nonvanishing on $\mathbb{D}$, even $H^\infty(\mathbb{D})$ is not enough for

$$|f(0)| = \exp\biggl(\int \log|f(e^{i\theta})|\,\frac{d\theta}{2\pi}\biggr) \tag{2.6.3}$$

Rather, we need a condition that implies f is outer, for example, that $\log(f)$ is in some $H^p(\mathbb{D})$, $p \geq 1$. In fact, $\log(\delta D)(z)$ will be in $\cap_{p<\infty} H^p(\mathbb{D})$.

Of course, by (2.3.55),

$$\frac{w(\theta)}{w_1(\theta)} = \lim_{r\uparrow 1} \frac{\operatorname{Re} F(re^{i\theta})}{\operatorname{Re} F_1(re^{i\theta})} \tag{2.6.4}$$

One might hope to find $\delta D(z)$ so $|\delta D(z)|^2 = \operatorname{Re} F(z)/\operatorname{Re} F_1(z)$, but this cannot be because it would imply the right side of (2.6.2) is 1, which is inconsistent with (2.1.16). Instead, we have to expect an extra factor, which goes to 1 as $|z| \to 1$ nontangentially. Even this will not be quite true as we will see.

It will pay to rewrite $\operatorname{Re} F$ and $\operatorname{Re} F_1$ in terms of f and f_1.

From (2.3.27),

$$\operatorname{Re} F(z) = \frac{\operatorname{Re}(1+zf)(1-\bar{z}\bar{f})}{|1-zf|^2} = \frac{1-|zf|^2}{|1-zf|^2} \tag{2.6.5}$$

Our goal is to find the absolute square of an analytic function—and the $|1-zf|^2$ is good. We have to hope for some cancellations in $(1-|zf|^2)/(1-|zf_1|^2)$, so we forge ahead using (2.3.75) to rewrite $|zf_1|$:

$$\begin{aligned} 1-|zf_1|^2 &= 1 - \left|\frac{f-\alpha_0}{1-\bar{\alpha}_0 f}\right|^2 \\ &= \frac{|1-\bar{\alpha}_0 f|^2 - |f-\alpha_0|^2}{|1-\bar{\alpha}_0 f|^2} \\ &= \frac{(1-|f|^2)(1-|\alpha_0|^2)}{|1-\bar{\alpha}_0 f|^2} \end{aligned} \tag{2.6.6}$$

This suggests we define

$$(\delta_0 D)(z) = \frac{1-zf_1}{1-zf} \frac{1-\bar{\alpha}_0 f}{(1-|\alpha_0|^2)^{1/2}} \tag{2.6.7}$$

so that (2.6.5) for F and F_1 and (2.6.6) imply

$$\frac{\operatorname{Re} F(z)}{\operatorname{Re} F_1(z)} = |(\delta_0 D)(z)|^2 \frac{1-|zf|^2}{1-|f|^2} \tag{2.6.8}$$

The "correction term," $(1-|zf|^2)/(1-|f|^2)$, may not go to 1 as $|z| \uparrow 1$ nontangentially, but it does at points where $w(\theta) > 0$ since:

Lemma 2.6.1. *If* $F(e^{i\theta}) \equiv \lim_{r\uparrow 1} F(re^{i\theta})$ *exists (and is finite) and* $\operatorname{Re}(F(e^{i\theta})) > 0$, *then* $f(e^{i\theta}) = \lim_{r\uparrow 1} f(re^{i\theta})$ *exists and*

$$|f(e^{i\theta})| < 1 \tag{2.6.9}$$

In particular, for a.e. θ *with* $w(\theta) > 0$,

$$\lim_{r\uparrow 1} \frac{1-r^2|f(re^{i\theta})|^2}{1-|f(re^{i\theta})|^2} = 1 \tag{2.6.10}$$

Proof. $\operatorname{Re} F \geq 0$ implies $1+F$ is bounded away from zero so (2.3.29) implies f has a limit. If $w(\theta) = \operatorname{Re} F(e^{i\theta}) > 0$, then

$$1-|f(e^{i\theta})|^2 = 1 - \frac{|F-1|}{|F+1|} = \frac{4\operatorname{Re} F(e^{i\theta})}{|1+F(e^{i\theta})|^2} > 0 \tag{2.6.11}$$

proving (2.6.9). Since $F(re^{i\theta})$ has a limit for a.e. θ and $\lim|f(e^{i\theta})|^2 < 1$ implies (2.6.10), we have the final assertion. $\square$

This allows us to state the main properties of $(\delta_0 D)(z)$.

Theorem 2.6.2. *$(\delta_0 D)(z)$ is analytic in $\mathbb{D}$ and nonvanishing. Moreover,*

$$\log(\delta_0 D)(z) \in \bigcap_{p<\infty} H^p(\mathbb{D}) \tag{2.6.12}$$

and

$$|(\delta_0 D)(0)|^2 = 1 - |\alpha_0|^2 = \exp\biggl(\frac{1}{2\pi}\int \log|(\delta_0 D)(e^{i\theta})|^2\, d\theta\biggr) \tag{2.6.13}$$

We have that if $w(\theta) > 0$, then

$$\lim_{r\uparrow 1} |(\delta_0 D)(re^{i\theta})|^2 = \frac{w(\theta)}{w_1(\theta)} \tag{2.6.14}$$

and that up to sets of $d\theta$-measure 0*,*

$$\{\theta \mid w(\theta) > 0\} = \{\theta \mid w_1(\theta) > 0\} \tag{2.6.15}$$

so if this set is all of $\partial\mathbb{D}$ (up to sets of measure 0*), then*

$$1 - |\alpha_0|^2 = \exp\biggl(\frac{1}{2\pi}\int \log\biggl(\frac{w(\theta)}{w_1(\theta)}\biggr)\, d\theta\biggr) \tag{2.6.16}$$

Remarks. 1. (2.6.16) is the step-by-step sum rule.

2. $(\delta_0 D)(z)$ has a pole at a point z_0 in $\partial\mathbb{D}$ if z_0 is an isolated pure point of $d\mu$, so $\delta_0 D$ may not itself lie in $H^1(\mathbb{D})$. But since any Carathéodory function lies in $H^p(\mathbb{D})$ for all $p < 1$ (see the Notes to Section 2.3) and $(1 - zf)^{-1} = \frac{1}{2}(1 + F)$, we see $(1-zf)^{-1}$ is in all $H^p(\mathbb{D})$, $p < 1$. Since $|(\delta_0 D)(z)| \le 4(1-|\alpha_0|^2)^{-1/2}|(1-zf)^{-1}|$, we see $(\delta_0 D) \in \bigcap_{0<p<1} H^p(\mathbb{D})$. We really care about H^p properties of $\log(\delta_0 D)$, so this is just an aside.

3. More generally than (2.6.13), we have

$$(\delta_0 D)(z) = \exp\biggl(\frac{1}{4\pi}\int \frac{e^{i\theta}+z}{e^{i\theta}-z} \log|(\delta_0 D)(e^{i\theta})|^2\, d\theta\biggr) \tag{2.6.17}$$

since, as noted after Proposition 2.3.9, there is a Poisson formula for any function in H^1.

Proof. Since $|f(z)| < 1$, analyticity and the nonvanishing properties are immediate. Moreover, if $|w| < 1$, $(1-w)$ lies in the right half-plane, so $|\arg(1-w)| < \pi/2$. It follows that

$$|\operatorname{Im}\log(\delta_0 D)| = |\arg(\delta_0 D)| \le \frac{3\pi}{2} \tag{2.6.18}$$

so, by Riesz's theorem (Proposition 2.3.8), (2.6.12) holds. Thus, by (2.3.51),

$$\log(\delta_0 D)(0) = \int \log(\delta_0 D)(e^{i\theta})\, \frac{d\theta}{2\pi} \tag{2.6.19}$$

Taking real parts, multiplying by 2, and exponentiating yields (2.6.13). (2.6.14) is immediate from (2.6.8) and Lemma 2.6.1.

Lemma 2.6.1 also implies (2.6.15) if we note the following:

(i) By joint continuity of $(z, w) \to (\alpha_0 + zw)/(1 + \bar{\alpha}_0 zw)$ on $|z| \leq 1$, $|w| < 1$ implies that if $\lim f_1(re^{i\theta})$ exists and is less than 1, then $\lim f(re^{i\theta})$ exists.

(ii) By joint continuity of $(z, w) \to z^{-1}(\alpha_0 - w)/(1 - \bar{\alpha}_0 w)$ on $\frac{1}{2} \leq z \leq 1$, $|w| < 1$ implies that if $\lim f(re^{i\theta})$ exists and is less than 1, then $\lim f_1(re^{i\theta})$ exists.

(iii) For $|z_0| = 1$ and $|\alpha_0| < 1$, $(\alpha_0 + z_0 w)/(1 + \bar{\alpha}_0 zw) \in \mathbb{D}$ if and only if $w \in \mathbb{D}$.

Given these facts, (2.6.13) implies (2.6.16) if $w(\theta) > 0$ for a.e. θ. □

This is all we need for Szegő's theorem. For a higher-order Szegő theorem, it will be useful to have:

Theorem 2.6.3. *Suppose* $\{\theta \mid w(\theta) > 0\}$. *Then we have that*

$$\int e^{-i\theta} \log\left(\frac{w(\theta)}{w_1(\theta)}\right) \frac{d\theta}{2\pi} = \alpha_0 - \alpha_1 + \bar{\alpha}_0 \alpha_1 \tag{2.6.20}$$

Proof. By (2.3.52), if $h \in H^1(\mathbb{D})$,

$$\begin{aligned} \int e^{-i\theta} \operatorname{Re} h(e^{i\theta}) \frac{d\theta}{2\pi} &= \frac{1}{2} \int e^{-i\theta} h(e^{i\theta}) \frac{d\theta}{2\pi} + \frac{1}{2} \overline{\int e^{i\theta} h(e^{i\theta}) \frac{d\theta}{2\pi}} \\ &= \tfrac{1}{2} h'(0) \end{aligned} \tag{2.6.21}$$

Taking

$$h(z) = 2 \log(\delta_0 D)(z) \tag{2.6.22}$$

and using (2.6.14), we see that (2.6.20) is equivalent to

$$\tfrac{1}{2} h'(0) = \alpha_0 - \alpha_1 + \bar{\alpha}_0 \alpha_1 \tag{2.6.23}$$

By (2.6.7),

$$\tfrac{1}{2} h(z) = \log(1 - zf_1) - \log(1 - zf) + \log(1 - \bar{\alpha}_0 f) - \log(\rho_0) \tag{2.6.24}$$

and we need the $O(z)$ terms. Since $f(0) = \alpha_0$, $f_1(0) = \alpha_1$, and $\log(1 - zg) = -zg + O(z^2)$, the first two terms have $O(z)$ terms $-\alpha_1$ and α_0. The last term is z-independent. That leaves the third term. We note first that

$$f(z) = \frac{\alpha_0 + zf_1(z)}{1 + \bar{\alpha}_0 z f_1(z)} = \alpha_0 + (1 - |\alpha_0|^2)\alpha_1 z + O(z^2) \tag{2.6.25}$$

so

$$\begin{aligned} 1 - \bar{\alpha}_0 f &= (1 - |\alpha_0|^2) + \bar{\alpha}_0 \alpha_1 (1 - |\alpha_0|^2) z + O(z^2) \\ &= (1 - |\alpha_0|^2)(1 + \bar{\alpha}_0 \alpha_1 z + O(z^2)) \end{aligned} \tag{2.6.26}$$

so

$$\log(1 - \bar{\alpha}_0 f) = \log(1 - |\alpha_0|^2) + \bar{\alpha}_0 \alpha_1 z + O(z^2) \tag{2.6.27}$$

proving (2.6.23). □

Remarks and Historical Notes. The relative Szegő function was defined by Simon in [399, Section 2.9]. If it were not for the Killip–Simon analog for OPRL [225], he might not have found it. There the m-function was a natural object whose boundary values are $\log(\operatorname{Im} m/\operatorname{Im} m_1)$ (see (3.4.9)). For some OPUC issues, the analog of $m(z)$ is $F(z)$ but not here. Simon noted the basic step-by-step sum rule, (2.6.16), and also the higher-order (2.6.20).

The symbol $\delta_0 D$ is used because, making the α-dependence explicit, it is natural to define $\delta_n D$ inductively by

$$(\delta_n D)(z; \{\alpha_j\}_{j=0}^\infty) = (\delta_{n-1} D)(z; \{\alpha_j\}_{j=0}^\infty) + (\delta_0 D)(z; \{\alpha_{j+n}\}_{j=0}^\infty) \tag{2.6.28}$$

connected to stripping off n α's.

There are two alternate ways of writing $(\delta_0 D)(z)$. One, due to Killip–Nenciu (unpublished), is noted in [399, Proposition 2.9.4]:

$$(\delta_0 D)(z) = \frac{\rho_0}{1+\bar\alpha_0 z f_1}\,\frac{1-zf_1}{1-zf} \tag{2.6.29}$$

The other is in terms of

$$M(z) = z(1+\alpha_0)(1+F(z)) + (1+\bar\alpha_0)(1-F(z))$$

introduced in [400, Section 11.7] for

$$(\delta_0 D)(z) = (2z\rho_0)^{-1} M(z) \tag{2.6.30}$$

We note—as will be important later (see Section 5.14)—that if $d\mu$ has a gap in its essential spectrum, $(\delta_0 D)(z)$ has an analytic continuation to $[\mathbb{C}\setminus\partial\mathbb{D}]\cup$ gap with poles in the gap precisely at pure points of $d\mu$ and zeros at pure points of $d\mu_1$. This is also a property of m for OPRL. These properties hold since, by (2.3.57), z_0 is a pure point of $d\mu$ if and only if F has a pole at z_0, and by (2.3.27)/(2.3.29), this happens if and only if $1 - z_0 f(z_0) = 0$.

2.7 THE PROOF OF SZEGŐ'S THEOREM

We have all the pieces lined up and can follow the strategy of Section 2.1 to prove Theorem 2.1.1. We begin by iterating (2.6.16).

Theorem 2.7.1 (Iterated Step-by-Step Sum Rule)**.** *Let $d\mu$ obey* (2.1.1) *and let $d\mu_N$ be given by* (2.1.14) *and obey* (2.1.15). *Then up to sets of measure zero,*

$$\{\theta \mid w(\theta)\neq 0\} = \{\theta \mid w_N(\theta)\neq 0\} \tag{2.7.1}$$

and if this set is all of $\partial\mathbb{D}$, then

$$\log\biggl(\frac{w(\theta)}{w_N(\theta)}\biggr) \in \bigcap_{p<\infty} L^p\biggl(\partial\mathbb{D}, \frac{d\theta}{2\pi}\biggr) \tag{2.7.2}$$

and

$$\prod_{j=0}^{N-1}(1-|\alpha_j|^2) = \exp\biggl(\int \log\biggl[\frac{w(\theta)}{w_N(\theta)}\biggr]\frac{d\theta}{2\pi}\biggr) \tag{2.7.3}$$

Proof. Follows by induction by repeated use of Theorem 2.6.2 and

$$\log\left(\frac{w(\theta)}{w_N(\theta)}\right) = \sum_{j=1}^{N} \log\left(\frac{w_{j-1}(\theta)}{w_j(\theta)}\right) \tag{2.7.4}$$

(with $w_0 \equiv w$). $\square$

This implies Szegő's theorems for measures obeying (2.4.48):

Theorem 2.7.2 (Szegő's Theorem for Bernstein–Szegő Measures). *Suppose $d\mu$ is a measure with*

$$\alpha_j(d\mu) = 0 \qquad j \geq N \tag{2.7.5}$$

and let $d\mu$ obey (2.1.1). *then*

$$\prod_{j=0}^{N-1}(1 - |\alpha_j|^2) = \exp\left(\int \log(w(\theta)) \frac{d\theta}{2\pi}\right) \tag{2.7.6}$$

Proof. Since $w_N(\theta) = 1$, $\{\theta \mid w(\theta) \neq 0\} = \partial\mathbb{D}$ and (2.7.6) is just (2.7.3). $\square$

Remark. One can prove Theorem 2.7.2 directly without the need for the step-by-step sum rule by using the explicit formula (2.4.54) that holds when (2.7.5) is true. For it says

$$\log(w(\theta)) = -\log|\varphi_{N-1}^*(e^{i\theta})|^2 \tag{2.7.7}$$

Since $\varphi_{N-1}^*(z)$ is nonvanishing, the function $-\log|\varphi_{N_1}^*(z)|^2$ is harmonic on $\mathbb{D}$, so

$$\begin{aligned}\int \log(w(\theta)) \frac{d\theta}{2\pi} &= -\log|\varphi_{N-1}^*(0)|^2 \\ &= \log\left(\prod_{j=0}^{N-1}(1 - |\alpha_j|^2)\right)\end{aligned} \tag{2.7.8}$$

which is (2.1.5).

We now turn to step three of the strategy of Section 2.1 and use upper semicontinuity of the entropy:

Proposition 2.7.3. *For any nontrivial probability measure $d\mu$ on $\partial\mathbb{D}$,*

$$\prod_{j=0}^{\infty}(1 - |\alpha_j|^2) \leq \exp\left(\int \log(w(\theta)) \frac{d\theta}{2\pi}\right) \tag{2.7.9}$$

Proof. Let $d\mu^{(N)}$ be given by (2.1.27). By Theorem 2.7.2 and (2.2.5),

$$\prod_{j=0}^{N-1}(1 - |\alpha_j(d\mu)|^2) = \exp\left(S\left(\frac{d\theta}{2\pi} \,\middle|\, d\mu^{(N)}\right)\right) \tag{2.7.10}$$

Since $d\mu^{(N)} \xrightarrow{w} d\mu$, Theorem 2.2.3 implies

$$\prod_{j=0}^{\infty}(1-|\alpha_j(d\mu)|^2) \leq \exp\biggl(S\biggl(\frac{d\theta}{2\pi}\biggm| d\mu\biggr)\biggr) \tag{2.7.11}$$

which, by (2.2.5), is (2.7.9). □

We are now ready for step four in the strategy of Section 2.1.

Proposition 2.7.4. *For any nontrivial probability measure, $d\mu$, on $\partial\mathbb{D}$, and any N,*

$$\prod_{j=0}^{N-1}(1-|\alpha_j|^2) \geq \exp\biggl(\int \log(w(\theta))\,\frac{d\theta}{2\pi}\biggr) \tag{2.7.12}$$

Proof. If $\int \log(w(\theta))\frac{d\theta}{2\pi} = -\infty$, the inequality is trivial. If $\int \log(w(\theta))\frac{d\theta}{2\pi} > -\infty$, by (2.7.2), $\int \log(w_N(\theta))\frac{d\theta}{2\pi} > -\infty$ and

$$\int \log\biggl[\frac{w(\theta)}{w_N(\theta)}\biggr]\frac{d\theta}{2\pi} = \int \log(w(\theta))\,\frac{d\theta}{2\pi} - \int \log(w_N(\theta))\,\frac{d\theta}{2\pi} \tag{2.7.13}$$

so (2.7.3) becomes

$$\exp\biggl(\int \log(w_N(\theta))\,\frac{d\theta}{2\pi}\biggr)\prod_{j=0}^{N-1}(1-|\alpha_j|^2) = \exp\biggl(\int \log(w(\theta))\,\frac{d\theta}{2\pi}\biggr) \tag{2.7.14}$$

By (2.2.15) and (2.2.3),

$$\exp\biggl(\int \log(w_N(\theta))\,\frac{d\theta}{2\pi}\biggr) \leq 1 \tag{2.7.15}$$

(2.7.14) and (2.7.15) imply (2.7.12). □

Proof of Theorem 2.1.1. Taking $N \to \infty$ in (2.7.12), we get

$$\prod_{j=0}^{\infty}(1-|\alpha_j|^2) \geq \exp\biggl(\int \log(w(\theta))\,\frac{d\theta}{2\pi}\biggr) \tag{2.7.16}$$

This and (2.7.9) imply (2.1.11). □

Remarks and Historical Notes. This is the proof of Section 2.3 of [399] but is close to Verblunsky's proof [453] and is motivated by Killip–Simon [225].

2.8 A HIGHER-ORDER SZEGŐ THEOREM

In this section, we want to use the tools of this chapter to prove the following gem:

Theorem 2.8.1. *Let $d\mu$ obey* (2.1.1). *Then*

$$\int (1-\cos\theta)\log(w(\theta))\,\frac{d\theta}{2\pi} > -\infty \tag{2.8.1}$$

if and only if

$$\text{(i)} \qquad \sum_{n=0}^{\infty} |\alpha_n|^4 < \infty \tag{2.8.2}$$

$$\text{(ii)} \qquad \sum_{n=0}^{\infty} |\alpha_{n+1} - \alpha_n|^2 < \infty \tag{2.8.3}$$

Remarks. 1. Since $w \in L^1$, the integral in (2.8.1) is either absolutely convergent or diverges to $-\infty$.

2. If $\alpha_n = (n+1)^{-\beta}$, then

$$\sum_{n=0}^{\infty} |\alpha_n|^2 < \infty \tag{2.8.4}$$

if and only if $\beta > \frac{1}{2}$. (2.8.2)/(2.8.3) hold if and only if $\beta > \frac{1}{4}$, so $\beta \in (\frac{1}{4}, \frac{1}{2}]$ provides examples where (2.8.1) holds, but the integral is $-\infty$ if the $1 - \cos\theta$ is dropped. Thus, $\log(w(\theta))$ has a nonintegral divergence at $0 = 0$, but $(1 - \cos\theta)\log(w(\theta))$ is still integrable; indeed, it should be $w(\theta) \sim \exp(-c\theta^{-\gamma})$ for $1 < \gamma < 3$. The analogous weight for OPRL is studied in [248].

3. It is known (see the Notes) that if

$$\sum_{n=0}^{\infty} |\alpha_{n+1} - \alpha_n| < \infty \tag{2.8.5}$$

then $w(\theta)$ is continuous and positive on $(0, 2\pi)$ (but can vanish at $\theta = 0$ rather rapidly).

$$\alpha_n = (n+3)^{-1/3} + (-1)^n (n+3)^{-2/3}$$

is an example where (2.8.2)/(2.8.3) hold, but neither (2.8.4) nor (2.8.5) holds.

As with Szegő's theorem, the key is a step-by-step sum rule:

Theorem 2.8.2. *Let* $d\mu$ *obey* (2.1.1) *and suppose* $w(\theta) > 0$ *for a.e.* θ. *Then*

$$\int (1 - \cos\theta) \log\left(\frac{w(\theta)}{w_1(\theta)}\right) \frac{d\theta}{2\pi} = \log(1 - |\alpha_0|^2) - \operatorname{Re}(\alpha_0 - \alpha_1 + \bar{\alpha}_0 \alpha_1) \tag{2.8.6}$$

Proof. Immediate from (2.6.16), (2.6.20), and

$$1 - \cos\theta = 1 - \tfrac{1}{2}(e^{i\theta} + e^{-i\theta}) \tag{2.8.7}$$

□

In applying the strategy of the last section, a key role is played by "positivity," which in the context there meant $\log(1 - |\alpha_0|^2) \leq 0$. If $\alpha_0 = 0$, then the right-hand side of (2.8.6) $= -\operatorname{Re}\alpha_1$, which can have either sign, so there is not any strict positivity or negativity. However, the right side can be rewritten as something negative plus something that telescopes—and that we will see is enough. We will need the function

$$g(\alpha) = -\log(1 - |\alpha|^2) - |\alpha|^2 \tag{2.8.8}$$

Lemma 2.8.3. *Let $A < 1$. Then, if $|\alpha| \le A$, we have*

$$\left(\frac{1}{1-A^2}\right)^2 \frac{|\alpha|^4}{2} \ge g(\alpha) \ge \frac{|\alpha|^4}{2} \tag{2.8.9}$$

Proof. Let

$$G(y) = -\log(1-y) - y \tag{2.8.10}$$

so

$$G'(y) = \frac{1}{1-y} - 1 \qquad G''(y) = \frac{1}{(1-y)^2} \tag{2.8.11}$$

In particular, $G(0) = G'(0) = 0$, so by Taylor's theorem with remainder, if $0 \le y \le A^2$, then

$$G(y) = \frac{y^2}{2} G''(z(y)) = \frac{y^2}{2} \frac{1}{(1-z(y))^2} \tag{2.8.12}$$

for some $z(y) \in [0, A^2]$. Thus, since $(1-z)^{-2}$ is decreasing on $[0, A^2]$,

$$\frac{y^2}{2}\left(\frac{1}{1-A^2}\right)^2 \ge G(y) \ge \frac{y^2}{2}$$

which implies (2.8.9). □

By simple algebra,

$$\text{RHS of (2.8.6)} = -g(\alpha_0) - \tfrac{1}{2}|\alpha_0 - \alpha_1|^2 - \tfrac{1}{2}(|\alpha_0 + 1|^2 - |\alpha_1 + 1|^2) \tag{2.8.13}$$

This gives

Theorem 2.8.4 (Iterated Step-by-Step Sum Rule). . *Let $d\mu$ obey* (2.1.1) *and let $d\mu_N$ be given by* (2.1.14) *and obey* (2.1.15). *Then up to sets of measure zero,*

$$\{\theta \mid w(\theta) \ne 0\} = \{\theta \mid w_N(\theta) \ne 0\} \tag{2.8.14}$$

and if this set is all of $\partial\mathbb{D}$, then

$$\log\left(\frac{w(\theta)}{w_N(\theta)}\right) \in \bigcap_{p<\infty} L^p\left(\partial\mathbb{D}, \frac{d\theta}{2\pi}\right) \tag{2.8.15}$$

and

$$\int (1-\cos\theta)\log\left[\frac{w(\theta)}{w_N(\theta)}\right]\frac{d\theta}{2\pi}$$
$$= -\sum_{j=0}^{N-1} g(\alpha_j) - \tfrac{1}{2}\sum_{j=0}^{N-1}|\alpha_j - \alpha_{j+1}|^2 - \tfrac{1}{2}|\alpha_0 + 1|^2 + \tfrac{1}{2}|\alpha_N + 1|^2 \tag{2.8.16}$$

Proof. Rewrite (2.8.6) using (2.8.13) and iterate. □

Since

$$e^{-g(n)} = (1 - |\alpha|^2)e^{|\alpha|^2} \tag{2.8.17}$$

this suggests the sum rule:

$$e^{\frac{1}{2}(1-|1+\alpha_0|^2)}\prod_{j=0}^{\infty}(1-|\alpha_j|^2)e^{|\alpha_j|^2}e^{-\frac{1}{2}|\alpha_{j+1}-\alpha_j|^2}$$

$$=\exp\biggl(\int(1-\cos\theta)\log(w(\theta))\frac{d\theta}{2\pi}\biggr) \tag{2.8.18}$$

We just follow the strategy of the last section with two extra steps:

Proposition 2.8.5. *If $d\mu$ is a measure where* (2.7.5) *holds, then* (2.8.18) *holds.*

Proof. Follows immediately from exponentiating (2.8.16). □

Proposition 2.8.6. *For any nontrivial probability measure, $d\mu$, on $\partial\mathbb{D}$,*

$$\textit{LHS of } (2.8.18) \leq \textit{RHS of } (2.8.18) \tag{2.8.19}$$

Proof. We have just proven the equality for $d\mu^{(N)}$. As $N\to\infty$, the left-hand side converges since (2.6.11) says

$$(1-|\alpha_j|^2)e^{|\alpha_j|^2}e^{-\frac{1}{2}|\alpha_{j+1}-\alpha_j|^2}\leq 1 \tag{2.8.20}$$

implying monotonicity and so convergence (perhaps to 0).

On the other hand, the measure

$$d\tilde{\mu}_0=(1-\cos\theta)\,\frac{d\theta}{2\pi} \tag{2.8.21}$$

is a probability measure and

$$\int(1-\cos\theta)\log(w(\theta))\,\frac{d\theta}{2\pi}=S(d\tilde{\mu}_0\mid d\mu)+C \tag{2.8.22}$$

where

$$C=\int(1-\cos\theta)\log(1-\cos\theta)\,\frac{d\theta}{2\pi} \tag{2.8.23}$$

(which has $C=1-\log(2)$, but the precise value is unimportant). As in the proof of Proposition 2.7.3, upper semicontinuity of entropy implies (2.8.19). □

Proposition 2.8.7. *For any nontrivial probability measure, $d\mu$, on $\partial\mathbb{D}$ with $S(d\tilde{\mu}_0 \mid d\mu)>-\infty$, and any N,*

$$e^{\frac{1}{2}(1-|1+\alpha_0|^2)}\prod_{j=0}^{N-1}(1-|\alpha_j|^2)e^{|\alpha_j|^2}e^{-\frac{1}{2}|\alpha_j-\alpha_{j+1}|^2}$$

$$\geq e^{\frac{1}{2}(1-|1+\alpha_N|^2)}\exp(-L(w_N))\exp(L(w)) \tag{2.8.24}$$

where

$$L(w)=\int(1-\cos\theta)\log(w(\theta))\,\frac{d\theta}{2\pi} \tag{2.8.25}$$

Proof. As in the proof of Proposition 2.7.4, immediate from (2.6.29) since we can separate the $\log(w(\theta))$ and $\log(w_N(\theta))$ integrals. □

Theorem 2.8.8. *For any nontrivial measure,* (2.8.18) *holds.*

Proof. If $S(d\tilde{\mu}_0 \mid d\mu) = -\infty$, the right side of (2.8.18) is zero, and then by (2.8.19), the left side is zero and equality holds. Thus, we can suppose

$$S(d\tilde{\mu}_0 \mid d\mu) > -\infty \tag{2.8.26}$$

Then by (2.8.24) and (2.8.20) (which implies monotonicity of the left side), we have that

$$\text{LHS of (2.8.18)} \geq e^{a+b} \exp(L(w)) \tag{2.8.27}$$

where

$$a = \liminf \tfrac{1}{2}\,(1 - |1 + \alpha_N|^2) \tag{2.8.28}$$

$$b = \liminf(-L(w_N)) \tag{2.8.29}$$

Clearly, $a \geq \frac{1}{2}(1-4) = -\frac{3}{2}$ and by (2.8.22) and $S \leq 0$, $b \geq -C$, so (2.8.24) implies

$$\text{LHS of (2.8.18)} \geq e^{-3/2} e^{-C}\ \text{RHS of (2.8.18)} \tag{2.8.30}$$

Since we are supposing (2.8.26) holds, the left-hand side of (2.8.18) > 0, so

$$\sum_{j=0}^{\infty} g(\alpha_j) + \sum_{j=0}^{\infty} |\alpha_j - \alpha_{j+1}|^2 < \infty \tag{2.8.31}$$

and thus, by (2.8.9),

$$\sum_{j=0}^{\infty} |\alpha_j|^4 < \infty \tag{2.8.32}$$

This implies

$$\lim|\alpha_j| = 0 \tag{2.8.33}$$

and so, $a = 0$.

Moreover, by (2.8.33), $d\mu_n \overset{w}{\longrightarrow} d\mu_0$, so by (2.8.22) and the semicontinuity of the entropy,

$$\limsup L(w_n) \leq L(w_\infty \equiv 1) = 0 \tag{2.8.34}$$

so $b \geq 0$. Thus, by (2.8.27) and (2.8.19),

$$\text{RHS of (2.8.18)} \geq \text{LHS of (2.8.18)} \geq e^b \text{RHS of (2.8.18)} \tag{2.8.35}$$

which, with $b \geq 0$, implies $b = 0$, and (2.8.18) holds. □

Remark. The extra steps were needed to show $a = 0$ and $b = 0$ and so get the sum rule, but for the corollary, Theorem 2.8.1, (2.8.30) suffices and we can give a shorter argument.

Proof of Theorem 2.8.1. (2.8.1) holds if and only if the right-hand side of (2.8.18) > 0. (2.8.2)/(2.8.18) hold if and only if the left-hand side of (2.8.18) > 0. Thus, Theorem 2.8.8 implies Theorem 2.8.1. □

Remarks and Historical Notes. Theorem 2.8.1 and the sum rule (2.8.18) first appeared in Section 2.8 of [399]. Rather than using a relative Szegő function and a step-by-step sum rule, the proof there exploits Szegő's theorem. Our proof here is patterned after Simon–Zlatoš [411] who proved a more complicated result (see later).

My motivation in seeking those results was that the OPRL analog of Szegő's theorem was the C_0 sum rule of Section 3.6. I felt there had to be an OPUC analog of the P_2 sum rule of Killip–Simon [225] with positivity.

Even before the higher-order sum rule discussed here, there were higher-order sum rules without full positivity for OPRL. These are discussed in the Notes to Section 3.6.

In [399], Simon conjectured a generalization of Theorem 2.8.1, namely,

Conjecture 2.8.9. *Fix $\theta_1, \dots, \theta_k$ distinct in $[0, 2\pi)$ and $m_1, \dots, m_k$ strictly positive integers. Then*

$$\int \prod_{j=1}^{k} [1 - \cos(\theta - \theta_j)]^{m_j} \log(w(\theta)) \frac{d\theta}{2\pi} > -\infty \tag{2.8.36}$$

if and only if

$$\sum_{n=0}^{\infty} \left| \left\{ \prod_{j=1}^{k} [\delta - e^{-i\theta_j}]^{m_j} \alpha \right\}_n \right|^2 + |\alpha_n|^{2\ell} < \infty \tag{2.8.37}$$

where $\ell = 1 + \max_{j=1,\dots,k} m_j$ and where

$$(\delta\alpha)_n = \alpha_{n+1} \tag{2.8.38}$$

For $n = 1$, this can be obtained by rotation covariance from Theorem 2.8.1. For $n = 2$, it was proven by Simon–Zlatoš [411]. It is open for general n, but there are partial results in Golinskii–Zlatoš [178].

The argument used to get (2.8.35) is borrowed from Simon–Zlatoš [410]. As mentioned, if (2.8.5) holds, $w(\theta)$ is strictly positive and continuous on $(0, 2\pi)$, a result of Peherstorfer–Steinbauer [341]. For a history of related results and a proof, see Section 12.1 of [400]. It is a conjecture of Last [270] that if

$$\sum_{n=0}^{\infty} |\alpha_{n+1} - \alpha_n|^2 < \infty \tag{2.8.39}$$

then $w(\theta) > 0$ for a.e. θ. The OPRL analog of this result has been proven recently by Denisov [109].

2.9 THE SZEGŐ FUNCTION AND SZEGŐ ASYMPTOTICS

In his great 1920 paper [430], Szegő realized that his earlier result on Toeplitz determinants allowed very strong asymptotic results on the OPUC, $\varphi_n^*(z)$, in $\mathbb{D}$ (and then two years after [431], he realized he could use this to obtain asymptotics for

OPRL; see Section 3.7). While an aside from our main thrust, we would be remiss to ignore this beautiful and simple result.

Consider a sequence $\{x_n\}_{n=1}^\infty$ and three senses in which $x_n \sim \beta^n$ for some $\beta \in \mathbb{C} \setminus \{0\}$:

(i) *Root asymptotics*:

$$x_n^{1/n} \to \beta \tag{2.9.1}$$

If x_n is complex, we have an issue of phase, but can fix it by looking only at $|x_n|^{1/n}$.

(ii) *Ratio asymptotics*:

$$\frac{x_{n+1}}{x_n} \to \beta \tag{2.9.2}$$

(iii) *Power asymptotics* (also called *Szegő asymptotics*) for some $c \in \mathbb{C} \setminus \{0\}$:

$$\frac{x_n}{\beta^n} \to c \tag{2.9.3}$$

It is easy to see that

Proposition 2.9.1. *Power asymptotics $\Rightarrow$ ratio asymptotics $\Rightarrow$ root asymptotics.*

Roughly speaking, root asymptotics is connected with the theory of regular measures (discussed in Sections 2.15 and 5.9), ratio asymptotics with the Denisov–Rakhmanov–Remling theorem (discussed in Section 7.6), and Szegő asymptotics with Szegő's theorem.

Before turning to the subtle Szegő asymptotics for $\Phi_n^*(z)$, we want to discuss an elementary result on ratio asymptotics:

Theorem 2.9.2. *Let $\Phi_n(z)$ be the monic OPUC associated to a nontrivial probability measure on $\partial\mathbb{D}$ with Verblunsky coefficients $\{\alpha_n\}_{n=0}^\infty$. Then*

$$\lim_{n\to\infty} \frac{\Phi_{n+1}^*(z)}{\Phi_n^*(z)} = 1 \text{ uniformly on } \overline{\mathbb{D}} \Leftrightarrow \lim_{n\to\infty} \alpha_n = 0 \tag{2.9.4}$$

Moreover, if either side of (2.9.4) *holds, then uniformly on compact subsets of* $\mathbb{D}$:

$$\lim_{n\to\infty} \frac{\Phi_n(z)}{\Phi_n^*(z)} = 0 \tag{2.9.5}$$

Remarks. 1. If $\lim_{n\to\infty} \alpha_n = 0$, $\rho_n \to 1$ so $\varphi_{n+1}^*(z)/\varphi_n^*(z)$ also goes to 1. On the other hand, $\varphi_{n+1}^*(0)/\varphi_n^*(0) = \rho_n^{-1}$ so $\varphi_{n+1}^*(0)/\varphi_n^*(0) \to 1$ implies $\alpha_n \to 0$.

2. It is not true that $\Phi_{n+1}^*(z)/\Phi_n^*(z) \to 1$ uniformly on compact subsets of $\mathbb{D}$ implies $\alpha_n \to 0$; see the Notes.

3. The proof shows that if (2.9.4) holds for a single z in $\partial\mathbb{D}$, then $\alpha_n \to 0$.

Proof. By (1.8.5),

$$\left| \frac{\Phi_{n+1}^*(z)}{\Phi_n^*(z)} - 1 \right| = |\alpha_n|\,|z| \left| \frac{\Phi_n(z)}{\Phi_n^*(z)} \right| \tag{2.9.6}$$

By the lemma below, the right side is bounded for $z \in \overline{\mathbb{D}}$ by $|\alpha_n|$ and equal to $|\alpha_n|$ if $|z| = 1$. Thus, (2.9.4) holds.

Next suppose $|\alpha_j| \to 0$ and fix $z_0 \in \mathbb{D}$. Pick $n_j \to \infty$ so that

$$\frac{|\Phi_{n_j+1}(z_0)|}{|\Phi^*_{n_j+1}(z_0)|} \to \limsup \frac{|\Phi_n(z_0)|}{|\Phi^*_n(z_0)|} \equiv q(z_0) \tag{2.9.7}$$

By (1.8.5),

$$\left|\frac{\Phi_{n_j+1}(z_0)}{\Phi^*_{n_j+1}(z_0)}\right| \le |z_0| \left|\frac{\Phi_{n_j}(z_0)}{\Phi^*_{n_j}(z_0)}\right| \left|\frac{\Phi^*_{n_j}(z_0)}{\Phi^*_{n_j+1}(z_0)}\right| + |\alpha_{n_j+1}| \left|\frac{\Phi^*_{n_j}(z_0)}{\Phi^*_{n_j+1}(z_0)}\right| \tag{2.9.8}$$

Using (2.9.4) and taking $j \to \infty$, we get

$$q(z_0) \le |z_0| q(z_0)$$

so $q(z_0) = 0$. Thus, $\Phi_n(z)/\Phi^*_n(z) \to 0$ pointwise. So, by the lemma and Vitali's theorem, the convergence is uniform on compact sets. □

Lemma 2.9.3. *For $n \ge 1$,*

$$z \in \partial\mathbb{D} \Rightarrow \left|\frac{\Phi_n(z)}{\Phi^*_n(z)}\right| = 1 \tag{2.9.9}$$

$$z \in \mathbb{D} \Rightarrow \left|\frac{\Phi_n(z)}{\Phi^*_n(z)}\right| < 1 \tag{2.9.10}$$

$$z \in \mathbb{C} \setminus \overline{\mathbb{D}} \Rightarrow \left|\frac{\Phi_n(z)}{\Phi^*_n(z)}\right| > 1 \tag{2.9.11}$$

Remark. Indeed, up to phase, $\Phi_n(z)/\Phi^*_n(z)$ is the Blaschke product of the zeros of $\Phi_n(z)$, which lie in $\mathbb{D}$.

Proof. (2.9.9) is immediate from the definition of Φ^*_n, (2.9.10) then follows from analyticity (using Theorem 1.8.4) and the maximum principle, and (2.9.11) follows from

$$\left|\frac{\Phi^*_n(1/\bar{z})}{\Phi_n(1/\bar{z})}\right| = \left|\frac{\Phi_n(z)}{\Phi^*_n(z)}\right| \tag{2.9.12}$$

□

We now turn to Szegő asymptotics. A key role will be played by:

Definition. Let $d\mu$ be a nontrivial probability measure on $\partial\mathbb{D}$ of the form (2.1.1). If the Szegő condition holds, that is,

$$\int \log(w(\theta)) \frac{d\theta}{2\pi} > -\infty \tag{2.9.13}$$

then the *Szegő function*, $D(z)$, is defined by

$$D(z) = \exp\left(\int \frac{e^{i\theta} + z}{e^{i\theta} - z} \log(w(\theta)) \frac{d\theta}{4\pi}\right) \tag{2.9.14}$$

Note that Szegő's theorem says that

$$D(0) = \prod_{n=0}^{\infty} \rho_n = \lim_{n\to\infty} (\varphi_n^*(0))^{-1} \tag{2.9.15}$$

Note also the $1/4\pi$, not $1/2\pi$, in (2.9.14). It is responsible for:

Proposition 2.9.4. *Whenever the Szegő condition holds, $D(z) \in H^2(\mathbb{D})$ and is nonvanishing on $\mathbb{D}$. Indeed,*

$$\sup_{0\le r<1} \int |D(re^{i\theta})|^2 \frac{d\theta}{2\pi} \le 1 \tag{2.9.16}$$

and the boundary values $D(e^{i\theta})$ obey (for a.e. θ)

$$|D(e^{i\theta})|^2 = w(\theta) \tag{2.9.17}$$

Proof. Define

$$w_\varepsilon(\theta) = \min(w(\theta), \varepsilon^{-1}) \tag{2.9.18}$$

and $D_\varepsilon(z)$ by (2.9.14) with w replaced by w_ε. Since $\frac{1}{2}\log(w_\varepsilon(\theta)) \le \frac{1}{2}\log(\varepsilon^{-1})$,

$$\operatorname{Re}\biggl(\int \frac{e^{i\theta}+z}{e^{i\theta}-z} \log(w_\varepsilon(\theta)) \frac{d\theta}{4\pi}\biggr) \le \tfrac{1}{2} \log(\varepsilon^{-1})$$

because the Poisson kernel is positive and in L^1. Thus,

$$|D_\varepsilon(z)| \le \varepsilon^{-1/2}$$

so $D_\varepsilon \in H^\infty$. By (2.3.55), its boundary values obey

$$\operatorname{Re}\log(D_\varepsilon(e^{i\theta})) = \tfrac{1}{2} \log(w_\varepsilon(\theta))$$

or

$$|D_\varepsilon(e^{i\theta})|^2 = w_\varepsilon(\theta) \tag{2.9.19}$$

For any H^2 function, f, if

$$f(z) = \sum_{n=0}^{\infty} a_n z^n \tag{2.9.20}$$

then

$$\int_0^{2\pi} |f(re^{i\theta})|^2 \frac{d\theta}{2\pi} = \sum_{n=0}^{\infty} |a_n|^2 r^{2n} \tag{2.9.21}$$

so

$$\|f\|_{H^2}^2 = \sum_{n=0}^{\infty} |a_n|^2 = \int_0^{2\pi} |f(e^{i\theta})|^2 \frac{d\theta}{2\pi} \tag{2.9.22}$$

and thus, (2.9.19) implies

$$\|D_\varepsilon\|_{H^2}^2 = \int w_\varepsilon(\theta) \frac{d\theta}{2\pi} \le 1 \tag{2.9.23}$$

Since $D_\varepsilon(z) \to D(z)$ uniformly on compact subsets of $\mathbb{D}$,

$$\int_0^{2\pi} |D(re^{i\theta})|^2 \frac{d\theta}{2\pi} = \lim_{\varepsilon \downarrow 0} \int_0^{2\pi} |D_\varepsilon(re^{i\theta})|^2 \frac{d\theta}{2\pi} \leq 1$$

Thus, $D \in H^2$. (2.9.17) is immediate from (2.3.50). This implies $\|D\|_{H^2} = \int w(\theta) \frac{d\theta}{2\pi} \leq 1$ by (2.3.50). □

The following calculation is needed only for the theorem that follows and could be included in the proof, but it is so simple and elegant that we separate it out:

Lemma 2.9.5. *Let $d\mu$ have the form* (2.1.1) *and obey the Szegő condition. Then for any n,*

$$\int |\varphi_n^*(e^{i\theta}) D(e^{i\theta}) - 1|^2 \frac{d\theta}{2\pi} + \int |\varphi_n^*(e^{i\theta})|^2 \, d\mu_s = 2\left(1 - \prod_{j=n}^{\infty} \rho_j\right) \tag{2.9.24}$$

Proof. Expanding the square and using $|D|^2 \frac{d\theta}{2\pi} + d\mu_s = d\mu$, we see

$$\text{LHS of (2.9.24)} = \int |\varphi_n(e^{i\theta})|^2 \, d\mu + \int \frac{d\theta}{2\pi} - 2\,\text{Re}\left[\int D(e^{i\theta}) \varphi_n^*(e^{i\theta}) \frac{d\theta}{2\pi}\right] \tag{2.9.25}$$

$$= 1 + 1 - 2\,\text{Re}(D(0)\varphi_n^*(0)) \tag{2.9.26}$$

by (2.3.51) and the fact that $D(z)\varphi_n^*(z)$ is in H^2. By Szegő's theorem, $D(0) = \prod_{j=0}^{\infty} \rho_j$, while by (1.8.11), $\varphi_n^*(0) = \prod_{j=0}^{n-1} \rho_j^{-1}$. This proves (2.9.24). □

Theorem 2.9.6. *Let $d\mu$ have the form* (2.1.1) *and obey the Szegő condition. Then:*

(i) *Uniformly on compact subsets of $\mathbb{D}$, we have that*

$$\varphi_n^*(z) \to D(z)^{-1} \tag{2.9.27}$$

(ii) *Uniformly on $\{z \mid |z| \geq R\}$ for any $R > 1$, we have that*

$$z^{-n} \varphi_n(z) \to \overline{D(1/\bar{z})}^{-1} \tag{2.9.28}$$

(iii)

$$\int |\varphi_n(e^{i\theta})|^2 \, d\mu_s \to 0 \qquad \text{as } n \to \infty \tag{2.9.29}$$

(iv) *Define $D_{\text{ac}}(e^{i\theta})^{-1}$ in $L^2(\partial\mathbb{D}, d\mu)$ by*

$$D_{\text{ac}}(e^{i\theta})^{-1} = \begin{cases} D(e^{i\theta})^{-1} & \text{a.e. } \theta \text{ w.r.t. } \frac{d\theta}{2\pi} \\ 0 & \text{a.e. } d\mu_s(\theta) \end{cases} \tag{2.9.30}$$

Then

$$\varphi_n^*(e^{i\theta}) \to D_{\text{ac}}(e^{i\theta})^{-1} \quad \text{in} \quad L^2(\partial\mathbb{D}, d\mu) \tag{2.9.31}$$

(v)

$$\varphi_n^*(e^{i\theta}) D(e^{i\theta}) \to 1 \quad \text{in} \quad L^2\left(\partial\mathbb{D}, \frac{d\theta}{2\pi}\right) \tag{2.9.32}$$

(vi)

$$|\varphi_n(e^{i\theta})^2|\,d\mu(\theta) \to \frac{d\theta}{2\pi} \tag{2.9.33}$$

weakly as measures in $\partial\mathbb{D}$.

(vii) *We have uniformly on compact subsets of $\mathbb{D}$ that*

$$\varphi_n(z) \to 0 \tag{2.9.34}$$

Remarks. 1. We will prove later (see Theorem 2.13.5) that if the Szegő condition fails, then

$$|\varphi_n^*(z)|^{-1} \to 0 \tag{2.9.35}$$

uniformly on compact subsets of $\mathbb{D}$.

2. We have not been explicit about the results for φ_n, for example, $e^{-in\theta}\varphi_n(e^{i\theta}) \to \overline{D_{\rm ac}(e^{i\theta})}^{-1}$.

3. (2.9.33) holds in much greater generality than the Szegő condition. For example, Rakhmanov [358, 359] has proven (2.9.33) so long as $w(\theta) > 0$ and Khrushchev [221] has found necessary and sufficient conditions for (2.9.33), namely, (2.9.33) holds if and only if for all ℓ, $\alpha_n\alpha_{n+\ell} \to 0$. These issues are discussed in Chapter 9, explicitly Sections 9.3 and 9.7, of [400].

4. (2.9.28) is called *Szegő asymptotics.*

5. It is a result of Nevai–Totik [323] that if $|\alpha_n| \to 0$ exponentially, then $D(z)^{-1}$ has an analytic continuation beyond $\overline{\mathbb{D}}$ and (2.9.27) holds in a larger disk; see Chapter 7 of [399].

Proof. (i), (iii), (v). Since $\prod_{j=0}^\infty \rho_j$ converges, the right-hand side of (2.9.1) $\to 0$ so $\varphi_n^*(e^{i\theta})D(e^{i\theta}) \to 1$ in $H^2(\partial\mathbb{D})$ (proving (v)) and (2.9.29) holds. For any $f \in H^2$, one can extend (2.3.51) to get

$$f(re^{i\theta}) = \int \frac{1-r^2}{1+r^2-2r\cos(\theta-\varphi)}\, f(e^{i\varphi})\,\frac{d\varphi}{2\pi} \tag{2.9.36}$$

Thus, H^2-norm convergence implies uniform convergence on compact subsets of $\mathbb{D}$.

(ii) Immediate from (2.9.27) and (1.8.2).

(iv) Since $w(\theta) = |D(e^{i\theta})|^2$,

$$\int |\varphi_n^*(e^{i\theta}) - D_{\rm ac}(e^{i\theta})^{-1}|^2\,d\mu = \int |\varphi_n^*(e^{i\theta})|^2\,d\mu_{\rm s} + \int |D(e^{i\theta})\varphi_n^*(e^{i\theta}) - 1|\,\frac{d\theta}{2\pi}$$

goes to zero by (2.9.29) and (2.9.32).

(vi) $|\varphi_n(e^{i\theta})|^2\,d\mu_s \xrightarrow{w} 0$ by (2.9.29) and $|\varphi_n(e^{i\theta})|^2 w(\theta)\frac{d\theta}{2\pi} = |D(e^{i\theta})\varphi_n(e^{i\theta})|^2 \times \frac{d\theta}{2\pi} \to \frac{d\theta}{2\pi}$ by (2.9.32).

(vii) Immediate from (2.9.27) and (2.9.28) (which holds since the Szegő condition implies $|\alpha_n| \to 0$). □

Remarks and Historical Notes. Root asymptotics are discussed further in Theorem 2.15.3.

The key calculation (2.9.24) and its consequences in Theorem 2.9.6 are in Szegő's great paper [430]. Khrushchev [221] (see also Section 9.5 of [400]) has a general study of what kind of ratio asymptotics can occur for OPUC. In particular, these references discuss examples of Máté–Nevai [301] where one has ratio asymptotics uniformly on compact subsets of $\mathbb{D}$ for which $\alpha_n \nrightarrow 0$. For OPRL, the analog is studied by Simon [397].

In this section, we discussed pointwise limits in $\mathbb{D}$ but only L^2 limits on $\partial\mathbb{D}$. That is because one cannot prove pointwise limit theorems on $\partial\mathbb{D}$ if one assumes only the Szegő condition. Under stronger hypotheses, one can prove pointwise theorems; for example, see Szegő's book [434, Chapter XII], Freud's book [141, Section V.4 and Section V.5], and Section 2.5 of the planned second edition of [399], available online at `http://www.math.caltech.edu/opuc/newsections.html`.

2.10 ASYMPTOTICS FOR WEYL SOLUTIONS

Recall that in Section 2.4 we defined the Weyl solution $\binom{g_n(z)}{g_n^*(z)}$ for $z \in \mathbb{D}$ by (2.4.25)/(2.4.27) and proved that (see (2.4.38))

$$z^{-n} g_n^*(z) \to 0 \tag{2.10.1}$$

As an aside on the aside that was the last section, we will prove here that

Theorem 2.10.1. *If the Szegő condition holds, then uniformly on compact subsets of* $\mathbb{D}$,

$$z^{-n} g_n(z) \to 2D(z) \tag{2.10.2}$$

If the Szegő condition fails, then

$$z^{-n} g_n(z) \to 0 \tag{2.10.3}$$

uniformly on compacts.

Our proof will require the following result from Section 2.13 (see Theorem 2.13.5):

Proposition 2.10.2. *If the Szegő condition fails,*

$$\varphi_n^*(z)^{-1} \to 0 \tag{2.10.4}$$

uniformly on compact subsets of $\mathbb{D}$.

Proof of Theorem 2.10.1. We apply (2.4.43) with $a_0 = b_0 = -1$, so $a_n = -\varphi_n(z)$, $b_n = -\varphi_n^*(z)$, and

$$a_0 g_0^*(z) - b_0 g_0(z) = -(-1 + F) + (1 + F) = 2$$

and we have

$$\varphi_n^*(z)(z^{-n} g_n(z)) - \varphi_n(z)(z^{-n} g_n^*(z)) = 2 \tag{2.10.5}$$

Suppose the Szegő condition fails. By (2.4.47) and (2.9.10),

$$|\varphi_n(z) g_n^*(z)| \leq |z|\, |\varphi_n^*(z) g_n(z_0)|$$

so (2.10.5) implies

$$|\varphi_n^*(z)z^{-n}g_n(z)| \le 2 + |z|\,|\varphi_n^*(z)z^{-n}g_n(z)|$$

which implies

$$|z^{-n}g_n(z)| \le 2(1-|z|)^{-1}|\varphi_n^*(z)|^{-1} \tag{2.10.6}$$

so if the Szegő condition fails, (2.10.4) implies (2.10.3).

Now suppose the Szegő condition holds. By (2.10.1) and (2.9.34),

$$\lim_{n\to\infty} \varphi_n(z)z^{-n}g_n^*(z) = 0 \tag{2.10.7}$$

and (2.10.5) implies

$$\lim z^{-n}g_n(z) = 2\lim_{n\to\infty} \varphi_n^*(z)^{-1} = 2D(z)$$

by (2.9.27). □

Remarks and Historical Notes. Theorem 2.10.1 is due to Peherstorfer [336] and is related to OPRL results of Damanik–Simon [99] on the equivalence of Jost and Szegő asymptotics in that case. One can argue in this case that the "Jost function" is $(2D(z))^{-1}$ or $(2D(z))^{-1}(1+F)$. [336] also has some results on asymptotics on $\partial\mathbb{D}$ when $w(\theta)$ has regularity properties.

2.11 ADDITIONAL ASPECTS OF SZEGŐ'S THEOREM

In this section, we discuss several additional issues connected with Szegő's theorem.

Szegő's Theorem as a Nonlinear Plancherel Theorem

As mentioned in Section 1.5, there is a "small coupling" limit of Szegő's theorem in which it becomes the Plancherel theorem so that Szegő's theorem is a kind of nonlinear Plancherel theorem.

Suppose that $f \in L^\infty(\partial\mathbb{D}, \frac{d\theta}{2\pi})$ is real-valued and obeys

$$\int f(\theta)\,\frac{d\theta}{2\pi} = 0 \tag{2.11.1}$$

Then for $|\lambda| < \|f\|_\infty^{-1}$,

$$w_\lambda(\theta) = 1 + \lambda f(\theta) \tag{2.11.2}$$

is a weight for a probability measure

$$d\mu_\lambda = w_\lambda(\theta)\,\frac{d\theta}{2\pi} \tag{2.11.3}$$

Clearly,

$$\int \log(w_\lambda)\,\frac{d\theta}{2\pi} = \int \lambda f(\theta)\,\frac{d\theta}{2\pi} - \frac{1}{2}\int \lambda^2 f(\theta)^2\,\frac{d\theta}{2\pi} + O(\lambda^3)$$

$$= -\tfrac{1}{2}\lambda^2\|f\|_2^2 + O(\lambda^3) \tag{2.11.4}$$

by (2.11.1). On the other hand:

Proposition 2.11.1.

$$\alpha_{n-1}(d\mu_\lambda) = \lambda \int e^{-in\theta} f(\theta)\frac{d\theta}{2\pi} + O(\lambda^2) \tag{2.11.5}$$

Proof. We begin by proving that

$$\Phi_n(z) = z^n + O(\lambda) \tag{2.11.6}$$

For $\{\Phi_j\}_{j=0}^{n-1}$ is an orthogonal basis for the polynomials of degree $n-1$ so

$$\Phi_n(z) = z^n - \sum_{j=0}^{n-1} \langle \Phi_j, z^n\rangle \Phi_j \|\Phi_j\|^{-2} \tag{2.11.7}$$

(2.11.6) certainly holds for $n = 0$, so inductively we have $\|\Phi_j\|^2 = 1 + O(\lambda)$ for $j = 0, \dots, n-1$. Moreover, since

$$\int z^\ell\, d\mu_\lambda = O(\lambda) \qquad \text{for } \ell \neq 0 \tag{2.11.8}$$

we have for $j < n$ that $\langle \Phi_j, z^n\rangle = O(\lambda)$, proving (2.11.6) from (2.11.7).

Szegő recursion and $\int \Phi_{n+1}\, d\mu = \langle 1, \Phi_{n+1}\rangle = 0$ implies (for any $d\mu$)

$$\int e^{i\theta}\Phi_n(e^{i\theta})\, d\mu(\theta) = \bar{\alpha}_n \int \Phi_n^*(e^{i\theta})\, d\mu(\theta) \tag{2.11.9}$$

Now

$$\int \Phi_n^*\, d\mu = \langle 1, \Phi_n^*\rangle = \langle z^n, \Phi_n\rangle$$

$$= \|\Phi_n\|^2 = 1 + O(\lambda) \tag{2.11.10}$$

by (2.11.6). In (2.11.6), the $O(\lambda)$ term is a polynomial whose coefficients are $O(\lambda)$, so since $\int zz^m\, d\mu = O(\lambda)$ by (2.11.8),

$$\text{LHS of (2.11.9)} = \int e^{i(n+1)\theta}\, d\mu(\theta) + O(\lambda^2)$$

by (2.11.6). Thus, (2.11.9) implies (2.11.5). □

Thus,

$$\prod_{j=0}^{n}(1 - |\alpha_j|^2) = 1 - \lambda^2 \sum_{j=0}^{n} |\widehat{f}_j|^2 + O(\lambda^3) \tag{2.11.11}$$

where

$$\widehat{f}_j = \int e^{-ij\theta} f(\theta)\,\frac{d\theta}{2\pi} \tag{2.11.12}$$

Therefore, formally (i.e., ignoring the passage $n \to \infty$, which can be subtle), Szegő's theorem says

$$1 - \lambda^2 \sum_{j=0}^{\infty} |\widehat{f}_j|^2 + O(\lambda^3) = \exp(1 - \tfrac{1}{2}\lambda^2 \|f\|_2^2 + O(\lambda^3)) \tag{2.11.13}$$

Thus, the small coupling limit of Szegő's theorem is the Plancherel theorem, and one can think of Szegő's theorem as a nonlinear Plancherel theorem.

Szegő's Theorem and the Density of the Polynomials

In $L^2(\partial\mathbb{D}, \frac{d\theta}{2\pi})$, the closure of the polynomials is $H^2(\mathbb{D})$, that is, analytic functions on $\mathbb{D}$. The polynomials are not dense in L^2. One can ask for which measures the polynomials are dense in $L^2(\partial\mathbb{D}, d\mu)$ and we will answer that here. The initial steps have nothing to do with Szegő's theorem:

Proposition 2.11.2. *Let P be the projection onto the closure of the polynomials in $L^2(\partial\mathbb{D}, d\mu)$ for $d\mu$ a nontrivial probability measure on $\partial\mathbb{D}$. Then*

$$\|(\mathbf{1} - P)z^{-1}\| = \lim_{n\to\infty} \|\Phi_n\| = \prod_{j=0}^{\infty} (1 - |\alpha_j|^2)^{1/2} \tag{2.11.14}$$

Proof. The second equality is (1.8.11), so we focus on the first. Let $P_{\{f_1,\dots,f_n\}}$ be the projection onto the span of $f_1, \dots, f_n$. By Proposition 1.8.1(iii), Φ_n^* is the projection of 1 onto the span of $\{z, \dots, z^n\}$,

$$\begin{aligned} \|\Phi_n^*\| &= \|(\mathbf{1} - P_{\{z,\dots,z^n\}})1\| \\ &= \|[z^{-1}(\mathbf{1} - P_{\{z,\dots,z^n\}})z]z^{-1}\| && (2.11.15) \\ &= \|(\mathbf{1} - P_{\{1,\dots,z^{n-1}\}})z^{-1}\| && (2.11.16) \end{aligned}$$

since multiplication by z^{-1} is a unitary that maps $P_{\{z,\dots,z^n\}}$ to $P_{\{1,\dots,z^{n-1}\}}$. Since $\|\Phi_n\| = \|\Phi_n^*\|$, (2.11.14) follows from (2.11.16) by taking $n \to \infty$. □

Lemma 2.11.3. *If z^{-1} is in the closure of the span of the polynomials in $L^2(\partial\mathbb{D}, d\mu)$, so is $z^{-\ell}$ for all ℓ.*

Proof. We use induction in ℓ. So suppose for polynomials Q_n and R_n, $Q_n \to z^{-\ell}$ and $R_n \to z^{-1}$ in $L^2(\partial\mathbb{D}, d\mu)$. Then ($\|\cdot\| = L^2$-norm and $\|\cdot\|_\infty$ the $L^\infty(\partial\mathbb{D})$ norm)

$$\begin{aligned} \|z^{-\ell-1} - Q_n R_m\| &= \|z^{-\ell-1} - z^{-1}Q_n + z^{-1}Q_n - Q_n R_m\| \\ &\le \|z^{-\ell} - Q_n\| + \|Q_n\|_\infty \|z^{-1} - R_m\| && (2.11.17) \end{aligned}$$

since $\|z^{-1}\|_\infty = 1$. Given ε, pick n so $\|z^{-\ell} - Q_n\| < \frac{\varepsilon}{2}$. Having chosen n, pick m so $\|z^{-1} - R_m\| < \frac{\varepsilon}{2}\|Q_n\|_\infty^{-1}$. Then $z^{-\ell-1}$ has a polynomial within ε in L^2-norm and $z^{-\ell-1}$ is in the L^2-closure of the polynomials. □

Theorem 2.11.4. *Let $d\mu$ be a nontrivial probability measure on $\partial\mathbb{D}$ with Verblunsky coefficients $\{\alpha_j\}_{j=0}^\infty$. The polynomials are dense in $L^2(\partial\mathbb{D}, d\mu)$ if and only if*

$$\prod_{j=0}^{\infty}(1-|\alpha_j|^2) = 0 \tag{2.11.18}$$

Proof. If (2.11.18) fails, Proposition 2.11.2 shows that z^{-1} is not in the closure and so the closure is not all of L^2. Conversely, if (2.11.18) holds, by Proposition 2.11.2 and Lemma 2.11.3, all Laurent polynomials (i.e., finite sums $\sum_{j=-k_1}^{k_2} c_j z^j$) are in the closure of the polynomials. By Weierstrass' theorem, the Laurent polynomials are dense in the continuous functions in $\|\cdot\|_\infty$, so in L^2, and so the polynomials are dense. □

Because of Szegő's theorem, we have

Theorem 2.11.5 (Kolmogorov's Density Theorem). *Let $d\mu$ be a probability measure on $\partial\mathbb{D}$ of the form* (2.1.1). *Then the polynomials are dense in $L^2(\partial\mathbb{D}, d\mu)$ if and only if*

$$\int \log(w(\theta))\, \frac{d\theta}{2\pi} = -\infty \tag{2.11.19}$$

As a final remark about the density result, if the Szegő condition holds so the span of $\{z^n\}_{n=1}^\infty$ is not dense, one can ask for explicit functions in $L^2(\partial\mathbb{D}, d\mu)$ in the orthogonal complement. One can take

$$g(\theta) = e^{-i\theta}\, D(e^{i\theta})^{-1}\chi_S(\theta) \tag{2.11.20}$$

where S is a set of full $\frac{d\theta}{2\pi}$-measure whose complement supports $d\mu_s$. For then

$$\begin{aligned}\int e^{in\theta}\,\overline{g(\theta)}\, d\mu &= \int e^{i(n+1)\theta}\,\overline{D^{-1}}\,|D|^2\,\frac{d\theta}{2\pi}\\ &= \int e^{i(n+1)\theta}\, D(e^{i\theta})\,\frac{d\theta}{2\pi}\\ &= 0\end{aligned}$$

since $D \in H^2(\partial\mathbb{D}, \frac{d\theta}{2\pi})$ and $z^{n+1}D(z)$ vanishes at $z = 0$.

Szegő's Theorem and CMV Matrices

One of the important aspects of Jacobi matrices is that they all act on the same space, $\ell^2(\{1, 2, \dots\})$ so operator comparison and cancellation methods are available. Here we want to present a similar matrix representation for OPUC and show how it allows an expression of the Szegő function as a Fredholm determinant. CMV matrices will appear again later in Sections 6.10 and 8.7.

Definition. The *CMV basis*, $\{\chi_j\}_{j=0}^\infty$, is the orthonormal basis for $L^2(\partial\mathbb{D}, d\mu)$ obtained by applying Gram–Schmidt to the sequence $1, z, z^{-1}, z^2, z^{-2}, \dots$. The *alternate CMV matrix basis*, $\{x_j\}_{j=0}^\infty$, is obtained from $1, z^{-1}, z, z^{-2}, \dots$.

Remark. We saw before that the $\{\varphi_j\}_{j=0}^\infty$ may or may not be a basis because the polynomials might or might not be dense. Since Laurent polynomials are always dense, the $\{\chi_j\}_{j=0}^\infty$ and $\{x_j\}_{j=0}^\infty$ are always bases.

It is not hard to see that for $n = 0, 1, 2, \dots,$

$$\chi_{2n}(z) = z^{-n}\varphi_{2n}^*(z) \tag{2.11.21}$$

$$\chi_{2n+1}(z) = z^{-n}\varphi_{2n+1}(z) \tag{2.11.22}$$

$$x_{2n}(z) = z^{-n}\varphi_{2n}(z) \tag{2.11.23}$$

$$x_{2n+1}(z) = z^{-n-1}\varphi_{2n+1}^*(z) \tag{2.11.24}$$

$$x_j(z) = \overline{\chi_j(1/\bar{z})} \tag{2.11.25}$$

The *CMV matrix*, $\mathcal{C}$, is just multiplication by z in the $\{\chi_j\}_{j=0}^\infty$ basis, and the *alternate CMV matrix*, $\tilde{\mathcal{C}}$, is multiplication by z in the $\{x_j\}_{j=0}^\infty$ basis. Thus,

$$\mathcal{C}_{k\ell} = \langle \chi_k, z\chi_\ell\rangle \qquad \tilde{\mathcal{C}}_{k\ell} = \langle x_k, zx_\ell\rangle \tag{2.11.26}$$

By (2.11.25) and unitarity of $\mathcal{C}$,

$$\tilde{\mathcal{C}}_{k\ell} = \mathcal{C}_{\ell k} \tag{2.11.27}$$

that is, $\tilde{\mathcal{C}} = \mathcal{C}^t$.

$\mathcal{C}$ is a five-diagonal matrix with the form

$$\mathcal{C} = \begin{pmatrix} \bar\alpha_0 & \bar\alpha_1\rho_0 & \rho_1\rho_0 & 0 & 0 & \dots \\ \rho_0 & -\bar\alpha_1\alpha_0 & -\rho_1\alpha_0 & 0 & 0 & \dots \\ 0 & \bar\alpha_2\rho_1 & -\bar\alpha_2\alpha_1 & \bar\alpha_3\rho_2 & \rho_3\rho_2 & \dots \\ 0 & \rho_2\rho_1 & -\rho_2\alpha_1 & -\bar\alpha_3\alpha_2 & -\rho_3\alpha_2 & \dots \\ 0 & 0 & 0 & \bar\alpha_4\rho_3 & -\bar\alpha_4\alpha_3 & \dots \\ \dots & \dots & \dots & \dots & \dots & \dots \end{pmatrix} \tag{2.11.28}$$

with a 2×3 block at the top and then 2×4 blocks clustered about the diagonal. The easiest way to see this is to use:

Proposition 2.11.6. *Define*

$$\mathcal{L}_{k\ell} = \langle \chi_k, z\chi_\ell\rangle \qquad \mathcal{M}_{k\ell} = \langle x_k, \chi_\ell\rangle \tag{2.11.29}$$

Then

$$\mathcal{C} = \mathcal{L}\mathcal{M} \tag{2.11.30}$$

and

$$\mathcal{L} = \Theta_0 \oplus \Theta_2 \oplus \Theta_4 \oplus \dots \tag{2.11.31}$$

$$\mathcal{M} = \mathbf{1}_{1\times 1} \oplus \Theta_1 \oplus \Theta_3 \oplus \dots \tag{2.11.32}$$

where $\mathbf{1}_{1\times1}$ *is the* 1×1 *identity and* Θ_j *is the* 2×2 *matrix*

$$\Theta_j = \Theta(\alpha_j) = \begin{pmatrix} \bar\alpha_j & \rho_j \\ \rho_j & -\alpha_j \end{pmatrix} \tag{2.11.33}$$

Sketch of Proof. (2.11.30) follows from the fact that $\{x_j\}_{j=0}^{\infty}$ is a basis. (2.11.31)/(2.11.32) are a restatement of the Szegő recursion for the φ's:

$$z\varphi_n(z) = \rho_n \varphi_{n+1}(z) + \bar{\alpha}_n \varphi_n^*(z) \tag{2.11.34}$$

$$\varphi_n^*(z) = \rho_n \varphi_{n+1}^*(z) + \alpha_n z\varphi_n(z) \tag{2.11.35}$$

□

$\mathcal{C}_0$ is the CMV matrix associated to $\frac{d\theta}{2\pi}$. In terms of the trace ideals, $\mathcal{I}_p$ ([170, 390]), one can show

$$\sum_{j=0}^{\infty} |\alpha_j|^p < \infty \Leftrightarrow \mathcal{C} - \mathcal{C}_0 \in \mathcal{I}_p \tag{2.11.36}$$

for $1 \leq p < \infty$. By the general theory of trace ideals, if $A \in \mathcal{I}_1$, one can define $\det(1 + A)$, and if $A \in \mathcal{I}_2$, $\det_2(1 + A)$, which is formally $\det(1 + A)e^{-\mathrm{Tr}(A)}$, and actually $\det((1 + A)e^{-A})$ (since $A \in \mathcal{I}_2 \Rightarrow (1 + A)e^{-A} - 1 \in \mathcal{I}_1$). One has

Theorem 2.11.7. *If* $\sum_{j=0}^{\infty}|\alpha_j| < \infty$, *then*

$$D(0)D(z)^{-1} = \det((1 - z\mathcal{C})(1 - z\mathcal{C}_0)^{-1})$$

If $\sum_{j=0}^{\infty}|\alpha_j|^2 < \infty$, *then*

$$D(0)D(z)^{-1} = \det_2((1 - z\mathcal{C})(1 - z\mathcal{C}_0)^{-1})e^{zw_1}$$

where

$$w_1 = \alpha_0 - \sum_{n=1}^{\infty} \alpha_n \bar{\alpha}_{n-1} \tag{2.11.37}$$

Remarks and Historical Notes. Tao and Thiele have emphasized the view of Szegő's theorem as a kind of nonlinear Plancherel theorem.

The density of polynomials results (i.e., Theorem 2.11.5) are due to Kolmogorov [239]. It was Krein [249] who realized the connection to OPUC. We return to the Kolmogorov density theorem in Section 3.9.

CMV matrices are named after [70], although the history is complicated due to early work in the numerical linear algebra community; see the discussion in [403]. For details of the proof of Proposition 2.11.6, see [399, Section 4.2].

(2.11.36) is a result of Golinskii–Simon that appeared in [399, Section 4.3]. Theorem 2.11.7 is due to Simon in that section where a proof is given. Note that (4.2.57) of [399] is wrong; see the erratum, which is available online at `http://www.math.caltech.edu/opuc.html`.

2.12 THE VARIATIONAL APPROACH TO SZEGŐ'S THEOREM

While we are emphasizing step-by-step sum rule approaches to Szegő's theorem, we should present Szegő's variational proof from his great 1920 paper [430]. On one technical point—which, as I will explain, Szegő did not address in 1920—we will provide an elegant resolution of Helson–Lowdenslager [197]. We begin with

Proposition 2.12.1. *Let $d\mu$ be a nontrivial probability measure on $\partial\mathbb{D}$ of the form* (2.1.1) *with Verblunsky coefficients $\{\alpha_n\}_{n=0}^\infty$. Then*

$$\prod_{n=0}^{\infty}(1-|\alpha_n|^2) = \lim_{n\to\infty} \|\Phi_n^*\|^2 \tag{2.12.1}$$

$$= \inf\{\|P\|^2 \mid P \text{ a polynomial},\ P(0)=1\} \tag{2.12.2}$$

and if $d\mu_s = 0$, this is

$$= \inf\{\|f\|^2 \mid f \in H^\infty(\mathbb{D}),\ f(0)=1\} \tag{2.12.3}$$

Remark. Here $\|g\|$ is given by

$$\|g\|^2 = \int |g(e^{i\theta})|^2\, d\mu(\theta) \tag{2.12.4}$$

and in (2.12.3), we use the $d\theta$-a.e. boundary values, which is why we suppose $d\mu_s = 0$.

Proof. (2.12.1) is just (1.8.11) and $\|\Phi_n\| = \|\Phi_n^*\|$. Since Φ_n^* is the projection of 1 onto the orthogonal complement of $\{z,\dots,z^n\}$,

$$\|\Phi_n^*\|^2 \le \|1 + a_1 z + \cdots + a_n z^n\|^2 \tag{2.12.5}$$

for any $a_1,\dots,a_n \in \mathbb{C}$, so

$$\|\Phi_n^*\| = \inf\{\|P\|^2 \mid P \text{ is a polynomial of degree at most } n,\ P(0)=1\} \tag{2.12.6}$$

which implies (2.12.2) by taking $n\to\infty$.

Since each polynomial, P, is in H^∞,

$$\text{RHS of (2.12.3)} \le \text{RHS of (2.12.2)}$$

On the other hand, if $f \in H^\infty$, $f(re^{i\theta}) \to f(e^{i\theta})$ pointwise for $d\theta$-a.e. θ (by (2.3.48)), and so a.e. $d\mu$ since we are assuming $d\mu_s = 0$. Since f is bounded, we have convergence in $L^2(\partial\mathbb{D}, d\mu)$. For $r<1$, the Taylor polynomials for $g(z) = f(rz)$ converge uniformly to $f(re^{i\theta})$, and so for any $f \in H^\infty$, we can find polynomials with P_n with $P_n(0) = f(0)$ and $P_n \to f$ in $L^2(\partial\mathbb{D}, d\mu)$. Thus,

$$\text{RHS of (2.12.3)} \ge \text{RHS of (2.12.2)}$$

□

One side of Szegő's theorem follows from Jensen's inequality that

$$\int e^{h(x)}\, d\gamma(x) \ge \exp\biggl(\int h(x)\, d\gamma(x)\biggr) \tag{2.12.7}$$

for any probability measure $d\gamma$.

Theorem 2.12.2. *For any polynomial P and any nontrivial probability measure $d\mu$ on $\partial\mathbb{D}$ of the form* (2.1.1), *we have*

$$\int |P(e^{i\theta})|^2\, d\mu \ge |P(0)|^2 \exp\biggl(\int \log(w(\theta))\,\frac{d\theta}{2\pi}\biggr) \tag{2.12.8}$$

In particular,

$$\prod_{n=0}^{\infty}(1-|\alpha_n|^2) \geq \exp\left(\int \log(w(\theta))\,\frac{d\theta}{2\pi}\right) \tag{2.12.9}$$

Proof. We have

$$\begin{aligned}
\int |P(e^{i\theta})|^2\, d\mu(\theta) &\geq \int |P(e^{i\theta})|^2 w(\theta)\,\frac{d\theta}{2\pi} \\
&= \int \exp(2\log|P(e^{i\theta})| + \log(w(\theta)))\,\frac{d\theta}{2\pi} \\
&\geq \left[\exp\left(\int 2\log|P(e^{i\theta})|\,\frac{d\theta}{2\pi}\right)\right]\left[\exp\left(\int \log(w(\theta))\,\frac{d\theta}{2\pi}\right)\right]
\end{aligned}$$

by Jensen's inequality. The lemma below completes the proof of (2.12.8). (2.12.9) then follows from (2.12.2). □

Lemma 2.12.3. *For any polynomial P,*

$$\int \log|P(e^{i\theta})|\,\frac{d\theta}{2\pi} \geq \log|P(0)| \tag{2.12.10}$$

Proof. Suppose first that P is nonvanishing on $\overline{\mathbb{D}}$. Then $\log(P(z))$ is analytic in $\mathbb{D}$, continuous on $\overline{\mathbb{D}}$, and so $\log|P(z)| = \operatorname{Re}\log(P(z))$ is harmonic. So (2.12.10) holds with equality. The same is true if P has zeros on $\partial\mathbb{D}$ by a limiting argument.

If $P(0) = 0$, (2.12.10) is trivial.

Let $\{z_j\}_{j=0}^{\ell}$ are all the zeros of P in $\mathbb{D}$ and no z_j is zero. Define

$$Q(z) = \prod_{j=1}^{\ell} \frac{1-\bar{z}_j z}{z - z_j}\, P(z) \tag{2.12.11}$$

Then $|Q(e^{i\theta})| = |P(e^{i\theta})|$ and Q is nonvanishing on $\mathbb{D}$. So by the special case at the start of the theorem,

$$\begin{aligned}
\int \log|P(e^{i\theta})|\,\frac{d\theta}{2\pi} &= \int \log|Q(e^{i\theta})|\,\frac{d\theta}{2\pi} \\
&= \log|Q(0)| \\
&= \log\left|\prod_{j=1}^{\ell}|z_j|^{-1} P(0)\right| \qquad (2.12.12) \\
&\geq \log|P(0)|
\end{aligned}$$

since $|z_j|^{-1} \geq 1$. □

Remark. (2.12.12) is Jensen's formula.

Theorem 2.12.4. *Suppose* $d\mu_s = 0$. *Then*

$$\textit{RHS of } (2.12.3) \leq \exp\left(\int \log(w(\theta))\,\frac{d\theta}{2\pi}\right) \tag{2.12.13}$$

Proof. For $\varepsilon > 0$, define

$$f_\varepsilon(z) = \exp\biggl(\int \biggl[1 - \frac{e^{i\theta} + z}{e^{i\theta} - z}\biggr] \log(w(\theta) + \varepsilon)\, \frac{d\theta}{4\pi}\biggr) \tag{2.12.14}$$

$$= L_\varepsilon g_\varepsilon(z)^{-1} \tag{2.12.15}$$

with

$$L_\varepsilon = \exp\biggl(\int \log(w(\theta) + \varepsilon)\, \frac{d\theta}{4\pi}\biggr) \tag{2.12.16}$$

and

$$g_\varepsilon(z) = \exp\biggl(\int \biggl(\frac{e^{i\theta} + z}{e^{i\theta} - z}\biggr) \log(w(\theta) + \varepsilon)\, \frac{d\theta}{4\pi}\biggr) \tag{2.12.17}$$

Clearly, $g_\varepsilon(0) = L_\varepsilon$, so $f_\varepsilon(0) = 1$. Moreover, $\log(w(\theta) + \varepsilon) \geq \log(\varepsilon)$, so $|g_\varepsilon(z)| \geq \varepsilon^{1/2}$, and so,

$$|f_\varepsilon(z)| \leq \varepsilon^{-1/2} L_\varepsilon < \infty$$

and $f \in H^\infty$. Thus, by (2.12.3) and $d\mu_{\rm s} = 0$,

$$\text{RHS of (2.12.3)} \leq \int |f_\varepsilon(e^{i\theta})|^2 w(\theta)\, \frac{d\theta}{2\pi} \tag{2.12.18}$$

But, by (2.3.62),

$$|f_\varepsilon(e^{i\theta})|^2 = L_\varepsilon^2 (w(\theta) + \varepsilon)^{-1} \tag{2.12.19}$$

so

$$\int |f_\varepsilon(e^{i\theta})|^2 w(\theta)\, \frac{d\theta}{2\pi} \leq L_\varepsilon^2 \tag{2.12.20}$$

so

$$\text{RHS of (2.12.3)} \leq \lim_{\varepsilon \downarrow 0} L_\varepsilon^2 = \text{RHS of (2.12.13)}$$

proving (2.12.13). □

This completes the discussion of what Szegő had in 1920. We want to end by saying something about allowing $d\mu_{\rm s} \neq 0$ in this proof, an issue first addressed in print by Szegő in 1958 [184] although by other means (close to our entropy arguments earlier in this chapter), Verblunsky [453] allowed $d\mu_{\rm s} \neq 0$ in 1934. The idea used by Szegő is to find suitable polynomials to "mask" $d\mu_{\rm s}$. Szegő did this by hand, others (e.g., Garnett [146]) use peak functions, and Section 2.5 of [399] has a construction using boundary values of Carathéodory functions of singular measures. Instead, we want to present here a simple and elegant argument due to Helson–Lowdenslager [197].

Given $d\mu$, a positive measure on $\mathbb{D}$, define $S_{d\mu} \subset L^2(\partial\mathbb{D}, d\mu)$ to be the closure of polynomials, P, with $P(0) = 0$. Let $H_{d\mu}$ be the orthogonal projection of the function 1 to $S_{d\mu}^\perp$, so

$$\|H_{d\mu}\|^2_{L^2(d\mu)} = \min_{\text{polynomials}} \biggl[\int |P(e^{i\theta})|^2\, d\mu(\theta) \biggm| P(0) = 1\biggr] \tag{2.12.21}$$

Moreover, since $e^{-ik\theta} = \overline{z^k}$ and $z^k \in S_{d\mu}$, H is the unique function, which is a norm limit of polynomials P with $P(0) = 1$ and so that

$$\int H_{d\mu}(\theta) e^{-ik\theta}\, d\mu(\theta) = 0 \qquad k = 1, 2, \ldots \tag{2.12.22}$$

Proposition 2.12.5. *Suppose that $d\mu$ has the form* (2.1.1). *We have that*

$$|H_{d\mu}|^2\, d\mu_s(\theta) = 0 \tag{2.12.23}$$

and

$$\int H_{d\mu}(\theta) e^{-ik\theta} w(\theta)\, \frac{d\theta}{2\pi} = 0 \tag{2.12.24}$$

Proof. $\{P \mid \text{polynomial},\ P(0) = 0\}$ is an ideal in the set of polynomials. Thus, for any $k > 0$, $H(1 + \alpha e^{ik\theta}) \in 1 + S_{d\mu}$, so if for $\alpha \in \mathbb{C}$ we define

$$F_k(\alpha) = \int |H_{d\mu}(\theta)(1 + \alpha e^{ik\theta})|^2\, d\mu \tag{2.12.25}$$

then

$$F_k(\alpha) \geq F_k(0) \tag{2.12.26}$$

Expanding

$$F_k(\alpha) = F_k(0) + \operatorname{Re}(\alpha c_k) + d_k|\alpha|^2 \tag{2.12.27}$$

we see (2.12.26) implies $c_k = 0$, that is,

$$\int |H_{d\mu}(\theta)|^2\, e^{ik\theta}\, d\mu(\theta) = 0 \tag{2.12.28}$$

for all $k > 0$. But taking complex conjugates, we conclude the measure $|H|^2\, d\mu$ has all $k \neq 0$ Fourier coefficients zero, from which we conclude that

$$|H|^2\, d\mu = c\, \frac{d\theta}{2\pi} \tag{2.12.29}$$

This immediately implies (2.12.23), and (2.12.23) plus (2.12.22) implies (2.12.24). □

This allows us to prove

Theorem 2.12.6. *Let $d\mu$ have the form* (2.1.1). *Then*

$$\inf\left\{\int |P(e^{i\theta})|^2\, d\mu(\theta) \;\middle|\; P(0) = 1\right\} = \inf\left\{\int |P(e^{i\theta})|^2 w(\theta)\, \frac{d\theta}{2\pi} \;\middle|\; P(0) = 1\right\} \tag{2.12.30}$$

for the inf *over all polynomials.*

Proof. Since $H_{d\mu}$ obeys (2.12.24) and, by

$$\|f\|_{L^2(w \frac{d\theta}{2\pi})} \leq \|f\|_{L^2(d\mu)} \tag{2.12.31}$$

we see that $H_{d\mu}$ is an $L^2(w\frac{d\theta}{2\pi})$ limit of polynomials with $P(0) = 1$. We have

$$H_{w\frac{d\theta}{2\pi}} = H_{d\mu} \tag{2.12.32}$$

Thus, by (2.12.23),

$$\begin{aligned}\text{RHS of (2.12.30)} &= \int |H_{d\mu}|^2 w(\theta)\, \frac{d\theta}{2\pi} \\ &= \int |H_{d\mu}|^2\, d\mu \\ &= \text{LHS of (2.12.30)}\end{aligned}$$

□

Thus, Szegő's theorem for $d\mu_s = 0$ implies the theorem for $d\mu_s \neq 0$, and we have a new proof of

Theorem 2.12.7. *If μ has the form* (2.1.1), *then*

$$\lim_{n\to\infty} \|\Phi_n\|^2 = \exp\biggl(\int \log(w(\theta))\, \frac{d\theta}{2\pi}\biggr) \tag{2.12.33}$$

Remarks and Historical Notes. As noted, the basic ideas when $d\mu_s = 0$ are in Szegő's 1920 paper [430]. The beautiful argument in Proposition 2.12.5, which relies on the fact that $\{P \mid P(0) = 0\}$ is an ideal, is due to Helson–Lowdenslager [197].

2.13 ANOTHER APPROACH TO SZEGŐ ASYMPTOTICS

In this section, we want to discuss another approach to Szegő asymptotics. Central is a formula of considerable interest as a tool in OPUC:

Theorem 2.13.1 (CD Formula). *Let $\{\varphi_n\}_{n=0}^\infty$ be the normalized OPUC for a nontrivial probability measure, $d\mu$. Then for any z, ζ with $z\bar{\zeta} \neq 1$,*

$$\sum_{j=0}^n \overline{\varphi_j(\zeta)}\, \varphi_j(z) = [\, \overline{\varphi_{n+1}^*(\zeta)}\, \varphi_{n+1}^*(z) - \overline{\varphi_{n+1}(\zeta)}\, \varphi_{n+1}(z)](1 - \bar{\zeta} z)^{-1} \tag{2.13.1}$$

Remarks. 1. The quantity in (2.13.1) is called the *CD kernel* (or Christoffel–Darboux kernel) and denoted $K_n(\zeta, z)$. We will study it further in Sections 2.14–2.17, 3.11, 3.12, and 5.11.

2. This is called the Christoffel–Darboux formula because they proved an analog for OPRL (see Subsection 1.2.9 of [399] and [407]). It is due to Szegő [434].

3. This result is an analog of the Wronskian relation for solutions of $-u'' + Vu = zu$, $-w'' + Vw = \zeta w$ with $u(0) = w(0) = 0$. Then,

$$(z - \bar{\zeta}) \int_0^a \overline{w(x)}\, u(x)\, dx = \overline{w'(a)}\, u(a) - \overline{u'(0)}\, w(0)$$

and the first proof is similar.

First Proof. Taking the conjugate of (1.8.15) for ζ and multiplying by (1.8.15) for z and subtracting the same for (1.8.14), we get

$$\overline{\varphi_{n+1}^*(\zeta)}\,\varphi_{n+1}^*(z) - \overline{\varphi_{n+1}(\zeta)}\,\varphi_{n+1}(z) = \overline{\varphi_n^*(\zeta)}\,\varphi_n^*(z) - \bar{\zeta}\,z\,\overline{\varphi_n(\zeta)}\,\varphi_n(z) \tag{2.13.2}$$

since the cross-terms cancel and $\rho_n^{-2}(1 - |\alpha_n|^2) = 1$. Thus,

$$\text{LHS of (2.13.2)} = (1 - \bar{\zeta} z)\,\overline{\varphi_n(\zeta)}\,\varphi_n(z) + [\,\overline{\varphi_n^*(\zeta)}\,\varphi_n^*(z) - \overline{\varphi_n(\zeta)}\,\varphi_n(z)] \tag{2.13.3}$$

which leads to (2.13.1) if we iterate. □

Second Proof. Fix first $\zeta \in \partial\mathbb{D}$. Then $\overline{\varphi_{n+1}^*(\zeta)}\,\varphi_{n+1}^*(z) - \overline{\varphi_{n+1}(\zeta)}\,\varphi_{n+1}(z)$ vanishes if $\zeta = z$ since $|\varphi_{n+1}^*(\zeta)| = |\varphi_{n+1}(\zeta)|$. Thus, for some polynomial h of degree n,

$$\overline{\varphi_{n+1}^*(\zeta)}\,\varphi_{n+1}^*(z) - \overline{\varphi_{n+1}(\zeta)}\,\varphi_{n+1}(z) = (\zeta - z)h(z) \tag{2.13.4}$$

Since $\varphi_{n+1}, \varphi_{n+1}^* \perp \{z^j\}_{j=1}^n$, we see

$$\langle z^j, (\zeta - z)h\rangle = 0 \qquad j = 1, \dots, n \tag{2.13.5}$$

or

$$\bar{\zeta}\,\langle z^{j-1}, h\rangle = |\zeta|^2\langle z^j, h\rangle = \langle z^j, h\rangle \tag{2.13.6}$$

so, by induction (with $C = \langle 1, h\rangle$),

$$\langle z^j, h\rangle = (\bar{\zeta})^j C \qquad j = 0, 1, \dots, n \tag{2.13.7}$$

and thus, for any polynomial Q of degree at most n,

$$\langle Q, h\rangle = C\,\overline{Q(\zeta)} \tag{2.13.8}$$

Since $\{\varphi_j\}_{j=0}^n$ is an orthonormal basis,

$$h(z) = \sum_{j=0}^n \langle\varphi_j, h\rangle\varphi_j(z) \tag{2.13.9}$$

we find

$$h(z) = C\sum_{j=0}^n \overline{\varphi_j(\zeta)}\,\varphi_j(z) \tag{2.13.10}$$

Equating powers of z^{n+1} in (2.13.4),

$$C\,\overline{\varphi_n(\zeta)} = \rho_n^{-1}[\,\overline{\varphi_{n+1}(\zeta)} + \alpha_n\,\overline{\varphi_{n+1}^*(\zeta)}\,] \tag{2.13.11}$$

$$= \bar{\zeta}\,\overline{\varphi_n(\zeta)} \tag{2.13.12}$$

by the inverse recursion relation (2.4.6). Thus, $C = \bar{\zeta}$, proving (2.13.1) for $\zeta \in \partial\mathbb{D}$. By analyticity (both sides are analytic in $\bar{\zeta}$), the formula holds for all ζ. □

Corollary 2.13.2. *For* $|z| < 1$,

$$|\varphi_n^*(z)| \geq (1 - |z|^2)^{1/2} \tag{2.13.13}$$

Proof. Taking $\zeta = z$ in (2.13.1),

$$|\varphi_n^*(z)|^2 = |\varphi_n(z)|^2 + (1-|z|^2)\sum_{j=0}^{n-1}|\varphi_j(z)|^2 \tag{2.13.14}$$

Since $\varphi_0(z) = 1$, (2.13.13) is immediate. □

Corollary 2.13.3. *Fix $z_0 \in \mathbb{D}$. Then either*

$$\sup_n |\varphi_n^*(z_0)| < \infty \tag{2.13.15}$$

or else

$$\lim_{n\to\infty} |\varphi_n^*(z_0)| = \infty \tag{2.13.16}$$

(2.13.15) *holds if and only if*

$$\sum_{j=0}^{\infty} |\varphi_j(z_0)|^2 < \infty \tag{2.13.17}$$

Proof. By (2.13.14),

$$(1-|z_0|^2)\sum_{j=0}^{n-1}|\varphi_j(z_0)|^2 \le |\varphi_n^*(z_0)|^2 \le \sum_{j=0}^{n}|\varphi_j(z_0)|^2 \tag{2.13.18}$$

This shows (2.13.15) ⇔ (2.13.17), and that if (2.13.17) fails, then (2.13.16) is true. □

Proposition 2.13.4. *Let $\{f_n(z)\}_{n=1}^{\infty}$ be a family of nonvanishing analytic functions on a connected open subset, Ω, of $\mathbb{C}$. Suppose that for each compact $K \subset \Omega$,*

$$C_K = \inf_{z\in K,n} |f_n(z)| > 0 \tag{2.13.19}$$

Then either

(a) *For every $z \in \Omega$,*

$$\lim_{N\to\infty} \sup_{n\le N} |f_n(z)| = \infty \tag{2.13.20}$$

with convergence uniform on compact $K \subset \Omega$, or

(b) *For every compact $K \subset \Omega$,*

$$D_K = \sup_{z\in K,n} |f_n(z)| < \infty \tag{2.13.21}$$

Proof. Let $g_n(z) = f_n(z)^{-1}$ so the g_n are uniformly bounded on compact subsets of Ω. Suppose (2.13.21) fails for some K. Then we can find $z_j \in K$ and n_j so

$$|f_{n_j}(z_j)| \ge j \tag{2.13.22}$$

By passing to a subsequence, we can suppose $z_j \to z_\infty \in K$ and g_{n_j} has a limit g_∞. By (2.13.22), $g_\infty(z_\infty) = 0$, so by Hurwitz's theorem and the fact that each g_n is nonvanishing, $g_\infty \equiv 0$, that is, (2.13.20) holds uniformly on compacts. □

Theorem 2.13.5. *Let $d\mu$ be a nontrivial probability measure on $\partial\mathbb{D}$ of the form* (2.1.1). *Then*

(i) *If*

$$\int \log(w(\theta))\,\frac{d\theta}{2\pi} = -\infty \tag{2.13.23}$$

then for each $R < 1$,

$$\inf_{n,|z|\le R} |\varphi_n^*(z)| \to \infty \tag{2.13.24}$$

(ii) *If*

$$\int \log(w(\theta))\,\frac{d\theta}{2\pi} > -\infty \tag{2.13.25}$$

then for each $R < 1$,

$$\sup_{n,|z|<R} |\varphi_n^*(z)| < \infty \tag{2.13.26}$$

$$\sup_{n,|z|<R} \sum_{j=0}^{\infty} |\varphi_j(z)|^2 < \infty \tag{2.13.27}$$

Proof. (i) By Szegő's theorem, if (2.13.23) holds, then $\varphi_n^*(0) = \prod_{j=0}^{n-1} \rho_j^{-1} \to \infty$. Thus, by (2.13.14) and Proposition 2.13.4, for each R,

$$\inf_{|z|<R}\, \sup_{n\le N} |\varphi_n^*(z)| \to \infty \tag{2.13.28}$$

By (2.13.18), if $k > n$,

$$|\varphi_k^*(z)|^2 \ge (1-|z|^2)|\varphi_n^*(z)|^2 \tag{2.13.29}$$

so (2.13.28) implies (2.13.24).

(ii) By Szegő's theorem, if (2.13.24) holds,

$$\sup_n |\varphi_n^*(0)| < \infty$$

so (2.13.26) follows from Proposition 2.13.4, and (2.13.27) then follows from (2.13.18). □

Theorem 2.13.6. *If the Szegő condition,* (2.13.25), *holds, then there is an analytic function, $D(z)$, nonvanishing on $\mathbb{D}$ so that uniformly on compact subsets of $\mathbb{D}$,*

$$\varphi_n^*(z) \to D(z)^{-1} \tag{2.13.30}$$

Moreover (*with* $\alpha_{-1} \equiv -1$),

$$D(z)^{-1} = -\sum_{j=0}^{\infty} \alpha_{j-1}\left[\prod_{m=j}^{\infty} \rho_m\right]\varphi_j(z) \tag{2.13.31}$$

converging uniformly on compacts.

Remark. At this point, we do *not* have a connection of D to $\log(w)$!

Proof. By (2.13.1) with $\zeta = 0$,

$$\varphi_n^*(z) = \varphi_n^*(0)^{-1} \sum_{j=0}^{n-1} \overline{\varphi_j(0)}\, \varphi_j(z) \tag{2.13.32}$$

Since

$$\varphi_n^*(0)^{-1} \to \prod_{m=0}^{\infty} \rho_m \tag{2.13.33}$$

and (2.13.27) holds, the sum converges uniformly to a limit, which is nonvanishing since $|\varphi_n^*(z)| \geq (1 - |z|^2)^{1/2}$. Thus, we can denote the inverse of the limit as $D(z)$. By the uniformity of convergence on compacts, D is analytic. (2.13.31) follows from (2.13.33) and

$$\overline{\varphi_j(0)} = -\alpha_j \prod_{m=0}^{j} \rho_j^{-1} \tag{2.13.34}$$

□

We are heading toward showing that $D \in H^2$ and has boundary values with $|D(e^{i\theta})|^2 = w(\theta)$:

Proposition 2.13.7. *Let* $P_r(\theta, \varphi)$ *be the Poisson kernel* (2.3.19). *Then*

$$|\varphi_n^*(re^{i\varphi})|^{-2} \leq \int P_r(\theta, \varphi) |\varphi_n^*(e^{i\theta})|^{-2} \frac{d\theta}{2\pi} \tag{2.13.35}$$

If the Szegő condition, (2.13.25), *holds, then*

$$|D(re^{i\theta})|^2 \leq \operatorname{Re} F(re^{i\theta}) \tag{2.13.36}$$

Proof. Since $\varphi_n^*(z)^{-1}$ is analytic in $\mathbb{D}$ and continuous on $\overline{\mathbb{D}}$,

$$\varphi_n^*(re^{i\varphi})^{-1} = \int P_r(\theta, \varphi) \varphi_n^*(e^{i\theta})^{-1} \frac{d\theta}{2\pi}$$

Since $P_r(\theta, \varphi)\frac{d\theta}{2\pi}$ is a positive measure, the Schwarz inequality implies that

$$\begin{aligned} \text{LHS of (2.13.35)} &\leq \left(\int P_r(\theta, \varphi) |\varphi_n^*(e^{i\theta})|^{-2} \frac{d\theta}{2\pi} \right) \left(\int P_r(\theta, \varphi) \frac{d\theta}{2\pi} \right) \\ &= \text{RHS of (2.13.35)} \end{aligned} \tag{2.13.37}$$

since the integral of the Poisson kernel is 1.

Taking $n \to \infty$, the left-hand side of (2.13.35) $\to$ $|D(re^{i\varphi})|^2$ while $\frac{d\theta}{2\pi}|\varphi_n^*(e^{i\theta})|^2 \to d\mu$ by (2.4.59), so the right-hand side of (2.13.36) $\to \operatorname{Re} F(re^{i\varphi})$. □

Theorem 2.13.8. *If the Szegő condition,* (2.13.25), *holds, then the function D of* (2.13.30) *lies in H^2 and its boundary values $D(e^{i\theta})$ obey*

$$|D(e^{i\theta})|^2 = w(\theta) \tag{2.13.38}$$

for a.e. θ.

Proof. By (2.13.36) and

$$\int \operatorname{Re} F(re^{i\theta})\, \frac{d\theta}{2\pi} = \operatorname{Re} F(0) = 1$$

we see that

$$\int |D(re^{i\theta})|^2\, \frac{d\theta}{2\pi} \le 1$$

so $D \in H^2$.

Taking limits in (2.13.36) using (2.3.55), we find

$$|D(e^{i\theta})|^2 \le w(\theta) \tag{2.13.39}$$

By Theorem 2.3.27, any H^2 function obeys

$$\log|D(0)|^2 \le \int \log|D(e^{i\theta})|^2\, \frac{d\theta}{2\pi} \tag{2.13.40}$$

so by (2.13.39),

$$\log|D(0)|^2 \le \int \log(w(\theta))\, \frac{d\theta}{2\pi} \tag{2.13.41}$$

By Szegő's theorem, we have equality in (2.13.41), so we must have equality in (2.13.39) (proving (2.13.38)) and (2.13.40) (proving D has no singular inner part). □

Remarks and Historical Notes. The first proof of the CD formula is a variant of one I learned from L. Golinskii. The second, due to Wong [462], is a variant of the original proof of Szegő [434]. For a review of the CD kernel and recent applications, see Simon [407].

The approach in this section is motivated by the paper of Delsarte, Genin, and Kamp [106].

2.14 PARAORTHOGONAL POLYNOMIALS AND THEIR ZEROS

Having introduced the CD kernel, $K_n(\zeta, z)$, by (2.13.1), we will turn to studying its asymptotics in the next three sections and further in Sections 3.11, 3.12, and 5.11. This section is a preliminary on some natural polynomials associated to OPUC. Zeros of OPRL have a number of attractive properties:

(i) They are real.
(ii) They are simple.
(iii) A gap (a, b) in $\operatorname{supp}(d\mu)$ has at most one zero of each p_n.
(iv) The zeros for p_n interlace those for p_{n+1}.
(v) They enter in Gaussian quadrature in best discrete approximations of $d\mu$.

We have seen (i) and (ii) in Proposition 1.3.1 and will provide a reference for (iii)–(v) in the Notes to this section. All these fail for OPUC—the case $d\mu = \frac{d\theta}{2\pi}$ where $\Phi_n(z; d\mu) = z^n$ shows zeros need not be simple and that they do not lie on $\partial\mathbb{D}$. Indeed (see Theorem 1.8.4), we have seen they lie on $\mathbb{D}$ and never on $\partial\mathbb{D}$.

(iii)–(v) do not really make sense unless the zeros lie on a one-dimensional set like $\partial\mathbb{D}$.

This dramatic difference between OPRL and OPUC is seen in Theorem 1.2.6 where π_n is the projection onto polynomials of degree at most $n-1$, $A = \pi_n M_z \pi_n$ (with $M_z f = zf$) and $\det(z - A)$ is the monic OP. In the OPRL case, M_z is selfadjoint and A remains selfadjoint. In the OPUC case, M_z is unitary, but A is only a contraction and definitely not unitary.

It pays to look at what destroys unitarity to figure out how to remedy the problem. We claim

$$\operatorname{Ran}(\pi_n) = \operatorname{Ran}(\pi_{n-1}) \oplus [\varphi_{n-1}] \tag{2.14.1}$$

as an orthogonal direct sum. This follows from the definition of φ_{n-1}. (Here $[\psi]$ means the subspace spanned by ψ.)

Notice that

$$A\psi = \begin{cases} z\psi & \text{if } \psi \in \operatorname{Ran}(\pi_{n-1}) \\ \bar{\alpha}_{n-1}\varphi_{n-1}^* & \text{if } \psi = \varphi_{n-1} \end{cases} \tag{2.14.2}$$

for $z\psi \in \operatorname{Ran}(\pi_n)$ if $\psi \in \operatorname{Ran}(\pi_{n-1})$ and

$$z\varphi_{n-1} = \rho_n \varphi_n + \bar{\alpha}_{n-1}\varphi_{n-1}^* \tag{2.14.3}$$

and

$$\pi_n(\varphi_{n-1}^*) = \varphi_{n-1}^* \qquad \pi_n(\varphi_n) = 0 \tag{2.14.4}$$

Thus, A is unitary on $\operatorname{Ran}(\pi_{n-1})$, but since $\|A\varphi_{n-1}\| = |\alpha_{n-1}| < 1$, it is not unitary overall. It is only this rank one failure of unitarity that prevents $\det(z - A)$ from having zeros on $\partial\mathbb{D}$. We also see:

Proposition 2.14.1. *There is a one-parameter family of unitaries $U_\beta^{(n)}$ on* $\operatorname{Ran}(\pi_n)$ *that agree with A on* $\operatorname{Ran}(\pi_{n-1})$ *and have the form*

$$U_\beta^{(n)} \restriction \operatorname{Ran}(\pi_{n-1}) = M_z \restriction \operatorname{Ran}(\pi_{n-1}) \tag{2.14.5}$$

$$U_\beta^{(n)} \varphi_{n-1} = \bar{\beta}\varphi_{n-1}^* \tag{2.14.6}$$

for some $\beta \in \partial\mathbb{D}$.

Proof. $A[\operatorname{Ran}(\pi_{n-1})] = [z^n, \dots, z]$ so, by Proposition 1.8.1, $(A[\operatorname{Ran}(\pi_{n-1})])^\perp = [\varphi_{n-1}^*]$. Thus, any unitary obeying (2.14.5) must obey (2.14.6) and conversely. □

Theorem 2.14.2. *The operator $U_\beta^{(n)}$ has the characteristic polynomial*

$$\det(z - U_\beta^{(n)}) = z\Phi_{n-1}(z; d\mu) - \bar{\beta}\Phi_{n-1}^*(z; d\mu) \tag{2.14.7}$$

First Proof. Define for $\alpha \in \mathbb{D}$, A_α to be given by (2.14.5) and (2.14.6) but with $\bar{\beta}$ replaced by $\bar{\alpha}$. Thus, A_α is the A associated to the OPUC for the measure $d\mu_\alpha$ with

$$\alpha_j(d\mu_\alpha) = \begin{cases} \alpha_j(d\mu) & j \neq n-1 \\ \alpha & j = n-1 \end{cases} \tag{2.14.8}$$

so, by Theorem 1.2.6,

$$\det(z - A_\alpha) = z\Phi_{n-1}(z; d\mu) - \bar{\alpha}\Phi_{n-1}^*(z; d\mu) \tag{2.14.9}$$

Now take $\alpha = \beta(1 - m^{-1})$ and note as $m \to \infty$, A_α converges, as an operator on $\text{Ran}(\pi_n)$, to $U_\beta^{(n)}$. □

Second Proof. Suppose z_0 is an eigenvalue of $U_\beta^{(n)}$ with eigenvector ψ. ψ cannot be in $\text{Ran}(\pi_{n-1})$ since $\pi_n(z - z_0)h = (z - z_0)h \neq 0$ for $h \in \text{Ran}(\pi_{n-1}) \setminus \{0\}$. Thus, we can renormalize ψ so

$$\psi = h + \varphi_{n-1} \tag{2.14.10}$$

with $h \in \text{Ran}(\pi_{n-1})$. By (2.14.5) and (2.14.6),

$$(U_\beta^{(n)} - z_0)\psi = (z - z_0)h + z\varphi_{n-1} - \bar{\beta}\varphi_{n-1}^* \tag{2.14.11}$$

For this to be the zero polynomial, it must vanish at $z = z_0$. Thus (since $\varphi_{n-1}/\varphi_{n-1}^* = \Phi_{n-1}/\Phi_{n-1}^*$),

$$z_0\Phi_{n-1}(z_0) - \bar{\beta}\Phi_{n-1}^*(z_0) = 0 \tag{2.14.12}$$

If z_0 has multiplicity larger than one, we can find an eigenvector in $\text{Ran}(\pi_{n-1})$, which we argued cannot happen. Thus, $U_\beta^{(n)}$ has n distinct eigenvalues, so by (2.14.12), both sides of (2.14.7) have the same n simple zeros. Since both are monic, they are equal. □

This motivates

Definition. The monic *paraorthogonal polynomial* (POPUC) of degree n associated to a nondegenerate probability measure, $d\mu$, on $\partial\mathbb{D}$ and parameter $\beta \in \partial\mathbb{D}$ is given by

$$\Phi_n(z; d\mu, \beta) = z\Phi_{n-1}(z; d\mu) - \bar{\beta}\Phi_{n-1}^*(z; d\mu) \tag{2.14.13}$$

The normalized POPUC is given by

$$\varphi_n(z; d\mu, \beta) = z\varphi_{n-1}(z; d\mu) - \bar{\beta}\varphi_{n-1}^*(z; d\mu) \tag{2.14.14}$$

We will sometimes just use $\Phi_n(z; \beta)$. (2.14.7) becomes

$$\det(z - U_\beta^{(n)}) = \Phi_n(z; d\mu, \beta) \tag{2.14.15}$$

If $\{z_j^{(n-1)}\}_{j=1}^{n-1}$ are the zeros of $\Phi_{n-1}(z)$, we define their Blaschke product by

$$b_{n-1}(z) = \prod_{j=1}^{n-1} \frac{z - z_j^{(n-1)}}{1 - \bar{z}_j^{(n-1)} z} \tag{2.14.16}$$

As we saw (see (2.3.67)), there is a phase factor in "true" Blaschke products but we use the symbol b_{n-1} without the phase factor. Here are two formulae for the zeros of $\Phi_n(z; \beta)$:

Proposition 2.14.3. (i) *We have*

$$b_{n-1}(z) = \frac{\Phi_{n-1}(z)}{\Phi_{n-1}^*(z)} = \frac{\varphi_{n-1}(z)}{\varphi_{n-1}^*(z)} \tag{2.14.17}$$

(ii) *The zeros of* $\Phi_n(z;\beta)$ *are the* n *solutions of*

$$zb_{n-1}(z) = \bar\beta \tag{2.14.18}$$

(iii) *Let* $w \in \partial\mathbb{D}$ *and let*

$$\beta(w) = \bar w\, \frac{\overline{\varphi_{n-1}(w)}}{\overline{\varphi_{n-1}^*(w)}} \tag{2.14.19}$$

Then (with K *the CD kernel* (2.13.1))

$$K_{n-1}(w,z) = -\beta(w)\,\overline{\varphi_{n-1}^*(w)}\,\frac{\varphi_n(z;\beta(w))}{1-z\bar w} \tag{2.14.20}$$

In particular, the zeros of $K_{n-1}(w,\,\cdot\,)$ *are precisely the zeros of* $\varphi_n(\,\cdot\,;\beta(w))$ *other than* w*, and the zeros of* $\varphi_n(\,\cdot\,,\beta(w))$ *are* w *plus the zeros of* $K_{n-1}(w,\,\cdot\,)$.

Proof. (i) We have

$$\Phi_{n-1}(z) = \prod_{j=1}^{n-1}(z - z_j^{(n-1)}) \tag{2.14.21}$$

so

$$\Phi_{n-1}^*(z) = \prod_{j=1}^{n-1}(1 - \bar z_j^{(n-1)} z) \tag{2.14.22}$$

and $b_{n-1} = \Phi_{n-1}/\Phi_{n-1}^*$ is immediate. Since $\|\Phi_{n-1}\| = \|\Phi_{n-1}^*\|$, the two polynomial ratios in (2.14.17) are equal.

(ii) $z\Phi_{n-1} - \bar\beta\Phi_{n-1}^* = 0$ if and only if $z\Phi_{n-1}/\Phi_{n-1}^* = \bar\beta$ so (2.14.17) implies (ii).

(iii) By the CD formula, (2.13.2), and (2.14.19),

$$\begin{aligned}(1-z\bar w)K_{n-1}(w,z) &= \overline{\varphi_{n-1}^*(w)}\,[\varphi_{n-1}^*(z) - z\varphi_{n-1}(z)\beta(w)]\\ &= -\beta(w)\,\overline{\varphi_{n-1}^*(w)}\,\varphi_n(z;\beta(w))\end{aligned}$$

proving (2.14.20). This implies the results on the zeros if we note that, by (2.14.19), $\varphi_n(w;\beta(w)) = 0$. □

Here is the main result on zeros of POPUC:

Theorem 2.14.4. (i) *All the zeros of* $\Phi_n(z;\beta)$ *lie on* $\partial\mathbb{D}$.

(ii) *The zeros of* $\Phi_n(z;\beta)$ *are all simple.*

(iii) *If* $\tilde\beta \neq \beta$*, both in* $\partial\mathbb{D}$*, the zeros of* $\Phi_n(z;\beta)$ *and* $\Phi_n(z;\tilde\beta)$ *strictly interlace.*

(iv) *For any* $\beta, \tilde\beta$*, if* z_1, z_2 *are two successive zeros of* $\Phi_n(z;\beta)$*, (i.e., going counterclockwise,* z_2 *is the first zero after* z_1*), then* $\Phi_{n+1}(z;\tilde\beta)$ *has a zero in* (z_1, z_2).

(v) *If* (w_0, w_1) *is an interval on* $\partial\mathbb{D}$ *disjoint from* μ*, then each* $\Phi_n(z;\beta)$ *has at most one zero in* (w_0, w_1).

Proof. (i) Each factor

$$\eta(z, z_0) = \frac{z - z_0}{1 - \bar z_0 z} \tag{2.14.23}$$

with $z_0 \in \mathbb{D}$ has $\eta(\,\cdot\,, -z_0)$ as the functional inverse, so it is a bijection of $\overline{\mathbb{D}}$ to $\overline{\mathbb{D}}$ and of $\mathbb{D}$ and $\partial\mathbb{D}$ to themselves. Thus, $z \mapsto zb_{n-1}(z)$ maps $\mathbb{D}$ to $\mathbb{D}$, which means, for $\beta \in \partial\mathbb{D}$, $zb_{n-1}(z) = \bar{\beta}$ has no solution in $\mathbb{D}$.

Since

$$\overline{\eta(1/\bar{z}, z_0)} = \eta(z, z_0)^{-1} \tag{2.14.24}$$

zb_{n-1} maps $\mathbb{C} \setminus \overline{\mathbb{D}}$ to itself, so $zb_{n-1}(z) = \bar{\beta}$ has no solution in $\mathbb{C} \setminus \overline{\mathbb{D}}$. Thus, all solutions lie in $\partial\mathbb{D}$.

(ii) As noted in (i), $e^{i\theta} \to \eta(e^{i\theta}, z_0)$ is a bijection of $\partial\mathbb{D}$ to itself. Since it maps $\mathbb{D}$ to itself, $\frac{\partial}{\partial r}|\eta(re^{i\theta}, z_0)| \geq 0$. So, by the Cauchy–Riemann equations, if

$$\eta(e^{i\theta}, z_0) = e^{i\psi(e^{i\theta}, z_0)} \tag{2.14.25}$$

(ψ determined mod 2π), then

$$\frac{d}{d\theta}\,\psi(e^{i\theta}, z_0) \geq 0 \tag{2.14.26}$$

(actually, one can show > 0). Since $\frac{1}{i}\frac{d}{d\theta}(e^{i\theta}) = 1 > 0$, we see, if

$$e^{i\theta} b_{n-1}(e^{i\theta}) = e^{i\gamma(\theta)} \tag{2.14.27}$$

then

$$\gamma'(\theta) > 0 \tag{2.14.28}$$

which implies that zeros are simple.

(iii) $\gamma(\theta)$ defined by (2.14.27) is strictly monotone, so if $\beta = e^{-i\gamma_0}$, then between two solutions of $\gamma(\theta) = \gamma_0 \pmod{2\pi}$, every value of $\gamma(\theta) \pmod{2\pi}$ is taken once.

(iv) By Szegő recursion,

$$\begin{aligned} b_n = \frac{\Phi_n}{\Phi_n^*} &= \frac{z\Phi_{n-1} - \bar{\alpha}_{n-1}\Phi_{n-1}^*}{\Phi_{n-1}^* - \alpha_{n-1} z\Phi_{n-1}} \\ &= \frac{zb_{n-1} - \bar{\alpha}_{n-1}}{1 - \alpha_n z b_{n-1}} = \eta(zb_{n-1}, \bar{\alpha}_{n-1}) \end{aligned} \tag{2.14.29}$$

In the interval (z_1, z_2), zb_{n-1} goes through a 2π change of phase. $e^{i\theta} \to \eta(e^{i\theta}, \bar{\alpha}_{n-1})$ is a monotone bijection of $\partial\mathbb{D}$ to $\partial\mathbb{D}$, so b_n also goes through a 2π change of phase. Thus, zb_n goes through more than a 2π change of phase, and so $zb_n = \bar{\tilde{\beta}}$ must have a solution.

(v) Suppose that $\Phi_n(z; \beta)$ has zeros at $e^{\pm i\varphi}$ and μ is supported in $(e^{-i\varphi}, e^{i\varphi})$. Let

$$P(z) = \frac{\Phi_n(z; \beta)z}{(z - e^{i\varphi})(z - e^{-i\varphi})}$$

This is a polynomial because of the assumed zeros, and since Pz^{-1} has degree $n-2$, P is a linear combination of $z, z^2, \dots, z^{n-1}$. This is orthogonal to Φ_{n-1}^* and to $z\Phi_{n-1}$, and so to $\Phi_n(z; \beta)$. Thus,

$$\int \frac{z}{(z - e^{i\varphi})(z - e^{-i\varphi})}\, |\Phi_n(z; \beta)|^2\, d\mu = 0 \tag{2.14.30}$$

But

$$\frac{(e^{i\theta} - e^{i\varphi})(e^{i\theta} - e^{-i\varphi})}{e^{i\theta}} = (e^{i\theta} + e^{-i\theta}) - (e^{i\varphi} + e^{-i\varphi})$$

so

$$\frac{z}{(z - e^{i\varphi})(z - e^{-i\varphi})}\bigg|_{z=e^{i\theta}} = \frac{1}{\cos\theta - \cos\varphi} > 0$$

for $|\theta| < \varphi$, that is, on supp$(d\mu)$. Thus, (2.14.30) cannot hold. By rotation covariance, any pair of zeros in a gap can be rotated to this case. □

We can make the consequences of (iv) explicit:

Corollary 2.14.5. *Let* $(z_0, \dots, z_{n-1})$ *and* $(w_0, \dots, w_n)$ *be the zeros of* $\Phi_n(z; \beta)$ *and* $\Phi_{n+1}(z; \tilde{\beta})$*, respectively, counted counterclockwise. Then one of the following happens:*

(a) Φ_n *and* Φ_{n+1} *have a single zero in common, which, by cyclic relabeling, we can suppose is* $z_0 = w_0$*. In that case, each of the n intervals* (z_0, z_1), $(z_1, z_2), \dots, (z_{n-2}, z_{n-1}), (z_{n-1}, z_0)$ *has exactly one* w.

(b) Φ_n *and* Φ_{n+1} *have no zeros in common, in which case among the n intervals,* $(z_0, z_1), \dots, (z_{n-1}, z_0)$*, one has exactly two* w*'s and each of the others has exactly one* w.

Proof. Follows from the fact that each of the n intervals $(z_0, z_1), \dots, (z_{n-1}, z_0)$ must contain at least one w. There is only one other w left. □

Remarks and Historical Notes. For properties (iii)–(v) for OPRL, see Section 1.2 of [399]. The gap property (property (iii)) comes as follows: If x_0, x_1 are two zeros of P_n in (a, b), which is disjoint from supp$(d\mu)$, then $P_n/(x - x_0)(x_1 - x)$ is of degree $n - 2$, so orthogonal to P_n, so

$$\int |P_n|^2 (x - x_0)^{-1}(x_1 - x)^{-1}\, d\mu(x) = 0$$

But $(x - x_0)^{-1}(x_1 - x)^{-1}$ is positive on supp$(d\mu)$. This classical argument motivated the final proof in the section.

For purposes of Gaussian quadrature on $\partial\mathbb{D}$, POPUC were introduced by Jones, Njåstad, and Thron [210]. Their zeros and other properties have been studied by Golinskii [174], Cantero–Moral–Velázquez [69], Wong [462], and Simon [405]. Our discussion here using CD kernels is influenced by Wong [462]. Most of Theorem 2.14.4 is from [69, 174] with parts from [405].

The use of b_n and of the recursion (2.14.29) is due to Khrushchev [219].

2.15 ASYMPTOTICS OF THE CD KERNEL: WEAK LIMITS

This is the first of three sections on the asymptotics of the CD kernel for OPUC, $K_n(w, z)$, especially when $|w| = |z| = 1$ and $w = z$ or $|w - z|$ is small. In this section, we will say something about limits of $\frac{1}{n+1} K_n(e^{i\theta}, e^{i\theta})\, d\mu(\theta)$ as a measure.

We start by relating it to limits of the zero counting measure for paraorthogonal polynomials.

Given a measure $d\mu$ on $\partial\mathbb{D}$, we let $d\nu_n$ be the zero counting measure for Φ_n, that is, ν_n is a pure point measure with

$$\nu_n(\{w\}) = n^{-1} \times (\text{multiplicity of } w \text{ as a zero of } \Phi_n) \tag{2.15.1}$$

Similarly, for any $\beta \in \partial\mathbb{D}$, we let $\nu_n^{(\beta)}$ be the zero counting measure for the POPUC $\Phi_n(z;\beta)$ (all multiplicities are one). Finally, we define

$$d\mu^{(N)}(\theta) = \frac{1}{N+1} K_N(e^{i\theta}, e^{i\theta})\, d\mu(\theta) \tag{2.15.2}$$

which is a probability measure on $\partial\mathbb{D}$, since $\int |\varphi_j|^2\, d\mu = 1$. $\nu_n^{(\beta)}$ is a probability measure on $\partial\mathbb{D}$ and ν_n on $\overline{\mathbb{D}}$. Here is a result that says they have the same weak limits:

Theorem 2.15.1. *For any $\ell = 1, 2, \dots$ and any β,*

$$\left| \int z^\ell\, d\mu^{(N)} - \int z^\ell\, d\nu_{N+1}^{(\beta)} \right| \le \frac{2\ell}{N+1} \tag{2.15.3}$$

$$\left| \int z^\ell\, d\nu_{N+1} - \int z^\ell\, d\nu_{N+1}^{(\beta)} \right| \le \frac{2\ell}{N+1} \tag{2.15.4}$$

In particular, for a subsequence, $N(1) < N(2) < \dots$, $d\nu_{N(j)+1}^{(\beta_j)} \xrightarrow{w} d\nu_\infty$ if and only if $d\mu^{(N(j))} \xrightarrow{w} d\nu_\infty$ (for one, and then for all choices of $\{\beta_j\}$), and in that case, for any $\ell = 1, 2, \dots$,

$$\lim_{j\to\infty} \int z^\ell\, d\nu_{N(j)+1} = \int z^\ell\, d\nu_\infty(z) \tag{2.15.5}$$

Conversely, if (2.15.5) *holds for some $d\nu_\infty$ on $\partial\mathbb{D}$, then $d\mu^{(N(j))} \xrightarrow{w} d\nu_\infty$.*

Proof. $\varphi_0, \dots, \varphi_N$ are a basis for $\operatorname{Ran}(\pi_{N+1})$, so with $A = \pi_{N+1} M_z \pi_{N+1}$,

$$\int z^\ell\, d\nu_{N+1} = \frac{1}{N+1} \operatorname{Tr}(A^\ell) = \frac{1}{N+1} \sum_{j=0}^{N} \langle \varphi_j, (A_j)^\ell \varphi_j \rangle \tag{2.15.6}$$

and similarly,

$$\int z^\ell\, d\nu_{N+1}^{(\beta)} = \frac{1}{N+1} \sum_{j=0}^{N} \langle \varphi_j, (U_\beta^{(N+1)})^\ell \varphi_j \rangle \tag{2.15.7}$$

By definition of K_N,

$$\int z^\ell\, d\mu^{(N)} = \frac{1}{N+1} \sum_{j=0}^{N} \langle \varphi_j, z^\ell \varphi_j \rangle \tag{2.15.8}$$

If $j \le N - \ell$, $(A_j)^\ell \varphi_j = (U_\beta^{(N+1)})^\ell \varphi_j = z^\ell \varphi_j$, so the terms in the sum cancel for such j's. Since $|\langle \varphi_j, z^\ell \varphi_j \rangle| \le 1$ and similarly for A and $U_\beta^{(N+1)}$ for any j, the remaining terms contribute at most $2\ell/(N+1)$ to the difference of the sums. This proves (2.15.3) and (2.15.4).

For $d\mu^{(N)}$ and $d\nu_{N+1}^{(\beta)}$, we have measures on $\partial\mathbb{D}$ so $\int z^{-\ell}\,d\eta = \overline{\int z^{\ell}\,d\eta}$. Polynomials in z and z^{-1} are dense in the continuous functions on $\partial\mathbb{D}$, so weak convergence is equivalent to convergence of $\int z^{\ell}\,d\eta$ (for all $\ell \geq 0$), which happens for one of $d\mu^{(N(j))}$ and $d\nu_{N(j)+1}^{(\beta_j)}$ if and only if it happens for both (by (2.15.3)). And convergence then implies (2.15.5). For the converse, note that (2.15.5) implies convergence of the moments of $d\nu_{N(j)+1}^{(\beta)}$ by (2.15.4). □

This is especially useful since there is a class of measures $d\mu$ for which w-lim $d\nu_n^{(\beta)}$ can be seen to be $\frac{d\theta}{2\pi}$.

Proposition 2.15.2. *Consider the conditions*

$$\text{(a)} \qquad \lim_{n\to\infty} (\rho_0 \dots \rho_{n-1})^{1/n} = 1 \tag{2.15.9}$$

$$\text{(b)} \qquad \lim_{n\to\infty} \frac{1}{n}\sum_{j=0}^{n-1} |\alpha_j|^2 = 0 \tag{2.15.10}$$

$$\text{(c)} \qquad \lim_{n\to\infty} \frac{1}{n}\sum_{j=0}^{n-1} |\alpha_j| = 0 \tag{2.15.11}$$

Then (a) ⇒ (b) ⇔ (c). *If*

$$\sup_n |\alpha_n| = R < 1 \tag{2.15.12}$$

then (b) ⇒ (a) *also.*

Proof. <u>(b) ⇔ (c).</u> Since $|\alpha_j| < 1$, we have that $|\alpha_j|^2 < |\alpha_j|$. This and the Schwarz inequality imply

$$\left(\frac{1}{n}\sum_{j=0}^{n-1} |\alpha_j|\right)^2 \leq \frac{1}{n}\sum_{j=0}^{n-1} |\alpha_j|^2 \leq \frac{1}{n}\sum_{j=0}^{n-1} |\alpha_j| \tag{2.15.13}$$

<u>(a) ⇒ (b).</u> We have that

$$-\log|\rho_j|^2 = |\alpha_j|^2 + \sum_{k=2}^{\infty} \frac{1}{k}\,|\alpha_j|^{2k} \geq |\alpha_j|^2 \tag{2.15.14}$$

so

$$\frac{1}{n}\sum_{j=0}^{n-1} |\alpha_j|^2 \leq -\log[(\rho_0 \dots \rho_{n-1})^{2/n}] \tag{2.15.15}$$

Thus, (a) ⇒ $\lim(-\log(\rho_0 \dots \rho_{n-1})^{2/n}) = 0$ ⇒ (b).

<u>(b) ⇒ (a) if (2.15.12) holds.</u> If (2.15.12) holds, then for some K (can be taken $-R^{-1}\log(1-R)$),

$$-\log|\rho_j|^2 \leq K|\alpha_j|^2$$

so

$$\frac{K}{n}\sum_{j=0}^{n-1}|\alpha_j|^2 \geq -\log[(\rho_0\dots\rho_{n-1})^{2/n}] \tag{2.15.16}$$

so (b) plus the fact that $\rho_j < 1$ implies (a). □

Definition. Let μ be a measure on $\partial\mathbb{D}$. If

$$\lim_{n\to\infty}(\rho_0\dots\rho_{n-1})^{1/n} = 1$$

we say μ is *regular*.

Regularity has two important consequences:

Theorem 2.15.3. *Let μ be a measure on $\partial\mathbb{D}$, which is regular. Then for any $z \in \mathbb{C}\setminus\mathbb{D}$, we have*

$$\lim_{n\to\infty}|\Phi_n(z)|^{1/n} = \lim_{n\to\infty}|\varphi_n(z)|^{1/n} = |z| \tag{2.15.17}$$

Remark. The proof shows the convergence is uniform on compact subsets of $\mathbb{C}\setminus\overline{\mathbb{D}}$.

Proof. Since $(\rho_1\dots\rho_n)^{1/n} \to 1$, we need only prove the result for Φ_n. Suppose $|z| > 1$. By Szegő recursion and $|\Phi_n(z)| \geq |\Phi_n^*(z)|$ if $|z| > 1$ (see (2.9.11)), we have

$$(|z| - |\alpha_n|)|\Phi_n(z)| \leq |\Phi_{n+1}(z)| \leq (|z| + |\alpha_n|)|\Phi_n(z)| \tag{2.15.18}$$

Since $|z| > 1$ holds, there is a $K(|z|)$ so that for all n,

$$1 - |\alpha_n|\,|z|^{-1} \geq \exp(-K|\alpha_n|) \tag{2.15.19}$$

Moreover, if $|z| > 1$,

$$1 + |\alpha_n|\,|z|^{-1} \leq \exp(|\alpha_n|) \tag{2.15.20}$$

Thus, (2.15.18) plus induction implies

$$\exp\left(-K\left(\sum_{j=0}^{n-1}|\alpha_j|\right)\right) \leq \frac{|\Phi_n(z)|}{|z|^n} \leq \exp\left(\sum_{j=0}^{n-1}|\alpha_j|\right) \tag{2.15.21}$$

(2.15.11) thus implies (2.15.17) for Φ_n.

This proves (2.15.17) for $|z| > 1$ and the limit is uniform in θ, for $z = re^{i\theta}$ with $r > 1$ fixed. By the maximum principle, for any $r > 1$,

$$|\Phi_n(e^{i\theta})| \leq \sup_{\varphi}|\Phi_n(re^{i\varphi})| \tag{2.15.22}$$

This plus the uniformity implies for any $r > 1$,

$$\limsup\left[\sup_{\theta}|\Phi_n(e^{i\theta})|^{1/n}\right] \leq r$$

Since r is arbitrary, the lim sup is at most 1.

Since the ρ's for the second kind polynomials are the same, we have

$$\limsup|\psi_n(e^{i\theta})|^{1/n} \leq 1 \tag{2.15.23}$$

But by (2.4.57),

$$|\varphi_n(e^{i\theta})|\,|\psi_n(e^{i\theta})| \geq 1 \tag{2.15.24}$$

This plus (2.15.22) implies

$$\liminf|\varphi_n(e^{i\theta})|^{1/n} \geq 1 \tag{2.15.25}$$

and so (2.15.17) for $|z| = 1$. □

Theorem 2.15.4. *Let μ be a measure on $\partial\mathbb{D}$, which is regular. Then*

$$\operatorname*{w-lim}_{n\to\infty} d\mu^{(n)} = \frac{d\theta}{2\pi} \tag{2.15.26}$$

and for any $\{\beta_j\} \in \partial\mathbb{D}$,

$$\operatorname*{w-lim}_{n\to\infty} d\nu_n^{(\beta_n)} = \frac{d\theta}{2\pi} \tag{2.15.27}$$

Proof. By Theorem 2.15.1, it suffices to prove for $\ell \geq 1$,

$$\int z^\ell \, d\nu_n(z) \to 0 \tag{2.15.28}$$

since $\frac{d\theta}{2\pi}$ is the unique measure on $\partial\mathbb{D}$ with $\int e^{i\ell\theta}\, d\eta(\theta) = 0$ for $\ell > 0$.

Let $d\nu_\infty$ be an arbitrary weak limit point of $d\nu_n$. For $|z| > 1$, $\log|z - w|$ is continuous for $w \in \overline{\mathbb{D}}$, so

$$\int \log|z - w|\, d\nu_n(w) \to \int \log|z - w|\, d\nu_\infty(w) \tag{2.15.29}$$

Since

$$\frac{1}{n} \log|\Phi_n(z)| = \int \log|z - w|\, d\nu_n(w) \tag{2.15.30}$$

(2.15.17) implies for $|z| > 1$,

$$\int \log\left|1 - \frac{w}{z}\right| d\nu_\infty(w) = 0 \tag{2.15.31}$$

In the region $|z| > 1$, uniformly in $|w| \leq 1$, $\log|1 - \frac{w}{z}|$ is the real part of an analytic function, so

$$\int \log\left(1 - \frac{w}{z}\right) d\nu_\infty(w) = 0 \tag{2.15.32}$$

since we first see it is an imaginary constant and then, by taking $|z| \to \infty$, we see the constant is zero. Now

$$\log\left(1 - \frac{w}{z}\right) = -\sum_{j=1}^\infty \frac{1}{j}\left(\frac{w}{z}\right)^j \tag{2.15.33}$$

uniformly in $|w| \leq 1$ and $|z| \geq 2$, so interchanging the sum and integral, we see

$$\int w^j \, d\nu_\infty(w) = 0 \tag{2.15.34}$$

for $j \geq 1$, proving (2.15.28). □

We have thus proven that if $d\mu$ is regular, then

$$\frac{1}{n+1}\, K_n(e^{i\theta}, e^{i\theta})\left[w(\theta)\, \frac{d\theta}{2\pi} + d\mu_s \right] \xrightarrow{w} \frac{d\theta}{2\pi} \tag{2.15.35}$$

When the Szegő condition holds, (2.9.30) says $\frac{1}{n+1} K_n\, d\mu_s \xrightarrow{w} 0$, and one might hope that this is true more generally (indeed, see Theorem 2.17.7), which leads us to a natural guess that under suitable hypotheses, pointwise in θ,

$$\frac{1}{n+1}\, K_n(e^{i\theta}, e^{i\theta}) w(\theta) \to 1 \tag{2.15.36}$$

It is precisely this surmise that we explore in the next two sections. Of course, it cannot hold at points with $w(\theta) = 0$. Note, however, if $d\mu_s = 0$, (2.15.35) implies that if the left side of (2.15.36) converges uniformly, the limit must be 1.

Remarks and Historical Notes. Theorem 2.15.1 is from Simon [409]. Regularity will be discussed more extensively in Section 5.9, mainly in the context of OPRL. In particular, its history is discussed in the Notes to that section. That regularity implies zeros are distributed according to an "equilibrium" measure (which is $\frac{d\theta}{2\pi}$ for $\partial\mathbb{D}$) is a major theme of that section. The proof of (2.15.28) is essentially potential theoretic—this is discussed in Section 5.5.

2.16 ASYMPTOTICS OF THE CD KERNEL: CONTINUOUS WEIGHTS

In this section, we will study the asymptotics of the CD kernel for continuous nonvanishing weights and apply this to obtain a refined estimate on the zeros of POPUC. We will call a function, f, on $\partial\mathbb{D}$ "continuous" on an interval $I = [\alpha, \beta]$ (i.e., $\alpha, \beta \in \partial\mathbb{D}$ and I is the set of points between α and β going counterclockwise from α to β) if, *as a function on* $\partial\mathbb{D}$, it is continuous at each $z \in [\alpha, \beta]$. This is stronger than saying the restriction of f to I is continuous on I; in particular, it says something if $\alpha = \beta$ and I is a single point. Here is the main theorem of this section:

Theorem 2.16.1 (Levin–Lubinsky [275]). *Let $d\mu$ be a regular probability measure on $\partial\mathbb{D}$ of the form*

$$d\mu = w(\theta)\, \frac{d\theta}{2\pi} + d\mu_s \tag{2.16.1}$$

Suppose for an interval $I = [\alpha, \beta] \subset \partial\mathbb{D}$,
(a) $\operatorname{supp}(d\mu_s) \cap I = \emptyset$
(b) *w is "continuous" on I and nonvanishing there.*

Then
(1) (Diagonal Asymptotics) *For any $A < \infty$, uniformly in $z_\infty \in I$, and sequences $z_n \in \partial\mathbb{D}$ with $n|z_n - z_\infty| \le A$ for all n,*

$$\frac{1}{n+1}\, K_n(z_n, z_n) \to w(z_\infty)^{-1} \tag{2.16.2}$$

(2) (Lubinsky Universality) *For any* $A < \infty$, *uniformly in* $z_\infty \in I$, *and* $a, b \in \mathbb{R}$ *with* $|a|, |b| \le A$, *we have*

$$\frac{K_n(z_\infty e^{ia/n}, z_\infty e^{ib/n})}{K_n(z_\infty, z_\infty)} \to \frac{e^{\frac{i}{2}(a-b)} \sin \frac{1}{2}(a-b)}{\frac{1}{2}(a-b)} \tag{2.16.3}$$

More generally, the limit of $K_n(z_n, w_n)/K_n(z_\infty, z_\infty)$ *is the right side of* (2.16.3) *so long as* $z_n, w_n \in \partial\mathbb{D}$, $|z_n - z_\infty| < A/n$, $|w_n - z_\infty| < A/n$, *and* $(z_n/w_n)^n \to e^{i(a-b)}$.

Remark. If $a = b$, $\sin(\frac{1}{2}(a-b))/\frac{1}{2}(a-b)$ is interpreted as 1.

As a most significant application of Lubinsky universality, we will analyze the fine structure of the spacing of the zeros of the POPUC (see Theorem 2.16.10). The most important tools in the proof will be a variational formula for $K_n(z, z)$ and Lubinsky's inequality, which relates off-diagonal asymptotics to diagonal asymptotics.

Here is the variational formalism. Define the *Christoffel function* by

$$\lambda_n(z_0; d\mu) = \inf_{P_n}\left(\int |P_n(e^{i\theta})|^2 \, d\mu(\theta) \,\middle|\, \deg P_n \le n;\ P_n(z_0) = 1 \right) \tag{2.16.4}$$

If $z = 0$, $\lambda_n(0) = \|\Phi_n^*\|^2$ and Szegő's theorem gives the asymptotics of λ_n. Here is the connection to the CD kernel:

Proposition 2.16.2. *The minimizer of* (2.16.4) *is given by*

$$P_n(z; z_0) = \frac{K_n(z_0, z)}{K_n(z_0, z_0)} \tag{2.16.5}$$

and

$$\lambda_n(z_0) = K_n(z_0, z_0)^{-1} \tag{2.16.6}$$

Proof. Expand any trial polynomial

$$P_n(z) = \sum_{j=0}^{n} a_j \varphi_j(z) \tag{2.16.7}$$

Then the normalization condition says

$$\sum_{j=0}^{n} a_j \varphi_j(z_0) = 1 \tag{2.16.8}$$

while

$$\| P_n \|^2 = \sum_{j=0}^{n} |a_j|^2 \tag{2.16.9}$$

(2.16.8) and the Schwarz inequality says

$$\left(\sum_{j=0}^{n} |a_j|^2 \right) K_n(z_0, z_0) \ge 1 \tag{2.16.10}$$

so

$$\lambda_n(z_0) \geq K_n(z_0, z_0)^{-1} \tag{2.16.11}$$

On the other hand, the choice

$$a_j = \frac{\overline{\varphi_j(z_0)}}{K_n(z_0, z_0)} \tag{2.16.12}$$

that is, P_n given by the right side of (2.16.5) has

$$\|P_n\|^2 = \frac{K_n(z_0, z_0)}{K_n(z_0, z_0)^2} = K_n(z_0, z_0)^{-1} \tag{2.16.13}$$

Thus, (2.16.6) is the minimum and (2.16.5) the minimizer. □

Remark. $\lambda_n(0) = \|\Phi_n^*\|^2$ and (2.16.6) is just the CD formula at $\zeta = z = 0$.

For comparison purposes, it will be useful to consider all positive but not necessarily normalized measures. If $\tilde{\mu} = \mu/\mu(\partial\mathbb{D})$, then the monic Φ_n are the same, that is,

$$\Phi_n(z; d\mu) = \Phi_n(z; d\tilde{\mu}) \tag{2.16.14}$$

so we define

$$\alpha_n(d\mu) = \alpha_n(d\tilde{\mu}) \qquad \rho_n(d\mu) = \rho_n(d\tilde{\mu}) \tag{2.16.15}$$

Thus,

$$\|\Phi_n(d\mu)\| = \rho_0 \dots \rho_{n-1}\mu(\partial\mathbb{D})^{1/2} \tag{2.16.16}$$

so

$$\lim \|\Phi_n(d\mu)\|^{1/n} = 1 \Leftrightarrow (\rho_0 \dots \rho_n)^{1/n} \to 1 \tag{2.16.17}$$

Thus, if we define regularity as

$$\lim \|\Phi_n(d\mu)\|_{L^2(d\mu)}^{1/n} = 1 \tag{2.16.18}$$

then μ regular $\Leftrightarrow$ $\tilde{\mu}$ regular.

It is also easy to see that Theorem 2.16.1 for probability measures implies the result for any positive μ by comparing w to $\tilde{w}$ and K to $\tilde{K}$.

From the definition (2.16.4) and (2.16.6), we immediately have (note that $\mu \leq \mu^*$ only makes sense because we allow nonnormalized measures):

Corollary 2.16.3. *For any two measures on $\partial\mathbb{D}$, for all n, $z \in \mathbb{C}$,*

$$\mu \leq \mu^* \Rightarrow \lambda_n(z, \mu) \leq \lambda_n(z, \mu^*) \tag{2.16.19}$$

$$\Leftrightarrow K_n^*(z, z) \leq K_n(z, z) \tag{2.16.20}$$

We will prove Theorem 2.16.1 by a comparison technique. We thus need one example where we can prove the theorem by calculation. The example will be $d\mu_0 = \frac{d\theta}{2\pi}$!

Theorem 2.16.4. *Fix any $A < \infty$. Let*

$$d\mu_0 = \frac{d\theta}{2\pi}$$

(so $w \equiv 1$) and let $K_n^{(0)}$ be its CD kernel. Then

(i)

$$\frac{1}{n+1} K_n^{(0)}(z_n, z_n) \to 1 \tag{2.16.21}$$

uniformly for all $z_n \in \partial\mathbb{D}$ for which $z_n \to z_\infty \in \partial\mathbb{D}$ and

$$n|z_n - z_\infty| \leq A \tag{2.16.22}$$

for all A.

(ii) *Uniformly for $z_\infty \in \partial\mathbb{D}$ and a, b real with $|a|, |b| \leq A$, we have*

$$\frac{K_n^{(0)}(z_\infty e^{ia/n}, z_\infty e^{ib/n})}{K_n^{(0)}(z_\infty, z_\infty)} \to \frac{e^{\frac{i}{2}(a-b)} \sin \frac{1}{2}(a-b)}{\frac{1}{2}(a-b)} \tag{2.16.23}$$

Proof. (i) Neither $z_n \to z_\infty$ nor (2.16.22) is needed (!) since

$$K_n^{(0)}(e^{i\theta}, e^{i\theta}) = n + 1 \tag{2.16.24}$$

for all $e^{i\theta} \in \partial\mathbb{D}$ since $|\varphi_n(e^{i\theta})| = 1$.

(ii) If $a = b$, this is immediate by (2.16.24). If $a \neq b$, since

$$\varphi_n^{(0)}(e^{i\theta}) = e^{in\theta}$$

we have, by summing a geometric series (or by using the CD formula),

$$K_n^{(0)}(e^{i\theta + ia/n}, e^{i\theta + ib/n}) = \frac{1 - e^{i(a-b)(n+1)/n}}{1 - e^{i(a-b)/n}} \tag{2.16.25}$$

Since

$$n(1 - e^{i(a-b)/n}) \to -i(a-b)$$

and $ie^{iu/2} 2 \sin \frac{u}{2} = (e^{iu} - 1)$, we get (2.16.23). □

That the measure is regular will provide a key estimate on the minimizer (2.16.5):

Lemma 2.16.5. *Let μ be a regular measure on $\partial\mathbb{D}$. Then for any $\varepsilon > 0$, there is a δ and C so that the minimizer, $P_n(z, z_0)$, of (2.16.5) obeys*

$$|P_n(z, z_0)| \leq Ce^{\varepsilon n} \tag{2.16.26}$$

for all z, z_0 with

$$|z|, |z_0| \in (1 - \delta, 1 + \delta) \tag{2.16.27}$$

Proof. Let $\delta = e^{\varepsilon/4} - 1$. By regularity and Theorem 2.15.3, for all m and $|z| = 1 + \delta$,

$$|\varphi_m(z)| \leq C_1 e^{3\varepsilon m/8} \tag{2.16.28}$$

for some C_1. By the maximum modulus, this holds for z, z_0 obeying (2.16.27). Thus, since $K_n(z_0, z_0) \geq 1$,

$$|P_n(z, z_0)| \leq nC_1^2 e^{3\varepsilon n/4} \tag{2.16.29}$$

so (2.16.26) holds for suitable C. □

Here is a main tool for shifting from a nice case like μ_0 to a less nice case. We state it in greater generality than used to prove Theorem 2.16.1 because of the needs of the next section.

Theorem 2.16.6 (Nevai Comparison Theorem). *Let μ, $\mu^\sharp$ be two regular measures on $\partial\mathbb{D}$ of the form*

$$d\mu = w\,\frac{d\theta}{2\pi} + d\mu_s \qquad d\mu^\sharp = w^\sharp\,\frac{d\theta}{2\pi} + d\mu_s^\sharp \tag{2.16.30}$$

Suppose $z_0 = e^{i\theta_0} \in \mathbb{D}$ obeys

(1) *$d\mu_s = d\mu_s^\sharp$ for $z \in (z_0 e^{-i\delta}, z_0 e^{i\delta})$ for some $\delta > 0$.*

(2) *For all ε sufficiently small, there is $a_\varepsilon > 1$ so for $|\theta - \theta_0| < \varepsilon$, we have*

$$a_\varepsilon^{-1} w(\theta) \le w^\sharp(\theta) \le a_\varepsilon w(\theta) \tag{2.16.31}$$

and

$$\lim_{\varepsilon\to 0} a_\varepsilon = 1 \tag{2.16.32}$$

(3) *For some $z_n \in \mathbb{D}$, $z_n \to z_0$, and every $\ell(n)$ with $\frac{n}{2} < \ell(n) < 2n$,*

$$\lim_{n\to\infty} \frac{1}{n+1} K_n(z_{\ell(n)}, z_{\ell(n)}) = B \neq 0 \tag{2.16.33}$$

Then

$$\lim_{n\to\infty} \frac{1}{n+1} K_n^\sharp(z_n, z_n) = B \tag{2.16.34}$$

Moreover, this is uniform in z_n in the sense that if (2.16.33) holds (with the same B) for all $z_n \to z_0$, there are, for any ε, a δ and N_0 so that if $n > N_0$ and $|z_n - z_0| < \delta$, then

$$\left| B - \frac{1}{n+1} K_n^\sharp(z_n, z_n) \right| < \varepsilon \tag{2.16.35}$$

There is also uniformity in z_0: If w and $w^\sharp$ are continuous and nonvanishing on an interval in $\partial\mathbb{D}$ and we have $d\mu_s = d\mu_s^\sharp$ in a neighborhood of I and (2.16.31) is replaced by

$$a_\varepsilon^{-1} \frac{w(\theta)}{w(\theta_0)} \le \frac{w^\sharp(\theta)}{w^\sharp(\theta_0)} \le a_\varepsilon \frac{w(\theta)}{w(\theta_0)}$$

for $|\theta - \theta_0| < \varepsilon$ (a_ε the same for all θ_0), and (2.16.33) holds uniformly in $z_0 \in I$ where $B(z_0)$ is z_0-dependent, then (2.16.34) holds with B in (2.16.34) replaced by $B(z_0)\frac{w(z_0)}{w^\sharp(z_0)}$.

Proof. We will leave the two uniformity statements to the reader and focus on the case of a single z_0 and a single sequence $z_n \to z_0$.

We will construct *Nevai trial functions* to put into the variational principle (2.16.4) for $\lambda_n^\sharp(z_0)$, the Christoffel functions for $\mu^\sharp$. Fix $\varepsilon > 0$ and write

$$n = n(\varepsilon) + m(\varepsilon)$$

where $|n\varepsilon - m(\varepsilon)| < 1$ and $n(\varepsilon) = n - m(\varepsilon)$. Let

$$q^{(n)}(z) = \frac{z + z_n}{2z_n} \tag{2.16.36}$$

which obeys ($z_n \equiv e^{i\theta_n}$)

$$q^{(n)}(z_n) = 1; \quad \sup_{z\in\partial\mathbb{D}} |q^{(n)}(z)| = 1; \quad \sup_{|\theta-\theta_n|\geq\eta} |q^{(n)}(e^{i\theta})| = \cos\biggl(\frac{\eta}{2}\biggr) < 1 \tag{2.16.37}$$

The Nevai trial function is

$$Q_n(z) = P_{n(\varepsilon)}(z, z_n)[q^{(n)}(z)]^{m(\varepsilon)} \tag{2.16.38}$$

It has three critical properties:

$$(1) \qquad \deg Q_n = n; \quad Q_n(z_n) = 1$$

$$(2) \qquad |Q_n(z)| \leq |P_{n(\varepsilon)}(z, z_n)| \tag{2.16.39}$$

$$(3) \qquad \sup_{|\theta-\theta_0|\geq\eta} |Q_n(e^{i\theta})| \leq C_{\eta,\varepsilon} e^{-K(\eta,\varepsilon)n} \tag{2.16.40}$$

where $K > 0$ and $C < \infty$. (2) follows from the second relation in (2.16.37) and (3) from the third relation in (2.16.37) and (2.16.26) (where the "ε" of (2.16.26) is picked so $e^{\text{“}\varepsilon\text{”}} < \cos(\eta)^{-\varepsilon/(1-\varepsilon)}$).

Use $Q_n(z)$ as a trial function for $\lambda_n^\sharp(z_n)$, breaking up the integral into $z = e^{i\theta}$ with $|\theta - \theta_0| \leq \eta$ and $> \eta$, writing the contributions as $\lambda_{n;\leq\eta}^\sharp$ and $\lambda_{n;\geq\eta}^\sharp$. By (2.16.40),

$$\lambda_{n;\geq\eta}^\sharp \leq C_{\eta,\varepsilon}^2 e^{-2K(\varepsilon,\eta)n} \tag{2.16.41}$$

For $\lambda_{n;\leq\eta}$, we use the fact that, by $d\mu_s = d\mu_s^\sharp$ and by (2.16.31),

$$d\mu^\sharp \restriction (\theta_0 + \eta, \theta_0 - \eta) \leq a_\eta d\mu \restriction (\theta_0 + \eta, \theta_0 - \eta) \tag{2.16.42}$$

if $\eta < \delta$, so

$$\lambda_{n;\leq\eta}^\sharp \leq a_\eta^2 \lambda_{n(\varepsilon)}(z_n) \tag{2.16.43}$$

since the contribution of $|\theta - \theta_0| > \eta$ to $\lambda_{n(\varepsilon)}$ is positive.

Note first that $\lim_{n\to\infty} ne^{-2Kn} = 0$. Thus, by (2.16.33) and $\lim(n+1)/(n(\varepsilon)+1) = (1-\varepsilon)^{-1}$,

$$\limsup(n+1)\lambda_n^\sharp(z_n) \leq a_\eta^2(1-\varepsilon)^{-1}B \tag{2.16.44}$$

Since ε and η are arbitrary, we can take them to zero and use (2.16.32) to see

$$\limsup(n+1)\lambda_n^\sharp(z_n) \leq B \tag{2.16.45}$$

In the other direction, we switch the roles of μ and $\mu^\sharp$. We define $m(\varepsilon)$ so $|n\varepsilon - m(\varepsilon)| < 1$ but now define

$$n(\varepsilon) = n + m(\varepsilon) \tag{2.16.46}$$

$q^{(n)}(z)$ is still defined as before, but now the Nevai trial function is

$$Q_{n(\varepsilon)}^\sharp(z) = P_n^\sharp(z, z_n)q^{(n)}(z)^{m(\varepsilon)} \tag{2.16.47}$$

Using $Q^\sharp_{n(\varepsilon)}(z)$ as a trial function of $\lambda_{n(\varepsilon)}(z_n)$, breaking the integral into two pieces, $\lambda_{n;\le\eta}$ and $\lambda_{n;\ge\eta}$, for $|\theta-\theta_0|\le\eta$ and $|\theta-\theta_0|>\eta$. By (2.16.40),

$$\lambda_{n;\ge\eta}\le C^2_{\eta,\varepsilon}e^{-2K(\eta,\varepsilon)n}$$

which

$$\lambda_{n;\le\eta}\le a^2_\eta\lambda^\sharp_n(z_n)$$

Multiply by n and use $ne^{-2Kn}\to 0$, plus $\lim n\lambda_{n(\varepsilon)}(z_n)=B\lim\frac{n}{n(\varepsilon)}=(1+\varepsilon)^{-1}B$ to see

$$(1+\varepsilon)^{-1}B\le a^2_\eta\liminf_{n\to\infty}n\lambda^\sharp_n(z_n) \tag{2.16.48}$$

Again, we take $\eta\downarrow 0$ and then ε to 0 and so, with (2.16.45), we obtain (2.16.34). □

Proof of Theorem 2.16.1, part (1). Let us denote μ by $\mu^\sharp$ and then take $\mu=w(z_\infty)\frac{d\theta}{2\pi}$. All the hypotheses of the Nevai comparison theorem hold with $B=w(z_\infty)^{-1}$ since $\frac{1}{n+1}K_n(z,z)=w(z_\infty)^{-1}$ for any z! Thus, by that theorem, (2.16.2) holds. The reader should check the uniformity statements. □

For the second part, the key is

Theorem 2.16.7 (Lubinsky's Inequality). *Let $\mu\le\mu^*$. Then for any $z,w\in\mathbb{C}$, we have*

$$|K_n(z,w)-K^*_n(z,w)|^2\le K_n(w,w)[K_n(z,z)-K^*_n(z,z)] \tag{2.16.49}$$

For this, we need a critical property of the CD kernel:

Theorem 2.16.8 (CD Reproducing Property). *For any polynomial Q_n of degree at most n and all $w\in\mathbb{C}$,*

$$\int K_n(\zeta,w)Q_n(\zeta)\,d\mu(\zeta)=Q_n(w) \tag{2.16.50}$$

In particular, for any $z,w\in\mathbb{C}$,

$$\int K_n(z,\zeta)K_n(\zeta,w)\,d\mu(\zeta)=K_n(z,w) \tag{2.16.51}$$

Remark. One way to understand this is that K_n is the integral kernel of π_{n+1}, the projection onto polynomials of degree at most n. (2.16.51) is then an expression that $\pi^2_{n+1}=\pi_{n+1}$.

Proof. If $Q_n(\zeta)=\sum_{j=0}^n a_j\varphi_j(\zeta)$, then (2.16.50) is just

$$\int K_n(\zeta,w)\varphi_j(\zeta)\,d\mu(\zeta)=\varphi_j(w) \tag{2.16.52}$$

which is immediate. Since $K_n(z,\zeta)$ is a polynomial of degree n in ζ, (2.16.50) implies (2.16.51). □

Proof of Theorem 2.16.7. Since $K_n(z,w) - K_n^*(z,w)$ is a polynomial of degree n in w, we have

$$\int K_n(\zeta,w)[K_n(z,\zeta) - K_n^*(z,\zeta)]\,d\mu(\zeta) = K_n(z,w) - K_n^*(z,w) \tag{2.16.53}$$

By the Schwarz inequality,

$$\text{LHS of (2.16.49)} \le \boxed{1}\cdot\boxed{2} \tag{2.16.54}$$

where

$$\boxed{1} = \int |K_n(\zeta,w)|^2\,d\mu(\zeta) \tag{2.16.55}$$

$$= \int K_n(w,\zeta)K_n(\zeta,w)\,d\mu(\zeta)$$

$$= K_n(w,w) \tag{2.16.56}$$

by (2.16.52), while

$$\boxed{2} = \int |K_n(z,\zeta) - K_n^*(z,\zeta)|^2\,d\mu(\zeta)$$

$$= \int |K_n(\zeta,z)|^2 - 2\,\text{Re}\int K_n(\zeta,z)K_n^*(z,\zeta)\,d\mu(\zeta) + \int |K_n^*(\zeta,z)|^2\,d\mu(\zeta)$$

$$\le K_n(z,z) - 2K_n^*(z,z) + K_n^*(z,z) \tag{2.16.57}$$

$$= K_n(z,z) - K_n^*(z,z) \tag{2.16.58}$$

The first $K_n(z,z)$ in (2.16.57) comes from the same calculation that went from (2.16.55) to (2.16.56), while the last term comes from first using $d\mu \le d\mu^*$ and then doing the same calculation for K^*, μ^*. The middle term in (2.16.57) is just (2.16.50) for $Q_n(\zeta) = K_n^*(z,\zeta)$. □

Lemma 2.16.9. *Let μ, ν be two positive measures on $\partial\mathbb{D}$. Suppose μ is regular. Then $\mu \vee \nu$, their sup, is also regular.*

Remark. For any two measures, μ, ν, one shows there is a smallest η larger than μ and ν. This is denoted $\mu \vee \nu$. It is discussed in [114, 207].

Proof. Since $\rho_j(\mu \vee \nu) \le 1$, we have

$$\limsup \|\Phi_n(\,\cdot\,, d(\mu \vee \nu))\|_{L^2(\mu\vee\nu)}^{1/n} \le 1$$

On the other hand, since $\Phi_n(\,\cdot\,, d\mu)$ is a minimizer and $\mu \le \mu \vee \nu$,

$$\|\Phi_n(\,\cdot\,, d\mu)\|_{L^2(\mu)} \le \|\Phi_n(\,\cdot\,, d(\mu \vee \nu))\|_{L^2(\mu)}$$

$$= \|\Phi_n(\,\cdot\,, d(\mu \vee \nu))\|_{L^2(\mu\vee\nu)}$$

so, by regularity,

$$\liminf \|\Phi_n(\,\cdot\,, d(\mu \vee \nu))\|_{L^2(\mu\vee\nu)}^{1/n} \ge 1$$ □

Completion of the Proof of Theorem 2.16.1. Let $\mu^* = \mu \vee (w(z_\infty)\,d\mu_0)$. Then w, w^* and $w_0 \equiv w(z_\infty)$ (the constant weight) are continuous and agree at z_∞. Thus, by part (1), uniformly in $|a|, |b| < A$,

$$\frac{1}{n+1} K_n^\sharp(e^{ic/n} z_\infty, e^{ic/n} z_\infty) \to w(z_\infty)^{-1} \tag{2.16.59}$$

for $c = 0$, a, or b, and for $K_n^\sharp$ associated to any of μ^*, μ, or $w(z_\infty)\,d\mu_0$ (where we used the lemma to assure μ^* is regular).

Apply (2.16.49) where K_n is associated to $w(z_\infty)\,d\mu_0$ and K_n^* is associated to μ^*. Here we divide by $K_n(z_\infty, z_\infty)$. By part (1) of the theorem,

$$\frac{K_n(z_\infty e^{ib/n}, z_\infty e^{ib/n})}{K_n(z_\infty, z_\infty)} - \frac{K_n^*(z_\infty e^{ib/n}, z_\infty e^{ib/n})}{K_n(z_\infty, z_\infty)} \to 0$$

so

$$\frac{K_n^*(z_\infty e^{ia/n}, z_\infty e^{ia/n})}{K_n(z_\infty, z_\infty)} \to \text{RHS of (2.16.3)} \tag{2.16.60}$$

where we used (2.16.25).

Now we use part (1) of the theorem again to replace $K_n(z_\infty, z_\infty)$ in (2.16.60) by $K_n^*(z_\infty, z_\infty)$.

Next, we use Lubinsky's inequality for $\mu \le \mu^*$ and, in the same way, transfer the limit to ratios of K_n for μ.

The more general assertion at the end of Theorem 2.16.1 follows by going through the proof and seeing it gives the stronger result. □

Finally, we want to turn to the zeros of POPUC.

Definition. Given any measure $d\mu$ on $\partial\mathbb{D}$ and $\beta, z_\infty \in \partial\mathbb{D}$, $z_n^{(j),(\beta)}(z_\infty)$ is defined for $j = 0, \pm 1, \dots, \pm[\sqrt{n}\,]$ to be "successive" zeros $\Phi(z;\beta)$ where $z_n^{(0),(\beta)}$ is the first one going counterclockwise from z_∞ (or z_∞ itself), $z_n^{(1),(\beta)}$ the next, and so on, and $z_n^{(-1),(\beta)}$ going clockwise. If, for each $j \in \mathbb{Z}$ and $\beta \in \partial\mathbb{D}$,

$$\lim\left[\frac{z_n^{(j+1),(\beta)}(z_\infty) - z_n^{(j),(\beta)}(z_\infty)}{e^{2\pi i/n}}\right] \to 1 \tag{2.16.61}$$

we say there is *clock behavior* at z_∞. If the limit in (2.16.61) is uniform in $z_\infty \in I \subset \partial\mathbb{D}$, we say there is *uniform clock behavior* on I.

Remark. The name comes from the fact that the zeros are spaced like numerals on a clock.

Theorem 2.16.10 (Freud–Levin Theorem)**.** *If the hypotheses of Theorem 2.16.1 hold, we have uniform clock behavior on I for each fixed β.*

Proof. Fix $z_\infty \in I$. Let β_n be defined so $\Phi_n(z_\infty;\beta_n) = 0$. By (2.14.20), the other zeros of $\Phi_n(\,\cdot\,;\beta_n)$ are the zeros of $K_n(z,\beta_n)$. By the uniformity of convergence of (2.16.3), K_n changes sign at points asymptotic to $e^{2\pi i/n} z_\infty$, and so Φ_n has a zero within a slightly larger interval. By Theorem 2.14.4(iii), $\Phi_n(z;\beta)$ has a zero within $\frac{2\pi i}{n}(1+o(1))$ of z_∞. Thus, $|z_n^{(0),(\beta)}(z_\infty) - z_\infty| \le \frac{2\pi i}{n}(1+o(1))$.

By (2.16.3), there are no zeros in $(z_n^{(0),(\beta)}(z_\infty)e^{-i\pi/n}, z_n^{(0),(\beta)}(z_\infty)e^{i\pi/n})$ and the next zero has argument $\frac{2\pi i}{n}(1+o(1))$ greater. Repeating this, we get clock behavior. □

Remarks and Historical Notes. That $\frac{1}{n+1}K_n(z_\infty, z_\infty) \to w(z_\infty)^{-1}$ for smooth w's (or its equivalent for Christoffel weights) goes back to the first half of the twentieth century. Nevai [321] describes the history, Freud's key role, and applications. Its great generality is a result of Máté–Nevai–Totik [302], discussed in the next section.

In the context of OPRL on $[-1, 1]$, the ability to wiggle z_∞ and the importance of doing so was noted by Lubinsky [288]. In the same paper, he noted what we call Lubinsky's inequality and used it to prove Lubinsky universality. For smooth w's, this universality result goes back at least to Freud [141] and was studied in the context of random matrices using Riemann–Hilbert techniques (see [102, 256]) but nothing like Lubinsky's generality. The extension to OPUC is in Levin–Lubinsky [275].

The idea of localizing trial functions for one problem using $[\frac{1}{2}(z + z_0)]^{n\varepsilon}$ (or its equivalent for OPRL) goes back to Nevai [320]. We name the Nevai trial functions and Nevai comparison theorem after this work.

At the very end of his book, Freud [141] noted that universality (or its OPRL analog) implied clock spacing for the zeros. It was Levin–Lubinsky [276] who applied this idea and Lubinsky's very general universality result to get clock behavior in general. Since it was Levin who rediscovered Freud's result that universality implies clock behavior, we call Theorem 2.16.10 the Freud–Levin theorem. Earlier Last–Simon [273], using very different methods, had the best clock behavior results for OPRL, and Simon [402] had some clock behavior results for (P)OPUC, but the two Levin–Lubinsky papers [275, 276] have the strongest results on clock behavior.

Lubinsky [287] has a second interesting approach to universality; it is discussed in the OPRL context in Section 3.12.

We note the clock behavior here is only local. At opposite ends of the circle, the zeros have about $n/2$ zeros in between and the errors can add up so that there is no result on, say, asymptotically opposite zeros for even n.

2.17 ASYMPTOTICS OF THE CD KERNEL: LOCALLY SZEGŐ WEIGHTS

In this final section on asymptotics of the CD kernel for OPUC supported on all of $\partial\mathbb{D}$, we consider the case of noncontinuous weights. To even state the main theorem, we need to recall some basic harmonic analysis.

Definition. Let $f \in L^1([a,b], dx)$ (or $L^1(\partial\mathbb{D}, \frac{d\theta}{2\pi})$). A point x in $[a,b]$ (or $\partial\mathbb{D}$) is called a *Lebesgue point* of f if and only if

$$\lim_{\varepsilon\downarrow 0}(2\varepsilon)^{-1}\int_{x-\varepsilon}^{x+\varepsilon}|f(y)-f(x)|\,dy = 0 \tag{2.17.1}$$

In particular, at a Lebesgue point, the maximal function, Mf, obeys

$$(Mf)(x) \equiv \sup_{a>0} (2a)^{-1} \int_{x-a}^{x+a} |f(y)|\, dy < \infty \tag{2.17.2}$$

Three fundamental results we will need (and discuss in the Notes) are:

Theorem 2.17.1. *For $f \in L^1(dx)$, a.e. x in $[a, b]$ (or $\partial\mathbb{D}$) is a Lebesgue point.*

We also need an analog of (2.17.1)/(2.17.2) for singular measures:

Theorem 2.17.2. *Let $d\mu_s$ be a singular measure on $[a, b]$ (or $\partial\mathbb{D}$). Then for Lebesgue almost every x, we have*

$$\lim_{\varepsilon \to 0} (2\varepsilon)^{-1} \mu_s(x - \varepsilon, x + \varepsilon) = 0 \tag{2.17.3}$$

Remark. It is also known for μ_s-a.e. x that the limit is infinite.

If $d\mu$ has the form (2.1.1), we say $e^{i\theta} \in \partial\mathbb{D}$ is a Lebesgue point for $d\mu$ if (2.17.3) holds for $x = e^{i\theta}$ and μ_s, the singular part of $d\mu$, and $e^{i\theta}$ is a Lebesgue point for the weight w.

Theorem 2.17.3 (Fatou's Theorem). *Let $f \in H^1(\mathbb{D})$ with boundary values denoted by $f(e^{i\theta})$. If $e^{i\theta_0}$ is a Lebesgue point of $f \restriction \partial\mathbb{D}$, then nontangential boundary values are given by $f(e^{i\theta_0})$, that is, for any $\varepsilon > 0$,*

$$\lim_{n\to\infty} \sup\biggl(|f(z) - f(e^{i\theta})| \biggm| |z| > 1 - \frac{1}{n},\ |\arg(1 - ze^{-i\theta_0})| \le (1 - \varepsilon)\frac{\pi}{2} \biggr) = 0 \tag{2.17.4}$$

Definition. We say $d\mu$ obeying (2.1.1) is *locally Szegő* on $I = [\alpha, \beta] \subset \partial\mathbb{D}$ if and only if

$$\int_I \log(w(e^{i\theta}))\, \frac{d\theta}{2\pi} > -\infty \tag{2.17.5}$$

If w obeys a local Szegő condition, we can find $\tilde{w}$ obeying a global Szegő condition with $\tilde{w} \restriction I = w$. Let $\tilde{D}$ be the Szegő function for $\tilde{w}$. If $w^\sharp$ is a second extension and $z_0 \in I^{\text{int}}$, then $\tilde{D} - D^\sharp$ is analytic near z_0, so z_0 is a Lebesgue point for $\tilde{D}$ if and only if it is for $D^\sharp$. Thus, being a Lebesgue point is independent of global Szegő extension. We will say z_0 is a Lebesgue point of the local Szegő function in that case.

Note. Being a Lebesgue point for $\log(w)$ is not sufficient to be a Lebesgue point of D. One needs to also be a Lebesgue point for the conjugate function of $\log(w)$.

Here are the main results of this section:

Theorem 2.17.4 (MNT Theorem [302]). *Let μ be a regular measure on $\partial\mathbb{D}$, which is locally Szegő on I. Let $e^{i\theta_0} \in I$ be a point where $w(\theta_0) \neq 0$ and be Lebesgue point for both μ and for the local Szegő function. Let $z_n \in \partial\mathbb{D}$ be a sequence obeying*

$$\sup_n n|z_n - e^{i\theta_0}| \equiv A < \infty \tag{2.17.6}$$

Then

$$\lim_{n\to\infty} \frac{1}{n+1} K_n(z_n, z_n) = w(e^{i\theta_0})^{-1} \tag{2.17.7}$$

Moreover, for each A, the limit is uniform in z_n obeying (2.17.6).

Theorem 2.17.5 (Findley's Theorem [133]). *Under the hypotheses of Theorem 2.17.4, we have that for any $A < \infty$, uniformly in $|a|, |b| < A$,*

$$\frac{K_n(e^{i\theta_0}e^{ia/n}, e^{i\theta_0}e^{ib/n})}{K_n(e^{i\theta_0}, e^{i\theta_0})} \to \frac{e^{\frac{i}{2}(a-b)} \sin \frac{1}{2}(a-b)}{\frac{1}{2}(a-b)} \tag{2.17.8}$$

More generally, the limit relation holds for $K_n(z_n, w_n)/K_n(e^{i\theta_0}, e^{i\theta_0})$ if $z_n, w_n \in \partial\mathbb{D}$, $|z_n - e^{i\theta_0}| < A/n$, $|w_n - e^{i\theta_0}| < A/n$, and $(z_n/w_n)^n \to e^{i(a-b)}$.

We will see later that Findley's theorem implies a local clock behavior for the zeros of POPUC. Two other theorems we will prove do not even require a local Szegő condition. We will use the first in the proof of Theorem 2.17.4:

Theorem 2.17.6 (Máté–Nevai Upper Bound [300]). *For any measure $d\mu$ on $\partial\mathbb{D}$ of the form* (2.1.1) *and any Lebesgue point, z_0, of $d\mu$,*

$$\limsup_{n\to\infty} (n+1)\lambda_n(z_n) \le w(z_0) \tag{2.17.9}$$

for any sequence $z_n \in \partial\mathbb{D}$ with

$$\sup_n n|z_n - z_0| < \infty \tag{2.17.10}$$

Remark. This includes points where $w(z_0) = 0$.

Theorem 2.17.7 (Simon [409]). *If $I = (\alpha, \beta)$ is an open interval in $\partial\mathbb{D}$, if μ is regular, and $w(z) > 0$ for a.e. $z \in (\alpha, \beta)$, then*

$$\text{(i)} \qquad \int_I \left| \frac{1}{n+1} K_n(e^{i\theta}, e^{i\theta}) w(\theta) - 1 \right| \frac{d\theta}{2\pi} \to 0 \tag{2.17.11}$$

$$\text{(ii)} \qquad \int_I \frac{1}{n+1} K_n(e^{i\theta}, e^{i\theta})\, d\mu_s(\theta) \to 0 \tag{2.17.12}$$

We now turn to the proof of these four theorems, starting with the third and fourth:

Lemma 2.17.8. *Let λ be a finite positive measure on $\mathbb{R}$. For $x_\infty \in \mathbb{R}$, define for $t > 0$*

$$L(t) = \frac{1}{2t} \lambda([x_\infty - t, x_\infty + t]) \tag{2.17.13}$$

Let $h(s)$ be a continuous, even $L^1(\mathbb{R}, dx)$ function on $\mathbb{R}$ with

$$0 \le s \le t \Rightarrow h(s) \ge h(t) \ge 0 \tag{2.17.14}$$

Suppose

$$\lim_{t\downarrow 0} L(t) = 0 \tag{2.17.15}$$

and $x_n \to x_\infty$ *with*

$$A = \sup n|x_n - x_\infty| < \infty \tag{2.17.16}$$

Then

$$\lim_{n\to\infty} \int nh(n(x - x_n))\, d\lambda(x) = 0 \tag{2.17.17}$$

Remarks. 1. Continuity of h is not needed.

2. (2.17.18) below is often called the Layer Cake Principle; see Lieb–Loss [282].

Proof. Let $d\nu = -dh$ as a Stieltjes measure on $(0, \infty)$, so $h(s) = \nu([s, \infty))$, or equivalently,

$$h(s) = \int_0^\infty \chi_{[-t,t]}(s)\, d\nu(t) \tag{2.17.18}$$

Thus,

$$\begin{aligned}\int nh(n(x - x_n))\, d\lambda(x) &= \int n\lambda\left(\left[x_n - \frac{t}{n}, x_n + \frac{t}{n}\right]\right) d\nu(t) \\ &\leq \int n\lambda\left(\left[x_\infty - \frac{t+A}{n}, x_\infty + \frac{t+A}{n}\right]\right) d\nu(t) \\ &= \int L\left(\frac{t+A}{n}\right)(t + A)\, d\nu(t) \end{aligned}\tag{2.17.19}$$

Since $\int d\nu(t) = h(0)$ and $\int 2t\, d\nu = \int h(s)\, ds$, we see $(t + A)\, d\nu(t)$ is a finite measure. By hypothesis, $\|L\|_\infty < \infty$ and $\lim_{n\to\infty} L(\frac{t+A}{n}) = 0$ for all t, so by the dominated convergence theorem, (2.17.19) goes to 0. □

Proof of Theorem 2.17.6. Let

$$Q_n(e^{i\theta}, e^{i\varphi}) = \frac{1}{n+1} \sum_{j=0}^{n} e^{ij(\theta-\varphi)} \tag{2.17.20}$$

precisely the $\lambda_n(e^{i\varphi})$ minimizer for $\frac{d\theta}{2\pi}$, that is, $K_n^{(0)}(e^{i\theta}, e^{i\varphi})/K_n^{(0)}(e^{i\theta}, e^{i\theta})$. Of course, one can sum the geometric series (essentially a special case of the CD formula!)

$$Q_n(e^{i\theta}, e^{i\varphi}) = \frac{1}{n+1} \frac{e^{i(n+1)(\theta-\varphi)} - 1}{e^{i(\theta-\varphi)} - 1} \tag{2.17.21}$$

$$= \frac{1}{n+1} \frac{e^{in(\theta-\varphi)/2} \sin(\frac{n+1}{2}(\theta - \varphi))}{\sin(\frac{1}{2}(\theta - \varphi))} \tag{2.17.22}$$

which is "essentially" the classical Dirichlet kernel, and

$$F_n(e^{i\theta}, e^{i\varphi}) \equiv (n+1)|Q_n(e^{i\theta}, e^{i\varphi})|^2$$
$$= \frac{1}{n+1} \frac{\sin^2(\frac{n+1}{2}(\theta-\varphi))}{\sin^2(\frac{1}{2}(\theta-\varphi))} \tag{2.17.23}$$

which is exactly the classical Fejér kernel. It has the following properties:

$$\text{(a)} \qquad \int [F_n(e^{i\theta}, e^{i\varphi})] \frac{d\theta}{2\pi} = 1 \tag{2.17.24}$$

$$\text{(b)} \qquad \sup_{\theta,\varphi} |F_n(e^{i\theta}, e^{i\varphi})| = n+1 \tag{2.17.25}$$

$$\text{(c)} \qquad |F_n(e^{i\theta}, e^{i\varphi})| \leq (n+1)G((n+1)(\theta-\varphi)) \tag{2.17.26}$$

where

$$G(x) = \min\left(\frac{\pi^2}{x^2}, \frac{\pi^2}{4}\right) \tag{2.17.27}$$

(a) is immediate from (2.17.20) and the orthogonality of $e^{ij\theta}$. (b) follows from $|Q| \leq 1$ and $Q(e^{i\theta}, e^{i\theta}) = 1$.

To get (c), we note that $(\sin x)/x$ is monotone decreasing for x in $[0, \pi/2]$, so for $|\theta - \varphi| < \pi$,

$$\left|\sin\left(\frac{1}{2}(\theta-\varphi)\right)\right| \geq \frac{2}{\pi} \frac{|\theta-\varphi|}{2} \tag{2.17.28}$$

Thus,

$$|F_n(e^{i\theta}, e^{i\varphi})| \leq (n+1)\tilde{G}((n+1)(\theta-\varphi)) \tag{2.17.29}$$

with

$$\tilde{G}(x) = \frac{\pi^2 \sin^2(\frac{x}{2})}{x^2} \tag{2.17.30}$$

$\tilde{G}(x) \leq \pi^2/x^2$ since $\sin^2(x/2) \leq 1$ and $\tilde{G}(x) \leq \pi^2/4$ since $\sin^2(x/2) \leq x^2/4$.

Since Q_n is a valid trial function in (2.16.4),

$$(n+1)\lambda_n(z_n) \leq \int (n+1)|Q_n(z, z_n)|^2 \, d\mu(z)$$
$$= \int F_n(z, z_n) \, d\mu(z) \tag{2.17.31}$$
$$\leq w(z_0) + \int (n+1)G((n+1)(\theta-\varphi_n)) \, d\lambda(\theta) \tag{2.17.32}$$

where $z_n = e^{i\varphi_n}$, $z_0 = e^{i\varphi_0}$ and

$$d\lambda(\theta) = |w(\theta) - w(\varphi_0)| \frac{d\theta}{2\pi} + d\mu_s(\theta) \tag{2.17.33}$$

Here (2.17.32) comes from (2.17.24) and (2.17.26).

Lemma 2.17.8 is applicable since z_0, being a Lebesgue point of $d\mu$, implies the $L(t)$ associated to λ obeys (2.17.13). □

Proof of Theorem 2.17.7. By compactness of the measures, η, on $\partial\mathbb{D}$, with $\eta(\partial\mathbb{D}) \le 1$, pick a subsequence $n(j)$ so

$$\frac{1}{n(j)+1} K_{n(j)}(e^{i\theta}, e^{i\theta}) w(\theta) \frac{d\theta}{2\pi} \to d\eta_1(\theta) \tag{2.17.34}$$

and

$$\frac{1}{n(j)+1} K_{n(j)}(e^{i\theta}, e^{i\theta})\, d\mu_s(\theta) \to d\eta_2(\theta) \tag{2.17.35}$$

By Theorem 2.15.4 and the regularity assumption,

$$d\eta_1 + d\eta_2 = \frac{d\theta}{2\pi} \tag{2.17.36}$$

On the other hand, by Theorem 2.17.6 and (2.16.6),

$$\liminf \frac{1}{n+1} K_n(e^{i\theta}, e^{i\theta}) w(\theta) \ge 1 \tag{2.17.37}$$

for a.e. $\theta \in I$ (this uses $w(\theta) > 0$ a.e. on I). Thus, by Fatou's lemma, for any positive continuous function, f, supported in I,

$$\begin{aligned}\int f(\theta)\, d\eta_1(\theta) &= \lim_{n\to\infty} \int_I \frac{1}{n+1} K_n(e^{i\theta}, e^{i\theta}) w(\theta) f(\theta) \frac{d\theta}{2\pi} \\ &\ge \int_I \liminf\biggl(\frac{1}{n+1} K_n(e^{i\theta}, e^{i\theta}) w(\theta)\biggr) f(\theta) \frac{d\theta}{2\pi} \\ &\ge \int_I f(\theta) \frac{d\theta}{2\pi}\end{aligned} \tag{2.17.38}$$

by (2.17.37). This means

$$d\eta_1 \restriction I \ge \frac{d\theta}{2\pi} \tag{2.17.39}$$

so, by (2.17.36),

$$d\eta_1 \restriction I = \frac{d\theta}{2\pi} \qquad d\eta_2 \restriction I = 0 \tag{2.17.40}$$

By compactness of the set of measures, (2.17.12) holds and

$$\int_I \frac{1}{n+1} K_n(e^{i\theta}, e^{i\theta}) w(\theta) \frac{d\theta}{2\pi} = \frac{|I|}{2\pi} \tag{2.17.41}$$

This and (2.17.37) implies (2.17.26). $\square$

We turn to the proof of Theorem 2.17.4. The Máté–Nevai upper bound provides half the result, so we only need a lower bound. We will suppose for now that a global Szegő condition holds. Throughout, z_n obeys (2.17.6). Since we will be using analytic continuation of $D(z)$ and $K_n(z, z_n)$ from $\partial\mathbb{D}$ to $\mathbb{D}$, the following lemma will be useful:

Lemma 2.17.9. *Let Q_n be a polynomial of degree at most n with no zeros in $\mathbb{D}$. Let $z_0 \in \partial\mathbb{D}$ and $0 < s < 1$. Then*

$$|Q_n(sz_0)| \ge \left(\frac{1+s}{2}\right)^n |Q(z_0)| > e^{-n(1-s)} |Q(z_0)| \tag{2.17.42}$$

Proof. For $0 < t < 1$,

$$-\log\left(1-\frac{t}{2}\right) = \frac{t}{2} + \frac{t^2}{4} + \cdots < t\left(\frac{1}{2} + \frac{1}{4} + \dots\right) = t \tag{2.17.43}$$

Let $t = (1-s)$ so $1 - \frac{t}{2} = \frac{1}{2}(1+s)$ and see

$$-\log\left(\frac{1+s}{2}\right) < 1 - s \tag{2.17.44}$$

which implies

$$\frac{1+s}{2} > e^{-(1-s)} \tag{2.17.45}$$

showing the second inequality in (2.17.42).

By rotation covariance, we can suppose $z_0 = 1$. Any such $Q_n(z) = c\prod_{j=1}^n (z - z_j)$ with $z_j \notin \mathbb{D}$, so it suffices to prove the case $n = 1$ and $c = 1$, that is, that for any $0 < s < 1$ and $z_1 \notin \mathbb{D}$, we have

$$|1 - sz_1| \geq \left(\frac{1+s}{2}\right)|1 - z_1| \tag{2.17.46}$$

To prove this, fix $s \in (0, 1)$ and let

$$g(w) = \frac{|1 - sw|}{|1 - w|} \tag{2.17.47}$$

g is harmonic on $(\mathbb{C} \setminus \overline{\mathbb{D}}) \cup \{\infty\}$, so its minimum on $\mathbb{C} \setminus \mathbb{D}$ is taken on $\partial\mathbb{D}$. If $x = \operatorname{Re} w$ and $w \in \partial\mathbb{D}$,

$$\begin{aligned} g(w)^2 &= \frac{(1-sx)^2 + s^2(1-x^2)}{(1-x)^2 + (1-x^2)} \\ &= \frac{(1+s^2) - 2sx}{2 - 2x} \equiv h(x) \end{aligned} \tag{2.17.48}$$

Since $h'(x) = 2(1-s^2)/(2-2x)^2 > 0$, we see on $[-1, 1]$, $h(x)$ takes its minimum at $x = -1$, that is, $g(w)$ is minimized at $w = -1$ where $g(-1) = \frac{1+s}{2}$. $\square$

We will be looking at points with $s \sim 1 - \varepsilon/n$, so define for $\varepsilon > 0$,

$$x_n(\varepsilon) = \left(1 + \frac{\varepsilon}{n}\right)^{-1} z_n \tag{2.17.49}$$

Without loss, we suppose $z_0 = 1$. Here is the key inequality that will prove the result:

Proposition 2.17.10. *Let $P_n(z)$ be the minimizer for $\lambda_n(z_n)$, that is,*

$$P_n(z) = \frac{K_n(z_n, z)}{K_n(z_n, z_n)} \tag{2.17.50}$$

Suppose we prove that

$$\lim_{\varepsilon\downarrow 0} \limsup_{n\to\infty} \left[\sum_{j=n+1}^{\infty} \int P_n(e^{i\theta}) D(e^{i\theta}) x_n(\varepsilon)^j e^{-ij\theta} \frac{d\theta}{2\pi}\right] = 0 \tag{2.17.51}$$

Then

$$\liminf (n+1)\lambda_n(z_n) \geq w(1) \tag{2.17.52}$$

Remarks. 1. The intuition about why (2.17.51) is reasonable comes from the following. Since P_n is concentrated near $e^{i\theta} = 1$ and D is "reasonable" near 1, we would expect to be able to replace $D(e^{i\theta})$ by $D(1)$. But then, since $\deg P_n \leq n$ and $j \geq n+1$, the integral is zero.

2. By Proposition 2.14.3 and Theorem 2.14.4, P_n has all its zeros on $\partial\mathbb{D}$, hence none in $\mathbb{D}$, so it obeys Lemma 2.17.9.

Proof. Since $P_n(z)D(z)$ lies in H^2, its Taylor coefficients are given by integrals over $\partial\mathbb{D}$, so

$$P_n(x_n(\varepsilon))D(x_n(\varepsilon)) = \sum_{j=0}^{\infty} \int P_n(e^{i\theta})D(e^{i\theta})x_n(\varepsilon)^j e^{-ij\theta}\,\frac{d\theta}{2\pi} \tag{2.17.53}$$

so if

$$H_n(e^{i\theta},\varepsilon) = \sum_{j=0}^{n} x_n(\varepsilon)^j e^{-ij\theta} \tag{2.17.54}$$

and if

$$E_n(\varepsilon) = P_n(x_n(\varepsilon))D_n(x_n(\varepsilon)) - \int P_n(e^{i\theta})D(e^{i\theta})H_n(e^{i\theta},\varepsilon)\,\frac{d\theta}{2\pi} \tag{2.17.55}$$

then (2.17.51) implies

$$\lim_{\varepsilon\downarrow 0}\limsup_{n\to\infty} |E_n(\varepsilon)| = 0 \tag{2.17.56}$$

Since $\{e^{-ij\theta}\}_{j=0}^n$ are $\frac{d\theta}{2\pi}$-orthonormal and $|x_n(\varepsilon)| \leq 1$,

$$\int |H_n(e^{i\theta},\varepsilon)|^2\,\frac{d\theta}{2\pi} = \sum_{j=0}^{n} |x_n(\varepsilon)|^{2j} \leq (n+1) \tag{2.17.57}$$

Thus, by the Schwarz inequality and $|D(e^{i\theta})|^2 = w(\theta)$,

$$\begin{aligned}\left|\int P_n(e^{i\theta})D(e^{i\theta})H_n(e^{i\theta},\varepsilon)\,\frac{d\theta}{2\pi}\right|^2 &\leq (n+1)\int |P_n(e^{i\theta})|^2 w(\theta)\,\frac{d\theta}{2\pi} \\ &\leq (n+1)\int |P_n(e^{i\theta})|^2\,d\mu \\ &= (n+1)\lambda_n(z_n) \end{aligned} \tag{2.17.58}$$

Thus, by (2.17.56),

$$\liminf (n+1)\lambda_n(z_n) \geq \lim_{\varepsilon\downarrow 0}\left[\liminf_{n\to\infty}[|P_n(x_n(\varepsilon))|^2|D(x_n(\varepsilon))|^2]\right] \tag{2.17.59}$$

Since $P_n(z_n) = 1$, Lemma 2.17.9 implies that

$$\begin{aligned} |P_n(x_n(\varepsilon))| &\geq \exp\biggl(-n\biggl[1 - \Bigl(1 + \frac{\varepsilon}{n}\Bigr)^{-1}\biggr]\biggr) \\ &= \exp\biggl(-\varepsilon\Bigl(1 + \frac{\varepsilon}{n}\Bigr)^{-1}\biggr) \\ &\geq \exp(-\varepsilon) \end{aligned} \tag{2.17.60}$$

Since 1 is a Lebesgue point of D, Fatou's lemma implies that $D(x_n(\varepsilon)) \to D(1)$. Moreover, $|D(1)|^2 = w(1)$. Thus, (2.17.59) becomes

$$\liminf (n+1)\lambda_n(z_n) \geq \lim_{\varepsilon \downarrow 0} [e^{-2\varepsilon} w(1)] = w(1) \tag{2.17.61}$$

proving (2.17.52). □

Thus, we need to prove (2.17.51). Once we have Theorem 2.17.5, we will know $|P_n(z)| = |K_n(z, z_n)|/K_n(z_n, z_n)$ is asymptotically less than 1 for $z \in \partial\mathbb{D}$ with $|z - 1| \leq B/n$. At this point, we only need a weaker bound.

Proposition 2.17.11. *Let $P_n(z)$ be the minimizer* (2.17.50). *Then for any B finite,*

$$\lim_{n\to\infty} \sup_{|\theta|<B/n} |P_n(e^{i\theta})| \leq \frac{e}{\sqrt{2}} \tag{2.17.62}$$

Proof. Let $\rho_n = (1 - \frac{1}{n})$. Then, since $P^2(z)D^2(z)$ is in H^1, the Cauchy integral formula implies

$$P_n^2(\rho_n e^{i\theta}) D^2(\rho_n e^{i\theta}) = \frac{1}{2\pi i(1-\rho_n^2)} \oint_{|z|=1} P_n^2(z) D^2(z) \biggl(\frac{1 - \rho_n z e^{-i\theta}}{z - \rho_n e^{i\theta}}\biggr) dz \tag{2.17.63}$$

Since the expression in parentheses in the integral has magnitude 1 and $1 - \rho_n^2 \geq 2/n$, we see

$$\begin{aligned} |P_n(\rho_n e^{i\theta}) D(\rho_n e^{i\theta})|^2 &\leq \frac{n}{2} \int |P_n(e^{i\varphi})|^2 |D(e^{i\varphi})|^2 \frac{d\theta}{2\pi} \\ &\leq \frac{n}{2} \int |P_n(e^{i\varphi})|^2 \, d\mu \\ &= \frac{n}{2} \lambda_n(z_n) \end{aligned} \tag{2.17.64}$$

On the other hand, by Lemma 2.17.9,

$$|P_n(\rho_n e^{i\theta})| \geq e^{-1} |P_n(e^{i\theta})| \tag{2.17.65}$$

Since $\limsup n\lambda_n(z_n) \leq w(1)$ (by Theorem 2.17.6), we see

$$\lim_{n\to\infty} \sup_{|\theta|\leq B/n} |P_n(e^{i\theta})|^2 \leq \tfrac{e^2}{2}\, w(1) \Big/ \liminf_{\substack{|\theta|\leq B/n \\ n\to\infty}} |D(\rho_n e^{i\theta})|^2 \tag{2.17.66}$$

Fatou's lemma says that the lim inf is exactly $|D(1)|^2 = w(1)$. □

Proposition 2.17.12. (2.17.51) *holds.*

Proof. Consider the integral in (2.17.51) where $D(e^{i\theta})$ is replaced by $\overline{D(e^{i\theta})}$. Since $\overline{D(e^{i\theta})} \in L^2$, we can expand in Fourier series in the integral. $P_n(e^{i\theta})$ contributes $\{e^{im\theta}\}_{m=0}^n$ terms and $\overline{D(e^{i\theta})}$ terms of the form $\{e^{-i\ell\theta}\}_{\ell=0}^\infty$. Since $j \geq n+1$, $\int e^{-ij\theta - i\ell\theta + im\theta}\, d\theta = 0$ for all allowed ℓ, m, j. Thus, the integral with $\overline{D(e^{i\theta})}$ is zero and we can replace $D(e^{i\theta})$ by

$$R(e^{i\theta}) = D(e^{i\theta}) - \frac{D(1)\, \overline{D(e^{i\theta})}}{\overline{D(1)}} \tag{2.17.67}$$

By summing the geometric series and noting

$$\left| \frac{x_n(\varepsilon)^{n+1} e^{-i(n+1)\theta}}{1 - x_n(\varepsilon) e^{-i\theta}} \right| \leq \left| \frac{1}{1 - x_n(\varepsilon) e^{-i\theta}} \right| \tag{2.17.68}$$

(2.17.62) reduces to

$$\lim_{\varepsilon \downarrow 0} \limsup_{n \to \infty} \int |R(e^{i\theta})|\, |P_n(e^{i\theta})|\, |1 - x_n(\varepsilon) e^{-i\theta}|^{-1} \frac{d\theta}{2\pi} = 0 \tag{2.17.69}$$

Fix $B \geq 2A$ and write the integral as $I_1 + I_2$ where I_1 is over the region $|\theta| < B/n$ and I_2 over the region $|\theta| \geq B/n$. For I_2, we note $|R| \leq 2|D|$, and for $n \geq 2$, $\varepsilon < 1$ and $|\theta| \geq B/n \geq 2A/n$,

$$|1 - x_n(\varepsilon) e^{-i\theta}|^{-1} \leq C|\theta| \tag{2.17.70}$$

so

$$\begin{aligned} I_2^2 &\leq \left(\int 4|D(e^{i\theta})|^2 |P_n(e^{i\theta})|^2 \frac{d\theta}{2\pi} \right) \int_{|\theta| \geq B/n} \frac{1}{C^2 |\theta|^2}\, d\theta \\ &= (4\lambda_n(z_n)) \frac{2n}{C^2 B} \end{aligned} \tag{2.17.71}$$

Since $\limsup n\lambda_n(z_n) = w(1) < \infty$, we see that as $B \to \infty$, $\limsup_{n\to\infty} |I_2| \to 0$.

Thus, we need only show that for each fixed B and ε,

$$\limsup_{n \to \infty} I_1 = 0 \tag{2.17.72}$$

Since

$$D(e^{i\theta}) - \frac{D(1)}{\overline{D(1)}} \overline{D(e^{i\theta})} = (D(e^{i\theta}) - D(1)) - \frac{D(1)}{\overline{D(1)}} (\overline{D(e^{i\theta})} - \overline{D(1)}) \tag{2.17.73}$$

and for n large,

$$|1 - x_n(\varepsilon) e^{i\theta}|^{-1} \leq |1 - x_n(\varepsilon)|^{-1} \leq 2\varepsilon^{-1} n \tag{2.17.74}$$

we have

$$I_1 \leq 4\varepsilon^{-1} n \int_{|\theta| \leq B/n} |D(e^{i\theta}) - D(1)|\, |P_n(e^{i\theta})| \frac{d\theta}{2\pi} \tag{2.17.75}$$

$$\leq 4e\varepsilon^{-1} B \left(\frac{2B}{n} \right)^{-1} \int_{|\theta| < B/n} |D(e^{i\theta}) - D(1)| \frac{d\theta}{2\pi} \tag{2.17.76}$$

by Proposition 2.17.11. Since 1 is a Lebesgue point for D, this goes to zero as $n \to \infty$. □

Proof of Theorem 2.17.4. If $d\mu$ obeys a global Szegő property, we get the lower bound on lim inf from Propositions 2.17.10 and 2.17.12 and the upper bound from Theorem 2.17.6. If μ obeys only a local Szegő property, let $\mu^\sharp$ be a global Szegő measure with $\mu^\sharp = \mu$ on I. The Nevai comparison theorem (Theorem 2.16.6) and the global case just proven implies the result. □

Proof of Theorem 2.17.5. Given Theorem 2.17.5 and Lubinsky's inequality (Theorem 2.16.7), the proof is identical to the proof of Theorem 2.16.1. □

From Findley's theorem and the argument used to prove Theorem 2.16.10, we obtain

Theorem 2.17.13. *Let μ be a regular measure on $\partial\mathbb{D}$, which is locally Szegő on $I = [\alpha, \beta]$ with $\alpha \neq \beta$. Let $e^{i\theta_0} \in I$ be a point where $w(\theta_0) \neq 0$ and be a Lebesgue point for both $w(\theta)$ and the for the local Szegő function. Then there is local clock behavior for the POPUC at $z_0 = e^{i\theta_0}$.*

Remarks and Historical Notes. Theorems 2.17.1–2.17.3 are discussed in standard texts in harmonic analysis; see, for example, Rudin [372] and Katznelson [217].

The Máté–Nevai upper bound (Theorem 2.17.6) is from Máté–Nevai [300], whose proof we follow. The lower bound that completes the MNT theorem (Theorem 2.17.4) and the method of proof we give of it is from Máté–Nevai–Totik [302]. These authors consider fixed z_n. The fact the one can extend the proofs to allow z_n's obeying (2.17.6) is due to Findley [133] who used this improvement and Lubinsky's method to get Theorem 2.17.5. Theorem 2.17.7 is from Simon [409].

Chapter Three

The Killip–Simon Theorem: Szegő for OPRL

In this chapter, we focus on OPRL whose essential support is $[-2, 2]$. See the end of Section 3.1 for a summary of the chapter.

3.1 STATEMENT AND STRATEGY

In this chapter, we turn to analogs of Szegő's theorem for OPRL that are close to a free case. The main theorem is Theorem 3.1.1.

Theorem 3.1.1 (Killip–Simon Theorem). *Let $\{a_n, b_n\}_{n=1}^\infty$ be the Jacobi parameters of a Jacobi matrix, J. Then*

$$\sum_{n=1}^\infty (a_n - 1)^2 + b_n^2 < \infty \tag{3.1.1}$$

if and only if

(a)

$$\sigma_{\text{ess}}(J) = \sigma_{\text{ess}}(J_0) \qquad \textit{(Blumenthal–Weyl)} \tag{3.1.2}$$

(b) *The eigenvalues $E_n \notin \sigma_{\text{ess}}(J_0)$ obey*

$$\sum_{n=1}^\infty \text{dist}(E_n, \sigma_{\text{ess}}(J_0))^{3/2} < \infty \qquad \textit{(Lieb–Thirring)} \tag{3.1.3}$$

(c) *The function f of* (1.4.3) *obeys*

$$\int_{\sigma(J_0)} \text{dist}(x, \mathbb{R} \setminus \sigma(J_0))^{1/2} \log f(x)\, dx > -\infty \qquad \textit{(Quasi-Szegő)} \tag{3.1.4}$$

Our proof will rely on a sum rule, namely, if F, g, and Q are given by (1.10.9), (1.10.10), and (1.10.16), then

$$Q(\rho) + \sum F(E_n) = \sum_{n=1}^\infty \left[\tfrac{1}{4} b_n^2 + \tfrac{1}{2} G(a_n)\right] \tag{3.1.5}$$

Theorem 3.1.1 will then result by noting that it is equivalent to saying one side of (3.1.5) is finite if and only if the other is.

The proof of the sum rule will follow the path trod in Sections 2.4–2.7. The analog of the function $(\delta_0 D)(z)$ needed there will be the m-function, so Section 3.2 will find the coefficient stripping formula for m. Since $\log m$ is not always in H^p—indeed, m can have zeros and poles—we will need a suitable Poisson–Jensen formula, which we prove in Section 3.3 and use to prove step-by-step sum rules in Section 3.4. In Section 3.5, we will then use the combination of semicontinuity of the entropy and positivity to go from step-by-step sum rules to the full sum rule (3.1.5), which will prove Theorem 3.1.1.

As a bonus, our machinery yields an improvement of the Shohat–Nevai theorem (Theorem 1.9.3) in Section 3.6: rather than supposing $d\rho$ is supported on $[-2, 2]$, we only need that its essential support is $[-2, 2]$ and a natural condition on the eigenvalues outside $[-2, 2]$, namely, that

$$\sum_{n=1}^{\infty} \text{dist}(E_n, \sigma_{\text{ess}}(J_0))^{1/2} < \infty \tag{3.1.6}$$

In Section 3.7, we turn to asymptotics of OPRL using these techniques. The next two sections discuss moment problems using a relation to Szegő's theorem but in a very different direction from the core, Sections 3.2–3.7, of this chapter. The final three sections continue the discussion in Sections 2.15–2.17 on asymptotics of the CD kernel, a subject to be continued in Section 5.11.

Remarks and Historical Notes. The strategy described here is from Killip–Simon [225], with improvements from Simon–Zlatoš [410] and Simon [396].

3.2 WEYL SOLUTIONS AND COEFFICIENT STRIPPING

In this section, we will obtain the analog of the Schur algorithm that relates the Schur functions for the measures with Verblunsky coefficients $\{\alpha_n\}_{n=0}^{\infty}$ and $\{\alpha_{n+1}\}_{n=0}^{\infty}$. Instead, we relate the m-functions for the OPRL measures with Jacobi parameters $\{a_n, b_n\}_{n=1}^{\infty}$ and $\{a_{n+1}, b_{n+1}\}_{n=1}^{\infty}$.

As in the OPUC case, our approach relies on the identification of Weyl solutions. We will use our basic analysis of Weyl solutions to derive coefficient stripping and end the section with a few asides that give more information about Weyl solutions.

We begin with the transfer matrix by writing the difference equation (1.2.17) for $n = 1, 2, \dots,$

$$\begin{pmatrix} u_{n+1} \\ a_n u_n \end{pmatrix} = A(a_n, b_n; z) \begin{pmatrix} u_n \\ a_{n-1} u_{n-1} \end{pmatrix} \tag{3.2.1}$$

$$A(a, b; z) = \frac{1}{a} \begin{pmatrix} z - b & -1 \\ a^2 & 0 \end{pmatrix} \tag{3.2.2}$$

(3.2.1) is solved by $(u_n, a_{n-1}u_{n-1}) = (p_{n-1}(z), a_{n-1}p_{n-2}(z))$ with initial conditions $(u_1, u_0) = (1, 0)$. Given Jacobi parameters $\{a_n, b_n\}_{n=1}^{\infty}$, we define the transfer matrix by

$$T_n(\{a_j, b_j\}_{j=1}^n, z) = A(a_n, b_n, z) A(a_{n-1}, b_{n-1}, z) \dots A(a_1, b_1, z) \tag{3.2.3}$$

Thus,

$$\begin{pmatrix} u_{n+1} \\ y_{n+1} \end{pmatrix} = T_n \begin{pmatrix} u_1 \\ y_1 \end{pmatrix} \tag{3.2.4}$$

is equivalent to

$$y_{n+1} = a_n u_n \qquad n \geq 1 \tag{3.2.5}$$

$$a_n u_{n+1} + (b_n - z)u_n + a_{n-1}u_{n-1} = 0 \qquad n \geq 1 \tag{3.2.6}$$

with initial conditions (a_0 thus drops out) u_0, u_1 with

$$a_0 u_0 \equiv y_1 \qquad u_1 = y_1 \tag{3.2.7}$$

In particular, for $n \geq 1$,

$$\begin{pmatrix} p_n(z) \\ a_n p_{n-1}(z) \end{pmatrix} = T_n \begin{pmatrix} 1 \\ 0 \end{pmatrix} \tag{3.2.8}$$

Note that (the reason we add a_n to u_n in (3.2.1)) $\det(A(a, b; z)) = 1$ so that

$$\det(T_n(\{a_j, b_j\}_{j=1}^n, z)) = 1 \tag{3.2.9}$$

Our first main goal will be to prove the OPRL analog of Theorem 2.4.1:

Theorem 3.2.1. *Fix $z \in \mathbb{C}_+ = \{z \mid \operatorname{Im} z > 0\}$ and let*

$$v_0 = \begin{pmatrix} m(z) \\ -1 \end{pmatrix} \tag{3.2.10}$$

Then

$$v_n = T_n(z)v_0 \tag{3.2.11}$$

obeys $v \in \ell^2$ (i.e., $\sum_{n=0}^{\infty} \|v_n\|^2 < \infty$), and for any $w_0 \in \mathbb{C}^2$ with $T_n(z)w_0 \in \ell^2$, we have that w_0 is a multiple of v_0.

To prove this, we introduce the analog of ψ_n:

Definition. The *second kind polynomials*, $q_n(z)$, are defined for $n = 0, 1, 2, \dots$ by

$$q_n(x) = \int \frac{p_n(x) - p_n(y)}{x - y}\, d\mu(y) \tag{3.2.12}$$

and for $n = -1$,

$$q_{-1}(x) = -1 \tag{3.2.13}$$

Theorem 3.2.2. *The vector*

$$w_n = \begin{pmatrix} q_n(z) \\ a_n q_{n-1}(z) \end{pmatrix} \tag{3.2.14}$$

solves

$$w_n = T_n(z)w_0 \tag{3.2.15}$$

where w_0 is $(0, -1)^t$. Moreover, for $n \geq 1$, $q_n(x)$ is a polynomial of degree $n - 1$. Indeed,

$$q_n(x) = a_1^{-1} p_{n-1}(x; \{a_{\ell+1}, b_{\ell+1}\}_{\ell=1}^{n-1}) \tag{3.2.16}$$

the OPRL for the once-stripped measure with Jacobi parameters $\{a_{\ell+1}, b_{\ell+1}\}_{\ell=1}^{\infty}$.

Proof. The recursion relation (1.2.15) obeyed by p_n implies

$$\begin{aligned} a_{n+1}q_{n+1}(x) + b_{n+1}q_n(x) + a_n\tilde{q}_{n-1}(x) &= \int \frac{xp_n(x) - yp_n(y)}{x-y}\,d\mu(y) \\ &= xq_n(x) + \int \frac{x-y}{x-y}\,p_n(y)\,d\mu \\ &= xq_n(x) + \delta_{n0} \end{aligned} \tag{3.2.17}$$

where $\tilde{q}_{-1} = 0$ and otherwise $\tilde{q}_j = q_j$. Since $a_0 q_{-1} = -1$, we see for $n \geq 0$,

$$a_{n+1}q_{n+1}(x) + (b_n - x)q_n(x) + a_n q_{n-1}(x) = 0$$

which, given (3.2.5)/(3.2.6), implies that (3.2.15) holds.

Since $a_1 q_1 + (b_1 - x)q_0 + a_0 q_{-1} = 0$, we see, using $q_0 = 0$, $a_0 q_{-1} = -1$, that $q_1 = 1/a_1$, so q_n obeys

$$\begin{pmatrix} q_{n+1} \\ a_{n+1} \end{pmatrix} = T_n(\{a_{j+1}, b_{j+1}\}_{j=1}^n; z)\begin{pmatrix} q_1 \\ a_1 \end{pmatrix} \tag{3.2.18}$$

with initial conditions $(q_1, q_0) = 1/a_1(1,0)$, which immediately implies (3.2.16). This in turn implies that q_n is a polynomial of degree $n-1$. □

Thus,

$$T_n(z) = \begin{pmatrix} p_n(z) & -q_n(z) \\ a_n p_{n-1}(z) & -a_n q_{n-1}(z) \end{pmatrix} \tag{3.2.19}$$

The analog of Proposition 2.4.6 is more powerful since there is no z^n factor:

Proposition 3.2.3. *If r_n and s_n solve*

$$\begin{pmatrix} t_{n+1} \\ t_n \end{pmatrix} = T_n \begin{pmatrix} t_1 \\ t_0 \end{pmatrix} \tag{3.2.20}$$

then

$$r_{n+1}s_n - r_n s_{n+1} = r_1 s_0 - r_0 s_1 \tag{3.2.21}$$

In particular,

$$a_n(q_n(x)p_{n-1}(x) - q_{n-1}(x)p_n(x)) = 1 \tag{3.2.22}$$

Proof. By (3.2.20), with $R_n = \left(\begin{smallmatrix} r_{n+1} & s_{n+1} \\ r_n & s_n \end{smallmatrix}\right)$, we have $R_n = T_n R_0$, so $\det(R_n) = \det(T_n)\det(R_0)$. By (3.2.9), (3.2.21) holds. □

Proof of Theorem 3.2.1. Clearly, $v_n = (g_n, a_n g_{n-1})$ where

$$g_n(z) = m(z)p_n(z) + q_n(z) \tag{3.2.23}$$

$$\begin{aligned} &= p_n(z)\int \frac{d\mu(x)}{x-z} - p_n(z)\int \frac{d\mu(x)}{x-z} + \int \frac{p_n(x)}{x-z}\,d\mu(x) \\ &= \int \frac{p_n(x)}{x-z}\,d\mu = \langle p_n, (\cdot - z)^{-1}\rangle \end{aligned} \tag{3.2.24}$$

Since $\{p_n\}_{n=0}^\infty$ is an orthonormal basis,

$$\sum_{n=0}^\infty |g_n(z)|^2 = \int \frac{d\mu(x)}{|x-z|^2} \le |\mathrm{Im}\, z|^{-2} \tag{3.2.25}$$

so $v \in \ell^2$ since $\{a_n\}_{n=1}^\infty$ is bounded.

On the other hand, if $w_n \equiv T_n(z)w_0 = (h_n, a_n h_{n-1})$, then

$$h_n g_{n-1} - g_n h_{n-1} = h_0 g_{-1} - h_{-1} g_0 \tag{3.2.26}$$

by (3.2.21). If $w_n \in \ell^2$, $h_n \in \ell^2$, so $h_n g_{n-1} - g_n h_{n-1} \in \ell^1$. Thus, the left-hand side of (3.2.26) is in ℓ^1. Since the right side is constant, the constant must be zero, which implies (h_0, h_{-1}) is a multiple of (g_0, g_{-1}). □

Remarks. 1. As we will see, g_n actually decays exponentially; see Proposition 3.2.6. This plus (3.2.26) shows that any other solution must grow exponentially.

2. We note that

$$\int \frac{d\mu(x)}{|x-z|^2} = \frac{1}{\mathrm{Im}\, z} \int \mathrm{Im}\left(\frac{1}{x-z}\right) d\mu(x) = \frac{\mathrm{Im}\, m(z)}{\mathrm{Im}\, z}$$

so (3.2.25) can be rewritten as

$$\sum_{n=0}^\infty |m(z)p_n(z) + q_n(z)|^2 = \frac{\mathrm{Im}\, m(z)}{\mathrm{Im}\, z} \tag{3.2.27}$$

3. We will call $g_n(z)$ the Weyl solution, although we note that [99, 400] define the Weyl solution by $w_n(z) = -g_{n-1}(z + \frac{1}{z})$. We will discuss the reasons for the differing conventions in the Notes to Section 3.7.

Given Theorem 3.2.1, coefficient stripping is immediate:

Theorem 3.2.4 (Coefficient Stripping for OPRL, aka Stieltjes Expansion). *Let $m(z)$ be the m-function of a Jacobi matrix with Jacobi parameters $\{a_n, b_n\}_{n=1}^\infty$ and let m_1 be the m-function of the once-stripped Jacobi matrix, that is, the one with parameters $\{a_{n+1}, b_{n+1}\}_{n=1}^\infty$. Then*

$$m(z) = \frac{1}{b_1 - z - a_1^2 m_1(z)} \tag{3.2.28}$$

Remarks. 1. To make the analog to the Schur algorithm precise, note that if $m(z)$ is any discrete m-function, $-m(z)^{-1}$ is also Herglotz and analytic on $\mathbb{R} \setminus I$ with (2.3.11). But (2.3.10) fails since $-m(z)^{-1} \sim z$. But it can be seen that for some $a > 0$ and b and discrete m-function $\tilde{m}$,

$$-m(z)^{-1} = z - b + a^2 \tilde{m}(z)$$

This is the analog of the Schur algorithm. This theorem says that a, b are the first two Jacobi parameters and $\tilde{m} = m_1$.

2. For a more streamlined proof, see Section 10.3.

Proof. By the uniqueness in Theorem 3.2.1,

$$\begin{pmatrix} z-b_1 & -1 \\ a_1^2 & 0 \end{pmatrix} \begin{pmatrix} m(z) \\ -1 \end{pmatrix} = c \begin{pmatrix} m_1(z) \\ -1 \end{pmatrix} \tag{3.2.29}$$

for some c since

$$T_{n-1}(\{a_{j+1}, b_{j+1}\}_{j=1}^{n-1}; z) A(a_1, b_1; z) = T_n(\{a_j, b_j\}_{j=1}^n; z) \tag{3.2.30}$$

implies that $T_{n-1}(\{a_{j+1}, b_{j+1}\}_{j=1}^{n-1}; z)$ [LHS of (3.2.29)] lies in ℓ^2. This means

$$-m_1(z) = \frac{[(z-b_1)m(z)+1]}{[a_1^2 m(z)]}$$

which is equivalent to (3.2.28). □

In Theorem 3.7.6, we will extend (3.2.28) and relate g_k/g_{k-1} to the m-function of a stripped J.

As for the OPUC case, we thus have a continued fraction expansion for $m(z)$,

$$m(z) = \cfrac{1}{b_1 - z - \cfrac{a_1^2}{b_2 - z - \cfrac{a_2^2}{b_3 - z - \cdots}}} \tag{3.2.31}$$

(3.2.31) provides a second way to go from $d\rho$ to $\{a_n, b_n\}_{n=1}^\infty$. This completes the results we need going forward, but we would like to make some additional remarks.

First, we want to note a connection to the spectral theorist's Green's function, and second, deduce exponential decay from that. Define for $z \in \mathbb{C}_+$,

$$G_{k\ell}(z) = \langle \delta_k, (J-z)^{-1}\delta_\ell \rangle \tag{3.2.32}$$

Proposition 3.2.5. *We have that for $z \in \mathbb{C}_+$,*

$$G_{n1}(z) = g_{n-1}(z) \tag{3.2.33}$$

where g is given by (3.2.23).

Remark. More generally, one can show that $G_{k\ell}(z) = G_{\ell k}(z)$, and for $k \le \ell$,

$$G_{k\ell} = p_{k-1}(z) g_{\ell-1}(z) \tag{3.2.34}$$

Proof. We have proven that $g_{\cdot-1}(z)$ is the unique ℓ^2 solution of $[(J-z)(u)]_n = 0$ for $n \ge 2$. But clearly, $u = (J-z)^{-1}\delta_1$ obeys the same equation, so (3.2.33) holds up to a single overall constant (a priori constant in n but not necessarily in z). But

$$G_{11}(z) = m(z) = g_0(z)$$

showing the constant is one. □

Proposition 3.2.6. *For any $Q > 2\sup_n |a_n|$, we have*

$$|G_{k\ell}(z)| \le C_{Q,z} \left(1 + \frac{|\mathrm{Im}\, z|}{Q}\right)^{-|k-\ell|} \tag{3.2.35}$$

In particular, for each $z \in \mathbb{C}_+$, $g_n(z)$ decreases exponentially in z.

We will prove this using the method of Combes–Thomas [92], which depends on:

Lemma 3.2.7. *Let A be a (possibly unbounded) selfadjoint operator and J a bounded operator. Suppose $z_0 \notin \sigma(J)$ and that*

$$J(s) = e^{isA} J e^{-isA} \tag{3.2.36}$$

originally defined for $s \in \mathbb{R}$, has an analytic continuation to $S_{K_0} \equiv \{s \mid |\mathrm{Im}\, s| < K_0\}$ with $z_0 \notin \sigma(J(s))$ for all $s \in S_{K_0}$. Let $\varphi \in D(e^{K_0|A|})$. Then $(J - z_0)^{-1}\varphi \in D(e^{KA})$ for all $K \subset (-K_0, K_0)$.

Proof. It is a simple general fact that follows from the spectral theorem that $\eta \in D(e^{KA})$ for all $K \subset (-K_0, K_0)$ if and only if $e^{isA}\eta$, defined initially for $s \in \mathbb{R}$, has a (Hilbert-space) analytic continuation to S_{K_0}.

For s real,

$$e^{isA}(J - z_0)^{-1}\varphi = (J(s) - z_0)^{-1} e^{isA}\varphi \tag{3.2.37}$$

Under the hypotheses of the theorem, the right side of (3.2.37) has an analytic continuation and thus, so does the left. □

Proof of Proposition 3.2.6. We will prove the result when $C_{Q,z}$ is also ℓ-dependent. That it can be chosen ℓ-independent follows from detailed estimates implicit in (3.2.37). Of course, our application is to a fixed ℓ, namely, $\ell = 0$.

Let A be multiplication by n on ℓ^2, that is, $A\delta_n = n\delta_n$. Then for s real, $e^{isA} J e^{-isA}$ is the tridiagonal matrix with b_n on the main diagonal and $e^{\pm is} a_n$ off-diagonal. This has an analytic continuation to all of $\mathbb{C}$. Moreover, for K real,

$$\|J(\pm iK) - J\| \le 2 \sup_n |a_n| \, |e^K - 1| \tag{3.2.38}$$

Since $\|(J - z)^{-1}\| = |\mathrm{Im}\, z|^{-1}$ and

$$(J(\pm iK) - z) = (J - z)^{-1}(1 + (J - z)^{-1}(J(\pm iK) - J)) \tag{3.2.39}$$

we see $z_0 \notin \sigma(J(\pm iK))$ so long as

$$|e^K - 1| Q |\mathrm{Im}\, z|^{-1} \le 1 \tag{3.2.40}$$

or

$$e^{|K|} \le e^{K_0} \equiv 1 + \frac{\mathrm{Im}\, z}{Q}$$

Since $\delta_\ell \in D(e^{K_0|A|})$, the lemma implies $(J - z)^{-1}\delta_\ell \in D(e^{|K|A})$, that is, $\sum_n |(e^{|K|})^n G_{n\ell}|^2 < \infty$, which implies (3.2.36). □

Finally, we want to prove that on $\mathbb{C}_+$,

$$m(z) = \lim_{n\to\infty} -\frac{q_n(z)}{p_n(z)} \tag{3.2.41}$$

something closely related to the fact that $q_n + mp_n \in \ell_2$. It can be derived from $q_n + mp_n \in \ell_2$ but only by getting lower bounds on p_n, which is tricky (but see the Notes). Instead, we will proceed with a result of independent interest:

Proposition 3.2.8. *Let $J_{n;F}$ be the truncated $n \times n$ Jacobi matrix of* (1.2.30). *Let*

$$G_{k\ell}^{n;F}(z) = \langle \delta_k, (J_{n;F} - z)^{-1}\delta_\ell \rangle \tag{3.2.42}$$

Then

$$G_{11}^{n;F}(z) = -\frac{q_n(z)}{p_n(z)} \tag{3.2.43}$$

Proof. By (1.2.31), $p_n(z) = (a_1 \dots a_n)^{-1} \det(z - J_{n;F})$. By (3.2.16), $q_n(z)$ is $(a_1 \dots a_n)^{-1}$ times the 11 minor of $z - J_{n;F}$. Taking into account that $G_{11}^{n;F}(z) = -\langle \delta_1, (z - J_{n;F})^{-1}\delta_1 \rangle$, we see that (3.2.43) is just Cramer's rule. □

Remark. By (2.3.15) and Proposition 1.3.4, $m(z) + q_n(z)p_n(z)^{-1} = O(z^{-2n-1})$ at infinity, which, given the degrees of p and q, implies that $(-q_n/p_n)$ are Padé approximants about infinity. This convergence of Padé approximants in all of $\mathbb{C}_+$ is a (special case of a) result of Stieltjes [422].

Theorem 3.2.9. *For k, ℓ fixed and $z \in \mathbb{C}_+$,*

$$G_{k,\ell}^{n;F}(z) \to G_{k,\ell}(z) \tag{3.2.44}$$

In particular, (3.2.41) *holds.*

Proof. View $J_{n;F}$ as acting on ℓ^2 by embedding it in a matrix with all zeros. Clearly, for any $\varphi \in \ell^2$, $J_{n;F}\varphi \to J\varphi$. Thus, for $z \in \mathbb{C}_+$,

$$\begin{aligned} \|[(J_{n;F} - z)^{-1} - (J - z)^{-1}]\varphi\| &\le \|(J_{n;F} - z)^{-1}(J - J_{n;F})(J - z)^{-1}\varphi\| \\ &\le |\operatorname{Im} z|^{-1} \|(J - J_{n;F})(J - z)^{-1}\varphi\| \to 0 \end{aligned}$$

Taking $\varphi = \delta_\ell$ yields (3.2.44). □

We remark that one can also prove coefficient stripping from (3.2.44) without using Theorem 3.2.1, and it is often done that way.

Remarks and Historical Notes. While we present this as the OPRL analog of OPUC results, the history is the opposite! Jacobi [206] essentially wrote down the finite matrix terminating continued fraction expansion (3.2.31). Stieltjes [422] wrote the infinite N case. Wall [455] calls them J-fractions. Similarly, Weyl solutions were first discussed (as Jost solutions) for OPRL; see, for example, Case [76].

We note that if $\operatorname{supp}(d\rho) = [-2, 2]$ so there is a $d\mu$ on $\partial\mathbb{D}$ via the Szegő mapping theorem, then there are formulae relating the Weyl solutions for $d\rho$ to those of $d\mu$.

While it may predate that, the use of the transfer matrix (3.2.1) with the extra a_n/a_{n-1} in the second factor that leads to $\det(T) = 1$ is borrowed from Damanik–Killip–Simon [97]. For other proofs of (3.2.35), see the use of CD kernels and/or potential theory in Stahl–Totik [417] and Simon [404]. These references (see also [98]) also provide direct proofs that $p_n(z)$ is bounded below as $n \to \infty$ for any $z \in \mathbb{C}_+$, and so direct proofs of (3.2.41) from (3.2.27).

3.3 MEROMORPHIC HERGLOTZ FUNCTIONS

In the proof of Szegő's theorem, a key role was played by the fact that a nonvanishing analytic function on $\mathbb{D}$ (in this case, $(\delta_0 D)(z)$), $f(z)$, with $\log f \in H^1(\mathbb{D})$ has a Poisson–Jensen representation

$$f(z) = \exp\biggl(\int \frac{e^{i\theta} + z}{e^{i\theta} - z} \log(f(e^{i\theta})) \frac{d\theta}{2\pi}\biggr) \tag{3.3.1}$$

if $f(0) > 0$.

For OPRL, the analog of $\delta_0 D$ will be the m-function moved to $\mathbb{D}$ by the map (1.9.1), that is,

$$M(z) = -m(z + z^{-1}) \tag{3.3.2}$$

(we will see the reason for the minus sign shortly). This function has zeros and poles in $\mathbb{D}$ so it cannot be represented in the form (3.3.1)!

There is a standard method for controlling zeros of H^p functions, namely, via Blaschke products that we discussed in Section 2.3. As we will see, one needs a variant on the products. To frame the change, we remark that one proves the convergence part of Proposition 2.3.16 by noting

$$b_{re^{i\theta}}(z) = b_r(e^{-i\theta} z) \tag{3.3.3}$$

and

$$b_r(z) - 1 = -\frac{(1-r)(z+1)}{1 - rz} \tag{3.3.4}$$

so that

$$\sup_{|z| \le R} \sum_j |b_{z_j}(z) - 1| \le \frac{1+R}{1-R} \sum_j (1 - |z_j|) \tag{3.3.5}$$

The absolute convergence of Blaschke products will sometimes be relevant so the $p = 1$ case of the following is important (as will $p = 3$):

Proposition 3.3.1. *Let* $\{E_j\}_{j=1}^\infty \subset \mathbb{R} \setminus [-2, 2]$ *and define* β_j *in* $\mathbb{R} \setminus [-1, 1]$ *by*

$$E_j = \beta_j + \beta_j^{-1} \tag{3.3.6}$$

Then for any $p > 0$,

$$\sum_{j=1}^\infty (|E_j| - 2)^{p/2} < \infty \Leftrightarrow \sum_{j=1}^\infty (1 - |\beta_j|^{-1})^p < \infty \tag{3.3.7}$$

Proof. Follows immediately from

$$|E_j| - 2 = |\beta_j|(|\beta_j|^{-1} - 1)^2 \tag{3.3.8}$$

□

The convergence result in Proposition 2.3.16 is thus an analog of $\sum_j |a_j| < \infty \Rightarrow \sum_{j=1}^J a_j$ convergent for numerical sums. We will instead need the analog of $(-1)^{j+1} a_j > 0$, $|a_j| \ge |a_{j+1}|$, and $|a_j| \to 0 \Rightarrow \sum_{j=1}^J a_j$ convergent, the result for alternating sums.

Theorem 3.3.2. *Let* $\{z_j\}_{j=1}^\infty$ *and* $\{p_j\}_{j=1}^\infty$ *be subsets of* $(-1, 1)$ *so that*

$$\text{(a)} \qquad |z_j| \to 1 \quad \text{as } j \to \infty \tag{3.3.9}$$

$$\text{(b)} \qquad \sum_{j=1}^\infty |z_j - p_j| < \infty \tag{3.3.10}$$

Then, with b *given by* (2.3.67),

$$\prod_{j=1}^N \left[\frac{b_{z_j}(z)}{b_{p_j}(z)}\right] \to B_\infty(z) \tag{3.3.11}$$

as $N \to \infty$ *uniformly on compact subsets of* $\mathbb{C} \setminus S$ *where*

$$S = \{p_j\}_{j=1}^\infty \cup \{z_j^{-1}\}_{j=1}^\infty \cup \{\pm 1\} \tag{3.3.12}$$

Moreover, on $\partial\mathbb{D} \setminus \{\pm 1\}$,

$$|z| = 1 \Rightarrow |B_\infty(z)| = 1 \tag{3.3.13}$$

If $z_j > p_j$, *let* $I_j = (p_j, z_j)$ *and* $\sigma_j = +1$, *and if* $z_j < p_j$, *let* $I_j = (z_j, p_j)$ *and set* $\sigma_j = -1$. *Define, for* $x \in (-1, 1)$,

$$N(x) = \sum_j \sigma_j \chi_{I_j}(x) \tag{3.3.14}$$

If

$$N_\infty \equiv \|N\|_\infty < \infty \tag{3.3.15}$$

then in $\mathbb{C}_+ \cap \mathbb{D}$,

$$|\arg B_\infty(z)| < \pi N_\infty \tag{3.3.16}$$

Remarks. 1. (3.3.9)/(3.3.10), of course, imply $|p_j| \to 1$ also.

2. The convergence as a function with values in $\mathbb{C} \cup \{\infty\}$ is uniform away from ± 1.

3. By (a), (b), any $x \in (-1, 1)$ lies in at most finitely many I_j and the sum in (3.3.14) is uniformly convergent on each $(-1+\varepsilon, 1-\varepsilon)$. N_∞ is an integer (if finite).

4. In (3.3.16), we mean the continuous branch of arg B_∞ with $\lim_{\varepsilon\downarrow 0} \arg B_\infty(x + i\varepsilon) = 0$ at points in $(0, \delta)$ for small δ, where $B_\infty(x) > 0$.

5. In the case where the z's and p's interlace so $N_\infty = 1$, it can happen that the set of values of arg $B_\infty(z)$ is either $(0, \pi)$ or $(-\pi, 0)$.

Before proving this theorem, we want to note that (3.3.15) implies (3.3.10) for suitable orderings of the z's and p's.

Lemma 3.3.3. *Suppose that*

$$0 \le z_1 \le z_2 \le \cdots \tag{3.3.17}$$

$$0 \le p_1 \le p_2 \le \cdots \tag{3.3.18}$$

Then

$$\sum_{j=1}^{\infty} |z_j - p_j| = \int_0^1 |N(x)|\, dx \tag{3.3.19}$$

so $N_\infty < \infty$ *implies* (3.3.10).

Proof. We claim that if $I_j \cap I_k \neq 0$, then $\sigma_j = \sigma_k$. Suppose that $j < k$ and $\sigma_j = 1$, $\sigma_k = -1$ (the other cases are similar). Then $z_j < p_j$ while $p_k < z_k$ so $I_j \cap I_k \neq \emptyset$ implies $p_k < p_j$, contrary to (3.3.18).

Thus,

$$|N(x)| = \#\{j \mid x \in I_j\} \tag{3.3.20}$$

since there are no $\chi_{I_j} - \chi_{I_k}$ cancellations. This in turn implies (3.3.19). □

Remarks. 1. The above proof also shows that if $x \in \{z_j\}_{j=1}^{\infty} \cup \{p_j\}_{j=1}^{\infty}$, then

$$N(x) = \#(j \mid z_j < x) - \#(j \mid p_j < x) \tag{3.3.21}$$

2. One can handle the situation where we consider $(-1, 1)$ instead of $(0, 1)$ with ± 1 limit points of the z's and p's by labeling $\{z_j\}_{j=-\infty}^{\infty}$ and $\{p_j\}_{j=-\infty}^{\infty}$ with

$$z_{-2} \leq z_{-1} < 0 \leq z_0 \leq z_1 < \cdots \tag{3.3.22}$$

One still has (3.3.19).

We begin the proof of Theorem 3.3.2 with two lemmas:

Lemma 3.3.4. *Let Q and K be two compact sets in $\mathbb{C}$ with Q a real interval and*

$$K \cap [Q \cup Q^{-1}] = \emptyset \tag{3.3.23}$$

For $z \in K$, $x \in Q$, define

$$\tilde{b}(z, x) = \frac{z - x}{1 - xz} \tag{3.3.24}$$

Then there is a constant C depending only on Q, K so for all $x, w \in Q$,

$$\left| 1 - \frac{\tilde{b}(z, x)}{\tilde{b}(z, w)} \right| \leq C|x - w| \tag{3.3.25}$$

Remarks. 1. Of course, if $z \in \mathbb{D}$, $\tilde{b}(z, x) = \operatorname{sgn}(-x) b(z, x)$. b is normalized by $b(0, x) > 0$ (for $x \neq 0$ and $b'(z, 0)|_{z=0} > 0$), which is convenient for products of b's not to oscillate, but for us here, smoothness of b in x is more important. The x, w cancellations control oscillations.

2. By (3.3.23), $\tilde{b}$ is analytic and nonvanishing in z, x for $z \in K$, $x \in Q$.

3. In our applications, we will take $Q = [-1, 1]$ and $K \subset \mathbb{C}_+$ or $K \subset \mathbb{C}_-$ or else $Q = [a, 1]$ or $[-1, a]$ and $K = \{z \mid |z| \leq a - \varepsilon\}$.

Proof. Clearly, (3.3.25) follows from

$$\inf_{\substack{z \in K \\ x \in Q}} |\tilde{b}(z, w)| > 0 \tag{3.3.26}$$

and

$$|\tilde{b}(z, w) - \tilde{b}(z, x)| \leq C_1|x - w| \tag{3.3.27}$$

In turn, since Q is connected, (3.3.27) is implied by

$$\sup_{\substack{z \in K \\ w \in Q}} \left| \frac{\partial}{\partial w} \tilde{b}(z, w) \right| < \infty \tag{3.3.28}$$

The required (3.3.26)/(3.3.28) are immediate by compactness, given analyticity and nonvanishing of $\tilde{b}$ on $K \times Q$. □

Lemma 3.3.5. *Fix* $z \in \mathbb{C}_+ \cap \mathbb{D}$. *Define* $\arg(\tilde{b}(z, x))$ *for* $x \in (-1, 1)$ *by requiring continuity and* $\arg(\tilde{b}(z, x = 0)) = \arg(z) \in (0, \pi)$. *Then*

$$\text{(i)} \qquad \arg(\tilde{b}(z, x)) \in (0, \pi)$$

$$\text{(ii)} \qquad 0 < \frac{\partial}{\partial x} \arg(\tilde{b}(z, x)) \tag{3.3.29}$$

$$\text{(iii)} \qquad \frac{\partial}{\partial x} \arg(\tilde{b}(z, x)) < 2\, \frac{\operatorname{Im} z}{|x - z|^2} \tag{3.3.30}$$

$$\text{(iv)} \qquad \int_{-1}^{1} \frac{\partial}{\partial x} \arg(\tilde{b}(z, x))\, dx = \pi \tag{3.3.31}$$

Proof. This depends on a remarkably simple formula for $\frac{\partial}{\partial x} \arg(\tilde{b}(z, x))$. We have

$$\begin{aligned} \frac{\partial}{\partial x} \arg(x - z) &= \operatorname{Im} \frac{\partial}{\partial x} \log(x - z) = \operatorname{Im}\left(\frac{1}{x - z}\right) \\ &= \frac{\operatorname{Im} z}{|x - z|^2} \end{aligned} \tag{3.3.32}$$

and

$$\begin{aligned} \frac{\partial}{\partial x} \arg(1 - xz) &= \operatorname{Im}\left(-\frac{z}{1 - xz}\right) = -\frac{\operatorname{Im} z}{|1 - zx|^2} \\ &= -\frac{\operatorname{Im} z}{|x - z|^2}\, |\tilde{b}(x, z)|^2 \end{aligned}$$

Thus,

$$\frac{\partial}{\partial x} \arg(\tilde{b}(z, x)) = \frac{\operatorname{Im} z}{|x - z|^2}\, (1 + |\tilde{b}(z, x)|^2) \tag{3.3.33}$$

(ii), (iii) are obvious from this formula.

By (3.3.32), (3.3.33), and $|\tilde{b}(z, x)| < 1$ for $x \in (-1, 1)$, we see

$$\frac{\partial}{\partial x} \arg(\tilde{b}(z, x)) \leq 2\, \frac{\partial}{\partial x} \arg(x - z) \tag{3.3.34}$$

By simple geometry,

$$\int_{-1}^{1} \frac{\partial}{\partial x} \arg(x - z)\, dx = \arg(1 - z) - \arg(-1 - z) < \pi \tag{3.3.35}$$

so

$$\int_{-1}^{1} \frac{\partial}{\partial x} \arg(\tilde{b}(z, w))\, dx \leq 2\pi \tag{3.3.36}$$

Note that $\tilde{b}(z, 1) = -1$, $\tilde{b}(z, -1) = 1$. Since $\arg(\tilde{b}(z, w))$ is monotone on $(-1, 1)$, the integral in (3.3.31) is $(2n + 1)\pi$ for some $n = 0, 1, 2, \dots$. By (3.3.35) and (3.3.36), we conclude (3.3.31). But this plus $\arg(\tilde{b}(z, 0)) \in (0, \pi)$ implies (i). □

Proof of Theorem 3.3.2. If z_j and p_j lie on the same side of 0, $b_{z_j}(z)/b_{p_j}(z) = \tilde{b}_{z_j}(z)/\tilde{b}_{p_j}(z)$ (there is a minus sign if they are on opposite sides). By hypotheses (a) and (b), only finitely many pairs are on opposite sides, so it suffices to prove convergence for b replaced by $\tilde{b}$. On $\mathbb{C} \setminus [Q \cup Q^{-1}]$ where $Q = \overline{\{z_j\}_{j=1}^{\infty}} \cup \overline{\{p_j\}_{j=1}^{\infty}}$ (which will include 1 and/or -1), this convergence is immediate by Lemma 3.3.4. The points at $\{z_j\} \cup \{p_j^{-1}\}$ are removable singularities. (3.3.13) is obvious from the uniform convergence.

To get (3.3.16), we use the fact that B_∞ is the product of $\tilde{b}$'s or its negative, so we need only prove (3.3.16) for the product of $\tilde{b}$'s. But then

$$\arg\left(\prod_{j=1}^{n} \frac{\tilde{b}(z, z_j)}{\tilde{b}(z, p_j)}\right) = \sum_{j=1}^{n} \sigma_j \int_{I_j} \frac{\partial}{\partial x} \arg(\tilde{b}(z, x))\, dx$$

so

$$\arg(\pm B_\infty) = \sum_{j=1}^{\infty} \sigma_j \int_{I_j} \frac{\partial}{\partial x} \arg(\tilde{b}(z, w))\, dx$$

and

$$\begin{aligned} |\arg(\pm B_\infty)| &\leq \int_{I_j} |N(x)| \frac{\partial}{\partial x} \arg(\tilde{b}(z, w))\, dx \\ &\leq N_\infty \pi \end{aligned}$$

by (3.3.31). □

We now turn to the main object of this section:

Definition. A *meromorphic Herglotz* (MH) function is a meromorphic function on $\mathbb{D}$ so

$$|z| < 1 \quad \text{and} \quad \operatorname{Im} z \neq 0 \Rightarrow \frac{\operatorname{Im} f(z)}{\operatorname{Im} z} > 0 \tag{3.3.37}$$

Example. Since $z \mapsto z + z^{-1} = E$ maps $\operatorname{Im} z > 0$ to $\operatorname{Im} E < 0$, the function M of (3.3.2) is an MH function.

Theorem 3.3.6. *Let f be an MH function. Then all its zeros, $\{z_j\}_{j=-N}^{M}$, and poles, $\{p_j\}_{j=-N}^{M}$, lie on $\mathbb{R} \cap \mathbb{D}$ and interlace. Their Blaschke product,*

$$\prod_{j=-K}^{K} \frac{b_{z_j}(z)}{b_{p_j}(z)} \tag{3.3.38}$$

converges as $K \to \infty$ to a function, B_∞, which obeys

(i) B_∞ *is analytic on* $\mathbb{C} \setminus (\{z_j^{-1}\} \cup \{p_j\} \cup \{\pm 1\})$.

(ii) $|B_\infty(e^{i\theta})| = 1$ *for* $e^{i\theta} \in \partial\mathbb{D} \setminus \{\pm 1\}$.

(iii)

$$|\arg B_\infty| \le 2\pi \tag{3.3.39}$$

For a.e. θ, $\lim_{r\uparrow 1} f(re^{i\theta})$ *exists and is nonzero with*

$$\int \big|\log|f(e^{i\theta})|\big|^p \frac{d\theta}{2\pi} < \infty \tag{3.3.40}$$

for all $p \in [1, \infty)$.

Moreover, f *has a representation*

$$f(z) = \sigma\, B_\infty(z) \exp\biggl(\int \frac{e^{i\theta} + z}{e^{i\theta} - z} \log|f(e^{i\theta})| \frac{d\theta}{2\pi}\biggr) \tag{3.3.41}$$

with $\sigma = \pm 1$. *Explicitly,*

$$f(0) \neq 0, \infty \Rightarrow \sigma = \operatorname{sgn}(f(0)) \tag{3.3.42a}$$

$$f(0) = 0 \Rightarrow \sigma = +1 \tag{3.3.42b}$$

$$f(0) = \infty \Rightarrow \sigma = -1 \tag{3.3.42c}$$

Remarks. 1. One can improve (3.3.39) to $\le \pi$.

2. In line with the discussion after Theorem 2.3.19, we will call (3.3.41) the Poisson–Jensen formula for MH functions.

Proof. In the neighborhood of any finite-order zero or pole of a meromorphic function, f takes values with all possible arguments, so (3.3.37) implies that all the zeros and poles lie on $(-1, 1)$.

As one goes around a circle centered on $(-1, 1)$, which intersects $\mathbb{R}$ at points in $(-1, 1)$, which are neither zeros nor poles, $\arg f$ can change by at most π in each half-plane, so at most 2π over all. Thus, by the argument principle, each such circle has

$$|\#\text{ of poles inside} - \#\text{ of zeros inside}| \le 1$$

This counts multiplicity. So zeros and poles are simple and must interlace.

Thus, the intervals (z_j, p_j) are disjoint and

$$\sum_{j=-N}^{M} |z_j - p_j| < 2 \tag{3.3.43}$$

(typically, $N = M = \infty$). Clearly, $|z_j| \to 1$ and $N_\infty = 1$. Theorem 3.3.2 is thus applicable and implies (i)–(iii).

Define

$$g(z) = \frac{f(z)}{B_\infty(z)} \tag{3.3.44}$$

g is nonvanishing, so $\log(g(z))$ defined with $\mathrm{Im}[\log(g(0))] = 0$ or π is analytic in $\mathbb{D}$. By (3.3.39) and $\arg|f(z)| \le \pi$ on $\mathbb{D} \cap \mathbb{C}_+$, we see

$$|\mathrm{Im}\log(g(z))| \le 3\pi \tag{3.3.45}$$

so by M. Riesz's theorem (Proposition 2.3.8), $\log(g(z)) \in \bigcap_{p<\infty} H^p$, so $\log(g(z))$ has boundary values in L^p. Since $|B_\infty(e^{i\theta})| = 1$, $\log|f(e^{i\theta})| = \mathrm{Re}\log(g(e^{i\theta})) \in \bigcap_p L^p$, that is, (3.3.40) holds.

Since $\log(g) \in H^1$ and the Poisson representation (2.3.17) holds for H^1 functions (by (2.3.50)), we immediately get (3.3.41) where

$$\sigma = \frac{g(0)}{|g(0)|} \tag{3.3.46}$$

Since $b_{z_0}(0) > 0$ for $z_0 \neq 0$, we see

$$\sigma = \frac{f(0)}{|f(0)|}$$

if $f(0) \neq 0, \infty$, which is (3.3.42a). If $f(0) = 0$, $B_\infty(z)$ has a factor z, so $\sigma = \mathrm{sgn}\, f'(0) > 0$ since $\frac{\partial\, \mathrm{Im}\, f}{\partial\, \mathrm{Im}\, z} \ge 0$ and simple zeros implies $f'(x) > 0$ on $(-1, 1)$. This proves (3.3.42b). This also implies residues of poles are negative, which implies (3.3.42c). □

Remarks and Historical Notes. Theorem 3.3.6 was first proven by Simon [396] as a tool for proving OPRL sum rules. Our proof follows his, given Theorem 3.3.2. He proved that theorem only for alternating z_j and p_j. Following a suggestion of Killip, the presentation in [400] (see Proposition 13.8.2 and Theorem 13.8.3) emphasized that only $\sum_j |z_j - p_j| < \infty$ was needed for convergence.

The extension of Theorem 3.3.2 essentially to the form we have it was needed by Damanik–Killip–Simon [97] to get sum rules for matrix-valued OPRL (see Section 4.4). Our proof here, by using Lemmas 3.3.4 and 3.3.5, is somewhat simpler than theirs.

It is worth emphasizing that there is some magic going on here and explaining where the magic comes from. In the usual analysis of Nevanlinna functions, f (or if we allow poles, functions of bounded characteristic), one assumes some weak bounds on $|f(re^{i\theta})|$ as $r \uparrow 1$. These bounds imply information on the number of zeros (by Jensen's inequality—essentially one goes from bounds on $|f(re^{i\theta})|$ to some control of $\arg(f(re^{i\theta}))$) and this allows construction and control of a Blaschke product, B. One proves that f/B has the same kind of growth property.

Here, we make no a priori assumptions on $|f(z)|$ but instead on $\arg f$, which it turns out implies bounds on $|f(z)|$.

The magic in both the usual analysis for Nevanlinna functions and the one here is, in essence, M. Riesz's duality. The difference is that in the usual case, one goes from $\mathrm{Re}\log|f|$ to $\mathrm{Im}\log|f|$, and here we go in the opposite direction. There is also a difference in how the Blaschke products are controlled.

3.4 STEP-BY-STEP SUM RULES FOR OPRL

At this point, we are ready to turn the crank: find the right function, write down a Poisson–Jensen formula for it, and obtain step-by-step sum rules as Taylor coefficients in the Poisson–Jensen formula. The magic, of course, is in picking the right function—it will be the m-function! The analog of (2.6.17) is:

Theorem 3.4.1 (Nonlocal Step-by-Step Sum Rule). *Let J be an infinite Jacobi matrix with $\sigma_{\text{ess}}(J) \subset [-2, 2]$ and J_1 the once-stripped matrix. Let $\{E_n^\pm\}_{n=1}^{N_\pm}$ (resp. $\{E_n^{(1)\pm}\}_{n=1}^{N_{1,\pm}}$) be the eigenvalues of J (resp. J_1) with $\pm E_n^\pm > 2$ and $|E_{n+1}^\pm| < |E_n^\pm|$. Let M (resp. M_1) be given by (3.3.2) for the m-function of J (resp. J_1). Then M is an MH function. Its poles in $\mathbb{D}$ are at*

$$\{\beta^{-1} \mid \beta + \beta^{-1} = E_n^\pm,\ |\beta| > 1\} \equiv P \tag{3.4.1}$$

and its zeros in $\mathbb{D}$ are at $z = 0$ and

$$\{\beta^{-1} \mid \beta + \beta^{-1} = E_n^{(1)\pm},\ |\beta| > 1\} \equiv Z \tag{3.4.2}$$

Moreover, $M(re^{i\theta})$, $M_1(re^{i\theta})$ have limits as $r \uparrow 1$ for a.e. θ, and up to sets of measure zero,

$$\{\theta \mid \operatorname{Im} M(e^{i\theta}) \neq 0\} = \{\theta \mid \operatorname{Im} M_1(e^{i\theta}) \neq 0\} \tag{3.4.3}$$

and

$$\log\left(\frac{\operatorname{Im} M(e^{i\theta})}{\operatorname{Im} M_1(e^{i\theta})}\right) \in \bigcap_{p<\infty} L^p\left(\partial\mathbb{D}, \frac{d\theta}{2\pi}\right) \tag{3.4.4}$$

and $a_1 M$ has the representation

$$\frac{a_1 M(z)}{z} = B_\infty(z) \exp\left(\frac{1}{4\pi} \int_0^{2\pi} \frac{e^{i\theta} + z}{e^{i\theta} - z} \log\left(\frac{\operatorname{Im} M(e^{i\theta})}{\operatorname{Im} M_1(e^{i\theta})}\right) d\theta\right) \tag{3.4.5}$$

Remark. As in the discussion of Theorem 2.6.2, $\operatorname{Im} M(e^{i\theta})/\operatorname{Im} M_1(e^{i\theta})$ is correct only if the set in (3.4.3) is all of $\partial\mathbb{D}$ (as will be true in applications). The more precise version replaces $\operatorname{Im} M(e^{i\theta})/\operatorname{Im} M_1(e^{i\theta})$ by a function (namely, $a_1^2|M(e^{i\theta})|^2$), which is always defined and in $\cap_p L^p(\partial\mathbb{D}, \frac{d\theta}{2\pi})$ and which equals $\operatorname{Im} M/\operatorname{Im} M_1$ on the set in (3.4.3).

Proof. Since m is meromorphic on $\mathbb{C} \setminus [-2, 2]$ with $\operatorname{Im} m/\operatorname{Im} z > 0$ if $\operatorname{Im} z \neq 0$, M is an MH function. Poles of m are at $\{E_n^\pm\}$ so M has poles precisely in P. By (3.2.28), m has zeros precisely at points where m_1 has poles, and so M has zeros precisely on $Z \cup \{0\}$ ($M(z) = 0$ since $m(E) \to 0$ as $|E| \to \infty$).

Since M is an MH function, Theorem 3.3.6 implies that $M(re^{i\theta})$ has a limit $M(e^{i\theta})$ as $r \uparrow 1$, and since $\log|M| \in \cap_{p<\infty} L^p(\partial\mathbb{D}, \frac{d\theta}{2\pi})$,

$$M(e^{i\theta}) \neq 0, \infty \qquad \text{for a.e. } e^{i\theta} \in \partial\mathbb{D} \tag{3.4.6}$$

(3.2.28) implies

$$-M(z)^{-1} = b_1 - z - z^{-1} + a_1^2 M_1(z) \tag{3.4.7}$$

so taking imaginary parts and then taking $re^{i\theta}$ with $r \uparrow 1$ (noting that $\operatorname{Im}(re^{i\theta} + r^{-1}e^{-i\theta}) \to 0$), we find

$$\operatorname{Im} M(e^{i\theta})|M(e^{i\theta})|^{-2} = a_1^2 \operatorname{Im} M_1(e^{i\theta}) \tag{3.4.8}$$

By (3.4.6), we obtain (3.4.3), and so, on the set in that equation,

$$|a_1 M(e^{i\theta})|^2 = \frac{\operatorname{Im} M(e^{i\theta})}{\operatorname{Im} M_1(e^{i\theta})} \tag{3.4.9}$$

Finally, by the last equation, (3.4.5) is just the Poisson–Jensen representation (3.3.41) for $a_1 M(z)$ if we note that B_∞ does not include the zero at $z = 0$. □

To get useful information from (3.4.5), we need to look at Taylor coefficients of its log. Since

$$\frac{e^{i\theta} + z}{e^{i\theta} - z} = 1 + 2\sum_{n=1}^\infty e^{-in\theta} z^n \tag{3.4.10}$$

the Taylor coefficients for the integral are easy.

For the $B_\infty(z)$ term, we need

Lemma 3.4.2. *We have for* $z_0 \in (0,1) \cup (-1,0)$ *and* $|z| < |z_0|$,

$$\log b_{z_0}(z) = \log(|z_0|) + \sum_{n=1}^\infty \frac{z^n}{n}(z_0^n - z_0^{-n}) \tag{3.4.11}$$

so

$$\log B_\infty(z) = \sum_{n=0}^\infty c_n z^n \tag{3.4.12}$$

where

$$c_n = \begin{cases} \sum_{j=1}^N [\log|z_j| - \log|p_j|] & n = 0 \\ \sum_{j=1}^N \frac{(z_j^n - p_j^n) - (z_j^{-n} - p_j^{-n})}{n} & n \geq 1 \end{cases} \tag{3.4.13}$$

Remark. In (3.4.13), N is often infinite. If $N < \infty$, then there could be two more poles than zeros, in which case one may need to modify the notation. If $N = \infty$, the poles and zeros interlace on $(0,1)$ and $(-1,0)$ starting with poles closest to $z = 0$ in each interval. Since $z_j^n - z_j^{-n} \to 0$, the interlacing implies the sums converge conditionally.

Proof. We have that

$$\begin{aligned} \log b_{z_0}(z) &= \log(|z_0|) + \log\left(1 - \frac{z}{z_0}\right) - \log(1 - z_0 z) \\ &= \log(|z_0|) - \sum_{n=1}^\infty \frac{z^n(z_0^{-n} - z_0^n)}{n} \end{aligned} \tag{3.4.14}$$

proving (3.4.11). (3.4.12) and (3.4.13) follow immediately. □

To identify the Taylor coefficients of $\log M(z)/z$, we will need some manipulations with Chebyshev polynomials of the first kind given by (1.2.34).

Lemma 3.4.3. *Define for* $x \in (-1, 1)$,

$$g(x, z) = -\tfrac{1}{2} \log[1 - 2xz + z^2] \tag{3.4.15}$$

Then, for $|z| < 1$,

$$g(x, z) = \sum_{n=1}^{\infty} T_n(x) \frac{z^n}{n} \tag{3.4.16}$$

Proof. If $x = \cos\theta$, then

$$1 - 2xz + z^2 = (1 - e^{i\theta} z)(1 - e^{-i\theta} z) \tag{3.4.17}$$

so

$$\begin{aligned} g(\cos\theta, z) &= -\tfrac{1}{2}[\log(1 - e^{i\theta} z) + \log(1 - e^{-i\theta} z)] \\ &= \sum_{n=1}^{\infty} (e^{in\theta} + e^{-in\theta}) \frac{z^n}{2n} \end{aligned}$$

which is (3.4.16) given (1.2.34). □

Lemma 3.4.4. *Fix* $h \in \mathbb{C}$. *Then for* $|z|$ *sufficiently small,*

$$\log\left(1 - \frac{h}{z + z^{-1}}\right) = \sum_{n=1}^{\infty} \frac{2}{n} [T_n(0) - T_n(\tfrac{1}{2}h)] z^n \tag{3.4.18}$$

Proof. Since $|z + z^{-1}| \to \infty$ as $z \to 0$, the left side of (3.4.18) is analytic in z near $z = 0$ for any h and the Taylor coefficients are clearly real analytic in h. Thus, we need only prove (3.4.18) for h small and real.

In terms of the function g of (3.4.15),

$$\begin{aligned} \log\left(1 - \frac{h}{z + z^{-1}}\right) &= \log\left(1 - \frac{hz}{1 + z^2}\right) \\ &= \log\left(\frac{1 - hz + z^2}{1 + z^2}\right) = 2g(0, z) - 2g(\tfrac{1}{2}h, z) \end{aligned}$$

so (3.4.18) for $0 < h < 1$ follows from (3.4.16). □

Proposition 3.4.5. *Let* J *be viewed as acting on* $\ell^2(\{1, 2, \dots\})$ *and* J_1 *on* $\ell^2(\{2, 3, \dots\})$. *Then for any* n,

$$T_n(\tfrac{1}{2}J) - 0 \oplus T_n(\tfrac{1}{2}J_1) \tag{3.4.19}$$

(where 0 *acts on* $\mathbb{C} = \ell^2(\{1\})$*) is trace class and*

$$\log\left(\frac{M(z)}{z}\right) = \sum_{n=1}^{\infty} \frac{2}{n} [\mathrm{Tr}(T_n(\tfrac{1}{2}J) - 0 \oplus T_n(\tfrac{1}{2}J_1))] z^n \tag{3.4.20}$$

Proof. J is tridiagonal, so the matrix elements of $(J^m)_{k\ell}$ are zero if $|k - \ell| > m$ and only depend on matrix elements of J, b_n, a_n, with $|n - \frac{1}{2}(k + \ell)| \le m + 1$. It follows that $T_n(\frac{1}{2}J) - 0 \oplus T_n(\frac{1}{2}J_1)$ has zero matrix element except in a block of size at most $(n + 2) \times (n + 2)$. It is thus trace class.

As we have seen (see Theorem 3.2.9),

$$
\begin{aligned}
M(z) &= \lim_{m\to\infty} \langle \delta_1, (E(z) - J_{m;F})^{-1}\delta_1\rangle \\
&= \lim_{m\to\infty} \frac{\det_{m-1}(E(z) - (J_1)_{m-1;F})}{\det_m(E(z) - J_{m;F})} && (3.4.21)\\
&= \lim_{m\to\infty} E(z)^{-1} \frac{\det_m(E(z) - 0 \oplus (J_1)_{m-1;F})}{\det_m(E(z) - J_{m;F})} && (3.4.22)\\
&= \lim_{m\to\infty} E(z)^{-1} \frac{\det_m(1 - [0 \oplus (J_1)_{m-1;F}]/E(z))}{\det_m(1 - J_{m;F}/E(z))} && (3.4.23)
\end{aligned}
$$

where $E(z) = z + z^{-1}$. (3.4.21) is Cramer's rule, (3.4.22) uses $\det_m(E(z) - 0 \oplus B) = E(z)\det_{m-1}(E(z) - B)$ for any $(m - 1) \times (m - 1)$ matrix B. Once we have two $\det_m$'s, we can use $\det_m(E(z) - B) = E(z)^m \det_m(1 - B/E(z))$.

By (3.4.18), if B has eigenvalues $h_1, \dots, h_m$ and $|z|$ is small,

$$
\begin{aligned}
\log\left(\det_m\left(1 - \frac{B}{E(z)}\right)\right) &= \sum_{j=1}^{m} \log\left(1 - \frac{h_j}{E(z)}\right) \\
&= \sum_{n=1}^{\infty} \frac{2}{n} [mT_n(0) - \operatorname{Tr}(T_n(\tfrac{1}{2}B))]z^n && (3.4.24)
\end{aligned}
$$

Thus, (3.4.23) implies

$$
\log(M(z)E(z)) = \sum_{n=1}^{\infty} \frac{2}{n} \operatorname{Tr}[T_n(\tfrac{1}{2}J) - T_n(0 \oplus \tfrac{1}{2}J_1)] \tag{3.4.25}
$$

where we took $\lim_{m\to\infty}$ by first noting, since $T_n(\frac{1}{2}J)$ and $T_n(0 \oplus \frac{1}{2}J_1)$ agree except on an $(n + 2) \times (n + 2)$ block, that the coefficients for fixed n are m-independent for m large, and then noting that the convergence in (3.4.21) is uniform in z for $|z|$ small.

Next we note that $E(z) = (1 + z^2)/z$ and that by Lemma 3.4.3,

$$
\log(1 + z^2) = -\sum_{n=1}^{\infty} \frac{2}{n} T_n(0)z^n
$$

Thus,

$$
\begin{aligned}
\log\left(\frac{M(z)}{z}\right) &= \log(M(z)E(z)) - \log(1 + z^2) \\
&= \sum_{n=1}^{\infty} \frac{2}{n} \{T_n(0) + \operatorname{Tr}(T_n(\tfrac{1}{2}J) - T_n(0 \oplus \tfrac{1}{2}J_1))\}z^n \\
&= \sum_{n=1}^{\infty} \frac{2}{n} \operatorname{Tr}(T_n(\tfrac{1}{2}J) - 0 \oplus T_n(\tfrac{1}{2}J_1))z^n
\end{aligned}
$$

since $T_n(0 \oplus \frac{1}{2}J_1) = T_n(0) \oplus T_n(\frac{1}{2}J_1)$. □

Let us write out (3.4.20) explicitly for $n = 1, 2$. $T_1(x) = x$, so

$$\mathrm{Tr}(T_1(\tfrac{1}{2}J) - 0 \oplus T_1(\tfrac{1}{2}J_1)) = \tfrac{1}{2}b_1$$

Next, $T_2(x) = 2x^2 - 1$. Thus, since

$$\mathrm{Tr}(\mathbf{1} - 0 \oplus \mathbf{1}) = 1$$

and

$$\mathrm{Tr}((\tfrac{1}{2}J)^2 - 0 \oplus (\tfrac{1}{2}J_1)^2) = \tfrac{1}{4}(b_1^2 + 2a_1^2)$$

(for the sum of the squares of all matrix elements is involved), we see that

$$\mathrm{Tr}(T_2(\tfrac{1}{2}J) - 0 \oplus T_2(\tfrac{1}{2}J_1)) = \tfrac{1}{2}b_1^2 + (a_1^2 - 1)$$

Thus, (3.4.20) says

$$\log\left(\frac{M(z)}{z}\right) = b_1 z + (\tfrac{1}{2}b_1^2 + a_1^2 - 1)z^2 + O(z^3) \tag{3.4.26}$$

Theorem 3.4.6 (Step-by-Step Case Sum Rules). *We have:*
(C_0) *Define*

$$Z(J \mid J_1) = \frac{1}{4\pi}\int_0^{2\pi} \log\left(\frac{\mathrm{Im}\, M_1(e^{i\theta})}{\mathrm{Im}\, M(e^{i\theta})}\right) d\theta \tag{3.4.27}$$

Then

$$-\log(a_1) = Z(J \mid J_1) + \sum_{j,\pm} [\log(|p_j|) - \log(|z_j|)] \tag{3.4.28}$$

(C_n) *For $n \geq 1$, we have*

$$\frac{2}{n}\,[\mathrm{Tr}(T_n(\tfrac{1}{2}J) - 0 \oplus T_n(\tfrac{1}{2}J_1))] = S_n + \tilde{\mathcal{E}}_n \tag{3.4.29}$$

where

$$S_n = -\frac{1}{2\pi}\int_0^{2\pi} \log\left(\frac{\mathrm{Im}\, M_1(e^{i\theta})}{\mathrm{Im}\, M(e^{i\theta})}\right) \cos n\theta\, d\theta \tag{3.4.30}$$

$$\tilde{\mathcal{E}}_n = \sum_{j,\pm} \frac{(z_j^n - p_j^n) - (z_j^{-n} - p_j^{-n})}{n} \tag{3.4.31}$$

Proof. Given (3.4.10), (3.4.13), and (3.4.20), these are just the Taylor coefficients of the logs of the two sides of (3.4.5) (we have defined C_0 as the negative of the zeroth coefficient). Here we need to note that since

$$\frac{\mathrm{Im}\, M_1(e^{i\theta})}{\mathrm{Im}\, M(e^{i\theta})} = \frac{\mathrm{Im}\, M_1(e^{-i\theta})}{\mathrm{Im}\, M(e^{-i\theta})}$$

we can replace $e^{-in\theta}$ by $\frac{1}{2}(e^{-in\theta} + e^{in\theta}) = \cos n\theta$. □

We want to note explicitly the C_1 formulae using (3.4.26):

$$b_1 = S_1 + \tilde{\mathcal{E}}_1 \tag{3.4.32}$$

and the combination $C_0 + \frac{1}{2}C_2$ called the P_2 sum rule:

Corollary 3.4.7 (Step-by-Step P_2 Sum Rule). *Define*

$$Q(J \mid J_1) = \frac{1}{2\pi} \int_0^{2\pi} \log\left(\frac{\operatorname{Im} M_1(e^{i\theta})}{\operatorname{Im} M(e^{i\theta})} \right) \sin^2 \theta \, d\theta \tag{3.4.33}$$

$$F(E) = \tfrac{1}{4}[\beta^2 - \beta^{-2} - \log \beta^4] \tag{3.4.34}$$

where $|\beta| > 1$ *and* $E = \beta + \beta^{-1}$,

$$G(a) = a^2 - 1 - \log(a^2) \tag{3.4.35}$$

Then

$$\tfrac{1}{4}b_1^2 + \tfrac{1}{2}G(a_1) = Q(J \mid J_1) + \sum_{j,\pm} [F(E_j^{\pm}(J)) - F(E_j^{\pm}(J_1))] \tag{3.4.36}$$

The P in P_2 is for "positive" and comes from the fact that the left side of (3.4.36) is positive and the right side is a difference of a positive term for J and a positive term for J_1. We will discuss this further in the next section.

Remarks and Historical Notes. The nonlocal step-by-step sum rule is from Simon [396], but he was motivated by the earlier step-by-step sum rule in Killip–Simon [225] and followup in Simon–Zlatoš [410].

The Case sum rules are named after Case [76, 77]. He did not have them in step-by-step form nor was he careful about conditions for them to hold, but he had the idea of looking at Taylor coefficients of a Poisson–Jensen representation of the Jost function, which is an iterated M-function; see (3.7.27). He did not have a general formula for the functions of the Jacobi parameters, but knew they were polynomials in the a's and b's and found the first few. The formula in terms of Chebyshev polynomials is due to Killip–Simon [225].

The positivity of P_2 is a discovery of Killip–Simon [225].

3.5 THE P_2 SUM RULE AND THE KILLIP–SIMON THEOREM

Our goal in this section is to prove (3.1.5) and use that to prove Theorem 3.1.1. As a preliminary, we need to study the functions Q, F, G of (3.1.5).

Lemma 3.5.1. *Let* ρ *have the form* (1.4.3) *and let* M *be its* M*-function, given by* (3.3.2). *Let* $Q(\rho)$ *be given by* (1.10.16). *Then*

$$Q(\rho) = \frac{1}{2\pi} \int_0^{2\pi} \log\left(\frac{\sin \theta}{\operatorname{Im} M(e^{i\theta})} \right) \sin^2 \theta \, d\theta \tag{3.5.1}$$

$$= -\tfrac{1}{2} S(\mu_0 \mid \mu) \tag{3.5.2}$$

where $d\mu = \mathrm{Sz}^{-1}(d\rho)$ *is given by* (1.9.5) *and*

$$d\mu_0 = \frac{1}{\pi} \sin^2 \theta \, d\theta \tag{3.5.3}$$

In particular,

$$Q(\rho) \geq 0 \tag{3.5.4}$$

and

$$d\rho_n \xrightarrow{w} d\rho \Rightarrow \liminf Q(\rho_n) \geq Q(\rho) \tag{3.5.5}$$

Remark. If $d\rho_0$ is given by (1.10.3), then $d\mu_0 = \text{Sz}(d\rho_0)$ and (3.5.2) is (1.10.17).

Proof. By (3.3.2), for $\theta \in [0, \pi]$, we have that

$$\operatorname{Im} M(e^{i\theta}) = \operatorname{Im} m(2\cos\theta) = \pi f(2\cos\theta) \tag{3.5.6}$$

by (2.3.56). If we first use $\theta \to -\theta$ symmetry to write the integral in (3.5.1) as $(2\pi)^{-1}\int_0^\pi$ and then make the change of variables $x = 2\cos\theta$ and use $\sqrt{4-x^2} = 2\sin\theta$, we obtain (3.5.1).

(3.5.2) is just the definition of entropy. (3.5.4) is then just (2.2.15) and (3.5.5) is Theorem 2.2.3. □

Lemma 3.5.2. *Let G be given by* (1.10.10). *Then*

$$G(a) > 0 \qquad \text{on } (0, \infty) \setminus \{1\} \tag{3.5.7}$$

and near $a = 1$,

$$G(a) = 2(a-1)^2 + O((a-1)^3) \tag{3.5.8}$$

Proof. We compute

$$G'(a) = 2(a - a^{-1}) \qquad G''(a) = 2 + 2a^{-2} \tag{3.5.9}$$

so $G(1) = G'(1) = 0$, $G''(a) \geq 0$, and $G''(1) = 4$. Since G is strictly convex, its minimum is at $a = 1$, proving (3.5.7). (3.5.8) is just Taylor's theorem at $a = 1$. □

Lemma 3.5.3. *Let F be given by* (1.10.9). *Then*

$$F(E) = \tfrac{1}{2}\int_2^{|E|} (E^2 - 4)^{1/2}\, dE \tag{3.5.10}$$

We have that

$$F(E) > 0 \qquad \text{on } \mathbb{R} \setminus [-2, 2] \tag{3.5.11}$$

and for $|E|$ *near* 2 *and in* $(2, \infty)$,

$$F(E) = \tfrac{2}{3}(|E| - 2)^{3/2} + O((|E| - 2)^{5/2}) \tag{3.5.12}$$

Proof. Differentiating (1.10.9) with respect to β yields

$$(1 - \beta^{-2})F'(\beta + \beta^{-1}) = \tfrac{1}{2}(\beta + \beta^{-3} - 2\beta^{-1})$$

so

$$F'(\beta + \beta^{-1}) = \frac{1}{2}\frac{(\beta - \beta^{-1})^2}{(\beta - \beta^{-1})} = \tfrac{1}{2}(\beta - \beta^{-1}) \tag{3.5.13}$$

If $E = \beta + \beta^{-1}$, then $(E^2 - 4)^{1/2} = |\beta| - |\beta|^{-1}$, so (3.5.13) says that if $E > 2$, then

$$F'(E) = \tfrac{1}{2}(E^2 - 4)^{1/2} \tag{3.5.14}$$

From (1.10.9), $\lim_{E \downarrow 2} F(E) = 0$, so (3.5.14) implies (3.5.10). This in turn implies (3.5.11) and (3.5.12) if we note that with $y = E - 2$,

$$(E^2 - 4)^{1/2} = (y(y+4))^{1/2} = 2y^{1/2} + O(y^{3/2}) \tag{3.5.15}$$

□

Proposition 3.5.4 (P_2 Sum Rule for Finite Rank $J - J_0$). *Suppose that for some N, $a_n = 1$, $b_n = 0$ for all $n \geq N$. Then the number of E_n outside $[-2, 2]$ is finite and (3.1.5) holds with each term finite.*

Proof. Define $d\rho_m$ to be the m-times stripped measure, that is, the measure with Jacobi parameters $\{a_{n+m}, b_{n+m}\}_{n=1}^{\infty}$ and J_m its Jacobi matrix. By iterating (3.4.36), we find $Q(J \mid J_m) < \infty$ and

$$Q(J \mid J_m) + \sum_{j,\pm} [F(E_j^{\pm}(J)) - F(E_j^{\pm}(J_m))] = \sum_{n=1}^{m} [\tfrac{1}{4}b_n^2 + \tfrac{1}{2}G(a_n)] \tag{3.5.16}$$

By hypothesis, $J - J_0$ is finite rank. Thus, by the min-max principle (see Subsection 1.4.9 of [399]), each of $(-\infty, -2)$ and $(2, \infty)$ has at most N eigenvalues, so $\sum F(E_j^{\pm}(J))$ is finite. (3.5.16) for $m = N$ is (3.1.5) since then $Q(J \mid J_m) = Q(\rho)$ (by Lemma 3.5.1) and there are no $E_j^{\pm}(J_N)$. □

Theorem 3.5.5 (P_2 Sum Rule). *Let J be a Jacobi matrix with $\sigma_{\text{ess}}(J) = [-2, 2]$. Then, with ρ its spectral measure,*

$$Q(\rho) + \sum_{E \notin \sigma_{\text{ess}}(J)} F(E) = \sum_{n=1}^{\infty} [\tfrac{1}{4}b_n^2 + \tfrac{1}{2}G(a_n)] \tag{3.5.17}$$

Each term is positive, including $+\infty$, and (3.5.17) *holds in the sense that either both sides are infinite or both are finite and equality holds.*

Proof. Define $J^{(m)}$ by

$$a_k^{(m)} = \begin{cases} a_k & k \leq m - 1 \\ 1 & k \geq m \end{cases} \tag{3.5.18}$$

$$b_k^{(m)} = \begin{cases} b_k & k \leq m \\ 0 & k \geq m + 1 \end{cases} \tag{3.5.19}$$

Then $J^{(n)} \xrightarrow{s} J$. It follows by the min-max principle that for ℓ fixed,

$$\limsup_{n\to\infty} \mp E_\ell^{\pm}(J^{(n)}) \leq \mp E_\ell^{\pm}(J) \tag{3.5.20}$$

Since $F \geq 0$, it follows that for any L,

$$\sum_{\ell=1,\pm}^{L} F(E_\ell^\pm(J)) = \liminf_{n\to\infty} \sum_{\ell=1,\pm 1}^{L} F(E_\ell^\pm(J^{(n)}))$$

$$\leq \liminf_{n\to\infty} \sum_{\ell=1,\pm 1}^{\infty} F(E_\ell^\pm(J^{(n)})) \tag{3.5.21}$$

Since the right side of (3.5.21) is L-independent, we can take $L \to \infty$. Moreover, since the spectral measure $\rho^{(n)}$ for $J^{(n)}$ converges weakly to that for J (i.e., $\int x^\ell \, d\rho^{(n)} = \langle \delta_1, (J^{(n)})^\ell \delta_1 \rangle \to \langle \delta_1, (J)^\ell \delta_1 \rangle = \int x^\ell \, d\rho$), (3.5.5) says

$$Q(\rho) \leq \liminf Q(\rho^{(n)})$$

By Proposition 3.5.4,

$$Q(\rho) + \sum_{\ell=1,\pm}^{\infty} F(E_\ell^\pm(J)) \leq \liminf \left[Q(\rho^{(n)}) + \sum_{\ell=1,\pm}^{\infty} F(E_\ell^\pm(J^{(n)})) \right]$$

$$\leq \liminf \left(\sum_{j=1}^{n} \tfrac{1}{4} b_j^2 + \sum_{j=1}^{n} \tfrac{1}{2} G(a_j) \right)$$

$$= \sum_{j=1}^{\infty} \left[\tfrac{1}{4} b_j^2 + \tfrac{1}{2} G(a_j) \right] \tag{3.5.22}$$

by the positivity of b_j^2 and $G(a_j)$.

Thus, we need only prove that

$$\sum_{j=1}^{\infty} \left[\tfrac{1}{4} b_j^2 + \tfrac{1}{2} G(a_j) \right] \leq Q(\rho) + \sum_{\ell,\pm}^{\infty} F(E_\ell^\pm(J)) \tag{3.5.23}$$

If the right side of (3.5.23) is ∞, there is nothing to prove, so suppose it is finite. Then $Q(\rho) < \infty$ and $Q(J \mid J_1)$ finite (which is always true) proves $Q(\rho_1) < \infty$ and

$$Q(J \mid J_1) = Q(\rho) - Q(\rho_1)$$

Similarly, $\sum_{\ell,\pm} F(E_\ell^\pm(J)) < \infty$ and interlacing proves that the sum for the right of (3.4.36) is a difference of separate J and J_1 sums. Thus,

$$Q(\rho) + \sum_{\ell,\pm 1} F(E_\ell^\pm(J)) = \tfrac{1}{4} b_1^2 + \tfrac{1}{2} G(a_1) + Q(\rho_1) + \sum_{\ell,\pm 1} F(E^\pm(J_1)) \tag{3.5.24}$$

Iterating this n times and noting $Q \geq 0$, $F \geq 0$, we get

$$Q(\rho) + \sum_{\ell,\pm} F(E_\ell^\pm(J)) \geq \sum_{j=1}^{n} \left[\tfrac{1}{4} b_j^2 + \tfrac{1}{2} G(a_j) \right]$$

Taking $n \to \infty$ yields (3.5.23). □

Proof of Theorem 3.1.1. If (3.1.1) holds, then $J - J_0$ is compact, so (3.1.2) holds by Weyl's theorem on invariance of the essential spectrum under compact perturbations (see Subsection 1.4.15 of [399]). By (3.1.5), (3.5.8), and (3.5.23),

$$Q(\rho) < \infty \Rightarrow (3.1.4)$$

and

$$\sum_{j,\pm} F(E_j^\pm(J)) < \infty \Rightarrow (3.1.3)$$

by (3.5.12).

Suppose (i)–(iii) hold. By (i), we have the sum rule (3.5.17). By (ii), (iii), and (3.5.12), the left-hand side of (3.5.17) is finite, so

$$\sum_{n=1}^{\infty} \left[\tfrac{1}{4}b_n^2 + \tfrac{1}{2}G(a_n)\right] < \infty \tag{3.5.25}$$

Thus, $G(a_n) \to 0$, $a_n \to 1$, and $\sum_{n=1}^{\infty} b_n^2 + (a_n - 1)^2 < \infty$ by (3.5.8). □

Remarks and Historical Notes. See the Notes to Section 3.1 for the history.

Just as for OPUC where, once one goes to slower than ℓ^2 decay, there might be no a.c. spectrum, one can use any of the methods described in the Notes to Section 2.1. In particular, one has the following theorem, which we will need later (see Section 10.2):

Theorem 3.5.6. *There exist Jacobi parameters so that each of the matrices J_m with parameters $\{a_{n+m}, b_{n+m}\}_{n=1}^{\infty}$ has only dense pure point spectrum in $[-2, 2]$ (and, in particular, no a.c. spectrum) and so that $a_n \equiv 1$ and*

$$|b_n| \le Cn^{-1/2} \tag{3.5.26}$$

For example, one can do this with decaying random potentials; see [227].

3.6 AN EXTENDED SHOHAT–NEVAI THEOREM

While it is missing positivity, the C_0 sum rule is useful and can be used to prove the following:

Theorem 3.6.1 (Extended Shohat–Nevai Theorem). *Let*

$$d\rho(x) = f(x)\,dx + d\rho_s(x)$$

with $\sigma_{\text{ess}}(J) = [-2, 2]$. Suppose that $\{E_n^\pm\}_{n=1}^{N_\pm}$ are the pure points of $d\rho$ in $\pm(2, \infty)$ and that

$$\sum_{n,\pm} (|E_n^\pm| - 2)^{1/2} < \infty \tag{3.6.1}$$

Then

$$\int_{-2}^{2} (4 - x^2)^{-1/2} \log f(x)\,dx > -\infty \tag{3.6.2}$$

if and only if

$$\limsup a_1 \dots a_n > 0 \tag{3.6.3}$$

If these conditions hold, then

$$\lim a_1 \dots a_n \tag{3.6.4}$$

exists in $(0, \infty)$ *and*

$$\sum_{n=1}^{\infty} (a_n - 1)^2 + b_n^2 < \infty \tag{3.6.5}$$

Moreover,

$$\sum_{n=1}^{N} (a_n - 1) \qquad \textit{and} \qquad \sum_{n=1}^{N} b_n \tag{3.6.6}$$

have limits in $(-\infty, \infty)$. *If* (3.6.3) *fails, the limit in* (3.6.4) *exists and is* 0.

As a first preliminary, we need

Lemma 3.6.2. *Define*

$$Z(\rho) = \frac{1}{4\pi} \int_0^{2\pi} \log\biggl(\frac{\sin\theta}{\operatorname{Im} M(e^{i\theta})}\biggr) d\theta \tag{3.6.7}$$

Then

$$Z(\rho) = -\tfrac{1}{2} S\biggl(\frac{d\theta}{2\pi} \biggm| d\mu\biggr) - \tfrac{1}{2}\log 2 \tag{3.6.8}$$

where $d\mu = \mathrm{Sz}^{-1}(d\rho)$.

Proof. Suppose $d\mu$ has the form (1.9.37). Then

$$-\tfrac{1}{2} S\biggl(\frac{d\theta}{2\pi} \biggm| d\mu\biggr) = \frac{1}{4\pi} \int_0^{2\pi} \log\biggl(\frac{1}{w(\theta)}\biggr) d\theta \tag{3.6.9}$$

On the other hand, by (1.9.38) and (3.5.6)

$$\frac{1}{w(\theta)} = \frac{\sin\theta}{2\sin^2\theta \operatorname{Im} M(e^{i\theta})} \tag{3.6.10}$$

and so, (3.6.8) is implied by

$$\frac{1}{4\pi} \int \log(2\sin^2\theta)\, d\theta = -\tfrac{1}{2}\log 2 \tag{3.6.11}$$

Let $f(z) = \log(|1 - z^2|)$, which is harmonic in $\mathbb{D}$ and continuous in the closure. Thus,

$$\begin{aligned} 0 = f(0) &= \frac{1}{2\pi} \int \log(|1 - e^{2i\theta}|)\, d\theta \\ &= \frac{1}{2\pi} \int \log(2|\sin\theta|)\, d\theta \\ &= \frac{1}{4\pi} \int \log(4\sin^2\theta)\, d\theta \end{aligned}$$

which is (3.6.11). □

Next we need approximation results for the eigenvalue sums:

Lemma 3.6.3. *Define*

$$\mathcal{E}_0(J) = \sum_{j,\pm} \log|\beta_j^\pm(J)| \tag{3.6.12}$$

which may be $+\infty$. *Define* $J^{(n)}$ *by* (3.5.18)/(3.5.19). *Then,*

(a) *For any* n,

$$\mathcal{E}_0(J^{(n)}) \le \mathcal{E}_0(J) + 2\sup_m |b_m| + 4\sup_m \{|a_m|, 1\} \tag{3.6.13}$$

(b) *If* $a_n \to 1$, $b_n \to 0$, *and* $\mathcal{E}_0(J) < \infty$, *then*

$$\lim_{n\to\infty} \mathcal{E}_0(J^{(n)}) = \mathcal{E}_0(J) \tag{3.6.14}$$

Proof. Let $\tilde{J}^{(n)}$ be $J^{(n)}$ with a_n replaced by 0. Then $\tilde{J}^{(n)}$ is a direct sum of $J_{n;F}$ and J_0. Since $J_{n;F}$ is a restriction of J, we have

$$\pm E_j^\pm(\tilde{J}^{(n)}) = \pm E_j^\pm(J_{n;F}) \le \pm E_j^\pm(J) \tag{3.6.15}$$

by the min-max principle (see Subsection 1.4.9 of [399]).

Since

$$\begin{pmatrix} 0 & a_n \\ a_n & 0 \end{pmatrix} = \tfrac{1}{2}a_n \left[\begin{pmatrix} 1 & 1 \\ 1 & 1 \end{pmatrix} - \begin{pmatrix} 1 & -1 \\ -1 & 1 \end{pmatrix} \right] \tag{3.6.16}$$

$J^{(n)} - \tilde{J}^{(n)}$ is the sum of a positive rank one and a negative rank one perturbation. For E_j^-, the positive term can only move eigenvalues up, while the negative one interlaces. We have the opposite for E_j^+. Thus,

$$\pm E_{j+1}^\pm(J^{(n)}) \le \pm E_j^\pm(J) \tag{3.6.17}$$

We get (3.6.13) from this if we note

$$\pm E_1(J^{(n)}) \le \|J^{(n)}\| \le \sup_m |b_m| + 2\sup_m\{|a_m|, 1\} \tag{3.6.18}$$

If $a_n \to 1$ and $b_n \to 0$, then $\|J^{(n)} - J\| \to 0$, which implies that for each fixed ℓ, $E_\ell^\pm(J^{(n)}) \to E_\ell^\pm(J)$. If $\mathcal{E}_0(J) < \infty$, we can use dominated convergence to get (3.6.14). □

Remark. Since $E_\ell^\pm(J^{(n)}) \to E_\ell^\pm(J)$ if $\|J^{(n)} - J\| \to 0$, (3.6.14) holds even if $\mathcal{E}_0(J)$ is infinite.

Lemma 3.6.4. *Define* J_m *to be the* m*th stripped Jacobi matrix, that is, the one with Jacobi parameters* $\{a_{n+m}, b_{n+m}\}_{n=1}^\infty$. *Then, with* $\mathcal{E}_0$ *given by* (3.6.12),

$$\mathcal{E}_0(J_m) \le \mathcal{E}_0(J) \tag{3.6.19}$$

and if $a_n \to 1$, $b_n \to 0$, *and* $\mathcal{E}_0(J) < \infty$, *then*

$$\lim_{m\to\infty} \mathcal{E}_0(J_m) = 0 \tag{3.6.20}$$

Proof. J_m is the restriction of J to $\ell^2(\{\delta_\ell\}_{\ell=m+1}^\infty)$, so by the min-max principle,

$$\pm E_j^\pm(J_m) \le \pm E_j^\pm(J) \tag{3.6.21}$$

so (3.6.19) follows by monotonicity of $\log|\beta(E)|$ in E.

If $a_n \to 1$, $b_n \to 0$, then $\|J_n\| \to 2$, so since $J_n - J_0$ is compact, $E_j^\pm(J_m) \to \pm 2$ for each j. By (3.6.21) and dominated convergence of $\mathcal{E}_0(J) < \infty$, we have (3.6.20). □

We are now ready to prove the relevant sum rule as two halves:

Proposition 3.6.5. *If $\mathcal{E}_0(J) < \infty$, then*

$$Z(\rho) \le \liminf_{n\to\infty} \left(-\sum_{j=1}^n \log(a_j) \right) + \mathcal{E}_0(J) + 2\sup_m |b_m| + 4\sup_m \{|a_n|, 1\} \tag{3.6.22}$$

and if $a_n \to 1$, $b_n \to 0$, then

$$Z(\rho) \le \liminf_{n\to\infty} \left(-\sum_{j=1}^n \log(a_j) \right) + \mathcal{E}_0(J) \tag{3.6.23}$$

Proof. Let $J^{(n)}$ be given by (3.5.18)/(3.5.19) and let $\rho^{(n)}$ be the corresponding measure. By (3.4.30), iterated $n+1$ times (so $(J^{(n)})_n = J_0$),

$$Z(\rho^{(n)}) \le \mathcal{E}(J^{(n)}) + \left(-\sum_{j=1}^{n-1} \log(a_j) \right) \tag{3.6.24}$$

We get (3.6.22) by using (3.6.13) and taking $n \to \infty$ along a sequence that takes $-\sum_{j=1}^{n-1} \log(a_j)$ to its lim inf. By (3.6.8), Z is lower semicontinuous, so we get (3.6.22).

For (3.6.23), we use (3.6.14) instead of (3.6.13). □

Proposition 3.6.6. *If $\sigma_{\text{ess}}(J) = [-2, 2]$, $Z(\rho) < \infty$, and $\mathcal{E}_0(J) < \infty$, then*

$$\limsup_{n\to\infty} \left(-\sum_{j=1}^n \log(a_j) \right) \le Z(\rho) - \mathcal{E}_0(J) \tag{3.6.25}$$

Proof. Since $Z(\rho) < \infty \Rightarrow Q(\rho) < \infty$ and $\mathcal{E}_0 < \infty \Rightarrow \sum_{j,\pm 1} F(E_j^\pm(J)) < \infty$, Theorem 3.1.1 implies

$$\sum_{n=1}^\infty (a_n - 1)^2 + b_n^2 < \infty \tag{3.6.26}$$

so, in particular,

$$a_n \to 1 \qquad b_n \to 0 \tag{3.6.27}$$

By (3.4.30), since $Z(\rho \mid \rho_1) < \infty$ and the $\mathcal{E}_0(J_1) \le \mathcal{E}_0(J)$ (see (3.6.19)),

$$-\log(a_1) = Z(\rho) - Z(\rho_1) - \mathcal{E}_0(J) + \mathcal{E}_0(J_1) \tag{3.6.28}$$

so iterating,

$$\sum_{j=1}^{n} -\log(a_j) = Z(\rho) - Z(\rho_n) - \mathcal{E}_0(J) + \mathcal{E}_0(J_n) \tag{3.6.29}$$

By (3.6.27) and (3.6.20), $\mathcal{E}_0(J_n) \to 0$. Moreover, since $\rho_n \xrightarrow{w} \rho_{J_0}$, the measure for J_0 and $Z(\rho_{J_0}) = 0$, we have

$$\liminf Z(\rho_n) \geq Z(\rho_{J_0}) = 0 \tag{3.6.30}$$

so (3.6.29) implies (3.6.25). □

Proof of Theorem 3.6.1. If (3.6.3) holds, then

$$\liminf \left(-\sum_1^n \log(a_j) \right) < \infty \tag{3.6.31}$$

Since (3.6.1) $\Rightarrow \mathcal{E}_0(J) < \infty$, $Z(\rho) < \infty$ by (3.6.22). But then, as in the last proof, we obtain (3.6.26) and so (3.6.27), and thus, (3.6.23) holds.

On the other hand, if $Z(\rho) < \infty$ and (3.6.11) holds, then by (3.6.25),

$$\limsup \left(-\sum_1^n \log(a_j) \right) < \infty \tag{3.6.32}$$

A fortiori, (3.6.31) holds, so (3.6.23) holds. Thus,

$$\limsup \left(-\sum_{j=1}^n \log(a_j) \right) \leq Z(\rho) - \mathcal{E}_0(J) \leq \liminf \left(-\sum_{j=1}^n \log(a_j) \right)$$

It follows that the limit exists and

$$\lim \left(-\sum_{j=1}^n \log(a_j) \right) = Z(\rho) - \mathcal{E}_0(J) \tag{3.6.33}$$

This proves (3.6.4), and (3.6.5) follows from Theorem 3.1.1.

(3.6.33) for J_n and (3.6.20) let us strengthen (3.6.30) to

$$\lim_{n\to\infty} Z(\rho_n) = 0 \tag{3.6.34}$$

Finally, we turn to the conditional convergence of $\sum_1^n b_j$. By iterating the step-by-step C_1 Case sum rule, (3.4.32), we get

$$\sum_1^n b_j = \mathcal{T}(J) - \mathcal{T}(J_n) + \mathcal{E}_1(J) - \mathcal{E}_1(J_n) \tag{3.6.35}$$

where

$$\mathcal{T}(J) = -\frac{1}{2\pi} \int \log\left(\frac{\sin\theta}{\operatorname{Im} M(e^{i\theta})} \right) \cos\theta \, d\theta \tag{3.6.36}$$

and

$$\mathcal{E}_1(J) = \sum_{j,\pm 1} [\beta_j^{\pm} - (\beta_j^{\pm})^{-1}] \tag{3.6.37}$$

Because $Z(\rho) < \infty$ and $\mathcal{E}_1(J)$ is convergent, we can separate out the terms in the step-by-step sum rule. Clearly, to prove $\lim \sum_1^n b_j$ exists, it is sufficient to prove that

$$\lim_{n\to\infty} \mathcal{T}(J_n) = 0 \tag{3.6.38}$$

$$\lim_{n\to\infty} \mathcal{E}_1(J_n) = 0 \tag{3.6.39}$$

The second result has a proof identical to (3.6.20). For the first, we define

$$\mathcal{T}^{\pm}(J) = \frac{1}{2\pi} \int_0^{2\pi} \log\left(\frac{\sin\theta}{\operatorname{Im} M(e^{i\theta})} \right) (1 \pm \cos\theta)\, d\theta \tag{3.6.40}$$

$$= 2Z(J) \pm \mathcal{T}(J) \tag{3.6.41}$$

As in the proof of (3.6.8), one sees that

$$\mathcal{T}^{\pm}(J) = -S\left((1 \pm \cos\theta) \frac{d\theta}{2\pi} \,\Big|\, d\mu \right) + c \tag{3.6.42}$$

for a constant c, so $\mathcal{T}^{\pm}$ is lower semicontinuous. Since $d\mu_{J_n} \to d\mu_{J_0}$ and $\mathcal{T}^{\pm}(J_0) = 0$, we see that

$$\liminf \mathcal{T}^{\pm}(J) \geq 0 \tag{3.6.43}$$

Remark. (3.6.33) is the C_0 sum rule.

(3.6.34), (3.6.41), and (3.6.43) imply (3.6.38). □

For one application, we need the following, which we state without proof since the application is peripheral (but see the Notes):

Theorem 3.6.7 (Hundertmark–Simon [201]). *For any J with $\sigma_{\text{ess}}(J) = [-2, 2]$,*

$$\sum_{j,\pm} ((E_j^{\pm})^2 - 4)^{1/2} \leq \sum_{n=1}^{\infty} (|b_n| + 4|a_n - 1|) \tag{3.6.44}$$

This implies that

Theorem 3.6.8. *If*

$$\sum_{n=1}^{\infty} |b_n| + |a_n - 1| < \infty \tag{3.6.45}$$

then the Szegő condition $Z(J) < \infty$ holds.

Remark. (3.6.45) is equivalent to saying $J - J_0$ is trace class.

Proof. Clearly, $\sum_{n=1}^{\infty} |a_n - 1| < \infty$ implies $\prod_{j=1}^{n} a_j$ has a nonzero limit. By (3.6.44), $\sum (|E_j^{\pm}| - 2)^{1/2} < \infty$, so Theorem 3.6.1 implies $Z(\rho_J) < \infty$. □

Remarks and Historical Notes. That $Z(\rho)$ and $\mathcal{E}_0(J) < \infty$ implies that $\prod_{j=1}^{J} a_j$ has a limit is a result of Peherstorfer–Yuditskii [343]. This result was rediscovered and the converse proven by Killip–Simon [225] as part of their analysis of sum rules. Our proof here follows [225] with some important refinements of Simon–Zlatoš [410].

Theorem 3.6.7 is due to Hundertmark–Simon [201] (see Section 13.8 of [400] for a historical discussion and an exposition of their proof). Theorem 3.6.8 is from Killip–Simon [225]. It settled a conjecture of Nevai [322].

There have been a variety of papers that attempt to find higher-order sum rules and associated gems: [111, 258, 259, 260, 268, 314, 465].

3.7 SZEGŐ ASYMPTOTICS FOR OPRL

Szegő [430] related his OPUC theorem to asymptotics of OPUC (and later to asymptotics of OPRL, as we will discuss in the Notes, but we want to go beyond that here). So it is natural to ask about the relation of OPRL asymptotics to the ideas of this chapter—and that is what we will do in this section.

To see what we are seeking, consider the case J_0 (i.e., $a_n \equiv 1$, $b_n \equiv 0$) where we have seen the OPRL are given by Chebyshev polynomials of the second kind, (1.10.2), which can be conveniently rewritten using the map $z + z^{-1}$ from $\mathbb{D}$ to $\mathbb{C} \setminus [-2, 2]$:

$$p_{n,0}\left(z + \frac{1}{z}\right) = \frac{z^{n+1} - z^{-(n+1)}}{z - z^{-1}} \tag{3.7.1}$$

so for all $z \in \mathbb{D} \setminus \{0\}$, since z^{-1} dominates z,

$$z^n p_{n,0}\left(z + \frac{1}{z}\right) \to 1 \tag{3.7.2}$$

This leads to

Definition. We say orthogonal polynomials, $\{p_n(x)\}_{n=0}^{\infty}$, have *Szegő asymptotics* at $z_0 \in \mathbb{D} \setminus \{0\}$ if and only if

$$\lim_{n\to\infty} z_0^n p_n\left(z_0 + \frac{1}{z_0}\right) \tag{3.7.3}$$

exists and is nonzero. We define the limit to be $(\sqrt{2}D(z))^{-1}$ (the reason for this choice of symbol and normalization will be explained in the Notes).

Our main theorem, proven below, is

Theorem 3.7.1 (Damanik–Simon [99]). *Suppose the Jacobi parameters, $\{a_n, b_n\}_{n=1}^{\infty}$, of a measure, $d\rho$, obey*

$$\text{(a)} \qquad \lim_{n\to\infty} \prod_{j=1}^{n} a_j \quad \textit{exists in } (0, \infty) \tag{3.7.4}$$

$$\text{(b)} \qquad \lim_{n\to\infty} \sum_{j=1}^{n} b_j \quad \textit{exists in } \mathbb{R} \tag{3.7.5}$$

$$\text{(c)} \qquad \sum_{j=1}^{\infty} (a_j - 1)^2 + b_j^2 < \infty \tag{3.7.6}$$

Then the limit in (3.7.3) *exists for all z in $\mathbb{D}\setminus\{0\}$ and is nonzero* (*precisely*) *for all z so that $z+z^{-1}\notin\sigma(J)$. The convergence is uniform on compact subsets of $\mathbb{D}$. Conversely, if there is $\varepsilon>0$ so that* (3.7.3) *holds uniformly for $|z|=r$ for all $r\in(0,\varepsilon)$, then* (a)–(c) *hold.*

Remarks. 1. Since (a)–(c) imply $\sigma_{\text{ess}}(J)=[-2,2]$, $\{z\mid z+z^{-1}\in\sigma(J)\}$ is a discrete set in $\mathbb{D}\cap\mathbb{R}$.

2. One can also ask about suitable L^2 convergence on $\partial\mathbb{D}$. This is true and proven in [99] (but not in the form (3.7.3); see the Notes), but the proof is much more involved than the proof in this section. For the stronger hypotheses of Theorem 3.6.1, we discuss L^2 convergence on $\partial\mathbb{D}$ at the end of this section.

The following predated [99]:

Corollary 3.7.2 (Peherstorfer–Yuditskii [343]). *Let*

$$d\rho(x)=f(x)\,dx+d\rho_{\text{s}}(x)$$

with $\sigma_{\text{ess}}(J)=[-2,2]$. Suppose that $\{E_n^\pm\}_{n=1}^{N_\pm}$ are the pure points of $d\rho$ in $\pm(2,\infty)$ and that

$$\sum_{n,\pm}(|E_n|-2)^{1/2}<\infty \tag{3.7.7}$$

and

$$\int_{-2}^{2}(4-x^2)^{-1/2}\log f(x)\,dx>-\infty \tag{3.7.8}$$

Then the limit (3.7.3) *exists for all $z\in\mathbb{D}\setminus\{0\}$ uniformly on compacts and the limit is nonzero* (*precisely*) *for z so that $z+z^{-1}\notin\sigma(J)$.*

Proof. Immediate from Theorems 3.6.1 and 3.7.1. □

In Section 2.10, we saw that for OPUC, asymptotics of Weyl solutions and polynomials are related and that is a theme we will use here but in the opposite direction from the OPUC case (where we went from asymptotics of polynomials to Weyl solutions). Since $T_n(z)$ is entire and $m(z)$ is analytic on $\mathbb{C}\setminus\sigma(J)$, the Weyl solution, $g_n(z)$, is defined via (3.2.23) on all of $\mathbb{C}\setminus\sigma(J)$.

Definition. We say the Weyl solutions, $g_n(z)$, have *Jost asymptotics* at $z_0\in\mathbb{D}\setminus[\{0\}\cup\{z\mid z+z^{-1}\in\sigma(J)\}]$ if

$$\lim_{n\to\infty}-z_0^{-(n+1)}g_n\left(z_0+\frac{1}{z_0}\right) \tag{3.7.9}$$

exists and is nonzero. We define the limit to be $1/u(z_0)$. u is called the *Jost function*.

Example 3.7.3. For J_0 (i.e., $a_n\equiv 1$, $b_n\equiv 0$), we have

$$p_n\left(z+\frac{1}{z}\right)=\frac{z^{n+1}-z^{-n-1}}{z-z^{-1}}\qquad q_n\left(z+\frac{1}{z}\right)=p_{n-1}\left(z+\frac{1}{z}\right) \tag{3.7.10}$$

and (recall $M(z) = z$ and $M(z) = -m(z + \frac{1}{z})$)

$$m\left(z + \frac{1}{z}\right) = -z \tag{3.7.11}$$

so by algebra,

$$g_n\left(z + \frac{1}{z}\right) = -z^{n+1} \tag{3.7.12}$$

which explains the reason for the minus sign and $n+1$ in (3.7.9). In [99] and [400], the term "Weyl solution" is used for

$$w_n(z) = -g_{n-1}\left(z + \frac{1}{z}\right) \tag{3.7.13}$$

□

The *Jost solution* is defined by

$$u_n(z) = -u(z)g_{n-1}\left(z + \frac{1}{z}\right) \tag{3.7.14}$$

so $u_n(z) \sim z^n$.

We will also prove the following below:

Theorem 3.7.4. *The conditions* (a)–(c) *of Theorem 3.7.1 imply that for all* $z \in \mathbb{D} \setminus [\{0\} \cup \{z \mid z + z^{-1} \in \sigma(J)\}]$, *one has Jost asymptotics uniformly on compact subsets of this set. Conversely, if one has Jost asymptotics uniformly on* $|z| = \varepsilon$ *for all sufficiently small* ε, *then* (a)–(c) *hold.*

We first show Jost asymptotics is equivalent to Szegő asymptotics because of the following lemma:

Lemma 3.7.5. *Suppose*

$$a_n \to 1 \qquad b_n \to 0 \tag{3.7.15}$$

Then the limit y_∞ *of* (3.7.3) *exists if and only if the limit* x_∞ *of* (3.7.9) *exists and*

$$x_\infty y_\infty = \frac{1}{1 - z^2} \tag{3.7.16}$$

Remark. We use the language of two-sided Jacobi matrices of Sections 5.2 and 5.4.

Proof. Let

$$G_{nm}(z) = \langle \delta_n, (J - (z + z^{-1}))^{-1}\delta_m \rangle \qquad G_{nm}^{(0)}(z) = \langle \delta_n, (J_0 - (z + z^{-1}))^{-1}\delta_m \rangle \tag{3.7.17}$$

where J_0 is the two-sided Jacobi matrix with $a_n \equiv 1$, $b_n \equiv 0$.

By (3.7.15), if $(u^{(k)})_n = u_{k+n}$, $Ju^{(k)} \to J_0 u$ for any u, so for any $z \in \mathbb{D}$,

$$\lim_{n\to\infty} G_{nn}(z) = G_{00}^{(0)}(z) = -(z - z^{-1})^{-1} \tag{3.7.18}$$

where we use (5.4.59) and $u_n^{(0)\pm} = z^{\pm n}$. By (5.4.41),

$$G_{nn}(z) = p_{n-1}(z + z^{-1})g_{n-1}(z + z^{-1}) \tag{3.7.19}$$

(since the Wronskian of $p_{\cdot-1}$ and $g_\cdot$ is 1).

Thus, if y_n is the quantity in (3.7.3) and x_n in (3.7.9), (3.7.18) says that

$$zy_{n-1}x_{n-1} \to \frac{z}{1-z^2} \tag{3.7.20}$$

proving that $x_n \to x_\infty$ if and only if $y_n \to y_\infty$ and that (3.7.16) holds. □

Proof of Theorem 3.7.1 given Theorem 3.7.4. (a)–(c) imply (3.7.15) by (3.7.6), so Jost asymptotics implies Szegő asymptotics. Conversely, since near $z = 0$ $(z + \frac{1}{z} = \infty)$,

$$z^n p_n\left(z + \frac{1}{z}\right) = \frac{1}{a_1 \dots a_n}\left(1 + z\left(\sum_{j=1}^n b_j\right) + O(z^2)\right) \tag{3.7.21}$$

so Szegő asymptotics implies (a), (b), and so (3.7.15), and thus, by the lemma, Jost asymptotics. □

Remark. By (3.7.16), we have

$$u(z) = \frac{(1-z^2)}{(D(z)\sqrt{2})} \tag{3.7.22}$$

To get from asymptotics of g_n to information on the Jacobi parameters, we need a relation between $\{g_n\}_{n=0}^\infty$ and $\{M(z, J_n)\}$.

Theorem 3.7.6. *We have for $k \geq 1$,*

$$m(x, J_k) = -\frac{g_k(x)}{a_k g_{k-1}(x)} \tag{3.7.23}$$

and

$$m(x, J) = g_0(x) \tag{3.7.24}$$

Proof. We have that for $k \geq 1$,

$$T_{n+k}(\{a_j, b_j\}_{j=1}^{n+k}, x) = T_n(\{a_{j+k}, b_{j+k}\}_{j=1}^n, x)T_k(\{a_j, b_j\}_{j=1}^k, x) \tag{3.7.25}$$

Thus, applying this to $\binom{m(z)}{-1}$, we find

$$\begin{pmatrix} g_k \\ a_k g_{k-1} \end{pmatrix} = c\begin{pmatrix} m(z, J_k) \\ -1 \end{pmatrix} \tag{3.7.26}$$

since there is a unique ℓ^2 solution. This implies (3.7.23). (3.7.24) is just the initial condition for g. □

Corollary 3.7.7. *We have that*

$$-z^{-(n+1)}g_n\left(z + \frac{1}{z}\right) = \left(\prod_{j=1}^n a_j\right)\left(\frac{M(z)}{z}\right)\prod_{k=1}^n \frac{M(z, J_k)}{z} \tag{3.7.27}$$

Proof. This follows from (3.2.27) and (3.2.28). The minus comes from n factors in (3.7.23) and $(n+1)$ factors in $M(z) = -m(z + z^{-1})$. □

Proof of the half of Theorem 3.7.4 that Jost asymptotics ⇒ (a)–(c). By (3.7.27), $-z^{-(n+1)}g_n(z+\frac{1}{z})$ has a removable singularity at $z=0$ and defines a function $\eta_n(z)$ analytic in $\mathbb{R}\setminus\{z\mid z+z^{-1}\in\sigma(J)\}$ (zeros of $M(z)$ include poles of M_1, etc.). Thus, convergence of η_n uniformly on $|z|=r$ implies convergence of the Taylor coefficients of η_n. In particular,

$$\eta_n(0)=(a_1\dots a_n) \tag{3.7.28}$$

has a finite limit. This limit is nonzero since η_n is nonvanishing on $\{z\mid |z|\le\varepsilon\}$ and the limit is not identically zero.

By (3.7.27) and (3.4.26),

$$\log(\eta_n(z))=\beta_n+\gamma_n z+\varphi_n z^2+O(z^2) \tag{3.7.29}$$

where

$$\beta_n=\sum_{j=1}^{n-1}\log(a_j)\qquad \varphi_n=\sum_{j=1}^{n}b_j\qquad \gamma_n=\sum_{j=1}^{n}(a_j^2-1+\tfrac{1}{2}b_j^2) \tag{3.7.30}$$

That γ_n has a finite limit is thus immediate. As usual, we combine

$$\varphi_n-2\beta_n=\sum_{j=1}^{n}[G(a_j)+\tfrac{1}{2}b_j^2] \tag{3.7.31}$$

so positivity and conditional convergence imply convergence. □

For the other direction, we need a general result about asymptotics of difference equations, which we state without proof (but see the Notes).

Theorem 3.7.8 (Discrete Hartman–Wintner Theorem). *Let B_0 be a $d\times d$ diagonal matrix whose diagonal elements, $\lambda_1,\dots,\lambda_d$, obey $|\lambda_i|\neq|\lambda_j|$ for all $i\neq j$. Let $\{(\delta B)(n)\}_{n=1}^{\infty}$ obey*

(i)

$$(\delta B)(n)_{kk}\to 0 \tag{3.7.32}$$

as $n\to\infty$ for $k=1,\dots,d$.

(ii)

$$\sum_n|(\delta B)(n)_{kj}|^2<\infty \tag{3.7.33}$$

for all $k\neq j$.

(iii) *Let*

$$B(n)=B_0+\delta B(n) \tag{3.7.34}$$

For all n, we suppose that

$$\det(B(n))\neq 0 \tag{3.7.35}$$

Then for each j, there exists $u^{(j)}$, so

$$\prod_{\ell=1}^{n}(\lambda_j+\delta B(\ell)_{jj})^{-1}[B(n)\dots B(1)u^{(j)}]\to\delta_j \tag{3.7.36}$$

the vector with $(\delta_j)_k=\delta_{jk}$.

End of the proof of Theorem 3.7.4. Suppose (a)–(c) hold. Let $z \in \mathbb{D}$ and define the 2×2 matrices

$$B(n) = \begin{pmatrix} 1 & 1 \\ z^{-1} & z \end{pmatrix}^{-1} A(a_n, b_n; z + z^{-1}) \begin{pmatrix} 1 & 1 \\ z^{-1} & z \end{pmatrix} \tag{3.7.37}$$

where A is given by (3.2.2). The conjugating matrix is chosen because

$$B_0 \equiv \begin{pmatrix} z & 0 \\ 0 & z^{-1} \end{pmatrix} = \begin{pmatrix} 1 & 1 \\ z^{-1} & z \end{pmatrix}^{-1} \begin{pmatrix} z + z^{-1} & -1 \\ 1 & 0 \end{pmatrix} \begin{pmatrix} 1 & 1 \\ z^{-1} & z \end{pmatrix} \tag{3.7.38}$$

is diagonal. By (c), if δB is given by (3.3.46), then

$$\sum_n \|(\delta B)(n)\|_2^2 < \infty$$

implying (3.7.33). (3.7.32) holds since (c) implies $a_n \to 1$, $b_n \to 0$. Since $|z| < 1$, $|z| \neq |z^{-1}|$.

(a) and (b) imply $\sum_{\ell=1}^n (\delta B(\ell))_{jk}$ has a limit and $\sum_{\ell=1}^n |\delta B(\ell)_{jk}|^2 < \infty$, so $\prod_{j=1}^n [1 + z^{-1}(\delta B(\ell))_{11}]$ has a limit. By (3.7.36), there exist some initial conditions so that

$$z^{-n} T_n(\{a_j, b_j\}_{j=1}^n, z + z^{-1}) \tilde{u}_0(1)$$

has a nonzero limit. This gives an ℓ^2 solution, and so the ℓ^2 solution has Jost asymptotics. □

The following is known to hold in the generality of Theorem 3.7.1 (see the Notes), but the proof is easier with the stronger hypotheses discussed in Theorem 3.6.1.

Theorem 3.7.9. *Let*

$$d\rho = f(x)\,dx + d\rho_s(x)$$

where (3.7.7) *and* (3.7.8) *hold. Then u has nontangential boundary values a.e. on $\partial\mathbb{D}$ and*

$$\int \left| p_n(x) - \frac{\operatorname{Im}[\bar{u}(e^{i\theta(x)}) e^{i(n+1)\theta(x)}]}{\sin(\theta(x))} \right|^2 f(x)\,dx \to 0 \tag{3.7.39}$$

and

$$\int |p_n(x)|^2 \,d\mu_s(x) \to 0 \tag{3.7.40}$$

Here $\theta(x) \in [0, \pi]$ is given, as usual, by

$$x = 2\cos(\theta(x)) \tag{3.7.41}$$

and one has $[\sin(\theta(x))]^{-1} u(e^{i\theta(x)}) \in L^2(\mathbb{R}, f(x)\,dx)$.

Proof. Let us begin by *defining* u on $\mathbb{D}$ by

$$u(z) = \prod_{j,\pm} b_{p_j^{\pm}}(z) \exp\left(\int \frac{e^{i\theta} + z}{e^{i\theta} - z} \log\left(\frac{\sin\theta}{\operatorname{Im} M(e^{i\theta})} \right) \frac{d\theta}{4\pi} \right) \tag{3.7.42}$$

where, as usual (see (3.5.6)),

$$\operatorname{Im} M(e^{i\theta(x)}) = \pi f(x) \tag{3.7.43}$$

and $p_j^\pm$ are the points in $\mathbb{D}$ with

$$E_j^\pm = (p_j^\pm) + (p_j^\pm)^{-1} \tag{3.7.44}$$

By (3.7.7), the Blaschke product in (3.7.42) converges, and by (3.7.8), the second factor exists.

As in the proof for $p(z)$ for OPUC (see Proposition 2.9.4), this second factor, $E(z)$, has $(1-z^2)E(z)^{-1}$ in $H^2(\mathbb{D})$, so $u(z)$ has boundary values obeying

$$f(x) = \frac{\sin(\theta(x))}{\pi\,|u(e^{i\theta(x)})|^2} \tag{3.7.45}$$

We will prove that $u(z)$, defined by (3.7.42), obeys (3.7.39) and then use that to prove it is also an inverse of the limit in (3.7.9), so it agrees with our previous definition.

To facilitate calculation, define

$$k^+(e^{i\theta}) = \frac{\bar{u}(e^{i\theta})}{2i\sin\theta} \tag{3.7.46}$$

$$k^-(e^{i\theta}) = \overline{k^+(e^{i\theta})} \tag{3.7.47}$$

$$u^{(m)}(z) = B_m(z)E(z) \tag{3.7.48}$$

$$B_m(z) = \prod_{1\le j\le m,\pm} b_{p_j^\pm}(z) \tag{3.7.49}$$

$$k_m^- = k^- \frac{u^{(m)}}{u} \tag{3.7.50}$$

We let $\|\cdot\|_f$ and $\langle\ ,\ \rangle_f$ be the $L^2(\mathbb{R}, f(x)\,dx)$ norm and inner product, and $\|\cdot\|_{\mathrm{s}}$, the $L^2(\mathbb{R}, d\mu_{\mathrm{s}})$ inner product. Clearly, (3.7.39)/(3.7.40) is equivalent to

$$\|p_n - z^{n+1}k^+ - \overline{z^{n+1}}\,k^-\|_f^2 + \|p_n\|_{\mathrm{s}}^2 \to 0 \tag{3.7.51}$$

where $z(x) = e^{i\theta(x)}$. This will follow from

$$\|p_n\|_f^2 + \|p_n\|_{\mathrm{s}}^2 = 1 \tag{3.7.52}$$

$$\|k^\pm\|_f^2 = \tfrac{1}{2} \tag{3.7.53}$$

$$\langle z^{n+1}k^+, z^{-(n+1)}k^-\rangle_f \to 0 \tag{3.7.54}$$

$$\lim_{n\to\infty} \langle z^{-n-1}k^-, p_n\rangle_f = \tfrac{1}{2} \tag{3.7.55}$$

(3.7.52) is the normalization condition. To get (3.7.53), use (3.7.45) and (3.7.46) to see

$$\|k^+\|_f^2 = \int_{-2}^{2} \frac{|u(e^{i\theta(x)})|^2}{4\sin^2(\theta(x))} \frac{\sin(\theta(x))}{\pi|u(e^{i\theta(x)})|^2}\, d(2\cos\theta) \tag{3.7.56}$$

$$= \int_0^\pi \frac{1}{2}\frac{d\theta}{\pi} = \frac{1}{2} \tag{3.7.57}$$

To prove (3.7.54), note that a calculation similar to one just done shows

$$\langle z^{n+1}k^+, z^{-(n+1)}k^-\rangle_f = \int_0^\pi e^{-(2n+2)i\theta}\left[\frac{u(e^{i\theta})}{|u(e^{i\theta})|}\right]^2 \frac{d\theta}{2\pi} \tag{3.7.58}$$

goes to zero since $[u/\bar{u}]^2 \in L^2(\partial\mathbb{D}, \frac{d\theta}{2\pi})$.

Finally, note

$$\begin{aligned} |\langle z^{-n-1}(k^- - k_m^-), p_n\rangle_f| &\le \|k^- - k_m^-\|_f \\ &= \|k^-(B_\infty - B_m)\|_f \\ &\to 0 \end{aligned} \tag{3.7.59}$$

as $m \to \infty$, by the dominated convergence theorem. Thus, uniformly in n,

$$\langle z^{-n-1}k_m^-, p\rangle_f \to \langle z^{-n-1}k^-, p\rangle_f \tag{3.7.60}$$

so, to prove (3.7.55), it suffices to prove

$$\lim_{m\to\infty}\Big(\lim_{n\to\infty} \operatorname{Re}(\langle z^{-n-1}k_m^-, p\rangle_f)\Big) = \tfrac{1}{2} \tag{3.7.61}$$

We compute (using $|u^{(m)}| = |u|$ on $\partial\mathbb{D}$)

$$\begin{aligned} \operatorname{Re}\langle z^{-n-1}k_m^-, p_n\rangle &= \operatorname{Re}\int_0^\pi \frac{u^{(m)}e^{i(n+1)\theta}}{2i\sin\theta}\, p_n(2\cos\theta)\, \frac{\sin\theta}{\pi|u^{(m)}|^2}\, 2\sin\theta\, d\theta \\ &= \int_0^{2\pi} z^{n+1}\frac{z^{-1}-z}{2}\, p_n\Big(z+\frac{1}{z}\Big)\frac{dz}{2\pi i z u^{(m)}(z)} \\ &= \frac{1}{2}\int_0^{2\pi} (1-z^2)z^n\, p_n\Big(z+\frac{1}{z}\Big)\frac{dz}{2\pi i z u^{(m)}(z)} \end{aligned} \tag{3.7.62}$$

$(zu^{(m)})^{-1}$ has poles at $z = 0$ and at $\{p_j^{\pm}\}_{j=1}^m$. Using the fact that $z_n p_n(z+\frac{1}{z})\big|_{z=0} = (a_1\dots a_n)^{-1}$, we see that

$$\langle k_m^-, p_n\rangle = \frac{1}{2}(a_1\dots a_n)^{-1}u(0)^{-1}\frac{B_\infty(0)}{B_m(0)} + \varepsilon_n^{(m)} \tag{3.7.63}$$

where

$$\varepsilon_n^{(m)} = \frac{1}{2}\sum_{j=1,\dots,m;\pm} [1-(p_j^{\pm})^2](p_j^{\pm})^{n-1}\, p_n(E_j^{\pm})\, \frac{1}{(u^{(m)})'(p_j^{\pm})} \tag{3.7.64}$$

This is a finite sum. Since

$$\Big(\sum_n |p_n(E_j^{\pm})|^2\Big)\, \rho_s(E_j^{\pm}) \le 1 \tag{3.7.65}$$

we see

$$\sup_{n,j=1,\dots,m,\pm} |p_n(E_j^\pm)| < \infty \tag{3.7.66}$$

so, since $\sup_{j=1,\dots,m,\pm} |p_j^\pm|^{n-1} \to 0$ as $n \to \infty$, we have $\varepsilon_n^{(m)} \to 0$. Thus,

$$\lim_{n\to\infty} \langle k_m^-, p_n\rangle = \frac{1}{2}\,\frac{B_\infty(0)}{B_m(0)}\,u(0)^{-1} \lim_{n\to\infty} (a_1 \dots a_n)^{-1} \tag{3.7.67}$$

One can rewrite (3.6.37) as

$$\lim_{n\to\infty} (a_1 \dots a_n)^{-1} = u(0) \tag{3.7.68}$$

Since $B_m(0) \to B_\infty(0)$ as $m \to \infty$, (3.7.67) implies (3.7.61). This completes the proof of (3.7.51) and so of (3.7.39) and (3.7.40) for u defined by (3.7.42). All that remains is to prove that this u is the same as the previously defined Jost function.

One can rewrite (3.7.39) as saying (with $z = e^{i\theta(x)}$) that

$$z^n p_n\left(z + \frac{1}{z}\right) - \frac{u(z)}{1-z^2} + \frac{\overline{u(z)}}{1-z^2}\, z^{2n+2} \to 0 \tag{3.7.69}$$

in $\|\cdot\|_f$.

Since

$$u(z) = E(z) B_\infty(z) \tag{3.7.70}$$

and (3.7.45) says

$$f(x)\,dx = \frac{2\sin^2\theta}{|E(z)|^2}\,\frac{d\theta}{\pi} = \left|\frac{1-z^2}{E(z)}\right|^2 \frac{d\theta}{2\pi} \tag{3.7.71}$$

we see that (3.7.69) is the same as

$$(1-z^2)E(z)^{-1} z^n p_n\left(1 + \frac{1}{z}\right) - B_\infty(z) - z^{2n+2}\,\frac{\overline{E(z)}}{E(z)}\,\overline{B_\infty(z)} \to 0 \tag{3.7.72}$$

in $L^2(\partial\mathbb{D}, \frac{d\theta}{2\pi})$ norm. Since $\overline{E(z)}/E(z)\overline{B_\infty(z)} \in L^\infty(\partial\mathbb{D}) \subset L^2(\partial\mathbb{D}, \frac{d\theta}{2\pi})$, we see that the last term goes to zero weakly in $L^2(\partial\mathbb{D}, \frac{d\theta}{2\pi})$. Since $(1-z^2)E(z)^{-1}$ and $B_\infty(z)$ are in $H^2(\partial\mathbb{D})$, we see that weakly in $H^2(\partial\mathbb{D})$,

$$(1-z^2)E(z)^{-1} z^n p_n\left(z + \frac{1}{z}\right) - B_\infty(z) \tag{3.7.73}$$

goes to zero, so the function goes to zero uniformly on compact subsets of $\mathbb{D}$.

Thus, we have Szegő asymptotics with

$$\left(\sqrt{2}\,D(z)\right)^{-1} = \frac{u(z)}{1-z^2} \tag{3.7.74}$$

By (3.7.22) and Lemma 3.7.5, we have Jost asymptotics with the Jost function given by the u of (3.7.42). $\square$

Remark. As a bonus, we get the explicit formula (3.7.42) for the Jost function.

Remarks and Historical Notes. The main theorems, Theorems 3.7.1 and 3.7.4, of this section are from Damanik–Simon [99]. There is earlier work. Szegő asymptotics for OPRL with $\mathrm{supp}(d\rho) = [-2, 2]$ (i.e., no bound states) is a result of Szegő [431]. He used the fact, (1.9.9), that

$$z^n p_n\left(z + \frac{1}{z}\right) = [2(1 - \alpha_{2n-1})]^{-1/2}[\varphi_{2n}(z) + \varphi_{2n}^*(z)]$$

Since $\alpha_{2n-1} \to 0$, $\varphi_{2n}(z) \to 0$, and $\varphi_{2n}^*(z) \to D(z)^{-1}$ (see Theorem 2.9.6), we see in this case that

$$z^n p_n\left(z + \frac{1}{z}\right) \to \frac{1}{(\sqrt{2}\, D(z))}$$

hence the definition of "D" in cases when there are bound states and D cannot be defined via the Szegő map. Nevai [320] extended this to allow finitely many $\{E_j^{\pm}\}$. Corollary 3.7.2 was then found by Peherstorfer–Yuditskii [343]. [99] had a different proof of Lemma 3.7.5; the proof here is from Christiansen–Simon–Zinchenko [88].

Jost asymptotics, Jost solution, and Jost function all come from an analogy to work of Jost [211] who studied solutions of the Schrödinger equation, $-u''(x) + Vu(x) = k^2u(x)$, with asymptotics $u(x) \sim e^{ikx}$. This Jost solution had a value at $x = 0$, called the Jost function. The use of Jost functions in OPRL was pioneered by Case [76, 77, 154].

If one has Jost asymptotics and the Jost function is a Nevanlinna function, then the hypotheses of Theorem 3.6.1 are valid, and conversely. That the Jost function is a Nevanlinna function with trivial singular inner factor is a result explicit in Killip–Simon [225] and implicit in Peherstorfer–Yuditskii [342].

Case labels his Jacobi parameters starting at $n = 0$, and as a result, he has various factors of z^{-1} in Jost function formulae. It is to avoid such factors that Killip–Simon used the now common convention to start labeling at $n = 1$. It is also why the solution in [99, 400] start at $n = 1$, not $n = 0$.

Theorem 3.7.8 is due to Coffman [91] and is an analog of a continuum result (ODE) of Hartman–Wintner [192]. A pedagogic presentation of this theorem and additional history will appear in the second edition of [399] and is available at `http://www.math.caltech.edu/opuc/newsection13-3.pdf`.

[99] has two other proofs of the direction (a)–(c) $\Rightarrow$ Jost asymptotics. One uses Fredholm determinant formulae for Jost functions and the other a renormalized inner-outer factorization.

Theorem 3.7.9 and the proof we give is due to Peherstorfer–Yuditskii [342]. The same theorem for the more general context of Theorem 3.7.1 is in Damanik–Simon [99], but the proof is different. One cannot define u by (3.7.42) because the Blaschke product and Szegő integrals may diverge. Instead, one needs to use "renormalized" Blaschke products and Poisson representations. Then the trick of replacing B_∞ by B_m with only finitely many poles in B_m^{-1} is not available and a different method is needed.

Notice that our proof of Theorem 3.7.9 also provides an independent proof of Szegő asymptotics for p_n on $\mathbb{D}$ when (3.7.2) and (3.7.8) hold. It also only needs

$$\limsup (a_1 \dots a_n)^{-1} \geq u(0)$$

which is the easier half of the proof of (3.7.66) (i.e., of (3.6.37)). It then implies the full (3.7.66).

Notice that (3.7.39) expands p_n in terms of $e^{\pm in\theta(x)}$, not $u_n(e^{i\theta(x)})$ and its conjugate. The product of u_n and u_0 is not necessarily L^2, but since $e^{\pm in\theta(x)}$ are in L^∞, their products with u_0 are in L^2.

3.8 THE MOMENT PROBLEM: AN ASIDE

In the next section, we will discuss an application of Szegő's theorem for OPUC to the moment problem on the real line. This section is background but also illustrates the use of OPRL and, in particular, transfer matrices to study the moment problem.

The moment problem in its primeval form is:

Moment Problem: First Form. Given a sequence $\{c_n\}_{n=0}^\infty$ of real numbers, when does there exist a nontrivial measure, $d\mu$, on $\mathbb{R}$ with

$$\int x^n \, d\mu(x) = c_n \tag{3.8.1}$$

When a solution exists, is it unique? If it is not unique, what is the structure of the set of solutions?

Of course, for (3.8.1) to make sense, one needs

$$\int |x|^n \, d\mu < \infty \tag{3.8.2}$$

By structure of the set of solutions, we mean is it closed in the weak topology? (This is not obvious since x^n is not bounded.) Is it of finite or infinite dimension? Among the solutions, are there any that are pure point or singular continuous or purely absolutely continuous?

If there exists a unique solution, we call the moment problem *determinate*, and if there are multiple solutions, *indeterminate*. Since we can replace c_n by c_n/c_0, we can and will always suppose that $c_0 = 1$.

Often the c_n are given by (3.8.1), so existence is trivial. The moment problem then becomes:

Moment Problem: Second Form. Suppose c_n is a sequence given by (3.8.1) for some nontrivial probability measure, $d\mu_0$, on $\mathbb{R}$ obeying (3.8.2). Is $d\mu_0$ the unique measure obeying (3.8.1) for the given c_n, or are there others? If there are others, what is the structure of the solutions?

Example 3.8.1. Fix $0 < \alpha$ real and let c_n be given by

$$c_n = N_\alpha^{-1} \int x^n \exp(-|x|^\alpha) \, dx \tag{3.8.3}$$

where $N_\alpha = \int \exp(-|x|^\alpha)\,dx$ is a normalization constant. Below (see later in this section and then in the next) we will show that this problem is determinate if $\alpha \geq 1$ and indeterminate if $0 < \alpha < 1$. □

There is an obvious necessary condition on the c_n's for there to be any nontrivial measure.

Proposition 3.8.2. *If a solution of the moment problem exists, then for each $n = 1, 2, \dots,$ the Hankel determinants*

$$H_m(\{c_n\}_{n=0}^\infty) = \det((c_{j+k-2})_{1\leq j,k\leq m}) \tag{3.8.4}$$

are strictly positive.

Proof. Let $\{\alpha_j\}_{j=1}^m$ lie in $\mathbb{C}^m$. Then

$$\sum_{j,k=1}^m \bar{\alpha}_j \alpha_k c_{j+k-2} = \int \left| \sum_{j=0}^{m-1} \alpha_j x^j \right|^2 d\mu \tag{3.8.5}$$

so H_m is positive as the determinant of a strictly positive matrix. □

We will see later (see Theorem 3.8.4) that, conversely, if $H_m > 0$ for all m, then the moment problem is soluble. For now, we note that it is easy to see that if each H_m is positive, there exists a unique nondegenerate inner product on polynomials with

$$\langle 1, x^m \rangle = c_m \tag{3.8.6}$$

This inner product defines OPs both monic and normalized and Jacobi parameters $\{a_n, b_n\}_{n=1}^\infty \in ((0,\infty)\times\mathbb{R})^\infty$. Thus, we have:

Moment Problem: Third Form. Given a set of Jacobi parameters, $\{a_n, b_n\}_{n=1}^\infty \in ((0,\infty)\times\mathbb{R})^\infty$, when does there exist a measure, $d\mu$, whose Jacobi parameters are $\{a_n, b_n\}_{n=1}^\infty$? If one exists, is it unique? If it is not unique, what is the structure of the set of solutions?

Existence is essentially Favard's theorem discussed in Section 1.3. Jacobi parameters determine moments, so an inner product on polynomials, and (3.8.4) holds. Thus, Problems 1 and 3 are equivalent. We will see (see Theorem 3.8.4) that in this form, the moment problem always has solutions, that is, any set of Jacobi parameters can occur.

Proposition 3.8.3. *Fix $k \geq 1$. Let $\{c_n\}_{n=0}^{2k}$ be a set of moments with* (3.8.4) *strictly positive for $m = 1, \dots, k+1$. Then the set of measures in $\mathbb{R}$ obeying* (3.8.1) *for $n = 0, \dots, 2k-1$ and*

$$\int x^{2k}\,d\mu \leq c_{2k} \tag{3.8.7}$$

is a nonempty set, compact in the topology of weak-$*$ convergence (i.e., $d\mu_\ell \to d\mu$ if and only if $\int f(x)\,d\mu_\ell \to \int f(x)\,d\mu$ for all bounded continuous functions on $\mathbb{R}$).

Proof. The $\{c_n\}_{n=0}^{2k}$ define an inner product on polynomials of degree up to k, so orthonormal polynomials $\{p_j\}_{j=0}^{k}$, and so Jacobi parameters $\{a_n, b_n\}_{n=1}^{k}$. Choose any value for b_{k+1} and so get a $(k+1) \times (k+1)$ finite Jacobi matrix, $J_{k+1;F}$. Let $d\mu$ be the spectral measure for this matrix and vector δ_1. Then $d\mu$ obeys (3.8.1) for $n = 0, \dots, 2k$, so there is a solution proving the set is nonempty; indeed, we can suppose equality in (3.8.7).

Using the fact that the probability measures on $[-R, R]$ are compact, it is easy to see that the set of probability measures on $\mathbb{R}$ obeying

$$\int_{|x|\geq R} d\mu(x) \leq c_{2k} R^{-2k} \tag{3.8.8}$$

for each R is compact in the topology of weak-$*$ convergence. Here we use $k \geq 1$ to assure weak limits are also probability measures.

(3.8.7) implies (3.8.8). Thus, we need only prove that the set, S, of μ's obeying (3.8.1) for $m \leq 2k-1$ and (3.8.7) is weakly closed.

Let

$$f_{n;R}(x) = \begin{cases} x^n & |x| \leq R \\ R^n & x \geq R \\ (-R)^n & x \leq -R \end{cases} \tag{3.8.9}$$

and suppose $d\mu_\ell \in S$ converges weakly to $d\mu$. Then

$$\int f_{2k;R}\, d\mu_\ell \leq c_{2k} \tag{3.8.10}$$

so, since $f_{2k;R} - R^{2k}$ has compact support,

$$\int f_{2k;R}\, d\mu \leq c_{2k} \tag{3.8.11}$$

and (3.8.7) holds by the monotone convergence theorem.

By dominated convergence, (3.8.7) implies that for any $m = 1, \dots, 2k-1$,

$$\lim_{R\to\infty} \int f_{m;R}\, d\mu = \int x^m\, d\mu \tag{3.8.12}$$

Moreover, for any finite ℓ,

$$\begin{aligned} \int |f_{m;R} - x^m|\, d\mu_\ell &\leq 2 \int_{|x|\geq R} |x|^m\, d\mu_\ell \\ &\leq 2 \int_{|x|\geq R} \left|\frac{x}{R}\right|^{2k-m} |x|^m\, d\mu \\ &\leq 2R^{-(2k-m)} c_{2k} \end{aligned} \tag{3.8.13}$$

so (3.8.12) converges for each ℓ uniformly in ℓ. This plus (3.8.12) plus $\int x^m\, d\mu = c_m$ implies $d\mu$ obeys (3.8.1) for $n = 0, \dots, 2k-1$. $\square$

We thus have existence:

Theorem 3.8.4. *A set, $\{c_n\}_{n=0}^\infty$, of real numbers with $c_0 = 1$ has solutions of the moment problem if and only if each $H_m(\{c_n\}_{n=0}^\infty)$ (given by (3.8.4)) is strictly positive. Any set of Jacobi parameters $\{a_n, b_n\}_{n=1}^\infty \in ((0, \infty) \times \mathbb{R})^\infty$ is the Jacobi parameter of some measure.*

Remark. The second sentence is essentially Favard's theorem in the general case; see Theorem 1.3.9.

Proof. Let S_k be the set of measures given by Proposition 3.8.3. Since S_k is compact and nonempty, and $S_{k+1} \subset S_k$, we see $\cap_k S_k$ is nonempty. This plus Proposition 3.8.2 proves the first sentence in this proposition. As noted, the first and third forms of the moment problem are equivalent, thus proving the second sentence. □

To go further and analyze uniqueness, we need to briefly study unbounded selfadjoint operators. A *densely defined operator*, A, on a Hilbert space, $\mathcal{H}$, has a domain $D(A) \subset \mathcal{H}$, a dense subspace, and is a linear map of $D(A)$ into $\mathcal{H}$. Associated to A is its *graph*, $\Gamma(A) \subset \mathcal{H} \times \mathcal{H}$, defined by

$$\Gamma(A) = \{(\varphi, A\varphi) \mid \varphi \in D(A)\} \tag{3.8.14}$$

$\Gamma(A)$ is always a subspace of $\mathcal{H} \times \mathcal{H}$. A is called *closed* if and only if $\Gamma(A)$ is closed. B is an *extension* of A if and only if $\Gamma(A) \subset \Gamma(B)$, that is, $D(A) \subset D(B)$ and $B \restriction D(A) = A$.

Given an operator, A, we define $D(A^*)$ to be those $\varphi \in \mathcal{H}$ for which there is an $\eta \in \mathcal{H}$ with

$$\langle \eta, \gamma \rangle = \langle \varphi, A\gamma \rangle \tag{3.8.15}$$

for all $\gamma \in D(A)$. η is uniquely determined if it exists since $D(A)$ is dense. We then set $\eta = A^*\varphi$, so

$$\langle A^*\varphi, \gamma \rangle = \langle \varphi, A\gamma \rangle \tag{3.8.16}$$

for all $\gamma \in D(A)$, $\eta \in D(A^*)$. A^* is called the *adjoint* of A. A^* is thus defined to be the maximal operator so that (3.8.16) holds. If $D(A^*)$ is dense, then it is easy to see that A^* is a closed operator. Note that there is a relation between extension and adjoint:

$$A \subset B \Rightarrow B^* \subset A^*$$

An operator is called

$$\textit{Hermitian} \Leftrightarrow A \subset A^*$$
$$\textit{Selfadjoint} \Leftrightarrow A = A^*$$
$$\textit{Essentially selfadjoint} \Leftrightarrow A \subset A^* = (A^*)^*$$

Notice that if A is Hermitian, then A^* is densely defined and we can define $(A^*)^*$.

Proposition 3.8.5. *Let A be a Hermitian operator and let $z = x + iy \in \mathbb{C} \setminus \mathbb{R}$. Then*

(i) *For all $\varphi \in D(A)$,*

$$\|(A - z)\varphi\|^2 = \|(A - x)\varphi\|^2 + y^2\|\varphi\|^2 \tag{3.8.17}$$

(ii) *A is closed $\Leftrightarrow$ $\text{Ran}(A - z)$ is closed.*

(iii) *A^{**} is the smallest closed extension of A, so we write*

$$\bar{A} = A^{**} \tag{3.8.18}$$

(iv) *$A^* = A^{***}$. Moreover, if A is Hermitian, so is $\bar{A}$.*

(v)

$$\overline{\text{Ran}(A - z)} = \text{Ran}(\bar{A} - z) \tag{3.8.19}$$

(vi)

$$D(A^*) = D(\bar{A}) + \ker(A^* - z) + \ker(A^* - \bar{z}) \tag{3.8.20}$$

(vii) *A is essentially selfadjoint if and only if*

$$\ker(A^* - z) = \ker(A^* - \bar{z}) = \{0\} \tag{3.8.21}$$

Remark. (3.8.20) holds in the sense of algebraic direct sum, that is, any $\psi \in D(A^*)$ is uniquely the sum of three vectors, one in each space.

Proof. (i) (3.8.17) follows from noting that the cross-term

$$\langle (A - x)\varphi, iy\varphi\rangle + \langle iy\varphi, (A - x)\varphi\rangle = 0 \tag{3.8.22}$$

by Hermiticity.

(ii) By (3.8.17),

$$(\varphi, A\varphi) \mapsto (A - z)\varphi \tag{3.8.23}$$

is a metric space equivalence of $\Gamma(A)$ and $\text{Ran}(A - z)$, so one space is complete if and only if the other is.

(iii) Let $J\colon \mathcal{H} \to \mathcal{H}$ by $J\langle\varphi, \psi\rangle = \langle\psi, -\varphi\rangle$. Then

$$\Gamma(A^*) = J[\Gamma(A)^\perp] = [J\Gamma(A)]^\perp \tag{3.8.24}$$

Since $J^2 = -\mathbf{1}$, we see $\Gamma(A^{**}) = [-\Gamma(A)]^{\perp\perp} = \overline{\Gamma(A)}$. Thus, A^{**} is closed and is the smallest closed extension.

(iv) A^* is closed by (3.8.24), so (3.8.18) implies $A^* = A^{***}$. Thus, $A \subset A^*$ implies $A^{**} \subset A^* = (A^{**})^*$.

(v) As noted in the proof of (ii), (3.8.23) is a metric space equivalence, so it takes closures to closures.

(vi) If $\psi \in D(\bar{A})$, $\varphi_+ \in \ker(A^* - z)$, $\varphi_- \in \ker(A^* - \bar{z})$, and

$$\varphi_+ + \varphi_- + \psi = 0 \tag{3.8.25}$$

Then applying $(A^* - z)$ and then $(A^* - \bar{z})$, we see

$$\varphi_- = i(2\,\text{Im}\, z)^{-1}\bar{A}\psi \tag{3.8.26}$$

$$\varphi_+ = -i(2\,\text{Im}\, z)^{-1}\bar{A}\psi \tag{3.8.27}$$

so $\varphi_+ = -\varphi_-$, which implies $\varphi_+ = \varphi_- = 0$, and then $\psi = 0$. This proves uniqueness. If $\eta \in D(A^*)$, since

$$\operatorname{Ran}(\bar{A} - z) + \operatorname{Ran}(\bar{A} - z)^\perp = \mathcal{H}$$

and $\operatorname{Ran}(\bar{A} - z)^\perp = \ker(A^* - \bar{z})$, we can find $\psi \in D(\bar{A})$, $\varphi_- \in \ker(A^* - \bar{z})$ so that

$$(A^* - z)\eta = (\bar{A} - z)\psi + (A^* - z)\varphi_-$$

Thus, $\varphi_+ = \eta - \psi - \varphi_- \in \ker(A^* - z)$.

(vii) By (3.8.20), $D(\bar{A}) = D(A^*)$ if and only if (3.8.21) holds. □

Given *any* sequence $\{u_n\}_{n=1}^\infty$, define $\mathcal{J}u$, a new sequence, by

$$(\mathcal{J}u)_n = a_n u_{n+1} + b_n u_n + a_{n-1} u_{n-1} \tag{3.8.28}$$

where $a_0 = 0$. Define an operator, A, by

$$D(A) = \{u \mid u_n = 0 \text{ for all large } n\} \qquad Au = \mathcal{J}u \tag{3.8.29}$$

Then $A\colon D(A) \to D(A) \subset \ell_2$ is a densely defined operator.

Theorem 3.8.6. (i) *We have that for any $u \in D(A)$ and any sequence v that (both sums are finite)*

$$\sum_{n=1}^\infty \bar{v}_n (Au)_n = \sum_{n=1}^\infty \overline{(\mathcal{J}v)_n}\, u_n \tag{3.8.30}$$

(ii) *We have that*

$$D(A^*) = \{u \in \ell^2 \mid \mathcal{J}(u) \in \ell^2\} \tag{3.8.31}$$

and

$$A^* u = \mathcal{J}(u) \tag{3.8.32}$$

(iii) *If $u, v \in D(A^*)$, then*

$$\langle u, A^* v\rangle - \langle A^* u, v\rangle = -\lim_{n\to\infty} W(\bar{u}, v)(n) \tag{3.8.33}$$

where

$$W_n(f, g) = a_n(f_{n+1} g_n - f_n g_{n+1}) \tag{3.8.34}$$

(iv) *If $u, v \in D(A^*)$ and*

$$\langle u, A^* v\rangle - \langle A^* u, v\rangle \neq 0 \tag{3.8.35}$$

then both

$$u, v \in D(A^*) \setminus D(\bar{A})$$

Remark. (iii) includes the assertion that the limit exists.

Proof. (i) is a simple summation by parts.

(ii) If $u \in \ell^2$ and $\mathcal{J}(u) \in \ell^2$, then (3.8.30) proves $u \in D(A^*)$ and $A^*u = \mathcal{J}(u)$. Conversely, if $u \in D(A^*)$ and $\eta \in A^*$, then by (3.8.30), $\eta - A^*u$ is a sequence with

$$\sum_{n=1}^{\infty} \overline{(\eta - \mathcal{J}(u))_n}\, w_n = 0$$

for all $w \in D(A)$. Picking $w_n = \delta_{kn}$ shows $\eta = \mathcal{J}(u)$, proving (3.8.32) and so $\mathcal{J}(u) \in \ell^2$.

(iii) By a direct calculation,

$$\sum_{n=1}^{N} [\bar{u}_n \mathcal{J}(v)_n - \overline{\mathcal{J}(u)_n}\, v_n] = W(\bar{u}, v)_N \tag{3.8.36}$$

from which (3.8.33) is immediate.

(iv) If $u \in D(\bar{A})$, then

$$\langle A^*u, v\rangle = \langle \bar{A}u, v\rangle = \langle u, A^*v\rangle \tag{3.8.37}$$

so (3.8.35) fails; similarly, if $v \in D(\bar{A})$. □

For each $z \in \mathbb{C}$, we define two sequences, $\pi(z), \xi(z)$, by

$$\begin{aligned} \pi(z)_n &= p_{n-1}(z) \\ \xi(z)_n &= q_{n-1}(z) \end{aligned} \tag{3.8.38}$$

Of course, $W(\pi, \xi)$ is constant and, by (3.2.22),

$$W(\pi, \xi)_n = -1 \tag{3.8.39}$$

Lemma 3.8.7. *If $d\mu$ solves the moment problem and*

$$m_\mu(z) = \int \frac{d\mu(x)}{x - z} \tag{3.8.40}$$

then $\xi(z) + m(z)\pi(z) \in \ell^2$ for any $z \in \mathbb{C} \setminus \mathbb{R}$.

Proof. By (3.2.24),

$$\xi_n(z) + m(z)\pi_n(z) = \langle p_{n-1}, (\cdot - z)^{-1}\rangle \tag{3.8.41}$$

So, by Bessel's inequality,

$$\sum_n |\xi_n(z) + m(z)\pi_n(z)|^2 \le \int \frac{d\mu(x)}{|x - z|^2} \tag{3.8.42}$$

$$= \frac{\operatorname{Im} m_\mu(z)}{\operatorname{Im} z} \tag{3.8.43}$$

□

Note that if $\{p_{n-1}\}_{n=1}^{\infty}$ is an orthonormal basis, we have that equality holds in (3.8.42)/(3.8.43). Here is one of the main results on the moment problem:

Theorem 3.8.8. *The following are equivalent:*

(i) *For one $z_0 \in \mathbb{C} \setminus \mathbb{R}$, $\pi(z_0) \in \ell^2$.*
(ii) *For one $z_0 \in \mathbb{C} \setminus \mathbb{R}$, $\xi(z_0) \in \ell^2$.*
(iii) *A is not essentially selfadjoint.*
(iv) *For all $z_0 \in \mathbb{C} \setminus \mathbb{R}$, $\pi(z_0) \in \ell^2$ and $\xi(z_0) \in \ell^2$.*
(v) *The moment problem is indeterminate.*

Remark. We will eventually show (see Theorem 3.8.15) that (iv) can be replaced by all of $\mathbb{C}$.

Proof. We will show that (i) $\Leftrightarrow$ (ii) $\Leftrightarrow$ (iii) so (iii) $\Leftrightarrow$ (iv) will be automatic. We will then prove (v) $\Rightarrow$ (i). We will postpone the proof that (iii) $\Rightarrow$ (v).

(i) $\Leftrightarrow$ (ii). By Theorem 3.8.4, the moment problem has solutions. So for some $m_\mu(z) \neq 0$, $\xi(z_0) + m_\mu(z_0)\pi(z_0) \in \ell^2$. This implies $\pi(z_0) \in \ell^2 \Leftrightarrow \xi(z_0) \in \ell^2$.

(i) $\Leftrightarrow$ (iii). There is a unique sequence solving

$$\mathcal{J} u = z_0 u \tag{3.8.44}$$

and

$$u_{n=1} = 1$$

and no solution with $u_{n=1} = 0$. This is given by $u = \pi$. Thus, by Theorem 3.8.6(ii),

$$\ker(A^* - z_0) \neq \{0\} \Leftrightarrow \pi(z_0) \in \ell^2 \tag{3.8.45}$$

Since $\pi(\bar{z}_0) = \overline{\pi(z_0)}$, we see

$$\ker(A^* - z_0) \neq \{0\} \Leftrightarrow \ker(A^* - \bar{z}_0) \neq \{0\}$$

By Proposition 3.8.5(vii),

$$A \text{ essentially selfadjoint} \Leftrightarrow \pi(z_0) \notin \ell^2 \tag{3.8.46}$$

proving (i) $\Leftrightarrow$ (iii).

(iii) $\Leftrightarrow$ (iv). Obviously, (iv) $\Rightarrow$ (i) $\Rightarrow$ (iii). But since (iii) $\Rightarrow$ (i) for any z_0, it implies it for all z_0.

Not (i) $\Rightarrow$ not (v). Since $\pi(z_0) \notin \ell^2$, there is at most one $m(z_0)$ with $\xi(z_0) + m(z_0)\pi(z_0) \in \ell^2$. So for any two μ's solving the moment problem and all $z_0 \in \mathbb{C} \setminus \mathbb{R}$, $m_{\mu_1}(z_0) = m_{\mu_2}(z_0)$, so $\mu_1 = \mu_2$, that is, we have not (v). □

The following depends only on (v) $\Rightarrow$ (i):

Corollary 3.8.9. *If*

$$\sum_{n=1}^{\infty} a_n^{-1} = \infty \tag{3.8.47}$$

then the moment problem is determinate. In particular, if a moment problem is indeterminate, then

$$\lim_{n\to\infty} a_n = \infty \tag{3.8.48}$$

Proof. If $\pi(z_0) \in \ell^2$, then so is $\xi(z_0)$, and thus,

$$a_n^{-1} = (q_n(z_0)p_{n-1}(z_0) - q_{n-1}(z_0)p_{n-1}(z)) \tag{3.8.49}$$

(by (3.2.22)) lies in ℓ^1. Therefore, (3.8.47) implies not (i) implies not (v). □

Lemma 3.8.10. *For any $\{a_j\}_{j=1}^n \in \mathbb{R}^n$, we have*

$$\sum_{j=1}^n (a_1 \dots a_j)^{-1/j} \le 2e \sum_{j=1}^n a_j^{-1} \tag{3.8.50}$$

Proof. We have $1 + x \le e^x$ so $(1 + \frac{1}{n})^n \le e$ and thus, inductively,

$$n^n \le e^n n! \tag{3.8.51}$$

It follows that

$$\begin{aligned}(a_1 \dots a_j)^{-1/j} &= [a_1^{-1}(2a_2^{-1}) \dots (ja_j^{-1})]^{1/j}(j!)^{-1/j} \\ &\le ej^{-2}\sum_{k=1}^j ka_k^{-1}\end{aligned} \tag{3.8.52}$$

by the arithmetic-geometric mean inequality.

Thus,

$$\begin{aligned}\sum_{j=1}^n (a_1 \dots a_j)^{-1/j} &\le e\sum_{k=1}^n a_k^{-1} \sum_{j=k}^n \frac{k}{j^2} \\ &\le 2e\sum_{k=1}^n a_k^{-1}\end{aligned} \tag{3.8.53}$$

since

$$\sum_{j=k}^n \frac{k}{j^2} \le 2k \sum_{j=k}^\infty \frac{1}{j(j+1)} = 2 \tag{3.8.54}$$

□

Corollary 3.8.11 (Carleman's criterion). *If*

$$\sum_{n=1}^\infty c_{2n}^{-1/2n} = \infty \tag{3.8.55}$$

then the moment problem is determinate.

Proof. Since $p_n(x) = (a_1 \dots a_n)^{-1} x^n +$ lower order,

$$\langle (a_1 \dots a_n)^{-1} x^n, p_n \rangle = 1 \tag{3.8.56}$$

and thus, by the Schwarz inequality,

$$c_{2n}^{-1/2n} \le (a_1 \dots a_n)^{-1/n} \tag{3.8.57}$$

By (3.8.50), we see (3.8.55) implies (3.8.47). □

Example 3.8.1, revisited. If $\alpha \ge 1$,

$$\begin{aligned}\int x^n \exp(-|x|^\alpha) &\le 2 + \int x^n \exp(-|x|^1) \\ &= 2 + 2n! \le 4n^n\end{aligned} \tag{3.8.58}$$

and

$$c_{2n}^{-1/2n} \ge \frac{1}{8n} \tag{3.8.59}$$

Thus, (3.8.55) holds, and so the moment problem is determinate. □

To get the last step in the proof of Theorem 3.8.8, we need to analyze selfadjoint extensions of A when $\bar{A} \ne A^*$, that is, operators B with $\bar{A} \subset B = B^*$. Since $\bar{A} \subset B$ implies $B^* \subset A^*$, we have

$$\bar{A} \subsetneq B = B^* \subsetneq A^* \tag{3.8.60}$$

where $B \ne \bar{A}$ and $B \ne A^*$ comes from $\bar{A} \ne A^* \ne A^{**}$. In our case where $D(A^*)/D(\bar{A})$ has dimension 2, we must thus have $\dim(D(B)/D(A)) = 1$, which simplifies the analysis.

Theorem 3.8.12. *Suppose $D(A^*)/D(\bar{A})$ has dimension 2. Then*

(i) $D(B) = D(A) + [\varphi]$ *with* $\varphi \in D(A^*) \setminus D(A)$ *is the domain of a selfadjoint extension (i.e.,* $A^* \restriction D(B)$ *is selfadjoint) if and only if*

$$\langle \varphi, A^* \varphi \rangle \in \mathbb{R} \tag{3.8.61}$$

(ii) *Suppose* $\varphi, \psi = D(A^*)$ *with* $\langle \varphi, A^*\psi \rangle$, $\langle \psi, A^*\varphi \rangle$, $\langle \varphi, A^*\varphi \rangle$, $\langle \psi, A^*\psi \rangle$ *all real. Let* $t \in \mathbb{R} \cup \{\infty\}$ *and let*

$$\varphi_t = \frac{\varphi + t\psi}{1 + |t|} \tag{3.8.62}$$

(where φ_∞ *is interpreted as* ψ*). Then*

$$D(B_t) = D(\bar{A}) + [\varphi_t] \qquad B_t = A^* \restriction D(B_t) \tag{3.8.63}$$

describes all the selfadjoint extensions of A.

Proof. (i) By (3.8.60), $D(B)/D(A)$ is of dimension 1, so for every selfadjoint extension, B, $D(B)$ always has the claimed form. Since $\varphi \in D(B)$,

$$\langle \varphi, A^*\varphi \rangle = \langle \varphi, B\varphi \rangle \tag{3.8.64}$$

is real.

Conversely, if (3.8.61) holds and $\eta \in D(A)$, then

$$\langle \varphi + \eta, A^*(\varphi + \eta) \rangle = \langle \varphi, A^*\varphi \rangle + \langle \eta, A\eta \rangle + \langle \varphi, A\eta \rangle + \langle A\eta, \varphi \rangle \tag{3.8.65}$$

is real, so $A^* \restriction D(A) + [\varphi]$ has real expectation values. By polarization, it is Hermitian. Since $\bar{A} \subsetneqq B \subset B^* \subsetneqq A^*$, we see that $D(B^*)$ must be $D(B)$ since every subspace between $D(B)$ and $D(A^*)$ is either $D(B)$ or $D(A^*)$. Thus, $B = B^*$.

(ii) We have, for all $\eta \in D(\bar{A})$,

$$\operatorname{Im}\langle \varphi + \alpha\psi + \eta, A^*(\varphi + \alpha\psi + \eta) \rangle = (\operatorname{Im}\alpha)[\langle \varphi, A^*\psi \rangle - \langle \psi, A^*\varphi \rangle]$$

Since there is $\alpha, \beta \in \mathbb{C}$ and an $\eta \in D(\bar{A})$ with $A^*(\beta\varphi + \alpha\psi + \eta) = i(\beta\varphi + \alpha\psi + \eta)$, we conclude that $\langle \varphi, A^*\psi \rangle - \langle \psi, A^*\varphi \rangle \neq 0$. It follows that (3.8.61) holds for $\varphi + \alpha\psi$ if and only if $\alpha \in \mathbb{R}$. Given (i), this proves (ii). □

Later (see Theorem 3.8.15), we will prove that if A is not selfadjoint for the concrete Jacobi matrix, then not only is $\pi(z_0), \xi(z_0) \in \ell^2$ for $z_0 \in \mathbb{C} \setminus \mathbb{R}$ but also for $z_0 \in \mathbb{R}$. We use that for now for $z_0 = 0$. We have

$$\mathcal{J}(\pi(0)) = 0 \qquad \mathcal{J}(\xi(0)) = \delta_{\cdot 1} \tag{3.8.66}$$

so if A is the operator of $\mathcal{J}$ restricted to finite sequences, by Theorem 3.8.6(ii), we have

$$\langle \xi(0), A^*(\pi(0)) \rangle = \langle \pi(0), A^*(\pi(0)) \rangle = \langle \xi(0), A^*(\xi(0)) \rangle = 0 \tag{3.8.67}$$

$$\langle \pi(0), A^*(\xi(0)) \rangle = 1 \tag{3.8.68}$$

By Theorem 3.8.6(iv), we have $\pi(0), \xi(0) \in D(A^*) \setminus D(A)$ and, by Theorem 3.8.12, there is a one-parameter family, $\{B_t\}_{t \in \mathbb{R} \cup \{\infty\}}$, of selfadjoint extensions with

$$D(B_t) = D(\bar{A}) + [\xi(0) + t\pi(0)] \tag{3.8.69}$$

Proposition 3.8.13. *Suppose $\bar{A}$ is not essentially selfadjoint.*

(i) *For each $z_0 \in \mathbb{C} \setminus \mathbb{R}$, we have $\pi(z_0), \xi(z_0) \in D(A^*) \setminus D(A)$. For each t, there is an $a_t(z_0) \in \mathbb{C}$ so that*

$$\xi(z_0) + a_t(z_0)\pi(z_0) \in D(B_t) \tag{3.8.70}$$

and for every such z_0, all $a_t(z_0)$ are distinct as t varies.

(ii)

$$\langle \delta_1, (B_t - z_0)^{-1}\delta_1 \rangle = a_t(z_0) \tag{3.8.71}$$

In particular, if $\bar{A}$ is not selfadjoint, there are multiple solutions to the moment problem.

Remark. The spectral measures for B_t, which solve the moment problem, are called the *von Neumann solutions* of the moment problem.

Proof. As noted in the proof of Theorem 3.8.8,

$$\ker(A^* - z_0) = [\pi(z_0)] \tag{3.8.72}$$

Moreover, as in (3.8.66),

$$(A^* - z_0)\xi(z_0) = \delta_{\cdot 1} \tag{3.8.73}$$

Thus, every solution of $(A^* - z_0)\eta = \delta_0$ has the form

$$\eta = \xi(z_0) + c\pi(z_0) \tag{3.8.74}$$

So for some $a_t(z_0) \in \mathbb{C}$,

$$(B_t - z_0)^{-1}\delta_1 = \xi(z_0) + a_t(z_0)\pi(z_0) \tag{3.8.75}$$

Let η_t be the right side of (3.8.75). By (3.8.33),

$$\langle \pi(\bar{z}_0), A^*\eta_t \rangle - \langle A^*\pi(\bar{z}_0), \eta_t \rangle = 1 \tag{3.8.76}$$

we conclude that $\eta_t \in D(A^*) \setminus D(A)$, so

$$D(B_t) = D(\bar{A}) + [\eta_t] \tag{3.8.77}$$

which implies that the η_t are distinct for distinct t.

Finally, by (3.8.75),

$$\langle \delta_1, (B_t - z_0)^{-1}\delta_1 \rangle = a_t(z_0) \tag{3.8.78}$$

proving (3.8.71). □

Next, we turn to the claim that in the indeterminate case, $\pi(z_0), \xi(z_0) \in \ell^2$ also for $z_0 \in \mathbb{R}$. We depend on a useful general perturbation theorem.

Theorem 3.8.14. *Suppose $\{A_j\}_{j=1}^\infty$ and $\{\tilde{A}_j\}_{j=1}^\infty$ are two sequences of bounded operators with bounded inverses, and define*

$$T_n = A_n \dots A_1 \tag{3.8.79}$$

$$\tilde{T}_n = \tilde{A}_n \dots \tilde{A}_1 \tag{3.8.80}$$

$$B_k = T_k^{-1}(\tilde{A}_k - A_k)T_{k-1} \tag{3.8.81}$$

where $T_0 = \tilde{T}_0 = \mathbf{1}$. Then

(i) *We have for each n,*

$$\|T_n^{-1}\tilde{T}_n\| \le \exp\left(\sum_{j=1}^n \|B_j\|\right) \tag{3.8.82}$$

(ii) *If*

$$\sum_{n=1}^\infty \|B_n\| < \infty \tag{3.8.83}$$

then

$$\lim_{n\to\infty} T_n^{-1}\tilde{T}_n \tag{3.8.84}$$

exists and is given by

$$\lim_{n\to\infty} T_n^{-1}\tilde{T}_n = \mathbf{1} + \sum_{j=1}^\infty B_j T_{j-1}^{-1}\tilde{T}_{j-1} \tag{3.8.85}$$

(iii) *If*

$$\sum_{n=1}^{\infty} \|T_n\|^2 < \infty \tag{3.8.86}$$

and (3.8.83) *holds, then*

$$\sum_{n=1}^{\infty} \|\tilde{T}_n\|^2 < \infty \tag{3.8.87}$$

Remark. By (3.8.81) and (3.8.85), we get

$$\lim_{n\to\infty} T_n^{-1}\tilde{T}_n = \mathbf{1} + \sum_{j=1}^{\infty} T_j^{-1}(\tilde{A}_j - A_j)\tilde{T}_{j-1} \tag{3.8.88}$$

Proof. Noticing that

$$T_k^{-1} A_k T_{k-1} = \mathbf{1} \tag{3.8.89}$$

we have

$$T_k^{-1} \tilde{A}_k T_{k-1} = \mathbf{1} + B_k \tag{3.8.90}$$

Therefore,

$$\begin{aligned} T_n^{-1}\tilde{T}_n &= (T_n^{-1}\tilde{A}_n T_{n-1})(T_{n-1}^{-1}\tilde{A}_{n-1}T_{n-2})\dots(T_1^{-1}\tilde{A}_1 T_0) \\ &= (\mathbf{1}+B_n)\dots(\mathbf{1}+B_1) \end{aligned} \tag{3.8.91}$$

(i) Thus,

$$\|T_n^{-1}\tilde{T}_n\| \le \prod_{j=1}^{n}(1+\|B_j\|) \le \exp\left(\sum_{j=1}^{n}\|B_j\|\right) \tag{3.8.92}$$

(ii) By (3.8.91), we have

$$T_n^{-1}\tilde{T}_n = \mathbf{1} + \sum_{j=1}^{n} B_j(\mathbf{1}+B_{j-1})\dots(\mathbf{1}+B_1) \tag{3.8.93}$$

$$= \mathbf{1} + \sum_{j=1}^{n} B_j T_{j-1}^{-1}\tilde{T}_{j-1} \tag{3.8.94}$$

By (3.8.82),

$$\|B_j T_{j-1}^{-1}\tilde{T}_{j-1}\| \le \|B_j\| \exp\left(\sum_{k=1}^{\infty}\|B_k\|\right) \tag{3.8.95}$$

so the sum is absolutely convergent, implying that the limit exists and is given by (3.8.85).

(iii) By (3.8.82),

$$\|\tilde{T}_n\| \le \|T_n\| \exp\left(\sum_{j=1}^{\infty}\|B_j\|\right) \tag{3.8.96}$$

so (3.8.86) implies (3.8.87). □

To apply this to moment problems, $T_n, A_n, \dots$ will be 2×2 transfer matrices, but we will want to modify from the definition in Section 3.2. There we added an a_n to the lower component of vectors to get a transfer matrix of determinant one. With a_n's bounded from above, this is normally harmless, but here our a_n's are unbounded so we will modify. Given Jacobi parameters $\{a_n, b_n\}_{n=1}^\infty$, we define (with $a_0 \equiv 1$) for this section only,

$$A_n(z) = \begin{pmatrix} \frac{z-b_n}{a_n} & \frac{-a_{n-1}}{a_n} \\ 1 & 0 \end{pmatrix} \tag{3.8.97}$$

so

$$\begin{pmatrix} p_n(z) \\ p_{n-1}(z) \end{pmatrix} = A_n(z) \begin{pmatrix} p_{n-2}(z) \\ p_{n-2}(z) \end{pmatrix} \tag{3.8.98}$$

and

$$T_n(z) = \begin{pmatrix} p_n(z) & -q_n(z) \\ p_{n-1}(z) & -q_{n-1}(z) \end{pmatrix} \tag{3.8.99}$$

to be compared with (3.2.19). Now $\det(T_n) \neq 1$ but rather

$$\det(T_n) = a_n^{-1} \tag{3.8.100}$$

and thus,

$$T_n(z)^{-1} = a_n \begin{pmatrix} -q_{n-1}(z) & q_n(z) \\ -p_{n-1}(z) & p_n(z) \end{pmatrix} \tag{3.8.101}$$

Our perturbation will be to change z to w, so

$$A_n(w) - A_n(z) = \begin{pmatrix} \frac{w-z}{a_n} & 0 \\ 0 & 0 \end{pmatrix} \tag{3.8.102}$$

and

$$B_n \equiv T_n(z)^{-1}(A_n(w) - A_n(z))T_{n-1}(z) \tag{3.8.103}$$

The a_n in (3.8.101) and the a_n^{-1} in (3.8.102) cancel! Thus, with

$$N_n(z) = (|p_n(z)|^2 + |p_{n-1}(z)|^2 + |q_n(z)|^2 + |q_{n-1}(z)|^2)^{1/2} \tag{3.8.104}$$

we obtain

$$\|B_n\| \leq |w - z|\, N_n(z) N_{n-1}(z) \tag{3.8.105}$$

and by the Schwarz inequality,

$$\sum_{n=1}^\infty N_n(z)^2 < \infty \Rightarrow \sum_{n=1}^\infty \|B_n\| < \infty \tag{3.8.106}$$

Thus, we can apply Theorem 3.8.14 and find

Theorem 3.8.15. *If $\pi(z), \xi(z)$ are both in ℓ^2 for a single z, then $\pi(w), \xi(w)$ are in ℓ^2 for any $w \in \mathbb{C}$ and*

$$\lim_{n\to\infty} T_n(z)^{-1} T_n(w) \tag{3.8.107}$$

exists.

One defines four functions, $A(z)$, $B(z)$, $C(z)$, and $D(z)$, by

$$\lim_{n\to\infty} T_n^{-1}(z)T_n(w=0) = \begin{pmatrix} -B(z) & -A(z) \\ D(z) & C(z) \end{pmatrix} \tag{3.8.108}$$

and the Nevanlinna matrix by

$$N(z) = \begin{pmatrix} A(z) & C(z) \\ B(z) & D(z) \end{pmatrix} \tag{3.8.109}$$

By (3.8.88), (3.8.99), (3.8.101), and (3.8.102), we get

Proposition 3.8.16. *The Nevanlinna matrix is given by*

$$A(z) = z\sum_{n=0}^{\infty} q_n(0)q_n(z) \tag{3.8.110}$$

$$B(z) = -1 + z\sum_{n=0}^{\infty} q_n(0)p_n(z) \tag{3.8.111}$$

$$C(z) = 1 + z\sum_{n=0}^{\infty} p_n(0)q_n(z) \tag{3.8.112}$$

$$D(z) = z\sum_{n=0}^{\infty} p_n(0)p_n(z) \tag{3.8.113}$$

These functions are entire functions obeying

$$|A(z)| \le C_\varepsilon \exp(\varepsilon|z|) \tag{3.8.114}$$

and similarly for B, C, D. Near $z = 0$,

$$B(z) = -1 + O(z) \tag{3.8.115}$$

$$D(z) = D_0 z + O(z^2) \tag{3.8.116}$$

where

$$D_0 > 0 \tag{3.8.117}$$

Proof. The formulae follow from the earlier equations. (3.8.115) is immediate, as is (3.8.116) where

$$D_0 = \sum_{n=0}^{\infty} p_n(0)^2 > 0 \tag{3.8.118}$$

To get (3.8.114), we note that

$$B_k(z) = zb_k \tag{3.8.119}$$

with b_k a constant matrix with $\sum_{k=1}^{\infty}\|b_k\| < \infty$. Thus,

$$\|(1+B_N)\dots(1+B_k)\| \le \prod_{j=1}^{n}(1+|z|\,\|b_j\|)\exp\left(|z|\sum_{j=n+1}^{N}\|b_j\|\right) \tag{3.8.120}$$

from which (3.8.114) follows. □

We can express the resolvent of the selfadjoint extensions, B_t, in terms of the Nevanlinna matrix:

Theorem 3.8.17. *Consider an indeterminate moment problem. For* $t \in \mathbb{R} \cup \{\infty\}$ *and* $z \in \mathbb{C} \setminus \mathbb{R}$, *the resolvent of the selfadjoint extensions,* B_t, *is given by*

$$(\delta_1, (B_t - z)^{-1}\delta_1) \equiv F(z, t) \tag{3.8.121}$$

where for $z, w \in \mathbb{C}$,

$$F(z, w) \equiv -\frac{C(z)w + A(z)}{D(z)w + B(z)} \tag{3.8.122}$$

Proof. Given a sequence, $\{s_n\}_{n=1}^{\infty}$, we let $R_n(s) \in \mathbb{C}^2$ be defined by

$$R_n(s) = (s_{n+1}, s_n) \tag{3.8.123}$$

and we define $w_n : \mathbb{C}^2 \times \mathbb{C}^2 \to \mathbb{C}$ by

$$w_n((\alpha, \beta), (\gamma, \delta)) = a_n(\alpha\delta - \beta\gamma) \tag{3.8.124}$$

so that

$$W_n(f, g) = w_n(R_n(f), R_n(g)) \tag{3.8.125}$$

Constancy of the Wronskian for solutions of the same difference equation shows that for any $z \in \mathbb{C}$ and $u, v \in \mathbb{C}^2$,

$$w_n(T_n(z)u, T_n(z)v) = w_0(u, v) \tag{3.8.126}$$

By (3.8.33), if $f, g \in D(B_t)$, then

$$\lim_{n\to\infty} w_n(\overline{R_n(f)}, R_n(g)) = 0 \tag{3.8.127}$$

since $\langle f, B_t g\rangle = \langle B_t f, g\rangle$ by Hermiticity of B_t.

Since

$$R_n(\xi(0) + t\pi(0)) = T_n(0)\begin{pmatrix} t \\ 1 \end{pmatrix} \tag{3.8.128}$$

$$R_n(\xi(z_0) + a_t(z_0)\pi(z_0)) = T_n(z_0)\begin{pmatrix} a_t(z_0) \\ 1 \end{pmatrix} \tag{3.8.129}$$

we see, by (3.8.127), that

$$\lim_{n\to\infty} w_n\left(T_n(0)\begin{pmatrix} t \\ 1 \end{pmatrix}, T_n(z_0)\begin{pmatrix} a_t(z_0) \\ 1 \end{pmatrix}\right) = 0 \tag{3.8.130}$$

So, by (3.8.126),

$$\lim_{n\to\infty} w_0\left(\begin{pmatrix} t \\ 1 \end{pmatrix}, T_n(0)^{-1}T_n(z_0)\begin{pmatrix} a_t(z_0) \\ 1 \end{pmatrix}\right) = 0 \tag{3.8.131}$$

By the existence of the limit, for some constant c,

$$\begin{pmatrix} a_t(z_0) \\ 1 \end{pmatrix} = cT_n(z_0)^{-1}T_n(0)\begin{pmatrix} t \\ 1 \end{pmatrix} \tag{3.8.132}$$

Given (3.8.108), this implies (3.8.121)/(3.8.122). □

Lemma 3.8.18. *For $z \in \mathbb{C}_+$, $\{F(z,t) \mid t \in \mathbb{R} \cup \{\infty\}\}$ is a circle in the upper complex plane. $F(z,\,\cdot\,)$ maps $\mathbb{C}_+$ to the interior of the disk bounded by the circle.*

Proof. By (3.8.121), F maps $\mathbb{R} \cup \{\infty\}$ to $\mathbb{C}_+$ and not to ∞, so the image is a circle in $\mathbb{C}$. Suppose $[F(z,\,\cdot\,)]^{-1}(\infty)$ lies in $\mathbb{C}_-$. Then $F(z,\,\cdot\,)$ maps $\mathbb{C}_-$ to the outside of the circle, and so $\mathbb{C}_+$ to the inside. Since, for $z \in \mathbb{C}_+$, it can never lie in $\mathbb{R}$, by continuity, $[F(z,\,\cdot\,)]^{-1}(\infty)$ is either always in $\mathbb{C}_-$ (or always in $\mathbb{C}_+$), so it suffices to show this for $z = i\varepsilon$, that is, that $\operatorname{Im}(-B(i\varepsilon)/D(i\varepsilon)) < 0$ for ε small and positive. This follows from (3.8.120)/(3.8.121). □

Next, we relate solutions of the moment problem to asymptotics of the Stieltjes transform.

Proposition 3.8.19. *Let μ be a probability measure on $\mathbb{R}$ solving* (3.8.1) *and let*

$$G_\mu(z) = \int \frac{d\mu(x)}{x - z} \tag{3.8.133}$$

Let

$$R_N(\mu; iy) = G_\mu(iy) + \sum_{n=0}^{N} (-i)^{n+1} y^{-n-1} c_n \tag{3.8.134}$$

Then

$$|R_N(\mu; iy)| \le \begin{cases} c_{N+1} y^{-N-2} & N \text{ odd} \\ \frac{1}{2}(c_N + c_{N+2}) y^{-N-2} & N \text{ even} \end{cases} \tag{3.8.135}$$

Conversely, if $G(z)$ is a Herglotz function, so R_N, given by (3.8.134), *is $O(y^{-N-2})$ for each N, then G is given by* (3.8.133) *for some measure μ solving* (3.8.1).

Proof. If (3.8.133) holds and μ obeys (3.8.1), write

$$(x - iy)^{-1} = \sum_{n=0}^{N} x^n (-i)^n y^{-n-1} + (-i)^{-N-1} x^{N+1} y^{-N-2} \left(1 - \frac{x}{iy}\right)^{-1} \tag{3.8.136}$$

to see that R_N, given by (3.8.134), is given by

$$R_N(\mu; iy) = (-i)^{N+1} y^{-N-2} \int x^{N+1} \left(1 - \frac{x}{iy}\right)^{-1} d\mu(x) \tag{3.8.137}$$

Since $|1 - \frac{x}{iy}| \ge 1$ for x, y real, the N odd case of (3.8.135) is immediate. For N even, use the fact that for such N,

$$|x|^{N+1} \le \tfrac{1}{2} x^N + x^{N+2} \tag{3.8.138}$$

For the converse, start with the Herglotz representation, (2.3.87). Since (3.8.134)/(3.8.135) imply

$$\lim_{y \to \infty} |y|^{-1} |G(iy)| = 0 \tag{3.8.139}$$

we see that $A = 0$. They also imply that

$$yG(iy) \to ic_0 \tag{3.8.140}$$

from which one first sees (with ρ replaced by μ)

$$\int d\mu(x) = c_0 \tag{3.8.141}$$

since

$$\operatorname{Im} yG(iy) = \int \frac{y^2}{x^2+y^2}\, d\mu(x) \tag{3.8.142}$$

and we can use the monotone convergence theorem, and then that there is a cancellation of real parts that implies (3.8.133).

From (3.8.134)/(3.8.135), one sees inductively, using (3.8.136), that

$$\int \frac{(iy)^2 x^{2n-1}}{x-iy}\, d\mu(x) + i\gamma c_{2n-1} \to c_{2n} \tag{3.8.143}$$

which implies, taking real and imaginary parts, that

$$c_{2n} = \lim_{y\to\infty} \int \frac{y^2 x^{2n}}{x^2+y^2}\, d\mu(x) \tag{3.8.144}$$

$$c_{2n-1} = \lim_{y\to\infty} \int \frac{y^2 x^{2n-1}}{x^2+y^2}\, d\mu(x) \tag{3.8.145}$$

Monotone convergence and the first of these implies $\int x^{2n}\, d\mu = c_{2n}$ and then dominated convergence and (3.8.145) implies $\int x^{2n-1}\, d\mu = c_{2n-1}$. $\square$

Corollary 3.8.20. *For $z \in \mathbb{C}_+$, let*

$$\mathbb{D}(z) = \{F(z,w) \mid \operatorname{Im} w > 0\} \tag{3.8.146}$$

be the disk of Lemma 3.8.18. If G has the form (3.8.134) *where μ solves* (3.8.1), *then*

$$G(z) \in \overline{\mathbb{D}(z)} \tag{3.8.147}$$

for all $z \in \mathbb{C}_+$. Conversely, if G is an analytic function on $\mathbb{C}_+$ obeying (3.8.150), *then G has the form* (3.8.133) *for some μ obeying* (3.8.1).

Proof. By Proposition 3.8.19, $G_\mu(iy)$ has an asymptotic series

$$G(iy) \sim -\sum_{n=0}^{\infty} (-i)^{n+1} y^{-n-1} c_n \tag{3.8.148}$$

uniformly in the von Neumann solutions. Since these solutions fill out the circle at the boundary of $\overline{\mathbb{D}(z)}$, the estimates hold in all on $\overline{\mathbb{D}(z)}$, so G solves the moment problem by Proposition 3.8.19.

Conversely, by (3.8.43), if μ solves the moment problem,

$$G_\mu(z) \in \overline{\Delta(z)} \tag{3.8.149}$$

where

$$\overline{\Delta(z_0)} = \left\{ w \,\middle|\, \|\xi(z_0) + w\pi(z_0)\|^2 \le \frac{\operatorname{Im} w}{\operatorname{Im} z_0} \right\}$$

This set is given by a quadratic inequality in $\operatorname{Re} w$, $\operatorname{Im} w$ whose quartic term is $|w|^2 \|\pi(z_0)\|^2$. Such a set always describes a disk or the empty set. Since equality holds in (3.8.43) for von Neumann solutions, $\partial\Delta(z) = \partial\mathbb{D}(z)$, so $\Delta(z) = D(z)$ and (3.8.149) is (3.8.147). $\square$

Here is the main result on the description of the solutions of the moment problem in the indeterminate case:

Theorem 3.8.21 (Nevanlinna's Parametrization). *Let $\{c_n\}_{n=1}^{\infty}$ be the moments of an indeterminate problem, and let A, B, C, D be the elements of the Nevanlinna matrix, and F given by (3.8.122). There is a one-one correspondence between $\mathcal{H}$, the set of all analytic functions, φ, of $\mathbb{C}_+$ to $\overline{\mathbb{C}}_+$ so that μ_φ is given by*

$$\int \frac{d\mu_\varphi(x)}{x-z} = F(z, \varphi(z)) \tag{3.8.150}$$

The von Neumann solutions correspond to $\varphi(z) \equiv t$ and all other solutions have $\operatorname{Ran}(\varphi) \subset \mathbb{C}_+$.

Proof. Any function of the form $G(z) \equiv F(z, \varphi(z))$ has G obeying (3.8.147) by Lemma 3.8.18. Conversely, if G obeys (3.8.150), then, because $F(z, \,\cdot\,)$ is a bijection of $\mathbb{C}$ taking $\mathbb{C}_+$ to $\mathbb{D}(z)$, there is a unique φ obeying $G(z) = F(z, \varphi(z))$ with φ analytic or infinite.

By the open mapping theorem, either $\varphi(z) = t \in \mathbb{R} \cup \{\infty\}$ or $\operatorname{Ran}(\varphi) \in \mathbb{C}_+$. Given Corollary 3.8.20, this proves the theorem. $\square$

This allows further analysis of solutions, of which the following is typical:

Theorem 3.8.22. (i) *The von Neumann solutions of an indeterminate moment problem are discrete pure point measures.*
(ii) *If φ is a rational Herglotz function, $d\mu_\varphi$ is pure point.*
(iii) *The positions of the pure points and weights of the von Neumann solutions are real analytic in t. The positions are nonconstant.*
(iv) *There are always purely a.c. and purely s.c. solutions of an indeterminate problem.*

Proof. (i), (ii) In these cases, G_μ has an analytic continuation to an entire meromorphic function.

(iii) This follows from analyticity of A, B, C, D and the form of $F(z, t)$.

(iv) If $d\mu_t$ is the von Neumann solution associated to B_t and $d\nu(t)$ is a probability measure, then

$$d\eta_\nu(x) = \int d\mu_t(x)\, d\nu(t) \tag{3.8.151}$$

is a solution of the moment problem. By (iii), $d\eta_\nu$ is a.c. (resp. s.c.) if $d\nu$ is a.c. (resp. s.c.). $\square$

Remarks and Historical Notes. The critical paper on the moment problem is by Stieltjes [422]. Earlier, Chebyshev had asked about uniqueness for Gaussian measures. The approach via selfadjoint operators was pioneered by Stone [423] and the transfer matrix connection was exploited especially by Simon [395], which we follow in much of this section. For other presentations, see Akhiezer [13] and Shohat–Tamarkin [385]. The name von Neumann solutions comes from Simon [395], after von Neumann's theory of selfadjoint extensions. Such solutions are called N-extremal in Akhiezer [13].

The Nevanlinna parametrization is from [325]. A further result (see [13, 395]) is that the polynomials are dense in $L^2(\mathbb{R}, d\mu)$ if and only if $d\mu$ is a von Neumann solution and their closure has finite codimension if and only if the Nevanlinna function, φ, is rational. All these solutions are extreme points in the convex set of solutions of the moment problem, proving that the extreme points are dense.

Carleman's criterion (Corollary 3.8.11) is due to Carleman [75].

The awkward terminology (at least in English) "determinate" and "indeterminate" comes from the French. While Stieltjes was Dutch, his paper [422] is in French.

There are actually two moment problems discussed in the next section: what we have called "the moment problem" (i.e., solution of (3.8.1) with the measure allowed to be supported anywhere on $\mathbb{R}$) is more properly the *Hamburger moment problem*. The *Stieltjes moment problem* is the problem one gets by restricting to measures supported on $[0, \infty)$.

There is a simple relation between the two problems. Let $d\rho_0$ be a probability measure on $[0, \infty)$ with moments c_n. Define $d\tilde{\rho}_0$ on $\mathbb{R}$ by

$$d\tilde{\rho}_0(x) = \tfrac{1}{2}[\chi_{[0,\infty)}(x)\, d\rho(x^2) + \chi_{(-\infty,0]}(x)\, d\rho(x^2)] \tag{3.8.152}$$

and let

$$\Gamma_n = \int x^n\, d\tilde{\rho}_0(x) = \begin{cases} 0 & n \text{ odd} \\ c_{n/2} & n \text{ even} \end{cases} \tag{3.8.153}$$

(3.8.152) sets up a one-one correspondence between all solutions of the Stieltjes moment problem with moments c_n and all solutions of the Hamburger moment problem with moments Γ_n symmetric under $x \to -x$. It is a basic fact that any indeterminate Hamburger moment problem with vanishing odd moments has multiple solutions that are invariant under $x \to -x$, namely, the von Neumann solutions with $t = 0$ and $t = \infty$. This implies immediately that

Theorem 3.8.23. *Let $(d\rho_0, c_n)$ be a measure and set of moments on $[0, \infty)$. Let $(d\tilde{\rho}_0, \Gamma_n)$ be given by* (3.8.152)/(3.8.153). *Then the Stieltjes moment problem for $\{c_n\}$ is determinate* (*resp. indeterminate*) *if and only if the Hamburger moment problem for $\{\Gamma_n\}$ is determinate* (*resp. indeterminate*).

Theorem 3.8.23 goes back at least to Chihara [83] and appears also in Berg [42] and Simon [395].

3.9 THE KREIN DENSITY THEOREM AND INDETERMINATE MOMENT PROBLEMS

If one sought a connection between measures on $\partial\mathbb{D}$ and measures on $\mathbb{R}$, one might not think first of $x = z + z^{-1}$, which is quadratic, but rather

$$x = i\left[\frac{1-z}{1+z}\right] \tag{3.9.1}$$

which is a fractional linear mapping of $\mathbb{D}$ to $\mathbb{C}_+$ and its inverse

$$z = \frac{i-x}{i+x} \tag{3.9.2}$$

For the version of Szegő's theorem that gives asymptotics of the leading term in OPs, this is not useful—it relates polynomials in z to polynomials in $\frac{i-x}{i+x}$—or what is the same time polynomials in $(i+x)^{-1}$ since

$$\frac{i-x}{i+x} = -1 + \frac{2i}{i+x} \tag{3.9.3}$$

Krein [250] realized it could be used to transfer the density theorem (Theorem 2.11.5), which gives criteria for when $\{e^{in\theta}\}_{n=0}^{\infty}$ span $L^2(\partial\mathbb{D}, d\mu)$, to a continuum analog:

Theorem 3.9.1 (Krein's Density Theorem [250]). *Let*

$$d\rho = F(x)\,dx + d\rho_s \tag{3.9.4}$$

be a finite measure on $\mathbb{R}$. *Then the span of* $\{e^{i\alpha x}\}_{\alpha\ge 0}$ *is dense in* $L^2(\mathbb{R}, d\rho(x))$ *if and only if*

$$\int_{-\infty}^{\infty} \frac{\log F(x)}{1+x^2}\,dx = -\infty \tag{3.9.5}$$

Remark. As usual, $F \in L^1$ implies that the integral with $\log_+ F$ is finite, so the integral is either convergent or it diverges to $-\infty$.

As a first preliminary for the proof, we need

Lemma 3.9.2. *For any finite measure* $d\rho$ *on* $\mathbb{R}$, *the span of* $\{(i+x)^{-n}\}_{n=0}^{\infty}$ *is the same as the span of* $\{e^{i\alpha x}\}_{\alpha=0}^{\infty}$.

Proof. Suppose that f is orthogonal to $(i+x)^{-n}$ for $n = 0, 1, 2, \dots$. Then since $1 - ix = -i(i+x)$, we see that if

$$F(w) = \int f(x)(1-wx)^{-1}\,d\rho(x) \tag{3.9.6}$$

which is analytic in $\mathbb{C}_+$, then $\frac{d^nF}{dw^n}(i) = 0$ for all n. So $F = 0$ and thus (taking derivatives of F), we have that f is orthogonal to $(1-wx)^{-n}$ for all $w \in \mathbb{C}_+$ and $n = 0, 1, 2, \dots$.

Since for $\alpha \ge 0$,

$$\left(1 - \frac{i\alpha x}{n}\right)^{-n} \to e^{i\alpha x} \tag{3.9.7}$$

pointwise in x with $|(1-\frac{i\alpha x}{n})^{-n}| \leq 1$, we have convergence in L^2 so f is orthogonal to $\{e^{i\alpha x}\}_{\alpha\geq 0}$.

Conversely, if f is orthogonal to $\{e^{i\alpha x}\}_{\alpha\geq 0}$, we have f orthogonal to $(1-i\beta x)^{-1}$ for all $\beta > 0$ since

$$\int_0^\infty e^{i\beta\alpha x}\, e^{-\alpha}\, d\alpha = (1-i\beta x)^{-1} \tag{3.9.8}$$

and the integral converges weakly in $L^2(\mathbb{R}, d\rho)$. But then, by analyticity of F (given by (3.9.6), F is zero on $\mathbb{C}_+$ so its derivatives at i are all zero and f is orthogonal to $\{(i+x)^{-n}\}_{n=0}^\infty$. □

As a second preliminary, we introduce an analog of the Szegő map, Sz, of Section 1.9. Notice that the boundary value of (3.9.1) on $\partial\mathbb{D}$ is

$$x(e^{i\theta}) = \tan\left(\frac{\theta}{2}\right) \tag{3.9.9}$$

Thus, we define the *Krein map* Kr: $\mathcal{M}_{+,1}(\partial\mathbb{D}) \to \mathcal{M}_{+,1}(\mathbb{R}\cup\{\infty\})$ by $d\rho = \text{Kr}(d\mu)$ if

$$\int g(\theta)\, d\mu(\theta) = \int g(2\arctan(x))\, d\rho(x) \tag{3.9.10}$$

Kr is a one-one correspondence between $\{\mu \in \mathcal{M}_{+,1}(\partial\mathbb{D}) \mid \mu(\{-1\}) = 0\}$ and measures $d\rho$ in $\mathcal{M}_{+,1}(\mathbb{R})$. Notice also that if

$$d\mu(\theta) = w(\theta)\,\frac{d\theta}{2\pi} + d\mu_{\text{s}} \tag{3.9.11}$$

and $d\rho$ is given by (3.9.4), then because (3.9.9) says

$$\frac{dx}{d\theta} = \frac{1+x^2}{2} \tag{3.9.12}$$

we have that

$$w(\theta) = \pi\sec^2\left(\frac{\theta}{2}\right) F\left(\tan\left(\frac{\theta}{2}\right)\right) \tag{3.9.13}$$

or

$$F(x) = \pi^{-1}\,\frac{w(2\arctan(x))}{1+x^2} \tag{3.9.14}$$

Proof of Theorem 3.9.1. By Lemma 3.9.2 and (3.9.3), $\{e^{i\alpha x}\}_{\alpha\geq 0}$ is dense in $L^2(\mathbb{R}, d\rho(x))$ if and only if polynomials in $\frac{i-x}{i+x}$ are dense in $L^2(\mathbb{R}, d\rho(x))$.

Pick $d\mu$ on $\partial\mathbb{D}$ so $d\rho = \text{Kr}(d\mu)$ and let

$$V\colon L^2(\partial\mathbb{D}, d\mu) \to L^2(\mathbb{R}, d\mu) \tag{3.9.15}$$

by

$$(Vf)(x) = f(2\arctan(x)) \tag{3.9.16}$$

By (3.9.10), V is unitary, and if M_g is multiplicative by g, we have ($z = e^{i\theta}$)

$$VM_zV^{-1} = M_{(i-x)/(i+x)} \tag{3.9.17}$$

It follows that $\{e^{i\alpha x}\}_{\alpha\ge 0}$ is dense in $L^2(\mathbb{R}, d\rho)$ if and only if polynomials in z are dense in $L^2(\partial\mathbb{D}, d\mu)$. By Theorem 2.11.5, this is true if and only if

$$\int_0^{2\pi} \log(w(\theta))\, \frac{d\theta}{2\pi} = -\infty \tag{3.9.18}$$

where w is given by (3.9.11).

Since

$$\int_0^{2\pi} \left|\log\left(\sec\left(\frac{\theta}{2}\right)\right)\right| \frac{d\theta}{2\pi} < \infty$$

(for there is only a single logarithmic singularity at $\theta = \pi$), (3.9.13) says that (3.9.18) is equivalent to

$$\int_0^{2\pi} \log\left(F\left(\tan\left(\frac{\theta}{2}\right)\right)\right) \frac{d\theta}{2\pi} = -\infty \tag{3.9.19}$$

Changing variables to $x = \tan(\frac{\theta}{2})$ so $d\theta = \frac{2dx}{1+x^2}$, we see that (3.9.19) is equivalent to

$$\frac{1}{\pi}\int_{-\infty}^{\infty} \log(F(x))\, \frac{dx}{1+x^2} = -\infty$$

which is (3.9.5). □

Corollary 3.9.3 (Krein). *Let $d\rho$ have the form* (3.9.4) *and suppose*

$$\int_{-\infty}^{\infty} \frac{\log F(x)}{1+x^2}\, dx > -\infty \tag{3.9.20}$$

and that

$$\int_{-\infty}^{\infty} |x|^n\, d\rho(x) < \infty \tag{3.9.21}$$

for all n so the polynomials lie in $L^2(\mathbb{R}, d\rho)$. Then the polynomials are not dense in $L^2(\mathbb{R}, d\rho)$.

Remarks. 1. We will soon see many examples where (3.9.20) holds.

2. It is known (see Theorem 3.8.22) that there are discrete measures ($F \equiv 0$ so (3.9.20) fails) with the polynomials not dense so the converse of Corollary 3.9.3 does not hold.

Proof. If (3.9.20) holds, then the span of $\{e^{i\alpha x}\}_{\alpha\ge 0}$ is not dense by Krein's density theorem. Find a nonzero $f \in L^2$ with

$$\int \overline{f(x)}\, e^{i\alpha x}\, d\rho(x) = 0 \tag{3.9.22}$$

for all $\alpha \ge 0$. By (3.9.21), for any f in L^2, the integral in (3.9.22) is C^∞ with derivatives given by $(i)^n \int \overline{f(x)} x^n e^{i\alpha x}\, d\rho(x)$. In particular, taking derivatives at $\alpha > 0$ and taking $\alpha \downarrow 0$,

$$\int \overline{f(x)}\, x^n\, d\rho(x) = 0 \tag{3.9.23}$$

that is, f is orthogonal to the polynomials. □

This has applications to the theory of moments.

Corollary 3.9.4 (Krein). *If $d\rho_0$ has the form* (3.9.5), (3.9.20), *and* (3.9.21), *then the moment problem is indeterminate.*

Proof. By Corollary 3.9.3, the polynomials are not dense in $L^2(\mathbb{R}, d\rho_0)$. By Theorem 3.8.8, if the moment problem is determinate, then the unique solution of the moment problem has the polynomials dense. Thus, if ρ_0 exists, the problem is indeterminate. □

Example 3.8.1, revisited. If $\alpha < 1$,

$$\int \frac{\log(e^{-|x|^\alpha})}{1+x^2}\, dx > -\infty$$

so, by Krein's result, the problem is indeterminate. Thus, we see $d\rho_\alpha$ is determinate (resp. indeterminate) if $\alpha \geq 1$ (resp. $0 \leq \alpha < 1$). □

Example 3.9.5. Stieltjes [422] showed that the log normal measure

$$\pi^{-1/2}\chi_{(0,\infty)}(x)\, e^{-(\log x)^2}\, dx$$

is indeterminate. One can see this from Krein's criteria since

$$\int \frac{(\log x)^2}{1+x^2}\, dx < \infty \tag{3.9.24}$$

(in fact, Stieltjes showed the Stieltjes moment problem is indeterminate—this follows from a translation of Krein's criterion that we discussed in the Notes to the last section). In this case, the moments can be written down explicitly

$$c_n = \exp(\tfrac{1}{4}(n+1)^2)$$

and one can even write down explicit measures with the same moments: For $\theta \in [-1, 1]$,

$$d\rho_\theta(x) = \pi^{-1/2}\chi_{(0,\infty)}(x)[1 + \theta\sin(2\pi\log x)]\, e^{-(\log x)^2}\, dx$$

also solves the moment problem. This moment problem is further discussed in Christiansen [85] and references therein. □

Example 3.9.6. Hamburger [190] showed that the Stieltjes moment problem for

$$\chi_{[0,\infty)}(x)\exp\left(-\frac{\pi\sqrt{x}}{[\log x]^2 + \pi^2}\right) dx$$

is indeterminate. This follows from the Krein criterion for that case (see the Notes). □

Remarks and Historical Notes. The Krein density theorem (Theorem 3.9.1) appeared in Krein [250] with a proof essentially identical to the one here. He refers to Kolmogorov [239] for the density theorem on the disk with no mention of the connection of the entropy integral to Szegő, although earlier in 1945, he wrote a paper [249] on extensions of Kolmogorov's density theorem that discusses Szegő's work. Interestingly enough, probably a sign of World War II solidarity, both 1945 papers were in English!

L^p versions of the Krein density theorem are due to Akhiezer [11].

Corollary 3.9.4 seems to have appeared first in Akhiezer's book on the moment problem (see p. 87 of [13]) and is attributed to Krein (without any reference). The proof he gives first shows Corollary 3.9.3—we follow his arguments for both corollaries.

Theorem 3.8.23 allows Corollary 3.9.4 to be translated to:

Corollary 3.9.7. *If* $d\rho_0$ *has the form* (3.9.4) *and* (3.9.21) *holds, and if* $d\rho_0$ *is supported on* $[0, \infty)$ *and*

$$\int_0^\infty \frac{\log(F(x))}{1+x} \frac{dx}{\sqrt{x}} > -\infty \tag{3.9.25}$$

then the Stieltjes problem is indeterminate.

This shows the Hamburger example of Example 3.9.6 is borderline for indeterminacy.

The orthogonal polynomials associated to various explicit indeterminate problems are included in what is called the Askey scheme. Among them are the Stieltjes–Wigert polynomials associated to the measure of Example 3.9.5 [85] and the q-Laguerre and $1/q$-Hermite (see, e.g., Christiansen [84]).

While there is no strict converse to Corollary 3.9.3, there is a weak variant of the converse: If the polynomials are not dense, there is always a measure with the same moments for which (3.9.20) holds. Indeed, there is—among all measures solving the moment problem—a unique one that maximizes the integral in (3.9.20); see Berg [42] or Gabardo [144].

As mentioned after Theorem 2.11.5, when the Szegő condition holds, one can use the Szegő function, D, to find an explicit function orthogonal to all polynomials. One can also do this directly in the case of $\mathbb{R}$, providing a "direct" proof of Corollary 3.9.3. In fact, by using an analog of D^2, one gets G analytic in the upper plane whose boundary values obey $|G(x+i0)| = F(x)$ and $\int x^n G(x+i0)\,dx = 0$. Then $d\rho - \mathrm{Re}(G(x+i0)\,dx$ gives an explicit second measure with the same moments (since $F - \mathrm{Re}\,G \geq 0$, it is a positive measure). This is discussed in Simon [395].

3.10 THE NEVAI CLASS AND NEVAI DELTA CONVERGENCE THEOREM

Recall a measure on $\mathbb{R}$ is said to lie in the Nevai class for $[-2, 2]$ if and only if its Jacobi parameters obey

$$a_n \to 1 \qquad b_n \to 0 \tag{3.10.1}$$

In preparation for carrying over the limit theorems for CD kernels of Sections 2.15–2.17 from $\partial\mathbb{D}$ to $[-2, 2]$, we focus here on two theorems of Nevai [320] and a consequence. Here are the three results:

Theorem 3.10.1. *Let $p_n(x; d\rho)$ be the normalized OPRL for a measure in the Nevai class. Then for any $x \in [-2, 2]$, we have*

$$\lim_{n\to\infty} |p_n(x; d\rho)|^2 K_n(x, x)^{-1} = 0 \tag{3.10.2}$$

Theorem 3.10.2 (Nevai's Delta Convergence Theorem). *Let $Q_n(x, x_0)$ be the minimizer in the Christoffel function*

$$\lambda_n(x_0; d\rho) = \min\left(\int |X_n(x, x_0)|^2\, d\rho(x) \,\middle|\, \deg X \le n;\ X_n(x_0) = 1\right) \tag{3.10.3}$$

for a measure $d\rho$ in the Nevai class. Then for all $x_0 \in [-2, 2]$, the probability measure

$$d\xi_n(x) \equiv \lambda_n(x_0)^{-1} |Q_n(x, x_0)|^2\, d\rho(x) \tag{3.10.4}$$

converges weakly to a point mass at x_0.

Theorem 3.10.3. *Let $d\mu$ and $g\,d\mu$ be two measures in the Nevai class where g is such that there are polynomials R_0, R_1 so that $R_0^2 g$ and $R_1^2 g^{-1}$ are bounded continuous functions in some bounded open interval containing* supp$(d\rho)$. *Then for any $x_0 \in [-2, 2]$,*

$$\lim_{n\to\infty} \frac{\lambda_n(x_0; g\,d\rho)}{\lambda_n(x_0; d\rho)} = g(x_0) \tag{3.10.5}$$

Remarks. 1. All limits are uniform on $[-2, 2]$ as our proofs show.

2. These results also hold at point masses in supp$(d\mu)$.

3. The R_0, R_1 condition says g and g^{-1} have finitely many zeros and the vanishing is of finite order in that $|g(x)| \ge C|x - x_0|^\ell$ for some integer ℓ and x near x_0.

4. We will need Theorem 3.10.3 for the case $g(x) = \frac{1}{4}(4 - x)^2$ in connection with Theorem 3.11.9.

We will show Theorem 3.10.1 is equivalent to Theorem 3.10.2 and the two together imply Theorem 3.10.3, and then we will turn to the more subtle proof of Theorem 3.10.1. The *Christoffel–Darboux kernel* (aka CD kernel) is defined by

$$K_n(x, y) = \sum_{j=0}^{n} p_n(x) p_n(y) \tag{3.10.6}$$

for $x, y \in \mathbb{R}$.

Theorem 3.10.4 (CD Formula). *For all $x \neq y$,*

$$K_n(x, y) = \frac{a_{n+1}(p_{n+1}(x)p_n(y) - p_{n+1}(y)p_n(x))}{x - y} \tag{3.10.7}$$

Proof. Let

$$L_n(x, y) = a_{n+1}(p_{n+1}(x)p_n(y) - p_{n+1}(y)p_n(x)) \tag{3.10.8}$$

Take

$$xp_n(x) = a_{n+1}p_{n+1}(x) + b_{n+1}p_n(x) + a_n p_n(x) \tag{3.10.9}$$

multiply by $p_n(y)$ and subtract the expression obtained by interchanging x and y. Then

$$(x-y)\,p_n(x)\,p_n(y) = L_n(x,y) - L_{n-1}(x,y) \tag{3.10.10}$$

This plus induction (starting with $p_{-1}(x) = 0$ and $K_{-1} = 0$) yields (3.10.7). □

As in the proof of Proposition 2.16.2, one immediately obtains that

$$\lambda_n(x_0) = K_n(x,x_0)^{-1} \tag{3.10.11}$$

and that the minimizer is

$$Q_n(x,x_0) = \frac{K_n(x,x_0)}{K_n(x_0,x_0)} \tag{3.10.12}$$

As in the proof of Theorem 2.16.8,

$$\int K_n(x,y)K_n(y,z)\,d\rho(y) = K_n(x,z) \tag{3.10.13}$$

which, in particular, implies that

$$\int \lambda_n(x_0)^{-1}[Q_n(x,x_0)]^2\,d\rho(x) = 1 \tag{3.10.14}$$

First, we will show that Theorem 3.10.2 implies Theorem 3.10.1 and is equivalent if

$$\inf_n a_n > 0 \tag{3.10.15}$$

Proposition 3.10.5. *If* (3.10.2) *holds for $x = x_0$, then the measure $d\xi_n$ converges weakly to δ_{x_0}, a point mass at x_0. Conversely, if* (3.10.15) *holds and if $d\xi_n \xrightarrow{w} \delta_{x_0}$, then* (3.10.2) *holds.*

Proof. We begin with three preliminaries. Since the $d\xi_n$ are probability measures with supports inside a fixed compact set, it is easy to see that

$$d\xi_n \xrightarrow{w} \delta_{x_0} \Leftrightarrow \int (x-x_0)^2\,d\xi_n(x) \to 0 \tag{3.10.16}$$

Second, by the CD formula (3.10.7) and orthogonality of $\{p_k\}$ in $L^2(\mathbb{R},d\rho)$,

$$\int (x-x_0)^2\,d\xi_n = \frac{a_{n+1}^2[p_n(x_0)^2 + p_{n+1}(x_0)^2]}{K_n(x_0,x_0)} \tag{3.10.17}$$

Finally, we claim

$$|p_n(x_0)|^2 K_n(x_0,x_0)^{-1} \to 0 \Rightarrow |p_{n+1}(x_0)|^2 K_n(x_0,x_0)^{-1} \to 0 \tag{3.10.18}$$

For

$$p_{n+1}(x_0)^2 + K_n(x_0,x_0) = K_{n+1}(x_0,x_0) \tag{3.10.19}$$

so

$$\begin{aligned}\frac{|p_{n+1}(x_0)^2|}{K_{n+1}(x_0,x_0)} \to 0 &\Rightarrow \frac{K_n(x_0,x_0)}{K_{n+1}(x_0,x_0)} \to 1\\ &\Rightarrow \frac{|p_{n+1}(x_0)|^2}{K_n(x_0,x_0)} \to 0 \qquad \text{(by (3.10.19) again)}\end{aligned}$$

Now we turn to the theorem. If (3.10.2) holds, then since $\sup_n |a_{n+1}| < \infty$, (3.10.18) and (3.10.17) imply $\int (x - x_0)^2 \, d\xi_n \to 0$, which implies the weak convergence.

Conversely, the weak convergence plus (3.10.16), (3.10.17), and (3.10.15) imply (3.10.2). □

Proof of Theorem 3.10.3 given Theorems 3.10.1 and 3.10.2. Since $d\mu = g^{-1}(g\,d\mu)$, there is a symmetry in hypothesis and it suffices to prove that

$$\limsup \frac{\lambda_n(x_0, g\,d\rho)}{\lambda_n(x_0, d\rho)} \leq g(x_0) \tag{3.10.20}$$

Let ℓ be the degree of R_0. By (3.10.2),

$$\lim_{n\to\infty} \frac{\lambda_n(x_0, d\rho)}{\lambda_{n-\ell}(x_0, d\rho)} = 1 \tag{3.10.21}$$

so it suffices to prove that

$$\limsup \frac{\lambda_n(x_0, g\,d\rho)}{\lambda_{n-\ell}(x_0, d\rho)} \leq g(x_0) \tag{3.10.22}$$

Let $Q_n(x, x_0)$ be the minimizer of $d\mu$ and take as $g\,d\mu$ trial function $Q_{n-\ell}(x, x_0) R_0(x)/R_0(x_0)$, which is 1 at $x = x_0$. Thus,

$$\frac{\lambda_n(x_0, g\,d\rho)}{\lambda_{n-\ell}(x_0, d\rho)} \leq \frac{1}{\lambda_{n-\ell}(x_0, d\rho)} \int g(x) \left(\frac{R_0(x)}{R_0(x_0)}\right)^2 [Q_{n-\ell}(x, x_0)]^2 \, d\rho(x) \tag{3.10.23}$$

Since $g(x)R_0(x)^2$ is continuous, Theorem 3.10.2 implies (3.10.22). □

Finally, we turn to the proof of Theorem 3.10.1. We begin by stating a general inequality, which is a uniform form of (3.10.2) for the free case and whose proof we defer:

Proposition 3.10.6 (Nevai–Totik–Zhang [324]). *For any* $(r, \rho) \in [0, \infty) \times [0, \infty)$, $\theta_1, \theta_2, \alpha_1, \alpha_2$, *and* $L = 1, 2, \dots,$ *we have*

$$|\rho e^{i((L-1)\theta_1 + \alpha_1)} - r e^{i((L-1)\theta_2 + \alpha_2)}|^2 \leq \frac{12}{L} \sum_{j=0}^{L-1} |\rho e^{i(j\theta_1 + \alpha_1)} - r e^{i(j\theta_2 + \alpha_2)}|^2 \tag{3.10.24}$$

Proof of Theorem 3.10.1. Let $\{u_j\}_{j=0}^{\infty}$ solve

$$u_{j+1} + u_{j-1} = \lambda u_j \qquad j = 1, 2, \dots \tag{3.10.25}$$

for some $\lambda \in [-2, 2]$. Then for any $k = 0, 1, \dots,$

$$|u_{k+L-1}|^2 \leq \frac{12}{L} \sum_{j=0}^{L-1} |u_{k+j}|^2 \tag{3.10.26}$$

To see this, note that by continuity and the fact that the constant $12/L$ is λ-independent, it suffices to prove this for $\lambda \in (-2, 2)$. In that case, u has the form

$$u_{j+k} = a e^{ikj} + b e^{-ikj} \tag{3.10.27}$$

for some $a, b \in \mathbb{C}$ and $2\cos k = \lambda$. Thus, (3.10.26) is just (3.10.24).

Given ε, pick L_0 so that

$$\frac{12}{L_0} < \frac{\varepsilon}{4} \tag{3.10.28}$$

For this fixed L, let $T_0(j, k; \lambda)$ be the transfer matrix for (3.10.25) and $T(j, k; \lambda; \{a_m, b_m\})$ for the transfer matrix of some Jacobi matrix in the Nevai class. Using the fact that T_0 and L_0 are fixed, we see that for any ε_1, there exists δ so that for any k,

$$\sup_{\lambda\in[-2,2]} \sup_{k\le m\le k+L_0-1} \|T(k, m; \lambda; \{a, b\}) - T_0(k, m; \lambda)\| < \varepsilon_1 \tag{3.10.29}$$

if

$$\sup_{k\le m\le k+L_0-1} (|a_k - 1| + |b_k|) < \delta \tag{3.10.30}$$

and so, for some δ, (3.10.30) implies

$$|u_{k+L-1}|^2 \le \varepsilon \sum_{j=0}^{L-1} |u_{k+j}|^2 \tag{3.10.31}$$

for any solution of (3.2.6).

Because we are in Nevai class, there is L_1 so (3.10.30) holds for $k > L_1$. Thus, for $q > L_0 + L_1$,

$$|u_q|^2 \le \varepsilon \sum_{j=q-L+1}^{q} |u_j|^2 \le \varepsilon \sum_{j=0}^{q} |u_j|^2 \tag{3.10.32}$$

This proves (3.10.2). □

We need the following in the proof of Proposition 3.10.6:

Lemma 3.10.7. *For all $r \in [0, 1]$, $\gamma \in [0, \pi]$, and all $\beta \in [\frac{\gamma}{2}, \pi]$, we have*

$$|1 - re^{i\beta}|^2 \ge \tfrac{1}{4}|1 - re^{i\gamma}|^2 \tag{3.10.33}$$

Remark. The worst case is $r = 1$, $\beta = \frac{\gamma}{2}$, and $\gamma \to 0$, in which case (3.10.33) approaches equality.

Proof. Since

$$|1 - re^{i\beta}|^2 = 1 + r^2 - 2r\cos\beta \tag{3.10.34}$$

is decreasing as $\beta \in [0, \pi]$ decreases, we need only consider the case $\beta = \frac{\gamma}{2}$. For $a > b$, both in $(-1, 1)$ and $r \in [0, 1]$,

$$\frac{d}{dr}\left[\frac{1 + r^2 - 2ar}{1 + r^2 - 2br}\right] < 0 \tag{3.10.35}$$

so without loss, we can suppose $r = 1$. In that case (i.e., $r = 1$, $\beta = \frac{\gamma}{2}$), (3.10.33) is equivalent to

$$\sin^2\left(\frac{\gamma}{4}\right) \ge \tfrac{1}{4}\sin\left(\frac{\gamma}{2}\right)^2$$

which is immediate from $\sin(\frac{\gamma}{2}) = 2\cos(\frac{\gamma}{4})\sin(\frac{\gamma}{4})$. □

Proof of Proposition 3.10.6. Without loss, we can suppose first that $\rho \neq 0$ (since (3.10.24) is trivial if $\rho = 0$), then replacing r by $\rho(\frac{r}{\rho})$ that $\rho = 1$, then that $\theta_1 = \alpha_1 = 0$ (by writing $\theta_2 = (\theta_2 - \theta_1) + \theta_1$, and taking out θ_1) and then by periodicity and $|\bar{u}| = |u|$ that

$$0 \le \alpha \le \pi \qquad 0 \le |\theta| \le \pi \tag{3.10.36}$$

and we need (also taking $j \to L - 1 - j$)

$$\frac{12}{L} \sum_{j=0}^{L-1} |1 - re^{i(j\theta+\alpha)}|^2 \ge |1 - re^{i\alpha}|^2 \tag{3.10.37}$$

Finally, by symmetry, we can take

$$0 \le r \le 1 \tag{3.10.38}$$

We consider two cases:

$$L|\theta| \ge 2\pi \tag{3.10.39}$$

and

$$L|\theta| \le 2\pi \qquad L \ge 12 \tag{3.10.40}$$

since (3.10.37) is trivial if $L \le 12$.

In case 1, that is, (3.10.39), we note

$$\sum_{j=0}^{L-1} |1 - re^{i(\alpha+j\theta)}|^2 = L(1+r^2) - r\left[\frac{e^{i\alpha}[e^{i(L-1)\theta} - 1]}{e^{i\theta} - 1} + \frac{e^{-i\alpha}[e^{-i(L-1)\theta} - 1]}{e^{-i\theta} - 1}\right] \tag{3.10.41}$$

$$\ge L(1+r^2) - 2r\,\frac{1}{|e^{i\theta} - 1|}$$

$$= L(1+r^2) - r\left|\sin\left(\frac{\theta}{2}\right)\right|^{-1} \tag{3.10.42}$$

Now for $\eta \in [0, \frac{\pi}{2}]$, $\sin(\eta) \ge \eta\frac{2}{\pi}$ and $|\theta| < \pi$, so $\eta = \frac{|\theta|}{2} \in [0, \frac{\pi}{2}]$ and (3.10.41) becomes

$$\begin{aligned}
\sum_{j=0}^{L-1} |1 - r^{i(\alpha+j\theta)}|^2 &\ge L(1+r^2) - r\,\frac{\pi}{|\theta|} \\
&\ge L(1+r^2) - \frac{Lr}{2} \qquad \text{(by (3.10.39))} \\
&\ge \tfrac{3}{4}L(1+r^2) \qquad \text{(since } (1 + r^2 - 2r) > 0\text{)}
\end{aligned}$$

But

$$|1 - re^{i\alpha}|^2 \le 1 + 2r + r^2 \le 2(1+r^2) = \frac{8}{3L}\,[\tfrac{3}{4}L(1+r^2)]$$

so (3.10.37) holds (since $\frac{8}{3} < 12$).

On the other hand, if (3.10.40) holds, since $L|\theta| \leq 2\pi$, the points $e^{i(j\theta+\alpha)}$ are equally spaced and only fill part of the circle.

If there are k points starting at $j = 0$ with $|j\theta + \alpha| \geq \frac{\alpha}{2}$, then at most $2k$ further points can have $|j\theta+\alpha| \leq \frac{\alpha}{2}$ since $(-\frac{\alpha}{2}, \frac{\alpha}{2})$ is only twice as big as $(\frac{\alpha}{2}, \alpha)$, that is, at least $\frac{1}{3}$ points have $|j\theta + \alpha| \geq \frac{\alpha}{2}$. For such points, by the lemma, $|1 - re^{i(j\theta+\alpha)}| \geq \frac{1}{4}|1 - re^{i\alpha}|^2$. So, again, (3.10.37) holds (since $\frac{L}{3}\frac{1}{4} = \frac{L}{12}$). □

Remarks and Historical Notes. For further discussion of the CD kernel, including an operator theoretic proof of the CD formula, see Simon [407].

Theorems 3.10.1–3.10.3 are from Nevai's AMS Memoir [320] whose proofs we follow for the implications of one theorem to another. His proof of Theorem 3.10.1 is different.

Our proof of Theorem 3.10.1 follows Nevai–Totik–Zhang [324] who prove (3.10.24) with a larger constant than 12 (but for $|\cdot|^p$ not just $|\cdot|^2$).

Theorem 3.10.1 is proven on $\mathfrak{e}$ for asymptotically zero perturbations of a periodic Jacobi matrix with essential spectrum $\mathfrak{e}$ by Lubinsky–Nevai [289], for all of $\mathfrak{e}$ by Zhang [464], and for more general situations by Breuer–Last–Simon [60]. In particular, [60] has a different approach to uniform estimates motivated by Szwarc [435] that is illuminating.

[60] provides an example of a regular measure on $[-2, 2]$ (see Section 5.9 for the definition of regular) where Theorem 3.10.1 fails for many x's in $[-2, 2]$. They also extend the theorems of this section beyond $[-2, 2]$.

3.11 ASYMPTOTICS OF THE CD KERNEL: OPRL ON [−2, 2]

In Sections 2.15–2.17, we studied asymptotics of the CD kernel for OPUC regular on all of $\partial\mathbb{D}$ with additional conditions on the weight. In this section, we will carry these over to OPRL on $[-2, 2]$ (and in Section 5.11 to more general OPRL). Most arguments will either be a straightforward analog or the use of the Szegő map of Section 1.9 to directly relate the CD kernel for OPUC to the CD kernel for OPRL. There are, however, three interesting twists:

(a) When $\text{supp}(d\mu)$ was $\partial\mathbb{D}$, there was no place outside to put point masses, but now the natural hypothesis is $\sigma_{\text{ess}}(d\mu) = [-2, 2]$ and there can be pure points outside. This will require an extension, albeit a simple one, in the Nevai comparison theorem (Theorem 2.16.6).

(b) For OPUC, the natural limit for the density of zeros was $\frac{d\theta}{2\pi}$. It is not so obvious what the analog is for $[-2, 2]$. It is, in fact, the measure on $[-2, 2]$:

$$d\rho_{[-2,2]}(x) = \frac{1}{\pi}\,\frac{dx}{\sqrt{4 - x^2}} \tag{3.11.1}$$

The right way to understand this is potential theoretic, and we will defer this part of the story to Section 5.9.

(c) When the Szegő map is used, the CD kernel for OPUC will be related to two measures on $[-2, 2]$: $d\mu$ and $\frac{1}{4}(4 - x^2)\, d\mu$. Theorem 3.10.3 will overcome this difficulty.

We begin with an analog of Theorem 2.15.1:

Theorem 3.11.1. *Let $d\mu$ be a measure of compact support on $\mathbb{R}$. Let $K_n(x, y)$ be its CD kernel and define*

$$d\mu^{(N)} = \frac{1}{N+1} K_N(x, x)\, d\mu(x) \tag{3.11.2}$$

and let $d\nu_n$ be the zero counting measure for $P_n(x; d\mu)$. Then for $\ell = 0, 1, 2, \dots,$

$$\left| \frac{1}{N+1} \int x^\ell\, d\mu^{(N)}(x) - \frac{1}{N+1} \int x^\ell\, d\nu_{N+1}(x) \right| \le \frac{2\ell}{N+1} \tag{3.11.3}$$

In particular, for any subsequence $N(1) < N(2) < \dots$, $d\nu_\infty$ is a weak limit of $d\nu_{N(j)+1}$ if and only if it is a weak limit of $d\mu_{N(j)}$.

Proof. By Theorem 1.2.6, the zeros of P_{N+1} are identical to the eigenvalues of $\pi J \pi \restriction \operatorname{Ran}(\pi)$, so the proof is identical to the proof of Theorem 2.15.1. □

Corollary 3.11.2. *Let μ_1, μ_2 be two (not necessarily normalized) measures of compact support on $\mathbb{R}$. Suppose*

(1)

$$\mu_1 \ge \mu_2 \tag{3.11.4}$$

(2) *For some open interval $I = (a, b)$,*

$$\mu_1 \restriction (a, b) = \mu_2 \restriction (a, b) \tag{3.11.5}$$

(3) *For some subsequence $N(1), N(2), \dots,$ and density of zeros $d\nu_n^{(j)}$ $(j = 1, 2)$ of μ_j, we have*

$$d\nu_{N(k)}^{(j)} \xrightarrow{w} d\nu_\infty^{(j)} \tag{3.11.6}$$

Then

$$d\nu_\infty^{(2)} \restriction I \ge d\nu_\infty^{(1)} \restriction I \tag{3.11.7}$$

Remark. An example is $\mu_2 = \mu_1 \restriction (a, b)$.

Proof.

$$\begin{aligned}
(3.11.4) &\Rightarrow \lambda_n(x, \mu_2) \le \lambda_n(x, \mu_1) \\
&\Rightarrow K_n(x, x; \mu_2) \ge K_n(x, x; \mu_1) \\
&\Rightarrow K_n(x, x; \mu_2)\, d\mu_2 \restriction I \ge K_n(x, x; \mu_1)\, d\mu_1 \restriction I \\
&\Rightarrow (3.11.7)
\end{aligned}$$

by Theorem 3.11.1. □

We next turn to what typical limits of $d\nu_n$ are

Example 3.11.3. Let $d\mu_1, d\mu_2$ be given by

$$d\mu_1(x) = d\rho_{[-2,2]}(x) \tag{3.11.8}$$

with $d\rho_{[-2,2]}$ given by (3.11.1) and

$$\begin{aligned} d\mu_2(x) &= 2\,\tfrac{1}{4}(4-x^2)\,d\mu_1 \\ &= \frac{2}{\pi}\sqrt{4-x^2}\,dx \end{aligned} \tag{3.11.9}$$

In terms of the change of variable ($\theta \in [0,\pi]$),

$$x = 2\cos\theta \tag{3.11.10}$$

so $dx = 2\sin\theta\,d\theta = \sqrt{4-x^2}\,d\theta$, we see

$$d\mu_1 = \frac{d\theta}{\pi} \qquad d\mu_2 = 2\sin^2\theta\,\frac{d\theta}{\pi} \tag{3.11.11}$$

Thus, the normalized OPRL (essentially Chebyshev polynomials of the first and second kind) are given by

$$p_n(2\cos\theta; d\mu_1) = \sqrt{2}\,\cos(n\theta) \qquad (n \geq 1) \tag{3.11.12}$$

$$p_n(2\cos\theta; d\mu_2) = \frac{\sin((n+1)\theta)}{\sin\theta} \qquad (n \geq 0) \tag{3.11.13}$$

Thus,

$$d\nu_n(x; d\mu_1) = \frac{1}{n}\sum_{j=0}^{n-1} \delta_{\theta,\,\frac{j+1/2}{n}\pi} \tag{3.11.14}$$

$$d\nu_n(x; d\mu_2) = \frac{1}{n}\sum_{j=0}^{n-1} \delta_{\theta,\,\frac{j+1}{n+1}\pi} \tag{3.11.15}$$

In both cases,

$$d\nu_n \to \frac{d\theta}{2\pi} = d\rho_{[-2,2]}(x) \tag{3.11.16}$$

□

Definition. A measure $d\mu$ on $\mathbb{R}$ is called *regular* for $[-2,2]$ if and only if $\sigma_{\rm ess}(\mu) = [-2,2]$ and

$$\lim_{n\to\infty} (a_1 \dots a_n)^{1/n} = 1 \tag{3.11.17}$$

Remark. By (3.11.12)/(3.11.13), the $d\mu$'s of Example 3.11.3 have $b_n(d\mu_1) = 0$, $a_1(d\mu_1) = \sqrt{2}$, $a_n(d\mu_1) = 1$ ($n \geq 2$) and $b_n(d\mu_2) = 0$, $a_n(d\mu_2) = 1$. Thus, they are regular.

We will prove a generalization of the following as Theorem 5.9.2:

Theorem 3.11.4. *Let $d\mu$ be regular for* $[-2, 2]$. *Then*
(i)

$$d\nu_n \to d\rho_{[-2,2]} \tag{3.11.18}$$

(ii) *For any $\varepsilon > 0$, there is a δ so*

$$\limsup_{n\to\infty} \Big[\sup_{\operatorname{dist}(x,[-2,2])<\delta} |p_n(x; d\mu)|^{1/n} \Big] \le e^{\varepsilon} \tag{3.11.19}$$

Letting

$$\rho_{[-2,2]}(x) = \frac{1}{\pi} \frac{1}{\sqrt{4 - x^2}} \tag{3.11.20}$$

we see that if μ is regular and, say, $d\mu_s = 0$, then

$$w(x) \frac{1}{n+1} K_{n+1}(x, x)\, dx \to \rho_{[-2,2]}(x)\, dx \tag{3.11.21}$$

So now the normal expectation becomes

$$\frac{1}{n+1} K_{n+1}(x, x) \to \frac{\rho_{[-2,2]}(x)}{w(x)} \tag{3.11.22}$$

Example 3.11.3, revisited. For OPUC, we used $\frac{d\theta}{2\pi}$ and its OPs as a model in a Nevai comparison theorem. Here we will use $d\mu_1$, although we could just as well use $d\mu_2$. We begin by computing $K_n^{(1)}(x, x)$ for $x \in (-2, 2)$:

$$\begin{aligned} K_n^{(1)}(2\cos\theta, 2\cos\theta) &= 1 + 2\sum_{j=1}^{n} \cos^2(j\theta) \\ &= (n+1) + \sum_{j=1}^{n} \cos(2j\theta) \\ &= n + \frac{1}{2} + \frac{1}{2}\frac{\sin((2n+1)\theta)}{\sin\theta} \end{aligned} \tag{3.11.23}$$

so if $x_n \to x_\infty \in (-2, 2)$, then (since $w = \rho_{[-2,2]}$)

$$\frac{1}{n+1} K_n^{(1)}(x_n, x_n) \to 1 = \frac{\rho_{[-2,2]}(x_\infty)}{w(x_\infty)} \tag{3.11.24}$$

uniformly in $x_n \in K$, any compact in $(-2, 2)$.

As for slightly off-diagonal, we note that

$$\cos a \cos b = \tfrac{1}{2}\cos(a+b) + \tfrac{1}{2}\cos(a-b) \tag{3.11.25}$$

Thus, for $n \ge 1$, by the CD formula and (3.11.12),

$$\begin{aligned} K_n(2\cos\theta, 2\cos\varphi) &= \frac{[\cos(n+1)\theta\cos n\varphi - \cos(n+1)\varphi\cos n\theta]}{[\cos\theta - \cos\varphi]} \\ &= \frac{\cos[n(\theta-\varphi)+\theta] - \cos[\varphi - n(\theta-\varphi)]}{2[\cos\theta - \cos\varphi]} \end{aligned} \tag{3.11.26}$$

Take θ, φ to θ_n, φ_n where

$$\theta_n = \theta_\infty + O\left(\frac{1}{n}\right) \qquad \varphi_n = \theta_\infty + O\left(\frac{1}{n}\right) \tag{3.11.27}$$

$$n(\theta_n - \varphi_n) \to C - D \tag{3.11.28}$$

Notice that

$$n[\cos\theta_n - \cos\varphi_n] \to -(\sin\theta_\infty)(C - D) \tag{3.11.29}$$

while, using

$$\tfrac{1}{2}\cos(a+b) - \tfrac{1}{2}\cos(a-b) = -\sin a \sin b \tag{3.11.30}$$

we see

$$\begin{aligned} &\tfrac{1}{2}\cos[n(\theta-\varphi)+\theta] - \tfrac{1}{2}\cos[\varphi - n(\theta-\varphi)] \\ &\qquad \to \tfrac{1}{2}\cos[\theta_\infty + (C-D)] - \tfrac{1}{2}\cos[\theta_\infty - (C-D)] \\ &\qquad = -\sin(\theta_\infty)\sin(C-D) \end{aligned} \tag{3.11.31}$$

so

$$\frac{1}{n+1} K_n(2\cos\theta_n, 2\cos\varphi_n) \to \frac{\sin(C-D)}{C-D} \tag{3.11.32}$$

Thus, if

$$x_n = x_\infty + O\left(\frac{1}{n}\right) \qquad y_n = x_\infty + O\left(\frac{1}{n}\right) \qquad n(x_n - y_n) \to A - B$$

using (with $x_\infty = 2\cos(\theta_\infty)$)

$$C - D = (2\sin(\theta_\infty))^{-1}(A-B) = \pi\rho_{[-2,2]}(x_\infty)(A-B) \tag{3.11.33}$$

with $\rho_{[-2,2]}$ given by (3.11.20), then

$$\frac{K_n(x_n, y_n)}{K_n(x_\infty, x_\infty)} \to \frac{\sin(\pi\rho_{[-2,2]}(x_\infty)(A-B))}{\pi\rho_{[-2,2]}(x_\infty)(A-B)} \tag{3.11.34}$$

This is Lubinsky universality for this measure.

This has a form similar to (2.16.3) where the analog $\rho_{[-2,2]}$ is in many ways $1/2\pi$. Of course, we have lost the leading $e^{\frac{i}{2}(a-b)}$. □

As remarked, there is one extra aspect of the Nevai comparison theorem:

Theorem 3.11.5 (Nevai Comparison Theorem). *Let μ, $\mu^\sharp$ be two regular measures on $[-2,2]$ of the form*

$$d\mu = w\,dx + d\mu_\mathrm{s} \qquad d\mu^\sharp = w^\sharp\,dx + d\mu_\mathrm{s}^\sharp \tag{3.11.35}$$

Suppose $x_0 \in (-2,2)$ obeys

(i) *For some $\delta > 0$, $d\mu_\mathrm{s} = d\mu_\mathrm{s}^\sharp$ on $(x_0 - \delta, x_0 + \delta)$.*

(ii) *For all ε sufficiently small, there is $\alpha_\varepsilon > 1$, so for $|x - x_0| < \varepsilon$, we have*

$$\alpha_\varepsilon^{-1} w(x) \le w^\sharp(x) \le \alpha_\varepsilon w(x) \tag{3.11.36}$$

(iii) *For* $\alpha_\varepsilon \to 1$ *and any* $x_n \in (-2,2)$ *with* $x_n \to x_0$ *and every* $\ell(n)$ *with* $n/2 < \ell(n) < 2n$, *we have that*

$$\lim_{n\to\infty} \frac{1}{n+1} K_n(x_{\ell(n)}, x_{\ell(n)}) = B \neq 0 \tag{3.11.37}$$

Then

$$\lim_{n\to\infty} \frac{1}{n+1} K_n^\sharp(x_n, x_n) = B \tag{3.11.38}$$

Moreover, this is uniform in x_n *in the sense that if (with the same* B*) for all* $x_n \to x_0$, *there are, for any* ε, *a* δ *and an* N_0 *so if* $n > N_0$ *and* $|x_n - x_0| < \delta$, *then*

$$\left| B - \frac{1}{n+1} K_n^\sharp(x_n, x_n) \right| < \varepsilon \tag{3.11.39}$$

This is also uniform in x_0. *If* w *and* $w^\sharp$ *are continuous and nonvanishing in a closed interval in* $(-2,2)$ *and we have* $d\mu_{\rm s} = d\mu_{\rm s}^\sharp$ *in a neighborhood of* I *and* (3.11.36) *is replaced by*

$$\alpha_\varepsilon^{-1} \frac{w(x)}{w(x_0)} \le \frac{w^\sharp(x)}{w^\sharp(x_0)} \le \alpha_\varepsilon \frac{w(x)}{w(x_0)} \tag{3.11.40}$$

for $|x - x_0| < \varepsilon$ (α_ε *independent of* x_0) *and if* (3.11.36) *holds uniformly in* $x_0 \in I$ *where* $B(x_0)$ *is* x_0*-dependent, then* (3.11.38) *with* B *replaced by* $B(x_0)w(x_0)/w^\sharp(x_0)$.

Proof. The proof is the same as the proof of Theorem 2.16.6 with one extra step. Because we only have that $\sigma_{\rm ess}$ is $[-2,2]$, there can be pure points for $\mu^\sharp$ where the regularity does not imply the polynomials $p_n(x; d\mu)$ are bounded in n by $e^{\varepsilon n}$, so the choice Q_n in (2.16.38) may not be small at those points. However, for each δ, there are only finitely many pure points $\{x_j\}_{j=1}^{N_\delta}$ of $\mu^\sharp$ with $\text{dist}(x_j, [-2,2]) > \delta$. Adding a multiplicative factor $\prod_{j=1}^{N_\delta} (x - x_j)/(x_n - x_j)$ to Q_n (adjusting $n(\varepsilon)$ to be $n - m(\varepsilon) - N_\delta$) kills this finite number of points. With this adjustment, the proof extends with no other change. □

We then have:

Theorem 3.11.6 (Lubinsky [288]). *Let* $d\mu$ *be a regular probability measure on* $[-2,2]$ *of the form*

$$d\mu = w(x)\, dx + d\mu_{\rm s} \tag{3.11.41}$$

Suppose that, for any interval $[\alpha,\beta] \subset (-2,2)$,

(a) $\text{supp}(d\mu_{\rm s}) \cap I = \emptyset$

(b) w *is "continuous" on* I *and nonvanishing there.*

Then, with $\rho_{[-2,2]}$ *given by* (3.11.20), *we have*

(1) (Diagonal Asymptotics) *For any* $A < \infty$, *uniformly in* $x_\infty \in I$, *and sequence* $x_n \in [-2,2]$ *with* $n|x_n - x_\infty| \le A$ *for all* n, *we have*

$$\frac{1}{n+1} K_n(x_n, x_n) \to \frac{\rho_{[-2,2]}(x_\infty)}{w(x_\infty)} \tag{3.11.42}$$

(2) (Lubinsky Universality) *For any* $A < \infty$, *uniformly in* $x_\infty \in I$ *and* $a, b \in \mathbb{R}$ *with* $|a|, |b| \le A$, *we have*

$$\frac{K_n(x_\infty + \frac{a}{n}, x_\infty + \frac{b}{n})}{K_n(x_\infty, x_\infty)} \to \frac{\sin(\pi \rho_{[-2,2]}(x_\infty)(b-a))}{\pi \rho_{[-2,2]}(x_\infty)(b-a)} \tag{3.11.43}$$

More generally, the limit of $K_n(x_n, y_n)/K_n(x_\infty, x_\infty)$ *is the right side of* (3.11.43) *so long as* $|x_n - x_\infty| \le A/n$, $|y_n - x_\infty| \le A/n$, *and* $n(x_n - y_n) \to b - a$.

Remark. If $b - a = 0$, the right side of (3.11.43) is interpreted as 1.

Proof. Given the improved version of the Nevai comparison theorem and the model, $d\mu_1$, of Example 3.11.3, the proof is identical to that of Theorem 2.16.1. □

Theorem 3.11.7 (Máté–Nevai Upper Bound). *For any measure* $d\mu$ *with* $\sigma_{\text{ess}}(d\mu) \subset [-2, 2]$ *and any Lebesgue point* x_0 *of* $d\mu$ *in* $(-2, 2)$, *we have*

$$\limsup (n+1)\lambda_n(x_n) \le \frac{w(x_0)}{\rho_{[-2,2]}(x_0)} \tag{3.11.44}$$

for any sequence $x_n \in (-2, 2)$ *with* $\sup_n n|x_n - x_0| < \infty$.

Remarks. 1. Theorem 3.10.2 asserts that, under great generality, $|Q_n(x, x_0)|^2 d\mu(x)/\lambda_n(x)$ converge weakly as a measure to a point mass at x_0, that is, smeared with continuous functions. In essence, this proof relies on the fact that for a very nice $d\mu$ (namely, $d\mu_1$), the convergence is in a much stronger sense.

2. We emphasize that in this result $w(x_0)$ can be 0, in which case $\lim(n + 1)\lambda_n(x_n) = 0$.

Proof. Suppose first that $\sigma(d\mu) = [-2, 2]$. Define

$$F_n(x, y) = \left[\frac{K_n^{(1)}(x, y)}{K_n^{(1)}(y, y)} \right]^2 \rho_{[-2,2]}(y) [\lambda_n^{(1)}(y)]^{-1} \tag{3.11.45}$$

the objects associated to the $d\mu_1$ measure of Example 3.11.3. We will show that

$$\lim_{n\to\infty} \int F_n(x, x_n)\, d\mu(x) = w(x_0) \tag{3.11.46}$$

this implies (3.11.44) for this $\sigma(d\mu) = [-2, 2]$ case since $K_n^{(1)}(x, x_n)/K_n^{(1)}(x_n, x_n)$ can be used as a trial polynomial in (3.10.3) showing that

$$(n+1)\lambda_n(x_n, d\mu) \le (n+1)\lambda_n^{(1)}(x_n) \rho_{[-2,2]}(x_n)^{-1} \int F_n(x, x_n)\, d\mu(x) \tag{3.11.47}$$

and $(n + 1)\lambda_n^{(1)}(x_n) \to 1$ by (3.11.24).

To prove (3.11.44), we pick A (eventually, very large) and write the integral as a sum of three terms. First, the integral over $|x - x_n| \ge A/n$; second, what we get by taking $|x - x_n| < A/n$ and replacing $K_n^{(1)}(x, x_n)/K_n^{(1)}(x_n, x_n)$ in F by $\sin(\pi \rho_{[-2,2]}(x_0) n(x - x_n))/\pi \rho_{[-2,2]}(x_0) n(x - x_n)$; and third, the difference between the true F and this approximate F.

Because of the uniform convergence in (3.11.34), the third term is bounded by

$$Cn\mu\left(\left(x_n - \frac{A}{n}, x_n + \frac{A}{n}\right)\right) o(1) \tag{3.11.48}$$

$n\mu((x_n - \frac{A}{n}, x_n + \frac{A}{n}))$ is bounded since x_0 is a Lebesgue point, so this term goes to zero for each fixed A.

By the CD formula and the boundedness of $p_n(x; d\mu_1)$ on $[-2, 2]$,

$$|F_n(x, y)| \leq \frac{C}{n|x - y|^2}$$

for y in a compact subset of $(-2, 2)$, and thus, the first term is bounded by

$$Cn^{-1} \int_{|x-y|\geq A/n} \frac{dx}{|x - y|^2} = CA^{-1} \tag{3.11.49}$$

which can be made small by taking A large.

Thus, the main contribution is the second term, which we control with Lemma 2.17.8, as in the proof of Theorem 2.17.6. This completes the proof of (3.11.46) and so of (3.11.44) when $\sigma(d\mu) = [-2, 2]$.

If now $\sigma(d\mu) = [-2, 2] \cup F$ with F a finite set $\{x_j\}_{j=1}^m$ outside $[-2, 2]$, we can set $\tilde{\mu} = \mu \restriction [-2, 2]$ and use $P_{n,m}(x, x_0; d\tilde{\mu}(\prod_{j=1}^m (x - x_j)/(x_0 - x_n)$ as a trial function for $\lambda_n(x_0, d\mu)$ to get (3.11.44).

Finally, if $\sigma_{\text{ess}}([-2, 2])$, then for any ε, $\sigma(d\mu) = [-2 - \varepsilon, 2 + \varepsilon] \cup F_\varepsilon$ with F_ε finite. So by the above,

$$\limsup (n + 1)\lambda_n(x_0) \leq \frac{w(x_0)}{\rho_{[-2-\varepsilon,2+\varepsilon]}(x_0)} \tag{3.11.50}$$

so taking $\varepsilon \downarrow 0$, we obtain (3.11.44). □

Theorem 3.11.8 (Simon [409]). *If $I = (\alpha, \beta) \subset [-2, 2]$ is an open interval, if μ is regular for $[-2, 2]$ and $w(x) > 0$ for a.e. $x \in I$, then*

$$\text{(i)} \qquad \int_I \left| \frac{1}{n + 1} K_n(x, x)w(x) - \rho_{[-2,2]}(x_0) \right| dx \to 0 \tag{3.11.51}$$

$$\text{(ii)} \qquad \int_I \frac{1}{n + 1} K_n(x, x)\, d\mu_s(x) \to 0 \tag{3.11.52}$$

Proof. Given Theorems 3.11.4 and 3.11.7, the proof is the same as for Theorem 2.17.7. □

Theorem 3.11.9 (MNT Theorem [302]). *Let μ be a regular measure for $[-2, 2]$, which is locally Szegő on I, an open interval in $[-2, 2]$. Let $x_\infty \in I$ be a point with $w(x_\infty) \neq 0$ and which is a Lebesgue point for both w and for the local Szegő function. Let $x_n \in (-2, 2)$ be a sequence with*

$$\sup_n n|x_n - x_\infty| \equiv A < \infty \tag{3.11.53}$$

Then (3.11.42) holds. The limit is uniform in all x_n obeying (3.11.53).

Remark. By the local Szegő condition, we mean if $I = (\alpha, \beta)$ that for any $\varepsilon > 0$,

$$\int_{\alpha+\varepsilon}^{\beta-\varepsilon} \log(w(x))\, dx > -\infty \tag{3.11.54}$$

Given any such w and any x_∞, we can find $\tilde{w}$ equal to w near x_∞ with $\tilde{w}\, dx = d\tilde{\mu}$, the image under the Szegő map of a measure obeying the Szegő condition on $\partial\mathbb{D}$. By the local Szegő function, we mean the pull back to $[-2, 2]$ of the Szegő function for this measure on $\partial\mathbb{D}$.

Proof. First we use the Nevai comparison theorem (Theorem 3.11.5) to reduce to a case where $d\mu$ is supported on $[-2, 2]$ and obeys a global Szegő condition. Let $d\tilde{\mu}$ be on $\partial\mathbb{D}$ so that $d\mu = \mathrm{Sz}(d\tilde{\mu})$ and let $d\mu^* = \mathrm{Sz}_1(d\tilde{\mu})$. (Sz and Sz_1 are the Szegő mappings defined in Section 1.9.) By (1.9.27),

$$K_{2n}(e^{i\theta_n}, e^{i\theta_n}; d\tilde{\mu}) = K_n(x_n, x_n; d\mu) + \sin^2\theta_n\, K_{n-1}(x_n, x_n; d\mu^\sharp) \tag{3.11.55}$$

where

$$x_n = 2\cos\theta_n \tag{3.11.56}$$

By a slight extension of Theorem 3.10.3 with $g(x) = \frac{1}{4}(4 - x^2)$ and the relation (1.9.14) of $d\mu$ and $d\mu^\sharp$,

$$\sin\theta_n^2\, \frac{K_n(x_n x_n; d\mu)}{K_{n-1}(x_n, x_n; d\mu^\sharp)} \to 1 \tag{3.11.57}$$

Thus, (3.11.42) follows from (3.11.55) and (2.17.7). □

Theorem 3.11.10 (Findley's Theorem [133]). *Under the hypotheses of Theorem 3.11.9, we have* (3.11.43) *for each $A < \infty$, uniformly in a, b with $|a|, |b| < A$. More generally, the limit relation holds for $K_n(x_n, y_n)/K_n(x_\infty, x_\infty)$ for any x_n, y_n with $|x_n - x_\infty| \le A/n$, $|y_n - x_\infty| \le A/n$, and $n(x_n - y_n) \to a - b$.*

Proof. One has a Lubinsky inequality in the real case by the same proof as for $\partial\mathbb{D}$. This inequality plus the MNT theorem implies the off-diagonal result. □

Definition. We say the zeros of $p_n(x)$ have *clock behavior at x_0 with density $\rho(x)$* if for all j, $(x_{j+1}(x_0) - x_j(x_0))/(2\pi/\rho(x_0)) \to 1$. If the limit is uniform for $x_0 \in I$, we say there is uniform clock behavior in I.

As in the OPUC case (see Theorem 2.16.10), Lubinsky universality immediately proves:

Theorem 3.11.11. *Under the hypotheses of Theorem 3.11.6, one has clock behavior with density, $\rho(x_0)$, uniformly on I. Under the hypotheses of Theorem 3.11.9, one has uniform clock behavior at x_0.*

Remarks and Historical Notes. The history of the ideas in this section is discussed in the Notes to Sections 2.15, 2.16, and 2.17.

3.12 ASYMPTOTICS OF THE CD KERNEL: LUBINSKY'S SECOND APPROACH

Our previous discussion of the slightly off-diagonal CD kernel has depended on Lubinsky's inequality and a comparison measure. Remarkably, having revolutionized this subject with his elegant inequality, Lubinsky [287] presented an entirely different approach to universality that does not require a comparison model and that illuminates why the kernel $\sin(\pi x)/\pi x$ occurs. Here we will discuss this approach as extended by Avila–Last–Simon [30].

The main theorem is the following:

Theorem 3.12.1 ([30]). *Let*

$$d\mu(x) = w(x)\,dx + d\mu_{\rm s}(x) \tag{3.12.1}$$

be a nontrivial probability measure of compact support in $\mathbb{R}$. *Let* $\Sigma \subset \mathbb{R}$ *be a set of positive Lebesgue measure with*

$$\text{(a)} \qquad \frac{w(x)}{n+1}\, K_n(x,x) \to \rho_\infty(x) \in (0,\infty) \quad \textit{for a.e. } x \in \Sigma \tag{3.12.2}$$

$$\text{(b)} \qquad \sup_n \frac{1}{n+1} \sum_{j=0}^{n} |q_j(x)|^2 < \infty \qquad \textit{for a.e. } x \in \Sigma \tag{3.12.3}$$

Then for a.e. $x_0 \in \Sigma$ *and all* $z, w \in \mathbb{C}$, *we have*

$$\lim_{n\to\infty} \frac{K_n(x_0 + \frac{z}{n+1}, x_0 + \frac{w}{n+1})}{K_n(x_0,x_0)} = \frac{\sin(\pi\rho_\infty(x)(z-w))}{\pi\rho_\infty(x)(z-w)} \tag{3.12.4}$$

uniformly for z, w *with* $|z| < A$, $|w| < A$ *for any* A. *In particular, one has clock behavior of the zeros with density* $\rho_\infty(x)$ *for a.e.* $x \in \Sigma$.

Remark. (3.12.2) is a definition of ρ_∞, that is, the assertion is the existence and positivity of the limit.

Thus, control of the limit in (3.12.2) and the bound of (3.12.3) imply universality, and Totik's results on the diagonal kernel imply results for the off-diagonal. In the Notes, we will make this more precise.

The sin kernel enters via the following elegant result in complex function theory:

Theorem 3.12.2 ([287]). *Let* $f(z)$ *be an entire function that obeys*

(i)

$$f(0) = 1 \qquad |f(x)| \le 1 \quad \textit{for } x \in \mathbb{R} \tag{3.12.5}$$

(ii)

$$\int_{-\infty}^{\infty} |f(x)|^2\,dx \le 1 \tag{3.12.6}$$

(iii) *For some* C *and* A,

$$|f(z)| \le Ce^{A|z|} \tag{3.12.7}$$

(iv) *f is real on $\mathbb{R}$, all the zeros of f are real, and if*

$$\cdots < x_{-2} < x_{-1} < 0 < x_1 < x_2 < \cdots \tag{3.12.8}$$

are the zeros, then

$$|x_n| \geq (|n| - 1) \tag{3.12.9}$$

Then

$$f(z) = \frac{\sin(\pi z)}{\pi z} \tag{3.12.10}$$

Proof. ([30]) We will prove shortly that for any $\varepsilon > 0$, there is C_ε with

$$|f(x+iy)| \leq C_\varepsilon e^{(\pi+\varepsilon)|y|} \tag{3.12.11}$$

Assuming this for a moment, let us prove (3.12.10). Let $\widehat{f}(k)$ be the Fourier transform of f:

$$\widehat{f}(k) = (2\pi)^{-1/2} \int_{-\infty}^{\infty} e^{-ikx} f(x)\, dx \tag{3.12.12}$$

so

$$f(x) = (2\pi)^{-1/2} \int_{-\infty}^{\infty} e^{ikx} \widehat{f}(k)\, dk \tag{3.12.13}$$

(where the integrals are shorthand for distributional Fourier transform).

By the Paley–Wiener theorem (see the Notes), (3.12.11) implies

$$\widehat{f}(k) = 0 \qquad k \neq [-\pi, \pi] \tag{3.12.14}$$

By (3.12.6) and the Plancherel theorem,

$$\int_{-\infty}^{\infty} |\widehat{f}(k)|^2\, dk \leq 1 \tag{3.12.15}$$

and, by (3.12.5) and (3.12.13),

$$\int_{-\infty}^{\infty} \widehat{f}(k)\, dk = (2\pi)^{1/2} \tag{3.12.16}$$

Therefore,

$$\int_{-\infty}^{\infty} |\widehat{f}(k) - (2\pi)^{-1/2} \chi_{[-\pi,\pi]}(k)|^2 \leq 1 + 1 - 2 = 0 \tag{3.12.17}$$

so

$$\widehat{f} = (2\pi)^{-1/2} \chi_{[-\pi,\pi]} \tag{3.12.18}$$

and (3.12.10) follows from (3.12.13).

Thus, we are reduced to proving (3.12.11), to which we now turn. By (3.12.7), (3.12.5), and the Hadamard factorization theorem (see the Notes), for some B real,

$$f(z) = e^{Bz} \prod_{j=1}^{\infty} \left[\left(1 - \frac{z}{z_j}\right) e^{z/z_j}\right] \left[\left(1 - \frac{z}{z_{-j}}\right) e^{z/z_{-j}}\right] \tag{3.12.19}$$

from which we see for y real that

$$|f(iy)|^2 \le \prod_{j=1}^{\infty} \left(1 + \frac{y^2}{z_j^2}\right)\left(1 + \frac{y^2}{z_{-j}^2}\right) \tag{3.12.20}$$

so, by (3.12.9),

$$|f(iy)|^2 \le (1 + Cy^2)^2 \left[\prod_{n=1}^{\infty} \left(1 + \frac{y^2}{n^2}\right)\right]^2 \tag{3.12.21}$$

$$= (1 + Cy^2)^2 \left[\frac{\sinh(\pi y)}{\pi y}\right]^2 \tag{3.12.22}$$

by the Euler product formula (see the Notes). This implies (3.12.11) for $x = 0$.

(3.12.6) implies (3.12.11) for $y = 0$, so we have (3.12.11) on the axes. Since we have (3.12.7), we can apply the Phragmén–Lindelöf principle (see the Notes) in each quadrant to get (3.12.11) for all x and y. □

Lubinsky used this theorem to prove the following precursor of Theorem 3.12.1:

Theorem 3.12.3 ([287]). *Suppose $d\mu$ has the form* (3.12.1) *and x_0 is a Lebesgue point for $d\mu$ in the sense that*

$$\lim_{\varepsilon \downarrow 0} \frac{1}{2\varepsilon} \int |w(x) - w(x_0)|\, dx \to 0 \qquad \frac{1}{2\varepsilon}\, \mu_s(x_0 - \varepsilon, x_0 + \varepsilon) \to 0 \tag{3.12.23}$$

and suppose that

(a) *For some A and C and all $R < \infty$, there exists N so that for $n \ge N$ and all z complex with $|z| < R$,*

$$\left|K_n\left(x_0 + \frac{\bar z}{n}, x_0 + \frac{z}{n}\right)\right| \le Ce^{A|z|} \tag{3.12.24}$$

(b)

$$\liminf \frac{1}{n} K_n(x_0, x_0) > 0 \quad \textit{and} \quad w_0(x) > 0 \tag{3.12.25}$$

(c) *For all $B < \infty$,*

$$\lim_{n\to\infty} \frac{K_n(x_0 + \frac{a}{n}, x_0 + \frac{a}{n})}{K_n(x_0, x_0)} = 1 \tag{3.12.26}$$

uniformly for real a with $|a| \le B$.

Then

$$\lim_{n\to\infty} \frac{K_n(x_0 + \frac{z}{n\rho_n}, x_0 + \frac{w}{n\rho_n})}{K_n(x_0, x_0)} = \frac{\sin(\pi(z - w))}{\pi(z - w)} \tag{3.12.27}$$

where

$$\rho_n = \frac{w(x_0)}{n} K_n(x_0, x_0) \tag{3.12.28}$$

Remarks. 1. (3.12.23) holds for a.e. x with $w(x) > 0$ by standard harmonic analysis [372].

2. Lubinsky does not write (3.12.24) as a hypothesis, but instead demands $w(x) \geq \varepsilon > 0$ near x and then deduces (3.12.24).

3. The key is thus the hypothesis (3.12.26), which we call the *Lubinsky wiggle condition*. It is clearly also a key piece of Lubinsky's other argument, but here it is the only requirement (when w is bounded strictly away from zero). Alas, Lubinsky could only prove that (3.12.26) holds in cases where his first argument also works (but see the Notes).

The proof of Theorem 3.12.3 depends on a critical classical inequality (see the Notes):

Proposition 3.12.4 (Markov–Stieltjes Inequality). *Let $x_j^{(n)}(x_0)$ be defined by requiring*

$$p_{n-1}(x_0)\,p_n(x_j^{(n)}(x_0)) - p_n(x_0)\,p_{n-1}(x_j^{(n)}(x_0)) = 0 \tag{3.12.29}$$

where $x_j^{(n)}(x_0) < x_{j+1}^{(n)}(x_0)$ and $j = 1, \dots, n$ if $p_{n-1}(x_0) \neq 0$ and $j = 1, \dots, n$ if $p_{n-1}(x_0) \neq 0$. Then

$$\sum_{\{j \mid x_j^{(n)}(x_0) \leq x_0\}} \frac{1}{K_{n-1}(x_j^{(n)}(x_0), x_j^{(n)}(x_0))} \geq \mu((-\infty, x_0])$$

$$\geq \mu((-\infty, x_0)) \geq \sum_{\{j \mid x_j^{(n)}(x_0) < x_0\}} \frac{1}{K_{n-1}(x_j^{(n)}(x_0), x_j^{(n)}(x_0))} \tag{3.12.30}$$

Remark. If $p_n(x_0) = 0$ (resp. $p_{n-1}(x_0) = 0$), $x_j^{(n)}(x_0)$ list all the zeros of $p_n(x)$ (resp. $p_{n-1}(x)$).

Sketch of the Proof of Theorem 3.12.3. (See [30] for details.) Fix a real and let $f_n(z)$ be defined by

$$f_n(z) = \frac{K_n(x_0 + \frac{a}{n\rho_n}, x_0 + \frac{a+z}{n\rho_n})}{K_n(x_0, x_0)} \tag{3.12.31}$$

where ρ_n is given by (3.12.28).

By hypothesis (a) and Montel's theorem (see the Notes), $\{f_n\}$ is compact in the topology of uniform convergence on compact subsets of $\mathbb{C}$, so we need only show that each limit point is $\sin(\pi z)/\pi z$ to conclude that the limit is $\sin(\pi z)/\pi z$. By analyticity and the same compactness plus the Schwarz inequality for $K_n(x, y)$ (i.e., $|K_n(x, y)|^2 \leq K_n(\bar{x}; x) K_n(\bar{y}; y)$), we then get (3.12.27) for all complex z, w.

So let f be a limit point of f_n. We will show f obeys all the hypotheses of Theorem 3.12.2. By the Lubinsky wiggle condition and Schwarz inequality, (3.12.4) holds.

By the reproducing kernel property of K_n and a change of variables and definition of ρ_n,

$$\int |f_n(x)|^2 \left[\frac{w(x)}{w(x_0)}\right] dx \leq \frac{K_n(x_0 + \frac{a}{n\rho_n}, x_0 + \frac{a}{n\rho_n})}{K_n(x_0, x_0)} \tag{3.12.32}$$

Using that x_0 is a Lebesgue point and the wiggle condition, we get (3.12.6). From (3.12.24), we get (3.12.7).

Finally, the Markov–Stieltjes inequalities, the wiggle condition, and the Lebesgue point condition imply (3.12.9). □

Sketch of the Proof of Theorem 3.12.1. (See [30] for details.) We will establish the hypotheses of Theorem 3.12.3. (3.12.25) is immediate from (3.12.2).

(3.12.2), (3.12.4), and $\sup a_n < \infty$, $\inf a_n > 0$ (by $\Sigma_{\mathrm{ac}} \neq \emptyset$) imply that

$$\sup_n \frac{1}{n+1} \sum_{j=0}^{n+1} \|T_j(x_0)\|^2 \tag{3.12.33}$$

By Theorem 3.8.14, we conclude that

$$\begin{aligned}\left\| T_j\left(x_0 + \frac{z}{n+1}\right)\right\| &\leq \inf_n(|a_n|)^{-1} \|T_j(x_0)\| \\ &\quad \times \exp\left(\sum_{k=1}^{j} \frac{|z|}{j+1} \|T_k(x_0)\|\, \|T_{k-1}(x_0)\|\right)\end{aligned} \tag{3.12.34}$$

since $\|T_k^{-1}\| = \|T_k\|$. By the Schwarz inequality and (3.12.33), we conclude that

$$\frac{1}{n+1} \sum_{j=0}^{n} \left\| T_j\left(x_0 + \frac{z}{n+1}\right)\right\|^2 \leq C_1 \exp(C_2|z|) \tag{3.12.35}$$

which implies (3.12.24).

That leaves (3.12.26). By Egoroff's theorem for any ε, we can find $\Sigma_\varepsilon \subset \Sigma$, so $|\Sigma \setminus \Sigma_\varepsilon| < \varepsilon$, and (3.12.2) holds uniformly on Σ_ε. We will show (3.12.26) a.e. on each such Σ_ε and thus, a.e. on Σ.

Let $x_0 \in \Sigma_\varepsilon$ be a point of density of Σ_ε, that is,

$$\lim_{\delta \to 0} (2\delta)^{-1} |(x_0 - \delta, x_0 + \delta) \cap \Sigma_\varepsilon| \to 1 \tag{3.12.36}$$

and let

$$g_n(a) = \text{LHS of (3.12.26)}$$

By the uniform convergence of f_n on Σ_ε and the implied continuity of limit, for every $A < \infty$,

$$\sup_{\substack{|b| \leq A \\ x_0 + \frac{b}{n} \in \Sigma_\varepsilon}} |g_n(b) - 1| \to 0 \tag{3.12.37}$$

By (3.12.36),

$$\sup_{|a| \leq A} \left(\inf_{\substack{|b_n| \leq A \\ x_0 + \frac{b_n}{n} \in \Sigma_\varepsilon}} |a - b| \right) \to 0 \tag{3.12.38}$$

By (3.12.24), for every A,

$$\sup_{n, |z| \leq A} |g_n'(z)| < \infty \tag{3.12.39}$$

By (3.12.37), (3.12.38), and (3.12.39),

$$\sup_{|a|\leq A} |g_n(a) - 1| \to 0$$

as $n \to \infty$. It follows that we have proven (3.12.38) for a.e. $x_0 \in \Sigma_\varepsilon$. □

Remarks and Historical Notes. Lubinsky [287] had the wonderful idea of using Markov–Stieltjes inequalities and complex variable characterizations of the sinc (i.e., $\sin x/x$) kernel. He used special properties of the sinc kernel (see [420]); translating these properties into a direct proof using the Paley–Wiener theorem is from [30].

Lubinsky did not directly state (3.12.24) as a hypothesis. Rather, he assumed $w(x) > \delta$ in an interval and deduced (3.12.24) using the Christoffel variational principle. He was unable to prove the Lubinsky wiggle condition except in situations where Totik and Simon had already shown how to get universality using Lubinsky's first method. But he opened the portals to Avila–Last–Simon to handle ergodic Jacobi matrices.

Let Ω be a compact metric space with probability measures, $d\eta$, and $T: \Omega \to \Omega$ ergodic. If $A, B: \Omega \to \mathbb{R}$ are continuous with $\inf_{\omega\in\Omega} A(\omega) > 0$, one defines ergodic Jacobi matrices to be the ω-dependent matrix with Jacobi parameters $\{a_n, b_n\}_{n=1}^\infty$ given by

$$a_n(\omega) = A(T^n\omega) \qquad b_n(\omega) = B(T^n\omega) \tag{3.12.40}$$

[30] were able to prove that for a.e. ω and a.e. x_0 in the a.c. spectrum, one has universality and clock behavior with $\rho_\infty(x)$ given by the a.c. part of the density of zeros. The canonical example is the almost Mathieu equation (see Jitomirskaya [209]) where $\Omega = \partial\mathbb{D}$, $A(\omega) \equiv 1$, and $B(\omega) = 2\lambda\cos(\pi\alpha\theta)$. If $|\lambda| < 1$ and α is irrational, the spectrum is a Cantor set with purely a.c. spectrum.

For the Paley–Wiener theorem, see [362], and for the Hadamard factorization theorem, Phragmén–Lindelöf principle, Euler product formula, and Montel's theorem, see Titchmarsh [439].

The Markov–Stieltjes inequalities are due to Markov [296] and Stieltjes [421] who consider the case where $p_n(x_0) = 0$. The general form we use is due to Freud [141]; for a proof, see this book or [407]. There is a still more general version in Krein–Nudel'man [253].

Chapter Four

Sum Rules and Consequences for Matrix Orthogonal Polynomials

In this chapter, we will discuss matrix-valued orthogonal polynomials on the real line (aka MOPRL). These are based on a measure, $d\mu$, which, instead of assigning a nonnegative number to any set, assigns a nonnegative $\ell \times \ell$ matrix. From the Jacobi matrix point of view, the Jacobi parameters become $\ell \times \ell$ matrices.

4.1 INTRODUCTION

MOPRL is a strange subject. Most parts are straightforward extensions of the OPRL theory, but every so often, a subtlety arises. Fortunately, in our case of sum rules, the only subtlety concerns a possible coincidence of eigenvalues of J_0 and J_1. There is another place in our considerations where a subtlety arises that we will come to shortly.

The result is that the MOPRL theory is so close to the OPRL theory that much of the last three sections of this chapter, where we turn to sum rules, will say: "Now just follow the proof from Chapter 3." The only reason something so similar to OPRL occurs in these notes is because, remarkably, as we will see in Chapter 8, we can study perturbations of scalar periodic Jacobi matrices by relating it to a perturbation of an MOPRL with constant coefficients (indeed, $A_n \equiv \mathbf{1}$, $B_n \equiv \mathbf{0}$). It is for this reason that we consider MOPRL here and not MOPUC, the unit circle analog. It will turn out (see Section 8.7, especially (8.7.4)) that even perturbations of periodic CMV matrices relate to MOPRL, not MOPUC.

What we will be missing from our discussion is an analog of full Szegő asymptotics of the matrix orthogonal polynomials—very recently, this has been obtained (see the Notes), but its exposition would take up some space.

Section 4.2 discusses the basic MOPRL formalism—the one surprise is that there are actually two natural families of OPs. Section 4.3 discusses coefficient stripping and contains in Theorem 4.3.3 what is not a straightforward copying of what we did in Chapter 3. Section 4.4 then proves matrix nonlocal sum rules while Sections 4.5 and 4.6 present the by now standard applications to MOPRL analogs of the Shohat–Nevai and Killip–Simon theorems.

Remarks and Historical Notes. The theory of matrix orthogonal polynomials on the real line goes back to seminal papers by Krein [251] and Berezans'ki [41]. A lot of the rather large literature and survey of many of the analytic results can be found in the review article of Damanik–Pushnitski–Simon [98].

From the point of view of sum rules, the two most significant later papers are Aptekarev–Nikishin [23] and Damanik–Killip–Simon [97].

One place where there is recent progress is in the asymptotics of polynomials when a Szegő condition holds (i.e., the subject we studied in the scalar case in Sections 2.9, 2.13, and 3.7). For MOPRL, these asymptotics were studied by Delsarte–Genin–Kamp [106]. Indeed, they developed there the approach we discuss in Section 2.13. Aptekarev–Nikishin [23] were able to handle MOPRL with no bound states by a Szegő mapping and with finitely many bound states using an inductive argument (alternatively, one could use coefficient stripping). The result with only a Blaschke-type condition on the bound states was settled by Kozhan [245]. The key is a factorization result for matrix Herglotz functions (we will get around this by looking only at their determinants). There are factorization theorems for matrix-valued H^p functions (see Potapov [354], Gohberg–Sakhnovich [171]), whose techniques are part of Kozhan's work.

4.2 BASICS OF MOPRL

An $\ell \times \ell$ matrix-valued measure on $\mathbb{R}$ is the assignment of a nonnegative $\ell \times \ell$ matrix, $\mu(S)$, to each Borel set $S \subset \mathbb{R}$ and which is countably additive. We will usually normalize by

$$\mu(\mathbb{R}) = \mathbf{1} \tag{4.2.1}$$

but in any event, suppose $\mu(\mathbb{R})$ is a (finite) matrix.

Define a scalar measure, μ_t, by

$$\mu_t(A) = \mathrm{Tr}(\mu(A)) \tag{4.2.2}$$

Then, since $\langle \varphi, B\varphi \rangle \leq \mathrm{Tr}(B)$ for $B \geq 0$ and $\|\varphi\| = 1$, we see each measure $A \mapsto \mu(A)_{ij}$ is μ_t-a.c., so

$$d\mu_{ij}(x) = M_{ij}(x)\, d\mu_t(x) \tag{4.2.3}$$

μ positive and (4.2.2) imply that

$$M(x) \geq \mathbf{0} \qquad \mathrm{Tr}(M(x)) = 1 \tag{4.2.4}$$

We will postpone the definition of nontriviality of μ.

Throughout, we suppose μ has finite moments, that is, for all $n = 0, 1, 2, \dots,$

$$\int |x|^n \, d\mu_t(x) < \infty \tag{4.2.5}$$

Now suppose f, g are $\ell \times \ell$ matrix-valued functions on $\mathbb{R}$; we define an $\ell \times \ell$ matrix, $\int f(x)\, d\mu(x)\, g(x)$, in the obvious way, that is,

$$\left(\int f(x)\, d\mu(x)\, g(x) \right)_{ij} = \sum_{k,n} \int f(x)_{ik} M_{kn}(x) g(x)_{nj}\, d\mu_t(x) \tag{4.2.6}$$

We define two "inner products," that is, sesquilinear maps of the $\ell \times \ell$ matrix-valued functions to $\ell \times \ell$ matrices by (to avoid confusion with Szegő dual, we use †, not *, for adjoint)

$$\langle\langle f, g\rangle\rangle_{\rm L} = \int g(x)\, d\mu(x)\, f(x)^\dagger \tag{4.2.7}$$

and

$$\langle\langle f, g\rangle\rangle_{\rm R} = \int f(x)^\dagger\, d\mu(x)\, g(x) \tag{4.2.8}$$

initially on bounded f and g with bounded support, but eventually for suitable L^2-like spaces.

The symbols L, R (for "left" and "right") come from fact that for scalar multiplication by a matrix A,

$$\langle\langle f, Ag\rangle\rangle_{\rm L} = A\langle\langle f, g\rangle\rangle \tag{4.2.9}$$

$$\langle\langle Af, g\rangle\rangle_{\rm L} = \langle\langle f, g\rangle\rangle A^\dagger \tag{4.2.10}$$

$$\langle\langle f, gA\rangle\rangle_{\rm R} = \langle\langle f, g\rangle\rangle_{\rm R} A \tag{4.2.11}$$

$$\langle\langle fA, g\rangle\rangle_{\rm R} = A^\dagger\langle\langle f, g\rangle\rangle_{\rm R} \tag{4.2.12}$$

We will also define norms via

$$\|f\|_{\rm R} = ({\rm Tr}\langle\langle f, f\rangle\rangle_{\rm R})^{1/2} \tag{4.2.13}$$

$$\|f\|_{\rm L} = ({\rm Tr}\langle\langle f, f\rangle\rangle_{\rm L})^{1/2} \tag{4.2.14}$$

One has that

$$\|f^\dagger\|_{\rm R} = \|f\|_{\rm L} \qquad \langle\langle f, g\rangle\rangle_{\rm R}^\dagger = \langle\langle g, f\rangle\rangle_{\rm R} \tag{4.2.15}$$

and

$$\langle\langle f, g\rangle\rangle_{\rm R} = \langle\langle g^\dagger, f^\dagger\rangle\rangle_{\rm L} \tag{4.2.16}$$

Let $\mathcal{P}$ be the family of all polynomials and $\mathcal{P}_{\rm R} = \mathcal{P}/\{P \in \mathcal{P} \mid \|f\|_{\rm R} = 0\}$, and similarly for $\mathcal{P}_{\rm L}$. The completion of $\mathcal{P}_{\rm R}$ (resp. $\mathcal{P}_{\rm L}$) in $\|\cdot\|_X$ we call $\mathcal{H}_{\rm R}$ (resp. $\mathcal{H}_{\rm L}$). If μ has bounded support, then multiplication by x is a bounded selfadjoint operator. $f \to f^\dagger$ is an anti-unitary map of $\mathcal{H}_{\rm R}$ to $\mathcal{H}_{\rm L}$, leaving multiplication by x invariant. The spectrum of multiplication by x we will call $\sigma(d\mu)$—it is the support in the measure theoretic sense.

The following is elementary:

Proposition 4.2.1. *Let μ be a matrix-valued measure on $\mathbb{R}$. Then the following are equivalent:*

(i) *For every nonzero matrix-valued polynomial, f, $\|f\|_{\rm L} > 0$.*
(ii) *For every nonzero matrix polynomial, f, $\|f\|_{\rm R} > 0$.*
(iii) *For every $n = 0, 1, 2, \dots$, $\dim(\{P \in \mathcal{P}_{\rm L} \mid \deg(P) \le n\}) = \ell^2(n+1)$.*
(iv) *For every $n = 0, 1, 2, \dots$, $\dim(\{P \in \mathcal{R}_{\rm R} \mid \deg(P) \le n\}) = \ell^2(n+1)$.*

Proof. Since $\dim(\{P \in \mathcal{P} \mid \dim(P) \le n\}) = \ell^2(n+1)$, we see (i) $\Leftrightarrow$ (iii) and (ii) $\Leftrightarrow$ (iv). By (4.2.15), (i) $\Leftrightarrow$ (ii). □

Example 4.2.2. For $0 \le t \le 1$, let $M(t)$ be the orthogonal projection onto the vector $\binom{1}{-t}$ in $\mathbb{C}^2$ and let

$$d\mu = \chi_{[0,1]}(t) M(t)\, dt \tag{4.2.17}$$

Let $P(t)$ be the polynomial

$$P(t) = \begin{pmatrix} t & t \\ 1 & 1 \end{pmatrix} \tag{4.2.18}$$

so $\text{Ran}(P(t)) \subset \binom{1}{-t}^{\perp}$ and $P(t)^\dagger M(t) P(t) \equiv 0$, so $\|P\|_{\text{R}} = 0$. Thus, $d\mu$ does not obey (i)–(iv) of the proposition. However, for any fixed nonzero φ, $\langle \varphi, M(t)\varphi\rangle = 0$ if and only if $\langle \binom{1}{-t}, \varphi\rangle = 0$, which happens at most at one t in $[0, 1]$. Thus, $\langle \varphi, d\mu\, \varphi\rangle$ is nontrivial for any φ. This shows nontriviality of $\langle \varphi, d\mu\, \varphi\rangle$ for all φ does not suffice for (i)–(iv) to hold. $\square$

Definition. If (i)–(iv) of Proposition 4.2.1 hold, we say that μ is *nontrivial.*

Henceforth, we assume that μ is nontrivial.

Proposition 4.2.3. *A sufficient condition for μ to be nontrivial is that there is a Borel set S with* $\text{rank}(M(x)) = \ell$ *for $x \in S$, and for any finite set, F, $\mu_t(S \setminus F) \neq 0$.*

Proof. Let P be a nonzero polynomial in $\mathcal{P}$. On S, $\text{Tr}(P(x)^\dagger M(x) P(x))$ vanishes only at points where $P(x) = 0$ as a matrix—and this can only happen on the finite set where $\det(P(x)) = 0$. Thus, by hypothesis, $\|P\|_{\text{R}} > 0$. $\square$

Introduce monic MOPRL, P_n^{R}, P_n^{L}, $\ell \times \ell$ matrix polynomials of the form (we will use X as a generic for R or L)

$$P_n^X(x) = x^n + \text{lower order in } x \tag{4.2.19}$$

so that

$$\langle\langle x^j, P_n^X \rangle\rangle_X = 0 \quad \text{for } j = 0, 1, \dots, n-1 \tag{4.2.20}$$

Here x^j is shorthand for the matrix $x^j\mathbf{1}$. It is easy to see these determine P_n^X inductively and, if μ is nontrivial, that such P_n^X exist. Moreover, by (4.2.16),

$$P_n^{\text{R}}(x) = P_n^{\text{L}}(\bar{x})^\dagger \tag{4.2.21}$$

Indeed, we have $P_0^X(x) = \mathbf{1}$ and that, by (4.2.16),

$$\gamma_n \equiv \langle\langle P_n^{\text{L}}, P_n^{\text{L}} \rangle\rangle_{\text{L}} = \langle\langle P_n^{\text{R}}, P_n^{\text{R}} \rangle\rangle_{\text{R}} \tag{4.2.22}$$

is nonzero if μ is nontrivial so that

$$P_n^{\text{R}}(x) = x^n - \sum_{j=0}^{n-1} P_j^{\text{R}}(x) \gamma_j^{-1} \langle\langle P_j^{\text{R}}, x^n \rangle\rangle_{\text{R}} \tag{4.2.23}$$

To define orthonormal MOPRL, we pick $\ell \times \ell$ unitaries, $\sigma_0 = \mathbf{1}, \sigma_1, \sigma_2, \dots$ and $\tau_0 = \mathbf{1}, \tau_1, \tau_2, \dots$, and let

$$p_n^{\text{R}}(x) = P_n^{\text{R}}(x)\gamma_n^{-1/2}\sigma_n \qquad p_n^{\text{L}}(x) = \tau_n \gamma_n^{-1/2} P_n^{\text{L}}(x) \tag{4.2.24}$$

which obey

$$\langle\langle p_n^X, p_k^X \rangle\rangle_X = \delta_{nk} \tag{4.2.25}$$

$$p_n^X(x) = \kappa_n^X x^n + \text{lower order} \tag{4.2.26}$$

$$\kappa_n^{\mathrm{L}} = \tau_n \gamma_n^{-1/2} \qquad \kappa_n^{\mathrm{R}} = \gamma_n^{-1/2} \sigma_n \tag{4.2.27}$$

It is easy to see that if one demands $p_0 = \mathbf{1}$, then p_n^X, obeying (4.2.25), is determined up to precisely a choice of σ_n, τ_n.

In the scalar case, one picks $\sigma_n \equiv 1$ since it is reasonable to demand $\kappa_n > 0$. We will see below why one does not always demand that, but instead we associate a matrix-valued measure with an equivalence class of normalized MOPRL. Henceforth, we will always suppose that

$$\tau_n = \sigma_n^\dagger \tag{4.2.28}$$

so that, by (4.2.24) and (4.2.21),

$$p_n^{\mathrm{R}}(x) = p_n^{\mathrm{L}}(\bar{x})^\dagger \tag{4.2.29}$$

Note that the p_n^X are an orthonormal module basis in that if f is any matrix polynomial of degree n, we have

$$f(x) = \sum_{m=0}^{n} p_m^{\mathrm{R}}(x) \langle\langle p_m^{\mathrm{R}}, f \rangle\rangle_{\mathrm{R}} \tag{4.2.30}$$

$$= \sum_{m=0}^{n} \langle\langle p_m^{\mathrm{L}}, f \rangle\rangle_{\mathrm{L}}\, p_m^{\mathrm{L}}(x) \tag{4.2.31}$$

$$\| f \|_X^2 = \sum_{m=0}^{n} \mathrm{Tr}(\langle\langle p_m^X, f \rangle\rangle_X^\dagger \langle\langle p_m^X, f \rangle\rangle_X) \tag{4.2.32}$$

If one completes $\mathcal{P}_X$ in $\|\cdot\|_X$, these three formulae hold for any f in the completion if one takes $n = \infty$. Of course, as in the scalar case, we get a three-term recurrence relation:

Theorem 4.2.4. *Given a nontrivial $\ell \times \ell$ matrix measure μ with finite moments and choice of $\{\sigma_n\}_{n=0}^\infty$ with $\sigma_0 = \mathbf{1}$, there exist $\ell \times \ell$ matrices $\{B_n\}_{n=1}^\infty$ and $\{A_n\}_{n=1}^\infty$ with*

$$B_n^\dagger = B_n \tag{4.2.33}$$

so that

$$x p_n^{\mathrm{R}}(x) = p_{n+1}^{\mathrm{R}}(x) A_{n+1}^\dagger + p_n^{\mathrm{R}}(x) B_{n+1} + p_{n-1}^{\mathrm{R}}(x) A_n \tag{4.2.34}$$

and

$$x p_n^{\mathrm{L}}(x) = A_{n+1} p_{n+1}^{\mathrm{L}}(x) + B_{n+1} p_n^{\mathrm{L}}(x) + A_n^\dagger p_{n-1}^{\mathrm{L}}(x) \tag{4.2.35}$$

Moreover, each A_n is invertible and if

$$\sigma(d\mu) \subset [-R, R] \tag{4.2.36}$$

then

$$\|A_n\| \leq R \qquad \|B_n\| \leq R \tag{4.2.37}$$

If $\tilde{\sigma}_n$ is another choice of the σ's and

$$u_n = \sigma_{n-1}^{-1} \tilde{\sigma}_{n-1} \tag{4.2.38}$$

then

$$\tilde{B}_n = u_n^{-1} B_n u_n \qquad \tilde{A}_n = u_n^{-1} A_n u_{n+1} \tag{4.2.39}$$

Conversely, for any set of $\ell \times \ell$ matrices $\{A_n, B_n\}_{n=1}^{\infty}$ obeying A_n invertible, (4.2.33), *and* (4.2.37), *there is a matrix measure $d\mu$ obeying* (4.2.36) *(with a possible increase of R) and $\{\sigma_n\}_{n=0}^{\infty}$ with $\sigma_0 = \mathbf{1}$ so that the $\{p_n^{\mathrm{R}}\}_{n=0}^{\infty}$ obeying* (4.2.34) *and*

$$p_0^{\mathrm{R}}(x) = \mathbf{1} \qquad p_{-1}^{\mathrm{R}}(x) = 0 \tag{4.2.40}$$

are the MOPRL for μ. Two sets of $\{A_n, B_n\}$, $\{\tilde{A}_n, \tilde{B}_n\}$ generate the same $d\mu$ if and only if (4.2.39) *holds for some unitaries, $u_1, u_2, \dots$, with $u_1 = \mathbf{1}$.*

Proof. Given the scalar case, this is straightforward. By (4.2.29), (4.2.34) implies (4.2.35), so we need only do the former. As usual, for $j < n - 1$,

$$\langle\langle p_j^{\mathrm{R}}, x p_n^{\mathrm{R}} \rangle\rangle_{\mathrm{R}} = \langle\langle x p_j^{\mathrm{R}}, p_n^{\mathrm{R}} \rangle\rangle = 0 \tag{4.2.41}$$

since $\deg(x p_j^{\mathrm{R}}) < n$. Define

$$B_{n+1} = \langle\langle p_n^{\mathrm{R}}, x p_n^{\mathrm{R}} \rangle\rangle_{\mathrm{R}} \qquad A_n = \langle\langle p_{n-1}^{\mathrm{R}}, x p_n^{\mathrm{R}} \rangle\rangle_{\mathrm{R}} \tag{4.2.42}$$

so by (4.2.27),

$$B_{n+1}^{\dagger} = B_{n+1} \qquad A_{n+1}^{\dagger} = \langle\langle x p_{n+1}^{\mathrm{R}}, p_n^{\mathrm{R}} \rangle\rangle = \langle\langle p_{n+1}^{\mathrm{R}}, x p_n^{\mathrm{R}} \rangle\rangle \tag{4.2.43}$$

(4.2.41)–(4.2.43) and (4.2.30) imply (4.2.34).

By (4.2.42) and $\|p_j^{\mathrm{R}}\| = 1$, (4.2.37) is immediate. By definition,

$$\tilde{p}_n^{\mathrm{R}} = p_n^{\mathrm{R}} u_{n+1} \tag{4.2.44}$$

so (4.2.39) follows from (4.2.28) and (4.2.11)/(4.2.12).

By (4.2.24) and (4.2.26), we have

$$\kappa_n^{\mathrm{L}} = \tau_n \gamma_n^{-1/2} \tag{4.2.45}$$

and so is invertible. By looking at x^{n+1} coefficients in (4.2.34),

$$\kappa_n^{\mathrm{L}} = A_{n+1} \kappa_{n+1}^{\mathrm{L}} \tag{4.2.46}$$

so A_{n+1} is invertible.

For the converse, given $\{A_n, B_n\}_{n=1}^{\infty}$, form the block Jacobi matrix,

$$J = \begin{pmatrix} B_1 & A_1 & 0 & \dots \\ A_1^{\dagger} & B_2 & A_2 & \dots \\ 0 & A_2^{\dagger} & B_3 & \dots \\ \vdots & \vdots & \vdots & \ddots \end{pmatrix} \tag{4.2.47}$$

acting on $\ell^2(\{1,2,\dots\},\mathbb{C}^\ell)$, which is a bounded operator with $\|J\| \le 3R$. Let $\{\delta_j^{(k)}\}_{k=1,\dots,\ell;\ j=1,2,\dots}$ be the vector with a 1 in position $\ell(j-1)+k$. By the spectral theorem and multiplication, there are measures $\{\mu_{pk}\}_{p,k=1,\dots,\ell}$ with

$$\langle \delta_1^{(p)}, J^m \delta_1^{(k)} \rangle = \int x^m \, d\mu_{pk}(x) \tag{4.2.48}$$

These can be put together into a matrix measure and the polynomials defined inductively by (4.2.34) are normalized MOPRL and so have the requisite form. □

We have thus set up a one-one correspondence between nontrivial $\ell \times \ell$ matrix-valued measures of compact support on $\mathbb{R}$ and equivalence classes of uniformly bounded Jacobi parameters under the equivalence:

Definition. Two sets of Jacobi parameters are called *equivalent* if there exist $\ell \times \ell$ unitaries, $u_1 = \mathbf{1}, u_2, u_3, \dots$, so that (4.2.39) holds.

Notice that, by (4.2.34), (4.2.35), and (4.2.26),

$$\kappa_n^{\mathrm{L}} = (A_1 \dots A_{n+1})^{-1} \qquad \kappa_n^{\mathrm{R}} = (A_{n+1}^\dagger \dots A_1^\dagger)^{-1} \tag{4.2.49}$$

This partly explains why we consider equivalence classes and not a single set of Jacobi parameters. From the scalar case, one might like a choice with $A_j > 0$ and with $\kappa_n^{\mathrm{L}} > 0$. But since the A_j's may not commute, (4.2.49) shows those desires may conflict. There is actually a third natural choice:

Definition. A matrix, A, is called *lower triangular* if $A_{jk} = 0$ for $j < k$, that is, it has potentially nonzero elements only on and below the main diagonal. $\mathcal{L}$ will denote the set of lower triangular matrices, which are positive on diagonal, that is, $A_{jj} > 0$ for all j.

In Section 6.3, we discuss the upper triangular matrices and the associated QR decomposition. Note that a block Jacobi matrix, J, has all $A_n \in \mathcal{L}$ if and only if it is $2\ell + 1$ diagonal with positive elements on the two extreme diagonals.

Definition. A set of Jacobi parameters $\{A_n, B_n\}_{n=1}^\infty$ is said to be of *type 1* if and only if, for all n, $A_n > 0$. We say it is of *type 2* if and only if, for all n, $A_1 \dots A_n > 0$ (equivalently, $\kappa_n^{\mathrm{L}} = \kappa_n^{\mathrm{R}} > 0$). We say it is of *type 3* if and only if all $A_n \in \mathcal{L}$.

Using polar and QR decompositions, one can prove (see the Notes):

Theorem 4.2.5 (Damanik–Pushnitski–Simon [98]). *Each equivalence class of matrix Jacobi parameters has exactly one representative each of type 1, type 2, and type 3.*

Remarks and Historical Notes. We follow the notation of the review article on MOPRL and MOPUC of Damanik–Pushnitski–Simon [98], which, in particular, proves Theorem 4.2.5.

4.3 COEFFICIENT STRIPPING

In this section, we will define the m-function, find the coefficient stripping formula, and examine zeros and poles of $\det(m(z))$, an object we will study in Section 4.4.

Let μ be an $\ell \times \ell$ matrix-valued measure. We define the m-function on $\mathbb{C}_+$ by

$$m(z) = \int \frac{d\mu(x)}{x - z} \tag{4.3.1}$$

which is an $\ell \times \ell$ matrix ($(x-z)^{-1}$ is scalar so it does not matter where we put it).

We have seen that μ is associated with a Jacobi matrix, J, acting on $\mathcal{H} \equiv \ell^2(\{1, 2, \dots\}, \mathbb{C}^\ell)$. Let P_1 be defined from $\mathcal{H}$ to $\mathbb{C}^\ell$ by

$$(P_1 f) = f_1 \tag{4.3.2}$$

which we can think of as a projection on $\mathcal{H}$, but for now as a map from $\mathcal{H}$ to $\mathbb{C}^\ell$ so P_1^* (Hilbert space adjoint, not just on $\mathbb{C}^\ell$, so we use * rather than †) takes $\mathbb{C}^\ell$ to $\mathcal{H}$. By construction of μ from J, we have as $\ell \times \ell$ matrix-valued functions,

$$m(z) = P_1 \frac{1}{J - z} P_1^* \tag{4.3.3}$$

On $\mathbb{C}_+$, we have

$$\text{Im}(m(z)) \equiv \frac{1}{2i} (m(z) - m(z)^\dagger) > 0 \tag{4.3.4}$$

so $m(z)$ is a matrix-valued Herglotz function.

We will get coefficient stripping from Weyl solutions, so we start with second kind polynomials. We will only consider the R case. Define for $n \geq 0$,

$$q_n^{\text{R}}(x) = \int d\mu(y) \left[\frac{p_n^{\text{R}}(x) - p_n^{\text{R}}(y)}{x - y} \right] \tag{4.3.5}$$

and

$$q_{-1}^{\text{R}}(x) = -\mathbf{1} \tag{4.3.6}$$

Set $A_0 = \mathbf{1}$ and consider solutions of (u_n defined for $n = 0, 1, 2 \dots$)

$$z u_n = u_{n+1} A_n^\dagger + u_n B_n + u_{n-1} A_{n-1} \tag{4.3.7}$$

for $n = 1, 2, 3, \dots$. We will also need the Weyl solutions, for $z \in \mathbb{C}_+$,

$$\psi_n^{\text{R}}(z) = q_n^{\text{R}}(z) + m(z) p_n^{\text{R}}(z) \tag{4.3.8}$$

Theorem 4.3.1. *$p_{\cdot-1}^{\text{R}}(z)$ and $q_{\cdot-1}^{\text{R}}(z)$ both solve* (4.3.7) *and, therefore, so does $\psi_{\cdot-1}^{\text{R}}(z)$. Moreover, for fixed $z \in \mathbb{C}_+$, we have*

$$\sum_{n=0}^{\infty} \text{Tr}(\psi_n^{\text{R}}(z) \psi_n^{\text{R}}(z)^\dagger) < \infty \tag{4.3.9}$$

Any solution of (4.3.7) *obeying*

$$\sum_{n=0}^{\infty} \text{Tr}(u_n u_n^\dagger) < \infty \tag{4.3.10}$$

has the form

$$u_n = c\psi_{n-1}^{\text{R}}(z) \tag{4.3.11}$$

for some $\ell \times \ell$ matrix c.

Proof. That $p^{\mathrm{R}}_{\cdot-1}$ solves (4.3.7) is (4.2.34). From this and the argument that led to (3.2.17), we get that $q^{\mathrm{R}}_{\cdot-1}$ solves (4.3.7). For this, it is important that in (4.3.5), $d\mu$ multiplies on the left, while in (4.3.7), $A_n^\dagger$, B_n, A_{n-1} multiply on the right.

From (4.3.5), we get

$$\psi_n^{\mathrm{R}}(z) = \int d\mu(y)(y-z)^{-1} p_n^{\mathrm{R}}(y) \tag{4.3.12}$$

Since $(\cdot - z)^{-1} \in L^2(d\mu)$, we get (4.3.9) from completeness of the polynomials and (4.2.32).

We next claim that any $\ell \times \ell$ matrix solution, u_n, of (4.3.7) has the form

$$u_n = a p_{n-1}^{\mathrm{R}}(z) + b q_{n-1}^{\mathrm{R}}(z) \tag{4.3.13}$$

for some $\ell \times \ell$ matrices a and b. For let $\tilde{u}_n$ be given by (4.3.13) with

$$a = u_1 \qquad b = -u_0 \tag{4.3.14}$$

Thus, $\tilde{u}_n$ solves (4.3.7), and therefore, so does $\tilde{u}_n - u_n = d_n$. But, by (4.3.14) and the initial conditions,

$$p_{-1}^{\mathrm{R}}(z) = q_0^{\mathrm{R}}(z) = \mathbf{0} \qquad p_0^{\mathrm{R}}(z) = -q_{-1}^{\mathrm{R}}(z) = \mathbf{1} \tag{4.3.15}$$

$d_0 = d_1 = 0$. Since each A_n is invertible, the difference equation then implies $d_n \equiv 0$. This proves (4.3.13).

We next claim that if u_0 solves (4.3.7) and obeys (4.3.10), then

$$\operatorname{Im} z \sum_{n=1}^{\infty} \operatorname{Tr}(u_n u_n^\dagger) = -\operatorname{Im}(u_1 u_0^\dagger) \tag{4.3.16}$$

For define

$$s_n = \operatorname{Tr}(u_n A_{n-1}^\dagger u_{n-1}^\dagger) \tag{4.3.17}$$

We multiply (4.3.7) by $u_n^\dagger$ on the right and sum from $n = 1$ to N. We get

$$z \sum_{n=1}^{N} \operatorname{Tr}(u_n u_n^\dagger) = \sum_{n=1}^{N} s_{n+1} + \sum_{n=1}^{N} \operatorname{Tr}(u_n B_n u_n^\dagger) + \sum_{n=1}^{N} \bar{s}_n \tag{4.3.18}$$

since

$$\bar{s}_n = \operatorname{Tr}((u_n A_{n-1}^\dagger u_{n-1}^\dagger)) = \operatorname{Tr}(u_{n-1} A_{n-1} u_n^\dagger) \tag{4.3.19}$$

Now take imaginary parts of (4.3.18) using that

$$\operatorname{Im}(\operatorname{Tr}(u_n B_n u_n^\dagger)) = 0 = \operatorname{Im}(s_n + \bar{s}_n) \tag{4.3.20}$$

We get

$$\operatorname{Im} \sum_{n=1}^{N} \operatorname{Tr}(u_n u_n^\dagger) = -\operatorname{Im} s_1 + \operatorname{Im} s_{N+1} \tag{4.3.21}$$

Since (4.3.10) holds and A_n is bounded, $s_{N+1} \to 0$ and (4.3.16) holds by taking $N \to \infty$.

One consequence of (4.3.16) is that if $ap_{n-1}^{\mathrm{R}}(z)$ obeys (4.3.10), then $a = \mathbf{0}$. For $p_{-1}^{\mathrm{R}}(z) = \mathbf{0}$ means $u_0 = \mathbf{0}$, so by (4.3.16), $p_0^{\mathrm{R}} = \mathbf{1}$ and $\operatorname{Im} z > 0$, $\operatorname{Tr}(u_1 u_1^\dagger) = \operatorname{Tr}(aa^\dagger) = 0$, which implies $a = \mathbf{0}$.

Now let u_n obey (4.3.10). By (4.3.15),

$$u_n = ap_{n-1}^{\mathrm{R}}(z) + bq_{n-1}^{\mathrm{R}}(z) = \tilde{a}\, p_{n-1}^{\mathrm{R}}(z) + b\psi_n^{\mathrm{R}}(z) \tag{4.3.22}$$

with $\tilde{a} = a - bm(z)$. Since ψ_n^{R} obeys (4.3.9) and

$$\operatorname{Tr}(b\psi_n^{\mathrm{R}}(\psi_n^{\mathrm{R}})^\dagger b^\dagger) \le \|b\|^2 \operatorname{Tr}(\psi_n^{\mathrm{R}}(\psi_n^{\mathrm{R}})^\dagger) \tag{4.3.23}$$

we conclude that $\tilde{a}\, p_{n-1}^{\mathrm{R}}$ obeys (4.3.10), so $\tilde{a} = \mathbf{0}$ and u_n has the claimed form. □

Corollary 4.3.2. *If u_n solves* (4.3.7) *and obeys* (4.3.10) *and u_0 is invertible, then*

$$m(z) = -u_0^{-1} u_1 \tag{4.3.24}$$

Proof. By the theorem, u_n has the form (4.3.11), so

$$u_0 = -c \qquad u_1 = cm(z) \tag{4.3.25}$$

from which (4.3.24) is immediate. □

Theorem 4.3.3 (Aptekarev–Nikishin [23])**.** *Let J be a bounded $\ell \times \ell$ block Jacobi matrix with Jacobi parameters $\{A_n, B_n\}_{n=1}^\infty$ and let J_1 be the block matrix with parameters $\{A_{n+1}, B_{n+1}\}_{n=1}^\infty$. Let $m(z)$ and $m_1(z)$ be the m-functions for J and J_1. Then*

$$m(z) = (-z + B_1 - A_1 m_1(z) A_1^\dagger)^{-1} \tag{4.3.26}$$

for all z with $\operatorname{Im} z > 0$.

Remarks. 1. Since $\operatorname{Im}(m_1(z)) \ge 0$, $\operatorname{Im}(z - B_1 + A_1 m_1(z) A_1^\dagger) \ge \operatorname{Im}(z)\mathbf{1}$, so the object in $(\ )^{-1}$ in (4.3.26) is invertible.

2. See the Notes for other proofs.

Proof. Let (note n, not $n-1$, in ψ_n)

$$u_n = \begin{cases} \psi_n(z) & n \ge 1 \\ A_1 & n = 0 \end{cases} \tag{4.3.27}$$

for $n = 0, 1, \dots$. Then u_n solves (4.3.7) for the parameters of J_1 (with the convention that the $A_0^{(1)}$ for J_1 is $\mathbf{1}$ not A_1!). Moreover, $u_0 = m(z)$ is invertible and u_n obeys (4.3.10). By the above corollary,

$$m_1(z) = -A_1^{-1} m(z)^{-1} \psi_1(z) \tag{4.3.28}$$

By the equation (4.3.7) and initial conditions (4.3.15), we have

$$p_1^{\mathrm{R}}(z) = (z - B_1)(A_1^\dagger)^{-1} \qquad q_1^{\mathrm{R}}(z) = (A_1^\dagger)^{-1} \tag{4.3.29}$$

so (4.3.28) becomes

$$\begin{aligned} A_1 m_1(z) &= -m(z)^{-1}(q_1^{\mathrm{R}}(z) + m(z) p_1^{\mathrm{R}}(z)) \\ &= -(m(z)^{-1} + (z - B_1))(A_1^\dagger)^{-1} \end{aligned} \tag{4.3.30}$$

or

$$m(z)^{-1} = -z + B_1 - A_1 m_1(z) A_1^\dagger$$

which is (4.3.26). □

As a final topic, we want to discuss poles and zeros of $m(z)$. Fix $x_0 \in \mathbb{R} \setminus \sigma_{\text{ess}}(J)$. Let

$$\ell_0 = \dim(\ker(J - x_0)) \qquad \ell_1 = \dim(\ker(J_1 - x_0)) \tag{4.3.31}$$

We will prove

Theorem 4.3.4. (i) *$m(z)$ has a simple pole at $z = x_0$ and its residue is an $\ell \times \ell$ matrix of rank exactly ℓ_0.*

(ii)

$$\ell_0 + \ell_1 \leq \ell \tag{4.3.32}$$

(iii) $\det(m(z))$ *has a pole at x_0 of order $\ell_0 - \ell_1$.*

Remark. If $\ell_0 - \ell_1 < 0$, we mean $\det(m(z))$ has a zero of order $\ell_1 - \ell_0$. If $\ell_0 = 0$, (i) means that m is regular at x_0.

We need two preliminaries:

Lemma 4.3.5. *If $(J - x_0)u = 0$ and $u \not\equiv 0$, then $u_1 \neq 0$.*

Proof. If $u_1 = 0$, the recursion relation and A_1 invertible implies $u_2 = 0$. By induction and the second recursion relation with A_2 invertible implies $u \equiv 0$. □

Lemma 4.3.6. *Let Q and P be finite-dimensional projections and suppose*

$$\varphi \in \text{Ran}(Q) \Rightarrow P\varphi \neq 0 \tag{4.3.33}$$

Then

$$\text{rank}(PQP) = \text{rank}(Q) \tag{4.3.34}$$

Proof. First, by (4.3.33), P maps $\text{Ran}(Q)$ into a space of the same dimension, so

$$\text{rank}(PQ) = \text{rank}(Q) \tag{4.3.35}$$

Second, if A is any operator, $A^*A\varphi = 0 \Leftrightarrow \|A\varphi\| = 0 \Leftrightarrow A\varphi = 0$, so

$$\text{rank}(A^*A) = \text{rank}(A) = \text{rank}(A^*) \tag{4.3.36}$$

If $A = QP$, we conclude

$$\text{rank}(PQP) = \text{rank}(PQ) \tag{4.3.37}$$

(4.3.35) and (4.3.37) imply (4.3.33). □

Proof of Theorem 4.3.4. (i) By the spectral theorem, if Q is the projection onto the eigenspace for J with eigenvalue x_0, then

$$\frac{1}{J - z} = \frac{Q}{x_0 - z} + \text{analytic at } x_0$$

Thus, by (4.3.3), $m(z)$ has a simple pole at x_0 with residue $-P_1 Q P_1^*$. By the lemmas, $\operatorname{rank}(P_1 Q P_1^*) = \operatorname{rank}(Q)$ if we think of P_1 as UP, where P is the projection onto $\{u \mid u_n = 0 \text{ for } n \geq 1\}$ and U is a unitary map to $\mathbb{C}^\ell$.

(ii), (iii) Define the $\ell \times \ell$ matrix-valued function

$$G(z) = (z - x_0)m(z)$$

for z near x_0. This is analytic near x_0 and selfadjoint for z real. It follows by eigenvalue perturbation theory (see the Notes) that the eigenvalues of $G(z)$, call them $\tilde{\lambda}_1(z), \dots, \tilde{\lambda}_\ell(z)$, are analytic near $z = x_0$.

By (i), $G(x_0)$ is a rank ℓ_0 operator, so exactly ℓ_0 of $\tilde{\lambda}_j(z)$ are nonvanishing. The others vanish at least linearly in $(z - x_0)$ by analyticity. Thus, $m(z)$ has eigenvalues $\lambda_1(z), \dots, \lambda_\ell(z)$ near x_0 and exactly ℓ_0 have first-order poles at x_0 and the others are analytic there.

By (4.3.26), part (i), and A_1 invertible, $m(z)^{-1}$ has a pole of order one at x_0 with residue of rank exactly ℓ_1. So, as above, exactly ℓ_1 of the $\lambda_j(z)^{-1}$ have poles, all first-order, at x_0. It follows that exactly ℓ_1 of the $\lambda_j(z)$ have zeros at x_0 and they are order one.

Clearly, a fixed λ_j can have either a pole or a zero but not both, so (4.3.31) holds. Moreover,

$$\det(m(z)) = \prod_{j=1}^{\ell} \lambda_j(z) \tag{4.3.38}$$

has a product of ℓ_0 simple poles, ℓ_1 simple zeros, and $\ell - \ell_0 - \ell_1$ nonzero regular functions, and thus, has a pole of order $\ell_0 - \ell_1$. $\square$

Remarks and Historical Notes. Theorem 4.3.3 is from Aptekarev–Nikishin [23]. [98] has a proof relying on the method of "Schur complements" (due to Schur [380]) that for block operators (even of different square size)

$$\begin{pmatrix} A & B \\ C & D \end{pmatrix}^{-1} = \begin{pmatrix} (A - BD^{-1}C)^{-1} & -A^{-1}B(D - CA^{-1}B)^{-1} \\ -D^{-1}C(A - BD^{-1}C)^{-1} & (D - CA^{-1}B)^{-1} \end{pmatrix}$$

as can be checked by multiplication. One takes $A = B_1 - z$, $B = A_1$, $C = A_1^\dagger$, $D = J^{(1)} - z$. One can also use the proof in Section 10.3.

Theorem 4.3.4 is from [98] whose proof we follow. It is related to the fact, due to Durán–López-Rodríguez [117] and Sinap [412] (discussed also in [98]) that $\det(P_n^{\mathrm{R}}(z))$ has all real zeros precisely at the eigenvalues of the $n \times n$ block truncated matrix $J_{n;F}$ with multiplicities of the zeros equal to the multiplicities of the eigenvalues.

Matrix-valued Herglotz functions are discussed in [162, 167] and references therein. Eigenvalue perturbation for finite matrices is discussed in [215, 364].

4.4 STEP-BY-STEP SUM RULES OF MOPRL

In this section, our goal is to prove nonlocal step-by-step sum rules for $\det(m(z))$ where $m(z)$ is the m-function of some $\ell \times \ell$ nontrivial matrix-valued measure on

$\mathbb{R}$ with $\sigma_{\mathrm{ess}}(\mu) = [-2, 2]$. The zeros and poles may not interlace, so in using Theorem 3.3.2, we will not always have that $N_\infty = 1$ as it is for the scalar case. The key will be to prove that $N_\infty \leq \ell$.

Lemma 4.4.1. *Let J be any $\ell \times \ell$ block Jacobi matrix of the form* (4.2.47) *where*

$$A_n \to \mathbf{1} \qquad B_n \to \mathbf{0} \tag{4.4.1}$$

(called, not surprisingly, the matrix Nevai class). *Let J_1 be the once-stripped matrix. Let $\{E_j^\pm(J)\}_{j=1}^{N_\pm(J)}$ be the eigenvalues of J in* $\pm(2, \infty)$ *ordered by*

$$E_1^- \leq E_2^- \leq \cdots < -2 < 2 < \cdots \leq E_2^+ \leq E_1^+ \tag{4.4.2}$$

(counted up to multiplicity), and similarly for $E_j^\pm(J_1)$. Then

$$\text{(i)} \qquad N_\pm(J_1) \leq N_\pm(J) \tag{4.4.3}$$

$$\text{(ii)} \qquad E_j^-(J) \leq E_j^-(J_1) \leq E_{j+\ell}^-(J) \tag{4.4.4}$$

$$\text{(iii)} \qquad E_j^+(J) \geq E_j^+(J_1) \geq E_{j+\ell}^-(J) \tag{4.4.5}$$

Remark. One can prove that the inequalities in (4.4.4)/(4.4.5) are strict.

Proof. If one defines $E_j^\pm$ for all j by setting $E_j^\pm(J) = \pm 2$ if $j > N^\pm(J)$, then one has the min-max principle (see the Notes)

$$E_j^-(J) = \max_{V:\ \dim(V) \leq j-1} \left[\min_{\substack{\varphi \perp V \\ \|\varphi\|=1}} \langle \varphi, J\varphi \rangle \right] \tag{4.4.6}$$

Let $\mathbb{C}^\ell$ denote the ℓ-dimensional space spanned by $\{\delta_1^{(p)}\}_{p=1}^{\ell}$. Then for any V,

$$\min_{\substack{\varphi \perp V \\ \|\varphi\|=1}} \langle \varphi, J_1\varphi \rangle = \min_{\substack{\varphi \perp V \oplus \mathbb{C}^\ell \\ \|\varphi\|=1}} \langle \varphi, J\varphi \rangle \tag{4.4.7}$$

which immediately implies

$$E_j^-(J_1) \leq E_{j+\ell}^-(J) \tag{4.4.8}$$

since $\dim(V \oplus \mathbb{C}^\ell) \leq \ell + \dim(V)$, so we are taking a max over a restricted set of W's with $\dim(W) \leq j + \ell - 1$.

On the other hand, for any V with $\dim(V) \leq j - 1$, if π is the projection onto $\mathbb{C}^\ell$, then $\dim((1-\pi)(V)) \leq j - 1$ and

$$\min_{\substack{\varphi \perp (1-\pi)V \\ \|\varphi\|=1 \\ \pi\varphi=0}} \langle \varphi, J_1\varphi \rangle = \min_{\substack{\varphi \perp V \\ \|\varphi\|=1 \\ \pi\varphi=0}} \langle \varphi, J\varphi \rangle \geq \min_{\substack{\varphi \perp V \\ \|\varphi\|=1}} \langle \varphi, J\varphi \rangle \tag{4.4.9}$$

leading to

$$E_j^-(J_1) \geq E_j^-(J) \tag{4.4.10}$$

The proof for E^+ is similar, using max-min rather than min-max. $\square$

Define

$$M(z) = -m(z + z^{-1}) \tag{4.4.11}$$

Proposition 4.4.2. *Let $z_1^\pm, z_2^\pm, \dots$ and $p_1^\pm, p_2^\pm, \dots$ be the zeros and poles of* $\det(M(z))$ *where $m(\cdot)$ is the m-function of an $\ell \times \ell$ block Jacobi matrix in Nevai class and M is given by* (4.4.11). *Do not include the zero at $z = 0$, which is ℓ-fold, and label so* (*counting multiplicity*)

$$\dots \le z_2^- \le z_1^- < 0 < z_1^+ \le z_2^+ \le \dots \tag{4.4.12}$$

and similarly for $p_j^\pm$. Then

(i)

$$|z_j^\pm| \to 1 \qquad |p_j^\pm| \to 1 \qquad \text{as } j \to \infty \tag{4.4.13}$$

(ii)

$$z_j^- < p_j^- < 0 < p_j^+ < z_j^+ \tag{4.4.14}$$

(iii) *Let*

$$I_j^+ = (p_j^+, z_j^+) \qquad I_j^- = (z_j^-, p_j^-) \tag{4.4.15}$$

$\sigma_j^+ = -1$, $\sigma_j^- = 1$ *in the language of Theorem 3.3.2. Then for* $0 < \pm x < 1$,

$$0 \le \mp N(x) \le \ell \tag{4.4.16}$$

where N is given by (3.3.14), *so*

$$N_\infty \le \ell \tag{4.4.17}$$

In particular, on compact subsets of $\mathbb{C} \setminus S$ *where*

$$S = \{p_j^\pm\}_{j=1}^\infty \cup \{(z_j^\pm)^{-1}\}_{j=1}^\infty \cup \{\pm 1\} \tag{4.4.18}$$

we have uniformly

$$\prod_{j=1}^N \left(\frac{b_{z_j^+}(z)}{b_{p_j}^+(z)}\right)\left(\frac{b_{z_j^-}(z)}{b_{p_j}^-(z)}\right) \to B_\infty(z) \tag{4.4.19}$$

for a function analytic on $\mathbb{C} \setminus S$ and meromorphic on $\mathbb{C} \setminus \{\pm 1\}$ with poles at $S \setminus \{\pm 1\}$ and zeros at $S^{-1} \setminus \{\pm 1\}$. Moreover, on $\partial\mathbb{D} \setminus \{\pm 1\} \subset \mathbb{C} \setminus S$, we have

$$|B_\infty(z)| = 1 \tag{4.4.20}$$

and in $\mathbb{C}_+ \cap \mathbb{D}$,

$$\arg B_\infty(z) < 2\pi\ell \tag{4.4.21}$$

Remarks. 1. While we prove this for the m-functions that arise in our applications, a similar result holds for any function, which is the determinant of $M(z)$, which is an $\ell \times \ell$ matrix-valued meromorphic function of $\mathbb{D}$ with $\operatorname{Im} M(z) \equiv (M(z) - M(z)^\dagger)/2i > 0$ on $\mathbb{D} \cap \mathbb{C}_+$. In particular, a suitable labeling of the type (4.4.15) holds (although $\mp N(x) \ge 0$ for $\pm x > 0$ may not).

2. $\arg B_\infty(z)$ is defined with the branch defined near $z = 0$, with $\arg B_\infty(0) = 0$, for $B_\infty(0) > 0$.

3. We suppose $N^\pm(J) = \infty$ with simple modification of notation if one (or both) is finite.

Proof. Let $\tilde{p}_j^{\pm}$ and $\tilde{z}_j^{\pm}$ be the points in $(-1, 1)$ with

$$\tilde{p}_j^{\pm} + (\tilde{p}_j^{\pm})^{-1} = E_j^{\pm}(J) \qquad \tilde{z}_j^{\pm} + (\tilde{z}_j^{\pm})^{-1} = E_j^{\pm}(J_1) \tag{4.4.22}$$

By Lemma 4.4.1, (4.4.12) and (4.4.13) hold for the $\tilde{z}_j^{\pm}$ and $\tilde{p}_j^{\pm}$, and (4.4.14) is only modified by the inequalities having $\leq$, not $<$. Moreover, by (4.4.4)/(4.4.5),

$$\pm\tilde{z}_j^{\pm} \leq \pm\tilde{p}_{j+\ell}^{\pm} \tag{4.4.23}$$

By Theorem 4.3.4, the zeros and poles of $\det(M(z))$ are obtained from the $\tilde{z}_j^{+}$, $\tilde{p}_j^{+}$ by dropping those $\tilde{z}$, $\tilde{p}$, which happen to be equal in equal numbers. That leads to $\{z_j^{\pm}\}_{j=1}^{\infty}$, $\{p_j^{\pm}\}_{j=1}^{\infty}$ with strict inequality in (4.4.14). Moreover, we still have

$$\pm z_j^{\pm} \leq \pm p_{j+\ell}^{\pm} \tag{4.4.24}$$

which implies (4.4.16).

We have

$$\sum_{j=1}^{\infty} |z_j^{+} - p_j^{+}| = \bigcup_{j=1}^{\infty} |I_j^{+}| \tag{4.4.25}$$

$$= \int_0^1 N(x)\, dx \tag{4.4.26}$$

$$\leq \ell \tag{4.4.27}$$

and similarly,

$$\sum_{j=1}^{\infty} |p_j^{-} - z_j^{-}| \leq \ell \tag{4.4.28}$$

The rest of this theorem now follows from Theorem 3.2.2. We get 2ℓ in (4.4.21) because we separately consider the $\{z_j^{+}, p_j^{+}\}_{j=1}^{\infty}$ and the $\{z_j^{-}, p_j^{-}\}_{j=1}^{\infty}$ products although, by adding z^{ℓ}, one can get a global $\pi\ell$. □

Proposition 4.4.3. *Let* $M(z)$ *be defined as in* (4.4.11). *Then for* $x \in (0, \varepsilon)$ *with* ε *small,* $\det(M(x)) > 0$, *and if* $\arg(\det(M(z)))$ *is defined there to be* 0 *and analytically continued to* $\mathbb{C}_+ \cap \mathbb{D}$, *then*

$$0 < \arg(\det(M(z))) < \pi\ell \tag{4.4.29}$$

there.

Proof. Since

$$(J - z)^{-1} = -z^{-1} + O(z^{-2}) \tag{4.4.30}$$

and

$$M(z) = -(J - (z + z^{-1}))^{-1} \tag{4.4.31}$$

we see, at $z = 0$,

$$M(z) = z\mathbf{1} + O(z^2) \tag{4.4.32}$$

so

$$\det(M(z)) = z^\ell + O(z^{\ell+1}) \tag{4.4.33}$$

proving $\det(M(x)) > 0$ for x in $(0, \varepsilon)$ as claimed.

Eigenvalue perturbation theory (see [215, 364]) implies there is a discrete set $D \subset \mathbb{D} \setminus \mathbb{R}$ (i.e., the only limit points of D lie in $\partial\mathbb{D}$) so, for $z_0 \notin \mathbb{D} \setminus D$, there are ℓ eigenvalues $\lambda_1(z), \dots, \lambda_\ell(z)$ analytic near z_0, which are all the eigenvalues of $M(z)$ counting multiplicity. Continuing around at point $z_1 \in \mathbb{D}$ can permute the λ_j's. If $z \in (\mathbb{C}_+ \cap \mathbb{D}) \setminus D$ and

$$M(z)\psi_j = \lambda_j(z)\psi_j \tag{4.4.34}$$

for $\|\psi_j\| = 1$, then

$$\operatorname{Im} \lambda_j(z) = \operatorname{Im}\langle \psi_j, M(z)\psi_j\rangle > 0 \tag{4.4.35}$$

by (4.4.25).

Let $z \in (\mathbb{C}_+ \cap \mathbb{D}) \setminus D$ and let $\gamma(z)$, $0 \le z \le 1$, be a simple curve with $\gamma(0) = \varepsilon/2$, $\gamma(1) = z$, and $\gamma(t) \in (\mathbb{C}_+ \cap \mathbb{D}) \setminus D$ for $t \in (0, 1]$. By local analyticity, we can define $\lambda_j(z)$ for z in a neighborhood of $\{\gamma(t) \mid 0 \le t \le 1\}$, and by (4.4.35), $\arg \lambda_j(z) \in (0, \pi)$.

Thus,

$$\arg(\det(M(z))) = \sum_{j=1}^{\ell} \arg \lambda_j(z) \in (0, \ell\pi) \tag{4.4.36}$$

Since D is removable singularities of $\det(M(z))$, we have (4.4.29). □

With these preliminaries, following the proof of Theorem 3.3.6 leads directly to

Theorem 4.4.4. *Let $M(z)$ be defined in* (4.4.11). *Let B_∞ have the form* (4.4.19) *(and so obey* (4.4.21) *and the analyticity properties listed in Proposition 4.4.2). Then*

(i) *For a.e. θ, $\lim_{r\uparrow 1} \det(M(re^{i\theta}))$ exists and is nonzero with*

$$\int |\log|\det(M(e^{i\theta}))||^p \, \frac{d\theta}{2\pi} < \infty \tag{4.4.37}$$

for all $p \in [1, \infty)$.

(ii) *We have for all $z \in \mathbb{D}$,*

$$\det(M(z)) = z^\ell B_\infty(z) \exp\biggl(\int_0^{2\pi} \frac{e^{i\theta} + z}{e^{i\theta} - z} \log(|\det(M(e^{i\theta}))|) \, \frac{d\theta}{2\pi}\biggr) \tag{4.4.38}$$

By (4.3.26), for $z \in \mathbb{C} \setminus [\sigma(J) \cup \sigma(J_1)]$,

$$\det(m(z))^{-1} \det(m(z)^\dagger)^{-1} \det(\operatorname{Im} m(z)) = \det(\operatorname{Im} z + A_1 \operatorname{Im} m_1(z) A_1^\dagger) \tag{4.4.39}$$

By following the proof of Theorem 3.4.1, we obtain

Theorem 4.4.5. *Under the hypotheses of Theorem 4.4.4, up to sets of measure* 0,

$$\{\theta \mid \det(\operatorname{Im} M(e^{i\theta})) \neq 0\} = \{\theta \mid \det(\operatorname{Im} M_1(e^{i\theta})) \neq 0\} \tag{4.4.40}$$

and

$$\log\left(\frac{\det(\operatorname{Im} M(e^{i\theta}))}{\det(\operatorname{Im} M_1(e^{i\theta}))}\right) \in \bigcap_{p<\infty} L^p\left(\partial\mathbb{D}, \frac{d\theta}{2\pi}\right) \tag{4.4.41}$$

and

$$\det(|A_1|M(z)) = B_\infty(z)\exp\left(\frac{1}{4\pi}\int_0^{2\pi} \frac{e^{i\theta}+z}{e^{i\theta}-z}\log\left(\frac{\det(\operatorname{Im} M(e^{i\theta}))}{\det(\operatorname{Im} M_1(e^{i\theta}))}\right)d\theta\right) \tag{4.4.42}$$

Remark. As usual, the function log(...) in (4.4.41) and (4.4.42) is only as written on the set in (4.4.40), which is often a.e. θ. If it is not, the function is defined as a suitable boundary value.

Remarks and Historical Notes. The results of this section are due to Damanik–Killip–Simon [97].

Our proof follows theirs except, as in Section 3.3, our handling of alternating Blaschke products is new and somewhat simpler than theirs.

The min-max principle is discussed in Reed–Simon [364, Section XIII.1].

4.5 A SHOHAT–NEVAI THEOREM FOR MOPRL

Our goal in this section is to prove a matrix analog of Theorem 3.6.1, specifically

Theorem 4.5.1 ([97]). *Let J be an $\ell \times \ell$ block Jacobi matrix of the form* (4.2.47). *Suppose the associated measure has the form*

$$d\mu(x) = f(x)\,dx + d\mu_s(x) \tag{4.5.1}$$

with f $\ell \times \ell$ matrix-valued (and nonnegative) and $d\mu_s$ singular with respect to dx. Suppose

$$\sigma_{\text{ess}}(J) = [-2, 2] \tag{4.5.2}$$

Let $\{E_j^\pm\}_{j=1}^{N_\pm}$ be the eigenvalues of J outside $[-2, 2]$, *counting multiplicity. Suppose that*

$$\sum_{j,\pm} (|E_j^\pm| - 2)^{1/2} < \infty \tag{4.5.3}$$

Then

$$\int_{-2}^{2} (4-x^2)^{-1/2} \log(\det(f(x)) > -\infty \tag{4.5.4}$$

if and only if

$$\limsup \det(|A_1|\dots|A_n|) > 0 \tag{4.5.5}$$

and if either (and so both) happens, then the limit in (4.5.5) *exists and also*

$$\lim_{n\to\infty} \sum_{n=1}^{\infty} [\||A_n| - \mathbf{1}\|^2 + \|B_n\|^2] < \infty \tag{4.5.6}$$

Remarks. 1. If (4.5.3) and (4.5.4) hold, then the hypotheses of Theorem 4.6.1 hold, so (4.5.6) follows. Thus, in this section, we will focus entirely on (4.5.4) $\Leftrightarrow$ (4.5.5).

2. $\det(f(x)) \leq \|f(x)\|^\ell$, so $\mathrm{Tr}(f) \in L^1$ implies $\log_+(\det(f(x))) \in L^1$. Thus, (4.5.4) can only diverge to $-\infty$.

3. (4.5.4) implies $f(x)$ is invertible for a.e. $x \in [-2, 2]$, which implies that J has $\Sigma_{\mathrm{ac}} = [-2, 2]$ with uniform multiplicity ℓ.

4. If $\{\tilde{A}_j, \tilde{B}_j\}_{j=1}^\infty$ is equivalent to $\{A_j, B_j\}_{j=1}^\infty$, then for suitable unitary $\tilde{A}_j = u_j^{-1} A_j u_{j+1}$, so

$$|\tilde{A}_j| = u_{j+1}^{-1} |A_j| u_{j+1} \tag{4.5.7}$$

and

$$\det(|\tilde{A}_j|) = \det(|A_j|) \tag{4.5.8}$$

so (4.5.5) is equivalence class independent.

This will depend first on a step-by-step C_0 sum rule:

Theorem 4.5.2. *If M, J, J_1 are as in Theorem 4.4.5 hold, let $p_j^\pm(J)$ be the poles of $M(z; J)$. Then*

$$\begin{aligned} -\log(\det(|A_1|)) &= \frac{1}{4\pi} \int_0^{2\pi} \log\left(\frac{\det(\mathrm{Im}\, M^{(1)}(e^{i\theta}))}{\det(\mathrm{Im}\, M(e^{i\theta}))} \right) d\theta \\ &\quad - \sum_{j,\pm} [\log(|p_j^\pm(J)|) - \log(|p_j^\pm(J_1)|)] \end{aligned} \tag{4.5.9}$$

Remark. The sum in (4.5.9) has at most ℓ signs unbalanced with

$$\lim_{j\to\infty} \log(|p_j^\pm(J)|) = 0 \tag{4.5.10}$$

so the sum is at least conditionally convergent.

Proof. Take $z \to 0$ in (4.4.42) and take $-\log(\dots)$ of both sides. $\det(M(z))/z^\ell \to 1$, so the left-hand side is $-\log(\det(|A_1|))$. Since $-\log(b(z, \omega))|_{z=0} = -\log(\omega)$, the $\log B_\infty(z)|_{z=0}$ is the sum in (4.5.9). □

Theorem 4.5.3 (C_0 Sum Rule). *Suppose* (4.5.3) *holds. Define*

$$Z(\mu) = \frac{1}{4\pi} \int_0^{2\pi} \log\left(\frac{\sin^\ell(\theta)}{\det(\mathrm{Im}\, M(e^{i\theta}))} \right) d\theta \tag{4.5.11}$$

and

$$\mathcal{E}_0(J) = \sum_{j,\pm} \log(|z_j^\pm|) \tag{4.5.12}$$

Then, if $|A_n| \to 1$ and $|B_n| \to 0$,

$$Z(\mu) \leq \liminf_{n\to\infty} \left(-\sum_{j=1}^n \log(\det(|A_j|)) \right) + \mathcal{E}_0(J) \tag{4.5.13}$$

$$\limsup_{n\to\infty}\left(-\sum_{j=1}^{n}\log(\det(|A_j|))\right)\leq Z(\mu)-\mathcal{E}_0(J) \tag{4.5.14}$$

Remark. As with (3.6.8), we have for $\tilde{\mu}$ a pullback of μ to $\partial\mathbb{D}$,

$$Z(\mu)=-\frac{1}{2}\,S\biggl(\frac{d\theta}{2\pi}\,\bigg|\,d\tilde{\mu}\biggr)-\frac{\ell}{2}\log 2 \tag{4.5.15}$$

Proof. Given Theorem 4.5.2, we need only follow the proof in Section 3.6. □

Proof that when (4.5.3) *holds, then* (4.5.4) ⇔ (4.5.5). If (4.5.4) holds, then $Z(\mu)<\infty$ by (4.5.11) and (4.5.12). The limit exists and is finite. Conversely, (4.5.5) implies

$$\liminf_{n\to\infty}\left(-\sum_{j=1}^{n}\log(\det(|A_j|))\right)<\infty$$

so $Z(\mu)<\infty$ by (4.5.13), and the limit exists as above. □

Remarks and Historical Notes. These results are from Damanik–Killip–Simon [97].

4.6 A KILLIP–SIMON THEOREM FOR MOPRL

Our goal in this section is to prove a matrix analog of Theorem 3.1.1, specifically

Theorem 4.6.1 ([97]). *Let* $\{A_n, B_n\}_{n=1}^{\infty}$ *be the Jacobi parameters of an* $\ell\times\ell$ *block Jacobi matrix, J, whose matrix measure has the form* (4.5.1). *Then*

$$\sum_{n=1}^{\infty}\mathrm{Tr}((|A_n|-\mathbf{1})^2)+\mathrm{Tr}(B_n^2)<\infty \tag{4.6.1}$$

if and only if

(a)

$$\sigma_{\mathrm{ess}}(J)=[-2,2] \tag{4.6.2}$$

(b) *The eigenvalues* $\{E_n\}_{j=1}^{\infty}\notin\sigma_{\mathrm{ess}}(J)$ *obey*

$$\sum_{n=1}^{\infty}(|E_n|-2)^{3/2}<\infty \tag{4.6.3}$$

(c) *The* $\ell\times\ell$ *matrix function, f, of* (4.5.1) *obeys*

$$\int_{[-2,2]}(4-x^2)^{1/2}\log(\det(f(x))\,dx>-\infty \tag{4.6.4}$$

Remarks. 1. As with Theorem 4.5.1, the integral in (4.6.4) can only diverge to $-\infty$ and (4.6.4) implies $\Sigma_{\mathrm{ac}}=[-2,2]$ with uniform multiplicity ℓ.

2. Since the Hilbert–Schmidt norm on $\ell\times\ell$ matrices is equivalent to the operator norm, $\mathrm{Tr}(\cot^2)$ in (4.6.1) is equivalent to $\|\cdot\|^2$.

3. By (4.5.7), (4.6.1) is true for one element of the set of equivalent Jacobi parameters if and only if it is true for all.

The first step in the proof is, of course, a step-by-step sum rule:

Theorem 4.6.2 (Step-by-Step P_2 Sum Rule for MOPRL). *Define*

$$Q(J \mid J_1) = \frac{1}{4\pi} \int_0^{2\pi} \log\left(\frac{\det(\operatorname{Im} M_1(e^{i\theta}))}{\det(\operatorname{Im} M(e^{i\theta}))} \right) \sin^2\theta \, d\theta \tag{4.6.5}$$

Let F be given by (1.10.9) *and G by* (1.10.10). *Then*

$$\tfrac{1}{4}\operatorname{Tr}(B_1^2) + \tfrac{1}{2}\operatorname{Tr}(G(|A_1|)) = Q(J \mid J_1) + \sum_{j,\pm} [F(E_j^{\pm}(J)) - F(E_j^{\pm}(J_1))] \tag{4.6.6}$$

Proof. By (4.3.26) and $m_1(z) = -\frac{1}{z} + O(z^{-2})$, we have

$$\left(\frac{M(z)}{z} \right)^{-1} = \mathbf{1} - B_1 z - (A_1^* A_1 - \mathbf{1}) z^2 + O(z^3) \tag{4.6.7}$$

Since

$$\det(C) = \exp(\operatorname{Tr}(\log(C))) \tag{4.6.8}$$

if $\|C - \mathbf{1}\| < 1$, we have

$$\log\left(\det\left(\frac{M(z)}{z} \right) \right) = \operatorname{Tr}(B_1) z + \operatorname{Tr}\{[A_1^* A_1 - \mathbf{1}] + \tfrac{1}{2} B_1^2\} z + O(z^3) \tag{4.6.9}$$

Moreover,

$$A_1^* A_1 = |A_1|^2 \tag{4.6.10}$$

and

$$|\det(A_1)| = \det(|A_1|) \tag{4.6.11}$$

Given the analog of (3.4.26) and Theorem 4.4.5, we get (4.6.6) by following the proof of Theorem 3.4.6 and Corollary 3.4.7. □

We can now follow the argument in Section 3.5 to obtain the following, which immediately implies Theorem 4.6.1:

Theorem 4.6.3 (P_2 Sum Rule for MOPRL). *Let J be a block Jacobi matrix with $\sigma_{\text{ess}}(J) = [-2, 2]$. Let $d\mu$ be its spectral measure and*

$$Q(\mu) = \frac{1}{4\pi} \int_0^{2\pi} \log\left(\frac{\sin^{\ell}\theta}{\det(\operatorname{Im} M(e^{i\theta}))} \right) \sin^2\theta \, d\theta \tag{4.6.12}$$

Then

$$Q(\mu) + \sum_{E \notin \sigma_{\text{ess}}(J)} F(E) = \sum_{n=1}^{\infty} \left[\tfrac{1}{4}\operatorname{Tr}(B_n^2) + \tfrac{1}{2}\operatorname{Tr}(G(|A_n|)) \right] \tag{4.6.13}$$

As a final topic, we want to note that for the type 3 case (i.e., $A_n \in \mathcal{L}$, the lower triangular matrices), we can replace $\operatorname{Tr}((|A_n| - 1)^2)$ in (4.6.1) by $\operatorname{Tr}((A_n - 1)^2)$.

Lemma 4.6.4. *Let $C_n \in \mathcal{L}$ and suppose $|C_n| \to \mathbf{1}$. Then $C_n \to \mathbf{1}$.*

Proof. Since $|C_n| \to \mathbf{1}$, $C_n^* C_n \to \mathbf{1}$. Let $x_n^{(1)}, \dots, x_n^{(\ell)}$ be the rows of C_n. Then $C_n^* C_n \to 1$ implies

$$\langle x_n^{(j)}, x_n^{(k)} \rangle \to \delta_{jk} \tag{4.6.14}$$

Since $\mathcal{L}$ is lower triangular, $x_n^{(1)}$ has only its first component nonzero. Since this component is positive, (4.6.14) says $x_n^{(1)} \to \delta_1 = (1, 0, \dots, 0)$. Orthogonality then implies the first column of C_n goes to $(1, 0, \dots, 0)^t$. Thus, by (4.6.14) for $j = k_0 = 2$, $x_n^{(2)} \to \delta_2 = (0\,1\,0 \dots 0)$. Repeating this shows that $C_n \to \mathbf{1}$. $\square$

Lemma 4.6.5. *The map from $\mathcal{L}$ to strictly positive matrices given by $A \mapsto |A^\dagger|$ is a smooth diffeomorphism.*

Proof. $A \in \mathcal{L}$ means $\det(A) > 0$, so A is invertible. On strictly positive matrices, $\sqrt{\cdot}$ is smooth, so

$$A \to |A^\dagger| = \sqrt{AA^\dagger} \tag{4.6.15}$$

is smooth.

For the converse, given B strictly positive, the QR factorization of Section 6.3 implies that we can write

$$B = QR \tag{4.6.16}$$

with Q unitary and R upper triangular with $R_{jj} > 0$ and the map $B \to R$ is smooth from invertible B's by construction. Then $L = R^\dagger$ is lower triangular, and since B is Hermitian,

$$B = LQ^{-1} \tag{4.6.17}$$

so

$$B^2 = LQ^{-1}QL^\dagger = LL^\dagger \tag{4.6.18}$$

and

$$B = |L^\dagger|$$

so the smoothness of the QR algorithm shows the smoothness of the inverse map. $\square$

Lemma 4.6.6. *For any invertible A,*

$$\operatorname{Tr}((|A^\dagger| - \mathbf{1})^2) = \operatorname{Tr}((|A| - 1)^2) \tag{4.6.19}$$

Proof. There is a unitary with

$$|A^\dagger| = U|A|U^{-1} \tag{4.6.20}$$

so

$$|A^\dagger| - \mathbf{1} = U(|A^\dagger| - \mathbf{1})U^{-1} \tag{4.6.21}$$

from which (4.6.19) is immediate. $\square$

Theorem 4.6.7. *If* $\{A_n, B_n\}_{n=1}^{\infty}$ *are type 3 Jacobi parameters, then*

$$\sum_{n=1}^{\infty} \mathrm{Tr}[(|A_n| - \mathbf{1})^2 + |B_n|^2] < \infty \tag{4.6.22}$$

if and only if

$$\sum_{n=1}^{\infty} \mathrm{Tr}((A_n - \mathbf{1})^2 + B_n^2) < \infty \tag{4.6.23}$$

Proof. By Lemma 4.6.6, (4.6.22) is equivalent to

$$\sum_{n=1}^{\infty} \mathrm{Tr}[(|A_n^\dagger| - \mathbf{1})^2 + |B_n|^2] < \infty \tag{4.6.24}$$

If only one of the three conditions holds by Lemma 4.6.4, $A_n \to \mathbf{1}$, so by Lemma 4.6.5, for n large and some c_0, c_1 and all large n,

$$c_0 \| |A_n^*| - \mathbf{1}\| \le \|A_n - \mathbf{1}\| \le c_1 \| |A_n^*| - \mathbf{1}\|$$

which shows (4.6.24) is equivalent to (4.6.23). □

Remarks and Historical Notes. These results are from Damanik–Killip–Simon [97]. If one permutes the rows and columns of a matrix in $\mathcal{L}$ under $(1\,2\,3\ldots n) \to (n\,n-1\ldots 2\,1)$, one gets a matrix in $\mathcal{R}$ and vice versa. Thus, one can find an analog of the QR algorithm so $B = QL$ with L lower triangular and so show the map $A \mapsto |A|$ on $\mathcal{L}$ is a diffeomorphism. This allows a slightly more direct proof of Theorem 4.6.7, which is what [97] do. We use QR since we need it again in Chapter 6.

Chapter Five

Periodic OPRL

5.1 OVERVIEW

Thus far we have been looking at perturbations of OPUC and OPRL with constant Jacobi parameters; specifically, we looked at perturbations of

$$a_n^{(0)} = a \qquad b_n^{(0)} = b \tag{5.1.1}$$

where $b \in \mathbb{R}$, $a \in (0, \infty)$. By scaling and translation covariance, we focused on $a = 1, b = 0$. In this chapter, we will study the periodic case where

$$a_{n+p}^{(0)} = a_n^{(0)} \qquad b_{n+p}^{(0)} = b_n^{(0)} \tag{5.1.2}$$

for all n and some fixed p. (5.1.1) is, of course, $p = 1$. The perturbation theory will be the focus of Chapters 8 and 9; this chapter will study the surprisingly rich unperturbed case. We will drop $^{(0)}$ henceforth in this chapter since we are restricting to the periodic case.

For (5.1.1), the spectrum is $\mathfrak{e} = [b - 2a, b + 2a]$ and is purely a.c. In the period p case, generically, the essential spectrum of the Jacobi matrix associated to $\{a_n, b_n\}_{n=1}^\infty$ will be p closed intervals

$$\mathfrak{e} = [\alpha_1, \beta_1] \cup \cdots \cup [\alpha_p, \beta_p] \tag{5.1.3}$$

with

$$\alpha_1 < \beta_1 < \alpha_2 < \cdots < \beta_p \tag{5.1.4}$$

Naively, the parameter counting seems simple. For $p = 1$ (i.e., (5.1.1)), every interval $[\alpha_1, \beta_1]$ occurs (take $b = \frac{1}{2}(\alpha_1 + \beta_1)$, $a = \frac{1}{4}(\beta_1 - \alpha_1)$) and the map of $(0, \infty) \times \mathbb{R}$ to $\mathfrak{e}$'s is one-one and onto. For period p, there are $2p$ free Jacobi parameters since $\{a_n, b_n\}_{n=1}^p$ and periodicity determine all Jacobi parameters, and there are $2p$ free $\{\alpha_j, \beta_j\}_{j=1}^p$, so the simple expectation is that all $\mathfrak{e}$'s of the form (5.1.3)/(5.1.4) are allowed and the map is one-one or it might be finite-to-one.

In fact, this naive expectation is wrong! The set of $\mathfrak{e}$'s that occurs as essential spectrum of period p is not of dimension $2p$ but only a small subset of dimension $p + 1$. Not surprisingly, given this, the inverse image of a single $\mathfrak{e}$ is a manifold of dimension $p - 1$. The reason for this is a natural set of $p + 1$-dimensional objects lies between $\{a_n, b_n\}_{n=1}^p$ and $\mathfrak{e}$.

Let $T_n(\lambda)$ be the transfer matrix of (3.2.3). By periodicity, for any $k = 1, 2, \dots,$

$$T_{kp}(\lambda) = T_p(\lambda)^k \tag{5.1.5}$$

Since $\det(T_n) = 1$, all solutions will be bounded if and only if the eigenvalues of T_p have magnitude 1 and T_p is diagonalizable.

Since $T_p(\lambda)$ has determinant 1, its eigenvalues are completely determined by

$$\Delta(\lambda) = \mathrm{Tr}(T_p(\lambda)) \tag{5.1.6}$$

called the *discriminant*. Since each factor in T is linear in λ and T_p has p factors, $\Delta(\lambda)$ is a polynomial of degree p.

Since $\det(T_p) = 1$, its eigenvalues are distinct and of magnitude one if and only if they are $e^{\pm i\theta}$, $0 < \theta < \pi$, in which case $\Delta(\lambda) = 2\cos\theta$. It is thus not surprising that we will show

$$\mathfrak{e} = \Delta^{-1}([-2, 2]) \tag{5.1.7}$$

The parameter counting is now clearer. Δ as a real degree p polynomial has $p+1$ free parameters, so rather than think of

$$\{a_n, b_n\}_{n=1}^{p} \to \mathfrak{e} \tag{5.1.8}$$

we should think of

$$\{a_n, b_n\}_{n=1}^{p} \to \Delta \to \mathfrak{e} \tag{5.1.9}$$

Since Δ has only $p+1$ parameters, the set of $\{a_n, b_n\}_{n=1}^{p}$ with a given Δ should be a set of dimension $p-1$ $(= 2p - (p+1))$. And indeed, in the generic case, when $\mathfrak{e}$ has p pieces, the set will be a torus of dimension $p-1$.

For the other piece of the map, $\{\alpha_j, \beta_j\}_{j=1}^{p}$ will be the points where $\Delta(\lambda) = \pm 2$. Indeed, $\Delta(\lambda) = 2$ at $\beta_p, \alpha_{p-1}, \beta_{p-2}, \alpha_{p-3}, \dots$, and -2 at $\alpha_p, \beta_{p-1}, \alpha_{p-2}, \beta_{p-3}, \dots$. Clearly, Δ is determined by the p points where it is $+2$ and one of the points where it is -2, showing the rigidity in possibilities of $\mathfrak{e}$.

There are two other big themes in the analysis of this chapter: quadratic equations and potential theory. It is an eighteenth century result (see the Notes to Section 5.2) that a real number x has a continued fraction expansion with $\xi_{n+p}(x) = \xi_n(x)$ for some p and all $n \geq N_0$ for some N_0 if and only if x obeys a quadratic equation with integral coefficients. Given this and the fact that the Jacobi parameters appear in a continued fraction expansion for $m(z)$, it should not be surprising that if a Jacobi matrix has periodic Jacobi parameters, then its m-function obeys a quadratic equation with polynomial (in z) coefficients. This in turn implies $m(z)$ has a natural continuation to a two-sheeted Riemann surface. This surface will play a major role, especially in Sections 5.12 and 5.13 and in Chapter 9.

One can ask for a simple intrinsic criterion that determines whether a set $\mathfrak{e}$ of the form (5.1.3)/(5.1.4) is the essential spectrum of a periodic Jacobi matrix. Given any compact $\mathfrak{e} \in \mathbb{C}$, which is not too small, $\mathfrak{e}$ supports measures, ν, with

$$\mathcal{E}(\nu) = \int \log|x-y|^{-1}\, d\nu(x) d\nu(y) < \infty \tag{5.1.10}$$

For example, if $\mathfrak{e}$ has the form of (5.1.3), Lebesgue measure restricted to $\mathfrak{e}$ has $\mathcal{E}(\nu) < \infty$.

If there is at least one ν with $\mathcal{E}(\nu) < \infty$, there is a unique probability measure, $\rho_{\mathfrak{e}}$, on $\mathfrak{e}$, called the equilibrium measure for $\mathfrak{e}$, that minimizes $\mathcal{E}(\nu)$ among all probability measures ν on $\mathfrak{e}$. Remarkably, if $\mathfrak{e}$ comes from a period p problem and has

p disjoint pieces, then

$$\rho_{\mathfrak{e}}([\alpha_j, \beta_j]) = \frac{1}{p} \tag{5.1.11}$$

for all j, and conversely. Via potential theory, (5.1.11) provides the desired intrinsic criteria.

There are two extensions of the sketch so far to keep in mind. First, while a period p problem generically has an essential spectrum, $\mathfrak{e}$, with p connected components, it can have fewer—indeed, any number $\ell + 1$ from 1 to p. We use ℓ for the number of gaps and $\ell + 1$ for the number of components.

What is happening is that $\Delta^{-1}((-2, 2))$ always has p disjoint components, but the boundaries of these sets (i.e., $\Delta^{-1}(\{-2, 2\})$), while generically distinct, can overlap (essentially, if $\Delta^2 - 4$ has a double zero). The set between the closures of the components of $\Delta^{-1}((-2, 2))$ are called "gaps," and when there are fewer than $p - 1$ gaps, we say some gaps are closed.

We thus consider sets

$$\mathfrak{e} = [\alpha_1, \beta_1] \cup \cdots \cup [\alpha_{\ell+1}, \beta_{\ell+1}] \tag{5.1.12}$$

$$\alpha_1 < \beta_1 < \alpha_2 < \cdots < \beta_{\ell+1} \tag{5.1.13}$$

The condition that a set $\mathfrak{e}$ be the essential spectrum of a period p Jacobi matrix with perhaps some gaps closed is that there are integers, $k_1, \dots, k_{\ell+1}$, so that

$$\rho_{\mathfrak{e}}([\alpha_j, \beta_j]) = \frac{k_j}{p} \tag{5.1.14}$$

Thus, $\mathfrak{e}$ is the spectrum of some periodic problem if and only if each $\rho_{\mathfrak{e}}([\alpha_j, \beta_j])$ is rational.

The other extension that will appear in Sections 5.12 and 5.13 is that if some $\rho_{\mathfrak{e}}([\alpha_j, \beta_j])$ is irrational, then there are almost periodic Jacobi matrices whose spectrum is $\mathfrak{e}$. These sets will be studied further in Chapter 9.

Section 5.2 will discuss quadratic equations for m. Sections 5.3 and 5.4 will discuss Δ and related structures. Section 5.5 will provide background on potential theory and its relevance to periodic Jacobi matrices and, in particular, prove (5.1.14). Sections 5.12 and 5.13 will explore the Riemann surface associated to $\mathfrak{e}$ and its function theory to prove that if $\mathfrak{e}$ has ℓ gaps, then the family of $\{(a_n, b_n)\}_{n=1}^p$ with essential spectrum $\mathfrak{e}$ is an ℓ-dimensional torus, called appropriately the *isospectral torus*.

Sections 5.6–5.11 are a grand aside that approximate general compact sets in $\mathbb{R}$ by ones that are spectra of periodic problems, and use this as a tool to complete the discussion of the CD kernel begun in Sections 2.14–2.17 and 3.11.

Remarks and Historical Notes. The issue of a period p problem having p bands generically will not be discussed formally, so let us make a few remarks. As we will see in Theorem 5.3.4, closed gaps are equivalent to degenerate eigenvalues for $J(\theta = 0)$ or $J(\theta = \pi)$, the $p \times p$ truncated Jacobi matrices with periodic and antiperiodic boundary conditions. By using degenerate eigenvalue perturbation theory (see Reed–Simon [364]), it is easy to see that if these operators have a degenerate eigenvalue and a single b_j is changed slightly (with all other parameters fixed), then there are no degenerate eigenvalues. That implies the set of $\{\alpha_j, \beta_j\}_{j=1}^p$ with any closed gaps is of codimension 1 at most. In fact, using ideas of Wigner–von Neumann [454], it is to be expected that the codimension is there, but I am not aware of any proof of this ([454] consider all $n \times n$ real matrices, not the Jacobi ones with a single corner matrix element added).

In his work in the 1880s on the stability of the moon's orbit, Hill was led to look at the $-u''(z) + V(z)u(z) = \lambda u(z)$ where V is periodic, so the continuum analog of periodic Jacobi matrices is called Hill's equation. Many of the ideas of this chapter are analogs of ideas developed there. Along the way, we will relate to these continuum forebears.

5.2 *m*-FUNCTIONS AND QUADRATIC IRRATIONALITIES

If one iterates the Stieltjes expansion (3.2.28), one sees that m and the n-times stripped m-function, m_n, are related by

$$m = \frac{Am_n + B}{Cm_n + D} \tag{5.2.1}$$

where A, B, C, D are polynomials in z. But if the original Jacobi matrix is periodic, that is, obeys (1.11.1), then $J_p = J$ and so $m_p = m$, and (5.2.1) becomes a quadratic equation for m. This allows m to be meromorphically continued in z to a compact Riemann surface, which will play a big role later. Our goal in this section is to make this precise and find the relation of the coefficients A, B, C, D to OPs.

In fact, we will go through the inverse of (5.2.1), which we have found as (3.7.23).

Theorem 5.2.1. *Let* $\{a_n, b_n\}_{n=1}^\infty$ *obey* (1.11.1). *Then the m-function obeys*

$$\alpha(z)m(z)^2 + \beta(z)m(z) + \gamma(z) = 0 \tag{5.2.2}$$

where

$$\alpha(z) = a_p p_{p-1}(z) \qquad \beta(z) = p_p(z) + a_p q_{p-1}(z) \qquad \gamma(z) = q_p(z) \tag{5.2.3}$$

The quadratic equation "discriminant" is given by

$$\beta^2 - 4\alpha\gamma = \Delta(z)^2 - 4 \tag{5.2.4}$$

where

$$\Delta(z) = p_p(z) - a_p q_{p-1}(z) \tag{5.2.5}$$

is called the *discriminant.*

Remarks. 1. It is an unfortunate terminology clash that in analogy with an object in the study of Hill's equation, Δ, given by (5.2.5), is called the discriminant, so the object in (5.2.4), which is usually called the discriminant of the quadratic equation, cannot be given that name!

2. We will see (see (5.4.5)) that $\Delta(z)$ is the trace of a transfer matrix.

Proof. Given (3.2.23) and (3.7.23), we have that if $J_p = J$ (implied by (1.11.1)), then

$$m = -\frac{mp_p + q_p}{a_p(mp_{p-1} + q_{p-1})} \tag{5.2.6}$$

which implies (5.2.2)/(5.2.3).

By (5.2.3),

$$\beta^2 - 4\alpha\gamma = \Delta^2 - 4[a_p(q_p p_{p-1} - p_p q_{p-1})]$$

and, by (3.2.22),

$$a_p(q_p p_{p-1} - p_p q_{p-1}) = 1$$

which proves (5.2.4). □

As a quadratic equation, (5.2.2) has a second solution, and the remarkable fact is that the other solution is also related to an m-function. To describe it, we need to extend $\{a_n, b_n\}_{n=1}^\infty$ to a two-sided sequence $\{a_n, b_n\}_{n=-\infty}^\infty$ by requiring that (1.11.1) holds for all n. The two-sided sequence generates a two-sided Jacobi matrix, which acts on $\ell^2(\mathbb{Z})$ by

$$(Ju)_n = a_n u_{n+1} + b_n u_n + a_{n-1} u_{n-1} \tag{5.2.7}$$

so the matrix is

$$J = \begin{pmatrix} \ddots & \ddots & \ddots & & & \\ & a_{-2} & b_{-1} & a_{-1} & & \\ & & a_{-1} & b_0 & a_0 & \\ & & & a_0 & b_1 & a_1 \\ & & & & \ddots & \ddots & \ddots \end{pmatrix} \tag{5.2.8}$$

If we replace a_ℓ by zero, the matrix breaks into a direct sum via $\ell^2(\{j\}_{j=-\infty}^{\ell}) \oplus \ell^2(\{j\})_{j=\ell+1}^\infty)$. For $\ell \geq 0$, the second summand is the Jacobi matrix we have called J_ℓ (i.e., J_0 is the original Jacobi matrix; J_ℓ is ℓ-times stripped) and will now call J_ℓ^+. We will use J_ℓ^- for the Jacobi matrix obtained from the other half turned around to be a conventional Jacobi matrix. Thus,

$$J_\ell^+ = J(\{a_{n+\ell}, b_{n+\ell}\}_{n=1}^\infty) \qquad J_\ell^- = J(\{a_{\ell-n}, b_{\ell+1-n}\}_{n=1}^\infty) \tag{5.2.9}$$

We will use $m(z; J_\ell^\pm)$, $P_k(z; J_\ell^\pm)$, $\alpha(z; J_\ell^\pm)$, and so on when we want to emphasize the J dependence; so, for example, $m(z; J_0^+)$ solves

$$\alpha(z; J_0^+)m^2 + \beta(z; J_0^+)m + \gamma(z; J_0^+) = 0 \tag{5.2.10}$$

The second solution is given by

Theorem 5.2.2. *The second solution of* (5.2.10) *for* $z \in \mathbb{C} \setminus \mathbb{R}$ *is given by*

$$m^\sharp(z) \equiv (a_p^2 m(z; J_0^-))^{-1} \tag{5.2.11}$$

As we discuss in the Notes, we will give several proofs of this theorem—not so much for their own sakes as they give different ways of looking at the result. Our proof here depends on relations of OPs among the various $J_\ell^\pm$. Recall that P are monic and p normalized.

Lemma 5.2.3. *We have that*

$$\text{(i)} \qquad q_k(z; J_\ell^\pm) = (a_{\ell\pm 1} \dots a_{\ell\pm k})^{-1} P_{k-1}(z; J_{\ell\pm 1}^\pm) \tag{5.2.12}$$

For $k = 0, 1, 2, \dots, p-1$ *and any* ℓ,

$$\text{(ii)} \quad P_{p-k}(z; J_\ell^+) = P_{p-k}(z; J_{\ell-k}^-) \tag{5.2.13}$$

$$\text{(iii)} \quad q_p(z; J_0^\pm) = (a_p)^{-1} p_{p-1}(z; J_0^\mp) \tag{5.2.14}$$

$$\text{(iv)} \quad p_p(z; J_0^+) = p_p(z; J_0^-) \tag{5.2.15}$$

$$\text{(v)} \quad q_{p-1}(z; J_0^+) = q_{p-1}(z; J_0^-) \tag{5.2.16}$$

Proof. (i) This is just a restatement of (3.2.16).

(ii) Take first $\ell = 0$. By Theorem 1.2.10, $P_{p-k}(z; J_0^+)$ is the characteristic polynomial for

$$\begin{pmatrix} b_1 & a_1 & & \\ a_1 & b_2 & \ddots & \\ & \ddots & \ddots & \\ & & a_{p-k-1} & b_{p-k} \end{pmatrix}$$

and $P_{p-k}(z; J_{-k}^-)$ for

$$\begin{pmatrix} b_{-k} & a_{-k-1} & & \\ a_{-k-1} & b_{-k-1} & \ddots & \\ & \ddots & \ddots & \ddots \\ & & a_{1-p} & b_{1-p} \end{pmatrix}$$

By periodicity, these matrices are obtained from each other by inverting the order of rows and columns. The general ℓ case follows by translation covariance.

(iii) By (5.2.12), $q_p(z; J_0^\pm) = (a_1 \dots a_p)^{-1} P_{p-1}(z; J_{\pm 1}^\pm)$, and by (5.2.13) for $k = 1$, $P_{p-1}(z; J_{\pm 1}^\pm) = P_{p-1}(z; J_0^\mp)$, from which (5.2.14) follows.

(iv) This follows from (5.2.13) for $k = 0$.

(v) Since $a_1 \dots a_{p-1} = a_{-1} \dots a_{-(p-1)}$, this is equivalent, by (5.2.12), to $P_{p-2}(z; J_1^+) = P_{p-2}(z; J_{-1}^-)$, which is (5.2.13) for $k = 2$, $\ell = 1$. □

Proof of Theorem 5.2.2. By (5.2.3), (5.2.14), (5.2.15), and (5.2.14), we have

$$\gamma(z; J_0^-) = a_p^{-2}\alpha(z; J_0^+) \qquad \alpha(z; J_0^-) = a_p^2\gamma(z; J_0^+)$$
$$\beta(z; J_0^-) = \beta(z; J_0^+) \tag{5.2.17}$$

Use ˜ for the J_0^- objects and no ˜ for the J_0^+ objects. This means

$$a_p^2\gamma\tilde{m}^2 + \beta\tilde{m} + a_p^{-2}\alpha = 0$$

or multiplying by $a_p^{-2}\tilde{m}^{-2}$,

$$\alpha(a_p^{-2}\tilde{m}^{-1})^2 + \beta(a_p^{-2}\tilde{m}^{-1}) + \gamma = 0$$

which says $m^\sharp$ given by (5.2.11) obeys (5.2.10).

That $m^\sharp$ is distinct from m on $\mathbb{C} \setminus \mathbb{R}$ and so the second solution is immediate if we notice on $\mathbb{C}^+$ that $\operatorname{Im} m > 0$ while $\operatorname{Im} m^\sharp < 0$. □

By the quadratic equation formula and (5.2.3)/(5.2.4), the solutions of (5.2.2) are

$$m(z) = -\frac{\beta(z) \pm \sqrt{\Delta(z)^2 - 4}}{2a_p p_{p-1}(z)} \tag{5.2.18}$$

where one takes the branch of square root with $\sqrt{\Delta(z)^2 - 4} = \Delta(z) + O(1/\Delta(z))$ near $z = \infty$. As a check, we see this leads to

$$\begin{aligned} m(z) &= -\frac{2a_p q_{p-k}(z)}{2a_p p_{p-1}(z)} + O\left(\frac{1}{z^2}\right) \\ &= -\frac{1}{z} + O\left(\frac{1}{z^2}\right) \end{aligned} \tag{5.2.19}$$

near infinity.

We will see below that (see Theorem 5.4.2 and 5.4.16)

(i) $\Delta(z)^2 = 4$ has all its solutions on $\mathbb{R}$.

(ii) $p_{p-1}(z)$ (whose roots are all simple and real) has zeros in $\Delta^{-1}([-2, 2])$ only at those points where $\Delta(z) \mp 2$ has a double zero and at such points $\beta(z) = 0$ also.

This implies $m(z)$ has continuous boundary values on $\Delta^{-1}([-2, 2])$, is real off that set, and has poles at some of the points where $p_{p-1}(z)$ (some because the numerator might also vanish). Thus, by Proposition 2.3.12,

Theorem 5.2.4. *The Jacobi matrix associated to a sequence of Jacobi parameters obeying* (1.11.1) *has purely a.c. spectrum on* $\Delta^{-1}([-2, 2])$ *and at most* $p-1$ *additional pure points off that set and no other spectrum.*

The quadratic equation (5.2.2) defines a two-sheeted branched cover of $\mathbb{C} \cup \{\infty\}$, the Riemann surface of $\sqrt{\Delta^2 - 4}$. This will be the major theme of Sections 5.12 and 5.13. Since $m^\sharp$ will define the second sheet and zeros of $m(z; J_0^-)$ are poles of $m(z; J_{-1}^-)$, we will see that poles of m on the two-sheeted surface are precisely the eigenvalues of $J_{\pm 1}^\pm$.

Remarks and Historical Notes. The link between continued fractions, periodicity, and quadratic equations goes back to the study of continued fraction expansions of reals like (2.5.10). Euler noted that if the ξ_j in (2.5.10) are periodic (i.e., $\xi_{j+p} = \xi_j$ for fixed p and all $j > 0$), then x obeys a quadratic equation with integral coefficients (the proof of Theorem 5.2.1 is essentially his proof). Legendre proved continued fractions of x's obeying a quadratic equation with integral coefficients are eventually periodic (i.e., $\xi_{j+p} \equiv \xi_j$ for some p and $j \geq J$ for some J). Galois specified the set with strictly periodic ξ's. This is discussed, for example, in Koch [232] and Lang [265].

We will see several other proofs of Theorem 5.2.2 later. In Section 5.4, we will see its close relation to reality of the Green's function for the whole-line problem. Finally, in Section 5.13, we will see its relation to reflectionless operators.

5.3 REAL FLOQUET THEORY AND DIRECT INTEGRALS

In the last section, we saw that two-sided periodic Jacobi matrices as defined in (5.2.8) are naturally associated to periodic Jacobi parameters. We will concentrate on these two-sided matrices in this section (and the next), although we will briefly return to the one-sided case in the next section.

While we will do spectral analysis of J as an operator on $\ell^2(\mathbb{Z})$, it is useful to allow Ju to be defined via (5.2.7) for *any* sequence $\{u_n\}_{n=-\infty}^{\infty}$. Periodicity of the Jacobi parameters implies that J has a large commutant. Define S by

$$(Su)_n = u_{n+1} \tag{5.3.1}$$

so by periodicity,

$$JS^p = S^pJ \tag{5.3.2}$$

thought of as operators on $\ell^2(\mathbb{Z})$ or on all sequences or on any $\ell^p(\mathbb{Z})$ including $\ell^\infty(\mathbb{Z})$.

A physicist would say (5.3.2) means J and S^p can be simultaneously diagonalized, that is, have a common complete set of eigenvectors. Of course, we have to be prepared to consider "continuum eigenvectors," which the general theory (see the Notes to Section 5.4) says are polynomially bounded eigenvectors of

$$Ju = \lambda u \tag{5.3.3}$$

Since S^p is unitary, its eigenvalues should lie in $\partial\mathbb{D}$, so we look for solutions obeying

$$u_{n+p} = e^{i\theta}u_n \tag{5.3.4}$$

for all n and some real θ. Such solutions are called *Floquet solutions*. In this section, we consider θ real to do a spectral resolution. In the next, to consider Weyl solutions, we will consider θ complex. While this paragraph is motivation, it will not be used directly below.

We will need the following result about (5.3.3):

Proposition 5.3.1. *Let $\lambda \in \mathbb{C}$. The set of solutions of* (5.3.3) *among all two-sided sequences is at most two-dimensional.*

Proof. If the dimension is more than 2, there is a nonzero solution with $u_0 = u_1 = 0$, but then (5.3.3) and $a_n \neq 0$ for all n implies $u \equiv 0$. □

To study solutions obeying (5.3.4) for $\theta \in [0, 2\pi)$, we define

$$\ell_\theta^\infty = \{u \mid u \text{ obeys } (5.3.4)\} \tag{5.3.5}$$

As the notation suggests, such u's lie in ℓ^∞. Since $u \in \ell^\infty_\theta \to \{u_n\}_{n=0}^{p-1}$ is a bijection,

$$\dim(\ell^\infty_\theta) = p \tag{5.3.6}$$

Indeed, if we define $\delta^{(j)}(\theta)$ for $j = 1, 2, \dots, p$ by

$$[\delta^{(j)}(\theta)]_{n+p\ell} = e^{i\theta\ell}\delta_{jn} \qquad n = 1, \dots, p \tag{5.3.7}$$

then $\{\delta^{(j)}(\theta)\}_{j=1}^p$ is a basis for ℓ^∞_θ. We have

Proposition 5.3.2. *J leaves ℓ^∞_θ invariant. Its restriction to ℓ^∞_θ (call it $J(\theta)$) in the $\{\delta^{(j)}(\theta)\}_{j=1}^p$ basis has the matrix*

$$J(\theta) = \begin{pmatrix} b_1 & a_1 & 0 & \dots & \dots & e^{-i\theta}a_p \\ a_1 & b_2 & a_2 & \dots & \dots & 0 \\ 0 & a_2 & b_3 & \ddots & & 0 \\ & \ddots & \ddots & \ddots & \ddots & \\ 0 & & \ddots & \ddots & b_{p-1} & a_{p-1} \\ e^{i\theta}a_p & \dots & \dots & \dots & a_{p-1} & b_p \end{pmatrix} \tag{5.3.8}$$

Proof. By (5.3.2), J takes ℓ^∞_θ to itself since $\ell^\infty_\theta = \{u \mid S^p u = e^{i\theta}u\}$. The extra corner pieces come from $(J(\theta)\delta^{(1)}(\theta))_0 = a_0$, so by definition of ℓ^∞_θ and periodicity of a, $(J(\theta)\delta^{(1)}(\theta))_p = e^{i\theta}a_p$. ☐

We will need the following below:

Lemma 5.3.3. *If $u^{(j)} \in \ell^\infty_{\theta_j}$ for $j = 1, \dots, q$ are nonzero with the θ_j distinct, then $\{u^{(j)}\}_{j=1}^q$ are linearly independent in ℓ^∞.*

Proof. For each j, n, because the θ_j are distinct,

$$\lim_{L\to\infty} \frac{1}{2L+1} \sum_{\ell=-L}^{L} e^{-i\theta_j\ell} u^{(k)}_{n+\ell p} = \delta_{jk} u^{(k)}_n$$

Thus, if $\sum_{j=1}^q \gamma_j u^{(j)} = 0$, then

$$\gamma_j u^{(j)}_n = \lim_{L\to\infty} \frac{1}{2L+1} \sum_{\ell=-L}^{L} e^{-i\theta_j\ell} \left(\sum_{k=1}^q \gamma_k u^{(k)}_{n+\ell p} \right) = 0$$

so $\gamma_j = 0$. ☐

$J(\theta)$ is a selfadjoint $p \times p$ matrix, so it has (counting multiplicity) p eigenvalues,

$$e_1(\theta) \le e_2(\theta) \le \dots \le e_p(\theta) \tag{5.3.9}$$

Theorem 5.3.4. (i)

$$e_j(2\pi - \theta) = e_j(\theta) \qquad \textit{for } \theta \in (0, \pi) \tag{5.3.10}$$

(ii) *For $e^{i\theta} \neq \pm 1$, the $e_j(\theta)$ are simple, that is, $J(\theta)$ has simple spectrum for $\theta \in (0, \pi) \cup (\pi, 2\pi)$. Each $e_j(\theta)$ is real analytic on $(0, \pi)$.*

(iii) *For* $\theta \neq \theta'$, $J(\theta)$ *and* $J(\theta')$ *have disjoint spectra.*

(iv) *We have*

$$e_p(0) > e_p(\pi) \geq e_{p-1}(\pi) > e_{p-1}(0) \geq \dots \tag{5.3.11}$$

(v) *On* $(0, \pi)$, $(-1)^{p-j} e_j(\theta)$ *is strictly monotone decreasing.*

Remark. For now we will prove strict monotonicity of $e_j(\theta)$ in $(0, \pi)$. Eventually (see Theorem 5.4.2), we will prove $(-1)^{p-j} e_j'(\theta) < 0$.

Proof. (i) If $\overline{M}$ means the matrix with complex conjugates, then $\overline{J(\theta)} = J(2\pi - \theta)$, which, given that the eigenvalues are real, immediately implies (5.3.10).

(ii) If $J(\theta)$ has a degenerate eigenvalue, say λ, then $u^{(1)}, u^{(2)} \in \ell_\theta^\infty$ are linearly independent. By (i), λ is also an eigenvalue of $\ell_{2\pi-\theta}^\infty$, and so there is an eigenvector, $u^{(3)}$, of $J(2\pi - \theta)$ (could be chosen as $\overline{u^{(1)}}$). By Lemma 5.3.3, $u^{(1)}, u^{(2)}, u^{(3)}$ are linearly independent, so there is a violation of Proposition 5.3.1. The $e_j(\theta)$ are analytic as simple roots of a polynomial with analytic coefficients.

(iii) By the same argument in (ii), if $e_j(\theta) = e_\ell(\theta')$ and $\theta \in (0, \pi)$, then there are at least three linearly independent eigenvectors, violating Proposition 5.3.1. This handles all cases but $\{\theta, \theta'\} = \{0, \pi\}$, which follows from part (iv).

(iv) Let $a_n^{(0)} \equiv 1$, $b_n^{(0)} \equiv 0$, so the solutions of $J^{(0)} u = \lambda u$ are $u_n = e^{ikn}$ with $\lambda = 2\cos(k)$. $u_{n+p} = e^{i\theta} u_n$ with $\theta = kp \pmod{2\pi}$ and k is real and in $[-\pi, \pi)$. It follows that $e_p^{(0)}(0) = 2$, $e_{p-1}^{(0)}(0) = e_{p-2}^{(0)}(0) = 2\cos(\pm\frac{2\pi}{p})$, $e_{p-3}^{(0)}(0) = e_{p-4}^{(0)}(0) = 2\cos(\pm\frac{4\pi}{p})$; $e_p^{(0)}(\pi) = e_{p-1}^{(0)}(\pi) = 2\cos(\pm\frac{\pi}{p})$, and so on. We thus have (5.3.11) for $J^{(0)}$. Since eigenvalues are continuous in θ and we have proven nondegeneracy for $\theta \in (0, \pi)$, then $e_j(\theta) \neq e_j(0 \text{ or } \pi)$. We also have that (v) holds for $J^{(0)}$.

For $y \in [0, 1]$, let $J^{(y)} = (1 - y) J^{(0)} + yJ$. The $J^{(y)}(\theta)$ are continuous in y and θ. There is no way for an eigenvalue of $J^{(y)}(0)$ to cross an eigenvalue of $J^{(y)}(\pi)$ as y varies without going past the $J^{(y)}(\theta)$ eigenvalues, which cannot happen by the proof of (iii) we have given. Thus, (5.3.11) still holds at $y = 1$.

(v) As noted in the proof of (iv), (iii) + (iv) implies (v). □

Eigenvalue perturbation theory ([215, 364]) implies $e_j(\theta)$ is also real analytic at $\theta = 0, \pi$, although the continuation will sometimes be $e_j(2\pi - \theta)$ and sometimes $e_{j\pm1}(2\pi - \theta)$. If a gap is open, that is, strict inequality holds in (5.3.11) for e_j, then e_j has a maximum or minimum at 0 or π, so $\partial e_j / \partial\theta = 0$ at this endpoint.

For some purposes, an important issue is that if a gap is closed, $\partial e_j / \partial\theta \neq 0$ at an endpoint, so θ is an analytic function of e_j. This can be deduced from the exact form of $\partial J / \partial\theta$ (from (5.3.8)) and degenerate eigenvalue perturbation theory ([215, 364]), but we will prove it easily later using the discriminant (see Corollary 5.4.4).

We can now define the important notions of bands and gaps. We define

$$\mathfrak{e}_j = \operatorname{Ran}(e_j(\theta) \mid \theta \in [0, 2\pi)) \qquad j = 1, \dots, p \tag{5.3.12}$$

as the *bands* with

$$\mathfrak{e} = \bigcup_{j=1}^{p} \mathfrak{e}_j \tag{5.3.13}$$

By Theorem 5.3.4, we have

$$\mathfrak{e}_j^{\text{int}} = e_j[(0, \pi)] \qquad \mathfrak{e}_j^{\text{int}} \cap \mathfrak{e}_k^{\text{int}} = \emptyset \quad \text{for } j \neq k \tag{5.3.14}$$

so the $\mathfrak{e}_j$ can only intersect in their endpoints. Thus,

$$\mathfrak{e}_j = [\alpha_j, \beta_j]$$

with

$$\alpha_1 < \beta_1 \leq \cdots \leq \alpha_p < \beta_p \tag{5.3.15}$$

a rewriting of (5.3.11).

The gaps are the sets (β_j, α_{j+1}) (or sometimes those of these sets that are nonempty). If $\beta_j = \alpha_{j+1}$, we say the jth gap is *closed*; otherwise, we say it is open. We use ℓ for the number of open gaps, so

$$\ell \leq p - 1 \tag{5.3.16}$$

In the Notes to Section 5.1, we discussed that $\ell = p - 1$ generically, and in Section 5.12, we will see that ℓ is the genus of the Riemann surface defined by m.

We are heading toward a proof that for the full-line Jacobi matrix,

$$\sigma(J) = \mathfrak{e} \tag{5.3.17}$$

and that the spectrum is purely a.c. of multiplicity 2. We begin by putting the usual Fourier transform into a mod p setting. We define

$$\mathcal{F}\colon \ell^2(\mathbb{Z}) \to L^2\biggl(\partial\mathbb{D}, \frac{d\theta}{2\pi}; \mathbb{C}^p\biggr) \tag{5.3.18}$$

the L^2 functions with values in $\mathbb{C}^p$ by ($n = 0, 1, \dots, p-1$)

$$(\mathcal{F}u)_n(\theta) = \sum_{\ell=-\infty}^{\infty} u_{n+\ell p}\, e^{-i\ell\theta} \tag{5.3.19}$$

where, as usual with Fourier transform, we define this for $u \in \ell^1$ and extend by using

$$\int_{\partial\mathbb{D}} \|\mathcal{F}u.(\theta)\|^2 \frac{d\theta}{2\pi} = \sum_n |u_n|^2 \tag{5.3.20}$$

since $\{e^{i\ell\theta}\}_{\ell=-\infty}^{\infty}$ is a basis for $L^2(\partial\mathbb{D}, \frac{d\theta}{2\pi})$. Of course, we have the inverse

$$\mathcal{F}^{-1}\colon L^2\biggl(\partial\mathbb{D}, \frac{d\theta}{2\pi}; \mathbb{C}^p\biggr) \to \ell^2(\mathbb{Z})$$

by

$$(\mathcal{F}^{-1}f)_{n+\ell p} = \int e^{i\ell\theta} f_n(\theta)\, \frac{d\theta}{2\pi} \tag{5.3.21}$$

for $\ell \in \mathbb{Z}$ and $n = 0, 1, \dots, p-1$.

By the spectral theorem for finite matrices, there exist unitaries $U(\theta)\colon \mathbb{C}^p \to \mathbb{C}^p$ so

$$U(\theta)J(\theta)U(\theta)^{-1} = \begin{pmatrix} e_1(\theta) & & & \\ & \ddots & & \\ & & \ddots & \\ & & & e_p(\theta) \end{pmatrix} \tag{5.3.22}$$

It is easy to see that U can be picked measurably and, not much harder, using the simplicity to see that it can be chosen continuously on $(0,\pi) \cup (\pi, 2\pi)$. We fix $U(\theta)$ once and for all measurable so that (5.3.22) holds.

We define $\mathcal{U}\colon L^2(\partial\mathbb{D}, \frac{d\theta}{2\pi}; \mathbb{C}^p)$ to itself by

$$(\mathcal{U}f)(\theta) = U(\theta)f(\theta) \tag{5.3.23}$$

Theorem 5.3.5. *Let J be a two-sided periodic Jacobi matrix. Then*
(a) *As operators on $L^2(\partial\mathbb{D}, \frac{d\theta}{2\pi}; \mathbb{C}^p)$,*

$$[(\mathcal{F}J\mathcal{F}^{-1})f]_n(\theta) = (J(\theta)f)_n(\theta) \tag{5.3.24}$$

(b)

$$[(\mathcal{U}\mathcal{F})J(\mathcal{U}\mathcal{F})^{-1}f]_n(\theta) = e_n(\theta)f_n(\theta) \tag{5.3.25}$$

Proof. (a) Let $\delta_n \in \ell^2(\mathbb{Z})$ be a delta function at $n \in \mathbb{Z}$ and let $f_n^{(\ell)}$ for $\ell \in \mathbb{Z}$, $n \in \{0,\dots,p-1\}$, be the function with nonzero component n and value $e^{-i\ell\theta}$. Then $\mathcal{F}(\delta_{n+\ell p}) = f_n^{(\ell)}$ by (5.3.19). (5.3.24) is then an easy calculation.

(b) is immediate from (5.3.22), (5.3.23), and (5.3.24). □

Lemma 5.3.6. *Let F be strictly monotone and continuous on $[a,b]$ and let A be the selfadjoint operator*

$$(Af)(x) = F(x)f(x) \tag{5.3.26}$$

on $L^2([a,b], dx)$. Then A is unitarily equivalent to

$$(Bg)(y) = yg(y) \tag{5.3.27}$$

on $L^2([F(a), F(b)], dF^{-1})$.

Remark. F^{-1} is also a continuous and strictly monotone function, and dF^{-1} means its Stieltjes measure. Here F^{-1} is the functional inverse (not $1/F$).

Proof. Let $V\colon L^2([F(a),F(b)], dF^{-1}) \to L^2((a,b), dx)$ by

$$(Vg)(x) = g(F(x)) \tag{5.3.28}$$

Then V is unitary and $VBV^{-1} = A$. □

This lemma and Theorem 5.3.5 immediately imply

Theorem 5.3.7. *Let J be a two-sided period p periodic Jacobi matrix with bands $\{\mathfrak{e}_j\}_{j=1}^p$. Then $\sigma(J) = \mathfrak{e}$ and the spectrum is purely absolutely continuous with multiplicity 2.*

Proof. We get multiplicity 2 by separately considering e_j in $(0,\pi)$ and $(\pi, 2\pi)$. Since $e_j(\theta)$ is real analytic, its inverse is real analytic after a discrete set is removed, and so de_j^{-1} is an absolutely continuous measure. □

There is another way of writing this more explicitly. The proof just follows the various mappings above, so we will only provide a sketch. Let $\tilde{\mathfrak{e}} = \cup_{j=1}^{p} \mathfrak{e}_j^{\rm int}$. If $\lambda \in \tilde{\mathfrak{e}}$, there is a unique $\theta \in (0,\pi)$ and j so $\lambda = e_j(\theta)$. We write $\theta(\lambda)$. There are solutions $\varphi_n^{\pm}(\lambda)$ of

$$(J-\lambda)\varphi^{\pm}(\lambda) = 0 \tag{5.3.29}$$

with

$$\varphi_{n+kp}^{\pm}(\lambda) = e^{\pm ik\theta(\lambda)}\varphi_n^{\pm}(\lambda) \tag{5.3.30}$$

We can normalize $\varphi^{\pm}$ by requiring

$$\varphi_0^{\pm}(\lambda) > 0 \qquad \sum_{j=0}^{p-1} |\varphi_j^{\pm}|^2 = 1 \tag{5.3.31}$$

$\varphi_0^{+}(\lambda)$ cannot be zero since then $\varphi_0^{-}(\lambda) = \overline{\varphi_0^{+}(\lambda)}$ is also zero and there is a linear combination vanishing at 0 and 1, violating Lemma 5.3.3. Thus, the normalization in (5.3.31) is possible.

With this normalization,

$$\varphi^{-}(\lambda) = \overline{\varphi_{+}(\lambda)} \tag{5.3.32}$$

We define for $\{u_n\}_{n=-\infty}^{\infty}$ of finite support

$$\widehat{u}^{\pm}(\lambda) = \sum_{n=-\infty}^{\infty} \overline{\varphi_n^{\pm}(\lambda)}\, u_n \tag{5.3.33}$$

We define the measure $d\nu$ on $\tilde{\mathfrak{e}}$ by

$$d\nu(\lambda) = \frac{1}{p\pi}\left|\frac{d\theta}{d\lambda}(\lambda)\right| d\lambda \tag{5.3.34}$$

Then:

Theorem 5.3.8. $\widehat{\ }$ *extends to a unitary map of $\ell^2(\mathbb{Z})$ to $L^2(\mathfrak{e}, d\nu(\lambda); \mathbb{C}^2)$ with inverse*

$$(\check{f})_n = \frac{p}{2}\int [\varphi_n^{+}(\lambda) f^{+}(\lambda) + \varphi_n^{-}(\lambda) f^{-}(\lambda)]\, d\nu(\lambda) \tag{5.3.35}$$

Moreover,

$$\widehat{Ju}^{\pm}(\lambda) = \lambda\widehat{u}^{\pm}(\lambda) \tag{5.3.36}$$

Remarks. 1. In (5.3.35), we use $f^{\pm}(\lambda)$ for the two components of $\mathbb{C}^2$-valued function $f \in L^2(\mathfrak{e}, d\nu(\lambda); \mathbb{C}^2)$.

2. $d\nu$ will be the density of states discussed in Proposition 5.4.7.

3. The normalization of $d\nu$, which requires a $p/2$ in (5.3.35) is made so $d\nu$ is a probability measure. For θ' has a fixed sign on each $\mathfrak{e}_j$ so $\int_{\mathfrak{e}_j} |\frac{d\theta}{d\lambda}|\, d\lambda = \pi$ as θ runs from 0 to π or π to 0. Thus,

$$\int d\nu = \sum_{j=1}^{p} \int_{\mathfrak{e}_j} \frac{1}{p\pi} \left|\frac{d\theta}{d\lambda}\right| d\lambda = p\,\frac{1}{p\pi}\,\pi = 1$$

Sketch. $\tilde{\varphi}^+ \equiv \{\varphi_n^+\}_{n=0}^{\ell-1}$ is an eigenvector of $J(\theta)$ normalized because of (5.3.31), so if $\lambda_1, \dots, \lambda_p$ are the λ's with a given θ, $\{\tilde{\varphi}^+(\lambda_j)\}_{j=1}^{p}$ is an orthogonal basis for $\mathbb{C}^p$ and unitarity of $\widehat{\ }$ follows from that for $\mathcal{F}$. (5.3.35) comes from the fact that the inverse of $\widehat{\ }$ is its adjoint. (5.3.36) comes from (5.3.29). □

Example 5.3.9. Let $a_n \equiv 1$, $b_n \equiv 0$, and $p = 1$. Then $\theta(\lambda)$ is given by

$$\lambda = 2\cos(\theta(\lambda)) \tag{5.3.37}$$

for $\theta \in (0, \pi)$ and $\lambda \in (-2, 2)$. We have

$$\varphi_n^{\pm}(\lambda) = e^{\pm in\theta(\lambda)} \tag{5.3.38}$$

and $\frac{d\lambda}{d\theta} = 2\sin(\theta(\lambda))$, so

$$\frac{d\theta}{d\lambda} = \frac{1}{\sqrt{4-\lambda^2}} \tag{5.3.39}$$

and

$$d\nu = \frac{1}{2\pi}\,\frac{1}{\sqrt{4-\lambda^2}}\, d\lambda \tag{5.3.40}$$

the *free density of states*. (5.3.33) is just the ordinary Fourier transform. □

Remarks and Historical Notes. The space $L^2(\partial\mathbb{D}, \frac{d\theta}{2\pi}; \mathbb{C}^p)$ is often written as a direct integral and this is the language used in discussing eigenfunction expansions for periodic Schrödinger operators in arbitrary dimension. This section is essentially a discrete version of that theory specialized to one dimension. The ideas originated in the physics literature (as Bloch waves) and were expressed mathematically by Gel'fand [149]; see the historical background and exposition in Reed–Simon [364].

5.4 THE DISCRIMINANT AND COMPLEX FLOQUET THEORY

In this section, we mainly discuss periodic full-line Jacobi matrices, J, although some results will hold for general full-line matrices (with bounded Jacobi parameters). We will also say something about the half-line operators $J_\ell^{\pm}$ of (5.2.9). Except for the fact that we will use that $J(\theta)$ has only real eigenvalues (see the Notes for a way to avoid this), the discussion in this section will not use results from the last section although it will illuminate them.

We will be interested in solutions of

$$(J - \lambda)u = 0 \tag{5.4.1}$$

where $\lambda \in \mathbb{C}$ and u is an arbitrary sequence. We focus on solutions that obey

$$u_{n+p} = \eta u_n \tag{5.4.2}$$

for some η, all n (and p the period of J). Unlike the previous section, η need not be in $\partial\mathbb{D}$. η is called the *Floquet index* and u a *Floquet solution*. When we want to focus on the solutions of the last section where $|\eta| = 1$, we speak of *Floquet plane waves*.

A major role will be played by the transfer matrix (3.2.19) over p units

$$T_p(\lambda) = \begin{pmatrix} p_p(\lambda) & -q_p(\lambda) \\ a_p p_{p-1}(\lambda) & -a_p q_{p-1}(\lambda) \end{pmatrix} \tag{5.4.3}$$

Notice that (3.2.28) says that

$$\det(T_p(\lambda)) = 1 \tag{5.4.4}$$

We will define the *discriminant*

$$\Delta(\lambda) = \operatorname{Tr}(T_p(\lambda)) = p_p(\lambda) - a_p q_{p-1}(\lambda) \tag{5.4.5}$$

the object defined already in (5.2.5). Recall that if u solves (5.4.1), then (since $a_p = a_0$)

$$T_p(\lambda)\begin{pmatrix} u_1 \\ a_0 u_0 \end{pmatrix} = \begin{pmatrix} u_{p+1} \\ a_0 u_p \end{pmatrix} \tag{5.4.6}$$

Thus,

Theorem 5.4.1. *There is a Floquet solution of* (5.4.1) *with Floquet index η if and only if η is an eigenvalue of $T_p(\lambda)$ and the Floquet solution has $(u_1\ a_0u_0)^t$ as eigenvector. In particular,*

(i) *If η is a Floquet index, so is η^{-1}.*

(ii) *If $\Delta(\lambda) \neq \pm 2$, there are exactly two Floquet solutions* (*up to constant multiples*)*.*

(iii) *We have that for $\theta \in [0, 2\pi]$,*

$$\det(\lambda - J(\theta)) = (a_1 \dots a_p)[\Delta(\lambda) - 2\cos\theta] \tag{5.4.7}$$

(iv) *The eigenvalues, $e_j(\theta)$, of $J(\theta)$ solve*

$$\Delta(e_j(\theta)) = 2\cos\theta \tag{5.4.8}$$

Remarks. 1. Since $\deg(\Delta(\lambda)) = p$, $\Delta(\lambda) = +2$, and $\Delta(\lambda) = -2$, each has at most p solutions. So there are two Floquet solutions except for at most $2p$ points.

2. We explore below (see Proposition 5.4.3) when there are two Floquet solutions and when only one if $\Delta(\lambda) = \pm 2$.

3. $J(\theta)$ in (5.4.7) is given by (5.3.8). It is an interesting exercise to expand $\det(\lambda - J(\theta))$ in minors to get (5.4.7) using Theorem 1.2.10 and the definition (5.4.5) in terms of orthogonal polynomials.

4. By the spectral theorem for Hermitian matrices like $J(\theta)$, (5.4.8) immediately implies

$$\Delta(J(\theta)) = (2\cos\theta)\mathbf{1} \tag{5.4.9}$$

Proof. If u obeys (5.4.1) and (5.4.2), then

$$\begin{pmatrix} u_{p+1} \\ a_0 u_p \end{pmatrix} = \eta \begin{pmatrix} u_1 \\ a_0 u_0 \end{pmatrix} \tag{5.4.10}$$

so η is an eigenvalue of $T_p(\lambda)$. Conversely, if $(u_1 \; a_0 u_0)^t$ is an eigenvector, (5.4.10) holds, which means by periodicity of $\{a_n, b_n\}$ that (5.4.2) holds for the solution of (5.4.1) with (u_1, u_0) initial conditions. This verifies the first statement in the theorem.

To prove (i), we note $\det(T_p(\lambda)) = 1$ says that if η is an eigenvalue, so is η^{-1}. (ii) then follows since if $\eta \neq \pm 1$, then $\eta^{-1} \neq \eta$, and there are two eigenvalues. But since the algebraic eigenvalues have product 1, $\eta = \pm 1$ if and only if $\Delta(\lambda) = \text{Tr}(T_p(\lambda)) = \pm 2$.

To get (iii), suppose first that $\theta \neq 0, \pi$. We note λ is an eigenvalue of $J(\theta)$ if and only if $\eta = e^{i\theta}$ is a Floquet index, and that happens if and only if $\Delta(\lambda) = \eta + \eta^{-1} = 2\cos\theta$. It follows that the two sides of (5.4.7) have the same zeros. Since both are monic polynomials, they must be equal. $\theta = 0, \pi$ then follows by continuity.

To obtain (iv), note that if $\lambda = e_j(\theta)$, then by the Hamiltonian–Jacobi theorem, $\det(\lambda - J(\theta))|_{\lambda = e_j(\theta)} = 0$, so by (5.4.9), $\Delta(e_j(\theta)) = 2\cos\theta$. We note that, conversely, (5.4.9) shows any solution of $\Delta(\lambda) = 2\cos\theta$ is an eigenvalue of $J(\theta)$. □

We can now analyze Δ rather completely:

Theorem 5.4.2. *Δ has the following properties:*

(i) $\Delta^{-1}([-2, 2]) \subset \mathbb{R}$

(ii) *Let* $x_1^\pm \leq x_2^\pm \leq \cdots \leq x_p^\pm$ *be the zeros (counting multiplicity) of* $\Delta(\lambda) \mp 2$. *Then*

$$x_p^+ > x_p^- \geq x_{p-1}^- > x_{p-1}^+ \geq x_{p-2}^+ > x_{p-2}^- \geq \dots \tag{5.4.11}$$

(iii) $\Delta(\lambda)$ *is strictly monotone on each interval* (x_{p-2j}^-, x_{p-2j}^+) *and* $(x_{p-1-2j}^+, x_{p-1-2j}^-)$, $j = 0, 1, 2, \dots$. *Indeed,* $\Delta'(\lambda) > 0$ *on intervals of the first type and* $\Delta'(\lambda) < 0$ *on intervals of the second type.*

(iv) *If* $e_j(\theta)$ *are the eigenvalues of* $J(\theta)$ *for* $\theta \in (0, \pi)$, *then* $(-1)^{p-j} e_j'(\theta) > 0$.

Remark. (5.4.11) is equivalent to (5.3.11).

Proof. (i) If $\Delta(\lambda) = 2\cos\theta$, then $e^{\pm i\theta}$ are Floquet indices, so λ is an eigenvalue of $J(\theta)$, which is selfadjoint. Thus, λ is real.

(ii), (iii) We first claim that if $\Delta(\lambda_0) \in (-2, 2)$, then $\Delta'(\lambda_0) \neq 0$, for if $\Delta'(\lambda_0) = 0$, then $\lambda \mapsto \Delta(\lambda)$ is many to one near $\lambda = \lambda_0$ in $\mathbb{C}$, which implies, by the implicit function for analytic functions, that there are nonreal λ's near λ_0 with $\Delta(\lambda) \in (-2, 2)$, violating (i). This means that when $\Delta(\lambda)$ varies in $(-2, 2)$, it is strictly monotone.

Similarly, we see that if $\Delta(\lambda_0) = \pm 2$ and $\Delta'(\lambda_0) = 0$, then $\pm\Delta''(\lambda_0) < 0$ to avoid nonreal solutions of $\Delta(\lambda) \in (-2, 2)$.

Since $\Delta(\lambda) = (a_1 \dots a_p)^{-1} \lambda^p +$ lower order, $\Delta(\lambda) > 2$ near $+\infty$. Thus, the first zero, x_p^+, of $\Delta(\lambda)^2 - 4$ has $\Delta(\lambda_p^+) = 2$. By the result on points where $\Delta' = 0$, we have $\Delta'(\lambda_p^+) \neq 0$. Thus, as λ decreases, $\Delta(\lambda)$ runs from 2 down to -2. Either

$\Delta^2 - 4$ has a double zero at this point or else $\Delta(\lambda) < -2$ just below this point. As λ decreases, Δ must turn around (for $\Delta^2 - 4$ to have $2p$ zeros), and so we see $x_p^+ > x_p^- \geq x_{p-1}^-$. Repeating this analysis leads to the full string (5.4.11) and proves (iii) at the same time.

(iv) Since $\Delta(e_j(\theta)) = 2\cos\theta$,

$$\Delta'(e_j(\theta))e_j'(\theta) = 2\sin\theta \tag{5.4.12}$$

proving the result. □

By $\deg(q_\ell) = \ell - 1$, we see, by (1.2.13), that

$$\begin{aligned}\Delta(\lambda) &= p_p(\lambda) + O(\lambda^{p-2}) \\ &= (a_1 \dots a_p)^{-1}\left(\lambda^p + \left(\sum_{j=1}^p b_j\right)\lambda^{p-1}\right) + O(\lambda^{p-2})\end{aligned} \tag{5.4.13}$$

As in the last section, we define *bands*

$$\mathfrak{e}_p = [x_p^-, x_p^+] \qquad \mathfrak{e}_{p-1} = [x_{p-1}^+, x_{p-1}^-] \qquad \dots \tag{5.4.14}$$

and gaps.

If $\Delta(\lambda) \neq \pm 2$, there are two Floquet solutions since the eigenvalues of $T_p(\lambda)$ are distinct. As for points where $\Delta(\lambda) = \pm 2$:

Proposition 5.4.3. *Suppose* $\Delta(\lambda_0) = \pm 2$. *Then the following are equivalent:*

(i) *All solutions of* (5.4.1) *at* λ_0 *are periodic* (*if* $\Delta(\lambda_0) = 2$) *or antiperiodic* (*if* $\Delta(\lambda_0) = -2$).

(ii) $T_p(\lambda_0) = \pm\mathbf{1}$

(iii) $J(\theta = 0)$ (*if* $\Delta(\lambda_0) = 2$) *or* $J(\theta = \pi)$ (*if* $\Delta(\lambda_0) = -2$) *has an eigenvalue of multiplicity* 2.

(iv) $\Delta'(\lambda_0) = 0$

(v) *The gap at* λ_0 *is closed.*

Remarks. 1. *Antiperiodic* means $u_{n+p} = -u_n$.

2. If (i) fails, there is a unique Floquet solution (up to a constant).

3. If (ii) fails, $T_p(\lambda_0)$ has $\left(\begin{smallmatrix}\pm 1 & 1 \\ 0 & \pm 1\end{smallmatrix}\right)$ as Jordan normal form, which implies any solution of (5.4.1) independent of the (anti-)periodic solution grows so there are a $c_1 n$ upper bound and $c_2 n$ lower bound on $|u_n|$.

4. If the gap is open at the edges where $\Delta(\lambda_0) = \pm 2$, there is a unique (up to a constant) periodic (if $\Delta(\lambda_0) = 2$) or antiperiodic (if $\Delta(\lambda_0) = -2$) solution.

Proof. (i) $\Leftrightarrow$ (ii) is immediate from (5.4.6).

(i) $\Leftrightarrow$ (iii) Eigenvectors of $J(\theta = 0)$ with eigenvalue λ_0 are precisely periodic solutions of (5.4.1) for $\lambda = \lambda_0$. Since the set of potential solutions is two-dimensional, (i) is equivalent to there being a two-dimensional family of eigenvectors.

(iii) $\Leftrightarrow$ (iv) $\Delta'(\lambda_0) = 0$ if and only if $\Delta(\lambda) \mp 2$ has a double zero at $\lambda = \lambda_0$. By (5.4.7), this is true if and only if $\det(\lambda - J(\theta))$ has a double zero at $\lambda = \lambda_0$ for

$\theta = 0$ (or π). Since $J(\theta)$ is selfadjoint, the order of the zero is the multiplicity of the eigenvector.

(iv) $\Leftrightarrow$ (v) A gap is closed if and only if $\Delta(\lambda) \mp 2$ has a double zero, which happens if and only if $\Delta'(\lambda_0) = 0$. □

Corollary 5.4.4. *At closed gaps, $\partial e_j/\partial\theta \neq 0$ and at open gaps, $\partial e_j/\partial\theta = 0$.*

Proof. By (5.4.12), at $\theta = 0$ or π, $\Delta'(e_j(\theta))e_j'(\theta)$ has a simple zero. Thus, either $\Delta'(e)$ or $e_j'(\theta)$ has a zero—not both. Thus, by (iv) of the above proposition, $e_j'(\theta) = 0$ if and only if one is at an open gap. □

Remarks. 1. If $\partial e_j/\partial\theta \neq 0$, because we order e_j, the continuation of e_j is $e_{j\pm 1}$.

2. In some applications, this is important since it implies that θ is an analytic function of E, and so, for example, Floquet solutions are analytic in E.

Recall that the measure, $d\nu$, in the spectral representation (5.3.35) has the form (5.3.34). The formula (5.4.12) lets us compute $d\nu$ in terms of Δ:

Theorem 5.4.5. *The measure $d\nu$ of* (5.3.34) *can be written*

$$d\nu(\lambda) = \frac{1}{p\pi}\,\frac{|\Delta'(\lambda)|}{\sqrt{4-\Delta^2(\lambda)}}\,d\lambda \tag{5.4.15}$$

$$= \frac{1}{p\pi}\left|\frac{d}{d\lambda}\arccos\left(\frac{\Delta(\lambda)}{2}\right)\right| d\lambda \tag{5.4.16}$$

Remark. Again, we see (via (5.4.16)) that $\nu(\mathfrak{e}_j) = 1/p$ since $\Delta(\lambda)/2$ runs from 1 to -1 or -1 to 1, and so arccos from 0 to π or π to 0.

Proof. (5.4.12) can be rewritten

$$\Delta'(\lambda)\left(\frac{d\theta}{d\lambda}\right)^{-1} = 2\sin\theta = 2\sqrt{1-\left(\tfrac{\Delta(\lambda)}{2}\right)^2}$$

so (5.4.15) follows from (5.3.34). (5.4.16) is a direct calculation of the derivative of arccos. □

Since (5.4.15) is explicit, we see that $\frac{d\nu}{d\lambda}$ is real analytic on $\mathfrak{e}$ with square root divergence at the edges:

Corollary 5.4.6. *The Radon–Nikodym derivative $\frac{d\nu}{d\lambda}$ of ν is real analytic on $\mathfrak{e}^{\text{int}}$ and obeys*

$$c_1\,\text{dist}(\lambda, \mathbb{R}\setminus\mathfrak{e})^{-1/2} \le \frac{d\nu}{d\lambda} \le c_2\,\text{dist}(\lambda, \mathbb{R}\setminus\mathfrak{e})^{-1/2} \tag{5.4.17}$$

Remarks. 1. In fact, $\text{dist}(\lambda, \mathbb{R}\setminus\mathfrak{e})^{-1/2}\frac{d\nu}{d\lambda}$ has nonzero limits as one approaches an open gap edge.

2. There is an "explicit" formula for $\frac{d\nu}{d\lambda}$ in Corollary 5.4.20 that immediately shows the bounds in (5.4.17) are exact.

Proof. Except for points in $\mathfrak{e}^{\rm int}$ where $\Delta = \pm 2$, this is obvious from (5.4.15). Such points occur at closed gaps, λ, where $\Delta'(\lambda_0)$ has a simple zero and $4 - \Delta^2(\lambda_0)$ a double zero, so $\Delta'/\sqrt{4-\Delta^2}$ is regular. □

(5.4.16) allows us to reinterpret $d\nu$ as a density of states (aka density of zeros). Let $J_{m;F}$ be the truncated transfer matrix associated to $\{a_n, b_n\}_{n=1}^m$ (actually, a_m does not enter) and let $J_{m;F}^{(p)}$ be the matrix with periodic boundary conditions (i.e., (5.3.8) with p replaced by m and $e^{i\theta} = 1$). The eigenvalues $\{\lambda_j^{(m)}\}_{j=1}^m$ of $J_{m;F}$ are the zeros of $P_m(z)$ by (1.2.31). We will let $\{\lambda_j^{p_i(m)}\}_{j=1}^m$ be the eigenvalues of $J_{m;F}^{(p)}$ (which may be degenerate, so we count multiplicities). Define the normalized counting measures

$$d\nu_m(\lambda) = \frac{1}{m}\sum_{j=1}^m \delta_{\lambda,\lambda_j^{(m)}} \tag{5.4.18}$$

$$d\nu_m^{(p)}(\lambda) = \frac{1}{m}\sum_{j=1}^m \delta_{\lambda,\lambda_j^{p_i(m)}} \tag{5.4.19}$$

Proposition 5.4.7. *Suppose $\{a_n, b_n\}_{n=-\infty}^\infty$ is periodic. Then as $m \to \infty$, the measures $d\nu_m$ and $d\nu_m^{(p)}$ converge weakly to the same measure, $d\nu$, called the* density of states *or* density of zeros.

Remarks. 1. We use the same symbol, $d\nu$, since we will prove shortly that it is the $d\nu$ defined in (5.3.34).

2. We will identify this limit as a potential theoretic equilibrium density in Theorem 5.5.17.

3. The same proof works in more general situations; see [399, Section 8.2].

Proof. Since $\|J_{m;F}\|$ and $\|J_{m;F}^{(p)}\|$ are uniformly bounded, the $d\nu_m$ and $d\nu_m^{(p)}$ are supported on a fixed interval $[-A, A]$, so it suffices to prove for all ℓ that $\int \lambda^\ell \, d\nu_m(\lambda)$ and $\int \lambda^\ell \, d\nu_m^{(p)}(\lambda)$ converge to a limit, and the limit is the same (for polynomials are dense in $C([-A, A])$). Note that

$$\int \lambda^\ell \, d\nu_m(\lambda) = \frac{1}{m}\,\mathrm{Tr}((J_{m;F})^\ell) \tag{5.4.20}$$

(and similarly for $d\nu_m^{(p)}$). It is easy to see that for $\ell < m$ and $\ell < j < m - \ell$, the jj matrix element of $(J_{m;F})^\ell$ and $(J_{m;F}^{(p)})^\ell$ are equal and are independent of m (for $m > \ell + j$) and periodic in j. From this, the existence and equality of the limits follow. □

Theorem 5.4.8. *The measure $d\nu$ of* (5.3.34) *and* (5.4.15) *is the density of states.*

Proof. Consider $J_{rp;F}^{(p)}$ where m is a multiple of p. As we have seen, its eigenvalues are connected with when $T_{rp}(\lambda)$ has eigenvalue 1. But $T_{rp}(\lambda) = T_p(\lambda)^r$ by periodicity of the a's and b's, so we want to know when $T_p(\lambda)$ has an eigenvalue, η, with $\eta^r = 1$, that is, $\eta = e^{2\pi j/r}$, $j = 0, 1, \dots, r-1$. Thus, the eigenvalues of $J^{(p)}_{rp;F}$ are precisely the solutions of

$$\Delta(\lambda) = 2\cos\left(\frac{2j\pi}{r}\right) \qquad j = 0, 1, \dots, r-1 \tag{5.4.21}$$

Except perhaps when $r = 0$ or $r/2$ (if r is even), these zeros are all simple but involve (except for those values of r) a doubling of j and $r - j$. The doubling cancels the 2 in $(2\pi)^{-1}$.

The normalized counting measure thus converges to

$$p^{-1}2(2\pi)^{-1}\left[\frac{d}{d\lambda}\arccos\left(\frac{\Delta(\lambda)}{2}\right)\right]d\lambda$$

which is (5.4.16). □

The following can be viewed as a whole-line analog of Theorem 2.15.1:

Theorem 5.4.9. *Let f be a continuous function on $\mathfrak{e} = \sigma(J)$, the spectrum of a full-line period p periodic Jacobi matrix, J. Let $f(J)_{nm}$ be the matrix elements of $f(J)$ in the standard basis. Then $f(J)_{nm}$ is periodic, that is,*

$$f(J)_{n+p\, m+p} = f(J)_{nm} \tag{5.4.22}$$

and

$$\frac{1}{p}\sum_{n=1}^{p} f(J)_{nn} = \int f(\lambda)\, d\nu(\lambda) \tag{5.4.23}$$

where ν is the density of states.

Proof. As usual, we need only prove this for $f(\lambda) = \lambda^\ell$, $\ell = 0, 1, 2, \dots$. As in the proof of Theorem 2.15.1, we have

$$\lim_{n\to\infty}\left|\frac{1}{n}\sum_{j=1}^{n}(J^\ell)_{jj} - \frac{1}{n}\operatorname{Tr}(J^\ell_{n;F})\right| \to 0$$

so by Proposition 5.4.7,

$$\lim_{n\to\infty}\frac{1}{n}\sum_{j=1}^{n}(J^\ell)_{jj} = \int \lambda^\ell\, d\nu(\lambda) \tag{5.4.24}$$

By (5.4.22),

$$\frac{1}{kp}\sum_{j=1}^{kp}(J^\ell)_{jj} = \text{RHS of (5.4.23) for } f(\lambda) = \lambda^\ell$$

proving (5.4.23). □

Next we turn to the *Lyapunov exponent*:

Theorem 5.4.10. *For* $\lambda \in \mathbb{C}$,

$$\lim_{n\to\infty} \frac{1}{n} \log \|T_n(\lambda)\| = \gamma(\lambda) \tag{5.4.25}$$

exists and is given by

$$\gamma(\lambda) = \frac{1}{p} \log\left[\left|\frac{\Delta(\lambda)}{2} + \sqrt{\left(\tfrac{\Delta(\lambda)}{2}\right)^2 - 1}\right|\right] \tag{5.4.26}$$

Remarks. 1. (5.4.26) requires one to specify which branch of $\sqrt{}$ is intended. We place branch cuts on $\mathfrak{e} \subset \mathbb{R} \subset \mathbb{C}$ and take the branch, which is $\frac{\Delta(\lambda)}{2} + O(\lambda^{-p})$ near $\lambda = \infty$. There is a discontinuity of $\sqrt{}$ across $\mathfrak{e}$, but since $|\ldots| = 1$ there and the two branches are complex conjugates, the function in (5.4.26) is continuous there.

2. We will place the existence of the limit in (5.4.25) into a more general framework in Theorem 5.5.17.

3. γ is called the *Lyapunov exponent.*

4. If $\Delta(\lambda) \in [-2, 2]$, the square root in (5.4.26) is pure imaginary and $|\ldots| = 1$. Thus, on e, $\gamma(\lambda) = 0$.

Proof. Since

$$T_{rp+j} = T_j(T_p)^r$$

on account of periodicity, and since $\{\|T_j\|, \|T_j^{-1}\|\}_{j=0}^{p-1}$ are bounded, it is easy to see that it suffices to establish the limit exists for $n = rp$ and to note that limit is just $\lim_{r\to\infty} \frac{1}{p} \log \|T_p(\lambda)^r\|^{1/r}$, which exists by the spectral radius formula. Thus, γ exists and

$$\gamma(\lambda) = \frac{1}{p} \log\big[\max\{|\eta| \mid \eta \text{ an eigenvalue of } T_p(\lambda)\}\big] \tag{5.4.27}$$

Thus, eigenvalues are the solutions of

$$\eta^2 - 2\Delta(\lambda)\eta + 1 = 0$$

so (with the branch of $\sqrt{}$ given in Remark 1 above)

$$\eta_\pm(\lambda) \equiv \frac{\Delta(\lambda)}{2} \pm \sqrt{\left(\tfrac{\Delta(\lambda)}{2}\right)^2 - 1} \tag{5.4.28}$$

$\eta_\pm$ are analytic in $\mathbb{C} \setminus \mathfrak{e}$ and nonvanishing, so

$$t(\lambda) \equiv |\eta_+(\lambda)| - |\eta_-(\lambda)|$$

is harmonic. $t \to \infty$ as $\lambda \to \infty$ (since $|\eta_+| = O(|\lambda|^p)$ and $|\eta_-| = O(|\lambda|^{-p})$ and $t(\lambda) \to 0$ as $\lambda \to \mathfrak{e}$. Thus, by the minimum principle, $t > 0$ on $\mathbb{C} \setminus \mathfrak{e}$, that is, $|\eta_+| > |\eta_-|$, so (5.4.27) is (5.4.26). □

Next we turn to a remarkable relation between the density of states, $d\nu$, and the Lyapunov exponent, γ. We first need a lemma:

Lemma 5.4.11. $\Delta'(\lambda)$ *has exactly one zero in each gap (including closed gaps) and no other zeros.*

Proof. We first prove there is at least one zero in each gap. If a gap is closed, $\Delta(\lambda)$ has a double zero at the location λ_0 of this closed gap so $\Delta'(\lambda_0) = 0$. In an open gap (λ_0, λ_1), we have $\Delta(\lambda_0) = \Delta(\lambda_1)$, so Δ' has a zero in (λ_0, λ_1) by Snell's theorem. Thus, each gap has at least one zero.

There are $p-1$ gaps (counting closed gaps) and Δ' is a polynomial of degree $p-1$, so this accounts for all the zeros: one per gap and no others. □

Theorem 5.4.12. *For any* λ *in* $\mathbb{C}$,

$$\gamma(\lambda) = -\frac{1}{p}\log(a_1 \dots a_p) + \int \log|\lambda - x|\, d\nu(x) \tag{5.4.29}$$

Remarks. 1. (5.4.29) is called the *Thouless formula.* We will provide a proof in a more general context in Theorem 5.5.17.

2. This formula also plays a role in the potential theoretic analysis; see Section 5.5 and its Notes.

Proof. Consider the function on $\mathbb{C}_+$,

$$g(\lambda) = \frac{1}{p}\log\left[\frac{\Delta(\lambda)}{2} + \sqrt{\left(\tfrac{\Delta(\lambda)}{2}\right)^2 - 1}\right] \tag{5.4.30}$$

Pick the branch of the square root, which, near $i\infty$, is $\frac{\Delta(\lambda)}{2} + O(\lambda^{-p})$ and the branch of log that, near $i\infty$, has $\log(\lambda^p) = p\log|\lambda| + \frac{ip\pi}{2}$. As usual, we put a branch cut of the square root on $\mathfrak{e}$ so the quantity in $[\dots]$ in (5.4.30) is analytic in $\mathbb{C}_+$. Since

$$\left(\frac{\Delta(\lambda)}{2} + \sqrt{\left(\tfrac{\Delta(\lambda)}{2}\right)^2 - 1}\right)\left(\frac{\Delta(\lambda)}{2} - \sqrt{\tfrac{\Delta(\lambda)}{2} - 1}\right) = 1$$

the expression in $[\dots]$ is nonvanishing. So $g(\lambda)$ is analytic in $\mathbb{C}_+$.

Then on $\mathbb{C}_+$,

$$g'(\lambda) = \frac{1}{p}\,\frac{\Delta'(\lambda)}{\Delta(\lambda) + \sqrt{\Delta'(\lambda)^2 - 4}}\left[1 + \frac{\Delta(\lambda)}{\sqrt{\Delta^2(\lambda)^2 - 4}}\right] \tag{5.4.31}$$

$$= \frac{1}{p}\,\frac{\Delta'(\lambda)}{\sqrt{\Delta(\lambda)^2 - 4}} \tag{5.4.32}$$

since $\frac{d}{dx}(x + \sqrt{x^2 - 4}) = 1 + \frac{x}{\sqrt{x^2-4}}$.

$g'(\lambda)$ is thus analytic in $\mathbb{C}_+$ with boundary values on $\mathbb{R} \setminus \{\lambda \mid \Delta(\lambda) = \pm 2\}$. Δ' is real on $\mathbb{R}$, positive above the top band, and so also on the top band. By the lemma, it alternates sign from one band to the next.

$\sqrt{\Delta^2(\lambda) - 4}$ is real in the gaps and above and below the bands. Every time it moves from above a zero of $\Delta^2 - 4$ to below, its argument increases by $\frac{1}{2}\pi$, so $(\sqrt{\Delta^2(\lambda) - 4})^{-1}$ is pure imaginary on each band with negative imaginary part on the top band, positive on the next, and so on. Taking into account that Δ' also alternates sign, we see, by (5.4.15),

$$\operatorname{Im} g'(\lambda + i0) \le 0 \qquad \lambda \in \mathfrak{e} \tag{5.4.33}$$

$$\frac{1}{\pi}\operatorname{Im} g'(\lambda + i0)\, d\lambda = -d\nu(\lambda) \tag{5.4.34}$$

Near $\lambda = \infty$, $g'(\lambda) \sim \lambda^{-1}$, so $\operatorname{Im} g'(\lambda) < 0$ near $\lambda = \infty$ in $\mathbb{C}_+$. Since $\operatorname{Im} g$ is harmonic, $\operatorname{Im} g' \le 0$ on all of $\mathbb{C}_+$, and thus, $-\operatorname{Im} g'$ is an m-function. So, by (5.4.34),

$$g'(\lambda) = -\int \frac{d\nu(x)}{x - \lambda} \tag{5.4.35}$$

Therefore,

$$g(\lambda) = c + \int \log(x - \lambda)\, d\nu(x) \tag{5.4.36}$$

for some constant c, and so

$$\gamma(\lambda) = \operatorname{Re} g(\lambda) = \operatorname{Re} c + \int \log|\lambda - x|\, d\nu(x) \tag{5.4.37}$$

Since $\log|\lambda - x| = \log|\lambda| + \log|1 - \frac{x}{\lambda}|$, near $\lambda = i\infty$,

$$\text{RHS of (5.4.37)} = \log|\lambda| + \operatorname{Re} c + O\biggl(\frac{1}{|\lambda|}\biggr)$$

By (5.4.26) and

$$\Delta(\lambda) = (a_1 \dots a_p)^{-1} \lambda^p + \text{lower order} \tag{5.4.38}$$

we have

$$\gamma(\lambda) = \frac{1}{p}\, [p \log|\lambda| + \log|a_1 \dots a_p|^{-1} + O(|\lambda|^{-1})]$$

which implies (5.4.29). □

Next we turn to considering the connection of Floquet solutions and the spectral theorist's Green's function, aka matrix elements of the resolvent. Our first two results hold for any bounded two-sided Jacobi matrices.

Theorem 5.4.13. *Let J be a two-sided bounded Jacobi matrix. For any $\lambda \in \mathbb{C}_+$, there are solutions $u_n^{\pm}(\lambda)$ of (5.4.1), which are ℓ^2 at $\pm\infty$ unique up to constants. Their Wronskian,*

$$W(\lambda) = a_n(u_{n+1}^{+}(\lambda) u_n^{-}(\lambda) - u_n^{+}(\lambda) u_{n+1}^{-}(\lambda)) \tag{5.4.39}$$

is n-independent, and for $n \ge m$,

$$\langle \delta_n, (J - \lambda)^{-1} \delta_m \rangle = \frac{u_n^{+}(\lambda) u_m^{-}(\lambda)}{W(\lambda)} \tag{5.4.40}$$

Moreover, if $p_n(\lambda)$ are the orthonormal polynomials associated to J_0^+, we have for $n \ge m$,

$$\langle \delta_n, (J_0^+ - \lambda)^{-1} \delta_m \rangle = \frac{u_n^{+}(\lambda) p_{m-1}(\lambda)}{\widetilde{W}(\lambda)} \tag{5.4.41}$$

where

$$\widetilde{W}(\lambda) = a_n(u_{n+1}^{+}(\lambda) p_{n-1}(\lambda) - u_n^{+}(\lambda) p_n(\lambda)) \tag{5.4.42}$$

which is n-independent.

In particular, if

$$m(\lambda, J_n^+) = \langle \delta_{n+1}, (J_n^+ - \lambda)^{-1}\delta_{n+1}\rangle \tag{5.4.43}$$

$$m(\lambda, J_n^-) = \langle \delta_n, (J_n^- - \lambda)^{-1}\delta_n\rangle \tag{5.4.44}$$

then

$$G_{nn}(\lambda) = -\frac{1}{a_n^2 m(\lambda, J_n^+) - m(\lambda, J_n^-)^{-1}} \tag{5.4.45}$$

Remarks. 1. We will normally normalize $u^\pm$ by requiring $u_{n=0}^\pm = 1$. Normalization changes drop out of (5.4.40).

2. Since J and J_0^+ are symmetric and real, $\langle \delta_n, (J-\lambda)^{-1}\delta_m\rangle = \langle \delta_m, (J-\lambda)^{-1}\delta_n\rangle$, so (5.4.40)/(5.4.41) determine the full resolvent.

3. (5.4.40) is usually called the *Green's function* by spectral theorists.

4. In (5.4.41), if we take $n = m = 1$ and note that (since $p_0 = 1$, $p_{-1} = 0$)

$$\widetilde{W}(\lambda) = -a_0 u_0^+$$

we have

$$\langle \delta_1, (J_0^+ - \lambda)^{-1}\delta_1\rangle = -\frac{u_1^+(\lambda)}{a_0 u_0^+(\lambda)} \tag{5.4.46}$$

which is essentially (3.2.33) for $n = 1$. So (5.4.41) generalizes (3.2.33).

5. By (3.2.23) and (3.2.25), u_n^+ normalized by $u_{n=0}^+ = 1$ has the form

$$u_n^+ = -q_{n-1}(\lambda) - m(\lambda)p_{n-1}(\lambda) \tag{5.4.47}$$

for $n \geq 1$.

6. (5.4.45) has a disconcerting asymmetry in J^+ and J^-. There are two ways of restoring the symmetry. One is to note (by the symmetry or by mimicking the proof)

$$G_{nn}(\lambda) = -\frac{1}{a_{n-1}^2 m(\lambda, J_{n-1}^-) - m(\lambda, J_{n-1}^+)^{-1}} \tag{5.4.48}$$

The other is to use coefficient stripping

$$-m(\lambda, J_n^-)^{-1} = z - b_n + a_{n-1}^2 m(\lambda, J_{n-1}^-) \tag{5.4.49}$$

to get from (5.4.45) that

$$G_{nn}(\lambda) = -\frac{1}{z - b_n + a_{n-1}^2 m(\lambda, J_{n-1}^-) + a_n^2 m(\lambda, J_n^+)} \tag{5.4.50}$$

Proof. By using Theorem 3.2.1 on $J_0^\pm$, we find solutions $u_n^\pm$ for $\pm n \geq 1$, which are ℓ^2 at $\pm\infty$ and unique up to a constant. But any solution on $(1, \infty)$ can be uniquely extended to $(-\infty, \infty)$, so we get $u_n^\pm$. Independence in n of (5.4.39) follows by the same argument, using determinants of transfer matrices, that led to (3.2.21).

Define

$$G_{mn}(\lambda) = \langle \delta_m, (J - \lambda)^{-1}\delta_n\rangle \tag{5.4.51}$$

Fix m and note that

$$\{(J-\lambda)[G_{m\cdot}]\}_n = \delta_{mn} \tag{5.4.52}$$

and $G_{m\cdot}$ is ℓ^2 at $+\infty$. So for $n \ge m$,

$$G_{mn}(\lambda) = c_m^- u_n^+ \tag{5.4.53}$$

By the symmetry in m and n and looking at $-\infty$, for $m \le n$,

$$G_{mn}(\lambda) = c_n^+ u_m^- \tag{5.4.54}$$

It follows that

$$G_{mn}(\lambda) = c\, u_m^- u_n^+ \tag{5.4.55}$$

Evaluating (5.4.52) if $n = m$ shows

$$c[a_{m-1}u_{m-1}^- u_m^+ + (b_m - \lambda)u_m^- u_m^+ + a_m u_m^- u_{m+1}^+] = 1 \tag{5.4.56}$$

Since

$$(b_m - \lambda)u_m^+ + a_m u_{m+1}^+ = -a_{m-1}u_{m-1}^+$$

this says $cW(\lambda) = 1$, which proves (5.4.40).

If we note that any $\{v_j\}_{j=1}^\infty$ obeying $[(J_0^+ - \lambda)v]_j = 0$, $j = 1, 2, \dots, n$, has $v_j = cp_{j-1}$ for $j = 1, 2, \dots, n+1$ (see Proposition 1.3.1), the proof of (5.4.41) is identical to the proof of (5.4.40).

To prove (5.4.45), we note that, by (5.4.46),

$$m(\lambda, J_n^+) = -\frac{u_{n+1}^+(\lambda)}{a_n u_n^+(\lambda)} \tag{5.4.57}$$

$$m(\lambda, J_n^-) = -\frac{u_n^-(\lambda)}{a_n u_{n+1}^-(\lambda)} \tag{5.4.58}$$

and, by (5.4.40),

$$G_{nn}(\lambda)^{-1} = a_n\left[\frac{u_{n+1}^+(\lambda)}{u_n^+(\lambda)} - \frac{u_{n+1}^-(\lambda)}{u_n^-(\lambda)}\right] \tag{5.4.59}$$

from which (5.4.45) is immediate. □

Define $G_{mn}^D(\lambda)$ to be the resolvent of $J_{-1}^- \oplus \infty \oplus J_0^+$ on $\ell^2(-\infty, -1) \oplus \ell^2(\{0\}) \oplus \ell^2(1, \infty)$ (i.e., set $b_0 = a_{-1} = a_0 = 0$). Thus,

$$G_{mn}^D(\lambda) = \begin{cases} \langle \delta_m, (J_0^+ - \lambda)^{-1}\delta_n\rangle & \text{if } m, n \ge 1 \\ \langle \delta_m, (J_{-1}^- - \lambda)^{-1}\delta_n\rangle & \text{if } m, n \le -1 \\ 0 & \text{otherwise} \end{cases} \tag{5.4.60}$$

Theorem 5.4.14. *For any whole-line Jacobi matrix and all n, m and $\lambda \in \mathbb{C}_+$, $G_{00}(\lambda) \ne 0$ and*

$$G_{nm}^D(\lambda) = G_{nm}(\lambda) - G_{0n}(\lambda)G_{0m}(\lambda)[G_{00}(\lambda)]^{-1} \tag{5.4.61}$$

Remarks. 1. If one considers $J(\alpha)$, which is J with b_0 replaced by $b_0 + \alpha$, as $\alpha \to \infty$, then as $\alpha \to \infty$, $G_{nm}(\lambda; J(\alpha)) \to G^D_{nm}(\lambda)$. (5.4.61) can be viewed in terms of rank one perturbations at infinite coupling; see [164].

2. Since $G^D_{nm} = 0$ for $n \le 0 \le m$, we see that

$$n \le k \le m \Rightarrow G_{nm}(\lambda)G_{kk}(\lambda) = G_{nk}(\lambda)G_{km}(\lambda) \tag{5.4.62}$$

which follows directly from (5.4.40).

3. If $(a, b) \subset \mathbb{R} \setminus \sigma(J)$, (5.4.61) extends to all λ in (a, b) with $G_{00}(\lambda) \neq 0$. Points with $G_{00}(\lambda) = 0$ are (as we will see below) eigenvalues of $J^-_{-1} \oplus J^+_0$, and so poles of $G^D_{nm}(\lambda)$ for suitable n and m.

Proof. We have

$$\operatorname{Im}\langle\delta_0, (J - \lambda)^{-1}\delta_0\rangle \ge |\operatorname{Im}\lambda|^{-1}$$

so $G_{00}(\lambda) \neq 0$. Thus, $u^+_0 \neq 0 \neq u^-_0$. Let $n \ge m \ge 0$. Then, with $u^\pm$ normalized by $u^\pm_0 = 1$,

$$\begin{aligned} \text{RHS of (5.4.61)} &= W^{-1}(u^+_n u^-_m - u^+_n u^+_m) \\ &= W^{-1}(u^+_n(u^-_m - u^+_m)) \end{aligned} \tag{5.4.63}$$

$u^-_m - u^+_m$ is a solution of (5.4.1) vanishing at $m = 0$, so a multiple of p_{m-1}. Since the Wronskian of u^+ and u^- is the Wronskian of u^+ and $u^- - u^+$, the right-hand side of (5.4.62) $= G^D_{nm}$ by (5.4.41). □

We now specialize to the periodic case. We define the *integrated density of states*, $N(\lambda)$, for $\lambda \in \mathbb{R}$ by

$$N(\lambda) = \int_{-\infty}^{\lambda} d\nu(x) \tag{5.4.64}$$

We also define

$$k(\lambda) = ig(\lambda) \tag{5.4.65}$$

where g is given by (5.4.30). Thus, $e^{-ik(\lambda)}$ is a pth root of the eigenvalues, η, of $T_p(\lambda)$ of larger magnitude. So, in particular,

$$\gamma(\lambda)^{1/p} = |e^{-ik(\lambda)}| \tag{5.4.66}$$

We state a result about zeros of G_{00} in the next theorem but defer its proof slightly:

Theorem 5.4.15. *Let J be a two-sided periodic Jacobi matrix. Then $G_{00}(\lambda)$ vanishes precisely once in each gap at points we label $\{\mu_j\}_{j=1}^{p-1}$ and nowhere else. For each $\lambda \in \mathbb{C} \setminus \mathfrak{e} \cup \{\mu_j\}_{j=1}^{p-1}$, there exist solutions $u^\pm_n(\lambda)$ of (5.4.1), which are ℓ^2 at $\pm\infty$ and obey*

$$u^\pm_0(\lambda) = 1 \tag{5.4.67}$$

They are analytic on $\mathbb{C} \setminus \mathfrak{e} \cup \{\mu_j\}_{j=1}^{p-1}$. In addition, $u^\pm_n(\lambda)$ have limits as $\lambda \to \mathfrak{e}^{\text{int}}$ from above and below $u^\pm_n(\lambda \pm i0)$. Moreover,

$$\text{(i)}\quad u^\pm_n(\bar\lambda) = \overline{u^\pm_n(\lambda)} \qquad \text{all } \lambda \in \mathbb{C} \setminus \mathfrak{e} \cup \{\mu_j\}_{j=1}^{p-1} \tag{5.4.68}$$

$$\text{(ii)}\quad u^+_n(\lambda + i0) = \overline{u^-_n(\lambda + i0)} \qquad \text{all } \lambda \in \mathfrak{e} \tag{5.4.69}$$

(iii) $u_n^+(\lambda + i0) = u_n^-(\lambda - i0)$ *all* $\lambda \in \mathfrak{e}$ (5.4.70)

(iv) $k(\lambda) = \pi[N(\lambda) - 1]$ *for* $\lambda \in \mathfrak{e}$ (5.4.71)

We have

$$u_n^\pm = e^{\pm ink(\lambda)} v_n^\pm \tag{5.4.72}$$

with $v_n^\pm(\lambda)$ *periodic, that is,*

$$v_{n+p}^\pm = v_n^\pm \tag{5.4.73}$$

Remarks. 1. Except for a different normalization, $\{u^\pm(\lambda + i0) \mid \lambda \in \mathfrak{e}\}$ are the plane wave solutions discussed in (5.3.30).

2. For $\lambda \in \mathfrak{e}$, (5.4.72) shows that $u_n^\pm(\lambda + i0)$ are almost periodic in n (unless $N(\lambda)$ is rational, in which case they are periodic).

3. There is a slight misstatement in the theorem. It can happen that $G_{00}(\lambda)$ has no zero in some gap—that is the case where $G_{00}(\lambda) \to 0$ so one approaches the edge of a gap. This point will be explained below.

Proof. The existence of $u^\pm$ is Theorem 5.4.13 supplemented by the discussion of G on gaps in $\sigma(J)$. (i) is immediate from $\overline{T_p(\lambda)} = T_p(\bar\lambda)$. (ii) follows by noting that since $\lambda \in e$ has $T_p(e)$ real, if $T_p s = \eta s$, then $T_p \bar s = \bar\eta \bar s = \eta^{-1} \bar s$ since $|\eta| = 1$. (iii) follows from (i) and (ii).

(iv) follows from (5.4.35), which implies on $\mathbb{R}$

$$\frac{1}{\pi} \operatorname{Im} g'(\lambda) = -\frac{d\nu}{d\lambda} \tag{5.4.74}$$

Since $g(\lambda) \sim \log(\lambda)$ near $\lambda = i\infty$, $\operatorname{Im} g(\lambda) = \pi$ for $\lambda \in \mathbb{R}$ near $-\infty$, so

$$\operatorname{Im} g(\lambda) = \pi[1 - N(\lambda)]$$

Since $\operatorname{Re} g(\lambda) = 0$ on e (i.e., $|\eta| = 1$ there), we find (5.4.71). Since $e^{\pm ik(\lambda)p} = \eta^{\mp 1}$ and $u_{n+p}^\pm = (\eta^\mp u_n^\pm$, we get (5.4.72)/(5.4.73). □

We next turn to when $G_{00}(\lambda) = 0$.

Theorem 5.4.16. *Let J be a two-sided periodic Jacobi matrix. Then the $p-1$ zeros of $p_{p-1}(\lambda)$ lie one in each gap. At each such zero, λ_0, exactly one of the following holds:*

(i) *it is an eigenvalue of J_0^+, in which case λ_0 is in the interior of a gap;*
(ii) *it is an eigenvalue of J_{-1}^-, in which case λ_0 is in the interior of a gap;*
(iii) *λ_0 is at a gap edge, in which case there is a periodic or antiperiodic solution of $Ju = \lambda_0 u$, which vanishes at $n = 0$.*

The zeros of $G_{00}(\lambda)$ in $\mathbb{C} \setminus \mathfrak{e}$ are precisely the points in $\mathbb{C} \setminus \mathfrak{e}$ where $p_{p-1}(\lambda) = 0$.

Remarks. 1. There are many proofs that p_{p-1} has one zero per gap besides the one we will give here.

2. If p_{p-1} has a zero at a boundary point of an open gap, we say that J_0^+ has a *resonance* at λ_0.

3. In a sense we will make precise later (see Theorem 5.4.19), resonances are also zeros of G_{00}; we will prove that if λ_0 is the edge of an open gap, then

$$\lim_{\substack{\lambda\to\lambda_0 \\ \lambda\notin\mathfrak{e}}} G_{00}(\lambda) = 0$$

if λ_0 is a resonance and ∞ if it is not.

Proof. We first analyze the zeros of p_{p-1}. We use the same device used in the proof of (iv) in Theorem 5.3.4. Define $J_0^+(\mu)$, $0 \le \mu \le 1$, to be the half-line periodic Jacobi matrix with Jacobi parameters

$$a_n(\mu) = (1-\mu) + \mu a_n \tag{5.4.75}$$

$$b_n(\mu) = \mu b_n \tag{5.4.76}$$

which interpolates between the free Jacobi matrix and J_0^+. Let $p_{p-1}(\lambda;\mu)$ be the associated orthogonal polynomials.

Since p_{p-1} is a multiple of an orthogonal polynomial, its zeros are all real. We will show in a moment that it cannot have zeros in any $\mathfrak{e}_j^{\text{int}}$.

At $\mu = 0$, p_{p-1} is a Chebyshev polynomial of the second kind, that is,

$$p_{p-1}(2\cos\theta, \mu = 0) = c\,\frac{\sin(p\theta)}{\sin\theta}$$

which vanishes precisely at the points $2\cos(j\pi/p)$, $j = 1, 2, \dots, p-1$ with each zero simple. These are the locations of the closed gaps of $J_0^+(\mu = 0)$, which, viewed as a period p Jacobi matrix, has all gaps closed.

As μ varies, the zeros move continuously, must stay on $\mathbb{R}$ and cannot go into the interiors of bands. Thus, they stay trapped, one in each gap. This concludes the proof of the first statement in the theorem.

By (5.4.2),

$$T_p(\lambda)\begin{pmatrix}1\\0\end{pmatrix} = \begin{pmatrix}p_p(\lambda)\\ a_p p_{p-1}(\lambda)\end{pmatrix}$$

so

$$p_{p-1}(\lambda_0) = 0 \Leftrightarrow \begin{pmatrix}1\\0\end{pmatrix} \text{ is an eigenvector of } T_p(\lambda_0) \tag{5.4.77}$$

First of all, this means $\lambda_0 \notin \mathfrak{e}_j^{\text{int}}$, for there $T_p(\lambda_0)$ has eigenvalues $e^{\pm i\theta}$ $(\theta \in (0,\pi))$, and so linearly independent nonreal eigenvectors.

If $\binom{1}{0}$ is an eigenvector, let η be the eigenvalue. If $\eta = \pm 1$, then $\Delta(\lambda_0) = \pm 2$, and we are at a band edge. If $|\eta| < 1$, $T_n(\lambda)\binom{1}{0}$ defines a solution of (5.4.1) that goes to zero like $|\eta|^{n/p}$ as $n \to +\infty$, and so is an eigenvector of J_0^+. Similarly, if $|\eta| > 1$, $T_n(\lambda)\binom{1}{0}$ defines a solution decaying as $|\eta|^{-|n|/p}$ as $n \to -\infty$, and so is an eigenvector of J_{-1}^-. This proves the second assertion of the theorem.

Finally, $G_{00}(\lambda_0) = 0$ for $\lambda_0 \in \mathbb{C}\setminus\mathfrak{e}$ if and only if either $\tilde{u}_0^+ = 0$ or $\tilde{u}_0^- = 0$ where $\tilde{u}^\pm$ are normalized by $\tilde{u}_1^\pm = 1$. As we have seen, that happens if and only if $\binom{1}{0}$ is an eigenvector with eigenvalue $|\eta| < 1$ or $|\eta| > 1$, and so is a zero of $p_{p-1}(\lambda)$. $\square$

Next we turn to the significance of (5.4.69):

Theorem 5.4.17. *For any* $\lambda \in \mathfrak{e}^{\text{int}}$ *and any* n, $\lim_{\varepsilon\downarrow 0} G_{nn}(\lambda + i\varepsilon) \equiv G_{nn}(\lambda + i0)$ *exists and*

$$\text{Re}[G_{nn}(\lambda + i0)] = 0 \tag{5.4.78}$$

Remarks. 1. For reasons we discuss in the Notes, either (5.4.69) or (5.4.78) is described by saying that J is *reflectionless*.

2. Reflectionless Jacobi matrices will be a major theme in Sections 7.4 and 7.6.

Proof. By (5.4.69),

$$W = a_0(u_1^+ u_0^- - u_0^+ u_1^-) = -\overline{W}$$

so W is pure imaginary. Thus, using (5.4.69) again and (5.4.40),

$$G_{nn} = \frac{|u_n^+|^2}{W} \in i\mathbb{R} \tag{5.4.79}$$

□

We have just seen that (5.4.69) implies (5.4.78). Interestingly enough, the converse holds:

Theorem 5.4.18 (Gesztesy–Krishna–Teschl [163]; Sodin–Yuditskii [413]). *Suppose* J *is a two-sided Jacobi matrix, and that for some* $\lambda_0 \in \mathbb{R}$, *we have*

(a) $\lim_{\varepsilon\downarrow 0} u_n^\pm(\lambda_0 + i\varepsilon) = u_n^\pm$ *exists for all* n

(b) $w(\lambda_0 + i0) \equiv a_n(u_{n+1}^+ u_n^- - u_n^+ u_{n+1}^-) \neq 0$

(c) $\text{Re}\, G_{nn}(\lambda_0 + i0) = 0$ *for* $n = 0, -1, 1$

(d) $u_0^\pm \neq 0$, $u_1^\pm \neq 0$, $u_{-1}^\pm \neq 0$, $b_0 \neq \lambda_0$

Then

$$u_n^- = \overline{u_n^+} \tag{5.4.80}$$

Proof. Since $u_0^\pm \neq 0$, we can normalize so $u_0^\pm = 1$. Then (c) is equivalent to

$$\text{Re}\, w = 0 \qquad \text{Im}(u_n^+ u_n^-) = 0 \quad \text{for } n = \pm 1 \tag{5.4.81}$$

Define $v_1^\pm = a_1 u_1^\pm$, $v_{-1}^\pm = a_0 u_1^\pm$. Then (5.4.81) plus (5.4.1) implies

$$\text{Re}\, v_1^+ = \text{Re}\, v_1^- \qquad \text{Im}(v_{\pm 1}^+ v_{\pm 1}^-) = 0 \tag{5.4.82}$$

$$v_\pm^+ + v_{-1}^+ = v_1^- + v_{-1}^- = \lambda_0 - b_0 \tag{5.4.83}$$

Writing $v_j^\pm = |v_j^\pm| e^{i\varphi_j^\pm}$, the second equation in (5.4.82) and $u_1^\pm \neq 0$ $u_{-1}^\pm \neq 0$ implies

$$\varphi_1^- = -\varphi_1^+ \qquad \varphi_{-1}^- = -\varphi_{-1}^+ \tag{5.4.84}$$

By (5.4.83) and $b_0 - \lambda_0 \neq 0$, one of v_1^+ or v_1^- has nonzero real part, in which case (5.4.84) and the first equation in (5.4.82) implies either $v_1^- = \overline{v_1^+}$ or $v_{-1} = \overline{v_1^-}$. In either case, this plus $\overline{u_0^-} = \overline{u_0^+}$ implies (5.4.80). □

One consequence of the fact that G_{nn} is purely imaginary is a remarkable explicit formula of Craig [95]:

Theorem 5.4.19 (Craig [95]). *Suppose*

$$\alpha_1 < \beta_1 < \alpha_2 < \beta_2 < \cdots < \alpha_{\ell+1} < \beta_{\ell+1}$$

are distinct real numbers. Suppose that $G(z)$ *is analytic on* $\mathbb{C} \setminus \cup_{j=1}^{\ell+1} [\alpha_j, \beta_j]$ *with*

(a) $\operatorname{Im} G(z) > 0$ *for* $\operatorname{Im} z > 0$

(b)

$$G(\bar{z}) = \overline{G(z)} \tag{5.4.85}$$

(c) *For a.e.* $x \in \cup_{j=1}^{\ell+1} (\alpha_j, \beta_j)$,

$$\lim_{\varepsilon \downarrow 0} \frac{\operatorname{Re} G(x + i\varepsilon)}{\operatorname{Im} G(x + i\varepsilon)} = 0 \tag{5.4.86}$$

(d) *Near* ∞,

$$G(z) = -\frac{1}{z} + o\left(\frac{1}{z}\right) \tag{5.4.87}$$

Then there exist $x_j \in [\beta_j, \alpha_{j+1}]$ *for* $j = 1, 2, \dots, \ell$ *so that*

$$G(z) = -\prod_{j=1}^{\ell} (z - x_j) \left[\prod_{j=1}^{\ell+1} (z - \alpha_j)(z - \beta_j) \right]^{-1/2} \tag{5.4.88}$$

where the branch of square root, which is $O(z^{\ell+1})$ *near* ∞, *is taken.*

The only zeros of G *on* $\mathbb{C} \setminus \cup_{j=1}^{\ell+1} [\alpha_j, \beta_j]$ *are at those* x_j *in* (β_j, α_{j+1}). *If* $x_j = \beta_j$ *or* α_{j+1}, *then* $G(x) \to 0$ *as* $x \to x_j$ *from* (β_j, α_{j+1}). *If* $x_j \in (\beta_j, \alpha_{j+1})$, *then* $G(x) \to \infty$ *as* $x \to x_j$ *in* (β_j, α_{j+1}).

Remarks. 1. We emphasize that this theorem is not specific to the periodic case—the intervals are arbitrary disjoint closed intervals. We will call the set $\mathfrak{e} = \cup_{j=1}^{\ell+1} [\alpha_j, \beta_j]$ a finite gap set, or sometimes an ℓ-gap set. In Section 5.13, we will see when such intervals arise as the intervals associated to an almost periodic problem.

2. In the periodic case, only open gaps contribute. In fact, if we added both the edges and zero for a closed gap, they would cancel in (5.4.88) in any event.

Proof. By the Herglotz representation theorem (Theorem 2.3.6), there is a measure $d\eta$ on $\cup_{j=1}^{\ell+1} [\alpha_j, \beta_j]$ so that

$$G(z) = \int \frac{d\eta(x)}{x - z} \tag{5.4.89}$$

In particular,

$$G(z) > 0 \text{ on } (-\infty, \alpha_1) \qquad G(z) < 0 \text{ on } (\beta_{\ell+1}, \infty) \tag{5.4.90}$$

On any gap (β_j, α_{j+1}),

$$G'(z) = \int \frac{d\eta(x)}{(x - z)^2} > 0 \tag{5.4.91}$$

so G is strictly monotone. In particular, there is at most a single zero in each gap, say at x_j. If there is no zero in (β_j, α_{j+1}), we set $x_j = \beta_j$ if $G(x) > 0$ on (β_j, α_{j+1}) and $x_j = \alpha_{j+1}$ if $G(x) < 0$ on (β_j, α_{j+1}).

Define for $z \in \mathbb{C}_+$,

$$H(z) = \log(G(z)) \tag{5.4.92}$$

with the branch of log picked so that near ∞,

$$H(re^{i\theta}) = -\log(r) + i(\pi - \theta) + O(r^{-1}) \tag{5.4.93}$$

H is a Herglotz function since $\arg G \in (0, \pi)$ on $\mathbb{C}_+$.

For a.e. $x \in \mathbb{R}$ by definition of x_j, by (5.4.90) and (5.4.86),

$$\operatorname{Im} H(x + i0) = \begin{cases} \dfrac{\pi}{2} \text{ on } \cup_{j=1}^{\ell+1} (\alpha_j, \beta_j) \\ 0 \text{ on } (-\infty, \alpha_1) \\ \pi \text{ on } (\beta_{\ell+1}, \infty) \\ 0 \text{ on each } (x_j, \alpha_{j+1}) \\ \pi \text{ on each } (\beta_j, x_j) \end{cases} \tag{5.4.94}$$

Let $\widetilde{H}(z)$ be defined on $\mathbb{C}_+$ by

$$\widetilde{H}(z) = \log[\text{RHS of (5.4.88)}]$$

with the branch chosen so

$$\widetilde{H}(re^{i\theta}) = -\log(r) + i(\pi - \theta) + O(r^{-1}) \tag{5.4.95}$$

It is easy to see that $\operatorname{Im} \widetilde{H}$ also obeys (5.4.94). It follows by the general Herglotz representation theorem discussed in the Notes to Section 2.3 that for some A, B,

$$H(z) = \widetilde{H}(z) + Az + B$$

and (5.4.92)/(5.4.94) then imply $A = B = 0$.

The assertions about zeros follow from the definition of x_j and about the behavior of G as $x \downarrow \beta_j$ or $x \uparrow \alpha_{j+1}$ from the explicit form of G. $\square$

As a first consequence, we get an "explicit" formula for the density of states.

Corollary 5.4.20. *Let $d\nu$ be a density of states for a periodic Jacobi matrix, J, whose essential spectrum is $\cup_{j=1}^{\ell+1}[\alpha_j, \beta_j]$ with $\beta_j < \alpha_{j+1}$ $(j = 1, 2, \dots, \ell)$. Then for suitable $x_j \in (\beta_j, \alpha_{j+1})$, we have*

$$d\nu(x) = \frac{1}{\pi} \prod_{j=1}^{\ell} |x - x_j| \left[\prod_{j=1}^{\ell+1} \frac{1}{|x - \alpha_j|\,|x - \beta_j|} \right]^{1/2} \chi_{\cup_{j=1}^{\ell+1}[\alpha_j, \beta_j]}(x)\, dx \tag{5.4.96}$$

Proof. By the theorem, we need only prove that, in this case, x_j is in the interior of the jth gap. In a gap, $\frac{\partial \gamma}{\partial x} = -G(x)$ so, since $\gamma = 0$ at both ends,

$$\int_{\beta_j}^{\alpha_{j+1}} G(x)\, dx = 0 \tag{5.4.97}$$

which implies G does not have a definite sign in (β_j, α_{j+1}) and so it must have a zero. □

Remarks. 1. Basically, the x_j are determined by (5.4.97). That there is a solution follows from the existence of a density of states. Uniqueness as well as a direct proof of existence of solutions to (5.4.97) will be proven in Proposition 5.5.21.

2. This corollary also follows from (5.4.15). The x_j are precisely the zeros of $\Delta'(\lambda)$; see Lemma 5.4.11.

Another consequence of Theorem 5.4.19 is

Theorem 5.4.21 (Borg–Hochstadt Theorem). *Let J be a periodic Jacobi matrix all of whose gaps are closed. Then for some α and β, $a_n \equiv \alpha$ and $b_n \equiv \beta$.*

Remarks. 1. Periodicity is not used in this proof—only that it is reflectionless, that is, $\operatorname{Re} G_{nn}(\lambda) = 0$ on $\sigma(J)$, which is assumed to be an interval.

2. There are many other proofs of this theorem; see Section 11.14 of [400] for the OPUC analog. Also see Corollary 5.13.9 later.

Proof. Since all gaps are closed, $\sigma(J) = [\gamma, \delta]$ for some γ, δ. By replacing J by $\kappa J + \lambda$ for suitable κ, λ, we can arrange that $\sigma(J) = [-2, 2]$, which we do henceforth.

By Theorem 5.4.19 (with $\ell = 0$), for each n,

$$G_{nn}(\lambda) = -(\lambda^2 - 4)^{-1/2} \tag{5.4.98}$$

If $J^{(0)}$ is the Jacobi matrix with $a_n \equiv 1$, $b_n \equiv 0$, we see

$$\langle \delta_n, (J - \lambda)^{-1} \delta_n \rangle = \langle \delta_n, (J^{(0)} - \lambda)^{-1} \delta_n \rangle \tag{5.4.99}$$

for all n and all $\lambda \in \mathbb{C} \setminus [-2, 2]$. Looking at the Taylor series about $\lambda = \infty$, we see for all n and $\ell = 0, 1, 2, \dots$ that

$$\langle \delta_n, J^\ell \delta_n \rangle = \langle \delta_n, (J^{(0)})^\ell \delta_n \rangle \tag{5.4.100}$$

Taking $\ell = 1$, we find for all n,

$$b_n = 0 \tag{5.4.101}$$

Then, using $\ell = 2$, $\langle \delta_n, J^2 \delta_n \rangle = \|J\delta_n\|^2$ and $J\delta_n = a_{n-1}\delta_{n-1} + a_n \delta_{n+1}$, we find

$$a_n^2 + a_{n-1}^2 = 2 \tag{5.4.102}$$

which, given $a_n > 0$, implies

$$a_{2n} = a_0 \qquad a_{2n+1} = a_1 \tag{5.4.103}$$

Now take $\ell = 4$, $n = 0$ in (5.4.100) and use (5.4.103) plus $\langle \delta_0, J^4 \delta_0 \rangle = \|J^2 \delta_0\|^2$ and $J^2 \delta_0 = a_0 a_1 (\delta_{-2} + \delta_2) + (a_0^2 + a_1^2)\delta_0$ to find

$$2(a_0 a_1)^2 + (a_0^2 + a_1^2)^2 = 6 \tag{5.4.104}$$

Using (5.4.102), we see $(a_0 a_1)^2 = 1$, so

$$(a_0 - a_1)^2 = a_0^2 + a_1^2 - 2a_0 a_1 = 0$$

and thus, $a_0 = a_1 = 1$, that is, $J = J^{(0)}$. □

Our final topic is to provide a second proof of Theorem 5.2.2 based on (5.4.69) and the fact that for $\lambda \in \mathfrak{e}^{\rm int}$,

$$m(\lambda + i0, J_0^+) = -\frac{u_1^+(\lambda + i0)}{a_0 u_0^+(\lambda + i0)} \tag{5.4.105}$$

by taking limits of (3.7.23). Similarly,

$$m(\lambda + i0, J_0^-) = -\frac{u_0^-(\lambda + i0)}{a_0 u_1^-(\lambda + i0)} \tag{5.4.106}$$

Second Proof of Theorem 5.2.2. By (5.4.69), (5.4.105), and (5.4.106), we have (with $m^\sharp$ given by (5.2.11)) that for $\lambda \in \mathfrak{e}_j^{\rm int}$,

$$m^\sharp(\lambda + i0) = \overline{m(\lambda + i0, J_0^+)} \tag{5.4.107}$$

Since (5.2.10) has real coefficients for $\lambda \in \mathfrak{e}$, $m^\sharp$ also solves (5.2.10) and so, by analyticity, it solves (5.2.10) for all λ.

For a.e. $\lambda \in e_j^{\rm int}$, $\operatorname{Im} m(\lambda + i0, J_0^+) \neq 0$, so $m^\sharp$ is distinct so it is the second solution. □

Remarks and Historical Notes. As explained in the Notes to Section 5.1, much of the theory of periodic ODEs goes back to Hill's equation. In particular, the use of discriminants goes back to Lyapunov [291], Hamel [191], Haupt [193], and Kramers [246]. Magnus–Winkler [292] and Eastham [120] provide monograph presentations. The discussion for Jacobi and/or discrete Schrödinger equations can be found in Hochstadt [200], van Moerbeke [450], Toda [441], Last [269], and Teschl [436].

Instead of using selfadjointness of $J(\theta)$ to conclude $\Delta^{-1}([2, 2]) \subset \mathbb{R}$, one can proceed as follows: If $\Delta(\lambda) \in (-2, 2)$, all solutions of (5.4.1) are bounded, and by cutting off a bounded solution, one gets $w_n \in \ell_2$, so $\|(J - \lambda)w_n\|/\|w_n\| \to 0$, implying $\lambda \in \sigma(J)$. Thus, λ is real. By continuity and analyticity, $\Delta^{-1}([-2, 2]) = \overline{\Delta^{-1}((-2, 2))}$.

There is a more "physical" meaning to "reflectionless." It can be proven (in analogy with the Schrödinger case discussed by Davies–Simon [101]) that if $\mathcal{H}_{\rm ac}$ is the range of the projection onto the a.c. subspace for J, an arbitrary bounded two-sided Jacobi matrix, then there exist spaces $\mathcal{H}_{\ell,r}^\pm$ so

$$\mathcal{H}_{\rm ac} = \mathcal{H}_\ell^+ \oplus \mathcal{H}_r^+ = \mathcal{H}_\ell^- \oplus \mathcal{H}_r^-$$

so that $\varphi \in \mathcal{H}_\ell^\pm \Leftrightarrow$ for all $n \in \mathbb{Z}$, $\lim_{t \to \mp\infty} \|\chi_{[n,\infty)} e^{-itJ}\varphi\| = 0$ (and similarly for $\mathcal{H}_r^\pm$ with $\chi_{[n,\infty)}$ replaced by $\chi_{(-\infty,n]}$). Thus, for example, $\mathcal{H}_\ell^+$ is the set φ for which $e^{-itJ}\varphi$ move to $-\infty$ (the left) as $t \to -\infty$. For this point of view, reflectionless means

$$\mathcal{H}_\ell^+ = \mathcal{H}_r^- \tag{5.4.108}$$

so that there is no reflection back from where $e^{-itJ}\varphi$ came from!

In great generality, Breuer–Ryckman–Simon [61] have proven that this notion of reflectionless is equivalent to the spectral version we defined in this section.

Theorem 5.4.19 is due to Craig [95] who considered some situations with infinitely many gaps. His proof (and ours) depends on an exponential Herglotz representation (i.e., passing to the log and then writing down a Herglotz representation), first emphasized by Akhiezer–Krein [15] and used extensively by Aronszajn–Donoghue [27].

The continuum analog of what we have called the Borg–Hochstadt theorem is due to Borg [55]. The Jacobi matrix analog is due to Hochstadt [200]; see also Flaschka [135]. The proof we give here is closely related to a proof of Clark et al. [90].

5.5 POTENTIAL THEORY, EQUILIBRIUM MEASURES, THE DOS, AND THE LYAPUNOV EXPONENT

Because of (5.4.29) and $\gamma(z) = 0$ on $\mathfrak{e}$, there is a close connection between potential theory and the fundamental objects of the periodic theory—the density of states will be the potential theoretic equilibrium measure, γ will be the potential theoretic Green's function, and $(a_1 \dots a_p)^{1/p}$ will be the logarithmic capacity. This realization shows that $d\nu$ is intrinsic to $\mathfrak{e}$ and will be important when we discuss other finite gap situations in Chapter 9. We begin this section with a brief minicourse on two-dimensional potential theory.

Define on $\mathbb{C}$,

$$G_0(z) = \log(|z|^{-1}) \tag{5.5.1}$$

If μ is a measure on $\mathbb{C}$ of compact support, its logarithmic potential is defined by

$$\Phi_\mu(z) = \int G_0(z-w)\, d\mu(w) \tag{5.5.2}$$

This integral converges if $z \notin \operatorname{supp}(d\mu)$, and since $d\mu$ has compact support, $G_0(z-w)$ is uniformly bounded below for $(z,w) \in \operatorname{supp}(d\mu) \times \operatorname{supp}(d\mu)$, so the integral for each $z \in \operatorname{supp}(d\mu)$ either converges or diverges to $+\infty$, in which case we set $\Phi_\mu(z) = +\infty$.

The same semiboundedness lets us use Fubini's theorem to conclude that for any two (positive) measures of compact support,

$$\int \Phi_\mu(z)\, d\nu(z) = \int \Phi_\nu(z)\, d\mu(z) \tag{5.5.3}$$

Potentials enter naturally in studying growth of polynomials as $n \to \infty$. For if

$$P_n(x) = \prod_{j=1}^{n} (x - x_j^{(n)}) \tag{5.5.4}$$

then

$$\frac{1}{n} \log|P_n(x)| = -\Phi_{\nu_n}(x) \tag{5.5.5}$$

where

$$\nu_n = \frac{1}{n} \sum_{j=1}^{n} \delta_{x_j^{(n)}}$$

is the counting measure for the zeros. So if ν_n converges to ν_∞, one can hope that root asymptotics of P_n (i.e., the limiting behavior of $|P_n(x)|^{1/n}$) is connected to the potential of ν_∞.

$\Phi_\mu(z)$ is bounded below on $\text{supp}(d\mu)$, so

$$\mathcal{E}(\mu) = \int \Phi_\mu(z)\, d\mu(z) \tag{5.5.6}$$

$$= \int \log(|z-w|^{-1})\, d\mu(z) d\mu(w) \tag{5.5.7}$$

is either finite or diverges to $+\infty$. $\mathcal{E}(\mu)$ is called the *potential energy* of μ or, for short, the *energy* of μ.

Given a compact set $\mathfrak{e} \subset \mathbb{C}$, we consider all probability measures, $\mathcal{M}_{+,1}(\mathfrak{e})$, on $\mathfrak{e}$. We say $\mathfrak{e}$ has *capacity zero* if and only if $\mathcal{E}(\mu) = \infty$ for all $\mu \in \mathcal{M}_{+,1}(\mathfrak{e})$. Otherwise, we define the *capacity*, $C(\mathfrak{e})$, of $\mathfrak{e}$ by

$$C(\mathfrak{e}) = \exp(-\inf(\mathcal{E}(\mu) \mid \mu \in \mathcal{M}_{+,1}(\mathfrak{e}))) \tag{5.5.8}$$

and we say $\mathfrak{e}$ has *positive capacity*.

Remark. The use of exp in (5.5.8) is as an inverse for log. We will eventually show (with $[a,b] \subset \mathbb{R}$ a closed interval; see Example 5.5.20)

$$C([a,b]) = \tfrac{1}{4}(b-a) \tag{5.5.9}$$

We are heading toward a proof of

Theorem 5.5.1. *Let $\mathfrak{e} \subset \mathbb{C}$ be a compact set with positive capacity. Then there is a unique measure, $\rho_\mathfrak{e}$, in $\mathcal{M}_{+,1}(\mathfrak{e})$ (called the* equilibrium measure *for $\mathfrak{e}$) so that*

$$\mathcal{E}(\rho_\mathfrak{e}) = \min_{\mu \in \mathcal{M}_{+,1}(\mathfrak{e})} \mathcal{E}(\mu) = \log(C(\mathfrak{e})^{-1}) \tag{5.5.10}$$

Lemma 5.5.2. (i) *G_0 is harmonic on $\mathbb{C} \setminus \{0\}$.*

(ii) *We have, as a tempered distribution,*

$$(\Delta G_0)(x) = -2\pi\delta(x) \tag{5.5.11}$$

(iii) *For any x_0, and r,*

$$\begin{aligned} h(r,x_0) \equiv \int_0^{2\pi} G_0(x_0 + re^{i\theta})\, \frac{d\theta}{2\pi} &= G_0(x_0) \qquad r \le |x_0| \\ &\le G_0(x_0) \qquad r > |x_0| \end{aligned} \tag{5.5.12}$$

(iv) *For x_0 fixed, $h(r,x_0)$ is monotone decreasing as r increases.*

(v) *If $j \in C_0^\infty(\mathbb{R}^2)$, $j(Rx) = j(x)$ for rotations R about 0, $j \ge 0$,*

$$\int j(x)\, d^2x = 1 \tag{5.5.13}$$

and

$$j_\varepsilon(x) = \varepsilon^{-1} j(\varepsilon^{-1} x) \tag{5.5.14}$$

and if

$$G_0^{(\varepsilon)}(x) = (j_\varepsilon * j_\varepsilon * G_0)(x) \tag{5.5.15}$$

then $G_0^{(\varepsilon)}$ *is* C^∞,

$$G_0^{(\varepsilon)} \le G_0 \tag{5.5.16}$$

$$\lim_{\varepsilon \downarrow 0} G_0^{(\varepsilon)}(x) = G_0(x) \tag{5.5.17}$$

Indeed, for any $r > 0$*, there is* $A > 0$ *so*

$$|x| > r \text{ and } \varepsilon < A \Rightarrow G_0^{(\varepsilon)}(x) = G_0(x) \tag{5.5.18}$$

(vi) *For any (positive) measure* μ,

$$\mathcal{E}(\mu) = \lim_{\varepsilon \downarrow 0} \mathcal{E}(\mu * j_\varepsilon) \tag{5.5.19}$$

(vii) *For any (positive) measure* μ,

$$\mathcal{E}(\mu) = \lim_{\varepsilon \downarrow 0} \int G_0^{(\varepsilon)}(x-y)\, d\mu(x) d\mu(y) \tag{5.5.20}$$

Remark. The proof provides an explicit formula for $h(r, x_0)$.

Proof. (i), (ii) Since Δ in polar coordinates is given by

$$\Delta f = \frac{1}{r}\frac{\partial}{\partial r} r \frac{\partial}{\partial r} f + \frac{1}{r^2}\frac{\partial^2}{\partial \theta^2} f \tag{5.5.21}$$

we see

$$\Delta G_0 = 0 \tag{5.5.22}$$

for $z \neq 0$, first classically and then as distributions. For any $f \in C_0^\infty$, say $f(z) = 0$ if $|z| \ge R$,

$$\begin{aligned} \int G_0(\Delta f)\, d^2x &= \lim_{\varepsilon \downarrow 0} \int_{R>|r|>\varepsilon} [(G_0 \Delta f) - (f \Delta G_0)]\, d^2x \quad \text{by (5.5.22)} \\ &= \lim_{\varepsilon \downarrow 0} \int_{R>|r|>\varepsilon} \operatorname{div}[G_0 \vec{\nabla} f - f \vec{\nabla} G_0]\, d^2x \\ &= \lim_{\varepsilon \downarrow 0} \int_{|r|=\varepsilon} f(z)\left[-\frac{1}{r}\right] - (\vec{\nabla} f)(z) G_0(z)\, r d\theta \\ &= -2\pi f(0) \end{aligned}$$

by Gauss's theorem, continuity of f and $|z|G_0 \to 0$ as $z \to 0$. This proves (5.5.11).

(iii), (iv) By (5.5.11) and Gauss's theorem and $r \neq x_0$,

$$\begin{aligned} r\,\frac{\partial h}{\partial r} &= \frac{1}{2\pi}\int_0^{2\pi} \nabla G_0(x+re^{i\theta})\cdot\widehat{n}\,d\sigma \\ &= \frac{1}{2\pi}\int_{|y-x|\le r} \Delta G_0\,d^2x \\ &= \begin{cases} 0 & r<|x_0| \\ -1 & r>|x_0| \end{cases} \end{aligned} \tag{5.5.23}$$

Since h is continuous at $r = x_0$ and $h(r, x_0) \to G_0(x_0)$ as $r \downarrow 0$, we get (5.5.12), and monotonicity by (5.5.23).

(5.5.23) and $h(r, x_0) = G_0(x_0)$ at $r = 0$ leads to the explicit formula

$$h(r, x_0) = \log(\min(|x_0|, r)^{-1}) \tag{5.5.24}$$

The analog of this for potentials on $\mathbb{R}^3$ goes back to Newton!

(v) By (5.5.12), if $\operatorname{supp}(j(x)) \subset \{x \mid |x| \le \rho_0\}$ obeys (5.5.13), $j \ge 0$, and $j(Rx) = j(x)$ for rotations, then $j * G_0$ obeys

$$\begin{aligned} (j*G_0)(x) &= G_0(x) \qquad |x|>\rho_0 \\ &\le G_0(x) \qquad |x|\le\rho_0 \end{aligned} \tag{5.5.25}$$

Moreover, if j is C^∞, so is $j*G_0$ by general results on convolutions of distributions. (5.5.25) implies $(j_\varepsilon * G_0)(x) = G_0(x)$ if $x \ge 2\varepsilon\rho_0$ where ρ_0 is such that $\operatorname{supp}(j) \subset \{x \mid |x| \le \rho_0\}$.

(vi), (vii) This follows from (5.5.16) and (5.5.17). If $\mathcal{E}(\mu) < \infty$, then dominated convergence implies (5.5.19) since

$$\mathcal{E}(\mu * j_\varepsilon) = \int (G_0 * j_\varepsilon * j_\varepsilon)(x-y)\,d\mu(x)d\mu(y) \tag{5.5.26}$$

If $\mathcal{E}(\mu) = \infty$, it is obvious, by (5.5.17), that for any $\rho > 0$,

$$\liminf \mathcal{E}(\mu * j_\varepsilon) \ge \int_{|x-y|\ge\rho} G_0(x-y)\,d\mu(x)d\mu(y)$$

Taking $\rho \downarrow 0$ and using monotone convergence, we see $\mathcal{E}(\mu * j_\varepsilon) \to \infty$. □

As a consequence of this lemma:

Theorem 5.5.3. (i) *For any measure μ of compact support in $\mathbb{C}$, $\Phi_\mu(z)$ is lower semicontinuous in z and superharmonic. On $\mathbb{C} \setminus \operatorname{supp}(\mu)$, Φ_μ is harmonic.*
(ii) *For fixed z, $\Phi_\mu(z)$ is weakly lower semicontinuous in μ.*
(iii) *$\mu \mapsto \mathcal{E}(\mu)$ is weakly lower semicontinuous.*

Remarks. 1. Lower semicontinuity in (iii) means $\mu_n \to \mu \Rightarrow \liminf \mathcal{E}(\mu_n) \ge \mathcal{E}(\mu)$ (mnemonic: the value at the limit can be lower). Equivalently, $\mathcal{E}^{-1}((-\infty, a])$ is closed for all a. Equivalently, $\mathcal{E}^{-1}((a, \infty])$ is open for all a.

2. g, taking values in $(-\infty, \infty]$, is called superharmonic if it is lower semicontinuous, $\int |g(z_0 + re^{i\theta})| \frac{d\theta}{2\pi} < \infty$ for all $z_0 \in \mathbb{C}$, $r > 0$, and if

$$\int g(z_0 + re^{i\theta})\,\frac{d\theta}{2\pi} \le g(z_0) \tag{5.5.27}$$

This implies (one inequality comes from (5.5.27) and the other from lower semicontinuity)

$$\lim_{r\downarrow 0}\int g(z_0+re^{i\theta})\,\frac{d\theta}{2\pi}=g(z_0) \tag{5.5.28}$$

3. g is harmonic if it is continuous and equality holds in (5.5.27); equivalently, if g is C^∞ with $\Delta g=0$.

Proof. Let j_ε be as in the lemma and

$$\Phi_\mu^{(\varepsilon)}=j_\varepsilon*\Phi_\mu \qquad \mathcal{E}_\varepsilon(\mu)=\mathcal{E}(\mu*j_\varepsilon) \tag{5.5.29}$$

Then $\Phi_\mu^{(\varepsilon)}(z)$ is jointly continuous in μ and z. By the lemma and monotone convergence,

$$\Phi_\mu(z)=\sup_\varepsilon\,\Phi_\mu^{(\varepsilon)}(z) \qquad \mathcal{E}(\mu)=\sup_\varepsilon\,\mathcal{E}_\varepsilon(\mu) \tag{5.5.30}$$

which implies the claimed semicontinuity results (if $g=\sup_n g_n$, then $g^{-1}((a,\infty))=\cup_n g_n((a,\infty))$). The mean inequalities are immediate from (5.5.12) and averaging in x_0. □

Proposition 5.5.4. (a) *Let* $f\in C_0^\infty(\mathbb{R}^2)$ *with*

$$\int f(x)\,d^2x=0 \tag{5.5.31}$$

Then

$$\int f(x)f(y)\log(|x-y|^{-1})\,d^2x d^2y=\frac{1}{2\pi}\int\biggl(\int\frac{f(y)}{|x-y|}\,d^2y\biggr)^2 d^2x \tag{5.5.32}$$

(b) *Under the hypothesis of* (a),

$$\textit{LHS of } (5.5.32)=2\pi\int\frac{|\widehat{f}(k)|^2}{|k|^2}\,d^2k \tag{5.5.33}$$

(c) *Let* μ *be a (positive) measure of compact support. Then*

$$\mathcal{E}(\mu)<\infty\Leftrightarrow\int_{|k|\ge 1}\frac{|\widehat{\mu}(k)|^2}{|k|^2}\,d^2k<\infty \tag{5.5.34}$$

(d) *Let* μ,ν *be two probability measures with* $\mathcal{E}(\mu)<\infty$, $\mathcal{E}(\nu)<\infty$. *Then*

$$B(\mu,\nu)=\int d\mu(x)d\nu(x)\log|x-y|^{-1} \tag{5.5.35}$$

is finite.

(e) *Under the hypothesis of* (d), *define*

$$\mathcal{E}(\mu-\nu)\equiv\mathcal{E}(\mu)+\mathcal{E}(\nu)-2B(\mu,\nu) \tag{5.5.36}$$

Then

$$\mathcal{E}(\mu - \nu) = \frac{1}{2\pi} \int \left(\int \frac{d\mu(y) - d\nu(y)}{|x - y|} \right)^2 d^2x \tag{5.5.37}$$

$$= 2\pi \int \frac{|\widehat{\mu}(k) - \widehat{\nu}(k)|^2}{|k|^2} \, d^2k \tag{5.5.38}$$

(f) *Under the hypothesis of* (d), *if* $\mu \neq \nu$,

$$\mathcal{E}(\mu - \nu) > 0 \tag{5.5.39}$$

Remarks. 1. $f \in C_0^\infty$ implies the integral on the left side of (5.5.32) is absolutely convergent. Since

$$\int \frac{f(y)\, d^2y}{|x - y|} = \frac{\int f(y)\, d^2y}{|x|} + O\left(\frac{1}{|x|^2}\right)$$

the integral on the right side is finite if and only if (5.5.31) holds. So (5.5.32) only holds if (5.5.31) does.

2. If μ has compact support, $\widehat{\mu}(k)$ is defined by

$$\widehat{\mu}(k) = (2\pi)^{-1} \int e^{-ik\cdot x} \, d\mu(x) \tag{5.5.40}$$

and is an entire function of $\vec{k}$.

3. Because μ, ν have compact support, the integral in (5.5.35) is either convergent or it diverges to $+\infty$.

4. $B(\mu, \nu)$ may not be positive. For example, if $d\mu$ is the probability measure uniformly distributed in $\{z \mid |z| \leq 2\}$, then $B(\mu, \mu) = -\log 2$.

5. (5.5.39) is called "strict conditional positive definiteness."

6. One can understand parts of this proposition in terms of the distribution $G_0(x) = \log|x|^{-1}$. Since (5.5.11) holds,

$$k^2 \widehat{G}_0(k) = 1 \tag{5.5.41}$$

This does not imply $\widehat{G}_0(k) = 1/k^2$ as a distribution because $1/k^2$ is *not* a distribution since it is not L^1 at $k = 0$. Rather $\widehat{G}_0(k)$ is a distribution, which is a regularization of $1/k^2$. If $h \in \mathcal{S}(\mathbb{R}^2)$ and $h(0) = 0$, then

$$\widehat{G}_0(h) = \int h(k) k^{-2} \, d^2k \tag{5.5.42}$$

which explains (5.5.33).

Proof. (a) Let

$$h_\alpha(x) = |x|^{-1-\alpha} \tag{5.5.43}$$

for $\alpha > 0$. Then, by rotation invariance and scale covariance,

$$(h_\alpha * h_\alpha)(x) = C_\alpha |x|^{-2\alpha} \tag{5.5.44}$$

where

$$C_\alpha = \int \frac{1}{|y|^{1+\alpha}} \frac{1}{|y - (1,0)|^{1+\alpha}} \, d^2y \tag{5.5.45}$$

We can write C_α as a sum of three terms: $C_\alpha^{(1)}, C_\alpha^{(2)}, C_\alpha^{(3)}$, where the first is the integral over $|y| < 2$, the second the integral over $|y| > 2$ with integrand

$$\frac{1}{|y|^{1+\alpha}}\left[\frac{1}{|y-(1,0)|^{1+\alpha}} - \frac{1}{|y|^{1+\alpha}}\right] \tag{5.5.46}$$

and the third

$$C_\alpha^{(3)} = \int_{|y|>2} \frac{1}{|y|^{2+2\alpha}}\, d^2y = \frac{2\pi}{2\alpha}\,(2)^{-2\alpha} \tag{5.5.47}$$

Since $C_\alpha^{(1)}$ and $C_\alpha^{(2)}$ have finite limits as $\alpha \downarrow 0$, we see

$$\lim_{\alpha\downarrow 0} \alpha C_\alpha = \pi \tag{5.5.48}$$

By (5.5.44) and (5.5.31),

$$\int\left(\frac{f(y)}{|x-y|^{1+\alpha}}\, d^2y\right)^2 d^2x = C_\alpha \int f(x)f(y)[|x-y|^{-2\alpha} - 1]\, d^2x d^2y \tag{5.5.49}$$

Take $\alpha \downarrow 0$ in each side of (5.5.49). On the left side, since (5.5.31) holds,

$$\int \frac{f(y)}{|x-y|^{1+\alpha}}\, d^2y = O(|x|^{-2-\alpha}) \tag{5.5.50}$$

uniformly in α. So the integral converges to $(2\pi)\times$ the right-hand side of (5.5.32).

On the other hand,

$$(2\alpha)^{-1}(|x-y|^{-2\alpha} - 1) \to \log|x-y|^{-1} \tag{5.5.51}$$

as $\alpha \downarrow 0$. So by dominated convergence and (5.5.48), the right side of (5.5.49) converges to $(2\pi)\times$ the left-hand side of (5.5.32).

(b) By rotation and scale invariance of h, $\widehat{h}_0 = ch_0$ and $c = 1$ since $\widehat{\widehat{h}}_0 = h_0$. Thus, by $\widehat{f * g} = (2\pi)\widehat{f}\widehat{g}$, we have if

$$g(x) = \int \frac{f(x)}{|x-y|}\, d^2y \tag{5.5.52}$$

then

$$\widehat{g}(k) = (2\pi)|k|^{-1}\widehat{f}(k) \tag{5.5.53}$$

and so, by the Plancherel theorem,

$$\frac{1}{2\pi}\int |g(x)|^2\, d^2x = 2\pi \int \frac{|\widehat{f}(k)|^2}{|k|^2}\, d^2k \tag{5.5.54}$$

proving (5.5.33).

(c) Let f_0 be a fixed function in $C_0^\infty(\mathbb{R}^2)$ with

$$\int f_0(x)\, d^2x = 1 \tag{5.5.55}$$

Let j_ε be as in Lemma 5.5.2 and let f_n be defined by

$$j_{1/n} * d\mu = f_n(x)\, d^2x \tag{5.5.56}$$

By (5.5.19),

$$\mathcal{E}(\mu) < \infty \Leftrightarrow \lim_{n\to\infty} \mathcal{E}(f_n\, d^2x) < \infty \tag{5.5.57}$$

Define B by (5.5.35) and $\mathcal{E}((f_n - f_1)d^2x)$ by (5.5.36). Of course, $f_n - f_1$ obeys (5.5.31) and $\mathcal{E}((f_n - f_1)d^2x)$ is given by (5.5.32) with $f = f_n - f_1$. $\mathcal{E}(f\, d^2x) < \infty$ and fixed and

$$B(f_n\, d^2x,\, f_1\, d^2x) = \int f_n(x)\Phi_{f_1}(x)\, d^2x \tag{5.5.58}$$

$$\to \int \Phi_{f_1}(x)\, d\mu(x) \tag{5.5.59}$$

since $f_n\, d^2x \xrightarrow{w} d\mu$. Thus, (5.5.57) becomes

$$\mathcal{E}(\mu) < \infty \Leftrightarrow \lim_{n\to\infty} \mathcal{E}((f_n - f_1)d^2x) < \infty$$

$$\Leftrightarrow \lim_{n\to\infty} \int \frac{|\widehat{f_n} - \widehat{f_1}|^2}{|k|^2}\, d^2k < \infty \tag{5.5.60}$$

on account of (5.5.33).

For $|k| < \infty$, $\widehat{f_n}(k) \to \widehat{\mu}(k)$ uniformly in k, and since all are analytic and $\widehat{\mu}(0) = (2\pi)^{-1} = \widehat{f_n}(0) = \widehat{f_1}(0)$, we have

$$\sup_n \int_{|k|\le 1} \frac{|\widehat{f_n} - \widehat{f_1}|^2}{|k|^2}\, d^2k < \infty \tag{5.5.61}$$

(5.5.60) becomes

$$\mathcal{E}(\mu) < \infty \Leftrightarrow \lim_{n\to\infty} \int_{|k|>1} \frac{|\widehat{f_n} - \widehat{f_1}|^2}{|k|^2}\, d^2k < \infty \tag{5.5.62}$$

But

$$\int_{|k|<1} \frac{|\widehat{f_1}(k)|^2}{|k|^2}\, d^2k < \infty \tag{5.5.63}$$

and $\frac{1}{2}|a|^2 - \frac{1}{2}|b|^2 \le |a-b|^2 \le 2|a|^2 + 2|b|^2$, so (5.5.62) becomes

$$\mathcal{E}(\mu) < \infty \Leftrightarrow \lim_{n\to\infty} \int_{|k|>1} \frac{|\widehat{f_n}|^2}{|k|^2} < \infty \tag{5.5.64}$$

But $\widehat{f_n}(k) = (2\pi)^{1/2}\widehat{j_1}(nk)\widehat{\mu}(k)$, so by a simple use of dominated convergence,

$$\lim_{n\to\infty} \int_{|k|>1} \frac{|\widehat{f_n}(k)|^2}{|k|^2}\, d^2k = \int \frac{|\widehat{\mu}(k)|^2}{|k|^2}\, d^2k \tag{5.5.65}$$

(d) Let j_ε be as in Lemma 5.5.2. Let

$$\mu_\varepsilon = j_\varepsilon * \mu \qquad \nu_\varepsilon = j_\varepsilon * \nu \tag{5.5.66}$$

by the same arguments that led to (5.5.19)

$$B(\mu, \nu) = \lim_{\varepsilon\downarrow 0} B(\mu_\varepsilon, \nu_\varepsilon) \tag{5.5.67}$$

Since $\mu_\varepsilon = f_\varepsilon\, d^2x$, $\nu_\varepsilon = g_\varepsilon\, d^2x$ for suitable C_0^∞ f_ε, g_ε, we can directly define

$$\mathcal{E}(\mu_\varepsilon - \nu_\varepsilon) = \int \frac{|\widehat{f_\varepsilon} - \widehat{g_\varepsilon}|^2}{k^2}\, d^2k > 0$$

and so conclude

$$B(\mu_\varepsilon, \nu_\varepsilon) \le \tfrac{1}{2}\,(\mathcal{E}(\mu_\varepsilon) + \mathcal{E}(\nu_\varepsilon)) \tag{5.5.68}$$

so by (5.5.19) and (5.5.67),

$$B(\mu, \nu) \le \tfrac{1}{2}\,(\mathcal{E}(\mu) + \mathcal{E}(\nu)) < \infty \tag{5.5.69}$$

(e) As in the proof of (c), dominated convergence and the result for μ_ε, ν_ε implies (5.5.38) for μ, ν, and the Plancherel theorem and calculation in the proof of (b) ran backwards proves (5.5.37).

(f) is immediate from (e). □

Here is the proof of Theorem 5.5.1, one of the main theorems of potential theory.

Proof of Theorem 5.5.1. If $R = \{\sup|x - y| \mid x, y \in \mathfrak{e}\}$, then $\log|x - y|^{-1} \ge \log R^{-1}$. So $\mathcal{E}(\mu) \ge \log R^{-1}$ if $\mu \in \mathcal{M}_{+,1}(\mathfrak{e})$.

If $C(\mathfrak{e}) > 0$, find $\mu_n \in \mathcal{M}_{+,1}(\mathfrak{e})$ so

$$\mathcal{E}(\mu_n) \le \log(C(\mathfrak{e})^{-1}) + \tfrac{1}{n} \tag{5.5.70}$$

By compactness of $\mathcal{M}_{+,1}(\mathfrak{e})$, we can find $\mu \in \mathcal{M}_{+,1}(\mathfrak{e})$ and a subsequence $n(1) < n(2) < \dots$ so $\mu_{n(j)} \overset{w}{\longrightarrow} \mu$. By Theorem 5.5.3,

$$\mathcal{E}(\mu) \le \liminf_{j\to\infty} \mathcal{E}(\mu_{n(j)}) = \log(C(\mathfrak{e})^{-1}) \tag{5.5.71}$$

so μ minimizes $\mathcal{E}$.

It is easy to see that for $\mu, \nu \in \mathcal{M}_{+,1}(\mathfrak{e})$ with $\mathcal{E}(\mu) < \infty$, $\mathcal{E}(\nu) < \infty$,

$$\tfrac{1}{2}\,[\mathcal{E}(\mu) + \mathcal{E}(\nu)] - \mathcal{E}(\tfrac{1}{2}\,\mu + \tfrac{1}{2}\,\nu) = \tfrac{1}{4}\,\mathcal{E}(\mu - \nu) \tag{5.5.72}$$

so, by (5.5.39), one has strict convexity

$$\mu \ne \nu \Rightarrow \mathcal{E}(\tfrac{1}{2}\,\mu + \tfrac{1}{2}\,\nu) < \tfrac{1}{2}\,[\mathcal{E}(\mu) + \mathcal{E}(\nu)] \tag{5.5.73}$$

which implies uniqueness of the minimizer. □

If $\mathfrak{e} \subset \mathbb{C}$, we will use $\rho_\mathfrak{e}$ for the equilibrium measure. If $X \subset \mathfrak{e}$, we call $\rho_\mathfrak{e}(X)$ the *harmonic measure* of X.

We are heading toward looking at continuity properties of $\Phi_{\rho_\mathfrak{e}}(z)$ and related regularity of $d\rho_\mathfrak{e}$. Naively, one might guess Φ_μ is continuous in the extended sense (i.e., $\Phi_\mu(z_0) = \infty$ is allowed only if $z_n \to z_0 = \Phi_\mu(z_n) \to \infty$, and otherwise one has continuity in the usual sense), but that is false:

Example 5.5.5. Let $x_n = n^{-1}$ and let

$$d\mu = \sum_{n=1}^\infty n^{-2}\delta_{x_n} \tag{5.5.74}$$

Clearly,

$$\Phi_\mu(x_n) = \infty$$

and $x_n \to 0$, but

$$\Phi_\mu(0) = \sum_{n=1}^{\infty} n^{-2} \log n < \infty \tag{5.5.75}$$

so $\Phi_\mu(z)$ is not continuous at $z = 0$.

This phenomenon is not limited to infinite values nor to point measures—it can even happen for equilibrium measures.

For $k = 1, 2, \dots$ and $n = 1, 2, \dots,$

$$\mathfrak{e}_{k,n} = \left[\frac{1}{n} - \frac{1}{k}\, e^{-n^6}, \frac{1}{n} + \frac{1}{k}\, e^{-n^6}\right] \tag{5.5.76}$$

and let

$$\mathfrak{e}_k = \{0\} \cup \bigcup_{n=1}^{\infty} \mathfrak{e}_{k,n} \tag{5.5.77}$$

which is a closed set. Since for any $\mathfrak{e} \subset \mathbb{R}$ (see Theorem 5.5.24),

$$C(\mathfrak{e}) \geq \tfrac{1}{4}\, |\mathfrak{e}| \tag{5.5.78}$$

with equality for intervals

$$\log(C(\mathfrak{e}_k)^{-1}) \leq \log(d_+^{-1} k) \tag{5.5.79}$$

where $d_+ = \frac{1}{2} \sum_{n=1}^{\infty} e^{-n^6}$.

By Corollary 5.8.5, if $\rho_k = \rho_{\mathfrak{e}_k}$ and $\Phi_k = \Phi_{\rho_k}$, then

$$x \in \mathfrak{e}_k \setminus \{0\} \Rightarrow \Phi_k(x) = \log(C(\mathfrak{e}_k)^{-1}) \tag{5.5.80}$$

On the other hand (since $|x - y| \leq 2$ on $\mathfrak{e}_k$),

$$\begin{aligned} \log 2 + \log(C(\mathfrak{e}_k)^{-1}) &= \log 2 + \mathcal{E}(\rho_k) \\ &\geq \mathcal{E}(\rho_k \restriction \mathfrak{e}_{k,n}) \\ &\geq \rho_k(\mathfrak{e}_{k,n})^2 \log(C(\mathfrak{e}_{k,n}^{-1})) \end{aligned} \tag{5.5.81}$$

so

$$\log 2 - \log d_+ + \log k \geq \rho(\mathfrak{e}_{k,n})^2 [\log[\tfrac{1}{2k}\, e^{-n^6}]^{-1}]$$

or

$$\begin{aligned} \rho(\mathfrak{e}_{k,n}) &\leq \sqrt{\frac{c_1 + \log k}{c_2 + \log k + n^6}} \\ &\leq \begin{cases} c_3 & n \leq (\log k)^{1/3} \\ \frac{c_4}{n^3} & n \geq (\log k)^{1/2} \end{cases} \end{aligned} \tag{5.5.82}$$

By (5.5.82),

$$\Phi_k(0) \le c_5 \log(k)^{1/3} \log(\log(k)) \tag{5.5.83}$$

We will also need a lower bound on $\log(C(\mathfrak{e}_k)^{-1}) \equiv \mathcal{E}(\rho_{\mathfrak{e}_k})$ and will settle for the following weak one. As in (5.5.81) (with $\rho_{k,n} \equiv \rho_k(\mathfrak{e}_{k,n})$),

$$\log(C(\mathfrak{e}_k)^{-1}) + \log 2 \ge \sum_{n=1}^{\infty} \rho_{k,n}^2 [\log k + n^6] \tag{5.5.84}$$

We break the sum into $n \le (\log k)^{1/5}$ and $n > (\log k)^{4/5}$.

Since

$$\sum_{n=1}^{N} \rho_{k,n}^2 \ge \frac{1}{N} \left(\sum_{n=1}^{N} \rho_{k,n} \right)^2 \tag{5.5.85}$$

we have

$$\sum_{n=1}^{(\log k)^{1/5}} (\log k) \rho_{k,n}^2 \ge (\log k)^{4/5} \left(\sum_{n=1}^{(\log k)^{1/5}} \rho_{k,n} \right)^2 \tag{5.5.86}$$

On the other hand,

$$\begin{aligned} \left(\sum_{n=N}^{\infty} \rho_{k,n} \right)^2 &\le \left(\sum_{n=N}^{\infty} n^6 \rho_{k,n}^2 \right) \left(\sum_{n=N}^{\infty} n^{-6} \right) \\ &\le C^{-1} N^{-5} \sum_{n=N}^{\infty} n^6 \rho_{k,n}^2 \end{aligned}$$

so

$$\sum_{n=(\log k)^{1/5}}^{\infty} n^6 \rho_{k,n}^2 \ge C(\log k) \left(\sum_{n=(\log k)^{1/5}}^{\infty} \rho_{k,n} \right)^2 \tag{5.5.87}$$

Since $\sum_{n=1}^{\infty} \rho_{k,n}$ implies either $\sum_{n=1}^{(\log k)^{1/5}} \rho_{k,n} \ge \frac{1}{2}$ or $\sum_{(\log k)^{1/5}}^{\infty} \rho_{k,n} \ge \frac{1}{2}$, we see, by (5.5.85), that

$$\log(C(\mathfrak{e}_k)^{-1}) \ge C(\log k)^{4/5} \tag{5.5.88}$$

Thus, by (5.5.80),

$$\Phi_k(\tfrac{1}{n}) \ge C(\log k)^{4/5} \tag{5.5.89}$$

Thus, for k large,

$$\lim \Phi_k(\tfrac{1}{n}) > \Phi_k(0) \tag{5.5.90}$$

and Φ_k, which is bounded, is discontinuous at $x = 0 \in \mathfrak{e}_k$. □

Thus, continuity properties of potentials are not automatic and we need to prove something in nonpathological cases. Because we are interested in $\mathfrak{e} \subset \mathbb{R}$ and this case has some simplifications, we will study that case but mention the general situation in the Notes.

Theorem 5.5.6. *Let* $\text{supp}(\mu) \subset \mathbb{R}$. *Then* Φ_μ *is continuous on* $\mathbb{C}$ *if and only if* $\Phi_\mu \restriction \text{supp}(\mu)$ *is continuous on* $\text{supp}(\mu)$.

Proof. We will prove the contrapositive, that in, if Φ_μ is discontinuous on $\mathbb{C}$, its restriction to $\mathfrak{e} \equiv \text{supp}(\mu)$ is discontinuous. Since Φ_μ is lower semicontinuous, if it is discontinuous, there exist $z_n \to z_\infty$, so

$$\lim_{n\to\infty} \Phi_\mu(z_n) = a > \Phi_\mu(z_\infty) \tag{5.5.91}$$

Φ_μ is harmonic, hence continuous off $\mathfrak{e}$, so $z_\infty \in \mathfrak{e}$.

If x, y, w are real, $|x + iy - w|^{-1} \le |x - w|^{-1}$, so

$$\mathfrak{e} \subset \mathbb{R} \Rightarrow \Phi_\mu(x + iy) \le \Phi_\mu(x) \tag{5.5.92}$$

and thus, $\text{Re}\, z_n \to z_\infty$ and

$$\liminf_{n\to\infty} \Phi_\mu(\text{Re}\, z_n) \ge a > \Phi_\mu(z_\infty) \tag{5.5.93}$$

Thus, by passing to a subsequence, we can suppose (5.5.91) holds and $z_n \in \mathbb{R}$ with either $z_n > z_\infty$ or $z_n < z_\infty$ for all n. For notational simplicity, we will suppose $z_n > z_\infty$ for all n.

Suppose $(\alpha, \beta) \subset \mathbb{R} \setminus \mathfrak{e}$ with $\alpha, \beta \in \mathfrak{e}$. Since $x \to \log|x|^{-1}$ is convex on $(0, \infty)$ and we can use monotone convergence at the endpoints, Φ_μ is convex and continuous on $[\alpha, \beta]$ (continuous in the extended sense that ∞ is an allowed value at α or β). By convexity,

$$\sup_{x\in[\alpha,\beta]} \Phi_\mu(x) = \max(\Phi_\mu(\alpha), \Phi_\mu(\beta)) \tag{5.5.94}$$

The above continuity plus (5.5.91) implies z_∞ is not the lower end of an open interval in $\mathbb{R} \setminus \mathfrak{e}$. Thus, there are $z_n^\pm \in \mathfrak{e}$, with $z_n \in [z_n^-, z_n^+]$ and $z_n^+ \to z_\infty$.

By (5.5.94),

$$\liminf \max(\Phi_\mu(z_n^+), \Phi_\mu(z_n^-)) > \Phi_\mu(z_\infty)$$

so, since $z_n^+ \to z_\infty$, $\Phi_\mu \restriction \mathfrak{e}$ is not continuous. □

Theorem 5.5.7. *If* $\mathfrak{e} \subset \mathbb{R}$ *and* $C(\mathfrak{e}) > 0$, *there exists* $\nu \in \mathcal{M}_{+,1}(\mathfrak{e})$, *so* Φ_ν *is continuous on* $\mathbb{C}$.

Proof. Pick μ in $\mathcal{M}_{+,1}(\mathfrak{e})$ with $\mathcal{E}(\mu) < \infty$. Then $\Phi_\mu \in L^1(d\mu)$, so by Lusin's theorem (see the Notes), there are $K_n \subset \mathfrak{e}$ compact with $\mu(K_n) \to 1$ and $\Phi_\mu \restriction K_n$ continuous. Pick K_{n_0} with $\mu(K_{n_0}) > 0$ and let

$$\eta = \mu \restriction K_{n_0} \tag{5.5.95}$$

By the choice, Φ_μ is continuous on K_{n_0} and so,

$$\Phi_\eta = \Phi_\mu - \Phi_{\mu-\eta} \tag{5.5.96}$$

is upper semicontinuous on K_{n_0}. Of course, it is lower semicontinuous there, so Φ_η is continuous on K_{n_0}, and so on $\text{supp}(\eta)$. Thus, Φ_η is continuous by Theorem 5.5.6.

By $\mu(K_{n_0}) > 0$, $\eta \ne 0$, so $\nu = \eta/\eta(\mathfrak{e}) \in \mathcal{M}_{+,1}(\mathfrak{e})$ with a continuous potential. □

For any Borel subset, $X \subset \mathbb{C}$, we define

$$C(X) = \sup_{\substack{\mathfrak{e} \subset X \\ \mathfrak{e} \text{ compact}}} C(\mathfrak{e}) \tag{5.5.97}$$

$$= \exp(\inf\{\mathcal{E}(\mu) \mid \text{supp}(\mu) \text{ compact, supp}(\mu) \subset X,\ \mu(\mathbb{C}) = 1\}) \tag{5.5.98}$$

Thus, $C(X) = 0$ if and only if $\mathcal{E}(\mu) = \infty$ for any measure μ with compact support in X. If an event depends on z and fails on a Borel subset of capacity zero, we say the event holds *quasi-everywhere* (q.e.).

Corollary 5.5.8. *For any measure, μ, of compact support, $\Phi_\mu(z) < \infty$ q.e. In fact, $\{z \mid \Phi_\mu(z) = \infty\}$ is a G_δ of capacity zero.*

Remark. It can be shown (see Landkof [264]) that if X is any bounded G_δ of capacity zero, there is a measure, μ, of compact support so that $X = \{z \mid \Phi_\mu(z) = \infty\}$.

Proof. Since

$$X = \{z \mid \Phi_\mu(z) = \infty\} = \bigcap_{n=1}^\infty \{z \mid \Phi_\mu(z) > n\} \tag{5.5.99}$$

the set is a G_δ. Suppose X has positive capacity. Then it contains a compact K with $C(K) > 0$. Let $\nu \in \mathcal{M}_{+,1}(K)$ so Φ_ν is continuous. Then Φ_ν is uniformly bounded on $\text{supp}(d\mu)$, so

$$\int \Phi_\nu(z)\, d\mu(z) < \infty \tag{5.5.100}$$

But $\Phi_\mu(z) = \infty$ on K, so

$$\int \Phi_\mu(z)\, d\nu(z) = \infty \tag{5.5.101}$$

This contradicts (5.5.3), so $C(X) = 0$. □

Proposition 5.5.9. *Let η be a measure of compact support with $\mathcal{E}(\eta) < \infty$, and let $X \subset \mathbb{C}$ with $C(X) = 0$. Then $\eta(X) = 0$.*

Proof. Since any measure is inner regular, it suffices to prove this result when X is compact. If $\eta(X) \neq 0$, $\eta \restriction X$ (i.e., $A \mapsto \eta(X \cap A)$) is a nonzero measure. Moreover, if $r = \sup_{x,y \in \text{supp}(\eta)} |x - y|$, then

$$\int_{X \times X} \log(r|x-y|^{-1})\, d\eta(x) d\eta(y) \le \int_{\mathbb{C}\times\mathbb{C}} \log(r|x-y|^{-1})\, d\eta(x) d\eta(y)$$
$$= \log r[\eta(C)^2] + \mathcal{E}(\eta) < \infty$$

so $\mathcal{E}(\eta \restriction X) < \infty$, showing $C(X) > 0$. Thus, $C(X) = 0 \Rightarrow \eta \restriction X = 0 \Rightarrow \eta(X) = 0$. □

A second major theorem in potential theory is

Theorem 5.5.10 (Upper Envelope Theorem). *Let $\mathfrak{e} \subset \mathbb{R}$ be compact and let ν_n, $\nu_\infty \in \mathcal{M}_{+,1}(\mathfrak{e})$ with $\nu_n \xrightarrow{w} \nu_\infty$. Then*

(i)

$$\Phi_{\nu_n}(z) \to \Phi_{\nu_\infty}(z) \tag{5.5.102}$$

for all $z \in \mathbb{C} \setminus \mathfrak{e}$.

(ii)

$$\liminf \Phi_{\nu_n}(z) \geq \Phi_{\nu_\infty}(z) \tag{5.5.103}$$

for all $z \in \mathfrak{e}$.

(iii) *Equality holds in* (5.5.103) *for q.e. $z \in \mathfrak{e}$.*

Remark. If ν_n gives weight $\frac{1}{2^n+1}$ to $\{\frac{j}{2^n}\}_{j=0}^{2^n}$, $d\nu_n \xrightarrow{w} dx$, Lebesgue measure. At any dyadic rational, $\liminf \Phi_{\nu_n}(x) = \infty$ but $\Phi_{\nu_\infty}(x) < \infty$. So equality in (5.5.103) may not hold everywhere.

Proof. (i) For $z \in \mathbb{C} \setminus \mathfrak{e}$, $\log|z-w|^{-1}$ is continuous in $w \in \mathfrak{e}$ so (5.5.102) follows from the weak convergence.

(ii) Let $a < \infty$ and

$$\Phi_\nu^{(a)}(x) = \int \log(\min(a, |x-y|^{-1}))\, d\nu(y)$$

Since $\Phi_\nu^{(a)} \leq \Phi_\nu$ and $\Phi_\nu^{(a)}(x)$ is weakly continuous in ν,

$$\Phi_{\nu_\infty}^{(a)}(x) = \lim \Phi_{\nu_n}^{(a)} \leq \liminf \Phi_{\nu_n}(x) \tag{5.5.104}$$

Taking $a \to \infty$, using $\Phi_\nu^{(a)} \to \Phi_\nu$ (by monotone convergence), we obtain (5.5.103).

(iii) Let $X \subset \mathfrak{e}$ be the set where strict inequality holds in (5.5.103). If $C(X) > 0$, use Theorem 5.5.7 to find $\eta \in \mathcal{M}_{+,1}(\mathfrak{e})$ with $\operatorname{supp}(\eta) \subset X$ so that Φ_η is continuous. Then

$$\begin{aligned}
&\int \Phi_\eta(x)\, d\nu_\infty(x) \\
&\quad = \lim \int \Phi_\eta(x)\, d\nu_n(x) \\
&\quad = \lim \int \Phi_{\nu_n}(x)\, d\eta(x) && \text{(by (5.5.3))} \\
&\quad \geq \int \liminf \Phi_{\nu_n}(x)\, d\eta(x) && \text{(by Fatou's lemma)} \\
&\quad > \int \Phi_{\nu_\infty}(x)\, d\eta(x) && \text{(by definition of } X \text{ and } \operatorname{supp}(\eta) \subset X) \\
&\quad = \int \Phi_\eta(x)\, d\nu_\infty(x) && \text{(by (5.5.3))}
\end{aligned}$$

The strict inequality is a contradiction, so $C(X) = 0$ and equality holds q.e. □

A third major theorem in potential theory is

Theorem 5.5.11 (Frostman's Theorem). *Let $\mathfrak{e} \subset \mathbb{R}$ be a compact set and let $\rho_{\mathfrak{e}}$ be its equilibrium measure. Then*

(i) *For all $z \in \mathbb{C}$,*

$$\Phi_{\rho_{\mathfrak{e}}}(z) \le \log(C(\mathfrak{e})^{-1}) \tag{5.5.105}$$

(ii) *Equality holds in* (5.5.105) *for q.e. $z \in \mathfrak{e}$.*

(iii) *Strict inequality holds in* (5.5.105) *on $\mathbb{C} \setminus \operatorname{supp}(d\rho_{\mathfrak{e}})$.*

Remark. Equality may not hold everywhere on $\mathfrak{e}$. For example, if $\mathfrak{e} = [-1, 1] \cup \{2\}$ and $\tilde{\mathfrak{e}} = [-1, 1]$, then $\rho_{\mathfrak{e}} = \rho_{\tilde{\mathfrak{e}}}$, so $\Phi_{\rho_{\mathfrak{e}}}(2) = \Phi_{\rho_{\tilde{\mathfrak{e}}}}(2) > \log(C(\mathfrak{e})^{-1})$ by (iii).

Proof. (i) Let f be a bounded Borel function on $\mathfrak{e}$ so $\int f\, d\rho_{\mathfrak{e}} = 0$. Then for ε real with $|\varepsilon|$ small, $(1 + \varepsilon f)\, d\rho_{\mathfrak{e}}$ is a probability measure, so

$$\frac{d}{d\varepsilon}\, \mathcal{E}((1 + \varepsilon f)\, d\rho_{\mathfrak{e}}) = 2 \int f(x) \Phi_{\rho_{\mathfrak{e}}}(x)\, d\rho_{\mathfrak{e}}(x) = 0$$

This implies that there is a constant c so

$$\Phi_{\rho_{\mathfrak{e}}}(x) = c \qquad d\rho_{\mathfrak{e}}\text{-a.e. } x \tag{5.5.106}$$

Thus,

$$c = \int \Phi_{\rho_{\mathfrak{e}}}(x)\, d\rho_{\mathfrak{e}}(x) = \mathcal{E}(\rho_{\mathfrak{e}}) = \log(C(\mathfrak{e})^{-1}) \tag{5.5.107}$$

By lower semicontinuity, (5.5.105) holds everywhere on $\operatorname{supp}(d\rho_{\mathfrak{e}})$. As noted in the proof of Theorem 5.5.6, $\Phi_{\rho_{\mathfrak{e}}}(z)$ is convex and continuous on any interval $[\alpha, \beta] \subset \mathbb{R}$ with $(\alpha, \beta) \cap \mathfrak{e} = \emptyset$, which, together with $\lim_{|z|\to\infty} \Phi_{\rho_{\mathfrak{e}}}(z) = -\infty$, implies that (5.5.105) holds on $\mathbb{R}$, and then, by (5.5.92), on all of $\mathbb{C}$.

(ii) Let

$$X = \{x \in \mathfrak{e} \mid \Phi_{\rho_{\mathfrak{e}}}(x) < \log(C(\mathfrak{e})^{-1})\} \tag{5.5.108}$$

We need to prove $C(X) = 0$. If not, there is a measure $d\eta$ concentrated on X with $\mathcal{E}(\eta) < \infty$. In particular, $\mathcal{E}(t\eta + (1 - t)\rho_{\mathfrak{e}})$ is finite for all t and is a quadratic function of t with

$$\begin{aligned} \frac{d}{dt}\, \mathcal{E}(t\eta + (1 - t)\rho_{\mathfrak{e}})\bigg|_{t=0} &= 2 \int \Phi_{\rho_{\mathfrak{e}}}(x)[d\eta - d\rho_{\mathfrak{e}}] \\ &= 2 \int [\Phi_{\rho_{\mathfrak{e}}}(x) - \log(C(\mathfrak{e})^{-1})]\, d\eta(x) \qquad (5.5.109) \\ &< 0 \end{aligned}$$

by (5.5.108). Here (5.5.109) comes from $\int \Phi_{\rho_{\mathfrak{e}}}\, d\rho_{\mathfrak{e}} = \mathcal{E}(\rho_{\mathfrak{e}}) = \log(C(\mathfrak{e})^{-1})$. This contradicts minimality and proves that $C(X) = 0$.

(iii) Immediate from the maximum principle that $\sup_{z \notin \mathfrak{e}} \Phi_{\rho_{\mathfrak{e}}}(z) \le \log(C(\mathfrak{e})^{-1})$, and $\Phi_{\rho_{\mathfrak{e}}}(z) \to -\infty$ as $|z| \to \infty$ implies the maximum cannot be taken on $\mathbb{C} \setminus \mathfrak{e}$. $\square$

Corollary 5.5.12. *Let $\mathfrak{e} \subset \mathbb{R}$ be compact and $X \subset \mathbb{R}$ open. If $C(X \cap \mathfrak{e}) > 0$, then $\rho_{\mathfrak{e}}(X) > 0$.*

Proof. If $\rho_{\mathfrak{e}}(X) = 0$, then $X \subset \mathbb{C} \setminus \text{supp}(d\rho_{\mathfrak{e}})$ (since X is open), and thus, the inequality in (5.5.105) is strict. But (5.5.105) holds q.e. on $\mathfrak{e}$, so $C(X \cap \mathfrak{e}) = 0$. $\square$

A closed set $\mathfrak{e} \subset \mathbb{R}$ is called *potentially perfect* if for all $x_0 \in \mathfrak{e}$ and $\varepsilon > 0$, $C((x_0 - \varepsilon, x_0 + \varepsilon) \cap \mathfrak{e}) > 0$. It is easy to see that any compact $\mathfrak{e}$ in $\mathbb{R}$ can be decomposed $\mathfrak{e} = \mathfrak{e}_1 \cup \mathfrak{e}_2$ where $\mathfrak{e}_1$ is potentially perfect and $C(\mathfrak{e}_2) = 0$. The last corollary immediately implies:

Corollary 5.5.13. *Let $\mathfrak{e} \subset \mathbb{R}$ be compact. Then*

$$\text{supp}(d\rho_{\mathfrak{e}}) = \mathfrak{e}$$

if and only if $\mathfrak{e}$ is potentially perfect.

For purposes of solving the Dirichlet problem, one often defines the *potential theorist's Green's function* by

$$G_{\mathfrak{e}}(z) = \log(C(\mathfrak{e})^{-1}) - \Phi_{\rho_{\mathfrak{e}}}(z) \tag{5.5.110}$$

It is the unique function harmonic on $\mathbb{C} \setminus \mathfrak{e}$, subharmonic on $\mathbb{C}$ with

$$G_{\mathfrak{e}}(z) = \log(|z|) + O(1) \qquad \text{as } |z| \to \infty \tag{5.5.111}$$

$$G_{\mathfrak{e}}(x) = 0 \qquad \text{for q.e. } x \in \mathfrak{e} \tag{5.5.112}$$

It is unfortunate that spectral theorists use the term Green's function for a different object (namely, (5.4.40)) than $G_{\mathfrak{e}}$, which is why we add "potential theorist's"!

Notice that as $|z| \to \infty$,

$$G_{\mathfrak{e}}(z) = \log|z| - \log(C(\mathfrak{e})) + O(\tfrac{1}{z}) \tag{5.5.113}$$

Theorem 5.5.14 (Bernstein–Walsh Lemma). *Let $q_n(x)$ be a polynomial of degree x and let*

$$\|q_n\|_{\mathfrak{e}} = \sup_{x \in \mathfrak{e}} |q_n(x)|$$

for any compact $\mathfrak{e} \subset \mathbb{R}$. Then, for all z,

$$|q_n(z)| \le \|q_n\|_{\mathfrak{e}} \exp(nG_{\mathfrak{e}}(z)) \tag{5.5.114}$$

Proof. Fix $\varepsilon > 0$. Let

$$g_\varepsilon(z) = \log|q_n(z)| - \log\|q_n\|_e - (n+\varepsilon)G_e(z)$$

g_ε is harmonic on $\mathbb{C} \setminus \mathfrak{e} \cup \{z_j\}_{j=1}^n$ where z_j are the zeros of q. By $\varepsilon > 0$ and (5.5.113), $g_\varepsilon(z) \sim -\varepsilon \log|z| \to -\infty$ at ∞ and $g_\varepsilon(z) \to -\infty$ at the z_j. Thus, for any $\delta > 0$ and $\text{dist}(z, \mathfrak{e}) > \delta$, we have

$$g_\varepsilon(z) \le \max_{\text{dist}(z,\mathfrak{e})=\delta} g_\varepsilon(z)$$

By Frostman's theorem, $G_{\mathfrak{e}}(z) \ge 0$ for all z, so

$$\max_{\text{dist}(z,\mathfrak{e})=\delta} g_\varepsilon(z) \le \max_{\text{dist}(z,\mathfrak{e})=\delta} |q_n(z)| - \max_{z \in \mathfrak{e}} |q_n(z)|$$
$$\to 0$$

as $\delta \downarrow 0$. Thus,

$$g_\varepsilon(z) \leq 0$$

first on $\mathbb{C} \setminus \mathfrak{e}$ and then, by $g_\varepsilon(z) = \lim_{\delta\downarrow 0} \int g_\varepsilon(z + \delta e^{i\theta}) \frac{d\theta}{2\pi}$, on $\mathfrak{e}$. Taking $\varepsilon \downarrow 0$, we obtain (5.5.16). $\square$

For applications of potential theory to periodic Jacobi matrices, we state a converse of Frostman's theorem whose hypotheses can be weakened.

Theorem 5.5.15. *Let $\mathfrak{e} \subset \mathbb{R}$ be compact. Suppose $\eta \in \mathcal{M}_{+,1}(\mathfrak{e})$ obeys*

$$\Phi_\eta(x) = \alpha \qquad \textit{for all } x \in \mathfrak{e} \tag{5.5.115}$$

for some α. Then

$$\alpha = \log(C(\mathfrak{e})^{-1}) \tag{5.5.116}$$

and

$$\eta = \rho_\mathfrak{e} \tag{5.5.117}$$

Remark. By Remark 4 after Theorem 5.4.10, the potential theorist's Green's function, $G_\mathfrak{e}$, obeys

$$G_\mathfrak{e}(x) = 0 \qquad \text{for } x \in \mathfrak{e} \tag{5.5.118}$$

if $\mathfrak{e}$ is the spectrum of a two-sided Jacobi matrix.

Proof. By (5.5.115),

$$\mathcal{E}(\eta) = \int \Phi_\eta(x)\, d\eta(x) = \alpha < \infty \tag{5.5.119}$$

so, by Proposition 5.5.9, η gives zero weight to the subset of $\mathfrak{e}$ where equality fails in (5.5.105), that is,

$$\Phi_{\rho_\mathfrak{e}}(x) = \log(C(\mathfrak{e})^{-1}) \qquad \text{a.e. } d\eta \tag{5.5.120}$$

Thus, by (5.5.3),

$$\begin{aligned} \log(C(\mathfrak{e})^{-1}) &= \int \Phi_{\rho_\mathfrak{e}}(x)\, d\eta(x) \\ &= \int \Phi_\eta(x)\, d\rho_\mathfrak{e}(x) \\ &= \alpha \qquad \text{(by (5.5.115))} \end{aligned}$$

proving (5.5.116).

Therefore, by (5.5.119), $\mathcal{E}(\eta) = \log(C(\mathfrak{e})^{-1})$, so (5.5.117) holds by uniqueness of minimizers. $\square$

Theorem 5.5.16. *Let μ, ν be two probability measures of compact support in $\mathbb{R}$. Suppose $\Phi_\mu(z) \leq \Phi_\nu(z)$ near infinity. Then $\mu = \nu$. In particular, if $\Phi_\eta(z) \leq \Phi_{\rho_\mathfrak{e}}(z)$ for all $z \in \mathbb{C}_+$ or $\Phi_\eta(z) \geq \Phi_{\rho_\mathfrak{e}}(z)$ for all $z \in \mathbb{C}_+$ (where* $\operatorname{supp}(\eta) \subset \mathbb{R}$, *$\mathfrak{e} \subset \mathbb{R}$), then $\eta = \rho_\mathfrak{e}$.*

Proof. Since

$$\log|x - z|^{-1} = \log|z|^{-1} + O\left(\frac{|x|}{|z|}\right)$$

$\Phi_\mu - \Phi_\nu$ is harmonic at infinity and vanishes there. By the maximum principle, it is either identically zero off $\mathbb{R}$ or takes both signs near infinity.

If it is identically zero off $\mathbb{R}$, by averaging, it is zero on $\mathbb{R}$ and then $\mu = \nu$ since $\Delta\Phi_\mu = -2\pi\mu$ as distributions (by (5.5.11)). □

This completes our minicourse on potential theory, and we return to periodic Jacobi matrices:

Theorem 5.5.17. *Let $\mathfrak{e} = \cup_{j=1}^{p} \mathfrak{e}_j$ be the spectrum of a two-sided Jacobi matrix, J, of period p. Let Δ be its discriminant, let $d\nu$ be given by* (5.3.34) *(or* (5.4.15)*), and let $\gamma(z)$ be the Lyapunov exponent (given by* (5.4.25) *and* (5.4.26)*). Then*

(i) *$d\nu$ is $d\rho_{\mathfrak{e}}$, the equilibrium measure of $\mathfrak{e}$.*

(ii) *$C(\mathfrak{e})$ is the capacity of $\mathfrak{e}$ given by*

$$C(\mathfrak{e}) = (a_1 \dots a_p)^{1/p} \tag{5.5.121}$$

(iii) *$\gamma(z)$ is the potential theorist's Green's function for $\mathfrak{e}$; equivalently, $-\gamma(\lambda) - p^{-1}\log(a_1 \dots a_p)$ is the equilibrium potential for $\mathfrak{e}$.*

Proof. By (5.4.29) and (5.4.26) (which says $\gamma(\lambda) = 0$ for $\lambda \in \mathfrak{e}$), we have for $x \in \mathfrak{e}$,

$$\Phi_\nu(x) = -\frac{1}{p}\log(a_1 \dots a_p) \tag{5.5.122}$$

By Theorem 5.5.15, $\nu = \rho_{\mathfrak{e}}$ and $\log(C(\mathfrak{e})^{-1}) = -\frac{1}{p}\log(a_1 \dots a_p)$, proving (5.5.121). By (5.4.29), $\gamma(z)$ is the potential theorist's Green's function. □

This has two immediate corollaries about periodic problems:

Corollary 5.5.18. *If two two-sided periodic Jacobi matrices of period p have the same spectra, they have the same Δ, the same $d\nu$, and the same γ.*

Proof. Theorem 5.5.17 shows that ν and γ are intrinsic to $\mathfrak{e} = \operatorname{spec}(J)$. $d\nu$ determines Δ by (5.4.16) or γ by (5.4.26). □

Corollary 5.5.19. *If $\mathfrak{e} = \tilde{\mathfrak{e}}_1 \cup \dots \cup \tilde{\mathfrak{e}}_\ell$ is the spectrum of a two-sided periodic Jacobi matrix, J, with $\tilde{\mathfrak{e}}_j$ the connected components of $\mathfrak{e}$, then the harmonic measure of each $\tilde{\mathfrak{e}}_j$ is rational.*

Remark. We will discuss the converse of this shortly.

Proof. Each band $\mathfrak{e}_k$ has harmonic measure $1/p$ (see the remark after Theorem 5.4.5), so $\tilde{\mathfrak{e}}_j$, which is a union of $\mathfrak{e}_k$'s, has harmonic measure n_j/p, which is rational. □

Example 5.5.20. Let $\mathfrak{e} = [\alpha, \beta]$. This is the spectrum of the two-sided Jacobi matrix with constant parameters

$$b_n = \tfrac{1}{2}(\alpha + \beta) \qquad a_n = \tfrac{1}{4}(\beta - \alpha) \tag{5.5.123}$$

Thus,

$$C([\alpha, \beta]) = \tfrac{1}{4}(\beta - \alpha) \tag{5.5.124}$$

By translation and scaling (5.3.39), we see

$$d\rho_{[\alpha,\beta]}(x) = \frac{1}{\pi} \frac{1}{[(x-\alpha)(\beta - x)]^{1/2}}\, dx \tag{5.5.125}$$

consistent with (5.4.96). □

(5.4.96) thus gives a formula for the equilibrium measure (with $\{\lambda_j\}_{j=1}^{\ell}$ determined by (5.4.97)) of the essential spectrum of periodic Jacobi matrices. Our next immediate goal is to extend this to general finite gap sets

$$\mathfrak{e} = [\alpha_1, \beta_1] \cup \cdots \cup [\alpha_{\ell+1}, \beta_{\ell+1}] \tag{5.5.126}$$

where

$$\alpha_1 < \beta_1 < \alpha_2 < \cdots < \alpha_{\ell+1} < \beta_{\ell+1} \tag{5.5.127}$$

The function

$$R(z) = \prod_{j=1}^{\ell+1} (z - \alpha_j)(z - \beta_j) \tag{5.5.128}$$

will play a critical role here and later (see Section 5.12). Notice that each factor in the product is positive on $\mathbb{R} \setminus (\alpha_j, \beta_j)$ and negative on (α_j, β_j) so

$$R(x) > 0 \text{ if } x \in \mathbb{R} \setminus \mathfrak{e} \qquad R(x) \le 0 \quad x \in \mathfrak{e} \tag{5.5.129}$$

We want to define $\sqrt{R}$ as an analytic function on $\mathbb{C} \setminus \mathfrak{e}$, the branch with

$$\sqrt{R(x)} > 0 \quad \text{ if } x > \beta_{\ell+1} \tag{5.5.130}$$

This implies

$$\sqrt{R(x)} < 0 \qquad (\beta_\ell, \alpha_{\ell+1}) \cup (\beta_{\ell-2}, \alpha_{\ell-1}) \cup \dots \tag{5.5.131}$$

$$\sqrt{R(x)} > 0 \qquad (\beta_{\ell-1}, \alpha_\ell) \cup (\beta_{\ell-3}, \alpha_{\ell-2}) \cup \dots \tag{5.5.132}$$

$$(-1)^{\ell-1}\sqrt{R(x)} > 0 \text{ on } (-\infty, \alpha_1) \tag{5.5.133}$$

and ($\sqrt{R(x+i0)}$ means $\lim_{\varepsilon \downarrow 0}(\sqrt{R(x+i\varepsilon)})$)

$$(-i)\sqrt{R(x+i0)} > 0 \text{ on } (\alpha_{\ell+1}, \beta_{\ell+1}) \cup (\alpha_{\ell-1}, \beta_{\ell-1}) \cup \dots \tag{5.5.134}$$

$$i\sqrt{R(x+i0)} > 0 \text{ on } (\alpha_\ell, \beta_\ell) \cup (\alpha_{\ell-2}, \beta_{\ell-2}) \cup \dots \tag{5.5.135}$$

Following (5.4.96)/(5.4.97), we are interested in solutions of

$$\int_{\beta_j}^{\alpha_{j+1}} \left[\frac{P(x)}{\sqrt{|R(x)|}} \right] dx = 0 \tag{5.5.136}$$

where P is a monic polynomial of degree ℓ:

Proposition 5.5.21. (a) *If P is a nonzero polynomial of degree $\ell - 1$ or less, it cannot happen that* (5.5.136) *holds for $j = 1, \dots, \ell$.*

(b) *There is a unique monic polynomial, P, of exact degree ℓ so that* (5.5.136) *holds for $j = 1, \dots, \ell$. This P has all its zeros in the gaps, one each and simple in each (β_j, α_{j+1}), $j = 1, \dots, \ell$.*

Remark. (a) assures us the $\ell \times \ell$ matrix

$$Y_{jk} = \int_{\beta_j}^{\alpha_{j+1}} x^{k-1} |R(x)|^{-1/2}\, dx \qquad 1 \le j, k \le \ell \tag{5.5.137}$$

is invertible, and then the coefficients of P can be explicitly written in terms of the inverse of this matrix and the vector $Y_{jk=\ell+1}$.

Proof. (a) For any real polynomial, if (5.5.136) holds for some j_0, P must change sign on (β_j, α_{j+1}) so have a zero there. Since $\deg(P) \le \ell - 1$ means P has $\ell - 1$ zeros, it cannot have a zero in each gap, so (5.5.136) cannot hold for all $j = 1, \dots, \ell$. Thus, there is no solution with P real. But if P is any nonzero solution, both $P(z) + \overline{P(\bar z)}$ and $i(P(z) - \overline{P(\bar z)})$ solve the same equations and are real, and at least one must be nonzero.

(b) (5.5.136) for $j = 1, \dots, \ell$ and $\deg(P) \le \ell$ (not necessarily monic) represents ℓ linear conditions on $\ell + 1$ parameters, so there is always a solution. By (a), the solution must have a nonzero x^ℓ term, so there is a monic solution. If there were two monic solutions, their difference would violate (a), so this solution is unique. As in (a), P must have at least one and so exactly one zero in each of the ℓ gaps. □

Henceforth, we will use $P(z)$ or $P(z; \alpha_1, \beta_1, \dots, \alpha_{\ell+1}, \beta_{\ell+1})$ or $\prod_{j=1}^{\ell}(z - z_j)$ where $z_j \in (\beta_j, \alpha_{j+1})$. With the function R above and branch of square root, we define, initially on $\mathbb{C} \setminus \mathfrak{e}$,

$$H(z) = -\frac{P(z)}{\sqrt{R(z)}} \tag{5.5.138}$$

which is clearly analytic there, and at infinity where

$$H(z) = -\frac{1}{z} + O\left(\frac{1}{z^2}\right) \tag{5.5.139}$$

Since $R(z)$ is entire and nonvanishing on $\mathfrak{e}^{\text{int}}$, $H(x \pm i0)$ exist (and are complex conjugate). We prove:

Theorem 5.5.22. (i) *$H(x)$ is real on $\mathbb{R} \setminus \mathfrak{e}$ and $H(x + i0)$ is pure imaginary with strictly positive imaginary part on $\mathfrak{e}^{\text{int}}$.*

(ii) *$H(z)$ is a Herglotz function on $\mathbb{C}_+$ so that*

$$H(z) = \int \frac{d\nu(x)}{x - z} \tag{5.5.140}$$

for a probability measure on $\mathfrak{e}$.

(iii) *$d\nu$ is a purely a.c. measure with density given by* (5.4.96).

(iv) *$d\nu$ is the equilibrium measure for $\mathfrak{e}$.*

(v) *The potential Φ_ν is given, for $z \in \mathbb{C}_+$, by*

$$-\Phi_\nu(z) - \log|z| = \operatorname{Re}\left[\oint_z \left[\left(H(w) + \frac{1}{w}\right)\right] dw\right] \tag{5.5.141}$$

where the curve is the straight line from z to $z + i\infty$. In particular, for any $x \in \mathfrak{e}$, $x \neq 0$,

$$\log(C(\mathfrak{e})^{-1}) = \log|x| - \int_0^\infty \operatorname{Im}\left[\left(H(x+iy) + \frac{1}{x+iy}\right)\right] dy \tag{5.5.142}$$

Remark. In fact, (5.5.141) can have any curve in $\mathbb{C} \setminus \mathfrak{e} \cup \{0\}$. The imaginary part is curve dependent, but not the real part.

Proof. (i) P is a monic polynomial with real zeros, hence real coefficients. Reality in the gaps is thus immediate by (5.5.131), (5.5.132), and (5.5.133). By (5.5.134), $\operatorname{Im}(-1/\sqrt{R(x+i0)}) > 0$ on $(\alpha_{\ell+1}, \beta_{\ell+1})$. Since P is monic with all zeros below $\alpha_{\ell+1}$, P is positive on that interval, so $\operatorname{Im}(H(x+i0)) > 0$ on $(\alpha_{\ell+1}, \beta_{\ell+1})$.

The sign of $\operatorname{Im}(1/\sqrt{R(x+i0)})$ shifts from band to band, but because it has a single zero, so does the sign of P, so H is pure imaginary with $\operatorname{Im}(H(x+i0)) > 0$ on each band.

(ii) Fix $\varepsilon > 0$. Then since $|H(z)| = O(1/|z|)$ near infinity, $\operatorname{Im}(H(z) + i\varepsilon) > 0$ near infinity. On $\mathbb{R}$, $\operatorname{Im}(H(x+i0)) \geq 0$. So, by the maximum principle and the fact that $\operatorname{Im}(H(z))$ is harmonic on $\mathbb{C}_+$, continuous on $\mathbb{C}_+ \cup \{0\} \cup \mathbb{R}$, we see $\operatorname{Im}(H(x + i0) + i\varepsilon) > 0$. Since ε is arbitrary, H is Herglotz. Thus, by Theorems 2.3.6 and 2.3.7, (5.5.140) holds. Since $H(z) = -\frac{1}{z} + O(\frac{1}{z^2})$ at infinity, ν is a probability measure.

(iii) $H(z)$ is bounded and continuous on $\mathbb{R} \setminus \{\alpha_j, \beta_j\}_{j=1}^{\ell}$ so on that set, $d\nu$ is a.c. with density given by $\frac{1}{\pi}\operatorname{Im}(H(x+i0))$, that is, by (5.4.96). The only potential singular measure is on $\{\alpha_j, \beta_j\}_{j=1}^{\ell}$, which, as a finite set, can only support a pure point piece. Since $\lim_{\varepsilon\downarrow 0} \varepsilon|H(x+i\varepsilon)| = 0$ for all $x \in \mathbb{R}$, ν has no pure points by Proposition 2.3.12.

(iv), (v) Define Φ_ν by (5.5.141). We claim

$$\Phi_\nu(z) = -\int \log|z-x|\, d\nu(x) \tag{5.5.143}$$

for both sides have the same derivative (by (5.5.140)) and both are $-\log|z| + o(1)$ at infinity, so their difference goes to zero. Φ_ν is continuous on $\mathbb{C} \setminus \{\alpha_j, \beta_j\}_{j=1}^{\ell}$ with derivative $\operatorname{Re}\int \frac{d\nu(z)}{(x-z)^{-1}}$ off $\mathbb{R}$ and with continuous boundary values on $\mathfrak{e} \setminus \{\alpha_j, \beta_j\}_{j=1}^{\ell}$.

Thus, $\Phi_\nu(x)$ is constant on each band, since the derivative is 0 there. By (5.5.136), the integral of the derivative across each gap is 0, so $\Phi_\nu(x)$ is constant on $\mathfrak{e}$. It follows by Theorem 5.5.15 that ν is the equilibrium measure for $\mathfrak{e}$ and that the constant value of Φ_ν on $\mathfrak{e}$, given by the right side of (5.5.142), is $\log(C(\mathfrak{e})^{-1})$, proving (5.5.142). □

Proposition 5.5.23. *Let $\mathfrak{e}_1 \subset (-\infty, 0]$, $\mathfrak{e}_2 \subset [0, \infty)$. For $a \geq 0$, let*

$$\mathfrak{e}(a) = \mathfrak{e}_1 \cup (\mathfrak{e}_2 + a) \tag{5.5.144}$$

Then $C(a)$ is monotone increasing as a increases.

Remark. This is an expression of the repulsive nature of the Coulomb force.

Proof. Let $\mathcal{M}_a$ be $M_{+,1}(\mathfrak{e}(a))$. Map $\mathcal{M}_a$ to $\mathcal{M}_{a'}$ by

$$Q_{a',a}(\mu) \restriction \mathfrak{e}_1 = \mu \restriction \mathfrak{e}_1$$
$$Q_{a',a}(\mu) \restriction \mathfrak{e}_2 + a' = (\mu \restriction \mathfrak{e}_2 + a) + (a' - a)$$

If $a' > a$, $|x - y + a' - a|^{-1} < |x - y|^{-1}$ for $x \in \mathfrak{e}_1$, $y \in \mathfrak{e}_2 + a$, so

$$\mathcal{E}(Q_{a',a}(\mu)) < \mathcal{E}(\mu) \tag{5.5.145}$$

Since $Q_{a',a}$ is a bijection,

$$\log(C(a'))^{-1} \leq \log(C(a))^{-1} \tag{5.5.146}$$

Thus, $C(a) \leq C(a')$. □

Theorem 5.5.24. *For any* $\mathfrak{e} \subset \mathbb{R}$,

$$C(\mathfrak{e}) \geq \tfrac{1}{4}\,|\mathfrak{e}| \tag{5.5.147}$$

Proof. By the last proposition, $C(\mathfrak{e})$ decreases as gaps are shrunk to zero, leaving an interval $\tilde{\mathfrak{e}}$ with $|\mathfrak{e}| = |\tilde{\mathfrak{e}}|$ and $C(\mathfrak{e}) \geq C(\tilde{\mathfrak{e}}) = \frac{1}{4}|\tilde{\mathfrak{e}}|$ by (5.5.124). □

Finally, we want to discuss the converse of Corollary 5.5.19. Here is the key theorem, part of which will not be proven until later:

Theorem 5.5.25. *Let* $\mathfrak{e} = \cup_{j=1}^{\ell+1} \tilde{\mathfrak{e}}_j$ *be a union of* $\ell + 1$ *disjoint closed intervals in* $\mathbb{R}$. *Then the following are equivalent:*

(i) *There is a two-sided Jacobi matrix,* J, *of period* p *so that* $\sigma(J) = \mathfrak{e}$.
(ii) *Each* $\tilde{\mathfrak{e}}$ *has rational harmonic measure.*
(iii) *There is a polynomial* Δ *with real coefficients and leading positive coefficients so*
  (a) *All zeros of* Δ *lie in* $\mathbb{R}$ *and are simple.*
  (b) *All zeros of* Δ' *lie in* $\mathbb{R}$ *and*

$$\Delta'(x_0) = 0 \Rightarrow |\Delta(x_0)| \geq 2 \tag{5.5.148}$$

  (c)

$$\mathfrak{e} = \Delta^{-1}([-2, 2]) \tag{5.5.149}$$

Remarks. 1. The proof shows that the minimal p in (i) is the minimal integer, p, with $p\rho_{\mathfrak{e}}(\tilde{\mathfrak{e}}_j) \in \mathbb{Z}$.

2. The proof also shows that the minimal degree of the Δ in (iii) is the minimal p in (i).

3. The analog of (iii) $\Rightarrow$ (i) for OPUC is dubbed the "Quacks like a discriminant" theorem in [400].

We will prove this result as Theorem 5.13.8 later. (i) $\Rightarrow$ (iii) is a combination of Theorems 5.3.7 and 5.4.2. That (ii) $\Leftrightarrow$ (iii) is sometimes called Aptekarev's theorem, after its discoverer [21].

The condition (5.5.149) is intended as a map of $\mathbb{R}$ to $\mathbb{R}$. However, (a)–(c) are equivalent to (5.5.149) as a complex result.

Proposition 5.5.26. (a)–(c) *in Theorem 5.5.25 where* (5.5.149) *is intended in the sense*

$$\mathfrak{e} = \{x \in \mathbb{R} \mid \Delta(x) \in [-2, 2]\} \tag{5.5.150}$$

are equivalent to (5.5.149) *in the sense that*

$$\mathfrak{e} = \{z \in \mathbb{C} \mid \Delta(z) \in [-2, 2]\} \tag{5.5.151}$$

Proof. Suppose $\deg(\Delta) = p$. If (a)–(c) hold, then as we have seen, $\Delta^{-1}((-2, 2)) \cap \mathfrak{e}$ is p disjoint intervals on each of which Δ is one-one onto $(-2, 2)$. Thus, if $\lambda \in (-2, 2)$, $\Delta(z) - \lambda = 0$ has p roots in $\mathfrak{e}$ so, since $\deg(\Delta) = p$, all roots. Thus,

$$\mathfrak{e} \supset \{z \in \mathbb{C} \mid \Delta(z) \in (-2, 2)\} \tag{5.5.152}$$

so, by continuity, (5.5.151) holds.

Conversely, suppose (5.5.151) holds. Then $\Delta(\bar{z}) = \overline{\Delta(z)}$ for $z \in \mathfrak{e}$ so, by polynomial continuation for all z, so Δ is real. Clearly, all roots are real. If f is an analytic function with $f'(x) = 0$ for some x_0 in $\mathbb{R}$ and $f(x_0)$ real, there are nonreal z near x_0, so $f(z)$ is real and near $f(x_0)$ (by writing $f(z) = f(x_0) + c(z - x_0)^{\ell} + \dots$ with $\ell \geq 2$), so (5.5.151) implies (5.5.148) for real solutions of $\Delta'(x_0)$ and, in particular, all zeros of Δ are simple. Since Δ has p zeros on $\mathbb{R}$, by Snell's theorem, Δ' has all its zeros on $\mathbb{R}$ also. □

Remarks and Historical Notes. The use of potential theory in the study of orthogonal polynomials goes back to work of Faber [125] and Szegő [432] about 1920 and was rediscovered in the physics literature fifty years later [199, 438]. After important contributions by Erdős–Turán [123], Widom [459], and Ullman [447], it was raised to high art by Stahl–Totik [417]. Applications of potential theory to OPs are reviewed in [404].

For expositions of the mathematics of two-dimensional potential theory, see especially Ransford [360], Landkof [264], the appendices of [404], and also [19, 196, 298, 378, 417, 446]. In particular, [263, 360, 417] discuss the theory for $\mathfrak{e} \subset \mathbb{C}$ rather than just $\mathfrak{e} \subset \mathbb{R}$.

The result mentioned after Corollary 5.5.8 that $\{z \mid \Phi_\mu(z) = \infty\}$ can be an arbitrary bounded G_δ of capacity zero is proven in [264]. Theorem 5.5.15 is true if (5.5.115) is assumed to hold for $d\rho_{\mathfrak{e}}$-a.e. x; see the appendix to [404].

It is easy to prove Lusin's theorem that we need in the proof of Theorem 5.5.7, namely, if μ is a measure on a compact set, E, and $f \in L^1(E, d\mu)$, then there are compact K with $\mu(E \setminus K)$ arbitrarily small and $f \restriction K$ continuous. For pick f_n continuous with $\|f - f_n\|_1 \leq 2^{-n}$ and $\|f_{n+1} - f_n\|_1 \leq 2^{-n}$. Let U_n be the open set where $|f_{n+1}(x) - f_n(x)| \geq 2^{-n/2}$ so $\mu(U_n) \leq 2^{-n/2}$. If $K_m = E \setminus \cup_{n=m}^{\infty} U_n$, then $\mu(E \setminus K_m) \leq \sum_{n=m}^{\infty} 2^{-n/2}$ can be made arbitrarily small, and on K_m, $f_1 + \sum_{n=1}^{\infty} f_{n+1} - f_1$ is uniformly convergent, hence continuous. One can easily go from this case ($f \in L^1$) to the general case (f measurable and finite almost everywhere).

Our discussion of the equilibrium measure for arbitrary finite gap sets, that is, Proposition 5.5.21, follows Totik [444].

Theorem 5.5.17 is a well-known fact associated with work of Widom [460] and Aptekarev [21].

5.6 APPROXIMATION BY PERIODIC SPECTRA, I. FINITE GAP SETS

The next six sections are a grand aside from the main subject of this chapter, periodic Jacobi matrices, and represent an application of this theory. In this section and Section 5.8, we approximate general compact subsets of $\mathbb{R}$ by periodic spectra in two stages: finite gap sets here and general sets in Section 5.8.

Our main result in this section is:

Theorem 5.6.1 (Bogatyrëv–Peherstorfer–Totik Theorem). *Let $\mathfrak{e} = \cup_{j=1}^{\ell+1} \mathfrak{e}_j$ be an ℓ-gap set of the form* (5.5.126). *Then for all m large, there exist ℓ-gap sets $\mathfrak{e}^{(m)} = \cup_{j=1}^{\ell+1} \mathfrak{e}_j^{(m)}$ with*

(i)

$$\mathfrak{e}_j \subset \mathfrak{e}_j^{(m)} \tag{5.6.1}$$

(ii) *Each $\mathfrak{e}_j^{(m)}$ has harmonic measure in $\mathfrak{e}^{(m)}$ equal to $k_j^{(m)}/m$ with $k_j^{(m)} \in \{1, 2, \dots\}$.*

(iii) *For some C_1, C_2,*

$$|\mathfrak{e}_j^{(m)} \setminus \mathfrak{e}_j| \le C_1 m^{-1} \tag{5.6.2}$$

$$C(\mathfrak{e}) \le C(\mathfrak{e}^{(m)}) \le C(\mathfrak{e}) + C_2 m^{-1} \tag{5.6.3}$$

Remarks. 1. We will construct $\mathfrak{e}_j^{(m)}$ so $\mathfrak{e}_j \not\subset (\mathfrak{e}_j^{(m)})^{\text{int}}$ (only the right endpoints will move and $\mathfrak{e}_{\ell+1}^{(m)} = \mathfrak{e}_\ell$), but as we will explain in the Notes, it is easy to arrange that $\mathfrak{e}_j \subset (\mathfrak{e}_j^{(m)})^{\text{int}}$.

2. Only (i) and (ii) are in [50], [336], [444]. (iii) is a later refinement of Totik [445].

Because of our explicit construction in Theorem 5.5.22, we can prove regularity of harmonic measures and capacities in $\{\alpha_j, \beta_j\}_{j=1}^{\ell+1}$. The key will then be to prove that if we fix $\{\alpha_j\}_{j=1}^{\ell+1}$ and $\beta_{\ell+1}$ and only vary $\beta_1, \dots, \beta_\ell$, the map from these ℓ variables to $(\mu_\mathfrak{e}([\alpha_1, \beta_1]), \dots, \mu_\mathfrak{e}([\alpha_\ell, \beta_\ell])$ is nonsingular, hence invertible. First the regularity:

Proposition 5.6.2. *Let $\mathfrak{e}$ be given by* (5.5.126) *where $\{\alpha_j, \beta_j\}_{j=1}^{\ell+1}$ obey* (5.5.127). *Then*

(1) *The coefficients of the monic polynomial P of degree ℓ obeying* (5.5.136) *are real analytic functions of $\{\alpha_j, \beta_j\}_{j=1}^{\ell+1}$ in the region* (5.5.127).

(2) *Each of the $\ell + 1$ measures $\mu_\mathfrak{e}([\alpha_j, \beta_j])$ is a real analytic function of $\{\alpha_j, \beta_j\}_{j=1}^{\ell+1}$ in the region* (5.5.127).

(3) *The capacity $C(\mathfrak{e})$ is a real analytic function of $\{\alpha_j, \beta_j\}_{j=1}^{\ell+1}$ in the region* (5.5.127).

Proof. (1) Let Y_{jk} for $j = 1, \dots, \ell; k = 1, \dots, \ell + 1$ be given by (5.5.137). We will show this is real analytic in $\{\alpha_j, \beta_j\}_{j=1}^{\ell+1}$. For $j = j_0$, analyticity of Y_{j_0k} in $\{\alpha_j\}_{j\neq j_0+1}$ and $\{\beta_j\}_{j\neq j_0}$ is immediate since h is real analytic in these parameters uniformly on each $(\beta_j + \varepsilon, \alpha_{j+1} - \varepsilon)$ with uniform $O(\varepsilon^{-1/2})$ integrable bounds on derivatives.

But there appears to be an issue with $\partial Y_{j_0k}/\partial\beta_{j_0}$ since $|h(x)|^{-1/2} = \infty$ at β_{j_0} and with $\partial h/\partial\beta_{j_0}$, which is not integrable at β_j! These problems actually cancel. To see this, change variables from x to $y = (x-\beta_{j_0})/(\alpha_{j_0+1}-\beta_{j_0})$ so the integral goes over $[0, 1]$. There is no endpoint variation and all derivatives in *any* α_j or β_j is bounded by $|y(1-y)|^{-1/2}$. Put more succinctly, $h(x(y))x^{-1}(1-x)^{-1}$ is real analytic in $\{\alpha_j, \beta_j\}_{j=1}^{\ell+1}$ and nonvanishing uniformly in a neighborhood of $y \in [0, 1]$.

Once we have analyticity of Y, the fact that $\det((Y_{jk})_{j,k=1,\dots,\ell}) \neq 0$ and the resulting explicit formula for P in terms of Y yields the required analyticity.

(2) We have

$$\mu_{\mathfrak{e}}([\alpha_{j_0}, \beta_{j_0}]) = \frac{1}{\pi}\int_{\alpha_{j_0}}^{\beta_{j_0}} \frac{|P(x)|}{\sqrt{|h(x)|}}\,dx \tag{5.6.4}$$

By the change of variables,

$$y = \frac{x-\alpha_{j_0}}{\beta_{j_0}-\alpha_{j_0}}$$

the region of integration becomes one over $[0, 1]$ and, as above, is real analytic.

(3) This follows from (1) and (5.5.142). □

We now turn to monotonicity properties of the harmonic measures, heading toward a proof that for $k \neq j$, $\partial\mu_{\mathfrak{e}}([\alpha_k, \beta_k])/\partial\beta_j < 0$.

Proposition 5.6.3. *If $\mathfrak{e}$, $\mathfrak{e}'$ are two ℓ-gap sets with $\mathfrak{e} \subset \mathfrak{e}'$, then for $x \in \mathfrak{e}$,*

$$\rho_{\mathfrak{e}}(x) \geq \rho_{\mathfrak{e}'}(x) \tag{5.6.5}$$

Remarks. 1. We will prove this in a more general context in Theorem 5.8.6. We will also see below (see (5.6.9)) that the inequality is strict.

2. This is saying that if an extra material is added to a perfect conductor, charge flows out into the extra material, decreasing the charge density everywhere in the original conductor.

Proof. Let $G_{\mathfrak{e}}$ be the potential theorist's Green's function given by (5.5.126). We claim first that for all $z \in \mathbb{C}$, we have

$$G_{\mathfrak{e}}(z) \geq G_{\mathfrak{e}'}(z) \tag{5.6.6}$$

For $G_{\mathfrak{e}} - G_{\mathfrak{e}'}$ is harmonic on $(\mathbb{C}\cup\{\infty\})\setminus\mathfrak{e}'$ continuous on $\mathbb{C}\cup\{\infty\}$. Thus, it suffices to prove the result on $\mathfrak{e}'$ by the maximum principle. On $\mathfrak{e}$, $G_{\mathfrak{e}} = G_{\mathfrak{e}'} = 0$, so (5.6.6) is trivial. On $\mathfrak{e}'\setminus\mathfrak{e}$, $G_{\mathfrak{e}'} = 0 \leq G_{\mathfrak{e}}$, since (5.5.105) holds. We have thus proven (5.6.6).

In the case at hand where $G_{\mathfrak{e}}$ is real analytic in a neighborhood of $\mathfrak{e}$ and $G_{\mathfrak{e}}(x) = 0$ for $x \in \mathfrak{e}$, we have for $x \in \mathfrak{e}$ that

$$\rho_{\mathfrak{e}}(x) = \frac{1}{\pi}\lim_{\varepsilon\downarrow 0}\frac{G_{\mathfrak{e}}(x+i\varepsilon)}{\varepsilon} \tag{5.6.7}$$

so (5.6.6) implies (5.6.5). □

Proposition 5.6.4. *Let $\mathfrak{e}$ be given in the form (5.5.126). Fix j_0 and let $\mathfrak{e}(\beta)$ for $\alpha_{j_0} < \beta < \alpha_{j_0+1}$ (or infinity if $j_0 = \ell+1$) be the set with β_{j_0} changed and the other parameters fixed. Then for $x \in \mathfrak{e}^{\text{int}}$,*

$$\left.\frac{\partial\rho_{\mathfrak{e}(\beta)}(x)}{\partial\beta}\right|_{\beta=\beta_{j_0}} < 0 \tag{5.6.8}$$

In particular, if $k \neq j_0$,

$$\left.\frac{d}{d\beta}\,\rho_{\mathfrak{e}(\beta)}([\alpha_k,\beta_k])\right|_{\beta=\beta_{j_0}} < 0 \tag{5.6.9}$$

Proof. We have, by Theorem 5.5.22, that

$$\rho_{\mathfrak{e}(\beta)}(x) = \frac{|P(x,\beta)|}{\sqrt{|h(x,\beta)|}}$$

where the signs of objects in absolute value are constant on each component of $\mathfrak{e}(\beta)$. Thus, on each component,

$$\frac{\partial\rho_{\mathfrak{e}(\beta)}(x)}{\partial\beta} = \frac{Q(x,\beta)}{\sqrt{|h(x,\beta)|^3}} \tag{5.6.10}$$

where

$$Q(x,\beta) = \pm\frac{\partial P(x,\beta)}{\partial\beta}\,h(x,\beta) \pm \tfrac{1}{2}\,\frac{\partial h(x,\beta)}{\partial\beta}\,P(x,\beta) \tag{5.6.11}$$

with the two $\pm$'s potentially different, determined by the signs of P and h.

Q is a polynomial in x since P and h are polynomials in x with analytic coefficients and $Q(x=\beta,\beta) \neq 0$ since $P(x=\beta,\beta)\neq 0$ (P has zeros strictly in the gaps), $\frac{\partial h}{\partial\beta}(x,\beta)\Big|_{x=\beta} \neq 0$, and $h(x=\beta,\beta)=0$. Thus, Q is not identically zero.

The degree of Q is $\ell+2\ell+2-1$ since P is of degree ℓ, h of degree $2\ell+2$, and both are monic. So $\partial/\partial\beta$ has degree at least one less. Moreover, Q has zeros (since both h and $\partial h/\partial\beta$ do) at $\{\alpha_k\}_{k=1}^{\ell+1}$ and $\{\beta_k\}_{k\neq j_0}$, so there are at most ℓ additional zeros.

For each gap (β_k,α_{k+1}), $k\neq j_0$, by taking the derivative of (5.5.136), we get

$$\int_{\beta_k}^{\alpha_{k+1}} \frac{Q(x,\beta)}{\sqrt{|h(x,\beta)|^3}}\,dx = 0 \tag{5.6.12}$$

(since Q vanishes at α_{k+1} and β_k, this is integrable), so Q must have a zero in each such gap.

That accounts for $\ell-1$ zeros, leaving one zero remaining. Since the last proposition implies $Q(x,\beta)\le 0$ on the bands, any zero on $\mathfrak{e}^{\text{int}}$ has to be a double zero, which means $Q(x,\beta)<0$ on the bands. This proves (5.6.8).

Since

$$\frac{d}{d\beta}\,\rho_{\mathfrak{e}(\beta)}([\alpha_k,\beta_k]) = \int_{\alpha_k}^{\beta_k} \frac{Q(x,\beta)}{\sqrt{|h(x,\beta)|^3}}\,dx \tag{5.6.13}$$

we then obtain (5.6.9). □

As a final preliminary, we need a result about diagonally dominant matrices:

Definition. A finite $n \times n$ matrix M is called *diagonally dominant* if and only if for $j = 1, \dots, n$, we have

$$|M_{jj}| > \sum_{k \neq j} |M_{jk}| \tag{5.6.14}$$

Lemma 5.6.5. *Any diagonally dominant matrix, M, is invertible; indeed, any eigenvalue, λ, obeys*

$$|\lambda| \geq \min_j \left[|M_{jj}| - \sum_{k \neq j} |M_{jk}| \right] \tag{5.6.15}$$

Proof. Suppose λ is an eigenvalue and $\{x_j\}_{j=1}^n$ solves

$$(Mx)_j = \lambda x_j \tag{5.6.16}$$

Without loss, we can pick j_0 so that

$$|x_{j_0}| \geq \max_{k \neq j_0} |x_k| \tag{5.6.17}$$

Thus,

$$\begin{aligned} |\lambda| &= \left| M_{j_0 j_0} - \sum_{k \neq j_0} M_{j_0 k} \frac{x_k}{x_{j_0}} \right| \\ &\geq |M_{j_0 j_0}| - \sum_{k \neq j_0} |M_{j_0 k}| \frac{|x_k|}{|x_{j_0}|} \\ &\geq |M_{j_0 j_0}| - \sum_{k \neq j_0} |M_{j_0 k}| \end{aligned}$$

□

Proof of Theorem 5.6.1. Fix $\alpha_1^{(0)}, \dots, \alpha_{\ell+1}^{(0)}$ and $\beta_{\ell+1}^{(0)}$ but allow $\beta_1, \dots, \beta_\ell$ to vary within the condition (5.5.127). Define for $j = 1, \dots, \ell + 1$,

$$f_j(\beta_1, \dots, \beta_\ell) = \mu_{\mathfrak{e}(\beta_1, \beta_2, \dots, \beta_\ell)}([\alpha_j, \beta_j]) \tag{5.6.18}$$

and F: allowed $\beta_1, \dots, \beta_\ell$ to $\mathbb{R}^\ell$ by

$$F(\beta_1, \dots, \beta_\ell) = (f_1(\vec{\beta}), \dots, f_\ell(\vec{\beta}))$$

Since

$$\sum_{j=1}^{\ell+1} f_j = 1 \tag{5.6.19}$$

we have

$$\sum_{j=1}^{\ell} \frac{\partial f_j}{\partial \beta_k} = -\frac{\partial f_{\ell+1}}{\partial \beta_k} > 0 \tag{5.6.20}$$

by (5.6.9).

Also by (5.6.9), $\partial f_j/\partial\beta_k < 0$ for $j \neq k$, and thus, $\partial f_k/\partial\beta_k > 0$ by (5.6.20). It follows that

$$\left|\frac{\partial f_k}{\partial \beta_k}\right| - \sum_{\substack{j\neq k\\ j\leq \ell}} \left|\frac{\partial f_j}{\partial \beta_k}\right| = \sum_{j=1}^{\ell} \frac{\partial f_j}{\partial \beta_k} > 0 \tag{5.6.21}$$

by (5.6.20). So the derivative of F is diagonally dominant and so invertible.

Thus, F is a locally invertible C^1 (indeed, real analytic) map with C^1 local inverse.

Therefore, for any fixed initial set $\mathfrak{e}$ with parameters $\alpha^{(0)}$, $\beta^{(0)}$, those $\beta_1, \dots, \beta_\ell$ in $\mathbb{R}_+^\ell$ near $\beta_1^{(0)}, \dots, \beta_\ell^{(0)}$ map to a set S, which contains the intersection of an open ball about $F(\beta^{(0)})$ and an open cone with vertex $F(\beta^{(0)})$. Such an S for all large n contains balls with center s_n obeying $|s_n - F(\beta^{(0)})| \leq K_1/n$ and radius $r_n = \sqrt{\ell}/n$.

Such balls contain points of the form $(p_1/n, \dots, p_\ell/n)$ for integral p_j, so since F^{-1} is C^1, we obtain $\beta_1, \dots, \beta_\ell$. Hence,

$$\beta_j^{(0)} \leq \beta_j \leq \beta_j^{(0)} + \frac{C_1}{n} \tag{5.6.22}$$

with $F_j(\beta) = p_j/n$ for $j = 1, \dots, \ell+1$ ($\ell+1$ can be included by (5.6.19)). Thus, we have (i), (ii), and (5.6.2). (5.6.3) then follows from the fact that $C(\cdot)$ is a C^1 function of $(\beta_1, \dots, \beta_\ell)$ near $(\beta_1^{(0)}, \dots, \beta_\ell^{(0)})$. $\square$

As an application of Theorem 5.6.1, we study:

Definition. Let $\mathfrak{e}$ be a compact subset of $\mathbb{C}$. The *Chebyshev constants*, $t_n(\mathfrak{e})$, are defined by

$$t_n(\mathfrak{e}) = \min\{\|Q_n\|_{\mathfrak{e}} \mid Q_n \text{ monic of degree } n\} \tag{5.6.23}$$

where

$$\|f\|_{\mathfrak{e}} = \sup_{z\in\mathfrak{e}} |f(z)| \tag{5.6.24}$$

There are minimizing Q's, the Chebyshev polynomials studied in the next section. The Chebyshev constants are relevant to the theory of orthogonal polynomials because:

Theorem 5.6.6. *If μ is a measure supported by a compact set, $\mathfrak{e}\subset\mathbb{C}$, and $X_n(z, d\mu)$ are the monic OPs for μ, then*

$$\|X_n\|_{L^2(\mathbb{C},d\mu)} \leq t_n(\mathfrak{e})\mu(\mathfrak{e})^{1/2} \tag{5.6.25}$$

In particular, if $\mathfrak{e} \subset \mathbb{R}$ and $\{a_n, b_n\}_{n=1}^\infty$ are the Jacobi parameters for μ, then

$$a_1 \dots a_n \leq t_n(\mathfrak{e}) \tag{5.6.26}$$

Proof. Clearly, for any Q_n,

$$\|Q_n\|_{L^2}^2 \leq \|Q_n\|_{\mathfrak{e}}^2 \mu(\mathfrak{e}) \tag{5.6.27}$$

so minimizing using

$$\min \| Q_n \|_{L^2(\mathbb{C},d\mu)} = \| X_n \|_{L^2(\mathbb{C},d\mu)}$$

we get (5.6.25). □

Theorem 5.6.7 (Totik–Widom Theorem). *Let $\mathfrak{e}$ be a finite gap set in $\mathbb{R}$. Then there exists a constant w so*

$$t_n(\mathfrak{e}) \le wC(\mathfrak{e})^n \tag{5.6.28}$$

In particular, if $\operatorname{supp}(\mu) \subset \mathfrak{e}$ *and* $\mu(\mathfrak{e}) = 1$, *then*

$$\frac{a_1 \dots a_n}{C(\mathfrak{e})^n} \le w \tag{5.6.29}$$

Remarks. 1. To put this in context, we note we will prove that for any $\mathfrak{e} \subset \mathbb{C}$, one has (see Theorem 5.7.8)

$$t_n(\mathfrak{e}) \ge C(\mathfrak{e})^n \quad \text{and} \quad \lim_{n\to\infty} t_n(\mathfrak{e})^{1/n} = C(\mathfrak{e}) \tag{5.6.30}$$

and for $\mathfrak{e} \subset \mathbb{R}$, one has (see Corollary 5.7.7)

$$t_n(\mathfrak{e}) \ge 2C(\mathfrak{e})^n$$

2. We show later (see Example 5.7.3) that the polynomial $\widetilde{T}_m$ below is actually the minimizer of $\| Q_m \|_{\mathfrak{e}^{(m)}}$, so one has equality in (5.6.35).

Proof. Pick M so for $m \ge M$, we have sets $\mathfrak{e}^{(m)}$ obeying the conclusions of Theorem 5.6.1. Since $\mathfrak{e}^{(m)}$ is the spectrum of a periodic problem of period m (see Theorem 5.5.25), there are Jacobi parameters $\{a_j, b_j\}_{j=1}^\infty$ of period m, so (see (5.5.121))

$$a_1 \dots a_m = C(\mathfrak{e}^{(m)})^m \tag{5.6.31}$$

and discriminant $\Delta_m(x)$ for this Jacobi matrix.

Since $\mathfrak{e}^{(m)} = \Delta_m^{-1}([-2, 2])$ and

$$\Delta_m(x) = (a_1 \dots a_m)^{-1} x^m + \cdots \tag{5.6.32}$$

we have that

$$\widetilde{T}_m(x) = (a_1 \dots a_m)\Delta_m(x) \tag{5.6.33}$$

is a monic polynomial with

$$\| \widetilde{T}_m \|_{\mathfrak{e}^{(m)}} = 2C(\mathfrak{e}^{(m)})^m \tag{5.6.34}$$

which implies

$$t_m(\mathfrak{e}^{(m)}) \le 2C(\mathfrak{e}^{(m)})^m \tag{5.6.35}$$

But trivially $t_m(\mathfrak{e}) \le t_m(\mathfrak{e}^{(m)})$ since $\mathfrak{e} \subset \mathfrak{e}^{(m)}$ and by (5.6.3),

$$t_m(\mathfrak{e}) \le 2C(\mathfrak{e}^{(m)})^m \le 2C(\mathfrak{e})^m \left[1 + \frac{C_2}{mC(\mathfrak{e})}\right]^m \tag{5.6.36}$$

so

$$\limsup \frac{t_m(\mathfrak{e})}{C(\mathfrak{e})^m} \le 2\exp(C_2 C([\mathfrak{e}])^{-1}) < \infty \tag{5.6.37}$$

proving (5.6.28). □

Remarks and Historical Notes. Theorem 5.6.1(i), (ii) were obtained with very different proofs by Bogatyrëv [50] (using conformal mapping techniques), Peherstorfer [337, 339, 340] (using Chebyshev polynomials; see the Notes to Section 5.7), and Totik [444] (using methods close to ours here). Totik then noted (iii) (with a different proof) in [445]. The argument we use to get (iii) is new here.

Theorem 5.6.7 follows from a theorem of Widom [460] who proved $t_n(\mathfrak{e})/C(\mathfrak{e})^n$ is a bounded almost periodic function. The much simpler approach we use here is due to Totik [445].

As noted, our $(\mathfrak{e}^{(m)})^{\text{int}}$ does not include all of $\mathfrak{e}$. However, one can first increase all β's by $O(1/m)$ and decrease all α's by $O(1/m)$ and then use our construction on this larger set to get new $\mathfrak{e}^{(m)}$'s that also obey $\mathfrak{e} \subset (\mathfrak{e}^{(m)})^{\text{int}}$.

5.7 CHEBYSHEV POLYNOMIALS

> Chebyshev polynomials are everywhere dense in numerical analysis.
>
> —Mason and Handscomb [299], who say it is well known and might be due to Phillip Davis or to George Forsythe.

In an aside on our asides, we study in more detail the minimizers in the definition of Chebyshev constants.

Definition. Given a compact set $\mathfrak{e} \subset \mathbb{R}$, the *Chebyshev polynomials*, T_n (or $T_n^{(\mathfrak{e})}(x)$ if we need to make $\mathfrak{e}$ explicit), are the monic polynomials of degree n that minimize

$$\|f\|_{\mathfrak{e}} \equiv \sup_{x\in\mathfrak{e}} |f(x)| \tag{5.7.1}$$

that is,

$$t_n(\mathfrak{e}) \equiv \|T_n\|_{\mathfrak{e}} = \min\{\|Q_n\|_{\mathfrak{e}} \mid Q_n \text{ monic of degree } n\} \tag{5.7.2}$$

We will prove later (see Corollary 5.7.6) that if $\mathfrak{e}$ is not a finite set, T_n is unique. To see there is a minimum, suppose $\mathfrak{e}$ is infinite, pick any $Q_n^{(0)}$, and note $\{Q_n \mid \|Q_n\|_{\mathfrak{e}} \le \|Q_n^{(0)}\|_{\mathfrak{e}}\}$ is a nonempty set compact in the topology of convergence of coefficients. $Q_n \to \|Q_n\|_{\mathfrak{e}}$ is continuous in this topology, so the minimum value is taken. Since $\mathfrak{e}$ is infinite, $\|Q_n\|_{\mathfrak{e}}$ is never zero for a monic Q_n, so $\|T_n\| > 0$.

One can make this definition for any compact $\mathfrak{e} \subset \mathbb{C}$, and occasionally we will indicate results for that case.

Theorem 5.7.1 (Alternation Principle). *If Q_n is a monic polynomial with n simple zeros in $\mathfrak{e}$ so that each zero z_j lies in an interval $(z_j^-, z_j^+) \subset \mathfrak{e}$ where $Q_n'(x) \neq 0$ and*

$$|Q_n(z_j^\pm)| = \|Q_n\|_\mathfrak{e} \tag{5.7.3}$$

then Q_n is a Chebyshev polynomial for $\mathfrak{e}$ of degree n.

Proof. Suppose there is a monic polynomial T_n with

$$\|T_n\|_\mathfrak{e} < \|Q_n\|_\mathfrak{e} \tag{5.7.4}$$

Then $|T_n(z_j^\pm)| < |Q_n(z_j^\pm)|$, so

$$\operatorname{sgn}(Q_n(z_j^\pm) - T_n(z_j^\pm)) = \operatorname{sgn}(Q_n(z_j^\pm)) \tag{5.7.5}$$

Since $Q_n'(x) \neq 0$ on (z_j^-, z_j^+), $Q_n(z_j^+) = -Q_n(z_j^-)$, and thus, $Q_n - T_n$ has different signs at z_j^+ and z_j^-, and so a zero in (z_j^-, z_j^+).

Since $Q_n'(x)$ has a zero between any two zeros of Q_n, these intervals are disjoint, so $Q_n - T_n$ has at least n zeros. But Q_n and T_n are distinct monic polynomials, so $Q_n - T_n$ has at most $n - 1$ zeros. This contradiction shows (5.7.4) cannot occur, so Q_n has minimum norm. □

Example 5.7.2. Recall that the classical Chebyshev polynomials of the first kind are defined by

$$p_n(\cos\theta) = \cos(n\theta) \tag{5.7.6}$$

Since $\cos(n\theta) = (e^{in\theta} + e^{-in\theta})/2$ and $\cos^n\theta = [(e^{i\theta} + e^{-i\theta})/2]^n = 2^{-n}e^{in\theta}+$ lower order, p_n is not monic. Rather,

$$p_n(x) = 2^{n-1}x^n + \text{lower order} \tag{5.7.7}$$

We claim that

$$T_n(x) = 2^{-(n-1)}p_n(x) \tag{5.7.8}$$

are the Chebyshev polynomials for $[-1, 1]$. They are monic and the zeros of $p_n(x)$ occur at $x_\ell = \cos(\frac{2\pi(\ell+\frac{1}{2})}{n})$ for $\ell = 0, 1, \dots, n-1$ and each x_ℓ lies in an interval $[x_\ell^-, x_\ell^+]$ where $|p_n(x_\ell^\pm)| = 2^{-(n-1)} = \|T_n\|_{[-1,1]}$ and $p_n(x)$ is monotone on (x_ℓ^-, x_ℓ^+). This proves (5.7.8) is indeed the Chebyshev polynomial for the set.

Notice that (with $C(\cdot)$ = capacity)

$$\|T_n\|^{1/n} = 2^{-(n-1)/n} \to \tfrac{1}{2} = C([-1, 1]) \tag{5.7.9}$$

□

Example 5.7.3. Let $\mathfrak{e} = \cup_{j=1}^{\ell+1}\mathfrak{e}_j$ be an ℓ gap set, which is the spectrum of a two-sided Jacobi matrix J of period p. Let

$$\Delta(x) = (a_1 \dots a_p)^{-1}x^p + \text{lower order} \tag{5.7.10}$$

be its discriminant. Let $\tilde{\mathfrak{e}}_1, \dots, \tilde{\mathfrak{e}}_p$ be the closed bands. Each has a zero $z_j \in \tilde{\mathfrak{e}}_j$ of Δ, $\sup_{x\in\mathfrak{e}}|\Delta(x)| = 2$ since $\mathfrak{e} = \Delta^{-1}([-2, 2])$, and every $\tilde{\mathfrak{e}}_j$ is precisely the kind of interval required in Theorem 5.7.1. It follows that

$$T_p(x) = (a_1 \dots a_p)\Delta(x) \tag{5.7.11}$$

is the Chebyshev polynomial of $\mathfrak{e}$ and

$$\|T_p\|_{\mathfrak{e}} = 2(a_1 \dots a_p) \tag{5.7.12}$$

For each $k = 1, 2, \dots$, we can consider J as a matrix of period kp with discriminant $\Delta^{(k)}$. Indeed, if $p_k(\cos\theta) = \cos(k\theta)$, then $\Delta^{(k)}(x) = 2p_k(\frac{1}{2}\Delta(x))$. As above,

$$T_{pk}(x) = (a_1 \dots a_p)^k \Delta^{(p)}(x) \tag{5.7.13}$$

and

$$\|T_{pk}\|_{\mathfrak{e}} = 2(a_1 \dots a_p)^k$$

In particular, by (5.5.15),

$$\|T_{pk}\|_{\mathfrak{e}}^{1/pk} \to (a_1 \dots a_p) = C(\mathfrak{e}) \tag{5.7.14}$$

$\square$

Notice that if $T_n(z) = z^n + a_{n-1}z^{n-1} + \dots$, $\widetilde{T}_n(z) = z^n + \operatorname{Re}(a_{n-1})z^{n+1} + \dots$ has $\widetilde{T}_n(x) = \operatorname{Re} T_n(x)$ on $\mathfrak{e}$, so $\|\widetilde{T}_n\|_{\mathfrak{e}} \le \|T_n\|_{\mathfrak{e}}$, and thus, we can suppose T_n is a real polynomial, which we henceforth do.

Lemma 5.7.4. *Let $q_m(x)$ be a real polynomial and $a \le b$. For $\varepsilon > -\frac{1}{2}(b-a)$, let*

$$p_{m+2}^{(\varepsilon)}(x) = (x - (b+\varepsilon))(x - (a-\varepsilon))q_m(x)$$

Then for $\varepsilon > 0$ and any compact $K \subset \mathbb{R} \setminus [a-\varepsilon, b+\varepsilon]$,

$$\sup_{x\in K} |p_{m+2}^{(\varepsilon)}(x)| < \sup_{x\in K} |p_{m+2}^{(0)}(x)| \tag{5.7.15}$$

and for $\varepsilon < 0$ and any compact $K \subset \mathbb{R} \setminus [a, b]$,

$$\sup_{x\in K} |p_{m+2}^{(\varepsilon)}(x)| > \sup_{x\in K} |p_{m+2}^{(0)}(x)| \tag{5.7.16}$$

For $\varepsilon > 0$ and any compact $K \subset (a, b)$,

$$\sup_{x\in K} |p_{m+2}^{(\varepsilon)}(x)| > \sup_{x\in K} |p_{m+2}^{(0)}(x)| \tag{5.7.17}$$

and for $\varepsilon < 0$ and any compact $K \subset (a-\varepsilon, b+\varepsilon)$,

$$\sup_{x\in K} |p_{m+2}^{(\varepsilon)}(x)| < \sup_{x\in K} |p_{m+2}^{(0)}(x)| \tag{5.7.18}$$

Remark. In other words, if a pair of zeros are moved symmetrically apart, $|p|$ decreases outside the zeros and increases inside, and vice versa if the zeros are symmetrically moved together.

Proof. Without loss, we can suppose $a = -b$ with $b > 0$. Then $x^2 - (b+\varepsilon)^2$ is strictly decreasing as ε increases. So $|x^2 - (b+\varepsilon)^2|$ strictly increases in $|x| < (b+\varepsilon)$ and strictly decreases in $|x| > (b+\varepsilon)$. This holds for all x so remains true if we multiply by $|q_m(x)|$. $\square$

Theorem 5.7.5. *Let $\mathfrak{e} \subset \mathbb{R}$ be compact and let T_n be the Chebyshev polynomials for $\mathfrak{e}$. Then*

(i) *All zeros of T_n lie in $\mathbb{R}$.*
(ii) *All zeros of T_n are simple.*
(iii) *All zeros of T_n lie in* cvh($\mathfrak{e}$), *the convex hull of $\mathfrak{e}$.*
(iv) *If $(a, b) \cap \mathfrak{e} = \emptyset$, then T_n has at most one zero in (a, b).*
(v) *If $x_j < x_{j+1}$ are two successive zeros of T_n, then $T_n'(y)$ has exactly one zero y_j in $[x_j, x_{j+1}]$ and*

$$|T_j(y_j)| \geq \|T_n\|_{\mathfrak{e}} \tag{5.7.19}$$

with equality if $y_j \in \mathfrak{e}$.
(vi) *Moreover, there is $w_j \in (x_j, x_{j+1})$ so $w_j \in \mathfrak{e}$ and $|T(w_j)| = \|T_n\|_{\mathfrak{e}}$. Similarly, there is $w_0, w_n \in \mathfrak{e}$, $w_0 \in (-\infty, x_0)$, and $w_n \in (x_n, \infty)$ so that $|T_n(w_0)| = |T_n(w_n)| = \|T_n\|_{\mathfrak{e}}$.*

Proof. (i) As noted above, we can suppose T_n is real on $\mathbb{R}$, so $T_n(\bar{z}) = \overline{T_n(z)}$, and if $a + ib$ is a zero, so is $a - ib$. Since

$$|(x - (a - ib))(x - (a + ib))| = (x - a)^2 + b^2 \tag{5.7.20}$$

$T_n(x)$ would be decreased for all x if we replace $b \neq 0$ by $b = 0$. By the minimum norm definition, no zero can have $b \neq 0$.

(ii) By the lemma, if x_0 is a double zero, replace $(x - x_0)^2$ by $(x - (x_0 + \varepsilon))(x - (x_0 - \varepsilon))$ and decrease $\|T_n\|$ on $\mathfrak{e} \setminus (x_0 - \varepsilon, x_0 + \varepsilon)$. Since for ε small, $(x - (x_0 - \varepsilon))(x - (x_0 + \varepsilon))$ is small on $[x_0 - \varepsilon, x_0 + \varepsilon]$, we see we can decrease $\|T_n\|_{\mathfrak{e}}$. Thus, T_n cannot have double zeros.

(iii) If $a = \inf \mathfrak{e}$ and $x_0 < a$, then $|x - (x_0 + \varepsilon)| < |x - x_0|$ for all $x > a$. Thus, we can decrease $\|T_n\|_{\mathfrak{e}}$ by moving a zero below a upward. By the minimum definition, there can be no zeros on $(-\infty, a)$.

(iv) By the lemma, if (a, b) has two zeros $x_0 < x_1$, we can decrease $\|T_n\|_{\mathfrak{e}}$ by moving them slightly apart, violating the minimum property.

(v) By Snell's theorem, T_n' has at least one zero in each (x_j, x_{j+1}). Since T_n has n distinct zeros on $\mathbb{R}$, this accounts for $n - 1$ zeros of T_n' and so for all the zeros, so there is exactly one in each (x_j, x_{j+1}). If $|T_n(y_j)| < \|T_n\|_{\mathfrak{e}}$, $\sup_{[x_j, x_{j+1}]} |T_n(x)| < \|T_n\|_{\mathfrak{e}}$. Moving x_j, x_{j+1} apart, we decrease $|T_n(x)|$ on $\mathfrak{e} \setminus [x_j, x_{j+1}]$ and the increase on $[x_j, x_{j+1}]$ can be kept so small that we remain strictly less than $\|T_n\|_{\mathfrak{e}}$ there. This would decrease $\|T_n\|_{\mathfrak{e}}$, violating the minimum definition. Thus, (5.7.19) holds. Clearly, if $y_j \in \mathfrak{e}$, $|T_n(y_j)| \leq \|T_n\|_{\mathfrak{e}}$.

(vi) As in the proof of (iv), if $\sup_{w \in [x_j, x_{j+1}] \cap \mathfrak{e}} |T_n(w)| < \|T_n\|_{\mathfrak{e}}$, we can move the zeros slightly apart, so the new polynomial is still strictly less than $\|T_n\|_{\mathfrak{e}}$ on $[x_j, x_{j+1}]$ and the sup is decreased off $[x_j, x_{j+1}]$. Similarly, if $\sup_{w \in (-\infty, x_1] \cap \mathfrak{e}} |T_n(w)| < \|T_n\|_{\mathfrak{e}}$, we can move x_1 up and decrease $\|T_n\|_{\mathfrak{e}}$. □

Corollary 5.7.6. *The T_n minimizing $\|T_n\|_{\mathfrak{e}}$ is unique.*

Proof. Suppose P_n and Q_n are two distinct minimizers and let $T_n = \frac{1}{2}(P_n + Q_n)$. Since $\|T_n\|_{\mathfrak{e}} \leq \max(\|P_n\|_{\mathfrak{e}}, \|Q_n\|_{\mathfrak{e}})$, T_n is also a minimizer. Let $x_1 < \cdots < x_n$ be its simple zeros (by (ii) of the theorem) and let $x_j < w_j < x_{j+1}$ and

$w_0 \in (-\infty, x_1)$, $w_n \in (x_n, \infty)$ be such that $w_j \in \mathfrak{e}$ and $\|T_n(w_j)\| = \|T_n\|_\mathfrak{e}$ (which exist by (vi) of the last theorem). Since $|P_n(w_j)| \le \|T_n\|_\mathfrak{e}$, $|Q_n(w_j)| \le \|T_n\|_\mathfrak{e}$, and $\frac{1}{2}|P_n(w_j) + Q_n(w_j)| = |T_n(w_j)| = \|T_n\|_\mathfrak{e}$, we have $Q_n(w_j) = P_n(w_j) = T_n(w_j)$. Thus, $P_n - Q_n$ has at least $n+1$ zeros! Since $\deg(P_n - Q_n) \le n-1$, $P_n = Q_n$. $\square$

Corollary 5.7.7 (Schiefermayr's Theorem). *We always have*

$$t_n(\mathfrak{e}) \ge 2C(\mathfrak{e})^n \tag{5.7.21}$$

where $C(\mathfrak{e})$ is the capacity of $\mathfrak{e}$.

Remarks. 1. By Example 5.7.2, one has equality in (5.7.21) for all n if $\mathfrak{e} = [-1, 1]$. Thus, the number 2 in (5.7.21) cannot be increased.

2. For $\mathfrak{e} = \partial\mathbb{D}$, $T_n(z) = z^n$ and $t_n(\partial\mathbb{D}) = 1$, so (5.7.21) only holds for $\mathfrak{e} \subset \mathbb{R}$, not all $\mathfrak{e} \subset \mathbb{C}$.

Proof. Let

$$\Delta_n(x) = \frac{2T_n(x)}{\|T_n\|_\mathfrak{e}} \tag{5.7.22}$$

and let

$$\mathfrak{e}_n = \Delta_n^{-1}([-2, 2]) \tag{5.7.23}$$

By Theorem 5.7.5, Δ_n has all its zeros on $\mathbb{R}$, they are simple, and $\Delta'(x_0) = 0 \Rightarrow |\Delta(x_0)| \ge 2$. Thus, by Theorem 5.5.25, $\mathfrak{e}_n$ is the spectrum of a Jacobi matrix of period n and Δ_n is its discriminant.

By Example 5.7.3, $T_n^{(\mathfrak{e}_n)}$ is the monic multiple of Δ_n, and so $T_n = T_n^{(\mathfrak{e}_n)}$ and $\|T_n\|_\mathfrak{e} = \|T_n^{(\mathfrak{e}_n)}\|_{\mathfrak{e}_n} = 2C(\mathfrak{e}_n)^n \ge 2C(\mathfrak{e})^n$ since $\mathfrak{e} \subset \mathfrak{e}_n$, proving (5.7.21). $\square$

We are heading toward generalizing (5.7.9) and (5.7.14) and showing $\|T_n\|_\mathfrak{e}^{1/n} \to C(\mathfrak{e})$ for all $\mathfrak{e} \subset \mathbb{R}$, a result that holds for all $\mathfrak{e} \subset \mathbb{C}$ essentially by the same proof. It will be useful to have an additional notion:

Definition. Let $\mathfrak{e} \subset \mathbb{R}$. An n-point *Fekete set* is $x_1^{(0)}, \dots, x_n^{(0)} \in \mathfrak{e}$ so that if

$$q_n(x_1, \dots, x_n) = \prod_{i \ne j} |x_j - x_\nu| \tag{5.7.24}$$

then

$$q_n(x_1^{(0)}, \dots, x_n^{(0)}) = \sup_{(x_1, \dots, x_n) \in \mathfrak{e}} q_n(x_1, \dots, x_n) \tag{5.7.25}$$

We set

$$\zeta_n(\mathfrak{e}) = q_n(x_1^{(0)}, \dots, x_n^{(0)})^{1/n(n-1)} \tag{5.7.26}$$

Remark. The number of $i \ne j$ in (5.7.24) is $n(n-1)$, explaining the power in (5.7.26).

Notice that

$$q_{n+1}(x_1^{(0)}, \dots, x_{n+1}^{(0)})^{n-1} = \prod_{j=1}^{n+1} q_n(x_1^{(0)}, \dots, \widehat{x_j^{(0)}}, \dots, x_{n+1}^{(0)}) \tag{5.7.27}$$

(where $\widehat{x}_j^{(0)}$ means dropping $x_j^{(0)}$) since each pair (i,j) occurs on the right of (5.7.3) $n-1$ times. Thus,

$$[\zeta_{n+1}^{(n+1)n}]^{n-1} \le [\zeta_n^{(n-1)n}]^{n+1}$$

or

$$\zeta_{n+1}(\mathfrak{e}) \le \zeta_n(\mathfrak{e}) \tag{5.7.28}$$

and

$$\zeta_\infty(\mathfrak{e}) = \lim_{n\to\infty} \zeta_n(\mathfrak{e}) \tag{5.7.29}$$

exists. It is called the *transfinite diameter* of $\mathfrak{e}$.

Theorem 5.7.8 (Faber–Fekete–Szegő Theorem). *Let $\mathfrak{e} \subset \mathbb{R}$ be compact. Then, for all n,*

$$C(\mathfrak{e}) \le \|T_n\|_{\mathfrak{e}}^{1/n} \le \zeta_{n+1} \tag{5.7.30}$$

Moreover,

(i) *The normalized counting measure for Fekete sets converges to $d\rho_{\mathfrak{e}}$, the equilibrium measure for $\mathfrak{e}$.*

(ii) *$\zeta_\infty(\mathfrak{e}) = C(\mathfrak{e})$, so*

$$\lim_{n\to\infty} \|T_n\|_{\mathfrak{e}}^{1/n} = C(\mathfrak{e}) \tag{5.7.31}$$

(iii) *If $\mathfrak{e}$ is potentially perfect, then the zero counting measure for T_n converges to $d\rho_{\mathfrak{e}}$, the equilibrium measure of $\mathfrak{e}$.*

Remark. We proved in (5.7.21) a stronger statement than the first inequality in (5.7.30). We include (5.7.30) here because, unlike (5.7.21), it holds for all $\mathfrak{e} \subset \mathbb{C}$.

Proof. Let Q_n be any monic polynomial. By the Bernstein–Walsh lemma (5.5.114),

$$\frac{|Q_n(z)|}{|z|^n} \le \|Q_n\|_{\mathfrak{e}} \exp(n[G_{\mathfrak{e}}(z) - \log(|z|)]) \tag{5.7.32}$$

Take $|z| \to \infty$, $|Q_n(z)|/|z|^n \to 1$ by the fact that Q_n is monic. By (5.5.113), $G_{\mathfrak{e}}(z) - \log(|z|) \to -\log(C(\mathfrak{e}))$. Thus, (5.7.32) becomes

$$1 \le \|Q_n\|_{\mathfrak{e}} \exp(-n \log(C(\mathfrak{e}))) \tag{5.7.33}$$

or

$$\|Q_n\|_{\mathfrak{e}}^{1/n} \ge C(\mathfrak{e}) \tag{5.7.34}$$

which implies the first inequality in (5.7.30).

For $j = 1, \dots, n+1$, let

$$Q_j(z) = \prod_{k\ne j} (z - x_k^{(0)}) \tag{5.7.35}$$

for an $(n+1)$-point Fekete set. By (5.7.25),

$$\sup_{x\in\mathfrak{e}} |Q_j(x)| = \prod_{k\ne j} |x_j^{(0)} - x_k^{(0)}| \tag{5.7.36}$$

so

$$\prod_{j=1}^{n+1} \|Q_j\|_{\mathfrak{e}} = [\zeta_{n+1}^{n(n+1)}]$$

By $\|Q_j\|_{\mathfrak{e}} \geq \|T_n\|_{\mathfrak{e}}$, we have

$$\|T_n\|_{\mathfrak{e}}^{n+1} \leq \zeta_{n+1}^{n(n+1)}$$

which is the second inequality in (5.7.30).

(i), (ii) Let ν_∞ be a limit point of $\nu_{n(j)}$, where $\nu_{n(j)}$ is a normalized counting measure for Fekete sets with $n(j)$ points. Fix $m > 0$. Then, if $\{x_j\}_{j=1}^{n(j)}$ are the Fekete points,

$$m^{1/n} \prod_{j\neq k} |x_j - x_k|^{1/n^2} \leq \exp\biggl(-\int \log[\min(m^{-1}, |x-y|^{-1})]\, d\nu_{n(j)}(x) d\nu_{n(j)}(y)\biggr) \tag{5.7.37}$$

since

$$e^{-\log[\min(m^{-1}, |x-y|^{-1})]} = \max(|x-y|, m) \geq |x-y| \tag{5.7.38}$$

and we can use m for the n-terms with $x = y$ in (5.7.38). In that inequality, take $n \to \infty$, $m^{1/n} \to 1$, and since $n(n-1)/n^2 \to 1$, we get

$$\zeta_\infty(\mathfrak{e}) \leq \exp(-\mathcal{E}_{m^{-1}}(\nu_\infty)) \tag{5.7.39}$$

where

$$\mathcal{E}_a(\nu) = \int \log(\min(a, |x-y|^{-1}))\, d\nu(x) d\nu(y) \tag{5.7.40}$$

and we used that $\nu \to \mathcal{E}_a(\nu)$ is weakly continuous.

By the monotone convergence theorem, $\lim_{a\to\infty} \mathcal{E}_a(\nu) = \mathcal{E}(\nu)$, so taking $m \to 0$ in (5.7.39), we obtain

$$\zeta_\infty(\mathfrak{e}) \leq \exp(-\mathcal{E}(\nu_\infty)) \tag{5.7.41}$$

But, by (5.7.30), $C(\mathfrak{e}) \leq \zeta_\infty(\mathfrak{e})$, so

$$\mathcal{E}(\nu_\infty) \leq \log(C(\mathfrak{e})^{-1}) = \mathcal{E}(\rho_{\mathfrak{e}})$$

By the minimization property of $\rho_{\mathfrak{e}}$, $\nu_\infty = \rho_{\mathfrak{e}}$, proving convergence of ν_n to $\rho_{\mathfrak{e}}$ (by compactness of $\mathcal{M}_{+,1}(\mathfrak{e})$), and by (5.7.41),

$$\zeta_\infty(\mathfrak{e}) \leq C(\mathfrak{e}) \tag{5.7.42}$$

proving $\zeta_\infty(\mathfrak{e}) = C(\mathfrak{e})$.

(iii) By the Bernstein–Walsh lemma (5.5.114),

$$\frac{1}{n} \log|T_n(z)| \leq \frac{1}{n} \log\biggl[\frac{\|T_n\|_{\mathfrak{e}}^{1/n}}{C(\mathfrak{e})}\biggr] - \Phi_{\rho_{\mathfrak{e}}}(z) \tag{5.7.43}$$

If $\nu_{n(j)} \to \nu_\infty$ with ν_n the counting measures for zeros of T_n, then for $z \notin H = \text{cvh}(\mathfrak{e})$, (5.7.43), (5.5.52), (5.5.5), and (5.5.102) (see Theorem 5.5.10) imply that

$$-\Phi_{\nu_\infty}(z) \leq -\Phi_{\rho_\mathfrak{e}}(z)$$

since $\|T_n\|_\mathfrak{e}^{1/n} \to C(\mathfrak{e})$ by (5.7.31). By Theorem 5.5.16, we conclude that $\nu_\infty = \rho_\mathfrak{e}$. $\square$

Remarks and Historical Notes. Classical Chebyshev polynomials were introduced by him in two papers [80, 81], neither of which used the relation to $\cos(n\theta)$! He noted that they minimized $\|p_n\|_{[-1,1]}$ among all other polynomials with the same top order coefficients.

Classical Chebyshev polynomials have many applications to numerical analysis (see Mason–Handscomb [299] and Rivlin [370]).

The Faber–Fekete–Szegő theorem is named after their papers [125, 132, 432]. For other discussions of general Chebyshev polynomials, see [19, 179, 378, 446]. For a single interval, the Fekete points are known to be the zeros of a certain Jacobi polynomial; see Szegő [434, p. 382, Problem 37].

Corollary 5.7.7 is due to Schiefermayr [379].

Peherstorfer's proof [336] of Theorem 5.6.1(i)/(ii) looks at $\mathfrak{e}_n$ defined by (5.7.22)/(5.7.23), which, in general, has ℓ bands containing $\mathfrak{e}$ (if $\mathfrak{e}$ is an ℓ gap set) and $\ell - 1$ tiny bands around the at most $\ell - 1$ zeros in gaps of $\mathfrak{e}$. He showed one could remove all tiny bands by slightly enlarging $\mathfrak{e}$.

5.8 APPROXIMATION BY PERIODIC SPECTRA, II. GENERAL SETS

If $\mathfrak{e}$ is the spectrum of a two-sided periodic Jacobi matrix, we have several nice properties. We have Floquet solutions and we know $G_\mathfrak{e}(x)$ is zero on $\mathfrak{e}$. In this section, we want to approximate any compact $\mathfrak{e} \subset \mathbb{R}$ from the outside by periodic spectra and use this in one way and in Section 5.11 in a deeper way. The question, of course, is what we mean by approximate. While there are weaker notions, we will find approximants in the following strong sense:

$$\mathfrak{e} \subset \cdots \subset \mathfrak{e}_{n+1} \subset \mathfrak{e}_n \subset \cdots \subset \mathbb{R} \tag{5.8.1}$$

and

$$\bigcap_n \mathfrak{e}_n = \mathfrak{e} \tag{5.8.2}$$

and each $\mathfrak{e}_n$ is the spectrum of a periodic problem. Define $\tilde{\mathfrak{e}}_n = \{x \in \mathbb{R} \mid \text{dist}(x, \mathfrak{e}) \leq \frac{1}{n}\}$. These will obey (5.8.1) and (5.8.2) and we will prove each is a finite union of disjoint intervals, that is, an ℓ-gap set. Since $\tilde{\mathfrak{e}}_n \subset (\tilde{\mathfrak{e}}_{n-1})^{\text{int}}$, we will be able to use Theorem 5.6.1 to find $\mathfrak{e}_n$ a periodic spectrum with $\tilde{\mathfrak{e}}_n \subset \mathfrak{e}_n \subset (\tilde{\mathfrak{e}}_{n-1})^{\text{int}}$, and so find the required $\mathfrak{e}_n$. First, we give a few preliminaries.

To carry over $G_{\mathfrak{e}} = 0$ in intervals, I, in $\mathfrak{e}$, we will want the following:

Proposition 5.8.1. *Let $I = (a, b) \subset \mathfrak{e} \subset \mathbb{R}$ with $\mathfrak{e}$ compact. Suppose we know $G_{\mathfrak{e}}(x) = 0$ on I. Then*

(i) *$G_{\mathfrak{e}}$ is the real part of a function analytic in a neighborhood of I.*

(ii)

$$d\rho_{\mathfrak{e}} \restriction I = \rho_{\mathfrak{e}}(x)\, dx \tag{5.8.3}$$

where $\rho_{\mathfrak{e}}$ is a real analytic function of x.

(iii) *For each $k = 0, 1, 2, \dots$ and $\varepsilon > 0$,*

$$\sup_{x\in[a+\varepsilon,b-\varepsilon]} \left| \frac{d^k \rho_{\mathfrak{e}}(x)}{dx^k} \right| \tag{5.8.4}$$

is bounded with bounds depending only on ε, a, b, and $\text{diam}(\mathfrak{e})$.

Remarks. 1. We will eventually see (Corollary 5.8.5) that $G_{\mathfrak{e}}(x) = 0$ on I always holds.

2. Let $J = \text{cvh}(\mathfrak{e})$ so $I \subset \mathfrak{e} \subset J$. We will eventually show (see Corollary 5.8.7) that on I,

$$\rho_J(x) \le \rho_{\mathfrak{e}}(x) \le \rho_I(x) \tag{5.8.5}$$

where ρ_I, ρ_J are the equilibrium density for an interval given by (5.5.125). This will imply bounds on (5.8.4) depending only on I and not on $\text{diam}(\mathfrak{e})$.

Proof. (i) Let $z_n \to z_\infty \in I$. By upper semicontinuity of $G_{\mathfrak{e}}$, $G_{\mathfrak{e}}(z_\infty) \ge \limsup G_{\mathfrak{e}}(z_n)$. But $G_{\mathfrak{e}}(z_\infty) = 0$ and $G_{\mathfrak{e}}(z_n) \ge 0$ (by Frostman's theorem), so $\liminf G_{\mathfrak{e}}(z_n) \ge G_{\mathfrak{e}}(z_\infty)$. Thus, $G_{\mathfrak{e}}$ is continuous on $\mathbb{C}_+ \cup I$. Since $G_{\mathfrak{e}}$ is harmonic on $\mathbb{C}_+$, there is an analytic function $\widetilde{G}_{\mathfrak{e}}(z)$ with $\text{Re}\, \widetilde{G}_{\mathfrak{e}}(z) = G_{\mathfrak{e}}(z)$. By the Schwarz reflection principle (in the strong form that only requires continuity of $\text{Re}\, f$; see Ahlfors [7, Theorem 4.24], $\widetilde{G}_{\mathfrak{e}}$ has an analytic continuation to $\mathbb{C}_+ \cup \mathbb{C}_- \cup I$ with

$$\widetilde{G}_{\mathfrak{e}}(\bar{z}) = -\overline{\widetilde{G}_{\mathfrak{e}}(z)} \tag{5.8.6}$$

(ii) By the formula for the potential,

$$\frac{d\widetilde{G}_{\mathfrak{e}}(z)}{dz} = -\int \frac{d\rho_{\mathfrak{e}}(w)}{w - z} \tag{5.8.7}$$

since

$$\frac{d}{dx} G_{\mathfrak{e}}(x + iy) = \int d\rho_{\mathfrak{e}}(w)\ \text{Re}\left[\frac{1}{x + iy - w}\right] \tag{5.8.8}$$

From Propositions 2.3.11 and 2.3.12, $\rho_{\mathfrak{e}}$ is absolutely continuous on I, and for $x \in I$,

$$\rho_{\mathfrak{e}}(x) = \frac{1}{\pi} \lim_{y\downarrow 0} \left.\frac{\partial G_{\mathfrak{e}}(x + iy)}{\partial y}\right|_{y=0} \tag{5.8.9}$$

$$= \frac{1}{\pi} \lim_{y\downarrow 0} \frac{G_{\mathfrak{e}}(x + iy)}{y} \tag{5.8.10}$$

(iii) A Cauchy estimate shows that for any function f analytic in a neighborhood of $\{z \mid |z - z_0| \le \varepsilon\}$, we have f for $\ell = 1, 2, \dots,$

$$|f^{(\ell)}(z_0)| \le 2\varepsilon^{-\ell} \sup_{|z-z_0|=\varepsilon} |\text{Re}\, f(z)| \tag{5.8.11}$$

This follows from

$$f^{(\ell)}(z_0) = (2\pi)^{-1}\varepsilon^{-\ell} \int e^{-i\ell\theta} f(z_0 + \varepsilon e^{i\theta})\, d\theta$$

and

$$\int e^{-i\ell\theta}\, \overline{f(z_0 + \varepsilon e^{i\theta})}\, d\theta = 0$$

This, in turn, implies (using (5.8.6) and (5.8.9)) that

$$\sup_{x\in[a+\varepsilon,b-\varepsilon]} \left|\frac{d^k \rho_{\mathfrak{e}}}{dx^k}\right| \le \frac{2}{\pi}\,\varepsilon^{-k-1} \sup_{\substack{x\in[a,b]\\ 0\le y\le\varepsilon}} |G_{\mathfrak{e}}(x+iy)| \tag{5.8.12}$$

Since $C(\mathfrak{e}) \ge C(I) = \frac{1}{4}|b-a|$ and $\log|x+iy-w| \le \log(\text{diam}(\mathfrak{e}) + \varepsilon)$ for $0 \le y \le \varepsilon$ and $x \in I$, $w \in \mathfrak{e}$, we find

$$(5.8.4) \le \frac{2}{\pi}\,\varepsilon^{-\ell-1}\left[\log\left(\frac{4}{|b-a|}\right) + \log(\varepsilon + \text{diam}(\mathfrak{e}))\right] \tag{5.8.13}$$

$\square$

We first turn to approximations in the strong sense:

Proposition 5.8.2. *Let* $\mathfrak{e}$ *be compact so* (5.8.1) *and* (5.8.2) *hold for compact* $\mathfrak{e}_1, \mathfrak{e}_2, \dots$. *Then*

(i)

$$\rho_{\mathfrak{e}_n} \xrightarrow{w} \rho_{\mathfrak{e}} \tag{5.8.14}$$

(ii)

$$C(\mathfrak{e}_n) \downarrow C(\mathfrak{e}) \tag{5.8.15}$$

(iii) *If* $I = (a,b)$ *is an interval in* $\mathfrak{e}$ *so that* $G_{\mathfrak{e}_n} = 0$ *on* I, *then* $G_{\mathfrak{e}} = 0$ *on* I *and the densities* $\rho_{\mathfrak{e}_n}(x)$ *converge uniformly to* $\rho_{\mathfrak{e}}(x)$ *on each* $(a+\varepsilon, b-\varepsilon)$.

Proof. (i), (ii) Let ρ_∞ be a weak limit of $\rho_{\mathfrak{e}_{n(j)}}$. By hypothesis, $\rho_\infty \in \mathcal{M}_{+,1}(\mathfrak{e})$. By the obvious $C(\mathfrak{e}) \le C(\mathfrak{e}_{n(j)})$ and lower semicontinuity of the Coulomb energy, $\mathcal{E}$,

$$\log(C(\mathfrak{e})^{-1}) \le \mathcal{E}(\rho_\infty) \le \liminf \mathcal{E}(\rho_n) = \lim(\log(C(\mathfrak{e}_n)^{-1})) \le \log(C(\mathfrak{e})^{-1})$$

It follows that $\lim C(\mathfrak{e}_n) = C(\mathfrak{e})$ and $\rho_\infty = \rho_{\mathfrak{e}}$, so by compactness of $\mathcal{M}_{+,1}(\mathfrak{e})$, $\rho_{\mathfrak{e}_n} \xrightarrow{w} \rho_{\mathfrak{e}}$.

(iii) By Proposition 5.8.1, we have uniform bounds on $\frac{d\rho_{\mathfrak{e}_n}}{dx}\big|_{[a+\varepsilon,b-\varepsilon]}$ so equicontinuity of $\rho_{\mathfrak{e}_n}(x)$, so compactness in the topology of uniform convergence. But if $\rho_{\mathfrak{e}_n}(x) \to f(x)$ uniformly, $d\rho_{\mathfrak{e}_n} \restriction [a+\varepsilon, b-\varepsilon] \to f(x)\,dx$, so $d\rho_{\mathfrak{e}} = f(x)\,dx$ determining f, and so proving uniform convergence.

The uniform bounds on $\rho_{\mathfrak{e}_n}(x)$ and $\rho_{\mathfrak{e}}(x)$ imply uniform convergence of $\int_{|x-x_0|<\varepsilon} \log|x - x_0|^{-1}\, d\rho_{\mathfrak{e}_n}(x)$ (and for $\rho_{\mathfrak{e}}$) to zero as $\varepsilon \downarrow 0$ for $x_0 \in [a+\varepsilon, b-\varepsilon]$. This plus convergence of $\int_{|x-x_0|>\varepsilon} \log|x-x_0|^{-1}\, d\rho_{\mathfrak{e}_n}(x)$ to $\int_{|x-x_0|>\varepsilon} \log|x - x_0|^{-1}\, d\rho(x)$ implies $G_{\mathfrak{e}_n}(x_0) \to G_{\mathfrak{e}}(x_0)$ on (a,b), so $G_{\mathfrak{e}}(x_0) = 0$. □

Proposition 5.8.3. *Let $\mathfrak{e} \subset \mathbb{R}$ be compact. Let*

$$\tilde{\mathfrak{e}}_n = \{x \in \mathbb{R} \mid \operatorname{dist}(x, \mathfrak{e}) \le \tfrac{1}{n}\} \tag{5.8.16}$$

Then

(i) $\tilde{\mathfrak{e}}_n$ *obey* (5.8.1) *and* (5.8.2).
(ii) *Each $\tilde{\mathfrak{e}}_n$ is a finite union of disjoint closed (positive measure) intervals.*

Proof. (i) $\tilde{\mathfrak{e}}_{n+1} \subset \tilde{\mathfrak{e}}_n$ is trivial, and $\cap\tilde{\mathfrak{e}}_n = \tilde{\mathfrak{e}}$ by the compactness of $\mathfrak{e}$ (for $x \in \cap\tilde{\mathfrak{e}}_n$ implies there are $x_n \in \mathfrak{e}$ with $\operatorname{dist}(x_n, x) \le \frac{1}{n}$).

(ii) $\mathbb{R} \setminus \mathfrak{e}$ is an open set, so a disjoint union of maximal open intervals, two unbounded and the others $\{J_k\}_{k=1}^N$ in $\operatorname{cvh}(\mathfrak{e})$. Thus, $\sum_{k=1}^N |J_k| < \infty$, so for each n, $\#\{k \mid |J_k| > 2/n\}$ is finite. Thus, all but finitely many J_k lie in a given $\tilde{\mathfrak{e}}_n$, showing $\mathbb{R}\setminus\tilde{\mathfrak{e}}_n$ has finitely many open intervals. It is easy to see that each of the finite disjoint closed intervals in $\tilde{\mathfrak{e}}_n$ must have positive measure. □

By combining this with Theorem 5.6.1, we get:

Theorem 5.8.4. *Let $\mathfrak{e} \subset \mathbb{R}$ be compact. Then there exist $\mathfrak{e}_n$ so that* (5.8.1) *and* (5.8.2) *hold, and moreover,*

$$\mathfrak{e}_n \subset \mathfrak{e}_{n-1}^{\text{int}} \tag{5.8.17}$$

and each $\mathfrak{e}_n$ is a finite gap set with rational harmonic measures, that is, each $\mathfrak{e}_n$ is the spectrum of some two-sided periodic Jacobi matrix. Moreover,

(i) $\rho_{\mathfrak{e}_n} \xrightarrow{w} \rho_{\mathfrak{e}}$
(ii) $C(\mathfrak{e}_n) \to C(\mathfrak{e})$
(iii) *If $I = (a,b)$ is an interval in $\mathfrak{e}$, then $G_{\mathfrak{e}} = 0$ on I and*

$$\rho_{\mathfrak{e}_n}(x) \to \rho_{\mathfrak{e}}(x) \tag{5.8.18}$$

uniformly on each $[a+\delta, b-\delta]$.

Proof. Let $\tilde{\mathfrak{e}}_n$ be given by (5.8.16). Since $x \in \tilde{\mathfrak{e}}_n$ and $|x-y| \le [n(n-1)]^{-1}$ implies $y \in \tilde{\mathfrak{e}}_{n-1}$, we see

$$\tilde{\mathfrak{e}}_n \subset (\tilde{\mathfrak{e}}_{n-1})^{\text{int}}$$

By Proposition 5.8.3, $\tilde{\mathfrak{e}}_n$ is a finite gap set, so by Theorem 5.6.1, we can find $\mathfrak{e}_n$ a periodic spectrum with

$$\tilde{\mathfrak{e}}_n \subset \mathfrak{e}_n \subset (\tilde{\mathfrak{e}}_{n-1})^{\text{int}}$$

This implies (5.8.17), and (5.8.2) for $\tilde{\mathfrak{e}}_n$ implies it for $\mathfrak{e}_n$.

(i)–(iii) are immediate from Proposition 5.8.2 and the fact that $G_{\mathfrak{e}}(x) = 0$ on $\mathfrak{e}$ for periodic spectra. □

Corollary 5.8.5. *Let $I = (a, b) \subset \mathfrak{e} \subset \mathbb{R}$. Then $G_{\mathfrak{e}} = 0$ on I and $\rho_{\mathfrak{e}} \restriction I$ is absolutely continuous with real analytic $\rho_{\mathfrak{e}}(x)$.*

As an application of the approximation theorem, we can prove various comparison theorems:

Theorem 5.8.6. *Let $\mathfrak{e} \subset \mathfrak{e}'$ be compact subsets of $\mathbb{R}$. Then:*
(i) *For all $z \in \mathbb{C}$,*

$$G_{\mathfrak{e}'}(z) \leq G_{\mathfrak{e}}(z) \tag{5.8.19}$$

(ii)

$$d\rho_{\mathfrak{e}'} \restriction \mathfrak{e} \leq d\rho_{\mathfrak{e}} \tag{5.8.20}$$

(iii) *If $I = (a, b) \subset \mathfrak{e} \subset \mathfrak{e}'$, then*

$$x \in I \Rightarrow \rho_{\mathfrak{e}'}(x) \leq \rho_{\mathfrak{e}}(x) \tag{5.8.21}$$

Proof. Since our periodic approximations obey $\mathfrak{e} \subset (\mathfrak{e}_n)^{\text{int}}$ and $\mathfrak{e}_n \subset \{x \mid \text{dist}(\mathfrak{e}, x) < \frac{1}{n-1}\}$, it is easy to see we can find periodic approximations $\mathfrak{e}_n$, $\mathfrak{e}_n'$ of $\mathfrak{e}$, $\mathfrak{e}'$ with $\mathfrak{e}_n \subset \mathfrak{e}_n'$. By the convergence results in Theorem 5.8.4, it suffices to prove this theorem in case $\mathfrak{e}$, $\mathfrak{e}'$ are finite gap sets. We did this in Proposition 5.6.3 (the statement of that proposition required the set have the same number of gaps, but all that was used in the proof was continuity of $G_{\mathfrak{e}}$ on $\mathbb{C}$ and absolute continuity of $d\rho_{\mathfrak{e}}$). □

Corollary 5.8.7. *Let $I \subset \mathfrak{e} \subset \text{cvh}(\mathfrak{e}) = J \subset \mathbb{R}$. Then, on I,*

$$\rho_J(x) \leq \rho_{\mathfrak{e}}(x) \leq \rho_I(x) \tag{5.8.22}$$

Remarks and Historical Notes. Totik [443, 444] emphasized the approximation of general compact $\mathfrak{e} \subset \mathbb{R}$ by periodic spectra as a tool for extending not only results on CD kernels (we follow him in part in Section 5.10) but also classical polynomial inequalities like the Markov inequality.

For Totik, the periodic spectrum did not play a big role—rather, he exploited the existence of a polynomial Δ of the type in (c) of Theorem 5.7.5 with $\Delta^{-1}([-2, 2]) = \mathfrak{e}$. Objects like Floquet solutions are never used. His use of polynomial inverse images was motivated by Geronimo–Van Assche [155].

Standard work on potential theory [196, 264, 360] develops a "theory of barriers" to prove $(a, b) = I \subset \mathfrak{e} \subset \mathbb{R}$ implies $G_{\mathfrak{e}}$ is continuous and vanishing on I.

5.9 REGULARITY: AN ASIDE

This section has nothing to do with periodic Jacobi matrices—rather it provides a tool needed in Section 5.11, which has the deepest application of periodic approximations. We will address the issue of root asymptotics mentioned in Sections 2.9, 2.15, and 3.11.

Definition. A measure μ with compact support $\mathfrak{e} \subset \mathbb{R}$ is called *regular* if and only if its Jacobi parameters obey

$$\lim_{n\to\infty} (a_1 \dots a_n)^{1/n} = C(\mathfrak{e}) \tag{5.9.1}$$

To partly motivate this notion, we note

Proposition 5.9.1. *For any measure μ of compact support $\mathfrak{e} \subset \mathbb{R}$, we have*

$$\limsup_n \, (a_1 \dots a_n)^{1/n} \le C(\mathfrak{e}) \tag{5.9.2}$$

Remark. Below (see the remark after the proof of Theorem 5.9.2) we will provide a second proof of (5.9.2).

Proof. By (1.2.13) (assuming $\mu(\mathbb{R}) = 1$),

$$\limsup_n \, (a_1 \dots a_n)^{1/n} = \limsup_n \| P_n(\,\cdot\,, d\mu) \|_{L^2(d\mu)}^{1/n} \tag{5.9.3}$$

while, by (5.6.6),

$$\begin{aligned} \| P_n(\,\cdot\,, d\mu) \|_{L^2(d\mu)} &\le \| T_n \|_{L^2(d\mu)} \\ &\le \mu(\mathbb{R})^{1/2} \| T_n \|_{\mathfrak{e}} \end{aligned} \tag{5.9.4}$$

Since $\| T_n \|_{\mathfrak{e}}^{1/n} \to C(\mathfrak{e})$ (by Theorem 5.7.8), (5.9.3) and (5.9.4) imply (5.9.2). □

Here is the main result on regular measures:

Theorem 5.9.2. *Let μ be a measure supported by a compact set $\mathfrak{e} \subset \mathbb{R}$ so that μ is regular. Then*

(i) *The zero counting measures, ν_n, for the OPRL obey*

$$\nu_n \xrightarrow{w} \rho_{\mathfrak{e}} \tag{5.9.5}$$

the equilibrium measure for $\mathfrak{e}$.

(ii) *For any $z \notin \mathrm{cvh}(\mathfrak{e})$, the convex hull of $\mathfrak{e}$,*

$$\lim_{n\to\infty} |p_n(z, d\mu)|^{1/n} = \exp(G_{\mathfrak{e}}(z)) \tag{5.9.6}$$

(iii) *For any $z \in \mathrm{cvh}(\mathfrak{e})$,*

$$\limsup_{n\to\infty} |p_n(z, d\mu)|^{1/n} \le \exp(G_{\mathfrak{e}}(z)) \tag{5.9.7}$$

and for q.e. $z \in \mathfrak{e}$,

$$\limsup_{n\to\infty} |p_n(z, d\mu)| = 1 \tag{5.9.8}$$

We need one preliminary:

Lemma 5.9.3. *For $z \in \mathbb{C}_+$ and any measure μ of compact support in $\mathbb{R}$, we have*

$$\liminf |p_n(z, d\mu)|^{1/n} \ge 1 \tag{5.9.9}$$

Proof. If not, there exist $n(j) \to \infty$ with

$$\lim |p_{n(j)}(z)| = 0 \tag{5.9.10}$$

since $a \in [0, 1)$ implies $a^n \to 0$.

Let $\varphi_j(x) = \sum_{k=0}^{n(j)-1} p_k(x)p_k(z)$. By the recursion relation for the p's and $Jp_k = a_{k+1}p_{k+1} + b_{k+1}p_k + a_k p_{k-1}$, we have

$$((J-z)\varphi_j)(x) = a_{n(j)}(p_{n(j)}(x)p_{n(j)-1}(z) - p_{n(j)-1}(x)p_{n(j)}(z)) \tag{5.9.11}$$

(essentially the CD formula (3.10.7)). Thus,

$$\langle \varphi_j, (J-z)\varphi_j \rangle = -a_{n(j)}\, p_{n(j)}(z)\, \overline{p_{n(j)-1}(z)} \tag{5.9.12}$$

which implies, using $\|\varphi_j\|^2 = \sum_{k=1}^{n(j)-1} |p_k(z)|^2 \geq |p_{n(j)-1}(z)|^2$,

$$|\langle \varphi_j, (J-z)\varphi_j \rangle| \leq a_{n(j)} \|\varphi_j\|\, |p_{n(j)}(z)| \tag{5.9.13}$$

Since $\|\varphi_j\| \geq 1$ (from the p_0 term) and $\sup_n |a_n| < \infty$, this implies, given (5.9.10), that

$$\lim_{j\to\infty} \frac{|\langle \varphi_j, (J-z)\varphi_j \rangle|}{\|\varphi_j\|^2} = 0 \tag{5.9.14}$$

But

$$|\langle \varphi_j, (J-z)\varphi_j \rangle| \geq (\operatorname{Im} z)\|\varphi_j\|^2 \tag{5.9.15}$$

This contradiction shows that (5.9.10) cannot happen, so (5.9.9) holds. □

Proof of Theorem 5.9.2. (i) Suppose that $\nu_{n(j)} \to \nu_\infty$. By (5.5.37) and (5.5.5),

$$|P_{n(j)}(z, d\mu)|^{1/n(j)} \to \exp(-\Phi_{\nu_\infty}(z)) \tag{5.9.16}$$

for $z \in \mathbb{C} \setminus \text{cvh}(\mathfrak{e})$. By (5.9.1),

$$\lim_{j\to\infty} |p_{n(j)}(z, d\mu)|^{1/n(j)} = \exp(\widetilde{G}(z)) \tag{5.9.17}$$

where

$$\widetilde{G}(z) = -\Phi_{\nu_\infty}(z) + \log(C(\mathfrak{e})^{-1}) \tag{5.9.18}$$

By (5.9.9),

$$\Phi_{\nu_\infty}(z) \leq \log(C(\mathfrak{e})^{-1}) \tag{5.9.19}$$

for $\operatorname{Im} z \neq 0$.

By (5.5.28), (5.9.19) holds for $z \in \mathbb{R}$ also. Integrating $d\nu_\infty$, we find

$$\mathcal{E}(\nu_\infty) \leq \log(C(\mathfrak{e})^{-1}) = \mathcal{E}(\rho_\mathfrak{e}) \tag{5.9.20}$$

Since ν_n has at most weight $1/n$ in any gap of $\mathfrak{e}$, ν_∞ is supported on $\mathfrak{e}$. Thus, by (5.9.20), $\nu_\infty = \rho_\mathfrak{e}$, that is, $\rho_\mathfrak{e}$ is the only limit point of ν_n. By compactness of $\mathcal{M}_{+,1}(\mathfrak{e})$, $\nu_n \xrightarrow{w} \rho_\mathfrak{e}$.

(ii) $\nu_\infty = \rho_\mathfrak{e}$ implies $\widetilde{G} = G_\mathfrak{e}$, and thus, (5.9.17) is (5.9.6).

(iii) This is immediate from (ii) and (iii) of Theorem 5.5.10. □

Remark. If $\nu_{n(j)} \to \nu_\infty$ and $(a_1 \dots a_{n(j)})^{1/n(j)} \to A$, (5.9.20) becomes

$$\mathcal{E}(\nu_\infty) \leq \log(A^{-1}) \tag{5.9.21}$$

so

$$\log(A^{-1}) \geq \log(C(\mathfrak{e})^{-1}) \tag{5.9.22}$$

that is, $A \leq C(\mathfrak{e})$. This provides the promised second proof of (5.9.2).

Definition. A set $\mathfrak{e}$ is called *regular for the Dirichlet problem* if and only if $G_{\mathfrak{e}}(x) = 0$ for all $x \in \mathfrak{e}$.

There is a converse to part of Theorem 5.9.2 that we will need:

Theorem 5.9.4. *Let $\mathfrak{e} \subset \mathbb{R}$ be compact and regular for the Dirichlet problem and let μ be a measure with $\sigma_{\text{ess}}(\mu) = \mathfrak{e}$. Suppose*

$$\nu_n \xrightarrow{w} \rho_{\mathfrak{e}} \tag{5.9.23}$$

Then either μ is regular for $\mathfrak{e}$ or else there exists K of capacity zero so that $\mu(\mathbb{R} \setminus K) = 0$.

Proof. Let $(a_1 \dots a_{n(j)}^{1/n(j)}) \to A$ for some A. By the argument leading to (5.9.16) and the upper envelope theorem (Theorem 5.5.10), there is a set K of capacity zero so that for $z \in \mathbb{C} \setminus K$,

$$|p_{n(j)}(x, d\mu)|^{1/n(j)} \to A^{-1} \exp(-\Phi_{\rho_{\mathfrak{e}}}(z)) \tag{5.9.24}$$

In particular, for all $x \in \mathfrak{e} \setminus K$,

$$|p_{n(j)}(x, d\mu)|^{1/n(j)} \to \frac{C(\mathfrak{e})}{A} \tag{5.9.25}$$

On the other hand, since $\|p_n\|_{L^2(d\mu)} = 1$, we have

$$\int \sum_{n=0}^{\infty} (n+1)^{-2} |p_n(x)|^2 \, d\mu < \infty$$

so for μ a.e. x and an x-dependent constant, $B(x)$,

$$|p_n(x)| \le B(x)(n+1) \tag{5.9.26}$$

If $A < C(\mathfrak{e})$, the object on the right of (5.9.25) is larger than 1 and this is inconsistent with (5.9.26)! Thus, either $A \ge C(\mathfrak{e})$ or else μ is supported on the set where (5.9.25) fails, that is, $\mu(\mathbb{R} \setminus K) = 0$. Since $A \le C(\mathfrak{e})$ always, we see if the first case holds, then $C(\mathfrak{e})$ is the only limit point (and μ is regular). □

Corollary 5.9.5. *Let $\mathfrak{e} \in \mathbb{R}$ be a potentially perfect set, which is regular for the Dirichlet problem. Let μ be a measure on $\mathbb{R}$ with $\sigma_{\text{ess}}(\mu) = \mathfrak{e}$. Then for any $\delta > 0$, there exists a neighborhood K_δ of $\mathfrak{e}$ and constant C_δ so that for all n,*

$$\sup_{z \in K_\delta} |p_n(z, d\mu)| \le C_\delta e^{\delta n} \tag{5.9.27}$$

Proof. By hypothesis, $G_{\mathfrak{e}}$ is continuous on $\mathfrak{e}$ and so on $\mathbb{C}$ and vanishing on $\mathfrak{e}$. Let

$$K_\delta = G_{\mathfrak{e}}^{-1}([0, \tfrac{1}{2}\delta)) \tag{5.9.28}$$

∂K_δ is compact, disjoint from $\mathfrak{e}$, and $G_{\mathfrak{e}} = \frac{1}{2}\delta$ on ∂K_δ. Thus, uniformly in $z \in K_\delta$,

$$\lim |p_n(z, d\mu)|^{1/n} = e^{\frac{1}{2}\delta} \tag{5.9.29}$$

It follows that we can find C_δ so (5.9.27) holds for $z \in \partial K_\delta$, and thus, by the maximum principle, on K_δ. □

The following, which we state without proof (but see the Notes), provide criteria for regularity:

Theorem 5.9.6. *Let $\mathfrak{e} \subset \mathbb{R}$ be potentially perfect and let μ obey $\sigma_{\text{ess}}(\mu) = \mathfrak{e}$ and*

$$d\mu(x) = f(x)\, d\rho_{\mathfrak{e}}(x) + d\mu_{\text{s}}(x) \tag{5.9.30}$$

where $f(x) > 0$ for $d\rho_{\mathfrak{e}}$-a.e. x. Then μ is regular.

Theorem 5.9.7. *Let $\mathfrak{e}$ be a finite union of disjoint closed intervals. Let μ have $\sigma_{\text{ess}}(\mu) = \mathfrak{e}$. Suppose for every $\eta > 0$,*

$$\lim_{n\to\infty} |\{x \in \mathfrak{e} \mid \mu([x - \tfrac{1}{n}, x + \tfrac{1}{n}]) \le e^{-n\eta}\}| = 0 \tag{5.9.31}$$

where $|\cdot|$ is Lebesgue measure. Then μ is regular.

We will prove a special case of Theorem 5.9.6 (when $\mathfrak{e}$ has a large interior so $d\rho_{\mathfrak{e}}$ is dx-a.c.) later (see Theorem 5.11.3).

Remarks and Historical Notes. For $\mathfrak{e} = [-1, 1]$, the relation of $\|P_n\|^{1/n} \to \frac{1}{2}$, of the convergence of $d\nu_n$ to $\pi^{-1}(1 - x^2)^{-1/2}\, dx$ and positivity of the weight (i.e., the hypothesis of Theorem 5.9.6 in this case) go back to a 1940 paper of Erdős–Turán [123]. Systematic study of regularity on $[-1, 1]$ was begun by Ullman [447] (see the references in [404]).

The general theory was initiated and brought to fruition in a remarkable book of Stahl–Totik [417] who, in particular, prove Theorems 5.9.2 and 5.9.7. Theorem 5.9.6 appears implicitly in Widom [459] and explicitly in Van Assche [449]. Simon [404] has a review of the theory, including proofs of Theorems 5.9.6 and 5.9.7.

We also note the following due to Stahl–Totik [417] and proven also in [404].

Theorem 5.9.8. *Let $\mathfrak{e} = \mathfrak{e}_1 \cup \cdots \cup \mathfrak{e}_\ell$ be a union of ℓ disjoint closed intervals. Let μ be a measure with $\sigma_{\text{ess}}(\mu) = \mathfrak{e}$ and let $\mu_j = \mu \restriction \mathfrak{e}_j$. Then μ is regular for $\mathfrak{e}$ if and only if each μ_j is regular for $\mathfrak{e}_j$.*

5.10 THE CD KERNEL FOR PERIODIC JACOBI MATRICES

As we have seen in our analysis of CD kernel asymptotics in Sections 2.15–2.17 and 3.11, a key role is played by an example that we can analyze completely. For OPUC, this was $\frac{d\theta}{2\pi}$, and for OPRL on $[-2, 2]$, it was the measures $d\mu_1, d\mu_2$ of Example 3.11.3. In this section, as preparation for the next, we will study in detail the asymptotics of the CD kernel associated to the spectral measure of a periodic Jacobi matrix. (This is the analog of $d\mu_2$ in Example 3.11.3; while we used $d\mu_1$ more extensively in that section, we could have used $d\mu_2$.) Throughout this section, all results refer to a fixed periodic Jacobi matrix.

We let $\{a_n, b_n\}_{n=-\infty}^{\infty}$ be the Jacobi parameters of the half-line Jacobi matrix J_0^+ extended by periodicity. $\mathfrak{e} = \mathfrak{e}_1 \cup \cdots \cup \mathfrak{e}_{\ell+1}$ is the essential spectrum of J_0^+.

By Theorem 5.4.15, for $\lambda \in \mathfrak{e}^{\text{int}}$, we can define solutions

$$u_n^{\pm}(\lambda) \equiv u_n^{\pm}(\lambda + i0) \tag{5.10.1}$$

of (5.4.1)/(5.2.7) with

$$u_n^-(\lambda) = \overline{u_n^+(\lambda)} \qquad u_{n=0}^{\pm}(\lambda) = 1 \tag{5.10.2}$$

and (by (5.4.105) and Proposition 5.10.2(ii) below)

$$\operatorname{Im} u_1^+(\lambda) \neq 0 \tag{5.10.3}$$

which implies $u_n^{\pm}$ are linearly independent.

Moreover,

$$u_{n+mp}^{\pm}(\lambda) = e^{\pm im\theta(\lambda)} u_n^{\pm}(\lambda) \tag{5.10.4}$$

and θ is related to $\rho_{\mathfrak{e}}$, the density of the equilibrium measure, $d\rho_{\mathfrak{e}}$, by

$$\rho_{\mathfrak{e}}(\lambda) = \frac{1}{p\pi} \left| \frac{d\theta}{d\lambda} \right| \tag{5.10.5}$$

by (5.3.34).

Since $p_{\cdot-1}$ also solves (5.2.7), we have

$$p_{n-1}(\lambda) = \frac{[u_n^+(\lambda) - u_n^-(\lambda)]}{[u_1^+(\lambda) - u_1^-(\lambda)]} \tag{5.10.6}$$

since equality holds at $n = 0, 1$. Define

$$I(\lambda) = -2 \operatorname{Im} u_1^+(\lambda) \tag{5.10.7}$$

Theorem 5.10.1. *Let J_0^+ be a periodic Jacobi matrix.*

(i) *The weight $w(x)$ of the spectral measure for J_0^+ is given by*

$$w(x) = \frac{I}{2a_0\pi} \tag{5.10.8}$$

(ii) *The density, $\rho_{\mathfrak{e}}(x)$, of the equilibrium measure for $\mathfrak{e}$ is given by*

$$\rho_{\mathfrak{e}}(x) = \frac{1}{a_0 p\pi I} \sum_{n=1}^{p} |u_n^+(\lambda)|^2 \tag{5.10.9}$$

Proof. (i) By (5.4.41) and $u_0^+(\lambda) = 1$,

$$m(\lambda) = \langle \delta_1, (J_0^+ - \lambda)^{-1} \delta_1 \rangle = -\frac{u_1^+(\lambda)}{a_0} \tag{5.10.10}$$

Since $m(\lambda) = \int d\mu(x)(x-\lambda)^{-1}$, we have

$$w(\lambda) = \lim_{\varepsilon \downarrow 0} \frac{1}{\pi} \operatorname{Im} m(\lambda + i\varepsilon) \tag{5.10.11}$$

$$= -\frac{1}{2\pi a_0} (2 \operatorname{Im} u_1^+(\lambda)) \tag{5.10.12}$$

$$= \frac{I}{2a_0\pi}$$

(ii) By Theorems 5.4.9 and 5.5.17,

$$\frac{1}{p}\sum_{n=1}^{p}\langle\delta_n,(J-\lambda)^{-1}\delta_n\rangle=\int\frac{\rho_{\mathfrak{e}}(x)\,dx}{x-\lambda} \tag{5.10.13}$$

so as above,

$$\rho_{\mathfrak{e}}(\lambda)=\frac{1}{\pi p}\lim_{\varepsilon\downarrow 0}\sum_{n=1}^{p}\operatorname{Im}(G_{nn}(\lambda+i\varepsilon)) \tag{5.10.14}$$

$$=-\frac{1}{\pi p}\frac{\sum_{n=1}^{p}|u_n^+(\lambda)|^2}{\operatorname{Im}(W(\lambda))} \tag{5.10.15}$$

by (5.4.79). Here

$$\begin{aligned}W(\lambda)&=a_0(u_1^+(\lambda)u_0^-(\lambda)-u_1^-(\lambda)u_0^+(\lambda))\\&=-a_0I\end{aligned} \tag{5.10.16}$$

(5.10.15) and (5.10.16) imply (5.10.9). □

When we square (5.10.6), $|p_{n-1}(\lambda)|^2$ will have a cross-term $u_n^+(\lambda)\,\overline{u_n^-(\lambda)}=u_n^+(\lambda)^2$ and a key role will be played by the fact that uniformly on compact subsets of $\mathfrak{e}^{\text{int}}$, one has

$$\lim_{N\to\infty}\frac{1}{N}\sum_{j=1}^{N}u_j^+(\lambda)^2=0 \tag{5.10.17}$$

This is more subtle than it might appear at first.

By (5.10.4), what is relevant is

$$\sum_{n=1}^{M}e^{2im\theta(\lambda)}=\frac{e^{2i(M+1)\theta(\lambda)}}{e^{2i\theta(\lambda)}-1} \tag{5.10.18}$$

which easily yields (5.10.17) pointwise if

$$2\theta(\lambda)\neq 2\pi k \tag{5.10.19}$$

for an integer k. But if (5.10.19) fails, there is an issue and uniformity fails in (5.10.18) as $\theta(\lambda)\to$ some πk. Points in $\mathfrak{e}^{\text{int}}$ where (5.10.19) fails are precisely closed gaps, so we will need to look closely at what happens there. The key will be that at a closed gap,

$$\sum_{j=1}^{p}u_j^+(\lambda)^2=0 \tag{5.10.20}$$

As a warmup:

Proposition 5.10.2. *Let J_0^+ be a periodic Jacobi matrix.*

(i) *At any closed gap, λ_0, $\rho_{\mathfrak{e}}(\lambda)$ is continuous and nonvanishing.*

(ii) *At a closed gap, λ_0, $w(\lambda)$ is continuous and nonvanishing.*

Proof. (i), *first proof.* By Craig's formula (5.4.86), $\rho_{\mathfrak{e}}$ is continuous and strictly positive on any compact subset of $\mathfrak{e}^{\text{int}}$.

(i), *second proof.* By (5.10.5), we need to show $\frac{d\theta}{d\lambda}$ remains smooth a nonzero at a closed gap. θ solves

$$2\cos(\theta(\lambda)) = \Delta(\lambda) \tag{5.10.21}$$

where Δ is the discriminant, (5.4.6) (see Theorem 5.4.1). At a closed gap, λ_0,

$$2 \pm \Delta(\lambda) = c(\lambda - \lambda_0)^2$$

with $c > 0$ (see Proposition 5.4.3). So, by (5.10.18), we have

$$(\theta(\lambda) - \theta(\lambda_0))^2 = d(\lambda - \lambda_0)^2 + O((\lambda - \lambda_0)^3)$$

that is, $\frac{d\theta}{d\lambda}\big|_{\lambda=\lambda_0} \neq 0$.

(ii) Let λ_0 be the gap edge and let α, β, γ be the coefficients of (5.2.2). By Theorem 5.4.16 and (5.2.3), $\alpha(\lambda)$ has a simple zero at λ_0. By Proposition 5.4.3, $\Delta^2 - 4$ has a double zero at λ_0 so, by (5.2.4), β vanishes at λ_0. Since γ, like α, has simple zeros, (5.2.4) implies β and α have simple zeros at λ_0. At λ_0, β/α is real and $\sqrt{\Delta^2 - 4}/\alpha$ is pure imaginary in $(\lambda_0 - \varepsilon, \lambda_0 + \varepsilon)/\{\lambda_0\}$, and so nonvanishing and imaginary at λ_0. Thus, $\operatorname{Im} m$ is continuous and nonvanishing near λ_0. $\square$

Theorem 5.10.3. *At any closed gap, λ_0, we have*

$$\sum_{j=1}^{p} u_j^+(\lambda_0)^2 = 0 \tag{5.10.22}$$

Proof. We consider the case that $\Delta(\lambda_0) = 2$. The case $\Delta(\lambda_0) = -2$ is similar. Thus, $\theta(\lambda_0) = 0$. $J(\theta = 0)$ given by (5.3.8) thus has a doubly generated eigenvalue at λ_0 by Proposition 5.4.3. Let θ be small and positive. Then $J(\theta)$ has two eigenvalues $e_+(\theta) > \lambda_0 > e_-(\theta)$ near λ_0. By eigenvalue perturbation theory (see the Notes), the corresponding eigenvalues have limits.

Since $e_\pm(\theta)$ are distinct, the eigenvectors are orthogonal, so the limits are orthogonal. For $\theta \neq 0, \pi$, the only possible Floquet eigenfunctions are $u_\pm(e(\theta))$, so either $u_\pm(e_\pm(\theta))$ or $u_\mp(e_\pm(\theta))$ are the eigenvectors. It cannot be that $u_+(e_+(\theta))$ and $u_+(e_-(\theta))$ are the eigenvectors for $\theta > 0$ since u_+ is continuous and the limits are orthogonal.

Since the limits are orthogonal,

$$\sum_{j=1}^{p} \overline{u_j^-(\lambda_0)}\, u_j^+(\lambda_0) = 0 \tag{5.10.23}$$

which is (5.10.22) by (5.4.69). $\square$

Here is one of two main results of this section:

Theorem 5.10.4. *Fix a periodic Jacobi matrix with* $\sigma_{\text{ess}}(J) = \mathfrak{e}$. *Let* $I = [\alpha, \beta]$ *be a closed interval in* $\mathfrak{e}^{\text{int}}$. *Then uniformly in* I:

(i) *For any* $A > 0$ *and uniformly for all* $\lambda_n \to \lambda_0 \in I$ *with* $n|\lambda_n - \lambda_0| \le A$, *we have*

$$\lim_{n\to\infty} \frac{1}{n+1} K_n(\lambda_n, \lambda_n) = \frac{\rho_{\mathfrak{e}}(\lambda_0)}{w(\lambda_0)} \tag{5.10.24}$$

(ii) *Under the same conditions as* (i) *for all such* λ_n *and* $|a| \le A$, $|b| \le A$,

$$\lim_{n\to\infty} \frac{K_n(\lambda_n + \frac{a}{n}, \lambda_n + \frac{b}{n})}{K_n(\lambda_0, \lambda_0)} = \frac{\sin(\pi\rho_{\mathfrak{e}}(\lambda_0)(b-a))}{\pi\rho_{\mathfrak{e}}(\lambda_0)(b-a)} \tag{5.10.25}$$

Remark. See the Notes for an alternate proof of this theorem.

Proof. (i) We first claim that uniformly for $\lambda \in I$,

$$\frac{1}{n}\sum_{j=1}^{n} (u_j^+(\lambda))^2 \to 0 \tag{5.10.26}$$

Since $|u_{j+1}^+(\lambda)| = |u_j^+(\lambda)|$ and each u_j^+ is continuous on I, we have

$$\sup_{j,\lambda\in I} |u_j^+(\lambda)| < \infty \tag{5.10.27}$$

That implies it suffices to prove (5.10.26) for $n = kp$, $k = 1, 2, \dots$. But then, by (5.10.24),

$$\begin{aligned}\frac{1}{kp}\sum_{j=1}^{kp} (u_j^+(\lambda))^2 &= \frac{1}{kp}\left(\sum_{\ell=0}^{k-1} e^{2i\ell\theta(\lambda)}\right)\left(\sum_{j=1}^{p} (u_j^+(\lambda))^2\right) \\ &= \frac{1}{kp}\left[\frac{e^{2ik\ell\theta(\lambda)}-1}{e^{2i\theta(\lambda)}-1}\right]\sum_{j=1}^{p}(u_j^+(\lambda))^2\end{aligned} \tag{5.10.28}$$

where we interpret $[\dots]$ as k if $e^{2i\theta(\lambda)} = 1$.

By (5.10.23), $R(\lambda) \equiv \sum_{j=1}^{p} u_j^+(\lambda)^2$ vanishes at each λ in $\mathfrak{e}^{\text{int}}$ where $e^{2i\theta(\lambda)} = 1$ (since $e^{i\theta(\lambda)} = \pm 1$ and we are at a closed band edge). By eigenvalue perturbation theory, $u_1^+(\lambda)$ is analytic in $\theta(\lambda)$, including at closed band edges, and θ is invertible and so real analytic in λ on $\mathfrak{e}^{\text{int}}$. By the recursion relation (and $u_0^+(\lambda) = 1$), $u_j^+(\lambda)$ is analytic for each j, and so $R(\lambda)$ is real analytic on $\mathfrak{e}^{\text{int}}$. It follows for any compact $K \subset \mathfrak{e}^{\text{int}}$ that

$$\sup_{\lambda\in K}\left|(e^{2i\theta(\lambda)}-1)^{-1}\sum_{j=1}^{p} u_j^+(\lambda)^2\right| \equiv R_K < \infty \tag{5.10.29}$$

so

$$\text{LHS of (5.10.28)} \le \frac{2}{kp} D_K \to 0 \tag{5.10.30}$$

as $k \to \infty$, that is, (5.10.26) holds.

Thus, squaring (5.10.6), the $(u_n^+)^2$ and $(u_n^-)^2$ converge to zero and we see, by (5.10.6),

$$\begin{aligned} \frac{1}{n+1} K_n(\lambda_0, \lambda_0) &= \frac{2}{I(\lambda_0)^2(n+1)} \sum_{j=1}^{n+1} |u_j^+(\lambda_0)|^2 \\ &\to \frac{2\pi a_0}{I(\lambda_0)} \rho_{\mathrm{e}}(x) \qquad \text{by (5.10.9)} \\ &= \frac{\rho_{\mathrm{e}}(x)}{w(x)} \end{aligned} \tag{5.10.31}$$

by (5.10.8).

The above shows the convergence is uniform, and by going through the above, it is easy to accommodate the $\lambda_n \to \lambda_0$ extension.

(ii) Suppose first that $a \neq b$ and $\lambda_n \equiv \lambda_0$. Let

$$n = kp + j \qquad j = 0, 1, \dots, p-1 \tag{5.10.32}$$

Then, since $u_0^+, u_1^+, \dots, u_{p-1}^+$ are real analytic near λ_0 as is $\theta(\lambda)$, we have by (5.10.4) that, for $\ell = 1, 2$,

$$u_{n+\ell}^+\left(\lambda_0 + \frac{a}{n}\right) = e^{ik(\theta(\lambda_0)+\theta'(\lambda_0)a/n)+O(1/n^2)}\left[u_{j+\ell}^+(\lambda_0) + O\left(\frac{1}{n}\right)\right] \tag{5.10.33}$$

Plugging this into (5.10.6) yields

$$\begin{aligned} a_{n+1}&\left[p_{n+1}\left(\lambda + \frac{a}{n}\right)p_n\left(\lambda_0 + \frac{b}{n}\right) - p_{n+1}\left(\lambda_0 + \frac{b}{n}\right)p_n\left(\lambda_0 + \frac{a}{n}\right)\right] \\ &= W\left[2i \sin\left(\left[\frac{1}{p} + O\left(\frac{1}{n}\right)\right]\theta'(\lambda_0)(a-b)\right)\right] \end{aligned} \tag{5.10.34}$$

where

$$W = a_{n+1}[u_{n+2}^+(\lambda_0)\, \overline{u_{n+1}^+(\lambda_0)} - \overline{u_{n+2}^+(\lambda_0)}\, u_{n+1}(\lambda_0)] \tag{5.10.35}$$

$$= a_0[2i \operatorname{Im} u_1^+(\lambda_0)] \tag{5.10.36}$$

(5.10.34) used $\frac{k}{n} = \frac{1}{p} + O(\frac{1}{n})$, (5.10.35) that $u_n^-(\lambda_0) = \overline{u_n^+(\lambda_0)}$, and (5.10.36) the constancy of the Wronskian and $u_0^+(\lambda_0) = 1$.

The left side of (5.10.6) enters in the CD formula for $K_n(\lambda_0 + \frac{a}{n}, \lambda_0 + \frac{b}{n})$, so we obtain, using the definition of I, (5.10.7),

$$\lim_{n\to\infty} \frac{1}{n+1} K_m\left(\lambda_0 + \frac{a}{n}, \lambda_0 + \frac{b}{n}\right) = \frac{2a_0}{I} \sin[\pi \rho_{\mathrm{e}}(\lambda_0)(a-b)] \tag{5.10.37}$$

on account of (5.10.5).

Using (5.10.8) and (5.10.24), we obtain (5.10.25) for the case ($a \neq b, \lambda_n \equiv \lambda_0$). Next, we return to (5.10.24) and note it holds if $K_n(\lambda_n, \lambda_n)$ is replaced by $K_n(\lambda_n, \tilde{\lambda}_n)$ so long as $n(\lambda_n - \tilde{\lambda}_n) \to 0$ with $n|\lambda_n - \lambda_0| \leq A$. For uniformly in n, $p_n'(\lambda)$ near λ_0 is $O(n)$ by (5.10.33) and (5.10.6), which in the CD formula controls the change of $\frac{1}{n} K_n$.

With this in place, one can easily control $a = b$ and $\lambda_n \to \lambda_0$ in (5.10.25). □

Finally, as preparation for extending the Máté–Nevai bounds to general sets in $\mathbb{R}$, we note the following pair of results:

Theorem 5.10.5. *Fix a periodic Jacobi matrix with $\sigma_{\text{ess}}(J) = \mathfrak{e}$. For any compact set $K \subset \mathfrak{e}^{\text{int}}$, we have*

$$\sup_{n,\lambda\in K} |p_n(\lambda)| < \infty \tag{5.10.38}$$

Remark. This also follows from an analysis of transfer matrices.

Proof. By (5.10.28), (5.10.8), and Proposition 5.10.2(ii), we get a uniform bound on $|u_n^+(\lambda)|$ and on $|u_1^+(\lambda) - \overline{u_1^+(\lambda)}\,|^{-1}$. By (5.10.6), we obtain (5.10.38). □

Theorem 5.10.6. *Fix a periodic Jacobi matrix with $\sigma_{\text{ess}}(J) = \mathfrak{e}$.*

(a) *For any compact $K \subset \mathfrak{e}^{\text{int}}$, we have*

$$\frac{K_n(x,y)}{K_n(x,x)} \le \frac{C}{|x-y|} \qquad \textit{for all } x, y \in K \tag{5.10.39}$$

(b) *For any A and ε, there is N so for $n > N$ and all x, y in the region $|x-y| \le A/n$, $x, y \in K$, we have*

$$\left| \frac{K_n(x,y)}{K_n(x,x)} - \frac{\sin(\pi\rho_{\mathfrak{e}}(x)(x-y)n)}{n\pi\rho_{\mathfrak{e}}(x)(x-y)} \right| < \varepsilon \tag{5.10.40}$$

Proof. (a) follows from Theorem 5.10.5 and the CD formula. (b) follows from the uniformity of the convergence in Theorem 5.10.4. □

Remarks and Historical Notes. The use of Floquet solutions to study asymptotics of the CD kernel is due to Simon [408] who used a different approach, which has the advantage of also working for almost periodic isospectral tori of the type studied in Chapter 9. Because it is illuminating how the other proof uses the magic of the CD formula to avoid the need to prove (5.10.22), we sketch that approach here. Actually, we go slightly further than [408]. That paper did not compute directly a constant that we compute below, but instead relied on Theorems 3.11.1 and 3.11.4.

Define for $\lambda \in \mathfrak{e}^{\text{int}}$

$$f_n(\lambda) = e^{-in\theta(\lambda)/p} u_n^+(\lambda) \tag{5.10.41}$$

By (5.10.4), f_n has period p in n and it is real analytic on $\mathfrak{e}^{\text{int}}$. So, for any compact $K \subset \mathfrak{e}^{\text{int}}$,

$$\sup_{\lambda\in K,n} \left| \frac{df_n}{d\lambda} \right| = B < \infty \tag{5.10.42}$$

By (5.10.5), we see that for $\lambda \in K$,

$$\left| \frac{du_n^+}{d\lambda}(\lambda) - in\pi\rho_{\mathfrak{e}}(\lambda)u_n^+ \right| \le B \tag{5.10.43}$$

Next, take $x \to y$ in the CD formula (3.10.7) to get

$$K_n(x,x) = a_{n+1}(p_{n+1}'(x)p_n(x) - p_n'(x)p_{n+1}(x)) \tag{5.10.44}$$

Thus, by (5.10.43) and (5.10.6), we get

$$\sup_{x\in K}\left| K_n(x,x) - (n+1)(i\pi\rho_{\mathfrak{e}})a_{n+1}\frac{(s_{n+2}(\lambda)d_{n+1}(\lambda) - s_{n+1}(\lambda)d_{n+2}(\lambda))}{(u_1^+(\lambda) - u_1^-(\lambda))^2}\right| < \infty \tag{5.10.45}$$

where

$$s_n(\lambda) = u_n^+(\lambda) + u_n^-(\lambda) \tag{5.10.46}$$

$$d_n(\lambda) = u_n^+(\lambda) - u_n^-(\lambda) \tag{5.10.47}$$

(The s_n terms come from p'_{n-1} and d_n from p_n.)

Expanding out s_n and d_n in (5.10.45), the non-cross-terms vanish and the u_n^+/u_{n+1}^- cross-term is a Wronskian, which we evaluate at $n = 0$ to get

$$i\pi\rho_{\mathfrak{e}}a_{n+1}\frac{(s_{n+2}(\lambda)d_{n+1}(\lambda) - s_{n+1}(\lambda)d_{n+2}(\lambda))}{(u_1^+(\lambda) - u_1^-(\lambda))^2} = \frac{2\pi\rho_{\mathfrak{e}}a_0}{I} = \frac{\rho_{\mathfrak{e}}}{\sqrt{I}} \tag{5.10.48}$$

by (5.10.8). Thus, (5.10.45) gives another proof of (5.10.24).

For a discussion of eigenvalue perturbation theory, see [364, Section XII.1].

5.11 ASYMPTOTICS OF THE CD KERNEL: OPRL ON GENERAL SETS

In this section, we complete the discussion begun in Sections 2.14–2.17 and 3.11. Using the last three sections, we will consider general $\mathfrak{e} \subset \mathbb{R}$ with the sole restriction being that $\mathfrak{e}$ is regular for the Dirichlet problem in the sense defined before Corollary 5.9.5. Here are the lower bound asymptotics of $K_n(x,x)$.

Theorem 5.11.1 (Máté–Nevai Bounds for General $\mathfrak{e}$). *Let $\mathfrak{e} \subset \mathbb{R}$ be compact with $|\mathfrak{e} \setminus \mathfrak{e}^{\text{int}}| = 0$. Let μ be a measure with $\sigma_{\text{ess}}(\mu) \subset \mathfrak{e}$. Let μ have a decomposition*

$$d\mu = w\,dx + d\mu_s \tag{5.11.1}$$

with $d\mu_s$ Lebesgue singular. Then for Lebesgue a.e. x_0, we have

$$\limsup (n+1)\lambda_n(x_0, d\mu) \le \frac{w(x_0)}{\rho_{\mathfrak{e}}(x_0)} \tag{5.11.2}$$

If $I \subset \mathfrak{e}$ is an interval with w "continuous" on I and strictly positive there, then (5.11.2) holds uniformly on $x_0 \in I$.

Moreover, for each $A > 0$, we can replace $\lambda_n(x, d\mu)$ by $\lambda_n(x_n, d\mu)$ for $x_n \to x$ with $n|x - x_n| \le A$.

Remark. We define "continuous" in the sense of Section 2.16, that is, w is continuous as a function on $\mathfrak{e}$ for each $x \in I$.

Proof. First, pick $\mathfrak{e}_m$ according to Theorem 5.8.4. By (5.8.18), it suffices to prove (5.11.2) with $\rho_{\mathfrak{e}}$ replaced by $\rho_{\mathfrak{e}_m}$ and then take a limit as $m \to \infty$.

Since μ has support in $\mathfrak{e}_m^{\text{int}}$ and $\mathfrak{e}_m$ is associated with a periodic Jacobi matrix, by Theorem 5.10.6, we can use the Christoffel minimizer $K_n(x, x_0; d\mu_0)/K_n(x_0, x_0; d\mu_0)$ for $d\mu_0$ in the isospectral torus $\mathfrak{e}_m$ as a trial function in the Christoffel principle and have (5.10.35)/(5.10.36) available. For the a.e. result, the argument of the proof of Theorem 2.17.6 applies. For the uniform result, we use the same bounds and uniform continuity of w on I. □

Given this result, the weak convergence theorem for OPRL (Theorem 3.11.1) and the argument in Theorem 2.17.7, we immediately get

Theorem 5.11.2. *If $I = (\alpha, \beta) \subset \mathfrak{e}$ is an open interval, $\sigma_{\text{ess}}(\mu) = \mathfrak{e}$ and μ is regular for $\mathfrak{e}$ and $w(x) > 0$ on I, then*

$$\int_I \left| \frac{1}{n+1} K_n(x,x) w(x) - \rho_{\mathfrak{e}}(x) \right| dx \to 0 \tag{5.11.3}$$

$$\int_I \frac{1}{n+1} K_n(x,x)\, d\mu_{\text{s}}(x) \to 0 \tag{5.11.4}$$

Theorem 5.11.3. *Let $\mathfrak{e} \subset \mathbb{R}$ be compact so that $\mathfrak{e} \setminus \mathfrak{e}^{\text{int}}$ has capacity zero; in particular, $\mathfrak{e}$ can be a finite gap set. Let μ be a measure with $\sigma_{\text{ess}}(\mu) = \mathfrak{e}$ and*

$$d\mu(x) = w(x)\, dx + d\mu_{\text{s}}(x)$$

where $w(x) > 0$ for Lebesgue a.e. x. Then μ is regular.

Remark. This is a special case of Theorem 5.9.6.

Proof. By Theorem 5.9.4, it suffices to show that the zero density $d\nu_n \to d\rho_{\mathfrak{e}}$ (since μ is obviously not supported on a set of capacity zero). Suppose $\frac{1}{n(j)+1} \times K_{n(j)}(x,x) w(x)\, dx \to d\kappa_1$ and $\frac{1}{n(j)+1} K_{n(j)}(x,x)\, d\mu_{\text{s}} \to d\kappa_2$. By the argument in (2.17.38),

$$d\kappa_1 \geq \rho_{\mathfrak{e}}(x)\, dx = d\rho_{\mathfrak{e}}$$

But $\int d\kappa_1 + d\kappa_2 = 1 = \int d\rho_{\mathfrak{e}}$, so $d\kappa_2 = 0$ and $d\kappa_1 = d\rho_{\mathfrak{e}}$.

By compactness, $\frac{1}{n+1} K_n(x,x)\, d\mu \to d\rho_{\mathfrak{e}}$. So by Theorem 3.11.1, $d\nu_n \to d\rho_{\mathfrak{e}}$. □

To get lower bounds, we need a one-sided but extended Nevai comparison theorem (see Theorems 2.16.6 and 3.11.5):

Theorem 5.11.4 (Nevai Comparison Theorem)**.** *Let $\mathfrak{e}$ be a compact subset of $\mathbb{R}$, which is regular for the Dirichlet problem. Let $I \subset \mathfrak{e}^{\text{int}}$ be a closed interval. For every ε, there is a δ so that if $\mathfrak{e}' \subset \{x \mid \text{dist}(x, \mathfrak{e}) < \delta\}$ and μ, μ' are any measures on $\mathbb{R}$ with μ regular for $\mathfrak{e}$ and*

$$\sigma_{\text{ess}}(\mu) = \mathfrak{e} \qquad \sigma_{\text{ess}}(\mu') \equiv \mathfrak{e}' \tag{5.11.5}$$

and for some x_0,

$$(n+1)\lambda_n(x_0, d\mu') \to C > 0 \tag{5.11.6}$$

and

$$\mu \restriction I = \mu' \restriction I \tag{5.11.7}$$

then

$$\liminf (n+1)\lambda_n(x_0, d\mu) \geq C(1-\varepsilon) \tag{5.11.8}$$

Moreover, these results are unchanged if x_0 in (5.11.6) *and* (5.11.8) *are replaced by x_n obeying $x_n \to x_0$, and if* (5.11.6) *(with x-dependent C) holds uniformly in I, then so does* (5.11.8).

Proof. Pick D so that for $\text{dist}(x, \mathfrak{e}) < 1$ and $x_0 \in I$, we have

$$\left(1 - \frac{|x - x_0|}{D}\right) > 0 \tag{5.11.9}$$

and let Q be defined by

$$\sup_{\substack{x \notin \mathfrak{e} \\ x_0 \in I \\ \text{dist}(x,\mathfrak{e})<1}} \left(1 - \frac{|x - x_0|}{D}\right) = e^{-Q} \tag{5.11.10}$$

Given ε, pick $\delta < 1$ so

$$\sup_{\text{dist}(x,\mathfrak{e})<\delta} |G_{\mathfrak{e}}(x)| < \tfrac{1}{2} Q\varepsilon \tag{5.11.11}$$

Now just use Nevai trial functions built from minimizers of $\lambda_n(x, d\mu)$ as in (2.16.38) (modified to handle point mass outside $\mathfrak{e}'$ as in the proof of Theorem 3.11.5) to get

$$\lambda_n(x_0, d\mu') \leq \lambda_{n(\varepsilon)}(x_0, d\mu) + O(e^{-Q\varepsilon n/2}) \tag{5.11.12}$$

Multiply by $n(\varepsilon) + 1$ and take $n \to \infty$ using

$$\frac{n(\varepsilon)+1}{n+1} \to (1 - \varepsilon)$$

to get (5.11.8).

The uniformity results follow from the proof. $\square$

We also need a version of the Nevai theorem that has $\sigma_{\text{ess}}(\mu) = \sigma_{\text{ess}}(\mu')$ but allows μ and μ' to differ by a continuous weight:

Theorem 5.11.5 (Nevai Comparison Theorem). *Let $\mathfrak{e} \subset \mathbb{R}$ be compact and regular for the Dirichlet problem. Let $\mu, \mu^\sharp$ be two regular measures on $\mathfrak{e}$ of the form*

$$d\mu = w\,dx + d\mu_{\text{s}} \qquad d\mu^\sharp = w^\sharp\,dx + d\mu_{\text{s}}^\sharp \tag{5.11.13}$$

Suppose $x_0 \in \mathfrak{e}^{\text{int}}$ obeys

(i) *For some $\delta > 0$, $d\mu_{\text{s}} = d\mu_{\text{s}}^\sharp$ on $(x_0 - \delta, x_0 + \delta)$.*

(ii) *For all ε sufficiently small, there is $\alpha_\varepsilon > 1$, so for $|x - x_0| < \varepsilon$, we have*

$$\alpha_\varepsilon^{-1} w(x) \leq w^\sharp(x) \leq \alpha_\varepsilon w(x) \tag{5.11.14}$$

(iii) *That $\alpha_\varepsilon \to 1$ and any $x_n \in \mathfrak{e}^{\text{int}}$ with $x_n \to x_0$ and every $\ell(n)$ with $n/2 < \ell(n) < 2n$, we have that*

$$\lim_{n\to\infty} \frac{1}{n+1} K_n(x_{\ell(n)}, x_{\ell(n)}) = B \neq 0 \tag{5.11.15}$$

Then

$$\lim_{n\to\infty} \frac{1}{n+1} K_n^\sharp(x_n, x_n) = B \tag{5.11.16}$$

Moreover, this is uniform in x_n in the sense that if (with the same B) for all $x_n \to x_0$, there are, for any ε, a δ and an N_0 so if $n > N_0$ and $|x_n - x_0| < \delta$, then

$$\left| B - \frac{1}{n+1} K_n^\sharp(x_n, x_n) \right| < \varepsilon \tag{5.11.17}$$

This is also uniform in x_0. If w and $w^\sharp$ are continuous and nonvanishing in a closed interval in $\mathfrak{e}^{\text{int}}$ and we have $d\mu_s = d\mu_s^\sharp$ in a neighborhood of I and (5.11.14) *is replaced by*

$$\alpha_\varepsilon^{-1} \frac{w(x)}{w(x_0)} \le \frac{w^\sharp(x)}{w^\sharp(x_0)} \le \alpha_\varepsilon \frac{w(x)}{w(x_0)} \tag{5.11.18}$$

for $|x - x_0| < \varepsilon$ (α_ε independent of x_0) and if (5.11.14) *holds uniformly in $x_0 \in I$ where $B(x_0)$ is x_0-dependent, then* (5.11.16) *holds with B replaced by $B(x_0)w(x_0)/w^\sharp(x_0)$.*

Proof. Identical to the proof of Theorem 3.11.5. □

Next we generalize Lubinsky's theorem (Theorem 3.11.6):

Theorem 5.11.6. *Let $\mathfrak{e}$ be a compact subset of $\mathbb{R}$ regular for the Dirichlet problem. Let $d\mu$ be a regular probability measure on $\mathfrak{e}$ of the form*

$$d\mu = w(x)\,dx + d\mu_s \tag{5.11.19}$$

Suppose that, for some interval $[\alpha, \beta] \subset \mathfrak{e}^{\text{int}}$,

(a) $\text{supp}(d\mu_s) \cap I = \emptyset$

(b) *w is "continuous" on I and nonvanishing there.*

Then, with $\rho_\mathfrak{e}$ given by the equilibrium measure for $\mathfrak{e}$, we have

(1) (Diagonal Asymptotics) *For any $A < \infty$, uniformly in $x_\infty \in I$, and sequence $x_n \in \mathfrak{e}$ with $n|x_n - x_\infty| \le A$ for all n, we have*

$$\frac{1}{n+1} K_n(x_n, x_n) \to \frac{\rho_\mathfrak{e}(x_\infty)}{w(x_\infty)} \tag{5.11.20}$$

(2) (Lubinsky Universality) *For any $A < \infty$, uniformly in $x_\infty \in I$ and $a, b \in \mathbb{R}$ with $|a|, |b| \le A$, we have*

$$\frac{K_n(x_\infty + \frac{a}{n}, x_\infty + \frac{b}{n})}{K_n(x_\infty, x_\infty)} \to \frac{\sin(\pi\rho_\mathfrak{e}(x_\infty)(b-a))}{\pi\rho_\mathfrak{e}(x_\infty)(b-a)} \tag{5.11.21}$$

More generally, the limit of $K_n(x_n, y_n)/K_n(x_\infty, x_\infty)$ is the right side of (5.11.21) *so long as $|x_n - x_\infty| \le A/n$, $|y_n - x_\infty| \le A/n$, and $n(x_n - y_n) \to b - a$.*

Proof. If $\mathfrak{e}$ is the spectrum of a two-sided periodic Jacobi matrix, the proof follows that of Theorems 3.11.6 and 2.16.1; the upper bound comes from Theorem 5.11.1 and the lower bound uses Theorems 5.10.4 and 5.11.5.

In the general case, we approximate $\mathfrak{e}$ using Theorem 5.8.4. By Theorem 5.11.1, we see that

$$\liminf_{m\to\infty} \frac{1}{m+1} K_m(x_m, x_m) \geq \frac{\rho_{\mathfrak{e}}(x_0)}{w(x_0)} \tag{5.11.22}$$

and by the approximation, the special case above, and Theorem 5.11.4, we see that for each n,

$$\limsup_{m} \frac{1}{m+1} K_m(x_m, x_m) \leq (1-\varepsilon_n)^{-1} \frac{\rho_{\mathfrak{e}_n}(x_0)}{w(x_0)} \tag{5.11.23}$$

where $\varepsilon_n \to 0$ as $n \to \infty$. Taking $n \to \infty$ using (5.8.18) yields (5.11.20).

To get (5.11.21), we compare μ with a measure μ_n, which is μ on I and $\max(\mu, \rho_{\mathfrak{e}_n})$ off I. This is regular for $\mathfrak{e}_n$ (see the Notes). Putting this into Lubinsky's inequality and using (5.10.25) shows the absolute value of the difference of the left-hand side of (5.11.21) for μ and $\frac{\sin(\pi\rho_{\mathfrak{e}_n}(x_\infty)(b-a))}{\pi\rho_{\mathfrak{e}_n}(x_\infty)(b-a)}$ is asymptotically less than $\frac{\rho_{\mathfrak{e}}(x_\infty)}{\rho_{\mathfrak{e}_n}(x_\infty)} \left|\frac{\rho_{\mathfrak{e}}(x_\infty)-\rho_{\mathfrak{e}_n}(x_\infty)}{\rho_{\mathfrak{e}_n}(x_\infty)}\right|$, which goes to zero as $n \to \infty$ by (5.8.18). □

Finally, we turn to results on locally Szegő weights. By using approximation by periodic spectra, the key will be the extension to weights on periodic spectra. Here we will use the discriminant, Δ, to map $\mathfrak{e}$ to $[-2, 2]$ and we will be able to relate the Christoffel variational problems for such weights to ones on $[-2, 2]$. We will be able to do this initially for weights with a symmetry between bands. The localization intrinsic in Nevai trial functions will let us then go to nonsymmetric weights. We begin by studying the symmetry between bands, that is, solutions of $\Delta(x) = \lambda \in (-2, 2)$.

We take a polynomial, $Q(z)$, which we will eventually specialize to Δ. Suppose $\deg(Q) = N$. For any λ, we will look at the solutions of

$$Q(z) = \lambda \tag{5.11.24}$$

Q has a double (or higher-order) root at z_0 if and only if $Q'(z_0) = 0$, so this occurs at a maximum of $N-1$ points. The corresponding values of Q are a set, Λ, of at most $N-1$ points, the *critical values* of Q.

If $\lambda \notin \Lambda$, (5.11.24) has N solutions $z_1(\lambda), \dots, z_n(\lambda)$, which can be chosen analytically in the neighborhood of any $\lambda \notin \Lambda$. One cannot make a global choice since points in Λ are branch points—following a path around them will permute the $z_j(\lambda)$. Indeed, if Q is irreducible by following some path in the region $\mathbb{C}\backslash\Lambda$, one can go from any z_j to any z_k. However, analytic symmetric functions of $\{z_j(\lambda)\}_{j=1}^N$ will be analytic and singular-valued on $\mathbb{C}\backslash\Lambda$. Typically, there will be removable singularities at points in Λ. The following describes the special case of symmetrized polynomials:

Theorem 5.11.7. *Let $\ell = 0, 1, 2, \dots$. For*

$$\ell N \leq k < (\ell+1)N \tag{5.11.25}$$

we have that

$$\sum_{j=1}^{N} (z_j(\lambda))^k = R_k(\lambda) \tag{5.11.26}$$

where R_k is a polynomial with

$$\deg(R_k) = \ell \tag{5.11.27}$$

Proof. We use induction in ℓ starting with $\ell = 0$ (which will be the most subtle case!). For $\ell = 0$, we need to show the sum is constant. R_k is continuous in λ so it suffices to prove constancy for $\lambda \notin \Lambda$. Thus, $z_j(\lambda)$ is locally analytic and

$$\frac{dQ(z_j(\lambda))}{d\lambda} = 1 \Rightarrow \frac{dz_j(\lambda)}{d\lambda} = \frac{1}{Q'(z_j(\lambda))} \tag{5.11.28}$$

Therefore,

$$\frac{dR_\lambda}{d\lambda} = 0 \Leftrightarrow k\sum_{j=1}^{N} \frac{z_j(\lambda)^{k-1}}{Q'(z_j(\lambda))} = 0 \tag{5.11.29}$$

and we need only prove the right equality in (5.11.29).

Fix $\lambda \notin \Lambda$. Consider a circle, Γ, about zeros of radius R so large that

$$\sup_j |z_j(\lambda)| < R \tag{5.11.30}$$

and look at

$$\frac{1}{2\pi i}\oint_{|z|=R} \frac{kz^{k-1}}{Q(z)-\lambda}\, dz \tag{5.11.31}$$

Since $Q - \lambda$ has no zeros outside Γ, we can take $R \to \infty$. This integral is bounded by

$$\frac{1}{2\pi}(2\pi R)kR^{k-1}\left[\inf_\Gamma |Q(z)-\lambda|\right]^{-1} \tag{5.11.32}$$

and the inf goes like R^{-N}. The quantity in (5.11.32) is bounded for large R by $R^{k-N} \to 0$ since $\ell = 0$ (so $k < N$). This means that the integral in (5.11.31) is zero. It can also be evaluated in terms of the residues inside the circle, which is the sum on the right of (5.11.29). This completes the proof for $\ell = 0$.

For general ℓ, we use induction in k, assuming $k \geq N$. Then z_j^k can be written using $(Q(z_j) - \lambda)z_j^{k-N} = 0$ as a sum of constants times $\{z_j^{k-m}\}_{m=1}^N$ plus a constant times λz_j^{k-N}, and so write $R_k(\lambda)$ as a sum of $\{R_{k-m}\}_{m=1}^N$ and λR_{k-N}. This proves the result inductively. □

Remark. By the general theory of symmetric functions (see, e.g., [418]) and the fact that the sums of products of distinct $z_j(\lambda)$ are the coefficients of $Q(z) - \lambda$, R_k can be calculated "explicitly."

Corollary 5.11.8. *Let P be any polynomial of exact degree k obeying* (5.11.25). *Then there is a polynomial R of exact degree ℓ so that for z obeying $Q(w) = Q(z) \Rightarrow Q'(w) \neq 0$,*

$$\sum_{\{w \mid Q(w)=Q(z)\}} P(w) = R(Q(z)) \tag{5.11.33}$$

Remark. (5.11.33) holds (by continuity) at points where $Q'(w) = 0$ for some root so long as we count multiplicity in the sum.

Proof. Immediate given the theorem that handles monomials. □

Given any measure μ on $[-2, 2]$ and Δ, the discriminant from $\mathfrak{e}$ to $[-2, 2]$, we define a measure $S\mu$ on $\mathfrak{e}$ as follows: Write $\mathfrak{e} = \cup_{j=1}^{p} \mathfrak{e}_j$ the closed bands, that is, $\mathfrak{e}_j$ is the closure of one of the connected components of $\Delta^{-1}((-2, 2))$. If $A \subset \mathfrak{e}_j^{\text{int}}$ for some j, then

$$S\mu(A) = \tfrac{1}{p}\, \mu(\Delta(A)) \tag{5.11.34}$$

If $x_0 \in \Delta^{-1}(\{-2, 2\})$, we set

$$S\mu(\{x_0\}) = \begin{cases} \tfrac{1}{p}\, \mu(\Delta(x_0)) & \text{if } x_0 \text{ is an open gap edge} \\ \tfrac{2}{p}\, \mu(\Delta(x_0)) & \text{if } x_0 \text{ is a closed gap edge} \end{cases}$$

This definition is such that for any $f\colon [-2, 2] \to \mathbb{R}$,

$$\int f(\Delta(x))\, d(S\mu)(x) = \int f(x)\, d\mu(x) \tag{5.11.35}$$

Moreover, if $X_{jk}\colon \mathfrak{e}_j \to \mathfrak{e}_k$ is defined by demanding

$$\Delta(X_{jk}(x)) = \Delta(x) \tag{5.11.36}$$

then for any function $g\colon \mathfrak{e}_k \to \mathbb{R}$ and any j, k,

$$\int_{\mathfrak{e}_j} g(X_{jk}(x))\, d(S\mu)(x) = \int_{\mathfrak{e}_k} g(x)\, d(S\mu)(x) \tag{5.11.37}$$

It is not hard to see that (5.11.35)/(5.11.36) uniquely characterize $S\mu$.

Proposition 5.11.9. (a) *If* $d\mu = w\, dx + d\mu_{\text{s}}$ *and* $d(S\mu) \equiv \tilde{w}\, dx + d\tilde{\mu}_{\text{s}}$*, then*

$$\tilde{\mu}_{\text{s}} = S\mu_{\text{s}} \qquad \tilde{w}(x) = \frac{1}{p}\, w(\Delta(x))\, \frac{d\Delta}{dx} \tag{5.11.38}$$

(b) *The equilibrium measures and their densities on* $\mathfrak{e}$ *and* $[-2, 2]$ *are related by*

$$S\rho_{[-2,2]} = \rho_{\mathfrak{e}} \tag{5.11.39}$$

$$\rho_{\mathfrak{e}}(x) = \tfrac{1}{p}\, \rho_{[-2,2]}(\Delta(x)) \tag{5.11.40}$$

Proof. (a) The formula for $\tilde{w}$ is a standard change of variables, and $\tilde{\mu}_{\text{s}} = S\mu_{\text{s}}$ follows from $|\Delta(A)| = 0 \Leftrightarrow |A| = 0$ where $|\cdot|$ is Lebesgue measure.

(b) (5.11.40) is equivalent to (5.11.39), given (5.11.38). To see (5.11.40), we use the explicit formulae (5.4.15) and (5.5.125):

$$\rho_{\mathfrak{e}}(x) = \frac{1}{p\pi}\, \frac{\Delta'(x)}{\sqrt{4 - \Delta^2(x)}} \qquad \rho_{[-2,2]}(x) = \frac{1}{\pi}\, \frac{1}{\sqrt{4 - x^2}} \tag{5.11.41}$$

and (5.11.38). □

Remark. Lest (b) seem like a miracle, if $E(z) = \frac{1}{2}[z + \sqrt{z^2 - 4}]$ is the conformal map of $(\mathbb{C} \cup \{\infty\} \setminus [-2, 2])$ bijectively to $(\mathbb{C} \cup \{\infty\}) \setminus \overline{\mathbb{D}}$ (the inverse of $z \to z + \frac{1}{z}$), then $E \circ \Delta$ maps $(\mathbb{C} \cup \{\infty\}) \setminus \mathfrak{e}$ to $(\mathbb{C} \cup \{\infty\}) \setminus \overline{\mathbb{D}}$ conformally and bijectively, and since $\deg(\Delta) = p$, $\log|E - \Delta(x)| \sim p \log(z)$ at infinity, so the potential theorist's Green's functions are related by $G_{\mathfrak{e}}(z) = \frac{1}{p} G_{[-2,2]}(\Delta(z_0))$, which leads to another proof of (5.11.39)/(5.11.40).

Theorem 5.11.10. *Suppose μ is a measure on $[-2, 2]$ so that $S\mu$ is regular and $x_0 \in \cup_j \mathfrak{e}_j^{\text{int}}$. Then*

$$\limsup n\lambda_n(x_0, S\mu) \le \limsup n\lambda_n(\Delta(x_0), \mu) \tag{5.11.42}$$

and

$$\liminf n\lambda_n(x_0, S\mu) \ge \liminf n\lambda_n(\Delta(x_0), \mu) \tag{5.11.43}$$

In particular if $\lim_{n\to\infty} n\lambda_n(\Delta(x_0), \mu)$ *exists, then*

$$\lim_{n\to\infty} n\lambda_n(x_0, S\mu) = \lim_{n\to\infty} n\lambda_n(\Delta(x_0), \mu) \tag{5.11.44}$$

Moreover, (5.11.42)/(5.11.43) *hold if all x_0's are replaced by $x_n \to x_0$, and for each $A > 0$, this is uniform in x_n's with* $\sup_n n|x_n - x_0| \le A$.

Proof. We suppose throughout that $x_n = x_0$. The accommodations for $x_n \to x_0$ are straightforward.

We first prove (5.11.42). Since $\lambda_n(x_0; S\mu)$ is an inf, we will use a trial function built from the optimizers $Q_\ell(x, \Delta(x_0); \mu)$ for μ.

One might first try

$$S_{p\ell}(x) = Q_\ell(\Delta(x), \Delta(x_0); \mu) \tag{5.11.45}$$

This certainly obeys $S_{p\ell}(x_0) = 1$ and $\deg(S_{p\ell}) = p\ell$. By (5.11.35),

$$\int |S_{p\ell}(x)|^2 d(S\mu)(x) = \int |Q_\ell(x, \Delta(x_0))|^2 \, d\mu(x) = \lambda_\ell(\Delta(x_0); \mu) \tag{5.11.46}$$

so

$$\lambda_{p\ell}(x_0, S\mu) \le \lambda_\ell(x_0, \mu) \tag{5.11.47}$$

This is terrible! It will not give (5.11.42) but only an inequality with the right multiplied by p. The problem is that S is symmetric in the sense of (5.11.37) and that makes the integral too large because $S_{p\ell}(x)$ is 1 not only at x_0 but at all of the p elements of $\Delta^{-1}(\Delta(x_0))$. To kill the contributions from the other points, we use the localization idea behind Nevai trial functions.

Let $B = \sup\{|x - x_0| \mid x \in \mathfrak{e}\}$ and let ℓ, k be positive integers. Let

$$L(x) = 1 - \left[\frac{x - x_0}{B}\right]^2 \tag{5.11.48}$$

and

$$T_{\ell p+k}(x) = L(x)^k Q_\ell(\Delta(x), \Delta(x_0); \mu) \tag{5.11.49}$$

which has degree $\ell p + k$ and has $T_{\ell p+k}(x_0) = 1$. Let $x_0 \in \mathfrak{e}_j^{\text{int}}$

$$D = \sup_{x \in \mathfrak{e} \setminus \mathfrak{e}_j} 1 - \left(\frac{x - x_0}{B} \right)^2 < 1 \tag{5.11.50}$$

Since the first factor on the right of (5.11.49) is bounded by D^k on any $\mathfrak{e}_m, m \neq j$, and by 1 on $\mathfrak{e}_j$, by (5.11.35) and (5.11.37), we have that

$$\int |T_{\ell p+k}(x)|^2 \, d(S\mu)(x) \leq \left[\frac{1}{p} + \frac{p-1}{p} D^k \right] \int |Q_\ell(x, \Delta(x_0))|^2 \, d\mu(x) \tag{5.11.51}$$

so

$$\limsup n\lambda_n(x_0, S\mu) \leq (1 + (p-1)D^k) \limsup n\lambda_n(\Delta(x_0), \mu) \tag{5.11.52}$$

Since k is arbitrary and $D < 1$, we get (5.11.42) by taking $k \to \infty$.

For the opposite inequality, let $Q_n(x, x_0; S\mu)$ be the minimizer for $\lambda_n(x_0, S\mu)$. Let δ be given and let

$$\ell p = n + k \tag{5.11.53}$$

where

$$|k - \delta n| < p \tag{5.11.54}$$

and let

$$X_{n+k}(x) = L(x)^k Q_n(x, x_0; S\mu) \tag{5.11.55}$$

Finally, let R_ℓ be that polynomial of degree ℓ with

$$R_\ell(\Delta(z)) = \sum_{\{w \mid \Delta(w) = \Delta(z)\}} X_{n+k}(w) \tag{5.11.56}$$

By regularity, on $\mathfrak{e}_j$, for $K(\delta) > 0$,

$$|R_\ell(\Delta(z)) - X_{n+k}(z)| \leq Ce^{-K(\delta)n} \tag{5.11.57}$$

with n positive, so

$$\begin{aligned} \int_{[-2,2]} R_\ell(x)^2 \, d\mu(x) &= \int_{\mathfrak{e}_j} R_\ell(\Delta(x))^2 \, d(S\mu)(x) \\ &\leq \lambda_n(x_0, S\mu) + C_1 e^{-K(\delta)n} \end{aligned} \tag{5.11.58}$$

so by (5.11.35) and (5.11.37),

$$\int_{[-2,2]} R_\ell(x)^2 \, d\mu(x) \leq p[\lambda_n(x_0, S\mu) + C_1 e^{-K(\delta)n}] \tag{5.11.59}$$

Thus,

$$\lambda_\ell(\Delta(x_0), \mu) \leq R_\ell(\Delta(x_0))^{-2} p[\lambda_n(x_0, S\mu) + C_1 e^{-K(\delta)n}] \tag{5.11.60}$$

Picking $n(j)$ so $n(j)\lambda_{n(j)}(x_0, S\mu)$ goes to lim inf and using $\ell p/n \to 1 + \delta$, we get (5.11.43) with an extra $(1 + \delta)$ on the left. Since δ is arbitrary, (5.11.43) follows. □

To apply this to general measures, we need

Proposition 5.11.11. *Let* $\mathfrak{e} = \cup_{j=1}^{p} \mathfrak{e}_j$ *be the essential spectrum of a period* p *Jacobi matrix with discriminant* Δ. *Let* $x_0 \in \mathfrak{e}_j^{\text{int}}$ *for some* j *and let* $J \subset \mathfrak{e}_j^{\text{int}}$ *be a closed interval containing* x_0. *Let* μ *be a measure with* $\sigma_{\text{ess}}(\mu) = \mathfrak{e}$, *which is locally Szegő on* J. *Then there exists a measure* ν *on* $[2, 2]$ *so that*
(i) $S\nu \restriction J = \mu \restriction J$
(ii) ν *is regular for* $[-2, 2]$ *and* $S\nu$ *is regular for* $\mathfrak{e}$.

Remark. It can be proven that any measure ν on $[-2, 2]$ is regular for $[-2, 2]$ if and only if $S\nu$ is regular for $\mathfrak{e}$ (see Totik [443]), but we will not use this.

Proof. Let w be the weight for μ. Since μ is locally Szegő on J, $w > 0$ for a.e. x in J. Let ν_1 be a measure on $\mathfrak{e}_j$ so that

$$\nu_1 \restriction J = \mu \restriction J$$

and $d\nu_1/dx > 0$ for a.e. x in $\mathfrak{e}_j$. This is possible by the positivity of w on J.

Let ν be the unique measure on $[-2, 2]$ so that $S\nu \restriction \mathfrak{e}_j = \nu_1$ (ν is made by mapping ν_1 to $[-2, 2]$ using Δ and then mapping to $S\nu$ using Δ^{-1}). (i) holds by construction. ν and $S\nu$ are regular by Theorem 5.9.6 (or Theorem 5.11.13). □

Theorem 5.11.12. *Let* $\mathfrak{e} = \cup_{j=1}^{p} \mathfrak{e}_j$ *be the spectrum of a periodic Jacobi matrix. Let* $x_0 \in \mathfrak{e}_j^{\text{int}}$ *and let* μ *be a measure with* $\sigma_{\text{ess}}(\mu) = \mathfrak{e}$, μ *regular for* $\mathfrak{e}$, *and locally Szegő near* x_0. *Suppose* x_0 *is a Lebesgue point for* μ *with* $w(x_0) \neq 0$ *and for the locally Szegő function. Let* x_n *be a sequence with*

$$\sup_n n|x_n - x_\infty| \equiv A < \infty \tag{5.11.61}$$

Then

$$\lim_{n\to\infty} \frac{1}{n+1} K_n(x_n, x_n) = \frac{\rho_{\mathfrak{e}}(x_0)}{w(x_0)} \tag{5.11.62}$$

and the limit is uniform in x_n*'s obeying* (5.11.61).

Proof. Let $d\nu$ be the measure on $[-2, 2]$ given by Proposition 5.11.11 and $\tilde{w}$ its weight. $\Delta(x_0)$ is a Lebesgue point for $\tilde{w}$ and for its local Szegő function (since Δ is real analytic near x_0 with $\Delta'(x_0) \neq 0$) so Theorem 3.11.9 is applicable. Thus,

$$\lim_{n\to\infty} \frac{1}{n+1} K_n(\Delta(x_n), \Delta(x_n); \nu) = \frac{\rho_{[-2,2]}(\Delta(x_0))}{\tilde{w}(\Delta(x_0))} \tag{5.11.63}$$

By (5.11.38) and (5.11.40),

$$\frac{\rho_{[-2,2]}(\Delta(x_0))}{\tilde{w}(\Delta(x_0))} = \frac{\rho_{\mathfrak{e}}(x_0)}{w(x_0)} \tag{5.11.64}$$

and by Theorem 5.11.10,

$$\lim_{n\to\infty} \frac{1}{n+1} K_n(\Delta(x_n), \Delta(x_n); \mu) = \lim_{n\to\infty} \frac{1}{n+1} K_n(x_n, x_n; S\nu) \tag{5.11.65}$$

Finally, by the Nevai comparison theorem, Theorem 5.11.5, since $\mu \restriction J = S\nu \restriction J$,

$$\lim_{n\to\infty} \frac{1}{n+1} K_n(x_n, x_n; \mu) = \lim_{n\to\infty} \frac{1}{n+1} K_n(x_n, x_n; S\nu) \tag{5.11.66}$$

(5.11.61) follows from (5.11.63)–(5.11.66). Uniformity follows from the uniformity in Theorem 3.11.9. □

Once we have this result and the general Máté–Nevai upper bound (Theorem 5.11.1), by following the proof of Theorem 5.11.6, we get

Theorem 5.11.13. *Let $\mathfrak{e}$ be a compact subset of $\mathbb{R}$ regular for the Dirichlet problem. Let $d\mu$ be a regular probability measure on $\mathfrak{e}$ of the form*

$$d\mu = w(x)\, dx + d\mu_s \tag{5.11.67}$$

Suppose that, for some closed interval $I \subset \mathfrak{e}^{\text{int}}$, w obeys a local Szegő condition on I. Then for a.e. $x_\infty \in I$, with $\rho_{\mathfrak{e}}$ given by the equilibrium measure for $\mathfrak{e}$, we have

(1) (Diagonal Asymptotics) *For any $A < \infty$ and sequence $x_n \in \mathfrak{e}$ with $n|x_n - x_\infty| \le A$ for all n, we have*

$$\frac{1}{n+1} K_n(x_n, x_n) \to \frac{\rho_{\mathfrak{e}}(x_\infty)}{w(x_\infty)} \tag{5.11.68}$$

(2) (Lubinsky Universality) *For any $A < \infty$ and $a, b \in \mathbb{R}$ with $|a|, |b| \le A$, we have*

$$\frac{K_n(x_\infty + \frac{a}{n}, x_\infty + \frac{b}{n})}{K_n(x_\infty, x_\infty)} \to \frac{\sin(\pi\rho_{\mathfrak{e}}(x_\infty)(b-a))}{\pi\rho_{\mathfrak{e}}(x_\infty)(b-a)} \tag{5.11.69}$$

More generally, the limit of $K_n(x_n, y_n)/K_n(x_\infty, x_\infty)$ is the right side of (5.11.21) *so long as $|x_n - x_\infty| \le A/n$, $|y_n - x_\infty| \le A/n$, and $n(x_n - y_n) \to b - a$.*

As in the case $\mathfrak{e} = [-2, 2]$ (see Theorem 3.11.11), Theorems 5.11.6 and 5.11.13 imply clock behavior for zeros.

Remarks and Historical Notes. If μ, ν are two measures, they have a sup, that is, a smallest measure less than μ and ν. For let $f, g \in L^2(d\mu + d\nu)$ be defined by $d\mu = f(d\mu + d\nu)$, $d\nu = g(d\mu + d\nu)$. If $h = \max(f, g)$, the pointwise sup, then $h(d\mu + d\nu)$ is the sup of μ and ν. In Theorem 5.11.6, we used the fact that if μ and ν are two measures with the same essential support and each is regular, then so is their sup. For a proof, see [417, 404].

5.12 MEROMORPHIC FUNCTIONS ON HYPERELLIPTIC SURFACES

As explained in the overview section, the map from $\{a_n, b_n\}_{n=1}^p$ to Δ, a polynomial of degree p, maps $\mathbb{R}^{2p}$ to $\mathbb{R}^{p+1}$ so inverse images of points are generically of dimension $p-1$ and, in all cases, turn out to be a torus of dimension ℓ, the number of gaps.

Our proof of this in the next two sections will involve a two-step process. We have already seen (see Theorem 5.4.16) that each periodic Jacobi matrix has an

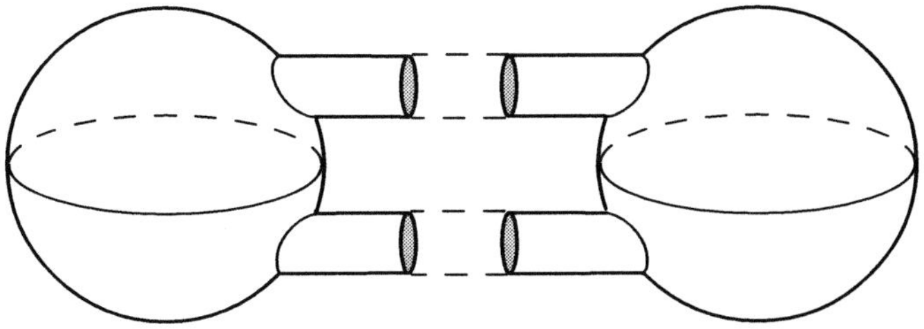

Figure 5.12.1. Two spheres with two cuts glued is a torus.

m-function with exactly one pole in each gap, although it may lie on either sheet. There will also be a pole at ∞ on the second sheet. Thus, m will be a function meromorphic on the two-sheeted Riemann surface associated to $\sqrt{\Delta^2 - 4}$ with exactly $\ell + 1$ poles (which we will see is minimal among all "nontrivial" meromorphic functions). We will prove that such minimal Herglotz functions (normalized to be $-\frac{1}{z} + O(1)$ at ∞ on the first sheet) are exactly in one-one correspondence to ℓ-tuples of points, one on each gap (on either part of the two-sheeted set associated to a gap), and thus, to a point on an ℓ-dimensional torus. This part of the argument, which shows the set of meromorphic m-functions is an ℓ-torus, will be discussed in the next section.

The second step will be to show that each such m-function is associated to a period p Jacobi matrix. This will involve coefficient stripping. Since the poles of the once-stripped m-function are the zeros of the unstripped m-function, we will care about the relation of zeros and poles of this meromorphic function. This is the subject of this section where we will also formally construct the Riemann surface that we study. We can do this in the context of general ℓ-gap sets, which is what we will do. The tools in these two sections will also help us to analyze perturbations of general finite gap Jacobi matrices with suitable properties; see Chapter 9.

So $\mathfrak{e} \subset \mathbb{R}$ has the form

$$\mathfrak{e} = [\alpha_1, \beta_1] \cup \cdots \cup [\alpha_{\ell+1}, \beta_{\ell+1}] \tag{5.12.1}$$

with

$$\alpha_1 < \beta_1 < \alpha_2 < \cdots < \beta_\ell < \alpha_{\ell+1} < \beta_{\ell+1} \tag{5.12.2}$$

Basic to what we do is the Riemann surface $\mathcal{S}$, which we will sometimes write as $\mathcal{S}_\mathfrak{e}$ to emphasize the set $\mathfrak{e}$. We start with an informal description: Take two copies, $\mathcal{S}_+$ and $\mathcal{S}_-$, of the Riemann sphere with $\mathfrak{e}$ removed, that is, $(\mathbb{C} \cup \{\infty\}) \setminus \mathfrak{e}$. Include the set $\mathfrak{e}$ as "top edges." $\mathcal{S}_+$ and $\mathcal{S}_-$ are glued together by the rule that when one passes through $\mathfrak{e}$ starting on $\mathbb{C}_+ \cap \mathcal{S}_+$, one winds up on $\mathbb{C}_- \cap \mathcal{S}_-$ and from $\mathbb{C}_+ \cap \mathcal{S}_-$ to $\mathbb{C}_- \cap \mathcal{S}_+$.

Two spheres with one cut, glued in this way, is topologically a sphere, two cuts are a torus (see Figure 5.12.1), ..., $\ell + 1$ cuts are a sphere with ℓ handles, so $\mathcal{S}$ is an orientable manifold of *genus* ℓ.

More formally, we begin without the points of infinity and think of $\mathcal{S} \subset \mathbb{C}^2$ as those points $\langle z, w\rangle$ with

$$w^2 = R(z) \equiv \prod_{j=1}^{\ell+1} (z - \alpha_j)(z - \beta_j) \tag{5.12.3}$$

Notice that

$$x \in \mathfrak{e} \Rightarrow R(x) \leq 0 \qquad x \in \mathbb{R} \setminus \mathfrak{e} \Rightarrow R(x) > 0 \tag{5.12.4}$$

In case $\mathfrak{e}$ is the essential spectrum of a periodic Jacobi matrix with all gaps open,

$$R(z) = (a_1 \dots a_p)^2[\Delta^2(z) - 4] \tag{5.12.5}$$

but double zeros are dropped if some gaps are closed and there is no Δ in general (i.e., if some band has irrational harmonic measure). $w^2 - R(z) = 0$ defines a Riemann surface (one-dimensional complex manifold) since $\nabla(w^2 - R(z)) \neq 0$ for all $\langle w, z\rangle \in \mathcal{S}$. If $z \notin \{\alpha_j, \beta_j\}_{j=1}^{\ell+1}$, then $\frac{\partial}{\partial w}(w^2 - R(z)) = 2w \neq 0$, so w is a smooth function of z and we can use z as a local coordinate. If $z \in \{\alpha_j, \beta_j\}_{j=1}^{\ell+1}$, $\frac{\partial}{\partial z}(w^2 - R(z)) = -R'(z)$ is nonzero, and we can use w as a local coordinate, but not z. This means functions defined on $\mathcal{S}$ near $\langle z_0, w_0\rangle \in \mathcal{S}$ are "analytic" if and only if they have convergent power series in $z - z_0$ if $z_0 \notin \{\alpha_j, \beta_j\}_{j=1}^{\ell+1}$, and if $z_0 \in \{\alpha_j, \beta_j\}_{j=1}^{\ell+1}$, we only need convergent power series in w, equivalently in $(z - z_0)^{1/2}$.

We obtain a compact surface by adding two points, ∞_+ and ∞_-, at infinity using $1/z$ local coordinates. ∞_+ and ∞_- are distinguished by the fact that nearby $w = \pm z^\ell(1 + O(z^{-1}))$. In more formal approaches, one embeds the finite part of $\mathcal{S}$ in two-dimensional complex projective space. There is now a single point at ∞ but it is a singularity, and that singularity is resolved by doubling the point; see, for example, Miranda [306].

The map $\pi : \langle z, w\rangle \to z$ is a two-to-one map over $\mathbb{C} \cup \{\infty\} \setminus \mathfrak{e}$. For any point $z \in \mathbb{C} \cup \{\infty\} \setminus \mathfrak{e}$, we use z_+ and z_- for the two points with $w > 0$ for z_+ and $z \in (\beta_{\ell+1}, \infty)$. We have labeled the two points at infinity ∞_+ and ∞_-.

We define

$$\tau : \mathcal{S} \to \mathcal{S}$$

by $\tau(z_+) = z_-$, where $\tau(z) = z$ if $\pi(z) \in \{\alpha_j, \beta_j\}_{j=1}^{\ell+1}$. We call this latter set *branch points*.

We will be interested in meromorphic functions f on $\mathcal{S}$, that is, maps from $\mathcal{S}$ to $\mathcal{S}_R$, the Riemann sphere, that are locally "analytic" as $\mathcal{S}_R$-valued maps, that is, locally meromorphic in the conventional sense.

We recall that if f is a meromorphic function (defined as being locally meromorphic at every point) on the entire Riemann sphere, then

$$f(z) = \frac{p(z)}{a(z)} \tag{5.12.6}$$

for polynomials p and a. For f has only finitely many poles by compactness. Take a to have zeros at the finite poles of order equal to the order of those poles. Then $f(z)a(z)$ is an entire function with a finite order pole at infinity, so a polynomial.

Proposition 5.12.1. *Every meromorphic function, f, on $\mathcal{S}$ has the form*

$$f(z) = \frac{p(z) + q(z)w}{a(z)} \tag{5.12.7}$$

where p, q, a are polynomials with no common zeros and with $a \not\equiv 0$, and conversely.

Remarks. 1. We will start writing

$$f = \frac{p \pm q\sqrt{R}}{a} \tag{5.12.8}$$

2. Here no common zeros means zeros of all three of p, q, a—not of just two.

Proof. If p, q, a have a common zero, we can factor it out, so we will ignore that condition henceforth.

Define

$$f_s(z) = \tfrac{1}{2}(f(z) + f(\tau z)) \tag{5.12.9}$$

f_s is symmetric under τ, so f_s is a function of $\pi(z)$ only which, by an abuse of notation, we will write as $f_s(z)$ also. f_s is obviously meromorphic in z at any nonbranch point since $f(z)$ and $f(\tau z)$ are. At a branch point z_0,

$$f(z) = \sum_{n=0}^{\infty} a_n (z - z_0)^{n/2}$$

$$f(\tau z) = \sum_{n=0}^{\infty} a_n (-1)^n (z - z_0)^{n/2}$$

so f_s is also analytic in z. Thus, by the remark before the theorem,

$$f_s(z) = \frac{p_1(z)}{a_1(z)}$$

Similarly, we define

$$f_a(z) = \frac{f(z) - f(\tau z)}{w}$$

and see it is also entire meromorphic, so $q_1(z)/a_2(z)$. Pick $a(z) = a_1(z)a_2(z)$ and get the required form after pulling out common zeros. $\square$

We want to note that meromorphic functions on $\mathcal{S}$ are all the solutions of quadratic equations.

Proposition 5.12.2. *Let f be a meromorphic function on $\mathcal{S}$ so that f has at most first-order poles at branch points and if $z \in \mathcal{S}$ is a pole but not a branch point, then τz is not a pole. Then the two values of f are the two solutions of*

$$\alpha(z) f(z)^2 + \beta(z) f(z) + \gamma(z) = 0 \tag{5.12.10}$$

Indeed, in terms of (5.12.7),

$$\alpha(z) = a(z) \qquad \beta(z) = -2p(z) \qquad \gamma(z) = \frac{p^2(z) - q^2(z)R(z)}{a(z)} \tag{5.12.11}$$

Remark. We claim γ given by (5.12.11) is a polynomial.

Proof. Clearly, (5.12.7) is equivalent to

$$(af - p)^2 = q^2 R \tag{5.12.12}$$

which is (5.12.10) if we prove γ is a polynomial. If $a(z)$ has a zero of order k at z_0, not a branch point, then by hypothesis, as an analytic function either $p + q\sqrt{R}$ or $p - q\sqrt{R}$ has a zero of order at least k. So $p^2 - q^2 R = (p + q\sqrt{R})(p - q\sqrt{R})$ as an analytic function, and so as a polynomial has a zero of order at least k.

At branch points, z_0, if $a(z_0)$, then for f to have a simple pole (given that a, p, q have no common zeros), we must have that a has a simple zero, $p(z) = 0$, $q(z) \neq 0$. So $p^2 - q^2 R$ has a simple zero. Thus, γ is a polynomial, as claimed. □

We count orders of zeros and poles in terms of local analytic coordinates. Thus, if z_0 is a branch point, we must use w or $(z - z_0)^{1/2}$ as local coordinates so $f(z) = z - z_0$ has a second-order zero at such a z_0.

Associated to any zero or pole, z_0, we associate a single integer $N(f; z_0)$, which is the order of the zero if z_0 is a zero and the negative of the order of the pole if a pole. That is, if ζ is a local analytic coordinate near z_0 with $\zeta(z_0) = 0$, then

$$f(z) = \zeta^{N(f;z)}(c + O(\zeta)) \tag{5.12.13}$$

with $c \neq 0$.

Theorem 5.12.3. *Let f be a meromorphic function on $\mathcal{S}$ with zeros/poles at $\{z_j\}_{j=1}^m$. Then*

$$\sum_{j=1}^{m} N(f; z_j) = 0 \tag{5.12.14}$$

Remarks. 1. This is usually stated by saying: "The number of zeros is equal to the number of poles." In case where all zeros and poles are simple (i.e., $|N(f; z_j)| = 1$ for all z_j), this is literally true. Otherwise, one needs to count with multiplicities.

2. There are two other ways to understand this result: first, a general connectedness argument (see [126, p. 12]), and second, in terms of homology theory.

Proof. Let Γ_+ be the curve in $\mathcal{S}_+$ that goes clockwise around a cut from α_1 to $\beta_{\ell+1}$, say, a distance ε around from the cut, and let Γ_- be the same curve on $\mathcal{S}_-$. We will consider

$$\xi = \frac{1}{2\pi i} \int_{\Gamma_+} \frac{f'(z)}{f(z)}\, dz + \frac{1}{2\pi i} \int_{\Gamma_-} \frac{f'(z)}{f(z)}\, dz \tag{5.12.15}$$

Suppose first that f has no zero or pole with $\pi(z_j) \in [\alpha_1, \beta_{\ell+1}]$. In that case, we claim

$$\xi = 0 \tag{5.12.16}$$

For by taking $\varepsilon \downarrow 0$, the contributions of the gaps cancel individually in Γ_+ and Γ_- (since the contours go in opposite directions).

Along the bands $[\alpha_j, \beta_j]$, f'/f is "continuous" across the band if we jump from $\mathcal{S}^+$ to $\mathcal{S}^-$, so the top piece of the Γ_+ contour cancels the bottom piece of the Γ_- contour, and we get (5.12.16).

On the other hand, one can evaluate the integrals by looking at residues at poles of f'/f (since infinity is either a regular point or a simple pole of f'/f) and get

$$\frac{1}{2\pi i}\int_{\Gamma_\pm} \frac{f'(z)}{f(z)}\,dz = -\sum_{\{z_j \mid z_j \in \mathcal{S}^\pm \setminus [\alpha_1, \beta_{\ell+1}]\}} N(f; z_j) \tag{5.12.17}$$

In this case, (5.12.16) and (5.12.17) yield (5.12.14). If there are zeros z_j with $\pi(z_j) \in [\alpha_1, \beta_{\ell+1}]$, we claim

$$\xi = \sum_{\{z_j \mid \pi(z_j) \in [\alpha_1, \beta_{\ell+1}]\}} N(f; z_j) \tag{5.12.18}$$

so, taking into account that (5.12.17) is always true, we get (5.12.14) in general.

For zeros $z_j \in [\alpha_1, \beta_{\ell+1}] \setminus \{\alpha_j, \beta_j\}_{j=1}^{\ell+1}$, (5.12.18) is immediate, since for zeros in gaps, the noncancelling parts of Γ_+ or Γ_- precisely surround the poles of f'/f, and for poles in $\mathfrak{e}^{\text{int}}$, the noncancelling parts of the contours that cancel surround the poles on $\mathcal{S}$.

If z_j is a branch point, $\arg(f)$ changes by $\pi N(f; z_j)$ (rather than $2\pi N(f; z_j)$) because of how orders are defined. But there are contributions for both Γ_+ and Γ_-, yielding a total change of $2\pi N(f; z_j)$, which proves (5.12.18). □

More generally, one can define an order of a value of f at any point as follows:

$$n(f; z_0, a) = \begin{cases} 0 & \text{if } f(z_0) \neq a \\ N(f - a; z_0) & \text{if } f(z_0) = a \neq \infty \\ -N(f; z_0) & \text{if } f(z_0) = a = \infty \end{cases} \tag{5.12.19}$$

so $n(f; z_0, a) \geq 0$ and is nonzero at only finitely many points.

Corollary 5.12.4. *For any meromorphic f on $\mathcal{S}$,*

$$\deg(f) \equiv \sum_{\{z \mid n(f; z, a) > 0\}} n(f; z, a) \tag{5.12.20}$$

is independent of a.

Proof. Call the right side of (5.12.20) $d(f; a)$. Then

$$d(f; a) - d(f; \infty) = \sum_{\{z_j \mid f(z_j) = a \text{ or } f(z_j) = \infty\}} N(f - a; z_j) = 0$$

by Theorem 5.12.3, which proves the a-independence. □

The number $\deg(f)$ is called the *degree of f*. As we will discuss in the Notes, degree and the formula (5.12.20) have a topological interpretation.

Definition. A meromorphic function, f, is called *root free* if $f(\tau z) = f(z)$. Equivalently,

$$f(z) = \frac{p(z)}{a(z)} \tag{5.12.21}$$

for polynomials p and a.

Theorem 5.12.5. (a) *Every root-free function has even order and all nonnegative even integers* $2, 4, 6, \dots$ *occur. Indeed, if f has the form* (5.12.21), *where p, a have no common zeros, then*

$$\deg(f) = 2\max(\mathrm{Deg}(p), \mathrm{Deg}(a)) \tag{5.12.22}$$

where $\mathrm{Deg}(\cdot)$ *is the conventional degree of a polynomial.*

(b) *If f is not root free, it has degree at least $\ell + 1$, and every degree larger than that occurs. In addition, if f has the form* (5.12.8),

$$\deg(f) \geq \max(\mathrm{Deg}(a), \ell + 1 + \mathrm{Deg}(q)) \tag{5.12.23}$$

Proof. (a) On the Riemann sphere, if f has the form (5.12.21) and $\mathrm{Deg}(p) \geq \mathrm{Deg}(a)$, f has a pole or nonzero value at ∞ and zeros (including multiplicity) at the zeros of p, so $\deg_{\mathrm{R.S.}}(f) = \mathrm{Deg}(p)$ (where $\deg_{\mathrm{R.S.}}$ means degree as a function on the Riemann sphere). If $\mathrm{Deg}(a) > \mathrm{Deg}(p)$, there are $\mathrm{Deg}(p)$ zeros on $\mathbb{C}$ and ∞ is a zero of degree $\mathrm{Deg}(a) - \mathrm{Deg}(p)$, so $\deg_{\mathrm{R.S.}}(f) = \mathrm{Deg}(p) + \mathrm{Deg}(a) - \mathrm{Deg}(p) = \mathrm{Deg}(a)$. Thus,

$$\deg_{\mathrm{R.S.}}(f) = \max(\mathrm{Deg}(p), \mathrm{Deg}(a))$$

This degree is doubled on $\mathcal{S}$ since nonbranch point values occur at both z_+ and z_- and branch point orders are doubled because of the change to $(z - z_0)^{1/2}$ counting. This proves (5.12.22), and that implies the allowed degrees of such functions are $2, 4, \dots$.

(b) We first prove (5.12.23). Let $z_1, \dots, z_A$ be the zeros of a (where $A = \mathrm{Deg}(a)$). If z_j is not a branch point, $R(z_j) \neq 0$ and so at least one of $p(z_j) + q(z_j)\sqrt{R(z_j)}$ or $p(z_j) - q(z_j)\sqrt{R(z_j)}$ is nonzero. (*Note:* If $q(z_j) = 0$, $p(z_j) \neq 0$ and both are nonzero.) f has a pole of order at least the order of the zeros z_j in a at either $(z_j)_+$ or $(z_j)_-$.

If z_j is a branch point, one of $p(z_j)$ or $q(z_j)$ is nonzero, so $p(z) + q(z)\sqrt{R(z)}$ is either $O(1)$ or $O((z - z_j)^{1/2})$, in which case if z_j is a zero of a of order n_j, f has a pole of order $2n_j$ or $2n_j - 1 \geq n_j$. We conclude

$$\deg(f) \geq \mathrm{Deg}(a)$$

If $Q = \mathrm{Deg}(q)$, $q\sqrt{R} \sim c\, z^{Q+\ell+1}$ near $\infty_\pm$. p can cancel $q\sqrt{R}$ or $-q\sqrt{R}$, but not both, that is, f has a pole of order $\mathrm{Deg}(q) + \ell + 1 - \mathrm{Deg}(a)$ (if positive), at least one of ∞_+ or ∞_-. So we have $\mathrm{Deg}(q) + \ell + 1 - \mathrm{Deg}(a) + \mathrm{Deg}(a)$ poles, that is,

$$\deg(f) \geq \mathrm{Deg}(q) + \ell + 1 \tag{5.12.24}$$

This proves (5.12.23) and shows that

$$\deg(f) \geq \ell + 1 \tag{5.12.25}$$

To see that every integer larger than or equal to $\ell = 1$ occurs, proceed as follows: Let $m \geq 0$ be an integer and define

$$g(z) = z^m \sqrt{R(z)} \tag{5.12.26}$$

where g is meromorphic near ∞ in $\mathbb{C}\cup\{\infty\}$. If we take the value of $\sqrt{R(z_j)}$, which is positive on $(\beta_{\ell+1}, \infty)$, it has a pole of order $m+\ell+1$. Let $p(z)$ be the "negative" order terms in the Laurent series at infinity—negative in z^{-1}. So $p(z)$ is that unique polynomial (or degree $\ell + 1 + m$) with

$$g(z) = p(z) + o(1) \tag{5.12.27}$$

near infinity.

Now let

$$f(z) = p(z) \pm z^m \sqrt{R(z)} \tag{5.12.28}$$

Clearly, f is meromorphic on $\mathcal{S}$. At finite points of $\mathcal{S}$, f is finite and, by (5.12.27),

$$f(z) = o(1)$$

near ∞_-, and so f has a zero there. Its only pole is at ∞_+ and there, $f(z) = 2z^{m+\ell+1} + O(z^{m+\ell})$. Thus, the pole is of order $m + \ell + 1$ and

$$\deg(f) = m + \ell + 1 \tag{5.12.29}$$

proving the claim. □

A compact Riemann surface that has meromorphic functions of degree 1 is conformally equivalent to the Riemann sphere since f is one-one and onto that sphere. A compact Riemann surface that has meromorphic functions of degree 2 is called *hyperelliptic* and this last theorem tells us that $\mathcal{S}$, the Riemann surface of $\sqrt{R}$, is hyperelliptic.

We now turn to the question of what sets can be the zeros/poles of a function on $\mathcal{S}$. By Theorem 5.12.3, if $\{z_j\}_{j=1}^{N_z}$ and $\{p_j\}_{j=1}^{N_p}$ are the zeros and poles of a meromorphic function (counting multiplicity and with no z_j equal to any p_k), then

$$N_z = N_p \tag{5.12.30}$$

so we will henceforth use N.

For the Riemann sphere, (5.12.30) is the only restriction on the zeros and poles. But we recall the situation for classical elliptic functions [7], that is, meromorphic functions on $\mathbb{C}$ that obey

$$f(z+1) = f(z) \qquad f(z+\tau) = f(z) \tag{5.12.31}$$

for some $\tau \notin \mathbb{R}$; by replacing τ by $-\tau$, we can suppose $\operatorname{Im}\tau > 0$. Let

$$\mathcal{L}_\tau = \{n + m\tau \mid n, m \in \mathbb{Z}\} \tag{5.12.32}$$

which is a discrete lattice in $\mathbb{C}$, and so let

$$\mathcal{S}_\tau = \frac{\mathbb{C}}{\mathcal{L}_\tau} \tag{5.12.33}$$

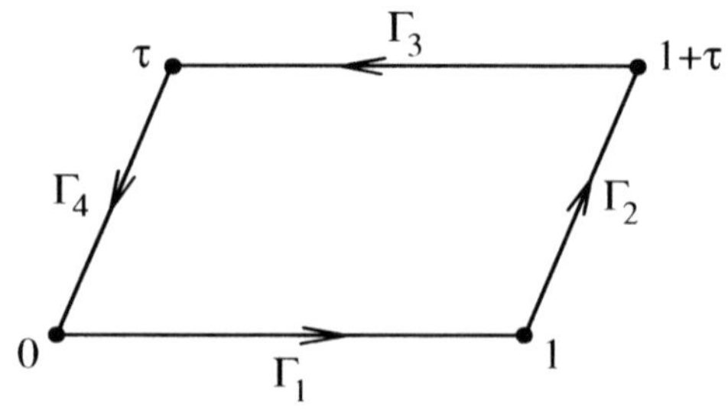

Figure 5.12.2. Contour for Liouville's second theorem.

equivalence classes in $\mathbb{C}$ mod $\mathcal{L}_\tau$. It can be shown $\mathcal{S}_\tau$ is conformal to $\mathcal{S}_{\tau'}$ if and only if for $c \in \mathbb{C} \setminus \{0\}$, $\mathcal{L}_\tau = c\mathcal{L}_{\tau'}$ if and only if there exists an $A \in \mathbb{SL}(2, \mathbb{Z})$ with $A\binom{\tau}{1} = c\binom{\tau'}{1}$ for $c \in \mathbb{C} \setminus \{0\}$. Moreover, every Riemann surface, which is topologically a torus, is conformal to some $\mathcal{S}_\tau$. In particular, our $\mathcal{S}$ is an $\mathcal{S}_\tau$ if $\ell = 1$ (with τ pure imaginary a function of $(\beta_2 - \alpha_2)/(\beta_1 - \alpha_1)$ and $(\alpha_2 - \beta_1)/(\beta_1 - \alpha_1)$).

Meromorphic functions on $\mathcal{S}_\tau$ are precisely the same as f's on $\mathbb{C}$ obeying (5.12.31). Liouville's second theorem on elliptic functions ((5.12.30) is his first theorem on elliptic functions) says that

$$\sum_{j=1}^{N} z_j - p_j \in \mathcal{L}_\tau \tag{5.12.34}$$

where, for example, one normalizes z_j, p_j by putting them in the fundamental region $\mathcal{F} = \{a + b\tau \mid 0 \le a < 1,\ 0 \le b < 1\}$.

To prove (5.12.34), one takes a contour, Γ, which is shown in Figure 5.12.2, that goes clockwise around the parallelogram with sides $\Gamma_1 = \{(a, 0) \mid 0 \le a < 1\}$, $\Gamma_2 = \{(1, b\tau) \mid 0 \le b < 1\}$, $\Gamma_3 = \{(a, \tau) \mid 0 \le a < 1\}$, $\Gamma_4 = \{(0, b\tau) \mid 0 \le b < 1\}$, and assuming f has no zeros or poles on Γ, one looks at

$$\xi = \frac{1}{2\pi i} \oint_\Gamma z \frac{f'}{f}\, dz \tag{5.12.35}$$

On the one hand, ξ is the left side of (5.12.34) by the residue calculus. On the other hand, since f'/f is periodic, the Γ_1 and Γ_3 contributions partially cancel to give $-\frac{1}{2\pi i}\int_{\Gamma_1} \tau \frac{f'}{f}\, dz$, and the Γ_2 and Γ_4 to give $\frac{1}{2\pi i}\int_{\Gamma_2} \frac{f'}{f}\, dz$. But by the argument principle and periodicity again, $\frac{1}{2\pi i}\int_{\Gamma_1} \frac{f'}{f}\, dz$ is an integer, so $\xi = n_1\tau + n_2 \in \mathcal{L}_\tau$.

In this argument, there are two main players: the function z and the contours Γ_1 and Γ_2. z enters because dz is an analytic one-form and z is its integral. In the torus $\mathcal{S}_\tau$, Γ_1 and Γ_2 are precisely homology generators for the homology of $\mathcal{S}_\tau$, closed curves that loop once about the two "holes" of $\mathcal{S}_\tau$.

Returning to our hyperelliptic surface, $\mathcal{S}$, its homology group has 2ℓ generators that loop about the two holes of each of the ℓ handles. We can realize these generators explicitly. For $j = 1, \dots, \ell$, let G_j^+ be the line on $\mathcal{S}_+$ from β_j to α_{j+1} and G_j^- the same on $\mathcal{S}_-$. $G_j = G_j^+ - G_j^-$ is a closed curve on $\mathcal{S}$ called $\Gamma(G_j)$. For $j = 1, \dots, \ell + 1$, let $\Gamma(B_j)$ be the closed curve that goes from α_j to β_j on $\mathcal{S}_+$ just below the cut and then returns from β_j to α_j just above the cut.

$\Gamma(B_1)+\Gamma(B_2)+\cdots+\Gamma(B_{\ell+1})$ is homologous to the curve Γ_+ used in the proof of Theorem 5.12.3 and that curve is even homotopic to 0 by "pulling it through ∞_+." Thus, $\{\Gamma(B_j)\}_{j=1}^{\ell+1}$ are not independent in homology, but $\{\Gamma(B_j)\}_{j=1}^{\ell}$ are, and $\{\Gamma(G_j)\}_{j=1}^{\ell} \cup \{\Gamma(B_j)\}_{j=1}^{\ell}$ are a set of homology generators.

As for that other player, analytic one-forms, consider

$$\omega_1 = \left(\sqrt{R(z)}\right)^{-1} dz \tag{5.12.36}$$

that is, $w^{-1}\,dz$ in $\langle w, z\rangle$ coordinates. Since R vanishes at each branch point, one might think w is singular there, but recall that the proper local coordinate there is w and $w \sim c_0(z-z_0)^{1/2}$, that is, $dz \sim c_1 w\,dw$ and $w^{-1}\,dz = c_1\,dw$ is nonsingular at z_0. Near ∞, we need to shift from z to $\zeta = z^{-1}$ and $dz = -z^2\,d\zeta$ is singular. But since $R(z) \sim O(z^{\ell+1})$ if $\ell \geq 1$, ω_1 is regular at $\infty_\pm$ also.

More generally, if $P(z)$ is a polynomial, then

$$\omega_P = P(z)\left(\sqrt{R(z)^{-1}}\right) dz \tag{5.12.37}$$

is regular at all finite points and is regular at infinity so long as

$$\deg(P) \leq (\ell+1) - 2 = \ell - 1 \tag{5.12.38}$$

We thus get an ℓ-dimensional family of analytic one-forms and, by de Rham's theorem and the fact that the homology is dimension 2ℓ, this is all of them (the 2ℓ-dimensional de Rham cohomology is spanned by ℓ analytic and ℓ anti-analytic forms).

It is natural to evaluate the cohomology elements on homology generators, and so define for $j = 1, \dots, \ell$,

$$\pi(P; B_j) = \int_{\Gamma(B_j)} \omega_P \tag{5.12.39}$$

$$\pi(P; G_j) = \int_{\Gamma(G_j)} \omega_P \tag{5.12.40}$$

called the *periods* of the one-form ω_P. The following is basic:

Theorem 5.12.6. *For P a real polynomial of degree at most $\ell - 1$, define vectors in $\mathbb{R}^\ell$ by*

$$\mathcal{B}(P)_j = -i\pi(P; B_j) \tag{5.12.41}$$

$$\mathcal{G}(P)_j = \pi(P; G_j)$$

Then $\mathcal{B}$ and $\mathcal{G}$ are bijections of real polynomials to $\mathbb{R}^\ell$.

Proof. Since the polynomials of degree at most $\ell-1$ are an ℓ-dimensional space, it suffices to prove $\ker\mathcal{B} = \ker\mathcal{G} = 0$. $i\sqrt{R(z)}$ has a definite sign on the top of each B_j and the contour has the opposite direction on the bottom and opposite signs, so if P has a definite sign on B_j, then $-i\pi(P; B_j) \neq 0$. It follows that if $\mathcal{B}(P) = 0$, the P has a zero on each of the ℓ sets $B_1, \dots, B_\ell$. But if P is nonzero, it can only have $\ell - 1$ zeros. Thus, $\ker\mathcal{B} = 0$.

A similar argument proves that $\ker\mathcal{G} = 0$. $\square$

Since $\mathcal{G}$ is a bijection, we can find polynomials $P_1, \dots, P_\ell$ so that

$$\pi(P_k; G_j) = \delta_{kj} \tag{5.12.42}$$

which we call the canonical basis. The *periods*, $\tau_{kj} \in \mathbb{R}$, of $\mathcal{S}$ are defined by

$$\pi(P_k; B_j) = i\tau_{kj} \tag{5.12.43}$$

In $\mathbb{C}^\ell$, we define the lattice $\mathcal{L}_\mathcal{S}$ of $\mathcal{S}$ by

$$\mathcal{L}_\mathcal{S} = \{\vec{n} + i\tau\vec{m} \mid n, m \in \mathbb{Z}^\ell\} \tag{5.12.44}$$

where

$$(\tau\vec{m})_k = \sum \tau_{kj} m_j \tag{5.12.45}$$

By the theorem, the vectors $\tau_{k\cdot}$ are independent, so $\mathcal{L}_\mathcal{S}$ is a discrete lattice in $\mathbb{C}^\ell$, which means that the *Jacobi variety*,

$$\mathcal{J}_\mathcal{S} = \frac{\mathbb{C}^\ell}{\mathcal{L}_\mathcal{S}} \tag{5.12.46}$$

is a torus of real dimension 2ℓ.

Given any rectifiable (not necessarily closed) contour, Γ, on $\mathcal{S}$, define $\mathbb{A}(\Gamma) \in \mathbb{C}^\ell$ by

$$\mathbb{A}(\Gamma)_k = \int_\Gamma \omega_{P_k}$$

If Γ is closed and homologous to zero, Cauchy's theorem implies $\mathbb{A}(\Gamma) = 0$. More generally, since $\{\Gamma(B_j\}_{j=1}^\ell$ and $\{\Gamma(G_j)\}_{j=1}^\ell$ are generators of homology, we have

$$\Gamma \text{ closed} \Rightarrow \mathbb{A}(\Gamma) \in \mathcal{L}_\mathcal{S}$$

That means that if $x, y \in \mathcal{S}$ is fixed and Γ_{xy} is any curve from x to y, $\mathbb{A}(\Gamma_{xy})$ has a value whose ambiguity is an element in $\mathcal{L}_\mathcal{S}$, and thus, $[\mathbb{A}(\Gamma_{xy})]_{\mathcal{L}_\mathcal{S}} \equiv \mathfrak{A}_x(y)$ is an element of $\mathcal{J}_\mathcal{S}$. Thus, once we fix a base point x in $\mathcal{S}$, we have a map

$$\mathfrak{A}_x : \mathcal{S} \to \mathcal{J}_\mathcal{S} \tag{5.12.47}$$

called Abel's map. $\mathcal{L}_\mathcal{S}$ is an abelian group, and it is easy to see that the change of base point is given by

$$\mathfrak{A}_{x_1}(y) = \mathfrak{A}_{x_0}(y) + \mathfrak{A}_{x_1}(x_0) \tag{5.12.48}$$

$$= \mathfrak{A}_{x_0}(y) - \mathfrak{A}_{x_0}(x_1) \tag{5.12.49}$$

While $\mathfrak{A}_{x_0}$ depends on the base point x_0, we will often just use $\mathfrak{A}$ with some fixed x_0 in mind.

The fundamental results about zeros and poles and meromorphic functions are:

Theorem 5.12.7 (Abel's Theorem, First Half). *Let f be a meromorphic function on $\mathcal{S}$ and let $\{z_j\}_{j=1}^{N_z}$ and $\{p_j\}_{j=1}^{N_p}$ be its zeros and poles counting multiplicity. Then*

(a) $N_z = N_p$

(b) *We have*

$$\sum_{j=1}^{N_z} \mathfrak{A}(z_j) = \sum_{j=1}^{N_p} \mathfrak{A}(p_j) \tag{5.12.50}$$

Theorem 5.12.8 (Abel's Theorem, Second Half). *Let $\{z_j\}_{j=1}^{N_z}$ and $\{p_j\}_{j=1}^{N_p}$ be points on $\mathcal{S}$ with no z_j equal to a p_k (although z's or p's can be repeated). Then there is a meromorphic f on $\mathcal{S}$ with zeros precisely at the z_j and poles precisely at the p_j if and only if* (a) *and* (b) *of Theorem 5.12.7 hold.*

Remarks. 1. This is a single result, which we state as two because we will prove and extensively use the first half below (and in the next section). We will only prove a special case of the second half in Section 9.11 and will use it once below in the first proof of a theorem (Theorem 5.12.10), for which we also provide a second proof below that does not use the second half.

2. Our use of Abel's theorem only requires the existence of a map $\mathcal{U}$ from $\mathcal{S}$ to $\mathcal{J}_{\mathcal{S}}$ with the required properties. Indeed, in Section 9.11, our $\mathcal{U}$ will map $\cup_{j=1}^{\ell} G_j \cup \{\infty_\pm\}$ to a natural torus group $(\partial\mathbb{D})^\ell$ and we will shift from additive notation for the group action to multiplicative.

3. Because of (5.12.49) and $N_z = N_p$, the equality (5.12.50) is base point independent.

4. The sum in (5.12.50) is in the abelian group $\mathcal{J}_{\mathcal{S}}$.

5. We emphasize that the sets in these theorems are really sets with multiplicity, and a zero of order k appears k times in $\{z_j\}_{j=1}^{N_z}$, and similarly for poles.

As a preliminary for the proof and because it is useful in further developments, we want to describe a specific realization of $\mathfrak{A}$ in $\mathbb{C}^\ell$ for $\mathcal{S}$ with suitable cuts. Remove from both $\mathcal{S}_+$ and $\mathcal{S}_-$ the interval $[\beta_1, \beta_{\ell+1}]$. The two halves are still connected by crossing (α_1, β_1), and the reader can convince himself/herself that the resulting set with $\infty_\pm$ included is simply connected. So taking the base point as α_1 for definiteness, one gets a single-valued map $\mathfrak{A}^\sharp$ with values in $\mathbb{C}^\ell$.

In each gap (β_j, α_{j+1}) on either sheet, $\mathfrak{A}^\sharp$ is discontinuous across the gap but only by a period (i.e., element of $\mathcal{L}_{\mathcal{S}}$), which, by discreteness, has to be constant on each gap. Similarly, for each band $[\alpha_j, \beta_j]$, $j = 2, \dots, \ell+1$, $\mathfrak{A}^\sharp$ is discontinuous if we approach $x \in (\alpha_j, \beta_j)$ from $\mathbb{C}_+ \cap \mathcal{S}_+$ or from $\mathbb{C}_- \cap \mathcal{S}_-$ (which are the same point in $\mathcal{S}$), and again we get a constant that is a period. For Theorem 5.12.7, all we need are these facts, but for Theorem 5.12.12 below, we need the precise constant:

Proposition 5.12.9. (a) *If $x \in (\beta_j, \alpha_{j+1})$, then*

$$\mathfrak{A}^\sharp(x_\pm + i0) - \mathfrak{A}^\sharp(x_\pm - i0) = \pm i\left[\sum_{m=1}^{j} \vec{\tau}_{\cdot m}\right] \tag{5.12.51}$$

(b) *If $x \in (\alpha_j, \beta_j)$, $j = 2, \dots, \ell+1$, then*

$$\mathfrak{A}^\sharp(x_\pm + i0) - \mathfrak{A}^\sharp(x_\mp - i0) = \pm \sum_{m=1}^{j-1} \vec{\delta}_{\cdot m} \tag{5.12.52}$$

Remark. $\vec{\tau}_{\cdot m}$ is the vector whose components are τ_{jm}. Similarly, $\vec{\delta}_{\cdot m}$ has components δ_{jm}, that is,

$$\sum_{m=1}^{j-1} \vec{\delta}_{\cdot m} = \left(\underbrace{1, \dots, 1}_{j-1}, \underbrace{0, \dots, 0}_{\ell-(j-1)} \right) \tag{5.12.53}$$

Proof. (a) A curve that goes from α_1 in $\mathcal{S}_+$ in the upper half-plane of $\mathcal{S}_+$ to $x \in (\beta_j, \alpha_{j+1})$ and returns to α_1 in the lower half-plane of $\mathcal{S}_+$ is homologous to $\Gamma(B_1) + \cdots + \Gamma(B_j)$, so (5.12.51) for $+$ is just (5.12.43). The minus sign is immediate if we note that with a base point α_1, the periods flip sign from $\sqrt{R}$ to $-\sqrt{R}$ in going from $\mathcal{S}_+$ to $\mathcal{S}_-$.

(b) Consider first $j = 2$. To get to $x_+ + i0$, we go above (α_1, β_1), then follow (β_1, α_2) in $\mathcal{S}_+$ and then go to $x + i0$. To get to $x_- + i0$, we do the same, but follow (β_1, α_2) in $\mathcal{S}_-$. The difference is just $\Gamma(G_1)$. For general j, the difference is $\Gamma(G_1) + \cdots + \Gamma(G_{j-1})$. This leads to (5.12.52). □

Proof of Theorem 5.12.7. (a) is Theorem 5.12.3. To prove (b), let $\Gamma_\pm$ be the contours used in the proof of that theorem and let

$$\xi_{\mathfrak{A}} = \frac{1}{2\pi i} \int_{\Gamma_+} \frac{f'(z)}{f(z)} \mathfrak{A}^\sharp(z)\, dz + \frac{1}{2\pi i} \int_{\Gamma_-} \frac{f'(z)}{f(z)} \mathfrak{A}^\sharp(z)\, dz \tag{5.12.54}$$

We will suppose no zeros or poles lie inside $\Gamma_\pm$—the change when some do is as with the proof of Theorem 5.12.3.

In that earlier theorem, the residues of f'/f outside $\Gamma_\pm$ at z_0 are just $\#(z_j = z_0) - \#(p_j = z_0)$. Now they are multiplied by $\mathfrak{A}^\sharp(z_0)$. Thus,

$$\xi_{\mathfrak{A}} = - \sum_{z_j, p_j \in \mathcal{S}^\pm \setminus [\alpha_j, \beta_{j+1}]} (\mathfrak{A}^\sharp(z_j) - \mathfrak{A}^\sharp(p_j)) \tag{5.12.55}$$

On the other hand, there is not the complete cancellation that caused $\xi = 0$ in (5.12.17) because $\mathfrak{A}^\sharp$ is discontinuous across cancelling curves. Rather, since $\mathfrak{A}^\sharp$ is constant on each B_j or G_j, we get

$$\begin{aligned} \xi_\Gamma = & \sum_{j=1}^{\ell} \left(i \sum_{m=1}^{j} \vec{\tau}_{\cdot m} \right) \frac{1}{2\pi i} \int_{\Gamma(G_j)} \frac{f'}{f}\, dz \\ & + \sum_{j=2}^{\ell+1} \left(\sum_{m=1}^{j-1} \vec{\delta}_{\cdot m} \right) \frac{1}{2\pi i} \int_{\Gamma(B_j)} \frac{f'}{f}\, dz \end{aligned} \tag{5.12.56}$$

Since $\frac{1}{i} \frac{f'}{f} = d(\arg f)$ (plus a change of $\log|f|$, which integrates to 0), which is a $2\pi i$ integer, so for integers n_j and m_j,

$$\frac{1}{2\pi i} \int_{\Gamma(G_j)} \frac{f'}{f}\, dz = n_j \qquad \frac{1}{2\pi i} \int_{\Gamma(B_j)} \frac{f'}{f}\, dz = m_j$$

and so, $\xi_{\mathfrak{A}} \in \mathcal{L}_{\mathcal{S}}$. Thus, the sum in (5.12.55) is 0 in $\mathbb{C}^\ell / \mathcal{L}_{\mathcal{S}}$. □

Next, we want to prove a result about sums of the type in Abel's theorem being one-one on certain special sets, whose relevance to m-functions of periodic problems should be evident. Let G_j be the set, which is the range of $\Gamma(G_j)$, that is,

$$G_j = \pi^{-1}([\beta_j, \alpha_{j+1}])$$

which is a circle formed from two lines between two branch points. Let

$$\mathbb{T}_{\mathfrak{e}} = G_1 \times \cdots \times G_\ell \tag{5.12.57}$$

Theorem 5.12.10. *Map $\mathbb{T}_{\mathfrak{e}}$ to $\mathcal{J}_S$ by*

$$\widetilde{\mathfrak{A}}(z_1, \dots, z_\ell) = \sum_{j=1}^{\ell} \mathfrak{A}(z_j) \tag{5.12.58}$$

Then $\widetilde{\mathfrak{A}}$ is one-one.

First Proof. If not, we can find $(z_1, \dots, z_\ell)$ and $(p_1, \dots, p_\ell)$ in $\mathbb{T}_{\mathfrak{e}}$, so

$$\sum_{j=1}^{\ell} \mathfrak{A}(z_j) = \sum_{j=1}^{\ell} \mathfrak{A}(p_j) \tag{5.12.59}$$

Drop those z's equal to p's, so we find $\{z_j\}_{j\in J}$ and $\{p_j\}_{j\in J}$, all distinct with $|J| \le \ell$, so

$$\sum_{j\in J} \mathfrak{A}(z_j) = \sum_{j\in J} \mathfrak{A}(p_j)$$

By the second half of Abel's theorem, there is a meromorphic function, f, with those zeros and poles. Clearly,

$$\deg(f) = |J| \le \ell \tag{5.12.60}$$

So, by Theorem 5.12.5, f is root free. But every such root-free function has zeros at $+/-$ pairs or double zeros at branch points, and this f does not. This is a contradiction. □

Second Proof. Let (5.12.59) hold. Then for integers $n_1, \dots, n_\ell$ and $m_1, \dots, m_\ell$, $j = 1, \dots, \ell$, and $k = 1, \dots, \ell$,

$$\sum_j \int_{\Gamma_j} \omega_k = n_j \delta_{kj} + i m_j \tau_{kj} \tag{5.12.61}$$

where Γ_j is that contour on $\Gamma(G_j)$ that goes clockwise from z_j to p_j. Since ω_k is real on $\Gamma(C_j)$ and the τ_k are linearly independent, we see $m_j = 0$ for all j. Moreover, since $\int_{\Gamma(G_j)} \omega_k = \delta_{jk}$, we can subtract n_j copies of $\Gamma(G_j)$ to Γ_j and get a contour $\Gamma_j^\sharp$ from z_j to p_j. So for all j, k,

$$\sum_j \int_{\Gamma_j^\sharp} \omega_k = 0 \tag{5.12.62}$$

Since $\{P_k\}_{k=1}^{\ell}$ is a basis, we conclude for any polynomial P of degree at most $\ell - 1$ and suitable $\tilde{\Gamma}_j$ and $\sigma_j = \pm 1$,

$$\sum_j \sigma_j \int_{\tilde{\Gamma}_j} P(x)\big(\sqrt{R(x)}\big)^{-1} dx = 0 \tag{5.12.63}$$

where $\tilde{\Gamma}_j$ is either $\Gamma_j^\sharp$ or $\Gamma_j^\sharp$ run backwards, and the choice is made so that

$$\int_{\tilde{\Gamma}_j} \bigl(\sqrt{R(x)}\,\bigr)^{-1}\, dx \geq 0 \tag{5.12.64}$$

Here σ_j are picked to accommodate the change of direction $\Gamma_j^\sharp$ if needed.

By multiplying all σ_j by -1 if necessary, we suppose $\sigma_1 = 1$. Pick P plus or minus a monic with zeros one in each band $(\alpha_{j+1}, \beta_{j+1})$ where $\sigma_j\sigma_{j+1} = -1$ and so P is positive on (β_1, α_2).

Thus,

$$\sigma_j P(x) > 0 \quad \text{on each } (\beta_j, \alpha_{j+1}) \tag{5.12.65}$$

and all terms in (5.12.63) are nonnegative and can only sum to zero if each $\tilde{\Gamma}_j$ is a single point, which implies $z_j = p_j$ for all j. □

It is a consequence of degree theory (see the Notes) that any one-one map between compact orientable manifolds of the same dimension is a bijection. Note that if $z_0, z \in \mathbb{T}_{\mathfrak{e}}$, then $\widetilde{\mathfrak{A}}(z) - \widetilde{\mathfrak{A}}(z_0)$ is real, that is,

$$\widetilde{\mathfrak{A}}(z) - \widetilde{\mathfrak{A}}(z_0) \in \frac{\mathbb{R}^\ell}{\mathcal{L}_S \cap \mathbb{R}^\ell} = \mathbb{T}^\ell \tag{5.12.66}$$

the standard ℓ torus $\mathbb{R}^\ell/\mathbb{Z}^\ell$. Thus:

Corollary 5.12.11. *Fix $z_0 \in \mathbb{T}_{\mathfrak{e}}$. then*

$$z \mapsto \widetilde{\mathfrak{A}}(z) - \widetilde{\mathfrak{A}}(z_0)$$

is a bijection of $\mathbb{T}_{\mathfrak{e}}$ and $\mathbb{T}^\ell$.

Finally, we want to find an explicit formula for $\mathfrak{A}(\infty_+) - \mathfrak{A}(\infty_-)$ in terms of harmonic measure. The analytic one-forms, ω_P, with $\deg(P) \leq \ell - 1$ played a critical role in defining $\mathfrak{A}$. If $\deg(P) = \ell$, then ω_P is no longer analytic at $\pm\infty$ but has simple poles at $\pm\infty$, so

$$\xi_H = \frac{1}{2\pi i} \int_{\Gamma_+ \cup \Gamma_-} H(z)\mathfrak{A}^\sharp(z)\, dz \tag{5.12.67}$$

will pick up $\mathfrak{A}^\sharp$ at the poles, that is, at $\pm\infty$. Here H is given by (5.5.138), that is,

$$H(z) = -\frac{P(z)}{\sqrt{R(z)}} \tag{5.12.68}$$

where P is the unique monic polynomial of degree ℓ with

$$\int_{\beta_j}^{\alpha_{j+1}} \frac{P(x)}{\sqrt{|R(x)|}}\, dx = 0 \tag{5.12.69}$$

and (see Theorem 5.5.22)

$$H(z) = \int -\frac{d\rho_{\mathfrak{e}}(x)}{x - z} \tag{5.12.70}$$

Theorem 5.12.12. *Let* $\mathfrak{e}_j = [\alpha_j, \beta_j]$. *Then*

$$\mathfrak{A}(\infty) - \mathfrak{A}(-\infty) = (\rho_\mathfrak{e}(\mathfrak{e}_1), \rho_\mathfrak{e}(\mathfrak{e}_1 \cup \mathfrak{e}_2), \dots, \rho_\mathfrak{e}(\mathfrak{e}_1 \cup \dots \cup \mathfrak{e}_\ell)) \tag{5.12.71}$$

Proof. $H(z)$ outside Γ has poles only at $\infty_\pm$ with residue 1 at ∞_+ and -1 at ∞_- (note if $w = 1/z$, $-dz/z = dw/w$, so $-dz/z$ has residue $+1$ at ∞!). Thus,

$$\xi_H = \mathfrak{A}^\sharp(-\infty) - \mathfrak{A}^\sharp(\infty) \tag{5.12.72}$$

On the other hand, $H(z)\,dz$ is regular inside Γ_+ and Γ_-, so there would be complete cancellation between the pieces if $\mathfrak{A}^\sharp$ was not there. Because $\mathfrak{A}^\sharp$ is discontinuous, these cancellations give constants times integrals of boundary values of H over each band and gap. The contributions over the gaps cancel and, by (5.12.52), (α_j, β_j) contributes (note $H(x_+ + i0) = \overline{H(x_- + i0)}$)

$$\left[\frac{2}{2\pi}\int_{\alpha_j}^{\beta_j} \operatorname{Im} H(x_+ + i0)\,dx\right] \sum_{m=1}^{j-1} \vec{\delta}_{\cdot m} \tag{5.12.73}$$

where the 2 comes from Γ_+ and Γ_- both contributing. But $\frac{d\rho_\mathfrak{e}}{dx} = \frac{1}{\pi}\operatorname{Im} H(x_+ + i0)$, so

$$\begin{aligned}(\xi_H)_k &= \sum_{j=1}^{\ell+1} \rho_\mathfrak{e}(\mathfrak{e}_j) \sum_{m=1}^{j-1} \delta_{km} \\ &= \sum_{j=k+1}^{\ell+1} \rho_\mathfrak{e}(\mathfrak{e}_j) = 1 - \sum_{j=1}^{k} \rho_\mathfrak{e}(\mathfrak{e}_j)\end{aligned} \tag{5.12.74}$$

which, given the fact that $\mathfrak{A}$ is measured modulo integers, yields (5.12.71). □

Remarks and Historical Notes. The theory of elliptic and hyperelliptic functions was a major theme in nineteenth century mathematics, with critical contributions by Abel, Liouville, Jacobi, Riemann, and Weierstrass. For the basic theory of meromorphic functions on Riemann surfaces, see Farkas–Kra [126], Griffiths–Harris [185], and Miranda [306].

The earliest realization that elliptic functions are connected to two-band problems is due to Akhiezer [12]. For finite gap Hill equations (a continuum analog of the Jacobi case), the relevance of hyperelliptic functions is a discovery of Dubrovin–Matveev–Novikov [115] and McKean–van Moerbeke [304] in the context of studying the KdV equation. Our use of P_j's obeying (5.12.42) and the resulting proof of Theorem 5.12.10 (given as the second proof) is motivated by Levitan's discussion [278] for the Hill equation.

The development of these ideas for Jacobi matrices is due to Flaschka–McLaughlin [136], Krichever [254, 255], and van Moerbeke [450]. For a list of the vast related literature, see [400].

Theorem 5.12.12 is motivated by the analogous result in [400], found following suggestions of Peherstorfer–Yuditskii. Our proof there is somewhat more complicated because it uses the potential $\int \log|z-x|\,d\rho_\mathfrak{e}(x)$ rather than $\int (x-z)^{-1}\,d\rho_\mathfrak{e}(x)$ and so has an extra logarithmic cut to cope with.

The degree theory result needed to obtain Corollary 5.12.11 runs as follows: Let M, N be two C^∞ orientable compact manifolds of the same dimension n so

$H^n(M) = H^n(N) = \mathbb{Z}$ for the homology groups. Any continuous $f\colon M \to N$ induces a map $H^n(f)\colon H^n(M) \to H^n(N)$, which is a group homomorphism, and so of the form $k \to Dk$ for some $D \in \mathbb{Z}$, called the *degree*, $\deg(f)$, of f.

Now let f be a C^∞ map. A point $m \in M$ is called a *regular point* if df_m, the derivative of f at m, is nonsingular. A point $n \in N$ is called a *regular value* if each point in $f^{-1}(n)$ is a regular point. In particular, if $f^{-1}(n)$ is empty, n is regular. By compactness and the inverse function theorem, each regular value has $f^{-1}(n)$, a finite set. Sard's theorem asserts the set of regular values is the complement of a set of measure zero.

If m is a regular value, the signature of f at m, $S_m(f)$, is the sign of $\det(f)$. (In general, this requires one to pick orientations on M and N as does determining the sign of $\deg(f)$; if $M = N$, making the two orientations the same fixes signs.) The fundamental theorem of degree theory says that for any regular value, n,

$$\sum_{m\in f^{-1}(n)} S_m(f) = \deg(f) \tag{5.12.75}$$

In particular, if $f^{-1}(n)$ is empty, $\deg(f) = 0$, and then regular points with $f^{-1}(n) \neq \emptyset$ must have an even number of points to get the sum of ± 1 to be 0. So if f is one-one, the degree is ± 1, and so f is onto, as claimed.

In the case studied in this section for f meromorphic on $\mathcal{S}$, f maps $\mathcal{S}$ to $\mathcal{S}_{\mathrm{R}}$, the Riemann sphere, and the topological degree is the degree as we have defined it. Analytic functions, f, where nonsingular, are conformal and so have signature $+1$ and (5.12.75) and (5.12.20) agree at points, a, for which $n(f; z, a) = 0$ or 1 for all z.

For expositions of degree theory for smooth maps, see Fonseca–Gangbo [137], Guillemin–Pollack [189], Krawcewicz–Wu [247], Lloyd [283], Milnor [305], and Spivak [416].

5.13 MINIMAL HERGLOTZ FUNCTIONS AND ISOSPECTRAL TORI

In Section 5.2, we saw the m-function, $m(z)$, for a periodic Jacobi matrix, J, with essential spectrum an ℓ-gap set, $\mathfrak{e}$, has a meromorphic continuation to $\mathcal{S}_{\mathfrak{e}}$. From the point of view of the last section, we will see m has some simple properties. And it will turn out that the study of all J's that lead to a fixed $\mathfrak{e}$ is related to the study of functions with these properties.

Theorem 5.13.1. *m is a meromorphic function on $\mathcal{S}_{\mathfrak{e}}$ with the following properties:*

(i) *m is Herglotz in the sense that if* $\operatorname{Im} z > 0$,

$$\operatorname{Im} m(z_+) > 0 \tag{5.13.1}$$

that is, $\operatorname{Im} m > 0$ *on* $\mathcal{S}_+ \cap \mathbb{C}_+$.

(ii) *On $\mathcal{S}_+$ near ∞_+,*

$$m(z) = -\frac{1}{z} + O\left(\frac{1}{z^2}\right) \tag{5.13.2}$$

(iii) *m has degree $\ell + 1$.*
(iv) *m has one zero and one pole on each set $\{G_j\}_{j=1}^{\ell}$ and, moreover, a zero at ∞_+ and a pole at ∞_-.*

Proof. (i) and (ii) hold for any m-function; see Example 2.3.1 and (2.3.10).

By Theorem 5.2.1, $m(z)$ obeys the quadratic equation

$$\alpha(z)m(z)^2 + \beta(z)m(z) + \gamma(z) = 0 \tag{5.13.3}$$

where

$$\alpha(z) = a_p p_{p-1}(z) \tag{5.13.4}$$

and the discriminant is $\Delta(z)^2 - 4$. Thus,

$$m(z) = -\frac{\beta(z) \pm \sqrt{\Delta^2(z) - 4}}{2\alpha(z)} \tag{5.13.5}$$

$m(z)$ clearly has a meromorphic continuation to all of $\mathcal{S}$ since $\sqrt{\Delta^2 - 4}$ has branch points precisely at the edges of open gaps (the double zeros of $\Delta^2 - 4$ at closed gaps are not branch points) with the only possible poles at $\infty_\pm$ and at the zeros of $p_{p-1}(x)$.

These zeros are analyzed in Theorem 5.4.16: one occurs in each gap. If the gap is closed, $\Delta^2 - 4$ has a double zero, and since that means $\beta^2 - \alpha\gamma = 0$ and $\alpha = 0$, we have $\beta = 0$. So, in (5.13.5), α has a simple zero and the numerator is also zero. So (as also remarked in Proposition 5.10.2), m has neither zero nor pole at the closed gaps.

If a gap is not closed and the zero is at the interior point of the gap, z_0, then $\alpha(z)$ has a simple zero at z_0. Since $\Delta(z_0) \neq \pm 2$, $\beta^2 - \alpha\gamma \neq 0$, so $\beta(z_0) \neq 0$. Thus,

$$-\beta(z) \pm \sqrt{\beta^2(z) - \alpha(z)\gamma(z)}$$

vanishes at one of $(z_0)_\pm$ and is nonzero (indeed, $-2\beta(z_0)$) at the other point. So m has a single pole on one sheet or the other, but not both.

If the pole is at a resonance, that is, at an edge, z_0, of a closed gap, $\Delta^2 - 4$ has a simple zero at z_0 and $\beta(z_0)^2 = \alpha(z_0)\gamma(z_0) + (\Delta^2(z_0) - 4) = 0$. Thus, $\beta(z_0) = c(z - z_0) + O((z - z_0)^2)$ while $\sqrt{\Delta^2 - 4} = c(z - z_0)^{1/2} + O((z - z_0)^{3/2})$ and $m(z) = c(z - z_0)^{-1/2} + O(1)$, so by the way poles are counted at branch points, m has a simple pole at z_0. We have thus proven $m(z)$ has exactly one pole in each G_j, $j = 1, \dots, \ell$.

By coefficient stripping (see (3.2.28)),

$$m(z)^{-1} = b_1 - z - a_1^2 m_1(z) \tag{5.13.6}$$

Since m_1 is also the m-function of a periodic Jacobi matrix, m_1 has one pole in each gap, and so m has exactly one zero in each two-sheeted gap.

Besides zeros of α, the only other possible poles of $m(z)$ are at $\infty_\pm$. At ∞_+, m is zero by (5.13.2). Thus, since $\alpha(z) \sim c_1 z^{p-1}$, $\beta(z) \sim c_2 z^p$, and $\Delta^2(z) \sim z^{2p}$, we must have $\beta(z)$ cancelling the z^p growth of $\sqrt{\Delta^2 - 4}$ at ∞_+. That means at ∞_-, the numerator is $-2c_2 z^p + O(z^{p-1})$ and so, $m(z)$ has a simple pole at ∞_-.

We have thus proven m has exactly $\ell + 1$ simple poles, so m has degree $\ell + 1$. Since we have accounted for $\ell + 1$ zeros of m, we have them all. $\square$

This leads to a natural definition in the context of general finite gap sets, not just those that are periodic spectra.

Definition. Let $\mathfrak{e}$ be a finite gap subset of $\mathbb{R}$ and let $\mathcal{S}_\mathfrak{e}$ be the associated Riemann surface. A *minimal Herglotz function* on $\mathcal{S}_\mathfrak{e}$ is a meromorphic function m on $\mathcal{S}_\mathfrak{e}$ obeying:

(i) m is Herglotz in the sense that (5.13.1) holds for $z \in \mathcal{S}_+ \cap \mathbb{C}_+$ and $\operatorname{Im} m(x_+ + i0)$ has compact support.
(ii) m obeys (5.13.2) (so m is a discrete m-function in the sense of Section 2.3).
(iii) $\deg(m) = \ell + 1$.
(iv) m has a pole at ∞_-.

Remark. The word minimal is used because m has minimal degree among non-square root-free functions.

The set of all minimal Herglotz functions on $\mathcal{S}_\mathfrak{e}$ will be denoted by $\mathcal{M}_\mathfrak{e}$. We will show first that $\mathcal{M}_\mathfrak{e}$ is a torus of dimension ℓ; indeed, naturally associated to the torus $\mathbb{T}_\mathfrak{e}$ of (5.12.57). We will then study the Jacobi matrix associated to an m in $\mathcal{M}_\mathfrak{e}$ and prove, for general $\mathfrak{e}$, it is almost periodic, and if $\mathfrak{e}$ comes from one periodic Jacobi matrix, then all the minimal Herglotz functions associated to $\mathfrak{e}$ have associated periodic Jacobi matrices and have the same Δ. This will provide the promised proof that the set of periodic J's with a given Δ is a torus.

Here is the general structure of minimal Herglotz functions:

Theorem 5.13.2. *Every minimal Herglotz function, m, in $\mathcal{M}_\mathfrak{e}$ has the form*

$$m(z) = \frac{p(z) \pm \sqrt{R(z)}}{a(z)} \tag{5.13.7}$$

where

$$\operatorname{Deg}(a) = \ell \tag{5.13.8}$$

$$\operatorname{Deg}(p) = \ell + 1 \tag{5.13.9}$$

and $-p$ is monic. Moreover,

(i) *p and a are real polynomials.*
(ii) *a has one simple zero in each gap.*
(iii) *m has exactly one simple pole in each gap plus the pole at ∞_-.*
(iv) *m has exactly one simple zero in each gap plus the zero at ∞_+.*

Remarks. 1. A polynomial is called *real* if all its coefficients are real.

2. In the periodic case with closed gaps, a is not the 2α of (5.13.5) but it has zeros at closed gaps that occur in the numerator removed. In addition, even if all gaps are open and $\Delta^2 - 4$ has simple zeros, it is not R, but rather $(a_1 \dots a_p)^{-2} R$.

Proof. As a rational function on $\mathcal{S}$, m has the form

$$m(z) = \frac{p(z) \pm q(z)\sqrt{R(z)}}{a(z)} \tag{5.13.10}$$

By (5.12.23) and $\deg(m) = \ell + 1$, we see $\deg(q) = 0$, so we can take $q = 1$. Also by (5.12.23), $\deg(a) \le \ell + 1$. Since (5.13.2) holds, and on $\mathcal{S}_+$,

$$+\sqrt{R(z)} = z^{\ell+1} + O(z^\ell) \tag{5.13.11}$$

near ∞_+, we must have that

$$p(z) = -z^{\ell+1} + O(z^\ell) \tag{5.13.12}$$

(since $\deg(a) \le \ell + 1$ means the $z^{\ell+1}$ term in the numerator must cancel). Thus, $-p$ is monic and (5.13.9) holds.

Since $-\sqrt{R(z)}$ (i.e., $\sqrt{R(z)}$ on $\mathcal{S}_-$) has the opposite sign, near ∞_-,

$$p(z) \pm \sqrt{R(z)} = -2z^{\ell+1} \tag{5.13.13}$$

so to have a pole at ∞_-, we must have

$$\mathrm{Deg}(a) \le \ell \tag{5.13.14}$$

Since $m(z)$ is real on $(\beta_{\ell+1}, \infty)$ and $\sqrt{R(z)}$ is real there, $p(z)/a(z)$ is real there. So, by analyticity, all its zeros and poles come in conjugate pairs or lie on $\mathbb{R}$. Since $-p$ is monic, we see p and then a is real.

On each band, p/a is real, so

$$\mathrm{Im}\, m(x_+ + i0) = \frac{\mathrm{Im}\sqrt{R(x_+ + i0)}}{a(x + i0)} \tag{5.13.15}$$

Since $\sqrt{R(x)}$ changes sign from one band to the next, a must change sign to keep $\mathrm{Im}\, m(x_+ + i0) \ge 0$. Thus, a has an odd number of zeros in each gap.

Since there are ℓ gaps and, by (5.13.14), at most ℓ zeros, we conclude each gap has precisely one zero and (5.13.8) holds.

As in the analysis in the proof of Theorem 5.13.1, if a has a zero at a point, z_0, in the interior of a gap where $R(z_0) \neq 0$, m must have a pole at either $(z_0)_+$ or $(z_0)_-$ (or both), and if a has a zero at a band edge, z_0, $p(z) \pm \sqrt{R(z)}$ vanishes at $(z - z_0)^{1/2}$ or approaches a constant. Thus, in that case also, m has a pole at z_0. Thus, m has at least one pole in each gap, and so since ∞_- is a pole and there are only $\ell + 1$ poles, we see each gap has exactly one simple pole.

Define m_1 by (5.13.6) where b_1, a_1 are picked so $m_1(z)$ obeys (5.13.2). By coefficient stripping, m_1 is a Herglotz function and clearly, m_1 is meromorphic on $\mathcal{S}$. m_1 has a pole at each finite zero of m and, by $\deg(m) = \ell + 1$ and the fact that ∞_- is not a zero, and by (5.13.2), ∞_+ is a simple zero, we know m has an ℓ finite zeros. Thus, m_1 has ℓ poles in $\mathcal{S}\setminus\{\infty_\pm\}$. At ∞_+, m_1 has a zero and, by (5.13.6) and $m(z)^{-1} \to 0$ at ∞_-, we see m_1 has a simple pole at ∞_-. Thus, $\deg(m_1) = \ell + 1$ and ∞_- is a pole, so m_1 is also in $\mathcal{M}_{\mathfrak{e}}$.

By the analysis above, m_1 has exactly one simple pole in each gap so, by (5.13.6), $m(z)$ has exactly one simple zero in each gap. □

Along the way, we have also proven:

Corollary 5.13.3. *If $m \in \mathcal{M}_{\mathfrak{e}}$, the coefficient stripped m_1 defined by* (5.13.16) *also lies in $\mathcal{M}_{\mathfrak{e}}$.*

Remark. The proof of this corollary did not use that m had a pole at ∞_-, only that m did not have a zero at ∞_-.

Example 5.13.4. This example shows that property (iv) in the definition of minimal Herglotz functions is not automatic. Let J be a periodic Jacobi matrix, and for $y \in \mathbb{R}$, let J_y be the matrix where only b_1 is changed from b_1 to $b_1 + y$. Let $m_y(z)$ be the associated m-function. By (5.13.6) and the fact that J_y and J once-stripped are the same, we see

$$m_y(z)^{-1} = y + m(z)^{-1} \tag{5.13.16}$$

Thus, m_y is also a meromorphic function of degree $\ell + 1$ and so obeys (i)–(iii) of the definition of $\mathcal{M}_\mathfrak{e}$. But, by (5.13.16),

$$m_y(\infty_-) = y^{-1} \tag{5.13.17}$$

so m_y fails to obey condition (iv) of the definition.

m_y still has a pole in each gap, but instead of a pole at ∞_-, there is one additional pole on $(-\infty, \alpha_1] \cup [\beta_{\ell+1}, \infty)$ whose location and sheet depend on the sign and magnitude of y. Also, now $\deg(a) = \ell + 1$ rather than $\deg(a) = \ell$.

Changing a_1 from the periodic value changes the degree of m. □

There is a natural map, $\mathcal{D}$, from $\mathcal{M}_\mathfrak{e}$ to $\mathbb{T}_\mathfrak{e}$, the torus described in (5.12.57). Namely, each $f \in \mathcal{M}_\mathfrak{e}$ has ℓ poles other than at ∞_-, one each in $G_1, G_2, \dots, G_\ell$. The set of these poles describes a point $(z_1, \dots, z_\ell) \in \mathbb{T}_\mathfrak{e}$. This is called the *Dirichlet data* for f. $\mathcal{D}$ is called the *Dirichlet map*. The reason for this name will be explained in the Notes.

Theorem 5.13.5. *$\mathcal{D}$ is a one-one continuous map of $\mathcal{M}_\mathfrak{e}$ onto $\mathbb{T}_\mathfrak{e}$. In particular, $\mathcal{M}_\mathfrak{e}$ is topologically a torus.*

Remark. Here $\mathcal{M}_\mathfrak{e}$ is topologized using the topology of uniform convergence (uniform as $\mathcal{S}_\mathrm{R}$-valued functions).

Proof. We will describe a point in $\mathbb{T}_\mathfrak{e}$ with coordinates

$$\mathcal{D}(f) = (z_1, \delta_1; z_2, \delta_2; \dots) \tag{5.13.18}$$

where $z_j \in [\beta_j, \alpha_{j+1}]$ and δ_j is ± 1, with the convention that we take $\delta_j = -1$ if z_j is at a band edge.

Any $f \in \mathcal{M}_\mathfrak{e}$ has the form

$$f(z) = \int_\mathfrak{e} \frac{g(x)\,dx}{x - z} + \sum_{\{j \mid \delta_j = 1\}} \frac{w_j}{x - z_j} \tag{5.13.19}$$

where

$$g(x) = \frac{1}{\pi} \operatorname{Im} f(x_+ + i0) \tag{5.13.20}$$

$$w_j = \lim_{\varepsilon \downarrow 0} (i\varepsilon) f((x_j)_+ + i\varepsilon) \tag{5.13.21}$$

This is just (2.3.7), (2.3.41), (2.3.54), and (2.3.58) where only the poles on $\mathcal{S}_+$ are relevant, since the measure is $\lim_{\varepsilon\downarrow 0} \frac{1}{\pi} \operatorname{Im} f(x_+ + i\varepsilon)\,dx$. Poles at branch points do not enter the sum because they only have $|x - z_j|^{-1/2}$ singularities. (They will affect g; at nonresonant gap edges, g vanishes as $(x - z_0)^{1/2}$, while at resonance edges, g diverges as $(x - z_0)^{-1/2}$.)

We know f has the form

$$f(z) = \frac{p(z) + \sqrt{R(z)}}{a(z)} \tag{5.13.22}$$

a has zeros at precisely the points $\{z_j\}_{j=1}^{\ell}$, so

$$a(z) = A \prod_{j=1}^{\ell} (z - z_j) \tag{5.13.23}$$

Since all $z_j < \alpha_{\ell+1}$ and $\operatorname{Im}(\sqrt{R(x_+ + i0)}) > 0$ on $[\alpha_{\ell+1}, \beta_{\ell+1}]$ (from $\sqrt{R(x_+ + i0)} > 0$ on $(\beta_{\ell+1}, \infty)$ and the branch of $(z - \beta_{\ell+1})^{1/2}$, which is positive on $(\beta_{\ell+1}, \infty) + i0$ has positive imaginary part on $(-\infty, \beta_{\ell+1}) + i0$), we have $A > 0$.

Thus, by (5.13.20), in (5.13.19) for $x \in \mathfrak{e}$,

$$g(x) = \frac{1}{\pi} \frac{\sqrt{|R(x)|}}{A \prod_{j=1}^{\ell} |x - z_j|} \tag{5.13.24}$$

while, by (5.13.21),

$$w_j = \frac{2\sqrt{|R(z_j)|}}{A \prod_{k \neq j} |z_k - z_j|} \tag{5.13.25}$$

for to avoid a pole on $\mathcal{S}_-$, we must have $p(z_j) - \sqrt{R(z_j)} = 0$, which yields to 2 in the numerator.

The normalization condition $f(z) = -z^{-1} + O(z^{-2})$ is equivalent to

$$\int_{\mathfrak{e}} g(x)\,dx + \sum_{\{j \mid \delta_j = 1\}} w_j = 1 \tag{5.13.26}$$

which determines A. Thus, knowing $\mathcal{D}(f)$ determines A and then g and w_j, and then f, which proves the map is one-one.

Conversely, given a set of Dirichlet data (i.e., a point in $\mathbb{T}_{\mathfrak{e}}$), define $a(z)$ by (5.13.23) where A is determined by (5.13.26), determine $p(z)$ by (since $(p(z) + \sqrt{R(z)})/a(z)$ is $O(z^{-1})$)

$$p(z) + \sqrt{R(z)} = O(z^{\ell-1}) \tag{5.13.27}$$

near ∞_+ (which determines the top two coefficients of $p(z)$) and the conditions (since m has no pole at $(z_j; \delta_j)$)

$$p(z_j) \mp \delta_j \sqrt{R(z_j)} = 0 \tag{5.13.28}$$

This defines f by (5.12.7). Tracking signs of a proves $\operatorname{Im} f(x_+ + i0) \geq 0$ on $\mathfrak{e}$ and that the residues of poles on $\mathcal{S}_+$ are positive. Thus, the Cauchy integral formula proves in $\mathbb{C}_+ \cap \mathcal{S}_+$

$$f(z) = \int_{\Gamma_+} \frac{f(w)}{w - z}\, dw \tag{5.13.29}$$

and then (5.13.19), which shows $\operatorname{Im} f > 0$ on $\mathcal{S}_+ \cap \mathbb{C}_+$.

In (5.13.29), Γ_+ is the contour in the proof of Theorem 5.12.3 and the fact that constructed f has $O(|z|^{-1})$ at ∞_+ means the contour at ∞_+ in the full Cauchy integral formula vanishes. This proves existence. □

Each $f \in \mathcal{M}_\mathfrak{e}$ is an m-function, so the m-function of a unique Jacobi matrix, J_f, which is determined either from the spectral measure $g(x)\, dx + \sum_{\{j \mid \delta_j = 1\}} w_j \delta_{z_j}$ or from the continued fraction expansion at ∞_+. The topology on $\mathcal{M}_\mathfrak{e}$ is equivalent to the topology of pointwise convergence on the parameters in J_f (once we prove J_f is periodic or almost periodic, this will be the same as uniform convergence in n). Note that f determines a_1, b_1 directly by

$$f(z)^{-1} = -z + b_1 + a_1^2 z^{-1} + O(z^{-2}) \tag{5.13.30}$$

at ∞_+.

We will study the n-dependence of the Jacobi parameters by studying the impact of coefficient stripping. We proved in Corollary 5.13.3 that $f \to f_1$, coefficient stripping given by (5.13.30) and (5.13.6) is a map of $\mathcal{M}_\mathfrak{e}$ to $\mathcal{M}_\mathfrak{e}$. We will also need a map of

$$\widetilde{\mathfrak{A}} \colon \mathcal{M}_\mathfrak{e} \to \mathbb{T}^\ell$$

the canonical ℓ-torus, $\mathbb{R}^\ell / \mathbb{Z}^\ell$, by mapping $\mathbb{T}_\mathfrak{e}$ to $\mathbb{T}^\ell$ by Corollary 5.12.11, and composing this with $\mathcal{D}$, that is, if

$$\mathcal{D}(f) = (z_1, \dots, z_\ell) \qquad (z_j \in G_j)$$

then

$$\widetilde{\mathfrak{A}}(f) = \sum_{j=1}^{\ell} \mathfrak{A}(z_j) - \mathfrak{A}(z_j^{(0)}) \tag{5.13.31}$$

where $z_j^{(0)}$ is some convenient point, say $z_j^{(0)} = \alpha_j$.

We can prove uniform (over the isospectral torus) bounds on the weight.

Theorem 5.13.6. *There are positive constants C, D so that uniformly over $\mathbb{T}_\mathfrak{e}$, one has for all $x \in \mathfrak{e}$,*

$$DR(x)^{1/2} \le g(x) \le CR(x)^{-1/2} \tag{5.13.32}$$

Proof. We have

$$\operatorname{dist}(x, \mathbb{R} \setminus \mathfrak{e}) \min_{j=1,\dots,\ell+1} \left(\tfrac{1}{2} |\beta_j - \alpha_j|\right)^{\ell-1} \le \prod_{j=1}^{\ell} |x - z_j| \le |\beta_{\ell+2} - \alpha_1|^\ell \tag{5.13.33}$$

so, by (5.13.24), for some C_1, D_1,

$$D_1 A^{-1} R(x)^{1/2} \le g(x) \le C_1 A^{-1} R(x)^{-1/2} \tag{5.13.34}$$

Also, we have, by (5.13.25),

$$0 \le w_j \le A^{-1} C_2 \tag{5.13.35}$$

where

$$C_2 = 2|\beta_{\ell+1} - \alpha_1|^{\ell+1} (\min|\beta_j - \alpha_j|)^{-\ell+1} \tag{5.13.36}$$

(5.13.26) and these bounds provide uniform (in $\mathcal{T}_{\mathfrak{e}}$) upper and strictly positive lower bounds on A and then (5.13.34) implies (5.13.32). □

Theorem 5.13.7. (a) *$\widetilde{\mathfrak{A}}$ is a bijection of $\mathcal{M}_{\mathfrak{e}}$ to $\mathbb{T}^\ell$.*

(b) *Coefficient stripping $f \to f_1$ obeys*

$$\widetilde{\mathfrak{A}}(f_1) - \widetilde{\mathfrak{A}}(f) = \mathfrak{A}(\infty_-) - \mathfrak{A}(\infty) \tag{5.13.37}$$

Proof. (a) $\widetilde{\mathfrak{A}}$ is the composition of $\mathcal{D}$ and the map of Corollary 5.12.11, each of which is a continuous bijection.

(b) f has poles at the points in $\mathcal{D}(f)$ plus at ∞_- and, by (5.13.6) (other than at $\infty_\pm$), zeros of f are precisely poles of f_1 plus the zeros at ∞_+. Thus, by the first half of Abel's theorem (Theorem 5.12.7),

$$\widetilde{\mathfrak{A}}(f) + \mathfrak{A}(\infty_-) = \widetilde{\mathfrak{A}}(f_1) + \mathfrak{A}(\infty_+)$$

which is (5.13.37). □

This is truly a remarkable theorem: $f \to f_1$ is a map of a torus to itself. In general, iterating maps on a torus is complicated, but if the map is just addition by a fixed group element, iteration n times is just adding n times that element! $x \to x + nx_0$ is an affine map (on $\mathbb{R}^\ell$), so (5.13.37) is sometimes summarized by the phrase: "Abel's map linearizes coefficient stripping." With this in place, we get some immediate consequences (they are corollaries, but so significant that we call them theorems!):

Theorem 5.13.8. *Let $\mathfrak{e} \subset \mathbb{R}$ be a finite gap set. Let $p \in \{1, 2, \dots\}$. The following are equivalent:*

(i) *One Jacobi matrix, J_f, associated to one $f \in \mathcal{M}_{\mathfrak{e}}$ is periodic of period p.*

(ii) *All Jacobi matrices, J_f, associated to all $f \in \mathcal{M}_{\mathfrak{e}}$ are of period p.*

(iii) *Each harmonic measure, $\rho_{\mathfrak{e}}(\mathfrak{e}_j)$ (where $\mathfrak{e}_j = [\alpha_j, \beta_j]$) is rational with*

$$p\rho_{\mathfrak{e}}(\mathfrak{e}_j) \in \mathbb{Z} \tag{5.13.38}$$

(iv) *There is a polynomial of degree p with*

$$\Delta^{-1}([-2, 2]) = \mathfrak{e} \tag{5.13.39}$$

(inverse as a map from $\mathbb{C}$).

Proof. Consider the statement

$$p(\mathfrak{A}(\infty_-) - \mathfrak{A}(\infty_+)) = 0 \tag{5.13.40}$$

that is, p times the element of the torus is the identity. By (5.13.37), if $f_1, f_2, \dots$ are what we get by coefficient stripping, (5.13.40) is equivalent to

$$\widetilde{\mathfrak{A}}(f_p) - \widetilde{\mathfrak{A}}(f) = 0 \tag{5.13.41}$$

for one f or for all f! Since $\widetilde{\mathfrak{A}}$ is a bijection, this is equivalent to $f_p = f$, that is, J is itself after stripping p times, that is, J is periodic!

By (5.12.71), (5.13.40) holds if and only if

$$p \sum_{j=1}^{k} \rho_{\mathfrak{e}}(\mathfrak{e}_j) \in \mathbb{Z}$$

for $k = 1, 2, \dots, \ell$, which is equivalent to (5.13.38).

Finally, we note that (i) $\Rightarrow$ (iv); just take Δ to be the discriminant. Conversely, (iv) implies (5.13.40). For let

$$F(z) = -\Delta(z) \pm \sqrt{\Delta^2(z) - 4}$$

Since $\Delta^{-1}([-2, 2]) = \mathfrak{e}$, $\Delta^2 - 4$ has double roots at internal points of $\mathfrak{e}$ and single roots at edges of $\mathfrak{e}$, so F is meromorphic on $\mathcal{S}_{\mathfrak{e}}$. Since

$$\pm\sqrt{\Delta^2 - 4} = \pm(\Delta(z) + O(\Delta(z)^{-1})) \tag{5.13.42}$$

we see at ∞_+, F has a zero of order p and at ∞_- a pole of order p. It thus has degree p (since there are no other poles) and so no other zeros (as can also be seen by noting that $F(z)^{-1} = \frac{1}{4}(-\Delta(z) \mp \sqrt{\Delta^2 - 4})$). Thus, (5.13.40) is just the first part of Abel's theorem for F. □

Notice that Theorem 5.13.8 implies Theorem 5.5.25 (given Proposition 5.5.26) and provides a proof of that theorem. Our proof of Aptekarev's theorem (i.e., (ii) $\Rightarrow$ (iii) in Theorem 5.5.25) is indirect: Rational harmonic measure implies (5.13.40) by the calculation in (5.12.71) and that implies there is a periodic J and then Δ is its discriminant. Peherstorfer's proof [338] is via a direct construction—its OPUC analog appears as Theorem 11.4.8 in [400].

The following generalizes the Borg–Hochstadt theorem (Theorem 5.4.21):

Corollary 5.13.9. *Let $\{a_n, b_n\}_{n=1}^{\infty}$ be a set of Jacobi parameters obeying*

$$a_{n+p} = a_n \qquad b_{n+p} = b_n \tag{5.13.43}$$

where $p = kq$ with k and q integral. Suppose all the gaps G_j are closed for $j \neq k, 2k, \dots, (q-1)k$. Then, a, b are periodic at period q, that is,

$$a_{n+q} = a_n \qquad b_{n+q} = b_n \tag{5.13.44}$$

Remark. The Borg–Hochstadt theorem is the case $q = 1$.

Proof. Each band has harmonic measure m/q. □

For general finite gap sets, the Jacobi matrices are quasiperiodic:

Theorem 5.13.10. *Let $\mathfrak{e}$ be a finite gap set and J_f a Jacobi matrix whose m-function is a minimal Herglotz function in $\mathcal{M}_{\mathfrak{e}}$. Then its Jacobi parameters are almost periodic. To be totally explicit, there are real analytic functions $A_{\mathfrak{e}}$ and $B_{\mathfrak{e}}$ on $\mathbb{T}^{\ell}$, the standard ℓ torus with values in $(0, \infty)$ and $\mathbb{R}$, respectively, so that for every such J_f, we have $t_0 \in \mathbb{T}^{\ell}$ so that*

$$a_n = A_{\mathfrak{e}}(t_0 - n\omega) \qquad b_n = B_{\mathfrak{e}}(t_0 - n\omega) \tag{5.13.45}$$

where ω is given in terms of the harmonic measures of $\mathfrak{e}$ by (5.12.71).

Proof. Define $\tilde{A}_{\mathfrak{e}}$ and $\widetilde{B}_{\mathfrak{e}}$ on $\mathcal{M}_{\mathfrak{e}}$ by

$$f(z)^{-1} = -z + \widetilde{B}_{\mathfrak{e}}(f) + \tilde{A}_{\mathfrak{e}}(f)^2 z^{-1} + O(z^{-2}) \tag{5.13.46}$$

which are clearly real analytic on $\mathcal{M}_{\mathfrak{e}}$. Define

$$A_{\mathfrak{e}} = \tilde{A}_{\mathfrak{e}} \circ \widetilde{\mathfrak{A}}^{-1} \qquad B_{\mathfrak{e}} = \widetilde{B}_{\mathfrak{e}} \circ \widetilde{\mathfrak{A}}^{-1}$$

where $\widetilde{\mathfrak{A}}$ is the bijection of $\mathcal{M}_{\mathfrak{e}}$ to $\mathbb{T}^{\ell}$ of Theorem 5.13.7. Then (5.13.45) is just (5.13.37) iterated. □

One can naturally use (5.13.45) to define (a_n, b_n) for all $n \in \mathbb{Z}$ and so get natural two-sided Jacobi matrices for any $\mathfrak{e}$. The set of such two-sided matrices is called the *isospectral torus*, $\mathcal{T}_{\mathfrak{e}}$, for $\mathfrak{e}$. In the periodic case, it is precisely the set of periodic J's with a given Δ. Just as Chapter 3 is the theory of special classes of perturbations of $\mathcal{T}_{\mathfrak{e}}$ for $\mathfrak{e} = [-2, 2]$, we want to understand the analogous perturbations for general $\mathfrak{e}$. For the rational harmonic measure case, this will be the subject of Chapter 8 and for general $\mathfrak{e}$'s, of Chapter 9.

Finally, we use these ideas to find another proof of (5.2.11) and show that for the general finite gap situation, the whole-line Jacobi matrices are reflectionless (i.e., have purely imaginary Green's functions).

Theorem 5.13.11. *Let $\mathfrak{e}$ be a finite gap set, m a minimal Herglotz function on $\mathcal{S}_{\mathfrak{e}}$, and J the two-sided Jacobi matrix given by* (5.13.45) *for $n \in \mathbb{Z}$, so that*

$$m(z) = m(z; J_0^+) \tag{5.13.47}$$

Then

$$m(z; J_0^-) = (a_0^2 m(\tau(z)))^{-1} \tag{5.13.48}$$

that is, one can recover $m(z; J_0^{-1})$ from the second sheet values of m.

Remark. In the periodic case, this provides another proof of (5.2.11).

Proof. By the fact that $m(z)$ has a pole at ∞_- and by (5.13.7), we see that $m_1(z) - (-a_1^{-2}z + a_1^{-2}b_1)$ has a zero at ∞_-, so near ∞_-,

$$m_1(z) = -a_1^{-2}z + a_1^{-2}b_1 + O(z^{-1}) \tag{5.13.49}$$

In particular, near ∞_- on $\mathbb{C}_+ \cap \mathcal{S}_+$, $\operatorname{Im} m_1(\tau(z)) \leq 0$. On the other hand, on $\mathfrak{e}$, $m_1(\tau(x + i0)) = \overline{m(x + i0)}$ also has a negative imaginary part. Finally, the same argument that showed poles on $\mathcal{S}_+$ have positive residues shows they have negative residues on $\mathcal{S}_-$ (for on $\mathcal{S}_-$, $p(z) + \sqrt{R(z)} = 0$ and $-2\sqrt{R(z)}/a(z)$ has positive sign). Thus, by the maximum principle for harmonic functions, $\operatorname{Im} m_1(\tau(z)) \leq 0$ on $\mathcal{S}_+ \cap \mathbb{C}_+$.

It follows that $(a_1^2 m_1(\tau(z)))^{-1}$ is a discrete m-function. Similarly, if we let

$$m_{+,n}(z) = m(z; J_n^+) \tag{5.13.50}$$

then

$$m_{-,n}(z) \equiv (a_n^2 m_{+,n}(\tau(z))^{-1} \tag{5.13.51}$$

is a discrete m-function.

With this definition, the recursion relation

$$m_{+,n}(z)^{-1} = b_{n+1} - z - a_{n+1}^2 m_{+,n+1}(z) \tag{5.13.52}$$

which initially holds on $\mathcal{S}_+ \cap \mathbb{C}_+$ extends by analytic continuation, and since $\tau(z) = z$ implies

$$a_n^2 m_{-,n}(z) = b_{n+1} - z - (m_{-,n+1}(z))^{-1} \tag{5.13.53}$$

which shows inductively that the Jacobi parameters associated to $m_{-,n}$ are $\{a_{j-2+n}, b_{j-1+n}\}_{j=1}^{\infty}$, that is, J_n^-. Thus,

$$m_{-,n}(z) = m(z; J_n^-) \tag{5.13.54}$$

which for $n = 0$ is (5.13.48). □

Theorem 5.13.12. *Let J be a two-sided Jacobi matrix in $\mathcal{T}_{\mathfrak{e}}$ where $\mathfrak{e}$ is a finite gap set. Then,*

(i) *The diagonal Green's function, $G_{nn}(z)$, is pure imaginary for $z = x + i0$ with $x \in \mathfrak{e}$. Thus, J is reflectionless on $\mathfrak{e}$.*
(ii) *$\sigma(J) = \mathfrak{e}$ and the spectrum is purely absolutely continuous of uniform multiplicity 2.*

Proof. (i) By (5.4.45),

$$G_{nn}(z) = -\frac{1}{a_n^2 m(z; J_n^+) - m(z; J_n^-)^{-1}} \tag{5.13.55}$$

On $\mathfrak{e}$,

$$\begin{aligned} m(x + i0, J_n^-) &= m(\tau(x - i0), J_n^-) \\ &= \overline{m(x + i0, J_n^-)} \end{aligned} \tag{5.13.56}$$

so, by translates of (5.13.48),

$$m(x + i0, J_n^-)^{-1} = \overline{a_n^2 m(x + i0, J_n)} \tag{5.13.57}$$

and, by (5.13.55), G_{nn} is pure imaginary.

(ii) By (5.13.55) and (5.13.48),

$$(-G_{nn}(z))^{-1} = a_n^2[m(z; J_n^+) - m(\tau(z); J_n^+)] \tag{5.13.58}$$

for all $z \in \mathbb{C} \setminus \mathfrak{e}$.

Consider a gap $[\beta_j, \alpha_{j+1}]$. Writing m in the form $(p \pm \sqrt{R})/a$, we see

$$(-G_{nn}(z))^{-1} = \frac{2a_n^2 \sqrt{R(z)}}{a(z)}$$

where $a(z)$ has a single zero in $[\beta_j, \alpha_{j+1}]$.

Suppose first that zero is in (β_j, α_{j+1}). Then $(-G_{nn}(z))^{-1}$ vanishes at β_j and α_{j+1}. Moreover, on $\mathbb{R} \setminus \sigma(J)$,

$$\frac{d}{dx} G_{nn}(x) > 0 \Rightarrow \frac{d}{dx} (-G_{nn}(x))^{-1} > 0$$

away from the zero of a. Thus, by monotonicity, $(-G_{nn}(z))^{-1}$ has no zero in (β_j, α_{j+1}).

If $(a(z))$ has a zero at β_j, then $(-G_{nn}(\beta_j))^{-1} = \infty$, $(-G_{nn}(\alpha_{j+1})) = 0$, and $(-G)^{-1}$ is finite and monotone in all of (β_j, α_{j+1}), so always strictly negative. Similarly, if $a(z)$ has a zero at α_j, $(-G_{nn}(z))^{-1}$ is strictly positive on (β_j, α_{j+1}).

In all cases, $(-G_{nn}(z))^{-1}$ is nonvanishing on (β_j, α_{j+1}), so no $G_{nn}(z)$ has a pole in those intervals, so $\sigma(J) \subset \mathfrak{e}$. By the fact that $G_{nn}(x + i0)$ is pure imaginary, Craig's theorem (Theorem 5.4.19) implies the spectrum is purely a.c. Since

$$\operatorname{Im}(a_n^2 m(x + i0, J_n^+)) = \operatorname{Im}((-m(x + i0, J_n^-))^{-1}) = \tfrac{1}{2} \operatorname{Im}((-G_{nn}(x + i0))^{-1})$$

we see that the a.c. spectrum is of multiplicity 2. □

Remarks and Historical Notes. This is the second half of the theory developed by Flaschka–McLaughlin–Krichever–van Moerbeke quoted (with background) in the Notes to the last section.

By the discussion in Example 5.13.4 and the remark after Corollary 5.13.3, if m obeys all the conditions for a function in $\mathcal{M}_{\mathfrak{e}}$, except it is finite and nonzero at ∞_- rather than a pole, then the once-stripped m_1 is in $\mathcal{M}_{\mathfrak{e}}$. So every such Jacobi matrix is an almost periodic one with b_1 modified.

In the periodic case, the Dirichlet data points are the roots of $p_{p-1}(z)$, which are eigenvalues of the truncated matrix $J_{p-1;F}$, so associated to solutions of $(J - \lambda)u = 0$ with $u_{n=0} = u_{n=p} = 0$, thus Dirichlet eigenvalues, which is the reason for the name. Alternatively, in terms of the operators $J_0^\pm$ of the truncated full-line problem, Dirichlet data in the interior of a gap are eigenvalues of J_0^+ if in $\mathcal{S}_+$ and of J_0^- if in $\mathcal{S}_-$.

There are basically two ways of thinking of the isospectral torus, $\mathcal{T}_{\mathfrak{e}}$: a set of whole-line Jacobi matrices or as their restrictions to the half-line (which, by almost periodicity, determine the whole-line matrix). The half-line objects are defined as the set of minimal Herglotz functions. The whole-line objects are the set of reflectionless whole-line J's with $\sigma_{\text{ess}}(J) = \Sigma_{\text{ac}}(J) = \mathfrak{e}$. That every such object lies in the isospectral torus, as we have defined it, will be the major theme in Section 7.5, which will also discuss the history of this point of view.

Among all almost periodic Jacobi matrices, the finite gap ones are unusual in that, generically, one expects infinitely many gaps and Cantor spectrum. For results on such generic Cantor spectrum, see [28, 29, 121, 172].

APPENDIX TO SECTION 5.13: A CHILD'S GARDEN OF ALMOST PERIODIC FUNCTIONS

As we have seen, Jacobi parameters induced by the minimal Herglotz functions associated to a general finite gap set are quasiperiodic, and so almost periodic. In this appendix, we discuss the general definition of quasiperiodic and almost periodic.

Given a function, f, on $\mathbb{Z}$ and $n \in \mathbb{Z}$, we define f_n on $\mathbb{Z}$ by

$$f_n(m) = f(n + m) \tag{5.13A.1}$$

Given a bounded function, f, on $\mathbb{Z}$, we define

$$\|f\|_\infty = \sup_n |f(n)| \tag{5.13A.2}$$

and let $C(\mathbb{Z})$ be the set of all bounded functions in this norm.

Definition. A function, f, from $\mathbb{Z}$ to $\mathbb{C}$ is called *almost periodic* (in Bochner sense) if and only if f is bounded and $\{f_n\}_{n\in\mathbb{Z}}$ has compact closure in $\|\cdot\|_\infty$.

Definition. A *Bohr almost periodic* function on $\mathbb{Z}$ is a bounded function, f, so that for any ε, there is an L so that for all $m \in \mathbb{Z}$, there is an n so that $|n - m| \le L$ and

$$\|f_n - f\|_\infty < \varepsilon \tag{5.13A.3}$$

Let $\mathbb{T}^1$ be the circle $\partial\mathbb{D} = \{z \mid |z| = 1\}$, $\mathbb{T}^n = \times_{j=1}^n \mathbb{T}^1$, the n-dimensional torus, and $\mathbb{T}^\infty$, the countably infinite product. We will think of $\mathbb{T}^n$ as $\partial\mathbb{D}^n$ and use $(z_1, \dots, z_n)$ as coordinates. Notice that we use additive notation for $\mathbb{Z}$ but multiplication for $\mathbb{T}$.

The main theorem at the center of the theory is:

Theorem 5.13A.1. *Let f be a bounded function on $\mathbb{Z}$. The following are equivalent:*

(1) *f is (Bochner) almost periodic.*
(2) *f is Bohr almost periodic.*
(3) *f is a uniform limit of finite sums of the form*

$$g_N(n) = \sum_{j=1}^N a_j e^{2\pi i \alpha_j^{(N)} n} \tag{5.13A.4}$$

for $\alpha_1, \dots, \alpha_N^{(N)} \in \mathbb{R}/\mathbb{Z}$.

(4) *There exists a continuous function F on $\mathbb{T}^\infty$ and $\{z_j\}_{j=1}^\infty$ in $\mathbb{T}^\infty$ so that*

$$f(n) = F(z^n) \tag{5.13A.5}$$

where $(z^n)_j = z_j^n$.

Remarks. 1. If F depends on only finitely many variables (equivalently, F can be viewed as a function of a finite-dimensional torus), f is called *quasiperiodic*.

2. In Theorem 5.13.10, we have functions of the form (5.13A.5) on a finite-dimensional torus, but only for $n \ge 0$. So the question comes up how to define almost periodic functions on $n \ge 0$. The answer is as restrictions to $n \ge 0$ of functions almost periodic on $\mathbb{Z}$, there is at most one such extension, for if there were two, their difference would be an almost periodic function vanishing for $n \ge 0$ and, by the Bohr definition, such a function is identically zero.

It is natural to prove this result in the general context of locally compact abelian groups. Let G be such a group, μ Haar measure, and $\widehat{G}$ the set of characters, that is, continuous homomorphisms of G to $\partial\mathbb{D}$. Besides $\mathbb{Z}$, the example to think about is $\mathbb{R}$.

Let $C(G)$ stand for bounded continuous functions on G with $\|\cdot\|_\infty$. For $f \in C(G)$ and $g \in G$, define f_g by

$$f_g(x) = f(x + g) \tag{5.13A.6}$$

f is (Bochner) almost periodic if $\{f_g\}_{g\in G}$ has compact closure in $\|\cdot\|_\infty$. f is called Bohr almost periodic if and only if for all ε, there is a compact set K so that for all g, there is h in $g+K$ so that

$$\|f_n - f\|_\infty \le \varepsilon \tag{5.13A.7}$$

The general form of Theorem 5.13A.1 is:

Theorem 5.13A.2. *Let G be a separable compact abelian group. Let $f \in C(G)$. Then the following are equivalent:*

(1) *f is (Bochner) almost periodic.*
(2) *f is Bohr almost periodic.*
(3) *f is a uniform limit of finite sums of the form*

$$g_N(x) = \sum_{j=1}^{N} a_j \chi_j^{(N)}(x) \tag{5.13A.8}$$

with $\chi_j^{(N)} \in \widehat{G}$.
(4) *There exists a continuous function F on $\mathbb{T}^\infty$ to $\mathbb{C}$ and a homomorphism $\zeta: G \to \mathbb{T}^\infty$ so*

$$f(x) = F(\zeta(x)) \tag{5.13A.9}$$

Theorem 5.13A.2 $\Rightarrow$ Theorem 5.13A.1. Only parts (2) and (4) look a little different. For (2), note compact sets in $\mathbb{Z}$ are finite and so contained in intervals. As for (4), note for $G = \mathbb{Z}$, homomorphisms $\zeta: G \to \mathbb{T}^\infty$ are given precisely by $\zeta(1)$ since $\zeta(n) = \zeta(1)^n$ (using a product rather than additive notation for $\mathbb{T}$). $\square$

(4) $\Rightarrow$ (3) in Theorem 5.13A.2. Let $z_1, z_2, \dots$ be coordinates on $\mathbb{T}^\infty$. Let $\chi_j: G \to \partial\mathbb{D}$ be $z_j \circ \varphi$. Then χ_j is a character on G, and thus, so is any finite product of χ_j's. By the Stone–Weierstrass theorem, polynomials in the z_j are dense in $C(\mathbb{T}^\infty)$, and so F is a uniform limit in polynomials in z_j. Thus, $F \circ \varphi$ is a uniform limit of finite linear combinations of characters. $\square$

(3) $\Rightarrow$ (1) in Theorem 5.13A.2. A set Q in a complete metric space, X, has compact closure if and only if for all ε, there are finitely many $q_1, \dots, q_\ell$ in X so that $\cup_{j=1}^{\ell}\{q \mid \rho(q, q_\ell) < \varepsilon\}$ contains Q. If f is a limit of f_N's of the form (5.13A.8), given ε, pick $\varepsilon/2$ so $\|f - f_N\|_\infty < \varepsilon/2$. Since

$$(f_N)_g = \sum_{j=1}^{N} a_j \chi_j(g) \chi_j \tag{5.13A.10}$$

$\{(f_N)_g\} \subset \{\sum_{j=1}^{N} a_j z_j \chi_j \mid |z_j| = 1\}$ is compact, and so covered by finitely many $\varepsilon/2$ balls. Thus, since $\|f_g - (f_N)_g\|_\infty = \|f - f_N\|_\infty$, $\{f_g\}$ is covered by finitely many ε balls. $\square$

(1) $\Rightarrow$ (2) in Theorem 5.13A.2. Given ε, pick $g_1, \dots, g_N$ in G so every f_g is within ε of some f_{g_j}. Let $K = \{-g_1, \dots, -g_N\}$, which is finite, and so compact. If $\|f_g - f_{g_j}\|_\infty < \varepsilon$, then $\|f_{g-g_j} - f\|_\infty < \varepsilon$ and $h = g - g_j \in g + K$. $\square$

Remark. Once we have (2) $\Rightarrow$ (1), this implies the compact K in Bohr almost periodic can be taken as a finite set!

Lemma 5.13A.3. *If f is Bohr almost periodic, then f is uniformly compact, that is, for any ε, there is a neighborhood N of the identity $e \in G$ so that if $x - y \in N$, then $|f(x) - f(y)| < \varepsilon$.*

Proof. Each f_y is continuous at e, so given ε, there is N_y, a neighborhood of e, so that $w \in N_y \Rightarrow |f_y(w) - f_y(e)| < \varepsilon/4$, so if $w, w' \in N_y$, then $|f_y(w) - f_y(w')| < \varepsilon/2$. By continuity of addition, we can find M_y, a neighborhood of e, so $M_y + M_y \subset N_y$. Thus, if

$$w, w', w'' \in M_y \Rightarrow |f_{y+w''}(w') - f_{y+w''}(w)| < \frac{\varepsilon}{2} \tag{5.13A.11}$$

If K is compact, we have $K \subset \cup_{y \in K}(y + M_y)$, so pick $y_1, \dots, y_\ell$ so $K \subset \cup_{j=1}^{\ell}(y_j + M_{y_j})$ and $M_K = \cap_{j=1}^{\ell} M_{y_j}$. Thus, by (5.13A.11),

$$y \in K,\ w, w' \in M_K \Rightarrow |f_y(w) - f_y(w')| < \frac{\varepsilon}{2} \tag{5.13A.12}$$

Given ε, let K compact be chosen so (5.13A.7) holds for $\varepsilon/4$ and pick M_K as above. Suppose $x - y \in M_K$. By Bohr almost periodicity, there is $h \in K$ so that $\|f_{h-y} - f\|_\infty < \varepsilon/4$. Thus, $\|f_h - f_y\|_\infty < \varepsilon/4$, so by (5.13A.12),

$$w, w' \in M_K \Rightarrow |f_y(w) - f_y(w')| < \varepsilon \tag{5.13A.13}$$

Taking $w = x - y$ and $w' = e$, we see

$$x - y \in M_K \Rightarrow |f(x) - f(y)| < \varepsilon \tag{5.13A.14}$$

which is uniform continuity. □

(2) $\Rightarrow$ (1) *in Theorem 5.13A.2.* By Lemma 5.13A.3, f is uniformly continuous, which implies $x \to f_x$ is continuous as a map of G to $C(G)$. Given ε, let K be the compact set so that (5.13A.7) holds for $\varepsilon/2$. Since $x \to f_x$ is continuous, $\{f_x\}_{x \in K}$ is compact, so we can find $x_1, \dots, x_\ell$ in K whose $\varepsilon/2$ balls cover this set of f's. Given any $y \in G$, there is $x \in K$ so $\|f_{-y+x} - f\|_\infty < \varepsilon/2$, so $\|f_y - f_x\| < \varepsilon/2$ and f_y is within ε of some f_{x_j}. Thus, $\{f_y\}_{y \in G}$ is covered by finitely many ε balls. Since ε is arbitrary, f is (Bochner) almost periodic. □

(1) $\Rightarrow$ (4) *in Theorem 5.13A.2.* This final step is the most elaborate and elegant. Let $H \subset C(G)$ be the closure of $\{f_x\}_{x \in G}$. H is called the *hull* of f. Define $\varphi_0 \colon G \to H$ by

$$\varphi_0(x) = f_x \tag{5.13A.15}$$

Since (1) $\Rightarrow$ (2) $\Rightarrow$ f is uniformly continuous, φ_0 is continuous. Since $\|p_x - q_x\|_\infty = \|p - q\|_\infty$, we see that

$$\|f_{x+y} - f_{x'+y'}\|_\infty \le \|f_x - f_{x'}\|_\infty + \|f_y - f_{y'}\|_\infty \tag{5.13A.16}$$

that is,

$$\|\varphi_0(x + y) - \varphi_0(x' + y')\| \le \|\varphi_0(x) - \varphi_0(x')\| + \|\varphi_0(y) - \varphi_0(y')\| \tag{5.13A.17}$$

Let $h, h' \in H$. Picking $x_n, y_n \in G$ so $\varphi(x_n) \to h$, $\varphi(y_n) \to h'$, we see, by (5.13A.17), that $\varphi(x_n + y_n)$ is Cauchy, which allows us to define $h + h'$ ("+" is map of $H \times H$ to H, not to be confused with adding the functions!). It is easy to see this turns H into a compact group. Since H is a metric space, compactness implies separability. By definition, φ is a homomorphism.

Now we need a fact about compact separable abelian groups (see the Notes): Such groups have characters that separate points, and by separability, there is a countable family, $\{\chi_j\}_{j=1}^{\infty} \subset \widehat{H}$, that separates points. Let $Q\colon H \to \mathbb{T}^{\infty}$ by $Q(h)_j = \chi_j(h)$ and $\varphi\colon G \to \mathbb{T}^{\infty}$ by $\varphi = Q \circ \widetilde{\varphi}$. Q is an injective map since $\{\chi_j\}$ separates points. φ is a group homomorphism.

Since H is compact, $Q[H]$ is closed in $\mathbb{T}^{\infty}$. Define $\widetilde{F}\colon H \to \mathbb{C}$ by $\widetilde{F}(h) = h(e)$. Then F is continuous and

$$\widetilde{F}(\varphi(x)) = \widetilde{F}(f_x) = f_x(e) = f(x) \tag{5.13A.18}$$

that is, $\widetilde{F} \circ \varphi = f$. Since Q is one-one, we can define a function F on $Q[H]$ so

$$F \circ Q = \widetilde{F} \tag{5.13A.19}$$

Since $Q[H]$ is closed, F has an extension to $\mathbb{T}^{\infty}$ by the Tietze extension theorem. We will still use F for this extension. Clearly, (5.13A.19) remains true; $F\colon \mathbb{T}^{\infty} \to \mathbb{C}$ and

$$F \circ \varphi = F \circ Q \circ \widetilde{\varphi} = \widetilde{F} \circ \widetilde{\varphi} = f \tag{5.13A.20}$$

by (5.13A.18). □

Remarks and Historical Notes. The definition of almost periodic functions on $\mathbb{R}$ and their properties is due to Harald Bohr [51, 52], using the definition we gave for Bohr almost periodic on $\mathbb{Z}$ (but for $\mathbb{R}$). The Bochner property (which we codified in the Bochner definition) is due to Bochner [47, 49].

Sometimes what we call "almost periodic" is called "uniformly almost periodic" since there are also Besicovitch almost periodic or L^2-almost periodic functions, which we will define below.

For book treatments of the theory, see Besicovitch [44], Bohr [53], Corduneanu [94], and Levitan–Zhikov [279].

We used the fact that any abelian separable compact group, G, has enough characters to separate points. This is essentially the Peter–Weyl theorem for such groups (see, e.g., Simon [394]); here is a sketch of the argument explicitly. Let f be a function on G with $f(-x) = \overline{f(x)}$. Define $T\colon L^2(G) \to L^2(G)$ by

$$(Th)(x) = \int f(x-y)h(y)\,d\mu(y)$$

where $d\mu$ is Haar measure. T is Hilbert–Schmidt (so compact) and selfadjoint.

Moreover, if $U_x\colon L^2 \to L^2$ by $(U_x f)(y) = f(y-x)$, then T commutes with $\{U_x\}$. Thus, $\{U_x\}$ leave each eigenspace invariant. If V is such an eigenspace and is finite-dimensional, the U_x are commuting unitaries on V, so they have a common eigenvector $\widetilde{\chi}(x)$. Thus,

$$\widetilde{\chi}(x+y) = (U_x\widetilde{\chi})(y) = \lambda_x\widetilde{\chi}(y)$$

and $U_{x+y} = U_x U_y$ implies $\lambda_{x+y} = \lambda_x \lambda_y$. Since $x \to U_x$ is continuous, this shows $\widetilde{\chi}$ is continuous and everywhere nonzero: $\chi(x) = \widetilde{\chi}(x)/\widetilde{\chi}(e)$ is thus a (continuous) character. So the characters span $\mathrm{Ran}(T)$. Since we can find f_n so $T_{f_n} \to \mathbf{1}$, we see the characters χ span L^2, which implies they separate points.

Further developments depend on the notion of the average of an almost periodic function. Given an almost periodic function, f, let H be its hull, $\widetilde{F}$ the function in (5.13A.18), and $d\nu$ normalized Haar on H. We define

$$\mathrm{Av}(f) = \int_H \widetilde{F}(x)\, d\nu(x) \tag{5.13A.21}$$

For $\mathbb{R}$ or $\mathbb{Z}$, one can prove that

$$\mathrm{Av}(f) = \lim_{T\to\infty} \frac{1}{2T} \int_{-T}^{T} f(x)\, dx \tag{5.13A.22}$$

(or $\frac{1}{2T+1} \sum_{-T}^{T} f(n)$ for $\mathbb{Z}$).

One defines the Fourier coefficients of f for $\chi \in \widehat{G}$ by

$$\widehat{f}(\chi) = \mathrm{Av}(\bar{\chi} f) \tag{5.13A.23}$$

noting that $\bar{\chi} f$ is also almost periodic. It is not hard to see that $\widehat{f}(\chi)$ is nonzero for only countably many χ's. Indeed, one has a Plancherel theorem

$$\sum_{\chi\in\widehat{G}} |\widehat{f}(\chi)|^2 = \mathrm{Av}(|f|^2) \tag{5.13A.24}$$

One also has an L^2 convergence of Fourier series; if $\{\chi_j\}_{j=1}^{\infty}$ is a numbering of those χ's with $\widehat{f}(\chi) \neq 0$, then

$$\mathrm{Av}\biggl(\biggl| f - \sum_{j=1}^{N} \widehat{f}(\chi_j)\chi_j \biggr|^2\biggr) \to 0 \tag{5.13A.25}$$

These results are all easy to prove by using the fact that if H is the hull, $\widehat{f}(\chi) \neq 0$ implies $\chi \in \widehat{H}$, that is,

$$\chi = \widetilde{\chi} \circ \widetilde{\varphi} \tag{5.13A.26}$$

where $\widetilde{\chi}$ is a character of H. (5.13A.24) and (5.13A.25) are then expressions of the fact that characters of H are a basis of $L^2(H, d\nu)$.

For $\mathbb{R}$, one defines Besicovitch almost periodic functions as functions on $\mathbb{R}$, for which there exists, for any z, a finite sum $f_N = \sum_{j=1}^{N} a_j^{(N)} e^{i w_j^{(N)} x}$ with

$$\limsup_{T\to\infty} \frac{1}{2T} \int_{-T}^{T} |f_n - f_N(x)|^2\, dx \leq \varepsilon \tag{5.13A.27}$$

The *frequency module* of f, an almost periodic function, is the set of characters of G that comes from H, the hull, via (5.13A.26). It is a countable subgroup of $\widehat{G}$. It is generated by $\{\chi \mid \widehat{f}(\chi) \neq 0\}$.

A function is called *limit periodic* if it is a uniform limit of periodic functions. Such functions are obviously almost periodic. A typical example is

$$f(x) = \sum_{n=1}^{\infty} 2^{-n} \cos(2\pi 2^{-n} x) \tag{5.13A.28}$$

We note that the term quasiperiodic is sometimes used for a very different notion from our use and that those quasiperiodic functions are not almost periodic.

The set of all almost periodic functions in $\|\cdot\|$ is Banach algebra. Its Gel'fand spectrum (see [150] for the theory of commutative Banach algebras) is called the Bohr compactification of G. It is huge, containing every hull as a subgroup. One can construct it by taking $\widehat{G}$ and putting the discrete topology in it and taking the dual of that.

5.14 PERIODIC OPUC

We have discussed OPRL with periodic Jacobi matrices in much of this chapter. The theory of OPUC whose Verblunsky coefficients obey

$$\alpha_{n+p} = \alpha_n \tag{5.14.1}$$

for all n and some fixed p is the subject of Chapter 11 of [400]. Our goal in this section is to sketch some parts of this theory, emphasizing the differences to the OPRL theory.

A major difference is that the transfer matrix for OPRL has determinant 1 since

$$\det\left[\frac{1}{a}\begin{pmatrix} z-b & -1 \\ a^2 & 0 \end{pmatrix}\right] = 1 \tag{5.14.2}$$

while in the OPUC case, the m step transfer matrix has determinant z^m since

$$\det\left[\frac{1}{\rho}\begin{pmatrix} z & -\bar{\alpha} \\ -\alpha z & 1 \end{pmatrix}\right] = z \tag{5.14.3}$$

(see (2.4.3)).

The natural discriminant is thus

$$\Delta(z) = z^{-p/2}\mathrm{Tr}(T_p(z)) \tag{5.14.4}$$

For this reason, it is natural to restrict to the case p even and control p odd by other means (e.g., by viewing it as period $2p$ instead of as period p). We shall do this henceforth.

$\Delta(z)$ is thus a Laurent polynomial (i.e., polynomial in z and z^{-1}). It is real on $\partial\mathbb{D}$, and one can show the associated measure is purely absolutely continuous on $\mathfrak{e} = \Delta^{-1}([-2,2]) \subset \partial\mathbb{D}$ with potentially one pure point per gap. The Carathéodory function obeys a quadratic equation and extends to a two-sheeted Riemann surface with branch points at the edges of connected components of $\mathfrak{e}$.

The most significant difference from OPRL comes from the following: If $\mathfrak{e}$ has $\ell+1$ connected components, in the OPRL case, there are ℓ significant gaps—the gap on $\mathbb{C}\setminus\mathfrak{e}$ that goes from $\beta_{\ell+1}$ to ∞ and then $-\infty$ to α_1 is not considered for

Dirichlet data. In some sense, the pole at $-\infty$ and zero at ∞_+ are fixed and only the zeros and poles in the finite gaps vary. But if $\mathfrak{e}$ has $\ell+1$ components, there are, on $\partial\mathbb{D}$, $\ell+1$ gaps and none is distinguished. The natural Dirichlet data is a torus of dimension $\ell+1$, one for each gap. On the other hand, $\mathcal{S}$ is still of genus ℓ, so the analog of our map $\widetilde{\mathfrak{A}}$ from $\mathcal{M}_\ell$ to $\mathbb{T}^\ell$ maps an $\ell+1$-dimensional torus to an ℓ torus and is no longer one-one. Instead, the inverse image $\widetilde{\mathfrak{A}}^{-1}(x)$ of a fixed point in $\mathbb{T}^\ell$ is a circle. Indeed, in the periodic case, $\{\alpha_n\}_{n=0}^{p-1}$ and $\{\alpha_n'\}_{n=0}^{p-1}$ have the same image under $\widetilde{\mathfrak{A}}$ if and only if $\alpha_n' = e^{i\theta}\alpha_n$ for some fixed θ. This means that the natural result of Abel's theorem is to show only that elements of $\mathcal{T}_\mathfrak{e}$ obey $\alpha_{n+p} = e^{i\theta}\alpha_n$ for some θ and, more generally, are almost periodic up to phase. Controlling this phase turns out to be simple in the periodic case and very involved in the almost periodic case.

Another significant difference is the function to be used in Abel's theorem. In the OPRL case, the m-function itself realized coefficient stripping, that is, the poles of the once-stripped Jacobi matrix were exactly the zeros of m. One might hope the Carathéodory function had this property, but that is not true. The zeros of the Carathéodory function associated to $\{\alpha_n\}_{n=0}^\infty$ are the poles for the Carathéodory function associated to $\{-\alpha_n\}_{n=0}^\infty$.

Instead, one needs to use the function $z(\delta_0 D)(z)$ of (2.6.9). It has poles at the poles of the Carathéodory function for $\{\alpha_n\}_{n=0}^\infty$ and zeros at the poles of the Carathéodory function for $\{\alpha_{n+1}\}_{n=0}^\infty$ (i.e., once-stripped). If these have a pole in common, the situation is slightly different. In addition, $z\delta_0 D(z)$ has a pole at ∞_- and a zero at 0_+, so in place of the $\mathfrak{A}(\infty_-) - \mathfrak{A}(\infty_+)$ of (5.13.37), we have $\mathfrak{A}(\infty_-) - \mathfrak{A}(0_+)$. This describes the major differences.

Remarks and Historical Notes. See [400] and its notes for the theory and history of periodic OPUC. [400] uses the function it calls $M(z)$ related to $\delta_0 D(z)$ by

$$M(z) = 2\rho_0 z(\delta_0 D)(z)$$

While [399] introduced $\delta_0 D$, the connection of $M(z)$ and $\delta_0 D$ was not realized in [399, 400].

Chapter Six

Toda Flows and Symplectic Structures

Having discussed periodic Jacobi matrices, we would be remiss if we did not discuss the closely related Toda lattice dynamical system. So even though it is definitely an aside, we provide the high points in this chapter.

6.1 OVERVIEW

The structure that the spectra of periodic Jacobi matrices induce on Jacobi parameters is striking. $[(0,\infty)\times\mathbb{R}]^p$, consisting of points $(a_n, b_n)_{n=1}^p$, is decomposed into its isospectral tori, generically of dimension $p-1$ with some degenerate tori of lower dimension. The fibration into tori is reminiscent of another structure, which we will discuss in Section 6.2. A completely integrable system is a manifold of dimension 2ℓ with ℓ Poisson commuting "independent" functions. If the sets where these functions have constant values are compact, then phase space is fibered into tori of dimension ℓ with some degenerate lower-dimensional tori. Of course, there is a dimension counting issue: $[(0,\infty)\times\mathbb{R}]^p$ has dimension $2p$ but the tori here are not of dimension p, but $p-1$. We will see shortly why that is not a problem.

Our main goal in this chapter will be to explore the completely integrable system on Jacobi matrices that helps "explain" the fibration into tori. Along the way, we will prove a technical fact about derivatives of coefficients of Δ with respect to Jacobi parameters that will be an important ingredient in the proof of the analog of the Killip–Simon theorem for periodic OPRL; see Section 8.5.

The *Toda lattice* was originally formulated in terms of the Hamiltonian

$$H(p_1,\dots,p_N,q_1,\dots,q_N)=\sum_{j=1}^{N}\tfrac{1}{2}\,p_j^2+\gamma\sum_{j=1}^{N-1}e^{q_j-q_{j+1}} \tag{6.1.1}$$

Here γ is a fixed positive coupling constant, which is usually set to 1, but which we will want to include. We will also consider the *periodic Toda lattice* where

$$H_P(p_1,\dots,p_N,q_1,\dots,q_N)=\sum_{j=1}^{N}\tfrac{1}{2}\,p_j^2+\gamma\left[\left(\sum_{j=1}^{N-1}e^{q_j-q_{j+1}}\right)+e^{q_N-q_1}\right] \tag{6.1.2}$$

To distinguish it from the periodic case, we will sometimes use the phrase "free Toda flow" for the solutions of the equation of motion associated to the Hamiltonian (6.1.1).

We consider the equations of motion in Poisson bracket (PB) form (discussed in Section 6.2)

$$\frac{df}{dt} = \{H, f\} \tag{6.1.3}$$

$$\{p_j, p_k\} = 0 \qquad \{q_j, q_k\} = 0 \qquad \{p_j, q_k\} = \delta_{jk} \tag{6.1.4}$$

so

$$\{f, g\} = \sum_{j=1}^{N} \left[\frac{\partial f}{\partial p_j} \frac{\partial g}{\partial q_j} - \frac{\partial f}{\partial q_j} \frac{\partial g}{\partial p_j} \right] \tag{6.1.5}$$

Thus, (6.1.3) says

$$\frac{dq_j}{dt} = p_j \qquad \frac{dp_j}{dt} = -\gamma(e^{q_j - q_{j+1}} - e^{q_{j-1} - q_j}) \tag{6.1.6}$$

with special formulae for $j = 1, N$.

Note that the setup is a bit unphysical. Particle j only interacts with particles $j-1$ and $j+1$ (as one might expect in a lattice), but the particle positions are not forced to be ordered. Moreover, the potential is highly nonsymmetric, minimum potential energies occur as $q_j - q_{j+1} \to -\infty$, and there is a kind of hard core: if the energy is E, then $q_j - q_{j+1} \leq \log(E/\gamma)$.

Flaschka [134] and Manakov [293] found a remarkable change of variables

$$b_j = -\tfrac{1}{2} p_j \qquad a_j = \frac{\sqrt{\gamma}}{2} e^{\frac{1}{2}(q_j - q_{j+1})} \tag{6.1.7}$$

In the free case, we only have $a_1, \dots, a_{N-1}$. In the periodic case, we have

$$a_N = \frac{\sqrt{\gamma}}{2} e^{\frac{1}{2}(q_N - q_1)} \tag{6.1.8}$$

not independent since

$$a_1 \dots a_N = \frac{\gamma^{N/2}}{2^N} \tag{6.1.9}$$

The Hamiltonians are

$$H = 2\sum_{j=1}^{N} b_j^2 + 4\sum_{j=1}^{N-1} a_j^2 \tag{6.1.10}$$

$$H_P = 2\sum_{j=1}^{N} b_j^2 + 4\sum_{j=1}^{N} a_j^2 \tag{6.1.11}$$

and the fundamental PB becomes

$$\{b_j, a_j\} = -\tfrac{1}{4} a_j \tag{6.1.12}$$

$$\{b_j, a_{j-1}\} = \tfrac{1}{4} a_{j-1} \tag{6.1.13}$$

γ drops out in the free case and only enters the periodic case through (6.1.9).

The equations of motion (6.1.3) (equivalent to (6.1.6)) become

$$\frac{da_j}{dt} = a_j(b_{j+1} - b_j) \tag{6.1.14}$$

$$\frac{db_j}{dt} = 2(a_j^2 - a_{j-1}^2) \tag{6.1.15}$$

with the proviso in the free case for (6.1.15): we interpret $a_0 = a_N = 0$, and in the periodic case, $a_0 = a_N$ in (6.1.15), and $b_{N+1} = b_1$ in (6.1.14) for $j = N$.

One can now understand the reason the tori are only dimension $N-1$, not N. The a, b variables have Poisson brackets that are degenerate. In the free case, $\sum_{j=1}^N b_j$ Poisson commutes (i.e., has zero Poisson bracket) with all a and b. In the periodic case, $\prod_{j=1}^N a_j$ also Poisson commutes (as does $\sum_{j=1}^N b_j$). Thus, in both cases, we need to restrict to $\sum_{j=1}^N p_j = \beta$, and in the periodic case to $\prod_{j=1}^N a_j = \alpha$ (which is no restriction to the q's; in p, q language, we are fixing $\sum_{j=1}^N p_j$ and $\sum_{j=1}^N q_j$). In either case, we get $2N-2$-dimensional manifolds with nondegenerate Poisson brackets. The natural completely integrable systems then have invariant tori of dimension $\frac{1}{2}(2N-2) = N-1$.

A hint of the connection to Jacobi matrices and invariance of the spectrum is seen in the fact that the free Hamiltonian (6.1.10) is given by

$$H = 2\,\mathrm{Tr}(J_{N;F}^2) \tag{6.1.16}$$

with $J_{N;F}$ given by (1.2.30) and the periodic Hamiltonian (6.1.11) by

$$H_P = 2\,\mathrm{Tr}(J_{N;P}^2) \tag{6.1.17}$$

where $J_{N;P}$ is the $J(\theta)$ of (5.3.8) with $\theta = 0$.

For complete integrability on our $2N-2$ phase space, one needs $N-1$ Poisson commuting functions and they will be $\mathrm{Tr}(J_{N;F}^\ell)$ and $\mathrm{Tr}(J_{N;P}^\ell)$, $\ell = 2, 3, \dots, N$, respectively. $\ell = 1$ is not included since it is $\sum_{j=1}^N b_j$ and constant on the manifold. We stop at $\ell = N$ since $\mathrm{Tr}(A^\ell)$, $\ell = 1, \dots, N$ for any $N \times N$ matrix determine the eigenvalues $\lambda_1, \dots, \lambda_N$, and so $\mathrm{Tr}(A^{N+1}), \mathrm{Tr}(A^{N+2}), \dots$.

Section 6.2 is a tutorial on symplectic manifolds and completely integrable systems, while Section 6.3 provides background on a piece of linear algebra (the QR factorization) needed later. Section 6.4 provides a first proof that in the free case, the PBs of the traces are zero: it goes from PBs of a, b to PBs of the orthogonal polynomials, and from there to PBs of eigenvalues and their spectral weights. Section 6.5 then solves the free Toda lattice in the eigenvalue-weight coordinates. Section 6.6 provides a second proof that the PBs of traces are zero, using Lax pairs, and Section 6.7 completes that analysis using the QR algorithm to link the two approaches. Section 6.7 also completes the proof that $\{\mathrm{Tr}(J_{N;F}^\ell), \mathrm{Tr}(J_{N;F}^k)\} = 0$ since Section 6.6 only does the calculation for $\ell = 2$ and general k. Section 6.8 turns to PBs for the periodic case and Section 6.9 proves an important independence result when all gaps are open. Finally, Section 6.10 has some remarks on the OPUC analog.

6.2 SYMPLECTIC DYNAMICS AND COMPLETELY INTEGRABLE SYSTEMS

In this section, we describe Hamiltonian dynamics on general manifolds and prove a key theorem about completely integrable systems. We suppose the reader is familiar with the basics of manifold theory, including the definition of tangent space, $T_p(M)$, cotangent space, $T_p^*(M)$, vector fields, forms, and flows; see [54, 93, 267, 284, 415, 416]. M will be a C^∞ manifold. We only sketch the proofs; for details, see [4, 26, 284, 297, 303, 437].

Two-forms can be viewed as functions on M with values at $p \in M$ in the anti-symmetric bilinear maps from $T_p(M) \times T_p(M)$, that is, given $p \in M$, a two-form, Ω, and $X, Y \in T_p(M)$, $\Omega_p(X, Y)$ is a number linear in each of X and Y with the other fixed, and

$$\Omega_p(X, Y) = -\Omega_p(Y, X) \tag{6.2.1}$$

If $\{x_j\}_{j=1}^n$ is a local coordinate system near p, then $(\frac{\partial}{\partial x_j})_{j=1}^n$ is a basis for $T_p(M)$ and $\{dx_j\}_{j=1}^n$ for $T_p^*(M)$. We normalize $dx_j \wedge dx_\ell$ $(j \neq \ell)$ by

$$(dx_j \wedge dx_\ell)\left(\frac{\partial}{\partial x_m}, \frac{\partial}{\partial x_q}\right) = \tfrac{1}{2}(\delta_{jm}\delta_{\ell q} - \delta_{\ell m}\delta_{jq}) \tag{6.2.2}$$

The half is there so we can write

$$\Omega_p = \sum_{k,\ell=1}^n \Omega_{k\ell}(p)\, dx_k \wedge dx_\ell \tag{6.2.3}$$

with

$$\Omega_{k\ell}(p) = -\Omega_{\ell k}(p) \tag{6.2.4}$$

and have

$$\Omega_p\left(\frac{\partial}{\partial x_m}, \frac{\partial}{\partial x_\ell}\right) = \Omega_{m\ell}(p) \tag{6.2.5}$$

Every form defines a map $\Omega_p^*\colon T_p(M) \to T_p^*(M)$ by

$$\Omega_p^*(X)(Y) = \Omega_p(X, Y) \tag{6.2.6}$$

equivalently,

$$\Omega_p^*\left(\sum_{j=1}^n a_j \frac{\partial}{\partial x_j}\right) = \sum_{j,k=1}^n a_j \Omega_{jk}(p)\, dx^k \tag{6.2.7}$$

A form is called *nondegenerate* at a point p if Ω_p^* is a bijection; equivalently, if $\det(\Omega_{m\ell}(p)) \neq 0$. Notice, by (6.2.4), that $\det(\Omega_{m\ell}) = (-1)^n \det(\Omega_{m\ell})$ is 0 if n is odd, so only even-dimensional manifolds can have nondegenerate forms. The key definition is

Definition. A *symplectic manifold* is a manifold, M, with distinguished two-form, Ω, nondegenerate at every point and closed, that is,

$$d\Omega = 0 \tag{6.2.8}$$

In (6.2.8), d is the canonical differential from ℓ-forms to $(\ell + 1)$-forms. Recall that a vector field is a smooth function on M taking values in $T_p(M)$, which, given that tangent vectors are equivalence classes of curves, are the same as first-order differential operators. Thus, vector fields map $C^\infty(M)$ to itself. In local coordinates, $X = \sum_{j=1}^n a_j \frac{\partial}{\partial x_j}$ and $Xf = \sum_{j=1}^n a_j \frac{\partial f}{\partial x_j}$. Vector fields define flows, and conversely. In good cases (always if M is compact), flows can be globally defined: There is $\varphi_t : M \to M$, C^∞ maps for all $t \in \mathbb{R}$ and

$$\varphi_{t=0} = \mathbf{1} \qquad \varphi_t \circ \varphi_s = \varphi_{t+s} \tag{6.2.9}$$

The relation to X is that

$$\frac{d}{dt}\, f(\varphi_t(x)) = (Xf)(\varphi_t(x)) \tag{6.2.10}$$

We will often write $\exp(tX)$ for φ_t.

In general, functions, f, define one-forms df but not vector fields. On a symplectic manifold, one can use $(\Omega^*)^{-1}$ to map one-forms to vector fields, and so associate functions to vector fields. The *Hamiltonian vector field*, X_f, associated to an arbitrary function f on a symplectic manifold is defined by

$$df = \Omega^*(X_f) \tag{6.2.11}$$

In local coordinates

$$X_f = \sum_{j=1}^n a_j \frac{\partial}{\partial x_j} \qquad a_j = -\sum_{k=1}^n (\Omega^{-1})_{jk} \frac{\partial f}{\partial x_k} \tag{6.2.12}$$

(the minus sign comes from antisymmetry of Ω^{-1} and the flip of order from (6.2.7)).

The *Poisson bracket* (aka PB), $\{f, g\}$, is defined by

$$\{f, g\} = X_f g \tag{6.2.13}$$

By (6.2.10), if φ_t is the flow defined by X_f (the *Hamiltonian flow*), then

$$\frac{d}{dt}\, g\bigg|_{t=0} = \{f, g\} \tag{6.2.14}$$

In particular, by (6.2.10), g is invariant under the Hamiltonian flow generated by f if and only if $\{f, g\} = 0$. By (6.2.12),

$$\{f, g\} = \sum_{j,k=1}^n (\Omega^{-1})_{jk} \frac{\partial f}{\partial x_j} \frac{\partial g}{\partial x_k} \tag{6.2.15}$$

which implies, by the antisymmetry of Ω, that

$$\{f, g\} = -\{g, f\} \tag{6.2.16}$$

An intrinsic way of seeing this is to note

$$\begin{aligned} \{f, g\} &= X_f g = dg(X_f) \\ &= \Omega^*(X_g)(X_f) \\ &= \Omega(X_g, X_f) \end{aligned} \tag{6.2.17}$$

from which the antisymmetry is obvious.

Note, in particular, (6.2.16) implies that

$$X_f f = 0 \tag{6.2.18}$$

So, by (6.2.14),

$$\frac{df}{dt} = 0$$

under the flow generated by f—this is *energy conservation*.

One advantage of the PB formalism is that it makes it easy to compute changes in the form of Hamiltonian equations under a change of variables. In this regard, two versions of the chain rule are invaluable. First,

$$\{f, G(g_1, \dots, g_\ell)\} = \sum_{j=1}^{\ell} \{f, g_j\} \frac{\partial G}{\partial g_j} \tag{6.2.19}$$

which follows from the chain rule for differential operators and $\{f, \cdot\} = X_f \cdot$. Using (6.2.16), one can iterate this to obtain a general formula for change of variables

$$\{F(f_1, \dots, f_m), G(g_1, \dots, g_\ell)\} = \sum_{k=1}^{m} \sum_{j=1}^{\ell} \{f_k, g_j\} \frac{\partial F}{\partial f_k} \frac{\partial G}{\partial g_j} \tag{6.2.20}$$

In particular, if M has coordinates $(p_j, q_j)_{j=1}^m$ with $\{p_j, p_k\} = \{q_j, q_k\} = 0$, then

$$\{F(p, q), G(p, q)\} = \sum_{k,j=1}^{m} \{p_j, q_k\} \left[\frac{\partial F}{\partial p_j} \frac{\partial G}{\partial q_k} - \frac{\partial G}{\partial p_j} \frac{\partial F}{\partial q_k} \right] \tag{6.2.21}$$

Recall that the Lie bracket of two vector fields is defined by

$$[X, Y] = XY - YX \tag{6.2.22}$$

as a composition of differential operators. It is also a vector field. The fact that $d\Omega = 0$ has the following important consequences:

Theorem 6.2.1. *On a symplectic manifold,*

(i)

$$[X_f, X_g] = X_{\{f,g\}} \tag{6.2.23}$$

(ii) (*Jacobi identity*)

$$\{f, \{g, h\}\} = \{\{f, g\}, h\} + \{g, \{f, h\}\} \tag{6.2.24}$$

(iii) *Hamiltonian flows preserve the symplectic form.*

Remarks. 1. Maps preserving the symplectic form are called *canonical transformations* or, in more modern discussions, *symplectomorphisms*.

2. (6.2.24) is often written in the more symmetric form

$$\{f, \{g, h\}\} + \{g, \{h, f\}\} + \{h, \{f, g\}\} = 0 \tag{6.2.25}$$

3. An invariant way to see (6.2.25) is to prove for general two-forms Ω,

$$\text{LHS of (6.2.25)} = c\, d\Omega(X_f, X_g, X_h)$$

for suitable constant c so that (6.2.25) is equivalent to $d\Omega = 0$.

Sketch. (i) Using (6.2.12), one easily computes $[X_f, X_g]$ and sees that it is $X_{\{f,g\}}$, plus some terms involving derivatives of $(\Omega^{-1})_{jk}$. Since $d\Omega = 0$, we have

$$\frac{\partial}{\partial x_k}(\Omega)_{j\ell} + \frac{\partial}{\partial x_j}(\Omega)_{\ell k} + \frac{\partial}{\partial x_\ell}(\Omega)_{kj} = 0 \tag{6.2.26}$$

and this plus $\frac{\partial}{\partial x_k}\Omega^{-1} = -\Omega^{-1}(\frac{\partial}{\partial x_k}\Omega)\Omega^{-1}$ (matrix multiplication) implies the terms involving derivatives of Ω cancel. This proves (6.2.23).

(ii) By (6.2.23),

$$X_g X_h f - X_h X_g f = X_{\{g,h\}} f \tag{6.2.27}$$

which, given (6.2.13), implies (6.2.24).

(iii) If φ_t^f is the flow generated by X_f and

$$(\varphi_t^f)^* g = g \circ \varphi_t^f \tag{6.2.28}$$

then, because $\{\cdot\,,\,\cdot\}$ determines $(\Omega^{-1})_{jk}$ and so Ω, invariance of Ω is equivalent to

$$\varphi_t^f(\{g,h\}) = \{\varphi_t^f(g), \varphi_t^f(h)\} \tag{6.2.29}$$

The derivative with respect to t of (6.2.29) is exactly (6.2.24). Since (6.2.29) holds at $t=0$ and derivatives are equal, we have (6.2.29) in general. $\square$

This has a number of important consequences. Suppose M is a symplectic manifold, so of even dimension $2m$. The m-fold wedge product $\Omega \wedge \cdots \wedge \Omega = \det(\Omega) \times dx_1 \wedge \cdots \wedge dx_{2m}$, called the *canonical volume form*, is, by nondegeneracy, an everywhere nonzero $2m$-form. Since Ω is invariant, so is this volume form. Thus,

Corollary 6.2.2 (Liouville's Theorem). *Any Hamiltonian flow preserves the canonical volume form.*

Secondly, we can set up several equivalences:

Theorem 6.2.3. *Let $f_1, \dots, f_\ell$ be ℓ functions on a symplectic manifold M. Then the following are equivalent:*

(i) *For all $1 \le j, k \le \ell$,*

$$\{f_j, f_k\} = 0 \tag{6.2.30}$$

(ii) *For all j, the flows $\exp(tX_{f_j})$ leave $\{f_k\}_{k=1}^\ell$ invariant.*

If these hold, then

(iii) *For $1 \le j, k \le \ell$,*

$$[X_{f_j}, X_{f_k}] = 0 \tag{6.2.31}$$

(iv) *The flows $\exp(tX_{f_j})$ and $\exp(sX_{f_k})$ commute.*

(v) *The map $\varphi^{\vec{f}}: (t_1, \dots, t_\ell) \mapsto \exp(\sum_{j=1}^\ell t_j X_{f_j})$ on $\mathbb{R}^\ell$ (assuming all flows are global) obeys*

$$\varphi_{t+s}^{\vec{f}} = \varphi_t^{\vec{f}} \varphi_s^{\vec{f}} \tag{6.2.32}$$

for all $t, s \in \mathbb{R}^\ell$.

Remarks. 1. Notice that if $\{f, g\}$ is a constant, $[X_f, X_g] = 0$, so (6.2.31) does not imply (6.2.30).

2. One can also show (iii), (iv), (v) are equivalent.

Sketch. (i) $\Leftrightarrow$ (ii) by (6.2.14). (ii) $\Rightarrow$ (iii) by (6.2.27). This in turn implies (iv) by standard results on Lie derivatives, and similarly, they imply (v). $\square$

Poisson commuting functions are said to be *in involution*. ℓ functions, $f_1, \dots, f_\ell$ are said to be *independent* at $p_0 \in M$ if and only if $(df_1)(p_0), \dots, (df_\ell)(p_0)$ are linearly independent. The implicit function theorem then implies

$$M_{p_0}^{\vec{f}} = \{p \mid f_j(p_0) = f_j(p),\ j = 1, \dots, \ell\} \tag{6.2.33}$$

intersected with a small neighborhood of p_0 is a submanifold of dimension $\dim(M) - \ell$. Put down a Riemann metric near p_0. grad(f) is the vector field associated to df under this metric. $\{\text{grad}(f_j)\}_{j=1}^{\ell}$ are all orthogonal to tangent vectors to $M_p^{\vec{f}}$. On the other hand, if $\{f_j, f_k\} = 0$, $\exp(tX_{f_j})$ leaves $M_{p_0}^{\vec{f}}$ invariant, and so the X_{f_j} are all tangent to $M_{p_0}^{\vec{f}}$, and so orthogonal to all grad(f_k). If $f_1, \dots, f_\ell$ are independent, the grad(f_k) are independent, as are the X_{f_j} since Ω and the Riemann metric are nondegenerate. Given the orthogonality, we get 2ℓ independent vectors at p_0. We have thus proven:

Proposition 6.2.4. *If M is a symplectic manifold of dimension $2m$ and $f_1, \dots, f_\ell$ are in involution and independent at some point $p_0 \in M$, then $\ell \le m$.*

If $\ell = m$, then $M_{p_0}^{\vec{f}}$ is of dimension m, $\{X_{f_j}\}_{j=1}^m$ span its tangent space and $\{\text{grad}(f_j)\}_{j=1}^n$ span the normal subspace to this tangent space.

Definition. A *completely integrable system* on a symplectic manifold, M, of dimension $2m$ is a set $f_1, \dots, f_m$ of functions in involution, which are linearly independent at almost all points in M.

Finally, here are the tori:

Theorem 6.2.5 (Arnold–Jost–Liouville Theorem). *Let $\{f_1, \dots, f_m\}$ be a completely integrable system on a symplectic manifold, M, of dimension $2m$. Let T be a connected compact set on which all f_j are constant and so that at each point of T, the f_j are independent. Then:*

(i) *For any $p_0 \in T$, $\{\exp(\sum_{j=1}^m t_j X_{f_j}) p_0 \mid t \in \mathbb{R}^m\} = T$.*

(ii) *T is diffeomorphic to an m-dimensional torus.*

Sketch. (i) Fix p_0. Let $\varphi\colon \mathbb{R}^m \to T$ by

$$\varphi(t) = \varphi_t^{\vec{f}}(p_0)$$

Clearly, φ maps to $M_{p_0}^{\vec{f}}$ since f_j are constant and so must lie in its connected component. Since T has a neighborhood N with $N \cap M_{p_0}^{\vec{f}} = T$ (by independence and the implicit function theorem), φ maps to T. By compactness, the flow is complete, that is, defined for all t.

Since $\{X_{f_j}\}_{j=1}^m$ are independent at any point, p_1, in T, $t \to \varphi_{\vec{t}}^{\vec{f}}(p_1)$ is a C^∞ bijection for $|t|$ small and so, by $\varphi_{\vec{t}+\vec{s}}^{\vec{f}} = \varphi_{\vec{t}}^{\vec{f}}\varphi_{\vec{s}}^{\vec{f}}$, the range of φ is open. Suppose $p_n \to p_\infty$ and $p_n \in \mathrm{Ran}(\varphi)$. Then, for n large, p_n lies in the image of $\varphi_t^{\vec{f}}$ on p_∞. So there is t with $\varphi_{-t}^{\vec{f}}(p_\infty) = p_n$. Since p_n is in $\mathrm{Ran}(\varphi)$, there is s with $\varphi(s) = p_n$. Thus, $\varphi(t+s) = p_\infty$, that is, $\mathrm{Ran}(\varphi)$ is closed. By connectedness, (i) holds.

(ii) Let $G \subset \mathbb{R}^m$ be $\{t \mid \varphi(t) = p_0\}$. Clearly, G is a closed subgroup. Since φ is a diffeomorphism of a small neighborhood of $0 \in \mathbb{R}^m$ to p_0 (by the independence and implicit function theorem), G is discrete, then $\tilde{\varphi}\colon \mathbb{R}^m/G \to M$ is a bijection, so compactness implies G is an m-dimensional lattice and $\mathbb{R}^m/G$ is an m-dimensional torus. □

Remarks and Historical Notes. In some ways, our symplectic manifolds will come via the ad hoc introduction of the PBs (6.1.12)/(6.1.13), but there are two natural classes of symplectic manifolds both related to the Toda lattices. First, the cotangent bundle of any manifold has a natural one-form, ω, and $\Omega = d\omega$ defines a closed nondegenerate two-form; see [4, 25, 26, 284].

Second, any Lie group defines an action on its Lie algebra and so on the dual of the Lie algebra. Orbits under this action are called *coadjoint orbits*. Using the Lie bracket, they have a natural symplectic form; see [202, 226, 355, 388, 463] for further discussion. As we will explain in the Notes to Section 6.7, this is also connected to Toda lattices.

Completely integrable systems were heavily studied in the nineteenth century with important contributions, via striking examples, by Jacobi, Neumann, and Kovalevskaya. They fell into the background after Poincaré's proof that celestial dynamics was not integrable and the focus on ergodicity to explain statistical mechanics. With the discovery by Gardner, Greene, Kruskal, and Miura [145] that KdV has an infinity of conserved quantities, there was an explosion of interest in the subject that has continued for the past forty years. The Lax formalism [274], which we discuss in Section 6.6, has been a central element of most of the examples found since then.

Missing from our discussion is the existence of angle variables. Under the hypotheses of Theorem 6.2.5, one can prove that there is a neighborhood, N, of T so that $N \cong M \times \mathbb{T}^m$ with $M \subset \mathbb{R}^m$ a neighborhood of 0 and $\mathbb{T}^m$ the m torus, so that if $(y_1, \dots, y_m)$ are coordinates on M and $\theta_1, \dots, \theta_m$ ($\theta \in [0, 2\pi)$) coordinates in $\mathbb{T}^m$, then $\{y_j, y_k\} = 0$, $\{\theta_j, \theta_k\} = 0$, $\{y_j, \theta_k\} = \delta_{jk}$ and the f's are functions only of y's. The y's are called *action variables* and the θ's *angle variables*. Angle variables are important in the study of perturbations, including KAM theory. For angle variables in free Toda, see [307], and in periodic Toda, see [33, 34, 198].

6.3 QR FACTORIZATION

In this section, we discuss an elementary piece of linear algebra that we will need later. While we will be mainly interested in finite matrices, the semi-infinite case presents no difficulty. The decomposition we will discuss is not about linear transformations but about matrices, that is, bases matter, and we will be talking about explicit $n \times n$ matrices and semi-infinite matrices, that is, operators on $\mathbb{C}^n$ and on $\ell^2(\mathbb{Z}_+)$. As we have done for Jacobi matrices, we label such vectors v_j, $j = 1, \dots, n$ or $j = 1, \dots$. Matrices have the form (a_{ij}), $i, j = 1, \dots, n$ or $i, j \in \mathbb{Z}_+ \equiv \{1, 2, \dots\}$.

Definition. An *upper triangular matrix* is one with $a_{jk} = 0$ if $j > k$, that is, it consists of diagonal elements and potentially nonzero elements above the diagonal. $\mathcal{R}$ will denote the set of upper triangular matrices that are strictly positive on diagonal, that is,

$$a_{jj} > 0 \tag{6.3.1}$$

Notice that $\mathcal{R}$ is closed under products and, at least in the finite matrix case, if $A \in \mathcal{R}$, it is invertible and $A^{-1} \in \mathcal{R}$, that is, in the case of $n \times n$ matrices, $\mathcal{R}$ is a subgroup of $\mathbb{GL}(n, \mathbb{R})$ or $\mathbb{GL}(n, \mathbb{C})$. We will let $\mathcal{U}$ stand for the group of unitary matrices.

Theorem 6.3.1 (QR Decomposition). *Let A be a bounded matrix with bounded inverse. Then there exist unique $Q \in \mathcal{U}$ and $R \in \mathcal{R}$ so that*

$$A = QR \tag{6.3.2}$$

Moreover,

$$Q\delta_1 = \frac{A\delta_1}{\|A\delta_1\|} \tag{6.3.3}$$

Proof. We begin by noting that

$$\mathcal{U} \cap \mathcal{R} = \{\mathbf{1}\} \tag{6.3.4}$$

for if $A \in \mathcal{U} \cap \mathcal{R}$, then unitarity of A and the fact that the first column is of the form $(a_{11}, 0, 0, \dots)^t$ implies $|a_{11}| = 1$ and then, since $a_{11} > 0$, we have $a_{11} = 1$. Unitarity of A means column j ($j \geq 2$) is orthogonal to $(1, 0, 0, \dots)^t$, so $a_{1j} = 0$, that is, A has the form

$$A = \begin{pmatrix} 1 & 0 & 0 & \dots & \dots \\ 0 & & & & \\ 0 & & \tilde{A} & & \\ \vdots & & & & \end{pmatrix} \tag{6.3.5}$$

where $\tilde{A} \in \mathcal{U} \cap \mathcal{R}$ so, by an obvious induction, $A = \mathbf{1}$, proving (6.3.4).

(6.3.4) implies uniqueness, for if

$$Q_1 R_1 = Q_2 R_2 \tag{6.3.6}$$

and R_1 is invertible, then

$$Q_2^{-1} Q_1 = R_2 R_1^{-1} \tag{6.3.7}$$

lies in $\mathcal{U} \cap \mathcal{R}$, showing $Q_1 = Q_2$ and $R_1 = R_2$.

For existence, $\{A\delta_j\}_{j=1}^N$ (N finite or infinite) are linearly independent since A is invertible, so we can use Gram–Schmidt to define $\{e_j\}_{j=1}^N$ inductively so the e's are orthonormal and

$$A\delta_j = \sum_{k=1}^{j} r_{kj} e_k \tag{6.3.8}$$

with

$$r_{jj} > 0 \tag{6.3.9}$$

and so

$$e_1 = \frac{A\delta_1}{\|A\delta_1\|} \tag{6.3.10}$$

Now define a unitary matrix, Q, by

$$Q\delta_j = e_j \tag{6.3.11}$$

Since A is invertible, the $\{e_k\}_{k=1}^N$ are a basis, for if $\psi \perp \{e_k\}_{k=1}^N$, by (6.3.8), $\langle \psi, A\delta_j \rangle = 0$, $A^*\psi = 0$, so $\psi = 0$. Define a matrix, R, by

$$(R)_{kj} = \begin{cases} r_{kj} & k \le j \\ 0 & R > j \end{cases} \tag{6.3.12}$$

Clearly, $R \in \mathcal{R}$ and

$$\begin{aligned} Q^{-1}(A\delta_j) &= Q^{-1}\left(\sum_{k=1}^{j} r_{kj} e_k\right) \\ &= \left(\sum_{k=1}^{j} r_{kj}\delta_k\right) = R\delta_j \end{aligned} \tag{6.3.13}$$

that is, $Q^{-1}A = R$, proving (6.3.2). (6.3.3) is (6.3.10) plus (6.3.11). □

Theorem 6.3.2. *If the matrix A of Theorem 6.3.1 is real, then Q is orthogonal (i.e., real and unitary) and R is real.*

Proof. Clearly, the vectors $\{e_j\}_{j=1}^N$ are real, so R is real and Q is orthogonal. □

One reason the QR factorization is important in numerical analysis is the QR algorithm. Given A invertible, write A by (6.3.2) and let

$$A_1 = RQ = Q^{-1}AQ \tag{6.3.14}$$

Since A and A_1 are unitarily equivalent, they have the same eigenvalues. The map $A \to A_1$ is called one step in the *QR algorithm*. It can be iterated, that is, one writes $A_1 = Q_1R_1$ and then $A_2 = R_1Q_1$. The remarkable fact is that in very many cases, one can prove that A_n converges to an upper triangular, sometimes even diagonal, matrix. Indeed, we will prove in Section 6.7:

Theorem 6.3.3. *Let J be an $n \times n$ Jacobi matrix that is strictly positive. Let $J^{(1)}, J^{(2)}, \dots$ be the results of repeatedly applying the QR algorithm to J. Then each $J^{(n)}$ is a Jacobi matrix and $J^{(n)}$ converges exponentially fast to a diagonal matrix whose eigenvalues are those of J.*

Thus, a practical method for effective numerical approximation of eigenvalues of a positive symmetric matrix is to first use Gram–Schmidt to find a basis in which the matrix is triangular and then to use this iterated QR algorithm.

Remarks and Historical Notes. The QR algorithm as a numerically convergent method goes back at least to Francis [140]. See the Notes to Section 6.7 for a discussion of works connecting it to the Toda lattice. Typical of generalizations of Theorem 6.3.3 are the following: If A is a finite symmetric matrix which is strictly positive with distinct eigenvalues, then the QR algorithm converges to a diagonal matrix; see, for example, Olver [329].

The QR algorithm is connected to the Iwasawa decomposition of the semisimple Lie group $\mathbb{GL}(n, \mathbb{C})$; see, for example, Helgason [195].

6.4 POISSON BRACKETS OF OPS, EIGENVALUES, AND WEIGHTS

As noted in Section 6.1, the free Toda lattice can be interpreted as the Hamiltonian equations of motion associated to the Poisson brackets (aka PBs)

$$\{b_k, a_k\} = -\tfrac{1}{4}\, a_k \qquad k = 1, \dots, N-1 \tag{6.4.1}$$

$$\{b_k, a_{k-1}\} = \tfrac{1}{4}\, a_{k-1} \qquad k = 2, \dots, N \tag{6.4.2}$$

(all other brackets are zero) and

$$H = 2\sum_{j=1}^{N} b_j^2 + 4\sum_{j=1}^{N-1} a_j^2 \tag{6.4.3}$$

Our goal here is to show, first of all, that there is a suitable symplectic structure in which these are the PBs and then to compute the PBs for the orthogonal polynomials generated by $\{a_j, b_j\}_{j=1}^{N-1} \cup \{b_N\}$ and for the eigenvalues and spectral weights of the associated Jacobi matrix, $J_{N;F}$. In particular, we will prove the Toda flow leaves the spectrum of $J_{N;F}$ invariant.

Of course, $\{(a_k)_{k=1}^{N-1}, (b_k)_{k=1}^{N}\}$ is an odd-dimensional space, so it cannot be a symplectic manifold. Related to this is that (6.4.1)/(6.4.2) imply that

$$\left\{ \sum_{j=1}^{N} b_j, a_k \right\} = 0 \qquad k = 1, \dots, N-1 \tag{6.4.4}$$

so that $\{\cdot\,,\,\cdot\}$ is not nondegenerate. In fact, we need to fix β and look at the submanifold where

$$\sum_{j=1}^{N} b_j = \beta \tag{6.4.5}$$

We will use $\mathbb{R}_+^{2N+1}$ for the set of $(a_j)_{j=1}^{N-1}$, $(b_j)_{j=1}^{N}$ with $a_j > 0$.

Proposition 6.4.1. *Fix β real. Let $X_\beta \subset \mathbb{R}_+^{2N+1}$ be the set of $(a,b) \in \mathbb{R}_+^{2N+1}$ obeying* (6.4.5). *Use $a_1, \dots, a_{N-1}, b_1, \dots, b_{N-1}$ for coordinates on X_β, and define*

$$\Omega = 4 \sum_{1\le \ell\le k\le N-1} (a_\ell^{-1} da_\ell) \wedge db_k \tag{6.4.6}$$

Then Ω is a closed nondegenerate two-form, which induces the PBs (6.4.1)/(6.4.2) *(where b_N is the function $\beta - \sum_{j=1}^{N-1} b_j$).*

Proof. If $\{x_j\}_{j=1}^{2L}$ are local coordinates and

$$\Omega = \sum_{j,k=1}^{2L} \Omega_{jk}(x)\, dx_j \wedge dx_k \tag{6.4.7}$$

where $\Omega_{jk} = -\Omega_{kj}$ is a symplectic form on a $2L$-dimensional manifold, then the Hamiltonian vector field H_{x_j} is given by

$$H_{x_j} = \sum_{k=1}^{2L} \alpha_k^{(j)} \frac{\partial}{\partial x_k} \tag{6.4.8}$$

where

$$\Omega\biggl(H_{x_j}, \frac{\partial}{\partial x_k}\biggr) = \delta_{jk} \Rightarrow \sum_{m=1}^{2L} \Omega_{mk}\alpha_m^{(j)} = \delta_{kj} \tag{6.4.9}$$

So

$$\alpha_m^{(j)} = (\Omega^{-1})_{jm} = -(\Omega^{-1})_{mj} \tag{6.4.10}$$

implies

$$\{x_j, x_k\} = H_{x_j}(x_k) = \alpha_k^{(j)} = -(\Omega^{-1})_{kj} \tag{6.4.11}$$

Thus, the coefficients of Ω are given by the negative of the inverse of the matrix of PBs. Next, we suppose the $2L$ coordinates are written in two blocks of L, say, $p_1, \dots, p_L$ and $q_1, \dots, q_L$. If (in our case, $U = 0$, but later we will want $U \neq 0$)

$$\Omega = \left(\begin{array}{c|c} U & W \\ \hline -W^t & 0 \end{array}\right) \tag{6.4.12}$$

then Ω is invertible if and only if W is, and

$$\Omega^{-1} = \left(\begin{array}{c|c} 0 & -(W^t)^{-1} \\ \hline W^{-1} & W^{-1}U(W^t)^{-1} \end{array}\right) \tag{6.4.13}$$

The PBs, (6.4.1)/(6.4.2), we are interested in have this form if $(p_1, \dots, p_{N-1}, q_1, \dots, q_{N-1}) = (b_1, \dots, b_{N-1}, a_1, \dots, a_{N-1})$ with $U \equiv 0$ and

$$(W^t)^{-1} = \begin{pmatrix} \frac{1}{4}a_1 & -\frac{1}{4}a_1 & 0 & \dots \\ 0 & \frac{1}{4}a_2 & -\frac{1}{4}a_2 & \dots \\ \dots & \dots & \dots & \dots \end{pmatrix} \tag{6.4.14}$$

$$= D(\mathbf{1} - M) \tag{6.4.15}$$

with D the diagonal matrix $D_{kj} = \delta_{kj}\frac{1}{4}a_k$ and M the standard nilpotent $\left(\begin{smallmatrix} 0 & 1 & 0 & \dots \\ 0 & 0 & 1 & \dots \\ \dots & \dots & \dots & \dots \end{smallmatrix}\right)$. Thus,

$$W = ([D(\mathbf{1} - M)]^t)^{-1} = D^{-1}(\mathbf{1} + M^t + (M^t)^2 + \dots + (M^t)^{N-1})$$

$$= \begin{pmatrix} 4a_1^{-1} & 0 & 0 \\ 4a_1^{-1} & 4a_2^{-1} & 0 \\ \dots & \dots & \dots \end{pmatrix} \tag{6.4.16}$$

Thus, Ω, given by (6.4.6), is nondegenerate and leads to the required PB. That Ω is closed follows from $a_\ell^{-1} da_\ell = (d\log(a_\ell))$ and $d^2 = 0$. $\square$

We are mainly interested in the PBs of the functions of $J_{N;F}$ given by the eigenvalues and weights. These are complicated functions of the a's and b's, so the key will be to use some intermediate functions, namely, the coefficients of the monic OPs.

Theorem 6.4.2. *Given Jacobi parameters in* $\mathbb{R}_+^{2N-1}$, *let* $(P_n, Q_n)_{n=1}^N$ *be the monic OPs and second kind monic (i.e., given by* (3.2.12) *with* p_n *replaced by* P_n*) polynomials. Then, for* $n = 1, \dots, N$,

$$\{P_n(x), P_n(y)\} = \{P_{n-1}(x), P_{n-1}(y)\} = 0 \tag{6.4.17}$$

$$2\{P_n(x), P_{n-1}(y)\} = \left[\frac{P_n(x)P_{n-1}(y) - P_n(y)P_{n-1}(x)}{x - y}\right] - P_{n-1}(x)P_{n-1}(y) \tag{6.4.18}$$

$$\{P_n(x), P_n(y)\} = \{Q_n(x), Q_n(y)\} = 0 \tag{6.4.19}$$

$$2\{P_n(x), Q_n(y)\} = -\left[\frac{P_n(x)Q_n(y) - P_n(y)Q_n(x)}{x - y}\right] + Q_n(x)Q_n(y) \tag{6.4.20}$$

Remarks. 1. While one tends to think of $P_n(x)$ as a function of a single variable, x, in fact, it is a function of x and also $(a_j)_{j=1}^{n-1} \cup (b_j)_{j=1}^n$. In (6.4.17), x and y are fixed and we mean PBs in the a's and b's! In essence, these PBs encode information on the PBs of the coefficients of P_n and P_{n-1}.

2. (6.4.18)/(6.4.20) hold for $x \neq y$ and then in a limit.

3. If S, T are polynomials, $[S(x)T(y) - S(y)T(x)]/(x - y)$ is called their *Bezoutian*.

Proof. We begin by proving (6.4.17)/(6.4.18) by induction. With $P_0(x) = 1$, $P_{-1}(x) = 0$, we see they hold when $n = 0$. So suppose they hold for n and let us check $\{P_{n+1}(x), P_j(y)\}$ for $j = n+1, n$.

As preliminary, we claim

$$\{a_n^2, P_n(x)\} = -\tfrac{1}{2}\, a_n^2 P_{n-1}(x) \tag{6.4.21}$$

for

$$P_n(x) = (x - b_n)P_{n-1}(x) - a_{n-1}P_{n-2}(x) \tag{6.4.22}$$

P_{n-1}, P_{n-2} are only functions of $\{b_j\}_{j=1}^{n-1}$ and $\{a_j\}_{j=1}^{n-2}$, and a_n only fails to Poisson commute with b_n and b_{n+1}, so

$$\{a_n^2, P_n(x)\} = \{a_n^2, -b_n\}P_{n-1}(x) \tag{6.4.23}$$

$$= -\tfrac{1}{2}\, a_n^2 P_{n-1}(x) \tag{6.4.24}$$

by (6.4.1).

Now use

$$P_{n+1}(x) = (x - b_{n+1})P_n(x) - a_n^2 P_{n-1}(x) \tag{6.4.25}$$

Since $P_n(x)$ is a function of $\{b_j\}_{j=1}^{n}$ and $\{a_j\}_{j=1}^{n-1}$, b_{n+1} Poisson commutes with $P_n(x)$, so by the induction hypothesis,

$$\{(x - b_{n+1})P_n(x), (y - b_{n+1})P_n(y)\} = 0 \tag{6.4.26}$$

Similarly, P_{n-1} is a function of $\{b_j\}_{j=1}^{n-1}$ and $\{a_j\}_{j=1}^{n-2}$, so a_n Poisson commutes with $P_{n-1}(x)$, and by induction,

$$\{a_n^2 P_{n-1}(x), a_n^2 P_{n-1}(y)\} = 0 \tag{6.4.27}$$

Thus, by (6.4.22),

$$-\{P_{n+1}(x), P_{n+1}(y)\} = \{(x - b_{n+1})P_n(x), a_n^2 P_{n-1}(y)\} - (x \leftrightarrow y) \tag{6.4.28}$$

Now $\{XY, WZ\} = XW\{Y, Z\} + XZ\{Y, W\} + YW\{X, Z\} + YZ\{X, W\}$, so (6.4.28) is a sum of four terms with one zero since $\{b_{n+1}, P_{n-1}(y)\} = 0$. The other terms are

$$\begin{aligned} 2t_1 &= 2(x - b_{n+1})a_n^2\{P_n(x), P_{n-1}(y)\} - (x \leftrightarrow y) \\ &= (x - y)a_n^2\{P_n(x), P_{n-1}(y)\} && (6.4.29) \\ &= (x - y)a_n^2\left[\frac{P_n(x)P_{n-1}(y) - P_n(y)P_{n-1}(x)}{x - y} - P_{n-1}(x)P_{n-1}(y)\right] && (6.4.30) \end{aligned}$$

where (6.4.29) comes from the symmetry of $\{P_n(x), P_{n-1}(y)\}$ under $x \leftrightarrow y$.

Next,

$$\begin{aligned} 2t_2 &= 2(x - b_{n+1})\{P_n(x), a_n^2\}P_{n-1}(y) - (x \leftrightarrow y) \\ &= a_n^2(x - y)P_{n-1}(x)P_{n-1}(y) && (6.4.31) \end{aligned}$$

by (6.4.21).

Finally,

$$\begin{aligned} 2t_3 &= -2\{b_{n+1}, a_n^2\}(P_n(x)P_{n-1}(y) - (x \leftrightarrow y)) \\ &= -a_n^2(P_n(x)P_{n-1}(y) - P_n(y)P_{n-1}(x)) \end{aligned} \tag{6.4.32}$$

which shows that $t_1 + t_2 + t_3 = 0$, proving (6.4.18) for $n + 1$.

Similarly, by (6.4.25), using (6.4.26),

$$\begin{aligned} \{P_{n+1}(x), P_n(y)\} &= \{-a_n^2 P_{n-1}(x), P_n(y)\} \\ &= -a_n^2\{P_{n-1}(x), P_n(y)\} - P_{n-1}(x)\{a_n^2, P_n(y)\} \end{aligned} \tag{6.4.33}$$

The first term is evaluated by induction and the second by (6.4.22)—it cancels one part of the first term, giving

$$\begin{aligned} 2\{P_{n+1}(x), P_n(y)\} &= a_n^2\left[\frac{P_n(y)P_{n-1}(x) - P_n(x)P_{n-1}(y)}{y-x}\right] \\ &= -\left[\frac{P_n(y)(P_{n+1}(x) - (x-b_{n+1})P_n(x)) - (x \leftrightarrow y)}{y-x}\right] \end{aligned} \tag{6.4.34}$$

$$= \frac{P_{n+1}(x)P_n(y) - P_{n+1}(y)P_n(x)}{x-y} - P_n(x)P_n(y)$$

proving (6.4.18) for $n + 1$. We get (6.4.34) using (6.4.25). This proves (6.4.17)/(6.4.18) inductively.

To get (6.4.19)/(6.4.20), we note that $P_n(x)$ is $\det(x - J_{n;F})$, while $P_{n-1}(x)$ is the minor of (n, n) while $Q_n(x)$ is the minor of $(1, 1)$. There is an obvious symmetry that says

$$P_n(x; a_1, \dots, a_{n-1}, b_1, \dots, b_n) = P_n(x; a_{n-1}, \dots, a_1, b_n, \dots, b_1) \tag{6.4.35}$$

$$Q_n(x; a_1, \dots, a_{n-1}, b_1, \dots, b_n) = P_{n-1}(x; a_{n-1}, \dots, a_1, b_n, \dots, b_1) \tag{6.4.36}$$

(P_{n-1} is not dependent on the last a and b nor Q_n on the first a and b).

Notice that under this reordering of variables, the signs of $\{a_j, b_k\}$ flip, so all Poisson brackets change signs. Thus, (6.4.19)/(6.4.20) follow by the change of variables. □

Finally, we turn to the spectral representation of $J_{n;F}$. We can write

$$\langle \delta_1, (J_{N;F} - z)^{-1}\delta_1 \rangle = \sum_{j=1}^{N} \frac{\rho_j}{\lambda_j - z} \tag{6.4.37}$$

where the ρ's are not independent since

$$\sum_{j=1}^{N} \rho_j = 1 \qquad \rho_j > 0 \tag{6.4.38}$$

The λ's are ordered by

$$\lambda_1 < \lambda_2 < \cdots < \lambda_N \tag{6.4.39}$$

and if (6.4.5) holds, then

$$\sum_{j=1}^{N} \lambda_j = \mathrm{Tr}(J_{N;F}) = \beta \tag{6.4.40}$$

Our analysis in Section 1.3 shows $\{(a,b) \in \mathbb{R}^{2N+1}\}\big|_{(6.4.5)\text{ holds}}$ is mapped bijectively to the set of (λ, ρ) obeying (6.4.38)/(6.4.39). The λ's and ρ's are functions on $\mathbb{R}_+^{2N+1}$ so we can ask about their PBs. But also $(a_1, \dots, a_{N-1}, b_1, \dots, b_{N-1}) \to (\rho_1, \dots, \rho_{N-1}, \lambda_1, \dots, \lambda_{N-1})$ is a coordinate change and we can ask about its Jacobian. We will be able to answer both!

Theorem 6.4.3. *We have*

$$\{\lambda_j, \lambda_k\} = 0 \qquad 1 \le j, k \le N \tag{6.4.41}$$

$$\{\lambda_j, \rho_k\} = \tfrac{1}{2}\,[\delta_{jk}\rho_j - \rho_j\rho_k] \qquad 1 \le j, k \le N \tag{6.4.42}$$

Remarks. 1. We will discuss $\{\rho_j, \rho_k\}$ in the Notes—we do not use it, so we do not make the calculation explicitly.

2. Notice the right side of (6.4.42) sums to zero, summed over either j or k, consistent with $\sum \rho_k$ and $\sum x_j$ being constant.

Proof. We have that

$$P_N(x) = \prod_{j=1}^{N} (x - \lambda_j) \tag{6.4.43}$$

and, by Cramer's rule, that

$$\frac{Q_N(x)}{P_N(x)} = \sum_{j=1}^{N} \frac{\rho_j}{x - \lambda_j} \tag{6.4.44}$$

so that

$$Q_N(x) = \sum_{j=1}^{N} \rho_j \prod_{k \neq j} (x - \lambda_k) \tag{6.4.45}$$

It follows that

$$\{P_N(x), P_N(y)\}|_{x=\lambda_j,\, y=\lambda_k} = \{\lambda_j, \lambda_k\} \prod_{\ell \neq j} (\lambda_j - \lambda_\ell) \prod_{m \neq k} (\lambda_k - \lambda_m) \tag{6.4.46}$$

where one makes the substitution only after evaluating all the PBs. Thus, $\{P_N(x), P_N(y)\} = 0$ implies (6.4.41).

Once one knows that (6.4.41) holds, we get from (6.4.45) that

$$\{P_N(x), Q_N(y)\}|_{x=\lambda_j,\, y=\lambda_k} = -\{\lambda_j, \rho_k\} \prod_{\ell \neq j} (\lambda_j - \lambda_\ell) \prod_{m \neq k} (\lambda_k - \lambda_\ell) \tag{6.4.47}$$

If $j \neq k$, since $P_N(x)|_{x=\lambda_j} = 0$, the first term in (6.4.20) vanishes. As for the second,

$$Q_N(x)Q_N(y)|_{x=\lambda_j,\, y=\lambda_k} = \rho_j\rho_k \prod_{\ell\neq j}(\lambda_j - \lambda_\ell) \prod_{m\neq k}(\lambda_k - \lambda_m) \tag{6.4.48}$$

which leads to (6.4.42) for $j \neq k$.

If $j = k$, the first term in (6.4.20) is $\frac{0}{0}$ if one sets $x = \lambda_j$, $y = \lambda_j$ directly. One needs to set $y = \lambda_j$ and take the limit as $x \to \lambda_j$. We get

$$-\rho_j \prod_{\ell\neq j}(\lambda_j - \lambda_\ell) \prod_{m\neq k}(\lambda_k - \lambda_m) \tag{6.4.49}$$

yielding the extra term in (6.4.42) when $j = k$. □

That the λ_j's all Poisson commute and there are $N - 1$ independent ones proves the promised complete integrability. However, as we will see, this is not on a compact set (we will need to pass to the periodic case to get compactness). As a final result from these calculations, we note:

Theorem 6.4.4. (a) *In terms of the coordinates* $\{\lambda_j, \rho_j\}_{j=1}^{N-1}$, *the symplectic form* Ω *has the form*

$$2\sum_{j=1}^{N} d\lambda_j \wedge \rho_j^{-1}\, d\rho_j + \sum_{i,j} U_{ij}\, d\lambda_i \wedge d\lambda_j \tag{6.4.50}$$

for suitable U_{ij}.

(b) *The Jacobian of the change of variables from* $\{a_j, b_j\}_{j=1}^{N-1}$ *to* $\{\lambda_j, \rho_j\}_{j=1}^{N-1}$ *is*

$$\left|\frac{\partial(a,b)}{\partial(x,p)}\right| = \frac{2^{-(N-1)}\prod_{j=1}^{N-1} a_j}{\prod_{j=1}^{N} \rho_j} \tag{6.4.51}$$

Remark. In (6.4.50), $d\lambda_N$ is shorthand for $-\sum_{j=1}^{N-1} d\lambda_j$ and $d\rho_N$ for $-\sum_{j=1}^{N-1} d\rho_j$. Alternatively, while not coordinate one-forms, $d\lambda_N$ and $d\rho_N$ are legitimate one-forms.

Proof. (a) We make a change of variables that helps "explain" the form of (6.4.42). Let

$$y_k = \log\left[\frac{\rho_k}{\rho_N}\right] \qquad k = 1, \dots, N-1 \tag{6.4.52}$$

Since $\sum_1^N \rho_k = 1$, one can invert this via

$$\rho_j = \frac{e^{y_j}}{[1 + \sum_{\ell=1}^{N-1} e^{y_\ell}]} \qquad j = 1, \dots, N-1 \tag{6.4.53}$$

$$\rho_N = \frac{1}{[1 + \sum_{\ell=1}^{N-1} e^{y_\ell}]} \tag{6.4.54}$$

mapping $\{(\rho_1, \dots, \rho_n) \mid \sum_{j=1}^{N} \rho_j = 1;\ \rho_j > 0\}$ to $\mathbb{R}^{N-1}$.

Moreover,

$$\begin{aligned}\{\lambda_j, y_k\} &= \{\lambda_j, \log(\rho_k)\} - \{\lambda_j, \log(\rho_N)\} \\ &= (\tfrac{1}{2}\delta_{jk} - \rho_j) - (-\rho_j) = \tfrac{1}{2}\delta_{jk}\end{aligned} \tag{6.4.55}$$

It follows by (6.4.13) (with $W^{-1} = \frac{1}{2}\mathbf{1}$) that

$$\Omega = 2\sum_{j=1}^{N-1} d\lambda_j \wedge dy_j + \sum_{i,j} U_{ij}\, d\lambda_j \wedge d\lambda_j \tag{6.4.56}$$

Since $dy_j = \rho_k^{-1}\, d\rho_k - \rho_N^{-1}\, d\rho_N$ and $\sum_{j=1}^{N-1} d\lambda_j = -d\lambda_N$, this implies (6.4.50).

(b) By (6.4.50), the $(N-1)$-fold wedge product

$$\Omega \wedge \cdots \wedge \Omega = 2^{N-1}(N-1)! \sum_{j=1}^{N} \bigwedge_{k\neq j} \rho_k^{-1}(d\lambda_k \wedge d\rho_k) \tag{6.4.57}$$

$$= 2^{N-1}(N-1)! \sum_{j=1}^{N} \left(\prod_{k\neq j} \rho_k^{-1}\right) d\lambda_1 \wedge d\rho_1 \wedge \cdots \wedge d\lambda_{N-1} \wedge d\rho_{N-1} \tag{6.4.58}$$

$$= 2^{N-1}(N-1)! \left(\prod_{k=1}^{N} \rho_k^{-1}\right) d\lambda_1 \wedge d\rho_2 \wedge \cdots \wedge d\lambda_{N-1} \wedge d\rho_{N-1} \tag{6.4.59}$$

since

$$\sum_{j=1} \prod_{k\neq j} \rho_k^{-1} = \prod_{k=1}^{N} \rho_k^{-1} \left(\sum_{j=1}^{N} \rho_j\right) = \prod_{k=1}^{N} \rho_k^{-1} \tag{6.4.60}$$

On the other hand, by (6.4.6),

$$\Omega \wedge \cdots \wedge \Omega = 4^{N-1}(N-1)! \left(\prod_{j=1}^{N-1} a_j^{-1}\right) da_1 \wedge db_1 \wedge \cdots \wedge da_{N-1} \wedge db_{N-1} \tag{6.4.61}$$

Since $|\frac{\partial(a,b)}{\partial(\lambda,\rho)}|$ is the absolute value of the coefficients in $da_1 \wedge \cdots \wedge db_{N-1} = C\, d\lambda_1 \wedge \cdots \wedge d\rho_{N-1}$, this proves (6.4.51). □

Remarks and Historical Notes. The analog of $\{\lambda_j, \lambda_k\} = 0$ for periodic boundary conditions under the symplectic form given by (6.4.1)/(6.4.2) is due to Flaschka [134], and (6.4.41)/(6.4.42) for $J_{n;F}$ is implicit in Moser [307]. The proof via the OP brackets in Theorem 6.4.2 is due to Cantero–Simon [71] with closely related calculations (brackets of m-functions) in Faybusovich–Gekhtman [131] and Gekhtman–Nenciu [147].

The Jacobian relation (6.4.51) appeared first in Dumitriu–Edelman [116] via an indirect calculation. A more direct proof using forms is in Forrester–Rains [139]. The idea we follow of getting it, via Poisson brackets, is due to Deift (unpublished).

For analogs of the results of this section for OPUC, see [71, 147] and Killip–Nenciu [223, 224].

The ρ_j are not angle variables conjugate to the λ's because we do *not* have $\{\rho_j, \rho_k\} = 0$. Instead, via $\{Q_N(x), Q_N(y)\} = 0$ and the computed $\{\rho_j, \lambda_k\}$, one obtains (see, e.g., [71])

$$\{\rho_j, \rho_k\} = \frac{\rho_j \rho_k}{\lambda_j - \lambda_k} - \sum_{m \neq j} \frac{\rho_j \rho_k \rho_m}{\lambda_j - \lambda_m} + \sum_{m \neq k} \frac{\rho_j \rho_k \rho_m}{\lambda_k - \lambda_m}. \tag{6.4.62}$$

6.5 SPECTRAL SOLUTION AND ASYMPTOTICS OF THE TODA FLOW

In this section, we will begin by noting that the PBs of Theorem 6.4.3 allow an immediate solution of the Toda equations of motion in (λ_j, ρ_j) coordinates, and then we will use OPs to deduce the asymptotics of the original Jacobi parameters. Here the main result will be

Theorem 6.5.1. *Let $a_n(t), b_n(t)$ solve the Toda equations in Flaschka form* (6.1.14) /(6.1.15) *and let $\lambda_1 < \lambda_2 < \cdots < \lambda_N$ be the eigenvalues of the Jacobi matrix $J(0)$ with parameters $a_n(0), b_n(0)$. Then*

$$\lim_{|t|\to\infty} a_j(t) = 0 \tag{6.5.1}$$

$$\lim_{t\to\infty} b_j(t) = \lambda_{N+1-j} \tag{6.5.2}$$

$$\lim_{t\to-\infty} b_j(t) = \lambda_j \tag{6.5.3}$$

Indeed, if

$$c = \min_{j=1,\dots,N-1} \lambda_{j+1} - \lambda_j \tag{6.5.4}$$

then $a_j(t)$ is $O(e^{-c|t|})$ and $|b_j(t) - b_j(\pm\infty)|$ is $O(e^{-2c|t|})$.

We begin with the equations for λ_j and ρ_j.

Theorem 6.5.2. *If $J(t)$ solves* (6.1.14)/(6.1.15) *for the Jacobi parameters, then its eigenvalues $\{\lambda_j(t)\}_{j=1}^N$ and weights $\{\rho_j(t)\}_{j=1}^N$ obey*

$$\lambda_j(t) = \lambda_j(0) \tag{6.5.5}$$

$$\rho_j(t) = \frac{e^{2t\lambda_j}\rho_j(0)}{\sum_{k=1}^N e^{2t\lambda_k}\rho_k(0)} \tag{6.5.6}$$

Proof. By (6.2.14), Hamilton's equation of motion for the values of an arbitrary smooth function on the manifold takes the form

$$\frac{df}{dt} = \{H, f\} \tag{6.5.7}$$

Shift to the coordinates $\lambda_1, \dots, \lambda_{N-1}, y_1, \dots, y_{N-1}$ of (6.4.52). Since, by (6.4.3),

$$H = 2\,\mathrm{Tr}(J^2) \tag{6.5.8}$$

$$= 2\sum_{j=1}^{N} \lambda_j^2 \tag{6.5.9}$$

where $\lambda_N = \beta - \sum_{j=1}^{N-1} \lambda_j$ and $\{\lambda_j, \lambda_k\} = 0$, $\{\lambda_j, y_k\} = \frac{1}{2}\delta_{jk}$ (by (6.4.55)), we first have

$$\{\lambda_N, y_j\} = -\tfrac{1}{2} \tag{6.5.10}$$

and thus,

$$\frac{d}{dt}\lambda_j = 0 \qquad \frac{d}{dt}y_j = 4(\lambda_j - \lambda_N)\tfrac{1}{2} \tag{6.5.11}$$

so

$$\lambda_j(t) = \lambda_j(0) \qquad y_j(t) = y_j(0) + 2t(\lambda_j(0) - \lambda_N(0)) \tag{6.5.12}$$

Plugging this into (6.4.53)/(6.4.54) leads to (6.5.6). □

Henceforth, since $\lambda_j(0)$ is constant, we write it as λ_j. We focus now on $t \to +\infty$; the analysis as $t \to -\infty$ is similar except for the ordering of λ's: the largest, $-\lambda_j$, that is, $-\lambda_1$, is relevant in place of λ_N. $d\rho$ has the form

$$d\rho_t = \frac{\sum_{j=1}^{N} e^{2t\lambda_j}\rho_j(0)\delta_{\lambda_j}}{\sum_{j=1}^{N} e^{2t\lambda_j}\rho_j(0)} \tag{6.5.13}$$

For t very large, the overwhelming largest weight is at λ_N since $e^{t\lambda_N} \gg e^{t\lambda_{N-1}} \gg \dots$, next largest at $\lambda_{N-1}, \dots$. Since $P_j(y, t)$, the orthogonal polynomial for $d\rho_t$, minimizes $\int |P(y)|^2\, d\rho_t(y)$ among all monic polynomials of degree j, the best strategy is to put its zeros very near the j largest weights. Proving this is the key to going from Theorem 6.5.2 to Theorem 6.5.1:

Proposition 6.5.3. *Let $P_j(\,\cdot\,, t)$ be the OPs for $d\rho_t$ and let $\{x_k^{(j)}(t)\}_{k=1}^{j}$ be the zeros ordered by*

$$x_1^{(j)} > x_2^{(j)} > \dots > x_j^{(j)} \tag{6.5.14}$$

Let c be given by (6.5.4). *Then*

(i)

$$\|P_j(\,\cdot\,, t)\|_{L^2(d\rho_t)}^2 \le e^{-2t(\lambda_N - \lambda_{N-j})}(\lambda_N - \lambda_1)^{2j}\rho_N(0)^{-1}$$
$$j = 1, 2, \dots, N-1 \tag{6.5.15}$$

(ii) *For t large, we have*

$$|x_k^{(j)}(t) - \lambda_{N+1-k}| \le \frac{c}{2} \tag{6.5.16}$$

(iii) *For large t,*

$$\|P_j(\,\cdot\,, t)\|_{L^2(d\rho_t)}^2 \ge \frac{(\frac{c}{2})^{2j} e^{-2t(\lambda_N - \lambda_{N-j})}\rho_{N-j}(0)}{\rho_N(0)} \tag{6.5.17}$$

(iv)

$$|x_k^{(j)}(t)-\lambda_{N+1-k}| \le \left(\frac{c}{2}\right)^{-(j-1)} (\lambda_N-\lambda_1)^j \rho_k(0)^{-1/2} e^{-t(\lambda_{N+1-k}-\lambda_{N-j})} \tag{6.5.18}$$

Proof. (i) P_j minimizes the norm, so picking $Q(x) = \prod_{k=1}^{j}(x - \lambda_{N+1-k})$, we see for $q = N - j, N - j - 1, \dots, 1$,

$$|Q(\lambda_q)| \le (\lambda_N - \lambda_1)^j$$

and thus (since Q vanishes at $\lambda_N, \dots, \lambda_{N-j+1}$),

$$\begin{aligned} \|P_j(\,\cdot\,,t)\|^2_{L^2(d\rho_t)} \le \|Q\|^2_{L^2(d\rho_t)} &\le (\lambda_N - \lambda_1)^{2j} \sum_{q=1}^{N-j} \rho_q(t) \\ &\le (\lambda_N - \lambda_1)^{2j} e^{-2t\lambda_N} \rho_N(0)^{-1} e^{2t\lambda_{N-j}} \end{aligned} \tag{6.5.19}$$

since

$$\sum_{j=1}^{N} e^{2t\lambda_j} \rho_j(0) \ge \rho_N(0) e^{2t\lambda_N} \tag{6.5.20}$$

and

$$\sum_{q=1}^{N-j} e^{2t\lambda_q} \rho_q(0) \le e^{2t\lambda_{N-j}} \sum_{q=1}^{N} \rho_q(0) \le e^{2t\lambda_{N-j}} \tag{6.5.21}$$

(ii) Let m_q be given by

$$m_q = \min_{k=1,\dots,j} |x_k^{(j)} - \lambda_q| \tag{6.5.22}$$

If we show for $q = N, N-1, \dots, N+1-j$ that $m_q \to 0$, we see that there is at least one zero in each $(\lambda_{N+1-k} - c/2, \lambda_{N+1-k} + c/2)$. This gives j disjoint intervals, so there has to be exactly one of j zeros in each interval, proving (6.5.16).

At λ_q,

$$P_j(\lambda_q) \ge (m_q)^j \tag{6.5.23}$$

so

$$\|P_j\|^2 \ge (m_q)^{2j} e^{-2t(\lambda_N - \lambda_q)} \rho_q(0) \rho_N(0)^{-1} \tag{6.5.24}$$

By (6.5.15),

$$(m_q)^{2j} \le (\lambda_N - \lambda_1)^{2j} e^{-2t(\lambda_q - \lambda_{N-j})} \rho_q(0)^{-1} \tag{6.5.25}$$

goes to zero since $\lambda_q > \lambda_{N-j}$.

(iii) By (6.5.16), we have no zero within $c/2$ of λ_{N-j}, so

$$m_{N-j} \ge \frac{c}{2} \tag{6.5.26}$$

and thus, (6.5.24) implies (6.5.17).

(iv) Since only one zero is within $c/2$ of any λ_q, we can improve (6.5.23) to

$$P_j(\lambda_q) \geq m_q \left(\frac{c}{2}\right)^{j-1} \tag{6.5.27}$$

and so improve (6.5.25) to

$$m_q^2 \left(\frac{c}{2}\right)^{2j-2} \leq (\lambda_N - \lambda_1)^{2j} e^{-2t(\lambda_q - \lambda_{N-j})} \rho_q(0)^{-1} \tag{6.5.28}$$

which implies (6.5.18). □

Proof of Theorem 6.5.1. Since

$$a_j = \frac{\|P_j\|}{\|P_{j-1}\|} \tag{6.5.29}$$

the upper and lower bounds in (6.5.15)/(6.5.17) imply

$$C_j e^{-t(\lambda_{N+1-j} - \lambda_{N-j})} \leq a_j \leq D_j e^{-t(\lambda_{N+1-j} - \lambda_{N-j})} \tag{6.5.30}$$

for nonzero constants C_j, D_j, which shows $a_j \to 0$ and is $O(e^{-ct})$ as $t \to \infty$.

Next, note that the x^{j-1} term in the recursion relation

$$\prod_{k=1}^{j} (x - x_k^{(j)}) = (x - b_j) \prod_{k=1}^{j-1} (x - x_k^{(j-1)}) + O(x^{j-2}) \tag{6.5.31}$$

implies that

$$b_j = \sum_{k=1}^{j} x_k^{(j)} - \sum_{k=1}^{j-1} x_k^{(j-1)} \tag{6.5.32}$$

By (6.5.18), as $t \to \infty$,

$$x_k^{(j)} \to \lambda_{N+1-k} \tag{6.5.33}$$

so (6.5.32) implies

$$b_j \to \lambda_{N+1-j} \tag{6.5.34}$$

proving (6.5.3).

Once we have $a_n = O(e^{-c|t|})$, the differential equation (6.1.15) implies that $db_n/dt = O(e^{-2c|t|})$, which implies $|b_n(t) - b_n(\pm\infty)| = O(e^{-2c|t|})$. □

In addition to the Toda flow, one can define generalized Toda flows. Pick a C^1 function, G, on $\mathbb{R}$. We want a Hamiltonian

$$H = \sum_{j=1}^{N} G(\lambda_j) = \mathrm{Tr}(G(J)) \tag{6.5.35}$$

which is a polynomial of degree ℓ in $\{a_j\}_{j=1}^{N-1}$ and $\{b_j\}_{j=1}^{N}$ if G is a polynomial of degree ℓ. Then Theorem 6.5.2 immediately extends:

Theorem 6.5.4. *If $J(t)$ solves the Hamiltonian equations of the generalized Toda flow associated to G, then the measure $d\rho_t$ associated to $J(t)$ is*

$$d\rho_t = \frac{\sum_{j=1}^{N} e^{\frac{1}{2}tG'(\lambda_j)}\rho_j(0)\delta_{\lambda_j}}{\sum_{j=1}^{N} e^{\frac{1}{2}tG'(\lambda_j)}\rho_j(0)} \tag{6.5.36}$$

The proof of Theorem 6.5.1 also extends so long as there is a nondegeneracy condition

$$G'(\lambda_j) \neq G'(\lambda_k) \tag{6.5.37}$$

for all $j \neq k$ (see the Notes for a discussion of the degenerate case).

Theorem 6.5.5. *If $J(t)$ solves the Hamiltonian equations of the generalized Toda flow associated to G and if the eigenvalues of $J(0)$ obey the nondegeneracy condition* (6.5.37)*, define $g_1, \dots, g_N$ to be the reordering of $\{G'(\lambda_j)\}_{j=1}^{N}$ obeying*

$$g_1 < g_2 < \cdots < g_N \tag{6.5.38}$$

Let $\tilde{\lambda}_j$ be defined by $g_j = G'(\tilde{\lambda}_j)$. Then with

$$c = \tfrac{1}{4} \min_{j=1,\dots,N-1} g_{j+1} - g_j \tag{6.5.39}$$

we have

$$a_j(t) = O(e^{-c|t|}) \tag{6.5.40}$$

$$b_j(t) \to \tilde{\lambda}_{N+1-j} \qquad \text{as } t \to \infty \tag{6.5.41}$$

$$b_j(t) \to \tilde{\lambda}_j \qquad \text{as } t \to -\infty \tag{6.5.42}$$

and

$$|b_j(t) - b_j(\pm\infty)| = O(e^{-2c|t|}) \tag{6.5.43}$$

Remarks and Historical Notes. Theorem 6.5.1 is due to Moser [307]. Our proof using zeros of OPs is taken from Simon [406]. The analysis of generalized Toda flows using critical points of the spectrum goes back to work of Deift–Li–Tomei [104] and Deift–Nanda–Tomei [105].

If one transfers the asymptotics back to p, q variables, one sees as $t \to \pm\infty$, $q_{j+1} - q_j \to \infty$, so as $t \to +\infty$, $p_1 < p_2 < \cdots < p_N$, and as $t \to -\infty$, $p_N < \cdots < p_2 < p_1$. The remarkable fact is that if $\lim_{t\to\pm\infty} p_j(t) = p_j^{\pm}$, then $p_j^+ = p_{N+1-j}^-$. Even though there are multiple particles, there is no momentum transfer, or put more precisely, the transfer preserves the set of values. This is, of course, a sign of the multiple conservation laws.

If G' has a degeneracy, the analysis is a little more complicated. One needs to break up λ_j's into groups with equal $G'(\lambda_j)$. If $\lambda_{j_1}, \dots, \lambda_{j_\ell}$ is a group with equal $G'(\lambda_j)$ and there are m λ_j's with larger $G'(\lambda)$, then as $t \to \infty$, the block

$\{b_j\}_{m+1}^{m+\ell} \cup \{a_j\}_{m+1}^{m+\ell-1}$ approaches the Jacobi parameters for the measure $\sum_{k=1}^{\ell} \rho_{t=0} \times (\{\lambda_{j_k}\})\delta_{\lambda_k}$. This is discussed in Simon [406].

Asymptotics of semi-infinite Toda and generalized Toda flows can be found in Deift–Li–Tomei [104], Deift–Nanda–Tomei [105], Golinskii [176], and Simon [406].

6.6 LAX PAIRS

One of the striking elements of our analysis so far is that the flow is isospectral—that is, the eigenvalues of $J_{N;F}$ are preserved. Lax [274] found a general class of isospectral flows, and we will note here that the Toda flow fits into this framework.

Let B be a smooth function from the set of all selfadjoint matrices to the set of skew adjoint (skew adjoint means $B^* = -B$) matrices or B might be defined on some manifold of selfadjoint matrices. Suppose $C(t)$ is a curve of selfadjoint matrices (lying in the domain of definition of B) solving the differential equation

$$\frac{dC(t)}{dt} = [B(C(t)), C(t)] \tag{6.6.1}$$

B and C are called a *Lax pair* and (6.6.1) is called a *Lax differential equation*. Note that there is no assumption that there is an underlying symplectic form or Hamiltonian. We will not address when (6.6.1) has a solution but suppose a solution is given. The point is

Theorem 6.6.1. *If $C(t)$ obeys* (6.6.1), *then there is a unitary family, $W(t)$, smooth in t, so $W(0) = \mathbf{1}$ and*

$$C(t) = W(t)^{-1}C(0)W(t) \tag{6.6.2}$$

In particular, for any m, $\mathrm{Tr}(C(t)^m)$ *is constant (equivalently, $C(0) \to C(t)$ is isospectral).*

Remark. $W(t)$ are called the *Lax unitaries*.

Proof. Consider the differential equation

$$\frac{dW(t)}{dt} = -W(t)B(C(t)) \tag{6.6.3}$$

Here $B(C(t))$ is given, so this is a linear differential equation with smooth coefficients, which has unique global solutions by standard techniques. We consider the solution with $W(0) = \mathbf{1}$.

Since $B^* = -B$, we see

$$\frac{dW^*(t)}{dt} = B(C(t))W^*(t) \tag{6.6.4}$$

so

$$\frac{d}{dt}WW^* = 0 \tag{6.6.5}$$

Since $WW^*|_{t=0} = \mathbf{1}$, we see $W(t)$ is unitary.

Let

$$D(t) = W(t)C(t)W^*(t) \tag{6.6.6}$$

Then, by (6.6.3)/(6.6.4),

$$\begin{aligned} \frac{d}{dt} D(t) &= W(t) \frac{dC}{dt} W^*(t) - W(t)[B(C(t)), C(t)]W^*(t) \\ &= 0 \end{aligned} \tag{6.6.7}$$

by (6.6.1). Thus, $D(t) = C(0)$, proving (6.6.2).

Since W is unitary, (6.6.2) says the flow preserves the spectrum. □

Given any Jacobi matrix, J, define $B(J)$ by

$$J = \begin{pmatrix} b_1 & a_1 & 0 & \cdots \\ a_1 & b_2 & a_2 & \cdots \\ 0 & a_2 & b_3 & \cdots \\ \vdots & \vdots & \vdots & \ddots \end{pmatrix} \qquad B(J) = \begin{pmatrix} 0 & a_1 & 0 & \cdots \\ -a_1 & 0 & a_2 & \cdots \\ 0 & -a_2 & 0 & \cdots \\ \vdots & \vdots & \vdots & \ddots \end{pmatrix} \tag{6.6.8}$$

Then, by a simple calculation (a special case of a more elaborate calculation in the next section!),

$$[B(J), J] = \begin{pmatrix} 2a_1^2 & a_1(b_2 - b_1) & 0 & \cdots \\ a_1(b_2 - b_1) & 2a_2^2 - 2a_1^2 & a_2(b_3 - b_2) & \cdots \\ 0 & a_2(b_3 - b_2) & & \cdots \\ \vdots & \vdots & \vdots & \ddots \end{pmatrix} \tag{6.6.9}$$

Thus, (6.6.1), the Lax differential equation, is just the Toda differential equation (6.1.14)/(6.1.15)! This provides a second proof of the isospectral nature of the Toda flow. In the next section, we will recover the action on the weights, ρ_j.

Remarks and Historical Notes. Lax pairs were introduced by Lax [274] in a seminal paper in the context of the KdV equation "explaining" the isospectral solution found by Gardner, Greene, Kruskal, and Miura [145]. Its relevance to Toda flows was discovered by Flaschka [134] and Moser [308].

6.7 THE SYMES–DEIFT–LI–TOMEI INTEGRATION: CALCULATION OF THE LAX UNITARIES

In this section, we will answer several hanging questions:

(a) From the Lax pair point of view, how can one get the dynamics of the weights?

(b) Can one find the Lax unitaries?

(c) What are the Lax pairs for the generalized Toda flow?

The key will be the QR factorization, and as a bonus, we will have a new understanding of the QR algorithm and its convergence to diagonal form under iteration.

Lemma 6.7.1. *If J is a Jacobi matrix and R_1, R_2 are in $\mathcal{R}$, then*

$$(R_1 J R_2)_{jk} = 0 \qquad \text{if } j > k+1 \tag{6.7.1}$$

Remark. A matrix H with $H_{jk} = 0$ if $j > k+1$ is called *upper Hessenberg*; it has only one possible nonzero diagonal below the main, precisely one diagonal below.

Proof. In

$$(R_1 J R_2)_{jk} = \sum_{\ell,m} (R_1)_{j\ell} J_{\ell m} (R_2)_{mk} \tag{6.7.2}$$

if the summand is nonzero, then $\ell \geq j$, $m \geq \ell - 1$, $k \geq m$, so $k \geq j-1$; that is, if $k < j-1$, then all terms in the sum are zero. □

Theorem 6.7.2 ([425, 426, 104]). *Let $J \equiv J(0)$ be a given Jacobi matrix. Let G be a C^1 function on $\mathbb{R}$. Let*

$$\exp(\tfrac{1}{4} t G'(J)) = Q_t R_t \tag{6.7.3}$$

be the QR factorization, and define

$$J(t) = Q_t^{-1} J Q_t \tag{6.7.4}$$

Then

(i) *$J(t)$ is a Jacobi matrix.*
(ii) *If $\lambda_j(t)$, $\rho_j(t)$ are eigenvalues and weights of $d\rho_t$, the spectral measure of $J(t)$, then*

$$\lambda_j(t) = \lambda_j(0) \equiv \lambda_j \tag{6.7.5}$$

$$\rho_j(t) = \frac{e^{\frac{1}{2} t G'(\lambda_j)} \rho_j(0)}{\sum_{k=1}^N e^{\frac{1}{2} t G'(\lambda_k)} \rho_k(0)} \tag{6.7.6}$$

Remarks. 1. Of course, given Theorem 6.5.4, this implies $J(t)$ solves the generalized Toda flow—we will say more about this later.

2. Q_t is, of course, the Lax unitary. For the Toda case, where $\frac{1}{4}G'(J) = J$, this calculation of the Lax unitary is due to Symes [425, 426]; for generalized Toda, it was discovered by Deift–Li–Tomei [104].

Proof. (i) Let

$$A_t \equiv \exp(\tfrac{1}{4} t G'(J)) \tag{6.7.7}$$

which is obviously invertible with $A_t J = J A_t$. Thus,

$$\begin{aligned} J(t) &= R_t R_t^{-1} Q_t^{-1} J Q_t R_t R_t^{-1} \\ &= R_t A_t^{-1} J A_t R_t^{-1} \\ &= R_t J R_t^{-1} \end{aligned} \tag{6.7.8}$$

R_t and R_t^{-1} lie in $\mathcal{R}$, so by the lemma,

$$J(t)_{jk} = 0 \tag{6.7.9}$$

if $j-k>1$. But $J(t)$ is symmetric, so (6.7.8) always holds if $j-k<-1$. Thus, $J(t)$ is tridiagonal and symmetric. Since (6.7.8) and (6.7.2) imply that

$$J(t)_{j\,j-1} = (R_t)_{jj} J_{j\,j-1} (R_t^{-1})_{j-1\,j-1} \tag{6.7.10}$$

we see $J(t)$ is positive off-diagonal, that is, $J(t)$ is a Jacobi matrix, as claimed.

(ii) Let $e_1, \dots, e_N$ be normalized eigenvectors of J, that is,

$$Je_j = \lambda_j e_j \tag{6.7.11}$$

Then (6.7.5) is immediate from (6.7.4) and unitarity of Q_t. Moreover,

$$\rho_j(t) = |\langle e_j, Q_t\delta_0\rangle|^2 \tag{6.7.12}$$

$$= \frac{|\langle e_j, A_t\delta_0\rangle|^2}{\|A_t\delta_0\|^2} \tag{6.7.13}$$

by (6.3.3). But $A_t e_j = e^{\frac{1}{4}tG'(\lambda_j)}e_j$ and $|\langle e_j, \delta_0\rangle|^2 = \rho_j(0)$, which leads to (6.7.6). □

This allows us to compute the other element of the Lax pair for general G and redo the calculation (6.6.9). Given any real symmetric matrix, A, define $\pi(A)$ anti-symmetric by

$$\pi(A)_{jk} = \begin{cases} A_{jk} & j<k \\ -A_{jk} & j>k \\ 0 & j=k \end{cases} \tag{6.7.14}$$

Theorem 6.7.3. *Let G be a C^1 function and let J, Q_t, R_t, $J(t)$ be given by Theorem 6.7.2. Then*

$$\frac{dQ_t}{dt} = -Q_t B_t \tag{6.7.15}$$

where

$$B_t = \pi(\tfrac{1}{4}\, G'(J(t))) \tag{6.7.16}$$

and

$$\frac{d}{dt} J(t) = [B_t, J(t)] \tag{6.7.17}$$

Remarks. 1. Since $\mathrm{Tr}(G(J))$ is the Hamiltonian that generates the generalized flow in question, (6.7.17) can be rewritten

$$\{\mathrm{Tr}(G(J)), J\} = [\tfrac{1}{4}\, G'(J), J] \tag{6.7.18}$$

intended in a matrix element equality, that is, for all k, ℓ,

$$\{\mathrm{Tr}(G(J)), J_{k\ell}\} = ([\tfrac{1}{4}\, G'(J), J])_{k\ell} \tag{6.7.19}$$

2. For the Toda lattice, $G'(\lambda) = 4\lambda$ and

$$B_t = \pi(J(t)) \tag{6.7.20}$$

is given by (6.6.8).

Proof. By (6.7.3),

$$\tfrac{1}{4}\, G'(J) = \left[\frac{d}{dt}\,(Q_t R_t)\right](Q_t R_t)^{-1} \tag{6.7.21}$$

$$= \frac{dQ_t}{dt}\, Q_t^{-1} + Q_t\, \frac{dR_t}{dt}\, R_t^{-1} Q_t^{-1} \tag{6.7.22}$$

Multiply by Q_t^{-1} on the left and Q_t on the right, using (6.7.4) to get

$$\tfrac{1}{4}\, G'(J(t)) = Q_t^{-1}\, \frac{dQ_t}{dt} + \frac{dR_t}{dt}\, R_t^{-1} \tag{6.7.23}$$

Since Q_t is orthogonal, $Q_t^{-1}\dot{Q}_t$ is skew symmetric and so 0 on diagonal. Since R_t is upper triangular, $\frac{dR_t}{dt}$ is also, and thus, so is $\frac{dR_t}{dt} R_t^{-1}$. Thus, below diagonal, that is, for $j > k$,

$$\left[\frac{1}{4}\, G'(J(t))\right]_{jk} = \left(Q_t^{-1}\, \frac{dQ_t}{dt}\right)_{jk}$$

so if B_t is given by (6.7.15), we have (6.7.16) for $j > k$ and then, by antisymmetry, for all j, k.

(6.7.17) is immediate from (6.7.15) and (6.7.4). □

We have thus found a Lax pair representation for the dynamics of generalized Toda flows.

Suppose now $J > 0$ and G is a C^1 function with $G'(x) = 4\log(x)$ for $x \geq \inf(\sigma(J))$. Then for $n = 1, 2, \ldots,$

$$A_{t=n} = J^n$$

In particular,

$$A_{t=1} = J = Q_{t=1} R_{t=1}$$

and

$$J_{t=1} = Q_t^{-1}(Q_t R_t) Q_t = R_t Q_t$$

The time one flow is just the QR algorithm! The time n flow can be seen to be the n times iterated flow. We thus have

Proof of Theorem 6.3.3. As just noted, the n times iterated QR algorithm is a generalized Toda flow at time n. Now use Theorem 6.5.5. □

Remarks and Historical Notes. Symes [425, 426] discovered the approach of this section for Toda lattices. Deift–Li–Tomei [104] developed this for generalized flows, emphasizing that it could be applied to convergence of the QR algorithm. See Simon [406] for the OPUC analog.

We already mentioned that the QR algorithm had a group theoretic interpretation in terms of the Iwasawa decomposition. There is a group theoretic version of the Toda chain found by Kostant [241] that, in part, motivated Symes [425]. The symplectic manifolds in this point of view are coadjoint orbits. This has spawned a huge literature, of which we mention [130, 148, 224, 242, 280, 290, 326, 328, 367, 383]. Some of them extend this approach to CMV matrices and Schur flows (see Section 6.9).

6.8 COMPLETE INTEGRABILITY OF PERIODIC TODA FLOW AND ISOSPECTRAL TORI

We turn now to the periodic case of greatest interest to us in this book. The main result in this case is

Theorem 6.8.1. *Let $\Delta(x, \{a_n, b_n\}_{n=1}^N)$ be the discriminant for period N Jacobi matrices. Then under the Poisson brackets* (6.1.12) *and* (6.1.13), *one has*

$$\{\Delta(x), \Delta(y)\} = 0 \tag{6.8.1}$$

If $\lambda_j(\theta)$ are the eigenvalues of $J(\theta)$ (given by (5.3.8)*), with $0 < \theta, \theta' < \pi$,*

$$\{\lambda_j(\theta), \lambda_k(\theta')\} = 0 \tag{6.8.2}$$

and this remains true for $\theta = 0$ or π at points of nondegeneracy. If ℓ, ℓ' are arbitrary,

$$\{\mathrm{Tr}(J(\theta)^\ell), \mathrm{Tr}(J(\theta')^{\ell'}) = 0 \tag{6.8.3}$$

Partial Proof. We discuss various proofs of one of these below—here we want to discuss their relation. For $\theta \neq 0, \pi$, $\lambda_j(\theta)$ are the roots of $\Delta(x) - 2\cos\theta = 0$ and are simple roots. Thus, (6.8.2) follows from $\{\Delta(x) - 2\cos\theta, \Delta(y) - 2\cos\theta'\} = 0$ by setting x to $\lambda_j(\theta)$ and y to $\lambda_k(\theta')$, as in the proof of Theorem 6.4.3 In this calculation, one uses $(a_1 \dots a_N)^{-1}$ is in the Poisson center since, unlike P, Δ is not monic.

One can go backwards using the fact that $\Delta(x) - 2\cos\theta = (a_1 \dots a_N)^{-1} \prod_{j=1}^N (x - \lambda_j(\theta))$.

Clearly, (6.8.2) implies (6.8.3). To get from (6.8.3), one uses a piece of combinatorics (see the Notes) that shows if $\lambda_1, \dots, \lambda_N$ are distinct, then $s_q = \sum_{i_j < \dots < i_q} \lambda_{i_j} \dots \lambda_{i_q}$ is a polynomial in $\{\sum_{j=1}^N \lambda_j^\ell \mid \ell = 1, \dots, q\}$ and that the s_q are coefficients of $(a_1 \dots a_N)(\Delta - 2\cos\theta)$. □

Remarks. 1. Note that to go from (6.8.2) to (6.8.1), we only used (6.8.2) for $\theta = \theta'$.

2. Since $J(\theta)$ is regular at $\theta = 0, \pi$ and continuous in θ, (6.8.3) holds also for $\theta, \theta' = 0$ or π.

We provide two proofs of the rest of Theorem 6.8.1 and mention a third in the Notes. The first depends on

Lemma 6.8.2 (Feynman–Hellman Theorem)**.** *Let $T(\beta)$ be an analytic family of matrices for β in $\mathbb{C}$ near 0, selfadjoint for β real. Suppose φ is a normalized eigenvector for $T(0)$ with eigenvalue $E(0)$, which is simple. Then for β small, $T(\beta)$ has exactly one eigenvalue, $E(\beta)$, near $E(0)$, and*

$$\left.\frac{dE(\beta)}{d\beta}\right|_{\beta=0} = \left\langle \varphi, \left.\frac{dT}{d\beta}\right|_{\beta=0} \varphi \right\rangle \tag{6.8.4}$$

Proof. By eigenvalue perturbation theory (see Kato [215] and Reed–Simon [364]), for β small, there is a unique eigenvalue $E(\beta)$ near $E(0)$ and one can choose an analytic eigenvector, $\varphi(\beta)$, so that $\varphi(0) = \varphi$ and

$$\beta \text{ real} \Rightarrow \|\varphi(\beta)\|^2 = 1 \tag{6.8.5}$$

Thus, by analyticity, $\langle \varphi(\bar{\beta}), \varphi(\beta)\rangle = 1$, and so for β real,

$$\left\langle \frac{d\varphi}{d\beta}, \varphi \right\rangle + \left\langle \varphi, \frac{d\beta}{d\varphi} \right\rangle = 0 \tag{6.8.6}$$

Since $T(\beta)\varphi(\beta) = E(\beta)\varphi(\beta)$ and $T^*(\beta) = T(\beta)$ for β real, (6.8.1) implies for β real that

$$\left\langle \frac{d\varphi}{d\beta}, T\varphi \right\rangle + \left\langle \varphi, T\,\frac{d\varphi}{d\beta} \right\rangle = 0 \tag{6.8.7}$$

On the other hand, since $\|\varphi(\beta)\|^2 = 1$ and $T(\beta)\varphi(\beta) = E(\beta)\varphi(\beta)$, for β real,

$$E(\beta) = \langle \varphi(\beta), T(\beta)\varphi(\beta)\rangle \tag{6.8.8}$$

so

$$\left.\frac{dE(\beta)}{d\beta}\right|_{\beta=0} = \left\langle \frac{d\varphi}{d\beta}, T\varphi \right\rangle + \left\langle \varphi, \frac{dT}{d\beta}\,\varphi \right\rangle + \left\langle \varphi, T\,\frac{d\varphi}{d\beta} \right\rangle \tag{6.8.9}$$

(6.8.7) and (6.8.9) imply (6.8.4). □

Proof of Theorem 6.8.1 (Flaschka [135]). Fix $\theta \neq 0, \pi$ so eigenvalues are simple. If λ_j is an eigenvalue of $J(\theta)$ with eigenvector $\varphi^{(j)}$ and components $\varphi_k^{(j)}$, then by the Feynman–Hellman theorem,

$$\frac{\partial \lambda_j}{\partial b_k} = |\varphi_k^{(j)}|^2 \qquad \frac{\partial \lambda_j}{\partial a_k} = 2\,\mathrm{Re}(\varphi_k^{(j)}\bar{\varphi}_{k+1}^{(j)}) \tag{6.8.10}$$

where

$$\varphi_{N+1}^{(j)} = e^{i\theta}\varphi_1^{(j)} \tag{6.8.11}$$

This is because

$$\begin{aligned} \left(\frac{\partial J}{\partial b_k}\right)_{\ell m} &= \delta_{k\ell}\delta_{\ell m} \\ \left(\frac{\partial J}{\partial a_k}\right)_{\ell m} &= (1 - \delta_{kN})[\delta_{k\ell}\delta_{m\,k+1} + \delta_{k+1\,\ell}\delta_{m\,k}] \\ &\qquad + \delta_{kN}[e^{i\theta}\delta_{k1}\delta_{\ell N} + e^{-i\theta}\delta_{kN}\delta_{\ell 1}] \end{aligned} \tag{6.8.12}$$

By the chain rule for PBs (see (6.2.21)),

$$\{f, g\} = \sum_{k,\ell}\{b_k, a_\ell\}\left[\frac{\partial f}{\partial b_k}\frac{\partial g}{\partial a_\ell} - \frac{\partial f}{\partial a_\ell}\frac{\partial g}{\partial b_k}\right] \tag{6.8.13}$$

so

$$\{\lambda_j, \lambda_\ell\} = \tfrac{1}{2}\sum_{k=1}^{N} a_{k-1}\big[|\varphi_k^{(j)}|^2 \operatorname{Re}(\varphi_k^{(\ell)}\bar{\varphi}_{k-1}^{(\ell)}) - |\varphi_k^{(\ell)}|^2 \operatorname{Re}(\varphi_k^{(j)}\bar{\varphi}_{k-1}^{(j)})\big]$$

$$-\tfrac{1}{2}\sum_{k=1}^{N} a_k\big[|\varphi_k^{(j)}| \operatorname{Re}(\varphi_k^{(\ell)}\bar{\varphi}_{k+1}^{(\ell)}) - |\varphi_k^{(\ell)}| \operatorname{Re}(\varphi_k^{(j)}\bar{\varphi}_{k+1}^{(j)})\big] \tag{6.8.14}$$

In the first term, for $k = 1$, $a_0 = a_N$, and

$$\varphi_0^{(j)} = e^{-i\theta}\varphi_N^{(j)} \tag{6.8.15}$$

so that $\operatorname{Re}(\varphi_1^{(\ell)}\bar{\varphi}_0^{(\ell)}) = \operatorname{Re}(\varphi_{N+1}^{(\ell)}\bar{\varphi}_N^{(\ell)})$. With this shift, we can combine terms to get

$$\{\lambda_j, \lambda_\ell\} = \tfrac{1}{2}\sum_{k=1}^{N} \operatorname{Re}(\bar{\varphi}_k^{(j)}\bar{\varphi}_k^{(\ell)}(W_k + W_{k-1})) \tag{6.8.16}$$

where W_k is the Wronskian

$$W_k = a_k(\varphi_{k+1}^{(j)}\varphi_k^{(\ell)} - \varphi_{k+1}^{(\ell)}\varphi_k^{(j)}) \tag{6.8.17}$$

Since φ obeys the equation

$$\lambda_j\varphi_k^{(j)} = a_k\varphi_{k+1}^{(j)} + b_k\varphi_k^{(j)} + a_{k-1}\varphi_{k-1}^{(j)} \tag{6.8.18}$$

multiplying this by $\varphi_k^{(\ell)}$ and subtracting the same equation with $j \leftrightarrow \ell$ yields

$$(\lambda_j - \lambda_\ell)\varphi_j^{(j)}\varphi_k^{(\ell)} = W_k - W_{k-1} \tag{6.8.19}$$

so (6.8.16) becomes

$$\{\lambda_j, \lambda_\ell\} = \tfrac{1}{2}\sum_{k=1}^{N} \frac{1}{\lambda_j - \lambda_\ell} \operatorname{Re}((\overline{W}_k - \overline{W}_{k-1})(W_k + W_{k+1}))$$

$$= \tfrac{1}{2}\sum_{k=1}^{N} \frac{1}{\lambda_j - \lambda_\ell}[|W_k|^2 - |W_{k-1}|^2] \tag{6.8.20}$$

$$= 0 \tag{6.8.21}$$

where (6.8.20) comes from $\operatorname{Re}((\bar{A}+\bar{B})(A-B)) = \operatorname{Re}(|A|^2 - |B|^2 + \bar{A}B - A\bar{B}) = |A|^2 - |B|^2$ and (6.8.21) comes from $W_N = e^{2i\theta}W_0$ given (6.8.11) and (6.8.15). $\square$

Our second proof is partly from Flaschka [134], who proved (6.8.3) for $\theta = 0$, $\ell = 2$ and arbitrary ℓ', and van Moerbeke [450], who developed the double period formalism we will use below to handle arbitrary ℓ, ℓ'. It depends on a Lax pair formalism.

Let us begin with case $\ell = 2$. (6.6.9) can be rewritten using the map π of (6.7.14) as

$$[\pi(J_{N;F}), J_{N;F}] = \{2\operatorname{Tr}(J_{N;F}^2), J_{N;F}\} \tag{6.8.22}$$

where, of course, if f is a scalar function and A a matrix-valued function, then $\{f, A\}$ is the matrix given by

$$\{f, A\}_{k\ell} = \{f, A_{k\ell}\} \tag{6.8.23}$$

From (6.8.22), we get

$$\begin{aligned}\{2\operatorname{Tr}(J_{N;F}^2), J_{N;F}^{\ell'}\} &= \sum_{j=0}^{\ell'-1} J_{N;F}^j \{2\operatorname{Tr}(J_{N;F}^2), J_{N;F}\} J_{N;F}^{\ell'-1-j} \\ &= \sum_{j=0}^{\ell'-1} J_{N;F}^j [\pi(J_{N;F}), J_{N;F}] J_{N;F}^{\ell'-1-j} \\ &= [\pi(J_{N;F}), J_{N;F}^{\ell'}]\end{aligned} \tag{6.8.24}$$

from which

$$\{\operatorname{Tr}(J_{N;F}^2), \operatorname{Tr}(J_{N;F}^{\ell'})\} = 0 \tag{6.8.25}$$

We want to note next that, at least if $N \geq 3$, then

$$[\pi(J_{N;P}), J_{N;P}] = \{2\operatorname{Tr}(J_{N;P}^2), J_{N;P}\} \tag{6.8.26}$$

This is true for the 21, 22, and 23 elements because they are the same as for the $J_{N;F}$ case since the different $1N$ and $N1$ matrix elements do not enter anywhere. But by the invariance of $J_{N;P}$ under cyclic change of variables, once we have it for the whole second row, we have it for all. From this and the argument leading to (6.8.25), we get

$$\{\operatorname{Tr}(J_{N;P}^2), \operatorname{Tr}(J_{N;P}^{\ell'})\} = 0 \tag{6.8.27}$$

The key fact in this calculation was $N \geq 3$. As ℓ increases, we need to increase the row number to be sure the terms in $J_{N;F}^{\ell+1}$ and $J_{N;P}^{\ell+1}$ are exactly the same; we can then get other rows by using periodicity. Basically, if we are at row r and $r > \ell$ and $N - r > \ell$, then we can only shift index by one and so not reach the rows (1 and N) where $J_{N;F}$ and $J_{N;P}$ differ. The optimal r is $N/2$ (or $(N-1)/2$ if N is odd), which limits ℓ to a lot less than the $\ell \leq N-1$ that we need. The clever solution of van Moerbeke [450] (following a similar KdV analysis in McKean–van Moerbeke [304]) is to look at the matrix with periodic boundary conditions but period $2N$.

So fix $\{a_j, b_j\}_{j=1}^N$ and let $J_{2N;2P}$ be the $2N \times 2N$ matrix, which is $J_{2N;P}$ for the Jacobi parameters of period $2N$ obtained by repeating $\{a_j, b_j\}_{j=1}^N$ two times (and is really of period N). Here is the key fact:

Theorem 6.8.3. *For $\ell = 2, 3, \dots, N$, we have*

$$[\tfrac{\ell}{4}\pi(J_{2N;P}^{\ell-1}), J_{2N;P}] = \tfrac{1}{2}\{\operatorname{Tr}(J_{2N;P}^{\ell}), J_{2N;P}\} \tag{6.8.28}$$

In particular, for $\ell, \ell' = 1, 2, \dots, N$, we have

$$\{\operatorname{Tr}(J_{2N;P}^{\ell}), \operatorname{Tr}(J_{2N;P}^{\ell'})\} = 0 \tag{6.8.29}$$

Proof. (6.7.18) for $G(x) = x^\ell$ says that for $2N \times 2N$ free boundary conditions with $2N - 1$ distinct a's and $2N$ distinct b's, we have

$$[\tfrac{\ell}{4}\pi(J_{2N;F}^{\ell-1}), J_{2N;F}] = \{\mathrm{Tr}(J_{2N;F}^{\ell}), J_{2N;F}\} \tag{6.8.30}$$

If we look at row N, since $\ell - 1 \le N - 1$, we have

$$[\tfrac{\ell}{4}\pi(J_{2N;P}^{\ell-1}), J_{2N;P}]_{Nq} = [\tfrac{\ell}{4}\pi(J_{2N;F}^{\ell-1}), J_{2N;F}]_{Nq}$$

since the differing matrix elements at $2N$, 1 cannot be linked in $\ell - 1$ steps to site N. Also, the difference of $\mathrm{Tr}(J_{2N;F}^{\ell})$ and $\mathrm{Tr}(J_{2N;P}^{\ell})$ Poisson commutes with $(J_{2N;P})_{Nq}$. However, in (6.8.30), we first compute $\{\ ,\ \}$ and then set $a_{N+j} = a_N$, $b_{N+j} = b_N$ ($j = 1, \dots, N$), while in (6.8.28), we have this equality and then compute $\{\ ,\ \}$. By the periodicity, this makes the PB in (6.8.28) twice as large, explaining the $\frac{1}{2}$.

For $\ell = 2, \dots, N$, the argument that led to (6.8.25) goes from (6.8.28) to (6.8.29). For $\ell = 1$, $\mathrm{Tr}(J_{2N;P}^{\ell}) = 2\sum_{j=1}^{N} b_j$ lies in the Poisson center, and so has zero PBs. □

At first sight, we have not helped the situation much because, while we now have N functions of the eigenvalues, we also have $2N$ rather than N eigenvalues! However, the new eigenvalues, which are the N periodic and the N antiperiodic eigenvalues, are not independent. This can be seen by the fact that $a_1 \dots a_N$ and the roots of $\Delta(x) - 2$ determine the roots of $\Delta(x) + 2$! So let $J_{N;A}$ refer to the antiperiodic boundary condition operator, that is, $J(\theta)$ of (5.3.8) with $\theta = \pi$. Notice, since $J_{2N;2P}$ has eigenvalues, which are the union of those of $J_{N;P}$ and $J_{N;A}$, that for any ℓ,

$$\mathrm{Tr}(J_{2N;2P}^{\ell}) = \mathrm{Tr}(J_{N;P}^{\ell}) + \mathrm{Tr}(J_{N;A}^{\ell}) \tag{6.8.31}$$

Proposition 6.8.4. *For $\ell = 1, 2, \dots, N - 1$, we have*

$$\mathrm{Tr}(J_{N;A}^{\ell}) = \mathrm{Tr}(J_{N;P}^{\ell}) \tag{6.8.32}$$

$$\mathrm{Tr}(J_{2N;2P}^{\ell}) = 2\,\mathrm{Tr}(J_{N;P}^{\ell}) \tag{6.8.33}$$

Moreover,

$$\mathrm{Tr}(J_{N;A}^{N}) = \mathrm{Tr}(J_{N;P}^{N}) - 4Na_1 \dots a_N \tag{6.8.34}$$

$$\mathrm{Tr}(J_{2N;2P}^{N}) = 2\,\mathrm{Tr}(J_{N;P}^{N}) - 4Na_1 \dots a_N \tag{6.8.35}$$

Proof. In computing $(J_{N;P}^{\ell})_{jj}$, we have products of ℓ matrix elements that change the index by either 0, ± 1 or $\pm N - 1$. Since $\ell \le N - 1$, the number that changes by $N - 1$ must equal the number that changes by $-(N - 1)$, if we are to return to j. Thus, a_N appears an even number of times and so is unchanged by the replacement $a_N \to -a_N$. This proves (6.8.32). This plus (6.8.31) implies (6.8.33).

Since $\Delta(J_{N;P}) = 2$ and $\Delta(J_{N;A}) = -2$, we have that

$$\mathrm{Tr}(\Delta(J_{N;P}) - \Delta(J_{N;A})) = 4N \tag{6.8.36}$$

Since

$$\Delta(x) = (a_1 \dots a_N)^{-1} x^N + \text{lower order}$$

by (6.8.32) in the left side of (6.8.36), all terms cancel but $(a_1 \dots a_n)\mathrm{Tr}(J_{N;P}^{N} - J_{N;A}^{N})$, so (6.8.36) implies (6.8.34). Then (6.8.34) implies (6.8.35). □

Second Proof of Theorem 6.8.1. For $\ell, \ell' \leq N-1$, (6.8.3) follows from (6.8.33) and (6.8.29). Since $a_1 \dots a_N$ is in the Poisson center, (6.8.33) yields (6.8.3) for $\ell = N$. Thus, $\lambda_1, \dots, \lambda_N$ Poisson commute, and so (6.8.2) holds for $\theta = 0$. □

Remarks and Historical Notes. Theorem 6.8.1 is due to Flaschka [134, 135]; the first proof we give is from [135] and the second from [134] and van Moerbeke [450].

There is a third proof in Cantero–Simon [71] who obtain it as a limit of free cases—the $\text{Tr}(J^\ell_{N;F})$ converge to moments of the density of states, not $\text{Tr}(J^\ell_{m;P}(\theta))$, so the argument is subtle. Basically, one can compute the θ-dependence using the fact that these are roots of $\Delta(\lambda) - 2\cos\theta$ and then relate moments of the density of states to integrals of $\text{Tr}(J^\ell_{N;P}(\theta))$ over $\frac{d\theta}{2\pi}$.

Our discussion of Lax pairs follows van Moerbeke [450].

For the fact that, given $\lambda_1, \dots, \lambda_N$, $s_q = \sum_{i_1<\dots<i_q} \lambda_{i_1} \dots \lambda_{i_q} \mid q = 1, \dots, N$ are polynomials in the $t_\ell = \sum_{i=1}^N \lambda_i^\ell \mid \ell = 1, \dots, N$, and vice versa, see a book on algebraic combinatorics such as Stanley [418] or see Cantero–Simon [71]. See also Lemma 6.9.3.

6.9 INDEPENDENCE OF TODA FLOWS AND TRACE GRADIENTS

In this section, we prove a technical result about the independence of the integrals of motion associated to the Toda flow. We define for $\ell = 1, \dots, N$,

$$t_\ell = \text{Tr}(J^\ell_{N;P}) \tag{6.9.1}$$

$$= \sum_{j=1}^N \lambda_j^\ell \tag{6.9.2}$$

and for $\ell = 1, \dots, N$,

$$s_\ell = \sum_{1\leq i_1<\dots<i_\ell\leq N} \lambda_{i_1} \dots \lambda_{i_N} \tag{6.9.3}$$

so that

$$\Delta(\lambda) = 2 + \left(\prod_{j=1}^N a_j\right)^{-1} [\lambda^N - s_1\lambda^{N-1} + s_2\lambda^{N-2} + \dots + (-1)^n s_n] \tag{6.9.4}$$

We will also let

$$\alpha = \prod_{j=1}^N a_j \tag{6.9.5}$$

Define $\{c_j\}_{j=1}^{N+1}$ by

$$\Delta(\lambda) = \sum_{j=1}^{N+1} c_j \lambda^{N+1-j} \tag{6.9.6}$$

so, by (6.9.4),

$$c_1 = \alpha^{-1}; \quad c_j = (-1)^{j+1}\alpha^{-1}s_{j-1}, \quad j = 2, \dots, N; \quad c_{N+1} = 2 + \alpha^{-1}s_N \tag{6.9.7}$$

Here is the main result of this section:

Theorem 6.9.1. *In $[(0,\infty)\times\mathbb{R}]^N$, let $(a_j^{(0)}, b_j^{(0)})_{j=1}^N$ be a point so that the corresponding periodic Jacobi matrix has all gaps open. Then*

(i) *$\{dt_\ell\}_{\ell=1}^N$ and $d\alpha$ are linearly independent.*
(ii) *$\{ds_\ell\}_{\ell=1}^N$ and $d\alpha$ are linearly independent.*
(iii) *$\{d\lambda_j\}_{j=1}^N$ and $d\alpha$ are linearly independent.*
(iv) *$\{dc_j\}_{j=1}^{N+1}$ are linearly independent.*

Remark. In $[(0,\infty)\times\mathbb{R}]^N$, when all gaps are open, the isospectral torus is a manifold of dimension $N-1$. This says certain sets of $N+1$ gradients orthogonal to the torus span the normal bundle.

This tells us that Theorem 6.2.5 is applicable. If $\{a_n^{(0)}, b_n^{(0)}\}_{n=1}^N$ corresponds to all gaps open, the connected component of its isospectral manifold is a torus. The theory of completely integrable systems does not explain why these manifolds are connected but sheds a lot of light on the fact that they are tori.

Clearly related to Theorem 6.9.1 is

Theorem 6.9.2 (van Moerbeke [450]). *Let $(a_j^{(0)}, b_j^{(0)})_{j=1}^N$ be a set of periodic Jacobi parameters with all gaps open. then the tangent vectors generated by the Hamiltonian flows of $\{\mathrm{Tr}(J_{N;P}^\ell)\}_{\ell=2}^N$ are linearly independent.*

We will prove that Theorem 6.9.2 $\Rightarrow$ Theorem 6.9.1 using the lemma below and then turn to the proof of Theorem 6.9.2.

Lemma 6.9.3. *We have for $\ell = 1, \dots, N$,*

$$ds_\ell = \sum_{j=1}^{\ell} f_{j\ell}\, dt_j \tag{6.9.8}$$

where $f_{\ell\ell} = (-1)^{\ell+1}/\ell$ and the f's are functions of the s's.

Proof. $s_1 = t_1$ so $ds_1 = dt_1$,

$$ds_2 = \sum_j \Big(\sum_{k\neq j} \lambda_k\Big)\, d\lambda_j = s_1\, dt_1 - \tfrac{1}{2}\, dt_2 \tag{6.9.9}$$

In general, by this same argument,

$$ds_\ell = s_{\ell-1}\, dt_1 - \tfrac{1}{2}\, s_{\ell-2}\, dt_2 + \tfrac{1}{3}\, s_{\ell-3}\, dt_3 + \cdots + \frac{(-1)^{\ell+1}}{\ell}\, dt_\ell \tag{6.9.10}$$

□

Proof of Theorem 6.9.2 $\Rightarrow$ Theorem 6.9.1. (i) Suppose

$$\gamma_0\, d\alpha + \sum_{j=1}^{N} \gamma_j\, dt_j = 0 \tag{6.9.11}$$

Then, since ω is nondegenerate on the space with α and t_1 fixed, we have

$$\sum_{j=2}^{N} \gamma_j X_{t_j} = 0 \tag{6.9.12}$$

so, by Theorem 6.9.2, $\gamma_2 = \gamma_3 = \cdots = \gamma_N = 0$. But α only depends on the a's and t_1 on the b's, so $d\alpha$ and dt_1 are independent. Thus, $\gamma_0 = \gamma_1 = 0$.

(ii) The matrix relating ds_ℓ to dt_ℓ is upper triangular and nonvanishing on diagonal, so it has nonzero determinant. Thus, $\{ds_\ell\}_{\ell=1}^N \cup \{d\alpha\}$ are independent if and only if $\{dt_\ell\}_{\ell=1}^N \cup \{d\alpha\}$ are.

(iii) Since dt_j is a linear combination of $\{d\lambda_j\}_{j=1}^N$ with a matrix whose determinant is $N!$ times a Vandermonde determinant, $\{\lambda_j\}$ distinct implies a nonzero determinant and that the $d\lambda_j$ are independent.

(iv) It is easy to see $\{dc_j\}_{j=1}^{N+1}$ and $\{ds_j\}_{j=1}^N \cup \{d\alpha\}$ are related by a nonsingular matrix (since $\alpha \neq 0$), and so (ii) implies (iv). □

Lemma 6.9.4. *Let A be a real symmetric matrix and B a real antisymmetric matrix. Suppose the eigenvalues of A are distinct and that $[B, A] = 0$. Then $B = 0$.*

Proof. Let $\lambda_1, \dots, \lambda_n$ be the eigenvalues of A and $\varphi_1, \dots, \varphi_n$ eigenvectors. By reality and simplicity of the eigenvalues, the φ_j are real, so by antisymmetry of B,

$$\langle \varphi_j, B\varphi_j \rangle = -\langle B\varphi_j, \varphi_j \rangle \tag{6.9.13}$$

Since A and B commute,

$$A(B\varphi_j) = B(A\varphi_j) = \lambda_j B\varphi_j \tag{6.9.14}$$

so by simplicity of the spectrum of A,

$$B\varphi_j = \mu_j \varphi_j \tag{6.9.15}$$

Since B and φ_j are real, μ_j is real. By (6.9.13), $\mu_j = -\mu_j$, that is, $\mu_j = 0$. Thus, $B\varphi_j = 0$ for all j, and so $B = 0$. □

Proof of Theorem 6.9.2. Suppose the flows are not linearly independent. Then for $\gamma_2, \dots, \gamma_N$ not all zero, $X_{\sum_{\ell=2}^N \gamma_\ell \operatorname{Tr}(J_{N;P}^\ell)} = 0$, and so for any function, f,

$$\left\{ \sum_{\ell=2}^{n} \gamma_\ell \operatorname{Tr}(J_{N;P}^\ell), f \right\} = 0 \tag{6.9.16}$$

Therefore, by (6.8.28),

$$\left[\sum_{\ell=2}^{N} \gamma_\ell \frac{\ell}{2} \pi(J_{2N;P}^{\ell-1}), J_{2N;P} \right] = 0 \tag{6.9.17}$$

Since all gaps are open, $J_{2N;P}$ has all simple eigenvalues, so by the lemma,

$$\sum_{\ell=2}^{N} \gamma_\ell \frac{\ell}{2} \pi(J_{2N;P}^{\ell-1}) = 0 \tag{6.9.18}$$

Notice that for $r = 1, 2, \dots, N-1$, we have

$$(J^q_{2N;P})_{N\,N+r} = 0 \qquad q = 1, \dots, r-1 \tag{6.9.19}$$

$$= a_1 a_2 \dots a_1 \qquad q = r \tag{6.9.20}$$

since q is not large enough to get a_N terms that jump to site 1.

Thus, the $(N, 2N-1)$ matrix element of (6.9.18) says $\gamma_N = 0$, after which the $(N, 2N-2)$ says $\gamma_{N-1} = 0, \dots$. Thus, all $\{\gamma_j\}_{j=2}^N$ are zero, proving independence. □

Remarks and Historical Notes. Theorem 6.9.2 and the proof we use is from van Moerbeke [450], based on a similar argument for KdV in [304]. Van Moerbeke also shows that if there are q closed gaps, $\{\mathrm{Tr}(J^\ell_{N;P})\}_{\ell=2}^{N-q}$ generate independent Hamiltonian flows.

6.10 FLOWS FOR OPUC

The connection of Toda flows to Jacobi matrices goes back forty years. Its OPUC analog only goes back to 2004–2005. The completely integrable flow, known as the defocusing Ablowitz–Ladik flow, was discovered in 1975 [1, 2, 3], but its connection to CMV matrices and OPUC is very recent. Nenciu–Simon [319] (reported on in Section 11.11 of [400]) found the symplectic form and proved Poisson commutation of Δ in the periodic case; Nenciu [316, 317] found the Lax pairs; and Killip–Nenciu [223] found the analog of the Symes–Deift–Li–Tomei unitaries; see the Notes for further references.

In this section, we want to summarize some of the differences with the Toda (OPRL) case:

(1) The natural "phase space" is complex for OPUC but Hamiltonians have to be real. Thus, if $\mathcal{C}$ is the CMV matrix, we have distinct flows generated by $\mathrm{Re}(\mathrm{Tr}(\mathcal{C}^\ell))$ and $\mathrm{Im}(\mathrm{Tr}(\mathcal{C}^\ell))$.

(2) The natural symplectic form in the OPUC case, unlike for Toda, is "decoupled," given by

$$\{\alpha_j, \alpha_k\} = \{\bar\alpha_j, \bar\alpha_k\} = 0 \qquad 1 \le j, k \le N \tag{6.10.1}$$

$$\{\alpha_j, \bar\alpha_k\} = -i\rho_j^2 \delta_{jk} \qquad 1 \le j, k \le N \tag{6.10.2}$$

The PBs are the "same" for free and periodic boundary conditions.

(3) The natural PBs for OPUC, unlike for periodic Toda, are nondegenerate. There is no Poisson center. The analog of $a_1 \dots a_N$ is $\rho_1 \dots \rho_N$—it generates a nontrivial flow that changes the phases of all the α's.

(4) The argument we gave, due to Flaschka [134], as the first proof of Theorem 6.8.1 does not extend. The dependence of the CMV matrix on the α's is more complicated than the dependence of J on the a's and b's ($\partial\mathcal{C}/\partial\alpha$ is rank two, not rank one) and the argument does not extend.

(5) The other element of the Lax pair is complex antisymmetric, and so it does not vanish on diagonal but only is pure imaginary there.

(6) Both because the CMV matrix is not tridiagonal and because Lemma 6.9.4 does not hold for complex antisymmetric B's, the argument used to prove Theorems 6.9.1 and 6.9.2 does not extend and it is an open question if these results are true in the CMV/OPUC case. The weaker result that for a dense open subset of $\{\alpha_j\}_{j=0}^{N-1}$, $\mathbb{C}^N$, the trace flows span the tangent space of the isospectral manifold at $\{\alpha_j\}_{j=0}^{N-1}$ is proven in Section 11.10 of [400].

Remarks and Historical Notes. The two simplest CMV flows are: the defocusing AL flow, generated by $2\,\mathrm{Re}(\mathrm{Tr}(\mathcal{C}))$, has

$$\dot{\alpha}_j = i\rho_j^2(\alpha_{j-1} + \alpha_{j+1}) \tag{6.10.3}$$

and the Schur flow, generated by $2\,\mathrm{Im}(\mathrm{Tr}(\mathcal{C}))$, has

$$\dot{\alpha}_j = \rho_j^2(\alpha_{j+1} - \alpha_{j-1}) \tag{6.10.4}$$

Although he did not realize the connection to OPUC or CMV matrices, Kulish [257] (see also [66]) had the symplectic form given by (6.10.1)/(6.10.2) for the AL equations.

The proof of $\{\Delta(z), \Delta(w)\} = 0$ for OPUC in [319] is quite involved. Simpler proofs can be found in Cantero–Simon [71], Gekhtman–Nenciu [147], and Nenciu [318].

For studies of the dynamics of these flows, see [18, 129, 176, 224, 406].

Chapter Seven

Right Limits

7.1 OVERVIEW

This chapter is an interlude. In our proof of Szegő asymptotics in Section 9.13, we will need the Denisov–Rakhmanov–Remling theorem, which we will prove below in Section 7.6. This leads us naturally to the notion of right limits. If J is a (one-sided) Jacobi matrix with parameters $\{a_n, b_n\}_{n=1}^{\infty}$ and J_r a two-sided matrix with parameters $\{a_n^{(r)}, b_n^{(r)}\}_{n=-\infty}^{\infty}$, we say J_r is a *right limit* of J if and only if for some subsequence $m_j \to \infty$, for all fixed $n \in \mathbb{Z}$, as $j \to \infty$,

$$a_{n+m_j} \to a_n^{(r)} \qquad b_{n+m_j} \to b_n^{(r)} \tag{7.1.1}$$

By compactness of product spaces, if J is bounded, there exist right limits. If we require Jacobi matrices to have $a_n > 0$, then this compactness argument requires the original J to have $\inf_n a_n > 0$. Instead, we will allow J_r to have some 0 a_n's.

Let $\mathcal{R}$ be the set of all right limits. The three major results on right limits are:

(1)

$$\sigma_{\text{ess}}(J) = \bigcup_{J_r \in \mathcal{R}} \sigma(J_r) \tag{7.1.2}$$

(2) For any $J_r \in \mathcal{R}$ (recall that $\Sigma_{\text{ac}}(J)$ is a set up to sets of Lebesgue measure zero, which supports the a.c. part of the spectral measures),

$$\Sigma_{\text{ac}}(J) \subset \Sigma_{\text{ac}}(J_r) \tag{7.1.3}$$

(3) For any $J_r \in \mathcal{R}$, J_r is reflectionless on $\Sigma_{\text{ac}}(J)$.

We will prove (7.1.2) in Section 7.2 and (7.1.3) in Section 7.3. We will define reflectionless and prove (3) in Section 7.4. In Section 7.5, we will show that for any finite gap set, the reflectionless operators are precisely the elements of the isospectral torus. In Section 7.6, we will then conclude easily, given what went before the Denisov–Rakhmanov–Remling theorem, that if $\mathfrak{e}$ is a finite gap set and $\Sigma_{\text{ac}}(J) = \sigma_{\text{ess}}(J) = \mathfrak{e}$, then $\mathcal{R}$ is a subset of the isospectral torus.

Remarks and Historical Notes. The importance of right limits in spectral theory is from Last–Simon [271] who defined right limits as one-sided Jacobi matrices. The proper emphasis on two-sided limits is from Last–Simon [272] and Remling [366]. See the Notes to Section 7.2 for more on the history of right limits.

The notion of reflectionless right limits has been applied by Breuer–Simon [63] to study power series with natural boundaries.

7.2 THE ESSENTIAL SPECTRUM

In this section, we will prove (7.1.2) and look at relevant examples. Let $\mathcal{R}(J)$ be defined by (7.1.1) where $m_j \to \infty$ and J is one-sided. We allow limits to have some a's equal to zero. We will also want to look at two-sided limits when J is two-sided. For such J's, $\mathcal{L}(J)$ will define the set of limits defined by (7.1.1), where now we only suppose $|m_j| \to \infty$. Finally, we will denote by $\sigma_{\infty,\text{pp}}(J)$ the ℓ^∞ point spectrum, that is, the set of λ so that there is u with

$$\sup_n |u_n| < \infty \tag{7.2.1}$$

and, as a difference equation,

$$Ju = \lambda u \tag{7.2.2}$$

Our main goal is to prove that

Theorem 7.2.1. *Let J be a one-sided Jacobi matrix with*

$$\sup_n |a_n| + |b_n| < \infty$$

Then

$$\sigma_{\text{ess}}(J) = \bigcup_{J_r \in \mathcal{R}(J)} \sigma(J_r) \tag{7.2.3}$$

$$= \bigcup_{J_r \in \mathcal{R}(J)} \sigma_{\infty,\text{pp}}(J_r) \tag{7.2.4}$$

Remarks. 1. Similarly, one shows a result for two-sided J's with $\mathcal{R}(J)$ replaced by $\mathcal{L}(J)$. Indeed, the proof below works in n dimensions.

2. It is not a priori obvious that the union in (7.2.3) is closed, but our result will prove that, since $\sigma_{\text{ess}}(J)$ is closed.

We will prove this theorem in three steps:

$$\text{(i)} \qquad \forall J_r \in \mathcal{R} \qquad \sigma(J_r) \subset \sigma_{\text{ess}}(J) \tag{7.2.5}$$

$$\text{(ii)} \qquad \sigma_{\infty,\text{pp}}(J_r) \subset \sigma(J_r) \tag{7.2.6}$$

$$\text{(iii)} \qquad \forall \lambda \in \sigma_{\text{ess}}(J), \qquad \exists J_r \quad \text{with} \quad \lambda \in \sigma_{\infty,\text{pp}}(J_r) \tag{7.2.7}$$

(i) and (ii) will be elementary; (iii) a little more involved. We begin with (i). The key is Weyl's trial function criterion—this is basic so we could assume the reader knows it. But in this context, it is so simple to prove, we begin with that:

Proposition 7.2.2 (Weyl's Principle). *For any selfadjoint operator, A,*

(a) *For any $\varphi \in \mathcal{H}$ and $\lambda \in \mathbb{R}$,*

$$\text{dist}(\lambda, \sigma(A))\|\varphi\| \le \|(A-\lambda)\varphi\| \tag{7.2.8}$$

Indeed,

$$\inf_{\varphi \ne 0} \frac{\|(A-\lambda)\varphi\|}{\|\varphi\|} = \text{dist}(\lambda, \sigma(A)) \tag{7.2.9}$$

(b) *For any sequence* $\varphi_n \in \mathcal{H}$, $\lambda \in \mathbb{R}$ *with* $\|\varphi_n\| = 1$ *and* $\varphi_n \xrightarrow{w} 0$, *we have*

$$\operatorname{dist}(\lambda, \sigma_{\text{ess}}(A)) \le \liminf \|(A-\lambda)\varphi_n\| \tag{7.2.10}$$

In addition, there exists such φ_n *with*

$$\operatorname{dist}(\lambda, \sigma_{\text{ess}}(A)) = \lim \|(A-\lambda)\varphi_n\| \tag{7.2.11}$$

Proof. (a) By the spectral theorem, for any φ with $\|\varphi\| = 1$, there is a probability measure $d\mu_\varphi$ supported on $\sigma(A)$ so

$$\begin{aligned}\|(A-\lambda)\varphi\|^2 &= \int (x-\lambda)^2\, d\mu_\varphi(x) \\ &\ge \operatorname{dist}(\lambda, \sigma(A))^2\end{aligned} \tag{7.2.12}$$

from which (7.2.8) is immediate. Again by the spectral theorem, for any $\alpha \in \sigma(A)$ and $\varepsilon > 0$, we can find φ with $\|\varphi\| = 1$, so $d\mu_\varphi$ is supported on $(\alpha - \varepsilon, \alpha + \varepsilon)$. Thus,

$$\|(A-\lambda)\varphi\| \le |\lambda - \alpha| + \varepsilon$$

from which (7.2.9) follows.

(b) By the definition of $\sigma_{\text{ess}}(A)$, given ε, $S = \{x \mid x \in \sigma(A),\ \operatorname{dist}(x, \sigma(A)) > \varepsilon\}$ is finite and there is a finite-dimensional projection, P, so $PA = AP$ and $\sigma(A(1-P)) \restriction \operatorname{Ran}(1-P)) = \sigma(A) \setminus S$. Since $\varphi_n \xrightarrow{w} 0$ and P is finite-dimensional, $\|P\varphi_n\| \to 0$, so $d\mu_{\varphi_n}(S) \to 0$ and, by the argument that led to (7.2.12),

$$\begin{aligned}\liminf \|(A-\lambda)\varphi_n\| &\ge \operatorname{dist}(\lambda, \sigma(A) \setminus S) \\ &\ge \operatorname{dist}(\lambda, \sigma_{\text{ess}}(A)) - \varepsilon\end{aligned}$$

Since ε is arbitrary, we get (7.2.10).

If $\alpha \in \sigma_{\text{ess}}(A)$, the spectral projection for $(\alpha - \varepsilon, \alpha + \varepsilon)$ is infinite-dimensional, so it contains an infinite orthonormal set $\{\varphi_n\}$. The set has $\varphi_n \xrightarrow{w} 0$, $\|\varphi_n\| = 1$, and $\liminf \|(A-\lambda)\varphi_n\| \le |\lambda - \alpha| + \varepsilon$. By picking from these sequences for $\operatorname{dist}(\alpha_n, \lambda) \to \operatorname{dist}(\sigma_{\text{ess}}(A), \lambda)$, and $\varepsilon_n \to 0$, we get $\varphi_n \xrightarrow{w} 0$, $\|\varphi_n\| = 1$ with

$$\lim \|(A-\lambda)\varphi_n\| \le \operatorname{dist}(\lambda, \sigma_{\text{ess}}(A))$$

so there is equality by (7.2.9). □

We can prove (7.2.5):

Proof of (7.2.5). Given $\lambda \in \sigma(J_r)$ and ε, first find (using (7.2.9)) φ with $\|\varphi\| = 1$ and $\|(J_r - \lambda)\varphi\| \le \frac{\varepsilon}{2}$. Then find $\tilde\varphi$ with $\|\tilde\varphi\| = 1$, $\tilde\varphi$ having finite support, and $\|\varphi - \tilde\varphi\| \le \frac{\varepsilon}{2} \frac{1}{\|J_r\| + |\lambda|}$, so

$$\|(J_r - \lambda)\tilde\varphi\| \le \varepsilon \tag{7.2.13}$$

Pick N so $\tilde\varphi_j = 0$ if $|j| > N$.

Now let m_j be chosen so (7.1.1) holds. For j such that $m_j > N$, define $\varphi^{(j)}$ by

$$\varphi^{(j)}(n) = \tilde\varphi(n - m_j) \tag{7.2.14}$$

Since $m_j > N$, $\|\varphi^{(j)}\| = 1$, and since $m_j \to \infty$, $\varphi^{(j)} \xrightarrow{w} 0$. By (7.1.1),

$$\limsup_{j\to\infty} \|(J-\lambda)\varphi^{(j)}\| = \|(J_r - \lambda)\tilde\varphi\| \le \varepsilon$$

By Proposition 7.2.2,

$$\operatorname{dist}(\lambda, \sigma_{\text{ess}}(J)) \le \varepsilon \tag{7.2.15}$$

Since ε is arbitrary, $\lambda \in \sigma_{\text{ess}}(J)$. □

The following is stronger than (7.2.6):

Theorem 7.2.3. *Let J be a half- or whole-line Jacobi matrix.*
(a) *If* (7.2.2) *holds for a u obeying*

$$|u_n| \le C(1+|n|)^k \tag{7.2.16}$$

for some k, then $\lambda \in \sigma(J)$.
(b) *For almost every λ in $\sigma(J)$, there is a solution of* (7.2.2) *obeying* (7.2.16) *with $k=1$.*

Remarks. 1. $k = 1$ in (b) can be replaced by any $k > \frac{1}{2}$.
2. (7.2.16) can be replaced by $\lim_{|n|\to\infty} |n|^{-1} \log(1+|u_n|) = 0$.
3. In (b), "almost every" means with respect to a spectral measure.

Proof. We will discuss the half-line results; the whole-line case is similar.
(a) We claim

$$\liminf_{n\to\infty} \frac{|u_{n+1}|^2 + |u_n|^2}{|u_1|^2 + \cdots + |u_{n-1}|^2} = 0 \tag{7.2.17}$$

for if not, for large n and some $\varepsilon > 0$,

$$(|u_1|^2 + \cdots + |u_{n+1}|^2) \ge (1+\varepsilon)(|u_1|^2 + \cdots + |u_{n-1}|^2) \tag{7.2.18}$$

which implies

$$\sum_{j=1}^{n} |u_j|^2 \ge C(1+|\varepsilon|)^{n/2} \tag{7.2.19}$$

violating (7.2.16).

Let $\tilde u^{(N)}$ be defined by

$$\tilde u_n^{(N)} = \begin{cases} u_n & n \ge N \\ 0 & n > N \end{cases} \tag{7.2.20}$$

Then

$$((J-\lambda)\tilde u^{(N)})_j = \begin{cases} 0 & j \ne N, N+1 \\ a_{n-1}u_n & j = N \\ -a_n u_{n+1} & j = N+1 \end{cases} \tag{7.2.21}$$

from which we get

$$\frac{\|(J-\lambda)\tilde{u}^{(N)}\|^2}{\|\tilde{u}^{(N)}\|^2} \le 2\sup_j |a_j| \left[\frac{|u_N|^2 + |u_{N+1}|^2}{|u_1|^2 + \cdots + |u_{N-1}|^2}\right] \tag{7.2.22}$$

So for a subsequence $N_j \to \infty$,

$$\frac{\|(J-\lambda)\tilde{u}^{(N_j)}\|}{\|u^{(N_j)}\|} \to 0 \tag{7.2.23}$$

which, by (7.2.8), implies $\lambda \in \sigma(J)$.

(b) Since $\int p_n^2(x)\,d\mu(x) = 1$, we have

$$\int \sum_{n=0}^{\infty} (n+1)^{-2} p_n^2(x)\,d\mu(x) < \infty \tag{7.2.24}$$

so for a.e. x,

$$\sum_{n=1}^{\infty} (n+1)^{-2} p_n^2(x) < \infty \tag{7.2.25}$$

which implies

$$(n+1)^{-1}|p_n(x)| \le C(x) \tag{7.2.26}$$

Use p_{n-1} for the u_n. □

As a preliminary for the final step, we need

Proposition 7.2.4. *If $J_r \in \mathcal{R}(J)$, then $\mathcal{L}(J_r) \subset \mathcal{R}(J)$.*

Proof. Let $\tilde{J} \in \mathcal{L}(J_r)$ be such that with $|m_j| > \infty$ and for $j \to \infty$,

$$a_{m_j+n}^{(r)} \to \tilde{a}_n \qquad b_{m_j+n}^{(r)} \to \tilde{b}_n \tag{7.2.27}$$

Pick $\ell_p \to \infty$ so that as $p \to \infty$,

$$a_{\ell_p+n} \to a_n^{(r)} \qquad b_{\ell_p+n} \to b_n^{(r)} \tag{7.2.28}$$

For each j, pick $p(j)$ so that

$$\text{(a)} \qquad \ell_{p(j)} > 2|m_j| \tag{7.2.29}$$

$$\text{(b)} \qquad |a_{\ell_{p(j)}+n} - a_n^{(r)}| + |b_{\ell_{p(j)}+n} - b_n^{(r)}| \le 2^{-j} \tag{7.2.30}$$

for $|n| \le |m_j| + |j|$. Then for all n fixed, as $j \to \infty$,

$$a_{\ell_{p(j)}+m_j+n} \to \tilde{a}_n \qquad b_{\ell_{p(j)}+m_j+n} \to \tilde{b}_n$$

and $m_j + \ell_{p(j)} \ge |m_j| \to \infty$ so $\tilde{J} \in \mathcal{R}(J)$. □

Example 7.2.5. Let $a_n \equiv 1$, $b_n = 1$ if $n = 1, 4, 9, 16, \dots$ and 0 otherwise. Then $\mathcal{R}(J)$ consists of J_r's with $a_n^{(r)} \equiv 1$ and either $b_n^{(r)} \equiv 0$ or $b_n^{(r)}$ has one n_0 for which $b_{n_0}^{(r)} = 1$ and all others are 0 (all n_0 occur). Each $\mathcal{L}(J_r)$ has only the element $\tilde{a}_n \equiv 1$, $\tilde{b}_n \equiv 0$. Thus, $\mathcal{L}(\mathcal{R}(J)) \subset \mathcal{R}(J)$ but strictly smaller. It can also happen that $\mathcal{L}(\mathcal{R}(J)) = \mathcal{R}(J)$. □

Proposition 7.2.6. *If there exist $J_k \in \mathcal{R}(J)$ and n_k and solutions $\varphi^{(k)}$ of $J_k\varphi^{(k)} = \lambda^{(k)}\varphi^{(k)}$ so that $\lambda^{(k)} \to \lambda_\infty$, and for some $C < \infty$,*

$$\max_{|j-n_k|\le k} |\varphi_j^{(k)}| \le C|\varphi_{n_k}^{(k)}| \tag{7.2.31}$$

then there exists $J_r \in \mathcal{R}(J)$ with $\lambda_\infty \in \sigma_{\infty,\mathrm{pp}}(J_r)$. In particular, $\cup_{J_r\in\mathcal{R}(J)}\sigma_{\infty,\mathrm{pp}}(J_r)$ is closed.

Proof. Let

$$u_j^{(k)} = \frac{\varphi_{n_k+j}^{(k)}}{\varphi_{n_k}^{(k)}} \tag{7.2.32}$$

and $\tilde{J}_k$ be J_k-translated by n_k units (which also lies in J_r). Then

$$\sup_{|j|\le k} |u_j^{(k)}| \le C \tag{7.2.33}$$

and $u_0^{(k)} = 1$. By compactness, we can pass to a subsequence so $\tilde{J}_{k(\ell)} \to J_r$ and $u^{(k)} \to u_\infty$. Clearly, $J_r u_\infty = \lambda_\infty u_\infty$ and $\|u_\infty\|_\infty \le C$ so $\lambda_\infty \in \sigma_{\infty,\mathrm{pp}}(J_r)$. □

We can now describe our strategy for completing the proof. By Theorem 7.2.3(b), if $\lambda \in \sigma_{\mathrm{ess}}(J)$, there exist distinct $\lambda_m \to \lambda$ and $u^{(m)}$ obeying (7.2.16) with $k = 1$. We will proceed as follows:

(1) If $u^{(m)}$ is unbounded, we will find $J_r \in \mathcal{R}(J)$ so $\lambda^{(m)} \in \sigma_{\infty,\mathrm{pp}}(J_r)$. Thus, by the last proposition, λ is in some $\sigma_{\infty,\mathrm{pp}}(J_r)$ or, for all m large, $u^{(m)}$ is bounded.
(2) If $u^{(m)}$ is not exponentially decaying, we find $J_r \in \mathcal{R}(J)$ so $\lambda^{(m)} \in \sigma_{\infty,\mathrm{pp}}(J_r)$.
(3) Thus, the only way to avoid $\lambda \in \cup\sigma_{\infty,\mathrm{pp}}(J_r)$ is for each λ_m to be an eigenvalue with exponentially decaying eigenvector, and we will prove they have to move to infinity in a way to get $\lambda \in \sigma_{\infty,\mathrm{pp}}(J_r)$ for some J_r.

Proposition 7.2.7. *If u obeys (7.2.2) and (7.2.16) for some k but $u \notin \ell^\infty$, then there is $J_r \in \mathcal{R}(J)$ and $\tilde{u} \ne 0$ in ℓ^∞ so*

$$J_r\tilde{u} = \lambda\tilde{u} \tag{7.2.34}$$

Proof. By the last proposition, it suffices, for each k, to find J_r and $\tilde{u}$ obeying (7.2.34) with $\tilde{u}_0 = 1$ and

$$\max_{|n|\le k} |\tilde{u}_n| \le 2 \tag{7.2.35}$$

Fix k and for $m = 1, 2, \dots,$ let

$$q_m = \max_{m(k-1)\le n<mk} |u_n| \tag{7.2.36}$$

By hypothesis, $q_m \to \infty$, so there are infinitely many m's, say $m \in M$, with

$$q_m \ge \max_{\ell\le m} q_\ell \tag{7.2.37}$$

If there are infinitely many $m \in M$ with

$$q_{m+1} \le 2q_m \tag{7.2.38}$$

then, by using compactness, we get a J_r and $\tilde{u}$ obeying (7.2.35). If $q_{m+1} \geq 2q_m$ with $m \in M$, then clearly, $m+1 \in M$. So if (7.2.38) fails to hold infinitely often, then there is m_0 with $m \in M$ for $m \geq m_0$ and $q_{m_0+\ell} \geq 2^\ell q_{m_0}$ for all ℓ, violating (7.2.16). Thus, (7.2.38) holds infinitely often and we get the required $\tilde{u}$. □

Proposition 7.2.8. *If* (7.2.2) *has a solution, $u \in \ell^\infty$, then either there is a $J_r \in \mathcal{R}(J)$ and a bounded nonzero $\tilde{u}$ obeying* (7.2.34) *or else, for some $A, B > 0$, we have*

$$|u_n| \leq A\exp(-Bn) \tag{7.2.39}$$

Proof. If u_n fails to go to zero, find $m_j \to \infty$ and $|u_{m_j}| \geq \varepsilon > 0$, and using a subsequence so u_{n+m_j}, a_{n+m_j}, and b_{n+m_j} all converge for J_r and $\tilde{u}$ obeying (7.2.34) with $|\tilde{u}_0| > \varepsilon$ and $\|\tilde{u}\|_\infty \leq \|u\|_\infty < \infty$. Thus, either there is a solution of (7.2.34) or $u_n \to 0$.

As above, we need only find for each k, J_r and $\tilde{u}$ obeying (7.2.35). Fix j and define q_m by (7.2.36). Let M be the set of m's with

$$q_m \geq \max_{\ell \geq m} q_\ell$$

Since $u_n \to 0$, M is infinite. If there are infinitely many m's with $q_{m-1} \leq 2q_m$, we can find a solution obeying (7.2.35). On the other hand, if $q_{m-1} \geq 2q_m$ for all large m, then u decays exponentially. □

Remark. This proposition proves that if $\lambda \in \sigma(J) \setminus \sigma_{\text{ess}}(J)$, then the corresponding eigenvectors decay exponentially; see the Notes for discussion.

Proof of Theorem 7.2.1. We need only prove (7.2.7). Since λ is not isolated, there exist $\lambda^{(m)} \to \lambda$ and solutions $u^{(m)}$ of $Ju^{(m)} = \lambda^{(m)} u^{(m)}$ obeying (7.2.16). By Proposition 7.2.6, if there are bounded solutions of some $J_r \tilde{u}^{(m)} = \lambda_m \tilde{u}^{(m)}$, then there is a J_r with $J_r \tilde{u} = \lambda \tilde{u}$ for $\tilde{u} \in \ell^\infty$. By Propositions 7.2.7 and 7.2.8, this happens, unless for all large m, $u^{(m)}$ obeys (7.2.39) so that each $\lambda^{(m)}$ is an eigenvalue. So we consider only that case. Pick $u^{(m)}$ so $\|u^{(m)}\|_2 = 1$.

Since $u_n^{(m)} \to 0$ as $n \to \infty$, $|u^{(m)}|$ takes its maximum value at some point—let n_m be picked so $|u_{n_m}^{(m)}| = \|u^{(m)}\|_\infty$ and so that n_m is the largest such point. If $n_m \to \infty$ as $m \to \infty$, we can use a limit point of $u^{(m)}/u_{n_m}^{(m)}$ to get a bounded solution of some $J_r \tilde{u} = \lambda \tilde{u}$. Thus, we need only consider the case $\sup_m |n_m| \equiv N < \infty$. Since $u^{(m)} \to 0$ weakly, $\sup_{|n| \leq n} |u_n^{(m)}| = C_m \to \infty$ and thus,

$$\|u^{(m)}\|_\infty = C_m \to 0 \tag{7.2.40}$$

As in the proof of the last proposition, we need only find, for each k, solutions of (7.2.34) obeying (7.2.35). By following the proof of that proposition, we see one cannot do this unless one has a B_1, an N_1, and an M_1 so that for all $m > M_1$ and $n > N_1$, we have

$$|u_n^{(m)}| \leq \|u^{(m)}\|_\infty e^{-B(n-N_1)} \tag{7.2.41}$$

But then, by (7.2.40), as $m \to \infty$,

$$\sum_{n \geq N_1} |u_n^{(m)}|^2 \to 0$$

By (7.2.40) again,

$$\sum_{n=1}^{N_1} |u_n^{(m)}|^2 \leq N_1 \|u^{(m)}\|_\infty^2 \to 0$$

so $\|u^{(m)}\|_2 \to 0$, violating the choice $\|u^{(m)}\|_2 = 1$. This contradiction shows that one can always construct the required bounded solutions. □

This completes the proof of Theorem 7.2.1.

Example 7.2.9. Let $S \equiv \{x_1, \dots, x_\ell\} \subset \mathbb{R}$. Let $P(\lambda) = \prod_{j=1}^{\ell}(\lambda - x_j)$. It is easy to see that $\sigma_{\text{ess}}(J) \subset S$ if and only if $P(J)$ is compact. For $\ell = 1$, this happens if and only if $a_n \to 0$, $b_n \to x_1$. But for $\ell \geq 2$, the conditions on the a's and b's are murky. Theorem 7.2.1 clarifies this. For example, we must have $a_n a_{n+1} \dots a_{n+\ell} \to 0$ as $n \to \infty$ so the J's have to be direct sums of finite matrices of size at most ℓ. The possible limits are those finite Jacobi matrices with spectrum in S. For example, if $\ell = 2$, the limits are the 1×1 matrices $b_1 = x_1$ or $b = x_2$ and the 2×2 matrices $\left(\begin{smallmatrix} b & a \\ a & x_1+x_2-b \end{smallmatrix}\right)$ where b and a are related by $b(x_1 + x_2 - b) - a^2 = x_1 x_2$. □

Example 7.2.10. If $\mathcal{T}_\mathfrak{e}$ is the isospectral torus associated to a finite gap set $\mathfrak{e}$ and if $\mathcal{R} \subset \mathcal{T}_\mathfrak{e}$, then by Theorem 7.2.1, $\sigma_{\text{ess}}(J) = \mathfrak{e}$. □

Remarks and Historical Notes. Last–Simon [271] looked at one-sided right limits $\tilde{J}_r$ and proved $\sigma_{\text{ess}}(\tilde{J}_r) \subset \sigma_{\text{ess}}(J)$, but their arguments prove (7.2.5). (7.2.3) and its proof are from Last–Simon [272]. The arguments in this section to get bounded eigenfunctions are new here.

In the spectral theory community, these ideas go back to geometric approaches to the HVZ theorem; see [272] for references.

There are three other independent threads that consider limits of differential or difference operators or some subclass, especially the almost periodic case. One thread using Fredholm operators goes back to Favard [127] with later developments by Muhamadiev [309, 310], Shubin [386, 387], Kurbatov [262], Rabinovich [356], and Chandler-Wilde–Lindner [78, 79]. In particular, Rabinovich has results close to (7.2.3) and Chandler-Wilde–Lindner have the result (7.2.4) on $\sigma_{\infty,\text{pp}}(J_r)$.

Two other threads involve C^*-algebras (see Georgescu–Iftimovici [152] and Măntoiu [294]) and what has been called "collectively compact operator approximation theory" (see Anselone [20] and references therein).

The result that $\sigma_{\text{ess}}(J) \subset S$ if and only if $P(J)$ is compact is due to Krein in Akhiezer–Krein [16]. The other parts of Example 7.2.9 are from [272]. Example 7.2.10, also from [272], answered a conjecture of Simon [400]. The history of this and related conjectures is discussed in the Notes to Section 8.1.

There is a huge literature on exponential decay of discrete eigenfunctions of Schrödinger operators, of which the seminal works are Combes–Thomas [92] and Agmon [6]. For CMV matrices, this is discussed in [400, Section 10.14], and the same analysis works for Jacobi matrices. This is superior to Proposition 7.2.8 since the Combes–Thomas method gives explicit positive lower bounds on the constant B in (7.2.39), while the proof of Proposition 7.2.8 does not. Still it is interesting to see this new approach to exponential decay.

7.3 THE LAST–SIMON THEOREM ON A.C. SPECTRUM

In this section, we will prove that

Theorem 7.3.1 (Last–Simon [271]). *For any $J_r \in \mathcal{R}$,*

$$\Sigma_{\rm ac}(J_r) \supset \Sigma_{\rm ac}(J) \tag{7.3.1}$$

Indeed, on $\Sigma_{\rm ac}(J)$, J_r has a.c. spectrum of multiplicity 2.

We will use the following characterization of $\Sigma_{\rm ac}(J)$:

Theorem 7.3.2 (Last–Simon [271]). *Let*

$$N^\sharp = \left\{ x_0 \;\middle|\; \liminf \frac{1}{n+1} K_n(x_0, x_0) < \infty \right\} \tag{7.3.2}$$

Then up to sets of measure zero,

$$\Sigma_{\rm ac}(J) = N^\sharp \tag{7.3.3}$$

Proof. Since $\int \frac{1}{n+1} K_n(x_0, x_0)\, d\mu = 1$, by Fatou's lemma,

$$\int \liminf \frac{1}{n+1} K_n(x_0, x_0)\, d\mu \le 1 \tag{7.3.4}$$

Thus, $\operatorname{supp}(d\mu) \subset N$, so up to sets of Lebesgue measure zero,

$$\Sigma_{\rm ac}(J) \subset N^\sharp \tag{7.3.5}$$

On the other hand, by Theorem 3.11.7 (scaling $(-2, 2)$ up to some interval I containing $\operatorname{supp}\mu$), up to sets of measure zero,

$$\{x \mid w(x) = 0\} \subset \left\{ x \;\middle|\; \liminf \frac{1}{n+1} K_n(x, x) = \infty \right\} = \mathbb{R} \setminus N^\sharp \tag{7.3.6}$$

which implies

$$N^\sharp \subset \Sigma_{\rm ac}(J) \tag{7.3.7}$$

□

Next, we need the invariance of $\Sigma_{\rm ac}$ under rank one perturbations. Suppose $\mathcal{H} = L^2(\mathbb{R}, d\mu)$ for μ a probability measure, $\varphi \equiv 1 \in \mathcal{H}$, $(A_0 f)(x) = x f(x)$, and for $\lambda \in \mathbb{R}$,

$$A_\lambda f = A_0 f + \lambda \langle \varphi, f \rangle f \tag{7.3.8}$$

Theorem 7.3.3. *φ is cyclic for A_λ and the spectral measure, $d\mu_\lambda$, for A_λ and φ, that is,*

$$F_\lambda(z) = \int \frac{d\mu_\lambda(x)}{x - z} = \left\langle \varphi, \frac{1}{A_\lambda - z} \varphi \right\rangle \tag{7.3.9}$$

$z \notin \mathbb{R}$, obeys

$$F_\lambda(z) = \frac{F_0(z)}{1 + \lambda F_0(z)} \tag{7.3.10}$$

In particular, if

$$d\mu_\lambda(x) = w_\lambda(x)\,dx + d\mu_{\lambda,s} \tag{7.3.11}$$

then for a.e. x,

$$w_\lambda(x) = \frac{w_0(x)}{|1+\lambda F_0(x+i0)|^2} \tag{7.3.12}$$

and

$$\Sigma_{\rm ac}(A_\lambda) = \Sigma_{\rm ac}(A) \tag{7.3.13}$$

Proof. We have that

$$(A_\lambda - z)^{-1} = (A_0 - z)^{-1} - (A_\lambda - z)^{-1}(A_\lambda - A_0)(A_0 - z)^{-1} \tag{7.3.14}$$

which leads to

$$F_\lambda(z) = F_0(z) - \lambda F_0(\lambda) F_\lambda(z) \tag{7.3.15}$$

which implies (7.3.10).

Also, by (7.3.14), if ψ is orthogonal to $(A_\lambda - z)^{-1}\varphi$ for all z, then ψ is orthogonal to $(A_0 - z)^{-1}\varphi$ for all z, so $\psi = 0$. Thus, φ is cyclic for A_λ.

By (2.3.55),

$$w_\lambda(x) = \frac{1}{\pi}\,\mathrm{Im}\,F_\lambda(x+i0) \tag{7.3.16}$$

Since (7.3.10) implies

$$\mathrm{Im}\,F_\lambda(z) = \frac{\mathrm{Im}\,F_0(z)}{|1+\lambda F_0(x)|^2} \tag{7.3.17}$$

we get (7.3.12), which implies (7.3.13) given that $\{x \mid F_0(x+i0) = -1/\lambda$ or $F_0(x+i0) = \infty\}$ has Lebesgue measure zero. $\square$

Corollary 7.3.4. *Let A be a bounded selfadjoint operator and F a finite rank selfadjoint operator. Then*

$$\Sigma_{\rm ac}(A+F) = \Sigma_{\rm ac}(A) \tag{7.3.18}$$

Indeed, the multiplicities are equal.

Remarks. 1. Traditionally, this is done via scattering theory; see the Notes.

2. By using cyclic sets of vectors when there is not a single cyclic vector (or by taking direct sums), one form of the spectral theorem is that any A is unitarily equivalent to multiplication by x on $L^2(\mathbb{R}, d\mu)$ where now $d\mu$ is a matrix- (or operator-) valued measure. One can still write

$$d\mu(x) = W(x)\,dx + d\mu_s \tag{7.3.19}$$

but now W is an operator. One shows

$$\Sigma_{\rm ac}^{(k)}(A) = \{x \mid \mathrm{rank}(W) = k\} \tag{7.3.20}$$

is independent (Lebesgue a.e.) of the representation and the equal multiplicities statement means

$$\Sigma_{\rm ac}^{(k)}(A+F) = \Sigma_{\rm ac}^{(k)}(A) \tag{7.3.21}$$

Proof. By diagonalizing F, we see any F is a sum of selfadjoint rank one operators, so it is sufficient to prove it for the case $F = \lambda(\varphi, \cdot\,)\varphi$ with $\|\varphi\| = 1$. Let $\mathcal{H}_1 =$ span of $\{(A-z)^{-1}\varphi \mid z \notin \mathbb{R}\}$. Then $\mathcal{H}_1$ and so $\mathcal{H}_1^\perp$ are invariant for A and $A+F$ and

$$A \restriction \mathcal{H}_1^\perp = (A+F) \restriction \mathcal{H}_1^\perp \tag{7.3.22}$$

By the theorem,

$$\Sigma_{\rm ac}((A+F) \restriction \mathcal{H}_1)) = \Sigma_{\rm ac}(A \restriction \mathcal{H}_1) \tag{7.3.23}$$

and are multiplicity 1. (7.3.22)/(7.3.23) imply (7.3.18) and (7.3.21). □

Closely related to the last theorem is

Theorem 7.3.5. *Let J be a Jacobi matrix and J_1 the once-stripped Jacobi matrix (see Theorem 3.2.4). Then*

$$\Sigma_{\rm ac}(J) = \Sigma_{\rm ac}(J_1) \tag{7.3.24}$$

Remarks. 1. In a sense, J_1 is the "rank one perturbation" with $b_1 = \infty$, so this result is a special or, at least, limiting case of Corollary 7.3.4. See the Notes for a discussion of this infinite coupling theory.

2. This is essentially an OPRL analog of (2.6.15).

Proof. By (3.2.28),

$$m_1(z) = (-z + b_1 - a_1^2 m_1(z))^{-1} \tag{7.3.25}$$

so, by (2.3.55),

$$w(z) = w_1(x)|-x + b_1 - a_1 m_1(x+i0)|^{-2} \tag{7.3.26}$$

so as in the proof of Theorem 7.3.3, up to sets of Lebesgue measure zero,

$$\{x \mid w(x) \neq 0\} = \{x \mid w_1(x) \neq 0\} \tag{7.3.27}$$

□

Now let μ be the measure for J and μ_1 for J_1. Just as there is a sup of measures discussed in Lemma 2.16.9, there is an inf

$$\eta = \mu \wedge \mu_1 \tag{7.3.28}$$

with

$$d\eta = n(x)\,dx + d\eta_{\rm s} \tag{7.3.29}$$

and one has

$$n(x) = \min(w(x), w_1(x))$$

and, in particular, by (7.3.27), up to sets of Lebesgue measure zero,

$$\{x \mid n(x) \neq 0\} = \Sigma_{\rm ac}(J) \tag{7.3.30}$$

By (3.2.16), the second kind polynomials obey

$$\int a_1^2 q_n(x)^2 \, d\mu_1(x) = 1 \tag{7.3.31}$$

Thus, by (3.2.19),

$$\int \|T_n(x)\|^2 \, d\eta(x) \leq (1 + a_n^2)(1 + a_1^{-2}) \tag{7.3.32}$$

Define

$$N = \left\{ x \,\middle|\, \liminf \frac{1}{n} \sum_{j=1}^{n} \|T_j(x)\|^2 < \infty \right\} \tag{7.3.33}$$

Then we have the following variant of Theorem 7.3.2:

Theorem 7.3.6 (Last–Simon [271]). *Up to sets of measure zero,*

$$N = \Sigma_{\text{ac}}(J) \tag{7.3.34}$$

Proof. Clearly,

$$N \subset N^\sharp = \Sigma_{\text{ac}}(J) \tag{7.3.35}$$

On the other hand, since $\sup(a_n) < \infty$, by (7.3.32) and Fatou's lemma,

$$\text{supp}(d\eta) \subset N \tag{7.3.36}$$

so, by (7.3.30),

$$\Sigma_{\text{ac}}(J) \subset N \tag{7.3.37}$$

$\square$

Recall that the transfer matrix, $T_{kj}(z; J)$, can be defined as mapping $\binom{u_{j+1}}{a_j u_j}$ to $\binom{u_{k+1}}{a_k u_k}$ for solutions of (3.2.6) and for J, a half-line Jacobi matrix ($k, j \geq 1$) or whole-line matrices. In terms of the transfer matrix T_n we discussed above,

$$T_n = T_{n1} \tag{7.3.38}$$

and since

$$T_{kj} T_{j\ell} = T_{k\ell} \tag{7.3.39}$$

we have

$$T_{kj} = T_k T_j^{-1} \tag{7.3.40}$$

Moreover, since $\det(T_j) = 1$,

$$\|T_j^{-1}\| = \|T_j\| \tag{7.3.41}$$

so, by (7.3.40),

$$\|T_{kj}\| \leq \|T_k\| \, \|T_j\| \tag{7.3.42}$$

Thus, by (7.3.32) and the Schwarz inequality, we get for half-line Jacobi matrices, J,

$$\int \|T_{kj}(x;\, J)\|\, d\eta_J(x) \le \Big[\sup_n(1+a_n^2)\Big](1+a_1^{-2}) \tag{7.3.43}$$

Let us denote the right side of (7.3.43) by $K(J)$. Suppose now that (7.3.2) holds. Then as $\ell \to \infty$,

$$T_{k+m_\ell\, j+m_\ell}(x;\, J) \to T_{kj}(x;\, J_r) \tag{7.3.44}$$

Theorem 7.3.7. *For any $J_r \in \mathcal{R}$, we have*

(i) $$\int \|T_{kj}(x;\, J_r)\|\, d\eta_J(x) \le K(J) \tag{7.3.45}$$

(ii) $$\int \left(\frac{1}{n}\sum_{k=1}^n \|T_{\pm k\, \pm 1}(x;\, J_r)\|^2\right)^{1/2} d\eta_J(x) \le K(J) \tag{7.3.46}$$

(iii) $$\int \liminf \left(\frac{1}{n}\sum_{k=1}^n \|J_{\pm k\, \pm 1}(x;\, J_r)\|^2\right)^{1/2} d\eta_J(x) \le K(J) \tag{7.3.47}$$

Proof. (i) This follows from (7.3.43), (7.3.44), and Fatou's lemma.

(ii) By (7.3.42), if S is a set with n elements,

$$\left(\frac{1}{n}\sum_{k\in S} \|T_{kj}\|^2\right)^{1/2} \le \|T_j\| \left(\frac{1}{n}\sum_{k\in S} \|T_k\|^2\right)^{1/2} \tag{7.3.48}$$

so, by the Schwarz inequality and (7.3.32),

$$\int \left(\frac{1}{n}\sum_{k\in S} \|T_{kj}\|^2\right)^{1/2} d\eta_J \le K(J) \tag{7.3.49}$$

By Fatou's lemma and (7.3.44), this leads to (7.3.46).

(iii) This follows from (7.3.46) and Fatou's lemma. □

Proof of Theorem 7.3.1. By (7.3.46), we see that if $(J_r)^\pm$ are the half-line Jacobi matrices obtained from J_r (i.e., $(J_r)^\pm$ have Jacobi parameters $(a_n^{(n)}, b_n^{(n)})_{n=1}^\infty$ and $(a_{-n}^{(n)}, b_{1-n}^{(n)})_{n=1}^\infty$), then

$$\int \left(\liminf \frac{1}{n}\sum_{k=1}^n \|T_k(x;\, J_r^\pm)\|^2\right)^{1/2} d\eta_J(x) < \infty \tag{7.3.50}$$

By (7.3.30), we see a.e. on $\Sigma_{\rm ac}(J)$, we have that the lim inf is finite so, by Theorem 7.3.6,

$$\Sigma_{\rm ac}(J_r^\pm) \subset \Sigma_{\rm ac}(J) \tag{7.3.51}$$

Since J_r and $J_r^+ \oplus J_r^-$ differ by a rank two operator (by replacing a_0 by 0), by Corollary 7.3.4, (7.3.51) implies (7.3.1) with the multiplicity 2 statement. □

Remarks and Historical Notes. As indicated, Theorems 7.3.1 and 7.3.6 are from [271] and the use of Fatou's lemma to get (7.3.5) is there. But the other direction, (7.3.7), is obtained there through the use of subordinacy theory; the idea we use here to exploit the Máté–Nevai variational principle seems to be new. The subordinacy theory yields more, namely, $\mu_s(N) = 0$ [271]. (*Note:* Sometimes $\mu_s(N^\sharp) \neq 0$; indeed, $\mu_s(\mathbb{R} \setminus N^\sharp) = 0$.)

The spectral theory of rank one perturbations goes back to Aronszajn and Donoghue [27, 113]. Implicit in their work is the invariance of a.c. spectrum. For further discussion, see Simon [390, Chapters 11 and 12].

Apparently without realizing the relevance of this work (even Kato's 1976 book [215] makes no mention of this work of Aronszajn and Donoghue!), invariance of the a.c. spectrum under finite rank perturbations was obtained by Kato [214] using scattering theory methods at about the same time as their work. The scattering approach also works for trace class perturbations; see Reed–Simon [363].

As mentioned, Theorem 7.3.5 can also be obtained using rank one perturbations at infinite coupling; see Gesztesy–Simon [164].

7.4 REMLING'S THEOREM ON A.C. SPECTRUM

We saw in the last section that J_r's in $\mathcal{R}(J)$ have a.c. spectrum of multiplicity 2 on $\Sigma_{\text{ac}}(J)$. In this section, we will prove they are actually reflectionless there, so we will have to begin with a definition of reflectionless. In Section 5.4, we proved a property called reflectionless for whole-line periodic Jacobi matrices. We proved Theorem 5.4.18, which had a number of conditions that turn out to hold for a.e. λ_0. We explored this further in Section 5.13. The natural notion is reflectionless on a set—indeed, on a *Lebesgue measure class*, that is, an equivalence class of Borel sets under the equivalence relation defined by $A \equiv B$ if and only if $|A \triangle B| = 0$.

Theorem 7.4.1. *Let Σ be a measure class and J a two-sided Jacobi matrix with bounded Jacobi parameters. The following are equivalent:*

(i) *For a.e. $\lambda \in \Sigma$ and all $n \in \mathbb{R}$,*

$$\operatorname{Re} G_{nn}(\lambda + i0) = 0 \tag{7.4.1}$$

(ii) *For a.e. $\lambda \in \Sigma$ and three successive n's,* (7.4.1) *holds.*

(iii) *For a.e. $\lambda \in \Sigma$ and all n,*

$$a_n^2 m(\lambda + i0, J_n^+) = \overline{m(\lambda + i0, J_n^-)^{-1}} \tag{7.4.2}$$

(iv) *For a.e. $\lambda \in \Sigma$ and one n,* (7.4.2) *holds.*

(v) *If $u_n^\pm(\lambda, J)$ are the Weyl solutions for $\lambda \in \mathbb{C}_+$ normalized by*

$$u_0^\pm(\lambda, J) = 1 \tag{7.4.3}$$

then for a.e. $\lambda \in \Sigma$ and all n,

$$u_n^-(\lambda + i0, J) = \overline{u_n^+(\lambda + i0, J)} \tag{7.4.4}$$

Proof. We recall some basic formulae we will need (see Theorem 5.4.13 and its proof):

$$J_\ell^+ \text{ has Jacobi parameters } \{a_{n+\ell}, b_{n+\ell}\}_{n=0}^\infty \tag{7.4.5}$$

$$J_\ell^- \text{ has Jacobi parameters } \{a_{\ell-n}, b_{\ell+1-n}\}_{n=1}^\infty \tag{7.4.6}$$

$$m(\lambda, J_\ell^+) = -\frac{u_{\ell+1}^+(\lambda)}{a_\ell u_\ell^+(\lambda)} \tag{7.4.7}$$

$$m(\lambda, J_\ell^-) = -\frac{u_\ell^-(\lambda)}{a_\ell u_{\ell+1}^-(\lambda)} \tag{7.4.8}$$

$$G_{\ell\ell}(\lambda) = \frac{u_\ell^+(\lambda)u_\ell^-(\lambda)}{a_\ell(u_{\ell+1}^+(\lambda)u_\ell^-(\lambda) - u_{\ell+1}^-(\lambda)u_\ell^+(\lambda))} \tag{7.4.9}$$

$$= -\frac{1}{a_\ell^2 m(\lambda, J_\ell^+) - m(\lambda, J_\ell^-)^{-1}} \tag{7.4.10}$$

initially for $\lambda \in \mathbb{C}_+$, then by taking boundary values for a.e. $\lambda \in \mathbb{R} + i0$. We will prove that (v) $\Rightarrow$ (iii) $\Rightarrow$ (iv) $\Rightarrow$ (v) and (iii) $\Rightarrow$ (i) $\Rightarrow$ (ii) $\Rightarrow$ (v).

<u>(v) $\Rightarrow$ (iii)</u>. (7.4.3), (7.4.7), and (7.4.8) imply (7.4.2).

<u>(iii) $\Rightarrow$ (iv)</u> is trivial.

<u>(iv) $\Rightarrow$ (v)</u>. By translation invariance, we can suppose $n = 0$, in which case by (7.4.1), we see (7.4.3), (7.4.7), and (7.4.8) imply (7.4.4). Here we use the fact that the difference equation is second-order, so it suffices for (7.4.4) to hold at $n = 0, 1$.

<u>(iii) $\Rightarrow$ (i)</u> is immediate from (7.4.10).

<u>(i) $\Rightarrow$ (ii)</u> is trivial.

<u>(ii) $\Rightarrow$ (v)</u>. By translation invariance, we can suppose that the three successive values are $n = 0, \pm 1$. Since boundary values of m-functions cannot vanish on sets of positive Lebesgue measure (see Theorem 2.3.21), for a.e. $\lambda \in \Sigma$, (a) and (d) of Theorem 5.4.18 hold. Similarly, since $w(\lambda) \neq 0$ for $\lambda \in \mathbb{C}_+$, for a.e. $\lambda \in \Sigma$, $w(\lambda + i0) \neq 0$, so (b) of that theorem holds for a.e. λ. By hypothesis, (c) holds, so by that theorem, (7.4.4) holds. □

Definition. Let J be a whole-line Jacobi matrix. If any and hence all of (i)–(v) hold on Σ, we say J is *reflectionless* on Σ. If $\Sigma \subset \mathfrak{e} \subset \mathbb{R}$ with $|\Sigma| > 0$ and $\mathfrak{e}$ compact, we let $\mathcal{R}(\Sigma, \mathfrak{e})$ denote those whole-line Jacobi matrices, which are reflectionless on Σ and have $\sigma(J) \subset \mathfrak{e}$.

Theorem 7.4.2. *If $J \in \mathcal{R}(\Sigma, \mathfrak{e})$, then $\Sigma \subset \Sigma_{\rm ac}(J)$. Indeed, J has multiplicity 2 on Σ.*

Proof. Let $m_\pm(z) = m(z, J_0^\pm)$ and let $G(z) = G_{00}(z)$. By hypothesis, $\operatorname{Re} G_{00}(\lambda + i0)$ for $\lambda \in \Sigma$. Since $|\{\lambda \mid G_{00}(\lambda + i0) = 0\}| = 0$, for a.e. $\lambda \in \Sigma$, we have $\operatorname{Im} G_{00}(\lambda + i0)) \neq 0$. Thus, for a.e. $\lambda \in \Sigma$, by (7.4.10),

$$\operatorname{Im} a_0^2 m_+(\lambda + i0) - m_-(\lambda + i0)^{-1} > 0 \tag{7.4.11}$$

But, by (7.4.2) on Σ,

$$\operatorname{Im} a_0^2 m_+(\lambda + i0) = \operatorname{Im} -m_-(\lambda + i0)^{-1} \tag{7.4.12}$$

so a.e. on Σ,

$$\operatorname{Im} m_\pm(\lambda + i0) > 0 \tag{7.4.13}$$

Thus, on Σ, $J_0^+ \oplus J_0^-$ has a.c. spectral multiplicity 2. By Corollary 7.3.4, J has also. □

In many cases, $\mu_{\text{sing}}(\Sigma) = 0$, but not for all possible Σ; see Theorem 7.4.8 below and the Notes. We can now state Remling's theorem whose proof we postpone until the end of the section:

Theorem 7.4.3 (Remling's Theorem [366]). *Let $J_r \in \mathcal{R}(J)$ be a right limit of a half-line Jacobi matrix, J. Then J_r is reflectionless on $\Sigma_{\text{ac}}(J)$.*

To apply this theorem, we need to know about reflectionless operators. The following is critical:

Theorem 7.4.4 (Kotani [243]). *Let $\Sigma \subset \mathfrak{e} \subset \mathbb{R}$ with $\mathfrak{e}$ compact and Σ a Borel set with $|\Sigma| > 0$. Put the product topology on $\{\{a_n, b_n\}_{n=-\infty}^0\}$. Then the set, $L(\Sigma, \mathfrak{e})$, of such Jacobi parameters obtained by restriction from $\mathcal{R}(\Sigma, \mathfrak{e})$ is compact, and there is a continuous function $F\colon L(\Sigma, \mathfrak{e}) \to (0, \infty) \times \mathbb{R}$ so that for any $\{a_n, b_n\}_{n=-\infty}^\infty \in \mathcal{R}(\Sigma, \mathfrak{e})$, we have*

$$(a_1, b_1) = F(\{a_n, b_n\}_{n=-\infty}^0) \tag{7.4.14}$$

Remark. By iteration, $\{a_n, b_n\}_{n=-\infty}^0$ then determine all $\{a_n, b_n\}_{n=1}^\infty$.

Proof. By (3.2.28),

$$(-m(z, J_0^-))^{-1} = z - b_0 - a_{-1}^2 m(z, J_{-1}^-) \tag{7.4.15}$$

so by the continuity of the map from half-line J's to $m \restriction \Sigma$ (in the weak topology discussed in (2.3.73)), we see that

$$\{a_n, b_n\}_{n=-\infty}^\infty \mapsto a_0^2 m(\lambda + i0, J_0^+) - \overline{m(\lambda + i0, J_0^-)^{-1}}$$

is continuous. Thus, the set on which it is 0 is compact. On this set, $-m(z, J_0^-)^{-1}$ is a continuous function of $\{a_n, b_n\}_{n=-\infty}^{-1} \cup \{b_0\}$ by (7.4.15). So by (7.4.2) on $\mathcal{R}(\Sigma, \mathfrak{e})$, $a_0^2 m(\lambda + i0, J_0^+) \restriction \Sigma$ is a continuous function of $\{a_n, b_n\}_{n=-\infty}^0$. Thus, by (3.2.31), F is continuous as claimed. □

Theorem 7.4.5 (Kotani [244]). *Let $\mathcal{F}$ be a finite subset of $(0, \infty) \times \mathbb{R}$. Let $\Sigma, \mathfrak{e}$ be given with $|\Sigma| > 0$ and $\Sigma \subset \mathfrak{e}$. Then there exists a p so that every $\{a_n, b_n\}_{n=-\infty}^\infty \in \mathcal{R}(\Sigma, \mathfrak{e})$ with $(a_n, b_n) \in \mathcal{F}$ for all n has period p.*

Remark. We are only claiming p is a period for (a_n, b_n), that is,

$$a_{n+p} = a_n \qquad b_{n+p} = b_n \tag{7.4.16}$$

not that p is the minimal period.

Proof. Let $\mathcal{R}(\Sigma, \mathfrak{e}, \mathcal{F})$ be the set of $\{a_n, b_n\}_{n=-\infty}^{\infty} \in \mathcal{R}(\Sigma, \mathfrak{e})$ with every $(a_n, b_n) \in \mathcal{F}$.

Pick $\varepsilon > 0$ so that for all $(\alpha, \beta) \neq (\alpha', \beta')$ both in $\mathcal{F}$,

$$|\alpha - \alpha'| + |\beta - \beta'| \geq \varepsilon \tag{7.4.17}$$

Then, with F given by (7.4.14), for each $(\alpha_0, \beta_0) \in \mathcal{F}$,

$$F^{-1}\left(\{(a', b')\} \mid |a' - \alpha_0| + |b' - \beta_0| < \tfrac{\varepsilon}{2}\right)$$

is open and so depends only on finitely many $\{a_j, b_j\}_{j \leq 0}$. Since $\mathcal{F}$ is finite, we see that there are k and $\widetilde{F}$ defined on $\{\{a_n, b_n\}_{n=-k+1}^{0}\}$ to $(0, \infty) \times \mathbb{R}$, so on $\mathcal{R}(\Sigma, \mathfrak{e}, \mathcal{F})$, $\widetilde{F}$ gives the value of (a_1, b_1). Thus, by iteration, the k block $(-k+1, 0)$ determines $(1, k)$, that is, there is a map H from allowed values in $\mathcal{F}^k$ to itself.

In the same way, there is a map from $(1, k')$ to $(-k' + 1, 0)$ for some k'. So by increasing k (if $k' > k$), we can suppose that H is invertible.

Since $\mathcal{F}^k$ is finite, for any allowed value $\alpha \in \mathcal{F}^k$, there must be a repeated entry among α, $H(\alpha)$, $H^2(\alpha), \dots$. But if $H^k(\alpha) = H^{k+q}(\alpha)$, then by invertibility, $H^q(\alpha) = \alpha$, that is, the corresponding J has period kq. Since $\mathcal{F}^k$ is finite, there is a maximal period, r, for all the J's in $\mathcal{R}(\Sigma, \mathfrak{e}, \mathcal{F})$. Then $p = r!$ is a common period of all such J's. □

For an alternate argument, see [63].

Theorem 7.4.6 (Remling [366]). *Let $\mathcal{F}$ be a finite subset of $(0, \infty) \times \mathbb{R}$. Let J be a half-line Jacobi matrix with each $(a_n, b_n) \in \mathcal{F}$. Suppose J has some a.c. spectrum, that is, $|\Sigma_{\text{ac}}(J)| > 0$. Then J is eventually periodic, that is, for some p and N, we have for all $n \geq N$,*

$$a_{n+p} = a_n \qquad b_{n+p} = b_n$$

Remarks. 1. Eventually, periodic matrices are finite rank perturbations of strictly periodic J's, so by Theorem 5.3.7 and Corollary 7.3.4, they have some a.c. spectrum.

2. This shows, for example, that if $a_n \equiv 1$ and $b_n = 0, 1, 0, 0, 1, 1, 0, 0, 0, 1, 1, 1, 0, \dots$, then J has purely singular continuous spectrum.

Proof. Put some metric on $\mathcal{F}^{\mathbb{Z}}$, say

$$d(\{a, b\}, \{a', b'\}) = \sum_{n=-\infty}^{\infty} 2^{-n}(|a_n - a'_n| + |b_n - b'_n|) \tag{7.4.18}$$

Extend J to $\mathbb{Z}$ by setting (a_n, b_n) to some fixed point in $\mathcal{F}$ for $n \leq 0$. Let $J^{(m)}$ have parameters $\{a_{n+m}, b_{n+m}\}_{n=-\infty}^{\infty}$. I claim that

$$\min_{J_r \in \mathcal{R}(J)} d(\{a^{(m)}, b^{(m)}\}, \{a_{(r)}, b_{(r)}\}) \to 0 \tag{7.4.19}$$

as $m \to \infty$. For if not, there is $m_j \to \infty$ and ε so for all J_r,

$$d(\{a^{(m_j)}, b^{(m_j)}\}, \{a_{(r)}, b_{(r)}\}) \geq \varepsilon \tag{7.4.20}$$

But, by compactness, there is a subsequence so $J^{(m_{j_k})}$ converges to some J_r. Thus, (7.4.20) fails and (7.4.19) must hold.

By Remling's theorem, each J_r is reflectionless on $\Sigma_{\text{ac}}(J)$. So, by Theorem 7.4.1, each J_r is periodic with period p. Pick ε so small that

$$d((a,b),(a',b')) < \varepsilon \Rightarrow \sup_{0\le k\le 2p} (|a_k - a'_k| + |b_k - b'_k|)$$

$$< \min_{(\alpha,\beta)\neq(\alpha',\beta')\in\mathcal{F}} |\alpha-\alpha'| + |\beta-\beta'| \qquad (7.4.21)$$

Pick M so $m > M \Rightarrow \min_{J_r\in\mathcal{R}(J)} d(J^{(m)}, J_r) < \varepsilon$. Thus, if $m > M$, J from $m \to m+2p$ must be equal a unique element of J_r. But from $m+p$ to $m+3p$, the same is true. On the overlap from $m+p$ to $m+2p$, they must agree. So, by periodicity of J_r, J must agree with a single J_r from M onward. □

As a second application, here is a very strong sparse potential theorem:

Theorem 7.4.7 (Remling [366]). *Let J be a bounded half-line Jacobi matrix so there exist M fixed and $m_j \to \infty$ with*
(a) $\liminf_{j\to\infty}(|a_{m_j} - 1| + |b_{m_j}|) > 0$
(b) $M \le |\ell| \le j \Rightarrow |a_{m_j+\ell} - 1| + |b_{m_j+\ell}| = 0$
Then J has no a.c. spectrum.

Proof. By passing to a subsequence of m_j, we find $J_r \in \mathcal{R}(J)$ with
(i) $|a_0^{(r)} - 1| + |b_0^{(r)}| \neq 0$
(ii) $a_\ell^{(r)} = 1$ and $b_\ell^{(r)} = 0$ for $|\ell| \ge M$
By (ii), J_r takes only finitely many values. By Theorem 7.4.5, if there is $|\Sigma| > 0$ on which J_r is reflectionless, then J_r is periodic. So by (ii), $a_\ell^{(r)} = 1$, $b_\ell^{(r)} = 0$ for all ℓ, but that violates (i). We conclude that J_r is not reflectionless on any Σ with $|\Sigma| > 0$.

By Remling's theorem, $|\Sigma_{\text{ac}}(J)| = 0$, that is, J has no a.c. spectrum. □

Clearly, reflectionless Jacobi matrices are important, so one defines a *reflectionless measure*, $\Sigma \subset \mathbb{R}$, as a measure, μ, on $\mathbb{R}$ so that

$$F_\mu(z) \equiv \int \frac{d\mu(x)}{x-z} \qquad (7.4.22)$$

has

$$\operatorname{Re} F_\mu(x+i0) = 0 \quad \text{for a.e. } x \in \Sigma \qquad (7.4.23)$$

As we have seen, if (here $d\mu_{\text{s}}$ is dx-singular)

$$d\mu = w(x)\,dx + d\mu_{\text{s}} \qquad (7.4.24)$$

then $w(x) > 0$ for a.e. $x \in \Sigma$. We want to explore when $d\mu_{\text{s}}(\Sigma) = 0$, which often holds. For finite gap sets, we will see this in Section 7.5, but there is a very general result:

Theorem 7.4.8 (Poltoratski–Remling [350]). *Let μ be a reflectionless measure. Let*

$$\Sigma_0 = \bigl\{x \in \Sigma \mid \limsup_{\delta\downarrow 0}(2\delta)^{-1}|\Sigma \cap (x-\delta, x+\delta)| > 0\bigr\} \qquad (7.4.25)$$

Then

$$\mu_s(\Sigma_0) = 0 \tag{7.4.26}$$

Note that in many cases $\Sigma_0 = \Sigma$. For example, if Σ is a finite gap set, the lim sup in (7.4.25) is $\frac{1}{2}$ at endpoints and 1 at interior points. There is a class of sets called homogeneous sets for which $\Sigma_0 = \Sigma$. See the Notes for further discussion.

The proof of the result depends on the following (whose proof is in the references in the Notes):

Theorem 7.4.9 (Poltoratski's Theorem [349]). *For any measure of compact support μ and any $f \geq 0$ with $f \in L^1(\mathbb{R}, d\mu)$,*

$$\lim_{\varepsilon \downarrow 0} \frac{F_{f\mu}(x+i\varepsilon)}{F_\mu(x+i\varepsilon)} = f(x) \tag{7.4.27}$$

for μ_s a.e. x.

Corollary 7.4.10. (i) *If ν is mutually singular to μ, then for μ_s a.e. x,*

$$\lim_{\varepsilon \downarrow 0} \frac{F_\nu(x+i\varepsilon)}{F_\mu(x+i\varepsilon)} = 0 \tag{7.4.28}$$

(ii) *For any measure ν and μ_s a.e. x,*

$$\lim_{\varepsilon \downarrow 0} \frac{F_\nu(x+i\varepsilon)}{F_\mu(x+i\varepsilon)} < \infty \tag{7.4.29}$$

Remark. (ii) includes that the limit exists.

Proof. (i) Pick $S \subset \mathbb{R}$, a Borel set, with $\mu(\mathbb{R} \setminus S) = 0$ and $\nu(S) = 0$. Let $\eta = \mu + \nu$ so $\mu = \chi_S \eta$. By the theorem, for η_s a.e. x,

$$\frac{F_\mu(x+i\varepsilon)}{F_\mu(x+i\varepsilon) + F_\nu(x+i\varepsilon)} = \frac{F_{\chi_S\eta}(x+i\varepsilon)}{F_\eta(x+i\varepsilon)} \to \chi_S(x) \tag{7.4.30}$$

Thus, for η_s a.e. $x \in S$,

$$\frac{F_\nu(x+i\varepsilon)}{F_\mu(x+i\varepsilon)} \to 0 \tag{7.4.31}$$

Since μ is η a.c. μ_s is η_s a.c. By $\mu(\mathbb{R} \setminus S) = 0$, $\chi_S(x) = 1$ for μ_s a.e. x. Thus, (7.4.28) holds.

(ii) Write

$$d\nu = f\,d\mu + d\tilde{\mu}_s \tag{7.4.32}$$

where $d\tilde{\mu}_s$ is $d\mu$ singular (as opposed to $d\nu_s$ which is dx singular!). By the theorem and (7.4.28), for μ_s a.e. x,

$$\frac{F_\nu(x+i\varepsilon)}{F_\mu(x+i\varepsilon)} = f(x) \tag{7.4.33}$$

is finite a.e. $d\mu_s$. □

Proof of Theorem 7.4.8. The general Herglotz representation theorem, (2.3.87), implies that if G is analytic in $\mathbb{C}_+$ with $\operatorname{Im} G$ bounded, then

$$\operatorname{Im} G(x+iy) = \frac{y}{\pi} \int \frac{\operatorname{Im} G(u+i0)\, d\mu}{(u-x)^2 + y^2} \tag{7.4.34}$$

where we used $\operatorname{Im} G$ bounded to set $A = 0$ (since $Ay \to \infty$ as $y \to \infty$) and then conclude that the boundary measure $d\mu$ in (2.3.87) is purely a.c.

Since F_μ is Herglotz, $0 < \operatorname{Im} \log(F_\mu(z)) < \pi$ on $\mathbb{C}_+$ so, by (7.4.34), if

$$A(z) = \frac{1}{\pi} \operatorname{Im} \log(F_\mu(z)) \tag{7.4.35}$$

then

$$A(x+iy) = y \int \frac{A(u+i0)}{(u-x)^2 + y^2} \tag{7.4.36}$$

That μ is reflectionless implies that for Lebesgue a.e. $x \in \Sigma$,

$$A(x+i0) = \tfrac{1}{2} \tag{7.4.37}$$

Let

$$X(z) = \int \frac{\chi_\Sigma(y)\, dy}{y - z} \tag{7.4.38}$$

Then $A(u+i0) > 0$ and (7.4.151) proves

$$A(z) \geq \tfrac{1}{2} \operatorname{Im} X(z)$$

Similarly, $A(u+i0) \leq 1$ implies

$$A(z) + \tfrac{1}{2} \operatorname{Im} X(z) \leq 1$$

Thus,

$$0 \leq \pi A(z) \pm \frac{\pi}{2} \operatorname{Im} X(z) \leq \pi$$

so

$$F^\pm(z) = F_\mu(z) \exp\Bigl(\pm \frac{\pi}{2} X(z)\Bigr) \tag{7.4.39}$$

are both Herglotz functions.

Since $X(z) \to 0$ at ∞,

$$F^\pm(z) = -\frac{1}{z} + o\left(\frac{1}{z}\right) \tag{7.4.40}$$

at ∞ and F is real outside $\operatorname{supp}(d\mu)$ since X is analytic and $\operatorname{Im} X = 0$ outside $\bar{\Sigma}$. Thus, F is a discrete m-function.

By Corollary 7.4.10, for μ_s a.e. x, we have

$$\lim_{\varepsilon \downarrow 0} \frac{F^\pm(x+i\varepsilon)}{F_\mu(x+i\varepsilon)} \equiv f^\pm(x) \tag{7.4.41}$$

exists and is finite. By (7.4.39),

$$\frac{F^+(z)F^-(z)}{F_\mu(z)^2} \equiv 1 \tag{7.4.42}$$

Thus, for μ_s a.e. x, $f^+(x)f^{-1}(x) = 1$, so for μ_s a.e. x, $f^\pm(x)$ is strictly positive.

It follows that for μ_s a.e. x,

$$\lim_{\varepsilon\downarrow 0} \frac{F^+(x+i\varepsilon)}{F^-(x+i\varepsilon)} = (0,\infty) \tag{7.4.43}$$

so for μ_s a.e. x,

$$\lim_{\varepsilon\downarrow 0} \operatorname{Im} X(x+i0) = 0 \tag{7.4.44}$$

Since

$$\begin{aligned}\operatorname{Im} X(x+i\varepsilon) &\geq \varepsilon \int_{|x-u|\leq\varepsilon} \frac{\chi_\Sigma\, d\mu}{(x-u)^2+\varepsilon^2} \\ &\geq \frac{1}{2\varepsilon} |(x-\varepsilon, x+\varepsilon)\cap\Sigma|\end{aligned}$$

(for $|x-u|\leq\varepsilon \Rightarrow [(x-u)^2+\varepsilon^2]^{-1} \geq (2\varepsilon^2)^{-1}$), we see for μ_s a.e. x,

$$\lim \delta^{-1}|(x-\delta, x+\delta)\cap\Sigma| = 0$$

so $\mu_s(\Sigma_0) = 0$. □

Finally, we turn to the proof of Remling's theorem. A key element in the proof will be a notion of convergence of Herglotz functions. We will consider general Herglotz functions rather than just m-functions because we will use (7.4.2) as the criterion for reflectionless, namely,

$$a_0^2 m(\lambda+i0, J_0^+) = \overline{m(\lambda+i0, J_0^-)}^{\,-1} \tag{7.4.45}$$

and the (negative of the complex conjugate of the) right side is Herglotz, but not an m-function since it is $z+O(1)$ at infinity not $-z^{-1}+O(z^{-2})$. If F is a Herglotz function and $s \in \mathbb{R}\cup\{\infty\}$, we define for $z\in\mathbb{C}_+$ and $s\neq\infty$ (and $F_\infty = F$)

$$F_s(z) \equiv \frac{1+sF(z)}{s-F(z)} = -s + \frac{1+s^2}{s-F(z)} \tag{7.4.46}$$

which is also Herglotz.

Since we are dealing with general Herglotz functions, we need the general representation (2.3.87), which says that for each s, there is $A_s \geq 0$ and $d\mu_F^{(s)}(x)$ obeying

$$\int \frac{d\mu_F^{(s)}(x)}{1+x^2} < \infty \tag{7.4.47}$$

so that

$$\frac{\operatorname{Im} F_s(z)}{\operatorname{Im} z} = A_s + \int \frac{d\mu_F^{(s)}(x)}{(x-\operatorname{Re} z)^2 + (\operatorname{Im} z)^2} \tag{7.4.48}$$

Using $A_s = \lim_{y\to\infty} F_s(iy)/y$, (7.4.46), and

$$F_t = (F_s)_{(\frac{1+st}{s-t})} \tag{7.4.49}$$

we see that there is at most one s in $\mathbb{R} \cup \{\infty\}$ with $A_s \neq 0$.

Define for Borel sets $A, S \subset \mathbb{R}$ with $|A| < \infty$, $|S| < \infty$,

$$\omega_F(A, S) = \int_{s\in S} \mu_F^{(s)}(A) \frac{ds}{1+s^2} \tag{7.4.50}$$

We will see below this is finite and can be defined for all Borel $S \subset \mathbb{R}$.

Definition. Let $\{F_n\}_{n=1}^\infty$ and F be Herglotz functions. We say that $F_n \to F$ in *Pearson sense* if and only if, for any A with $|A| < \infty$ and all $S \subset \mathbb{R}$,

$$\omega_{F_n}(A, S) \to \omega_F(A, S) \tag{7.4.51}$$

Here are the facts we will prove about Pearson convergence, which we will see imply Remling's theorem:

Proposition 7.4.11. *If* $\operatorname{Im} F(\lambda + i0) > 0$ *and* $\operatorname{Im} G(\lambda + i0) > 0$, *for a.e.* $\lambda \in \Sigma$, *a Borel set of* $\mathbb{R}$, *and*

$$\omega_F(A, S) = \omega_G(A, -S)$$

for all $S \subset \mathbb{R}$ *and all* $A \subset \Sigma$, *then for a.e.* $\lambda \in \Sigma$,

$$F(\lambda + i0) = -\overline{G(\lambda + i0)}$$

Proposition 7.4.12. *Let* $\{F_n\}_{n=1}^\infty$ *and* F *be Herglotz functions. Then* $F_n \to F$ *in Pearson sense if and only if* $F_n(z) \to F(z)$ *uniformly on all compact sets* $K \subset \mathbb{C}_+$.

As usual, given a half-line Jacobi matrix, J, we let J_n be the n-times stripped Jacobi matrix, and we define

$$m_{+,n}(z) = m(z, J_n) \tag{7.4.52}$$

We let $J_{n;F}$ be the $n \times n$ finite Jacobi matrix and

$$m_{-,n}(z) = \langle \delta_n, (J_{n;F} - z)^{-1} \delta_n \rangle \tag{7.4.53}$$

The core of the proof of Remling's theorem will be:

Theorem 7.4.13 (Breimesser–Pearson Theorem)**.** *Let* J *be a Jacobi matrix with bounded parameters. Then for* $A \subseteq \Sigma_{\mathrm{ac}}(J)$ *and any* $S \subset \mathbb{R}$,

$$\lim_{n\to\infty} \omega_{m_{+,n}}(A, S) - \omega_{(-a_n^2 m_{-,n})^{-1}}(A, -S) = 0 \tag{7.4.54}$$

Proof of Theorem 7.4.3 given the above results. Let J_r be a right limit of J and $n_k \to \infty$ so that

$$a_{n_k+m} \to a_m^{(r)} \qquad b_{n_k+m} \to b_m^{(r)} \tag{7.4.55}$$

Let $m_\pm^{(r)}$ be $m(\,\cdot\,, (J_r)_0^\pm)$.

By (7.4.55), $J_{n_k} \xrightarrow{s} J_0^+$ and $\tilde{J}_{n;F} \xrightarrow{s} J_0^-$ with $\tilde{J}_{n;F}$ the Jacobi matrix with parameters

$$\tilde{a}_\ell^{(n)} = \begin{cases} a_{n-\ell} & \ell \le n-1 \\ 0 & \ell \ge n \end{cases} \qquad \tilde{b}_\ell^{(n)} = \begin{cases} b_{n-\ell+1} & \ell \le n \\ 0 & \ell \ge n+1 \end{cases} \tag{7.4.56}$$

so, uniformly on compact sets $K \subset \mathbb{C}_+$,

$$m_{\pm,n_k} \to m_\pm^{(r)} \tag{7.4.57}$$

By Proposition 7.4.12 and $a_{n_k} \to a_0^{(r)}$,

$$\omega_{m_{+,n_k}}(A, S) \to \omega_{m_+^{(r)}}(A, S) \tag{7.4.58}$$

$$\omega_{(-a_{n_k}^2 m_{-,n})^{-1}}(A, -S) \to \omega_{(-a_0^{(r)} m_-^{(r)})^{-1}}(A, -S) \tag{7.4.59}$$

By Theorem 7.4.13, for $A \subset \Sigma_{\rm ac}(J)$,

$$\omega_{m_+^{(r)}}(A, S) = \omega_{(-a_0^{(r)2} m_-^{(r)})^{-1}}(A, -S) \tag{7.4.60}$$

By Theorem 7.3.1, $\operatorname{Im} m_+^{(r)}(\lambda+i0)$ and $\operatorname{Im}(\overline{m_-^{(r)}(\lambda+i0)})^{-1} > 0$ for a.e. $\lambda \in \Sigma_{\rm ac}(J)$, so, by Proposition 7.4.11 for a.e. $\lambda \in \Sigma_{\rm ac}(J)$,

$$m_+^{(r)}(\lambda + i0) = \big(a_0^{(r)2}\, \overline{m_-^{(r)}(\lambda + i0)}\big)^{-1} \tag{7.4.61}$$

that is, J_r is reflectionless on $\Sigma_{\rm ac}(J)$. □

Proof of Proposition 7.4.11. We begin by noting that

$$\omega_F(A, S) = \lim_{\varepsilon \downarrow 0} \int_{x \in A} dx \int_{t \in S} \frac{dt}{\pi} \frac{\operatorname{Im} F(x+i\varepsilon)}{(t - \operatorname{Re} F(x+i\varepsilon))^2 + (\operatorname{Im} F(x+i\varepsilon))^2} \tag{7.4.62}$$

Accepting this for a moment, we note that since

$$\int_{-\infty}^{\infty} \frac{dt}{\pi} \frac{y}{(t-x)^2 + y^2} = 1$$

for $y \neq 0$, we see that for all S,

$$0 \le \omega_F(A, S) \le |A| \tag{7.4.63}$$

proving that ω_F can be defined for all S, by (7.4.62), so long as $|A| < \infty$.

For $|S| < \infty$, define

$$H_S(z) = \frac{1}{\pi} \int_S \frac{dt}{t - z} \tag{7.4.64}$$

so (7.4.64) says

$$\omega_F(A, S) = \lim_{\varepsilon \downarrow 0} \int_{x \in A} dx \, \operatorname{Im} H_S(F(x+i\varepsilon)) \tag{7.4.65}$$

$$= \int_{x \in A} dx \, \operatorname{Im} H_S(F(x+i0)) \tag{7.4.66}$$

because $\operatorname{Im} H_S(z) \leq 1$, $|A| < \infty$, $F(x+i\varepsilon) \to F(x+i0)$ for a.e. $x \in A$, we can suppose $\operatorname{Im} F(x+i0)$ and $\operatorname{Im} G(x+i0)$ are a.e. strictly positive on A, and H_S is continuous on $\{z \mid \operatorname{Im} z > 0\}$. A similar formula holds for G.

Thus, for all $A \subset \Sigma$ and all S,

$$\int_{\substack{x\in A\\ t\in S}} \frac{\operatorname{Im} F(x+i0)}{(t-\operatorname{Re} F(x+i0))^2 + \operatorname{Im} F(x+i0)^2}\, dx\, dt$$
$$= \int_{\substack{x\in A\\ t\in S}} \frac{\operatorname{Im} G(x+i0)}{(-t-\operatorname{Re} G(x+i0))^2 + (\operatorname{Im} G(x+i0))^2} \tag{7.4.67}$$

By Lebesgue's theorem on differentiation of integrals for a.e. x, t,

$$\frac{\operatorname{Im} F(x+i0)}{(t-\operatorname{Re} F(x+i0))^2 + (\operatorname{Im} F(x+i0))^2}$$
$$= \frac{\operatorname{Im} G(x+i0)}{(t+\operatorname{Re} G(x+i0)^2) + (\operatorname{Im} G(x+i0)^2)} \tag{7.4.68}$$

By continuity in t, we have this for a.e. x and all t. Since $b/((t-a)^2+b^2)$ takes its maximum value at $t=a$ and the value there is b^{-1}, we conclude that for a.e. x, $F(x+i0) = -\overline{G(x+i0)}$ as required.

Thus, we only need (7.4.62). This comes from noting that, by (7.4.48) and then (7.4.46),

$$\begin{aligned}\mu_F^{(s)}(A) &= \lim_{\varepsilon\downarrow 0} \frac{1}{\pi} \int_{x\in A} \operatorname{Im} F_s(x+i\varepsilon)\, dx \\ &= \frac{1+s^2}{\pi} \lim_{\varepsilon\downarrow 0} \int_{x\in A} \frac{\operatorname{Im} F(x+i\varepsilon)}{(s-\operatorname{Re} F(x+i\varepsilon))^2 + (\operatorname{Im} F(x+i\varepsilon))^2}\, dx\end{aligned}$$

so (7.4.50) implies (7.4.62). □

If F is a Herglotz function, it will be convenient several places below to define $\omega_{\bar F}$ by $\omega_{\bar F}(A,S) \equiv \omega_F(A,-S)$ for all $A \subset \mathbb{R}$ with $|A| < \infty$ and all $S \subset \mathbb{R}$. This is consistent with (7.4.62) in that it holds for ω_F if we replace F by $\bar F$ on the right.

We need a number of preparatory lemmas:

Lemma 7.4.14. *For $S = (\alpha, \beta)$ and $z \in \overline{\mathbb{C}}_+$, let* $\operatorname{Im} H_S(z)$ *be given by* (7.4.64) *for* $\operatorname{Im} z > 0$ *and by*

$$\operatorname{Im} H_S(x) = \lim_{\varepsilon\downarrow 0} \operatorname{Im} H_S(x+i\varepsilon) = \begin{cases} 1 & \alpha < x < \beta \\ 0 & x < \alpha \text{ or } x > \beta \\ \frac{1}{2} & x = \alpha \text{ or } x = \beta \end{cases} \tag{7.4.69}$$

Suppose

$$\operatorname{Im} H_S(z) = \operatorname{Im} H_S(w) \tag{7.4.70}$$

for some $z, w \in \overline{\mathbb{C}}_+$ and all $S = (\alpha, \beta)$ with α, β rational. Then $z = w$.

Proof. If $z \in \mathbb{R}$, there are α, β rational, so $\operatorname{Im} H_{(\alpha,\beta)}(z) = 1$, and if $z \notin \mathbb{R}$, then $\operatorname{Im} H_{(\alpha,\beta)}(z) < 1$ for all α, β. Thus, z and w are either both real or both nonreal.

If z, w are real and distinct, say $z < w$, pick α, β rational so that $\alpha < z < \beta < w$. Then $\operatorname{Im} H_{(\alpha,\beta)}(z) = 1 \neq 0 = \operatorname{Im} H_{(\alpha,\beta)}(w)$, so reality plus (7.4.70) imply $z = w$.

If z is in $\mathbb{C}_+$,

$$\lim_{\varepsilon\downarrow 0} \varepsilon^{-1}\pi \operatorname{Im} H_{(\alpha,\alpha+\varepsilon)}(z) = \frac{\operatorname{Im} z}{(\alpha - \operatorname{Re} z)^2 + (\operatorname{Im} z)^2} \tag{7.4.71}$$

so if (7.4.70) holds for all rational (α, β), then

$$\frac{\operatorname{Im} z}{(\alpha - \operatorname{Re} z)^2 + (\operatorname{Im} z)^2} = \frac{\operatorname{Im} w}{(\alpha - \operatorname{Re} w)^2 + (\operatorname{Im} w)^2} \tag{7.4.72}$$

first for all rational α, and then for all real α by continuity. The maximum occurs at $\alpha = \operatorname{Re} z$ and the value is $(\operatorname{Im} z)^{-1}$, so z is determined by these values, that is, $z = w$ if (7.4.70) holds. □

Lemma 7.4.15. *If F, G are Herglotz functions so that for all $S \subset \mathbb{R}$ and A with $|A| < \infty$,*

$$\omega_F(A, S) = \omega_G(A, S) \tag{7.4.73}$$

then $F = G$.

Remarks. 1. The proof shows that we only need this for A, S finite intervals with rational end points.

2. This implies Pearson convergence is Hausdorff.

Proof. We first claim that for all Herglotz F, (7.4.66), which we previously proved for all S, but only when $\operatorname{Im} F(\lambda + i0)$ for a.e. $\lambda \in A$, holds for $S = (\alpha, \beta)$. For $\operatorname{Im} H_S(F(x + i\varepsilon))$ is uniformly bounded by 1, and if $F(x + i\varepsilon) \to F(x + i0)$ and $F(x + i0) \neq \alpha, \beta$, then $\operatorname{Im} H_S(F(x + i\varepsilon)) \to \operatorname{Im} H_S(F(x + i0))$. Since $F(x + i0)$ takes any given value on a set of Lebesgue measure zero and fails to have a limit on a set of measure zero, the integrals converge.

Thus, (7.4.73) implies that for any $\alpha, \beta \in \mathbb{R}$, for a.e. x, that (7.4.70) holds for $S = (\alpha, \beta)$. Since $\{\alpha, \beta \mid \alpha, \beta \text{ rational}\}$ is countable, for a.e. x, (7.4.71) holds for all rational α, β. So, by Lemma 7.4.14, for a.e. x, $F(x + i0) = G(x + i0)$. Since F is determined by its boundary value on any set of positive measure, (see Theorem 2.3.20), $F = G$. □

Define the Cauchy measure $\omega_z(\cdot)$ for $z \in \mathbb{C}_+$ and Borel $S \subset \mathbb{R}$ by,

$$\omega_z(S) = \frac{1}{\pi} \int_{t\in S} \frac{\operatorname{Im} z}{(t - \operatorname{Re} z)^2 + (\operatorname{Im} z)^2}\, dt \tag{7.4.74}$$

Lemma 7.4.16. *Let F be a Herglotz function. Then for any $z \in \mathbb{C}_+$ and Borel $S \subset \mathbb{R}$,*

$$\omega_{F(z)}(S) = \lim_{\varepsilon\downarrow 0} \int_{-\infty}^{\infty} \omega_{F(t+i\varepsilon)}(S)\, d\omega_z(t) \tag{7.4.75}$$

Proof. Suppose first that F is continuous on $\overline{\mathbb{C}}_+$ with $\operatorname{Im} F(x+i0) > 0$ for all x. Then since $d\omega_z$ is a probability measure, $\omega_{F(t+i\varepsilon)}(S) \le 1$, and we have convergence for all t, (7.4.75) becomes

$$\omega_{F(z)}(S) = \int_{-\infty}^{\infty} \omega_{F(t+i0)}(S)\, d\omega_z(t) \tag{7.4.76}$$

Both sides of (7.4.76) are bounded harmonic functions of z in $\mathbb{C}_+$, continuous on $\overline{\mathbb{C}}_+$, which agree on $\mathbb{R}$, so they are equal. Thus, (7.4.75) holds for continuous F's obeying $\operatorname{Im} F(x+i0) > 0$ for all x. That implies, applying to

$$F_\varepsilon(z) = F(z+i\varepsilon) \tag{7.4.77}$$

that

$$\int_{-\infty}^{\infty} \omega_{F(t+i\varepsilon)}(S)\, d\omega_z(t) = \omega_{F(z+i\varepsilon)}(S) \tag{7.4.78}$$

Picking $\varepsilon \downarrow 0$ yields (7.4.75). □

Lemma 7.4.17. *For any $A \subset \mathbb{R}$ with $|A| < \infty$,*

$$\lim_{\varepsilon\downarrow 0} \int_{t\in A} \omega_{t+i\varepsilon}(\mathbb{R}\setminus A)\, dt = 0 \tag{7.4.79}$$

Proof. For each $k \in (0,\infty)$,

$$\begin{aligned}(\text{LHS of (7.4.79)}) \le{}& \lim_{\varepsilon\downarrow 0} \int_{\substack{|s-t|\le \varepsilon k\\ s\in\mathbb{R}\setminus A\\ t\in A}} \varepsilon^{-1}\, ds dt \\ &+ \lim_{\varepsilon\downarrow 0} \int_{\substack{|s-t|> \varepsilon k\\ s\in\mathbb{R}\setminus A\\ t\in A}} \frac{\varepsilon}{(s-t)^2+\varepsilon^2}\, ds dt\end{aligned} \tag{7.4.80}$$

By letting $u = |s-t|/\varepsilon$, the second integral is bounded by $|A|\frac{1}{\pi}\int_{|u|\ge k}\frac{du}{1+u^2}$. The first integral is bounded by

$$\int_{t\in A} \varepsilon^{-1} |(t-\varepsilon k, t+\varepsilon k) \cap (\mathbb{R}\setminus A)|\, dt \tag{7.4.81}$$

By the Lebesgue theorem on integration of derivatives, the integrand goes to zero for each fixed k and a.e. x, and it is bounded by $2k$, so it goes to zero by dominated convergence.

Thus, for each k,

$$(\text{LHS of (7.4.79)}) \le \frac{1}{\pi}|A| \int_{|u|\ge k} \frac{du}{1+u^2}$$

so, taking $k\to\infty$, (7.4.79) holds. □

Lemma 7.4.18. *Let $A \subset \mathbb{R}$ be a Borel set with $|A| < \infty$. Then, uniformly in Borel $S \subset \mathbb{R}$ and Herglotz functions F,*

$$\lim_{\varepsilon\downarrow 0} \omega_{F_\varepsilon}(A,S) = \omega_F(A,S) \tag{7.4.82}$$

where F_ε is given by (7.4.77).

Remark. This is remarkable in that if F is Herglotz and $\alpha \in (0,\infty)$, the result holds for all αF, uniformly in α!

Proof. By (7.4.62) and (7.4.75),

$$|\omega_{F_\varepsilon}(A,S) - \omega_F(A,S)| = \lim_{\substack{\delta\downarrow 0\\ \delta<\varepsilon}} \left| \int_{-\infty}^{\infty} \omega_{F_\delta(t)}(S)(\omega_{t+i(\varepsilon-\delta)}(A) - \chi_A(t))\, dt \right| \tag{7.4.83}$$

Since $0 \le \omega_{F_\delta(t)}(S) \le 1$ for all F and all S and

$$\omega_{t+i(\varepsilon-\delta)}(A) - \chi_A(t) = \begin{cases} > 0 & \text{for } t \notin A \\ < 0 & \text{for } t \in A \end{cases} \tag{7.4.84}$$

we conclude that

$$\begin{aligned} |\omega_{F_\varepsilon}(A,S) - \omega_F(A,S)| &\le \max_{B=A,\mathbb{R}\setminus A} \int_B \omega_{t+i\varepsilon}(\mathbb{R}\setminus B)\, dt \\ &= \int_A \omega_{t+i\varepsilon}(\mathbb{R}\setminus A)\, dt \end{aligned} \tag{7.4.85}$$

this integral is independent of F and S, and goes to zero by Lemma 7.4.17. □

Lemma 7.4.19. *For any Herglotz function F and any A with $|A| < \infty$,*

$$\lim_{R\to\infty} \omega_F(A,(-R,R)) = |A| \tag{7.4.86}$$

and for all $a \in \mathbb{R}$,

$$\lim_{R\to\infty} \omega_F\left(A,(-R,R)\setminus\left[a-\tfrac{1}{R},a+\tfrac{1}{R}\right]\right) = |A| \tag{7.4.87}$$

Proof. By Lemma 7.4.18, it suffices to prove this for each F_ε. Since $\operatorname{Im} F_\varepsilon(x+i0) = \operatorname{Im} F(x+i\varepsilon) > 0$, (7.4.66) holds, that is,

$$\omega_{F_\varepsilon}(A,S) = \int_{x\in A} ds \int_{t\in S} \frac{dt}{\pi} \frac{\operatorname{Im} F(x+i\varepsilon)}{(t-\operatorname{Re} F(x+i\varepsilon))^2 + \operatorname{Im}|F(x+i\varepsilon)|^2}$$

and (7.4.86) and (7.4.87) are obvious for each fixed x, and then, by dominated convergence, for the integral. □

Lemma 7.4.20. *For any sequence F_n of Herglotz functions, at least one of the following is true:*

(a) *There is a Herglotz function F and $n_j \to \infty$ so that $F_{n_j} \to F$ uniformly on compact subsets $K \subset \mathbb{C}_+$.*
(b) *There is $n_j \to \infty$ so that $F_{n_j} \to a \in \mathbb{R}$ uniformly on compact subsets of $\mathbb{C}_+$.*
(c) *There is $n_j \to \infty$ so that $-F_n^{-1} \to 0$ uniformly on compact subsets $K \subset \mathbb{C}$.*

Remark. This is a standard normal families argument; see, for example, Ahlfors [7, Section 5.5].

Proof. Let $G_n(z) = (F_n(z) - i)/(F_n(z) + i) \in \mathbb{D}$. By Montel's theorem and the maximum principle, there is a subsequence G_{n_j} so that one of

(a) $G_{n_j}(z) \to G(z)$ uniformly on $K \subset \mathbb{C}_+$ with values in $\mathbb{D}$.

(b) $G_{n_j}(z) \to \alpha \in \partial\mathbb{D} \setminus \{-1\}$ uniformly on compacts.
(c) $G_{n_j}(z) \to -1$ uniformly on compacts.
Then F_{n_j} obeys one of the conditions in the theorem. $\square$

Lemma 7.4.21. *If $F_n \to a \in \mathbb{R}$ uniformly on compacts, then for all bounded $[\alpha, \beta] \not\ni a$ and all bounded $A \subset \mathbb{R}$, $\omega_{F_n}(A, (\alpha, \beta)) \to 0$. If $(F_n)^{-1} \to 0$ uniformly on compacts, then for all bounded (α, β) and all bounded $A \subset \mathbb{R}$, $\omega_{F_n}(A, (\alpha, \beta)) \to 0$.*

Proof. By (7.4.50), $F_s(i) \to 0$ uniformly for $s \in S$ an allowed (α, β). But, by (7.4.48), $\int \frac{d\mu_F^{(s)}(x)}{x^2+1} \leq \operatorname{Im} F_s(i)$. So, by (7.4.50), $\omega_F(A, S) \to 0$. $\square$

Proof of Proposition 7.4.12. If $F_n \to F$ on compacts, we claim first that for any $\varepsilon > 0$, $(F_n)_\varepsilon \to F_\varepsilon$ in Pearson sense. For given δ and A with $|A| < \infty$, pick R with $|\mathbb{R} \setminus A_R| < \delta/3$ where $A_R = A \cap [-R, R]$. Since $|\omega_G(A, S)| \leq |A|$ (by (7.4.63)),

$$|\omega_{(F_n)_\varepsilon}(A, S) - \omega_{F_\varepsilon}(A, S)| \leq \frac{2\delta}{3} + |\omega_{(F_n)_\varepsilon}(A_R, S) - \omega_{F_\varepsilon}(A_R, S)| \qquad (7.4.88)$$

By (7.4.66), which is valid for $(F_n)_\varepsilon$ and (F_ε) (since $\operatorname{Im}(F_n)_\varepsilon(x+i0) = \operatorname{Im} F_n(x+i\varepsilon) > 0$), the presumed uniform convergence on compacts, and the dominated convergence theorem, for n large, the last term in (7.4.88) is less than $\delta/3$. Since δ is arbitrary, we have the claimed Pearson convergence.

This plus Lemma 7.4.18 implies $F_n \to F$ in Pearson sense.

Conversely, let $F_n \to F$ in Pearson sense. If F_n does not converge to F uniformly on compacts, by Lemma 7.4.20, there is n_j so that on compacts, either $F_{n_j} \to G \neq F$ with G Herglotz, or $F_{n_j} \to a \in \mathbb{R} \cup \{\infty\}$. In the first case, $\omega_F(A, S) = \omega_G(A, S)$ by the first part of this proof, and then $F = G$ by Lemma 7.4.15.

In the latter case, by Lemma 7.4.21 and the presumed Pearson convergence,

$$\omega_F(A, (\alpha, \beta)) = 0 \qquad (7.4.89)$$

for all α, β and $[\alpha, \beta] \not\ni a$. This contradicts Lemma 7.4.19. $\square$

The remaining part of the proof of Remling's theorem is to prove the Breimesser–Pearson theorem, Theorem 7.4.13, which will require some machinery! We are going to make use of a natural pseudo-hyperbolic "metric" (which does not obey the triangle inequality) on $\mathbb{C}_+$:

$$\gamma(z, w) = \frac{|z - w|}{\sqrt{\operatorname{Im} z}\,\sqrt{\operatorname{Im} w}} \qquad (7.4.90)$$

We will see below that it is a monotone increasing function of the hyperbolic metric and locally equivalent to it. The important property for us is:

Theorem 7.4.22. *If F is a Herglotz function on $\mathbb{C}_+$, then for all $z, w \in \mathbb{C}_+$,*

$$\gamma(F(z), F(w)) \leq \gamma(z, w) \qquad (7.4.91)$$

Remark. In the Schwarz lemma, one has that if $|F(z)| = |z|$ for a single $z \neq 0$, $z \in \mathbb{D}$, then F is a rotation. Pushing this fact through the proof below shows equality for a single pair $z \neq w$ implies F is a fractional linear transformation that maps $\mathbb{C}_+$ to itself.

Proof. This will require several facts about the hyperbolic metric on $\mathbb{D}$ that we will discuss in Section 9.3. It can be written as (see Theorem 9.3.11)

$$\tanh(\rho(z,w)) = \frac{|z-w|}{|1-\bar{z}w|} \tag{7.4.92}$$

and it has the critical property that any fractional linear transformation, G, that maps $\mathbb{D}$ onto $\mathbb{D}$ is an isometry (Theorem 9.3.10), that is,

$$\rho(G(z),G(w)) = \rho(z,w) \tag{7.4.93}$$

The Schwarz lemma says that if $F\colon \mathbb{D} \to \mathbb{D}$ is analytic and $F(0) = 0$, then $|F(z)| \le |z|$. Since

$$\rho(z,0) = \text{arctanh}|z| \tag{7.4.94}$$

and arctanh is monotone, we conclude for such F,

$$\rho(F(z),F(0)) \le \rho(z,0) \tag{7.4.95}$$

Since for any $z \in \mathbb{D}$, there is a fractional linear transformation mapping $\mathbb{D} \to \mathbb{D}$ with $G(0) = z$ (and so $G^{-1}(z) = 0$), any F that maps $\mathbb{D}$ to $\mathbb{D}$ has the form $G_1 \circ \tilde{F} \circ G_2$ where G_1, G_2 obey (7.4.93) and $F(0) = 0$, so (7.4.95). Thus, any map $F\colon \mathbb{D} \to \mathbb{D}$ obeys (7.4.95), and so, using maps of 0 to w, we see that such F obey

$$\rho(F(z),F(w)) \le \rho(z,w) \tag{7.4.96}$$

The map $\varphi\colon \mathbb{C}_+ \to \mathbb{D}$ by

$$\varphi(z) = \frac{z-i}{z+i} \tag{7.4.97}$$

is a bijection and allows us to define a metric; we'll also call it ρ on $\mathbb{C}_+$ by

$$\rho_{\mathbb{C}_+}(z,w) = \rho_{\mathbb{D}}(\varphi(z),\varphi(w)) \tag{7.4.98}$$

(where we temporarily indicate the set). F is a Herglotz function on $\mathbb{C}_+$ if and only if $\varphi \circ F \circ \varphi^{-1}$ maps $\mathbb{D}$ to $\mathbb{D}$ analytically, so by (7.4.96), for any Herglotz function,

$$\rho_{\mathbb{C}_+}(F(z),F(w)) \le \rho_{\mathbb{C}_+}(z,w) \tag{7.4.99}$$

A straightforward calculation shows that

$$\frac{|\varphi(z)-\varphi(w)|}{|1-\overline{\varphi(z)}\,\varphi(w)|} = \frac{|z-w|}{|\bar{z}-w|} \tag{7.4.100}$$

so (7.4.92) implies

$$\tanh(\rho_{\mathbb{C}_+}(z,w)) = \frac{|z-w|}{|\bar{z}-w|} \tag{7.4.101}$$

Another straightforward calculation shows that

$$4\left(\frac{|z-w|}{\sqrt{\text{Im}\, z}\,\sqrt{\text{Im}\, w}}\right)^{-1} = \left(\frac{|z-w|}{|\bar{z}-w|}\right)^{-2} - 1 \tag{7.4.102}$$

so there is a strictly monotone function Ψ with

$$\gamma(z,w) = \Psi(\rho(z,w)) \tag{7.4.103}$$

Thus, (7.4.97) implies (7.4.91) for Herglotz functions, F. □

Proposition 7.4.23. *For any $w, z \in \mathbb{C}_+$ and $S \subset \mathbb{R}$ Borel*

$$|\omega_z(S) - \omega_w(S)| \le \gamma(z, w) \tag{7.4.104}$$

where $\omega_z(S)$ is given by (7.4.74).

Proof. $\omega_z(S) = \operatorname{Im} H_S(z)$ where H_S is the Herglotz function (7.4.64). Moreover, $0 < \omega_z(S) < 1$. Thus,

$$|\omega_z(S) - \omega_w(S)| \le \frac{|\omega_z(S) - \omega_w(S)|}{\omega_z(S)\omega_w(S)} \tag{7.4.105}$$

$$\le \gamma(H_S(z), H_S(w)) \tag{7.4.106}$$

$$\le \gamma(z, w) \tag{7.4.107}$$

by (7.4.91). □

We know for $\operatorname{Im} z > 0$, there is a unique solution $u_n^+(z)$ of (3.2.6). It obeys

$$\begin{pmatrix} u_{n+1}^+ \\ -a_n u_n^+ \end{pmatrix} = A_+(a_n, b_n; z) \begin{pmatrix} u_n^+ \\ -a_{n-1} u_{n-1}^+ \end{pmatrix} \tag{7.4.108}$$

where

$$A_\pm(a, b; z) = \begin{pmatrix} \frac{z-b}{a} & \pm\frac{1}{a} \\ \mp a & 0 \end{pmatrix} \tag{7.4.109}$$

A_- is the transfer matrix of (3.2.2). We use A_+ because of the minus sign in (7.4.108), whose presence we want for reasons that will be clear shortly. $A_\pm$ induce fractional linear transformations on $\mathbb{C}_+ \cup \{\infty\}$ by

$$\mathcal{A}_\pm(a, b; z)(w) = \frac{(z-b)w \pm 1}{\mp a^2 w} \tag{7.4.110}$$

Since $m_n^+(z) = -u_{n+1}^+(z)/a_n u_n^+(z)$ (see (5.4.57)) with

$$M_n^+(z) = m_n^+(z) \tag{7.4.111}$$

(7.4.108) implies

$$M_n^+(z) = \mathcal{A}_+(a_n, b_n; z)(M_{n-1}^+(z)) \tag{7.4.112}$$

equivalent to the familiar (5.4.49).

On the other hand, by Cramer's rule, with P_n the monic OPs,

$$m_{-,n}(z) = -\frac{P_{n-1}(z)}{P_n(z)} = -\frac{p_{n-1}(z)}{a_n p_n(z)} \tag{7.4.113}$$

and thus defining M_n^- by

$$M_n^-(z) = -(a_n^2 m_{-,n}(z))^{-1} = \frac{p_n(z)}{a_n p_{n-1}(z)} \tag{7.4.114}$$

The usual recursion relation for p_j:

$$\begin{pmatrix} p_n(z) \\ a_n p_{n-1}(z) \end{pmatrix} = A_-(a_n, b_n; z) \begin{pmatrix} p_{n-1}(z) \\ a_{n-1} p_{n-2}(z) \end{pmatrix} \tag{7.4.115}$$

implies that

$$M_n^-(z) = \mathcal{A}_-(a_n, b_n; z)(M_{n-1}^-(z)) \tag{7.4.116}$$

Define

$$\mathcal{J}_n^\pm(z) = \mathcal{A}_\pm(a_n, b_n; z) \circ \mathcal{A}_\pm(a_{n-1}, b_{n-1}; z) \circ \cdots \circ \mathcal{A}_\pm(a_1, b_1; z) \tag{7.4.117}$$

so we have

$$M_n^\pm(z) = \mathcal{J}_n^\pm(z)(M_0^\pm(z)) \tag{7.4.118}$$

where (m_J is the m-function of the Jacobi matrix, J)

$$M_0^+(z) = m_J(z) \qquad M_0^-(z) \equiv \infty \tag{7.4.119}$$

since $p_{-1}(z) \equiv 0$ and (7.4.114) holds.

$M_n^\pm$ are Herglotz functions and (7.4.54), the Breimesser–Pearson result we seek to prove, says that

$$\omega_{M_n^+}(A, S) - \omega_{M_n^-}(A, -S) \to 0 \tag{7.4.120}$$

for all $S \subset \mathbb{R}$ and $A \subseteq \Sigma_{\text{ac}}(J)$.

Proposition 7.4.24. (a) *For any $x \in \mathbb{R}$, $\mathcal{J}_n^\pm(x)(\cdot)$ are analytic bijections of $\mathbb{C}_+$.*
(b) *For any $z \in \mathbb{C}_+$, $a \in (0, \infty)$, and $b \in \mathbb{R}$, $\mathcal{A}_-(a, b; z)(\cdot)$, and so, $\mathcal{J}_n^-(z)(\cdot)$ are Herglotz functions.*

Proof. (a) Fractional linear transformations generated by real coefficients have

$$\operatorname{Im}\left(\frac{az+b}{cz+d}\right) = \frac{(ad-bc)\operatorname{Im} z}{|cz+d|^2} \tag{7.4.121}$$

and so if determinant 1 are bijections of $\mathbb{C}_+$ to $\mathbb{C}_+$.

(b)

$$\mathcal{A}_-(a, b; z)(w) = \frac{z-b}{a^2} - \frac{1}{a^2 w} \tag{7.4.122}$$

clearly takes $\operatorname{Im} w > 0$ to $\operatorname{Im} w > 0$ if $\operatorname{Im} z > 0$ and $a > 0$, $b \in \mathbb{R}$. □

Proposition 7.4.25. *Let*

$$Q = \sup_n |a_n| \tag{7.4.123}$$

(a) *If $w \in \overline{\mathbb{C}}_+ \cup \{\infty\}$, for any n,*

$$\operatorname{Im}[\mathcal{A}_-(a_n, b_n; z)(w)] \geq \frac{\operatorname{Im} z}{Q^2} \tag{7.4.124}$$

(b) *If w, v obey*

$$\operatorname{Im} w \geq \frac{\operatorname{Im} z}{Q^2} \qquad \operatorname{Im} v \geq \frac{\operatorname{Im} z}{Q^2} \tag{7.4.125}$$

then, for all n,

$$\gamma(\mathcal{A}_-(a_n, b_n; z)w, \mathcal{A}_-(a_n, b_n; z)v) \leq \frac{1}{1 + \left(\frac{\operatorname{Im} z}{Q}\right)^2}\, \gamma(w, v) \tag{7.4.126}$$

(c) *For any $w, v \in \mathbb{C}_+ \cup \{\infty\}$ and all n,*

$$\operatorname{Im}[\mathcal{J}_n^-(w)] \geq \frac{\operatorname{Im} z}{Q^2} \tag{7.4.127}$$

$$\gamma(\mathcal{J}_n^-(w), \mathcal{J}_n^-(v)) \leq \left[\frac{1}{1+\left(\frac{\operatorname{Im} z}{Q}\right)^2}\right]^{n-1} \gamma(\mathcal{J}_1^-(w), \mathcal{J}_1^-(v)) \tag{7.4.128}$$

(d) *Uniformly for $z \in K$, a compact subset of $\mathbb{C}_+$ and $w \in \overline{\mathbb{C}}_+ \cup \{\infty\}$,*

$$\lim_{n\to\infty} \gamma(M_n^-(z), \mathcal{J}_n^-(z)w) = 0 \tag{7.4.129}$$

Proof. (a) is immediate from (7.4.122) and $1/a_n \geq 1/Q$.

(b) For any $w \in \mathbb{C}_+$,

$$(\operatorname{Im} w)(\operatorname{Im}(-w^{-1})) = \frac{(\operatorname{Im} w)^2}{|w|^2} \leq 1 \tag{7.4.130}$$

so if w obeys (7.4.125),

$$\frac{\operatorname{Im} z}{\operatorname{Im}(-w^{-1})} \geq (\operatorname{Im} z)(\operatorname{Im} w) \geq \left(\frac{\operatorname{Im} z}{Q}\right)^2 \tag{7.4.131}$$

Rewrite (7.4.122) as

$$\mathcal{A}_-(a_n, b_n; z)(w) = c_n + g_n(w) \tag{7.4.132}$$

where

$$c_n = \frac{z - b_n}{a_n^2} \qquad g_n(w) = -\frac{1}{a_n^2 w} \tag{7.4.133}$$

Then

$$\begin{aligned} &\gamma(\mathcal{A}_-(a_n, b_n; z)w, \mathcal{A}_-(a_n, b_n; z)v) \\ &\quad = \gamma(c_n + g_n(w), c_n + g_n(v)) && (7.4.134) \\ &\quad = \gamma(g_n(w), g_n(v)) \left[\frac{1}{\sqrt{1+\frac{\operatorname{Im} c_n}{\operatorname{Im} g_n(w)}}\sqrt{1+\frac{\operatorname{Im} c_n}{\operatorname{Im} g_n(v)}}}\right] && (7.4.135) \end{aligned}$$

Since g_n is an isometry of $\mathbb{C}_+$,

$$\gamma(g_n(w), g_n(v)) = \gamma(w, v) \tag{7.4.136}$$

and, by (7.4.131),

$$\frac{\operatorname{Im} c_n}{\operatorname{Im} g_n(w)} = \frac{\operatorname{Im} z}{\operatorname{Im}(-w^{-1})} \geq \left(\frac{\operatorname{Im} z}{Q}\right)^2 \tag{7.4.137}$$

and similarly for v. (7.4.135)–(7.4.137) imply (7.4.126).

(c) (7.4.127) follows from (a), and that implies (7.4.128), given (b).

(d) is immediate from (c) and $M_n^-(z) = \mathcal{J}_n^-(z)(\infty)$. □

As a final lemma, we need

Lemma 7.4.26. *For any $w \in \mathbb{C}_+$, any n, and any $t \in \mathbb{R}$,*

$$-\overline{\mathcal{J}_n^+(t)w} = \mathcal{J}_n^-(t)(-\bar{w}) \tag{7.4.138}$$

Proof. Since the coefficients of $\mathcal{J}_n^{\pm}(t)$ are real, $\mathcal{J}_n^{\pm}(t)(\bar{\zeta}) = \overline{\mathcal{J}_n^{\pm}(t)\zeta}$, and therefore we need only prove (7.4.138) with the $^-$ dropped. For any $\alpha, \beta, \gamma, \delta$,

$$-\frac{\alpha w + \beta}{\gamma w + \delta} = \frac{\alpha(-w) + (-\beta)}{(-\gamma)(-w) + \delta} \tag{7.4.139}$$

so (7.4.138) follows if we prove $\mathcal{J}_n^+$ and $\mathcal{J}_n^-$ have the same diagonal elements and off-diagonal elements of the negatives of each other. That property is preserved under matrix products, so, since it is true of $A^{\pm}$, it is true of $\mathcal{J}^{\pm}$. □

Proof of Theorem 7.4.13. Fix $A \subset \Sigma_{\text{ac}}(J)$ and $\varepsilon > 0$. We claim there exists a decomposition $A = A_0 \cup \cdots \cup A_\ell$ and $m_1, \dots, m_\ell \in \mathbb{C}_+$ so that

$$|A_0| < \varepsilon \qquad \forall t \in A_j,\ \gamma(m_j(t+i0), m_j) < \varepsilon \tag{7.4.140}$$

For, first place in A_0 the set of measure zero, $\tilde{A}_0$, for which $\lim_{\varepsilon \downarrow 0} m_j(t+i\varepsilon)$ fails to exist or is real. Let

$$B_N = \left\{ t \in A \setminus \tilde{A}_0 \;\middle|\; |m_0(t+i0)| \le N,\ \operatorname{Im} m_j(t+i0) \ge \frac{1}{N} \right\}$$

Since $\cup_{n=1}^{\infty} B_N = A \setminus \tilde{A}_0$, we can find N so that $A_0 \equiv A \setminus B_N$ has $|A_0| < \varepsilon$. Since $\{z \mid |z| \le N,\ \operatorname{Im} z \ge N^{-1}\}$ is compact, we can cover it by finitely many ε-balls in γ, and then by finitely many disjoint sets, each contained in ε-balls. Let $A_1, \dots, A_\ell$ be the set of $t \in A$ with $m_j(t+i0)$ in these sets and m_j the centers of the ε-balls.

By Proposition 7.4.24(a) for each n and $t \in A_j$,

$$\gamma(M_n^+(t), \mathcal{J}_n^+(t)m_j) < \varepsilon \tag{7.4.141}$$

and then, by (7.4.104) and (7.4.53),

$$|\omega_{M_n^+}(A_j, S) - \omega_{\mathcal{J}_n^+(\cdot)m_j}(A_j, S)| \le \varepsilon |A_j| \tag{7.4.142}$$

where we can take ε to zero inside the integral in (7.4.62) because $\operatorname{Im} M_n^+(t) > 0$ on A (since $\operatorname{Im} m_J(t) > 0$ there) and $\operatorname{Im} m_j > 0$.

In (7.4.62), we can flip the sign of t and $\operatorname{Re} F$ to see that for any Herglotz function, F,

$$\omega_F(A, S) = \omega_{-\bar{F}}(A, -S) \tag{7.4.143}$$

and then use (7.4.138) to replace $-\overline{\mathcal{J}_n^+(\cdot)(m_j)}$ by $\mathcal{J}_n^-(t)(-\bar{m}_j)$, that is, we have

$$|\omega_{M_n^+}(A_j, S) - \omega_{\mathcal{J}_n^-(\cdot)(-\bar{m}_j)}(A_j, -S)| \le \varepsilon |A_j| \tag{7.4.144}$$

Next, appeal to Lemma 7.4.18 to find $y_0 > 0$, so for all Herglotz functions, F, and all Borel $S \subset \mathbb{R}$ and $j = 1, \dots, \ell$, we have that

$$|\omega_{F_{y_0}}(A_j, S) - \omega_F(A_j, S)| \le \varepsilon |A_j| \tag{7.4.145}$$

Then use J bounded and Proposition 7.4.25 to find n_0 so for all $n \geq n_0$, $t \in A_j$, and $j = 1, \dots, \ell$,

$$\gamma(M_n^-(t+iy_0), \mathcal{J}_n^-(t+iy_0)(-\bar{m}_j)) < \varepsilon \tag{7.4.146}$$

Use (7.4.104) and (7.4.62) to get for all $S \subset \mathbb{R}$ and $j = 1, \dots, \ell$,

$$|\omega_{(M_n^-)_{y_0}}(A_j, S) - \omega_{(\mathcal{J}_n^-)_{y_0}(-\bar{m}_j)}(A_j, -S)| \leq \varepsilon |A_j| \tag{7.4.147}$$

and (7.4.145) twice to conclude that

$$|\omega_{M_n^-}(A_j, -S) - \omega_{\mathcal{J}_n^-(-\bar{m}_j)}(A_j, -S)| \leq 3\varepsilon |A_j| \tag{7.4.148}$$

Since, for any S and F, $|\omega_F(A_0, S)| < \varepsilon$, we conclude from (7.4.148) and (7.4.144) that for $n \geq n_0$,

$$|\omega_{M_n^+}(A, S) - \omega_{M_n^-}(A, -S)| \leq 4\varepsilon |A| + 2\varepsilon \tag{7.4.149}$$

Since ε is arbitrary, this proves (7.4.120), and so the Breimesser–Pearson and Remling theorems. □

Remarks and Historical Notes. The main results of this section are from Remling [366]. The proof of his main theorem (Theorem 7.4.3) relies on ideas of Pearson [334, 335], especially two papers of Breimesser–Pearson [56, 57]. However, these earlier papers did not realize the link to reflectionless potentials nor the deep applications on finite-valued potentials and on the extensions of the Denisov–Rakhmanov theorem. It is unfortunate that we still do not have a proof of Remling's theorem less involved than his original proof. An extension of Remling's theorem to continuum Schrödinger operators was found by Remling [365] and to CMV matrices by Breuer, Ryckman, and Zinchenko [62].

What we have called Pearson convergence (following [62]) was introduced by Pearson in [334, 335], as "value distribution theory." That name has been taken by an unrelated set of ideas in the theory of meromorphic functions called Nevanlinna theory—hence the use of a different name here. The use of γ and the other techniques of the proof we give of the Breimesser–Pearson theorem are from their papers.

An important precursor which provided guideposts for the applications is the work of Kotani [243], who considered Schrödinger operators. Simon extended this work to Jacobi matrices with $a_n \equiv 1$ [391] and to OPUC [400, Section 10.11]. Kotani (or its Jacobi analog) considered stochastic Jacobi matrices, which involve certain families depending on a parameter, ω, in a probability measure space where many properties are a.e. constant. In this case, Kotani proved that a.e. J_ω was reflectionless on its a.c. spectrum. In [244], he proved Theorems 7.4.4 and 7.4.5.

Sparse potentials go back to a basic paper of Pearson [333]; see the notes to [400, Section 12.5] for further references.

Theorem 7.4.5 depended on two ideas:

(a) Because of the form of open sets in product spaces, the values of the function F of (7.4.14) are determined up to ε by $\{a_n, b_n\}_{-k+1}^{0}$ for some k.
(b) By a compactness argument, for any $\varepsilon > 0$, for n large, $J^{(n)}$ is within ε of $\mathcal{R}(J)$ for any fixed metric on the product space of Jacobi parameters.

Finiteness entered in Theorem 7.4.6 only because then ε bounds imply equality so long as ε obeys (7.4.20). Thus, without finiteness, one gets the following result, which Remling calls the Oracle theorem:

Theorem 7.4.27 (Remling [366]). *Let $\Sigma \subset \mathfrak{e} \subset \mathbb{R}$ with $|\Sigma| > 0$ and $\mathfrak{e}$ compact. Then for any $\varepsilon > 0$, there are k and function $F\colon [(0,\infty) \times \mathbb{R}]^k \to (0,\infty) \times \mathbb{R}$, so that for any half-line Jacobi matrix, J, with*

$$\Sigma \subset \Sigma_{\mathrm{ac}}(J) \subset \sigma(J) \subset \mathfrak{e} \tag{7.4.150}$$

we have for $n \geq N$, which can be J-dependent, that

$$|(a_n, b_n) - F(\{a_j, b_j\}_{j=n-k}^{n-1})| < \varepsilon \tag{7.4.151}$$

Theorem 7.4.9 is due to Poltoratski [349]. For a simple proof using rank one perturbation theory (starting with Theorem 7.3.3), see Jakšić–Last [208]. Theorem 7.4.9 does not require compact support but only that F_μ exists, for example, if $\int (x^2+1)^{-1}\, d\mu(x) < \infty$. It also holds for signed measures and general $f \in L^1(\mathbb{R}, d\mu)$.

Theorem 7.4.8 is due to Poltoratski–Remling [350]. For cases where Σ is a homogeneous set (i.e. $\exists \varepsilon$, $\delta_0 > 0$, so for all $\delta < \delta_0$ and all $x \in \Sigma$, $|(x-\delta, x+\delta) \cap \Sigma| \geq 2\varepsilon\delta$), and $\operatorname{supp}(d\mu) = \Sigma$, Sodin–Yuditskii [413] found all $d\mu$'s reflectionless on Σ (as isospectral torus) and found that $\mu_{\mathrm{s}} = 0$. Earlier extensions where $\operatorname{supp}(d\mu)$ are bigger than Σ are found in Gesztesy–Zinchenko [169]. Another proof that reflectionless measures on homogeneous sets have no singular part is due to Poltoratski–Simon–Zinchenko [351].

It is easy to construct reflectionless measures on $\{x_0\} \cup [-2,2]$ for any $x_0 \notin [-2,2]$, so if Σ has isolated points, μ_{s} can be nonzero. Much more subtle are examples where μ is reflectionless on Σ but μ has pure points or even singular continuous components on Σ (obviously, such examples must have $\Sigma_0 \neq \Sigma$). These are constructed in Nazarov–Volberg–Yuditskii [315].

For consideration of Jacobi matrices, our definition is adequate, but for continuum Schrödinger operators, one wants to allow measures, which are not of compact support and not finite weight, but only obey (2.3.86). One defines ρ to be reflectionless on Σ if for some choice of A and $\operatorname{Re} G(i)$, $\operatorname{Re} G(x+i0) = 0$ on Σ for G given by (2.3.87). [350] prove Theorem 7.4.8 in this context.

7.5 PURELY REFLECTIONLESS JACOBI MATRICES ON FINITE GAP SETS

Let $\mathfrak{e} \subset \mathbb{R}$ be a finite gap set. In Section 5.13, we defined the isospectral torus, $\mathcal{T}_{\mathfrak{e}}$, of whole-line Jacobi matrices and proved that if $J \in \mathcal{T}_{\mathfrak{e}}$, then J is reflectionless on $\mathfrak{e}$ and

$$\sigma(J) = \mathfrak{e} \tag{7.5.1}$$

We also showed that J had purely a.c. spectrum of multiplicity 2. Here we will prove a converse:

Theorem 7.5.1 (Sodin–Yuditskii [413]). *Let $\mathfrak{e} \subset \mathbb{R}$ be a finite gap set. Let J be a whole-line operator which obeys* (7.5.1) *and which is reflectionless on $\mathfrak{e}$. Then $J \in \mathcal{T}_{\mathfrak{e}}$.*

Proof. Let $\mathfrak{e}$ be given by (5.5.126) and (5.5.127). By Craig's theorem (Theorem 5.4.19), there exist $x_1, \dots, x_\ell$ with $x_j \in [\beta_j, \alpha_{j+1}]$ so

$$G_{00}(z) = -\frac{\prod_{j=1}^{\ell}(z - x_j)}{\left[\prod_{j=1}^{\ell+1}(z-\alpha_j)(z-\beta_j)\right]^{1/2}} \tag{7.5.2}$$

By this explicit formula, $(-G_{00}(z))^{-1}$ has real boundary values on $\mathbb{R} \setminus \mathfrak{e} \cup \{x_j\}_{j=1}^{\ell}$, is bounded on $\mathfrak{e}^{\text{int}}$, and has poles exactly at those x_j's, which are in the interior of gaps.

By (5.4.50),

$$(-G_{00}(z))^{-1} = z - b_n + \int \frac{d(\mu_+ + \mu_-)(x)}{x - z} \tag{7.5.3}$$

where

$$a_0^2 m(z, J_0^+) = \int \frac{d\mu_+(x)}{x - z} \tag{7.5.4}$$

$$a_{-1}^2 m(z, J_{-1}^-) = \int \frac{d\mu_-(x)}{x - z} \tag{7.5.5}$$

By the properties of $(-G_{00}(z_0))^{-1}$ above, we see $(\mu_+ + \mu_-)$ has pure points at x_j's in the gaps and otherwise is purely a.c. and supported on $\mathfrak{e}$.

If x_j is a pole of $m(z, J_0^+)$, there is a solution of $Ju = x_j u$ which is ℓ^2 at $+\infty$ with $u(0) = 0$. Similarly, if x_j is a pole of $m(z, J_{-1}^-)$, there is a solution ℓ^2 at $-\infty$ with $u(0) = 0$. If x_j were a pole of both, then $Ju = x_j u$ would have an ℓ^2 solution. But then $\sigma(J)$ would not be just $\mathfrak{e}$. Thus, each x_j in the interior of the gap is a pole of either $m(z, J_0^+)$ or of $m(z, J_0^-)^{-1}$, but not both.

By the reflectionless hypothesis, we can take $m(z, J_0^+)$ on $\mathcal{S}_+ \cap \mathbb{C}_+$ and $a_0^{-2} m(z, J_0^-)^{-1}$ on $\mathcal{S}_- \cap \mathbb{C}_+$ and extend to $\mathcal{S}_+$ and $\mathcal{S}_-$ and sew together to a single meromorphic function on $\mathcal{S}$. By the above, there is exactly one pole in each two-sheeted gap (if x_j is at β_j or α_{j+1}, there is a square root divergence, which is a pole the way poles at branch points on $\mathcal{S}$ are counted). Thus, m has $\ell + 1$ poles and so is a minimal Herglotz function on $\mathcal{S}$. The analysis of Theorem 5.13.11 shows that J is the corresponding point in the isospectral torus. □

Remark. While we used the theory of minimal Herglotz functions in the above, the fact that J is determined by the x_j's plus a left/right choice can be seen directly from reflectionless J's. For the reflectionless condition implies

$$\operatorname{Im} a_0^2 m(x + i0, J_0^+) = \tfrac{1}{2} \operatorname{Im} -G_{00}(x + i0)^{-1} \tag{7.5.6}$$

This plus the choice of poles and residues (which come from residues of $-G_{00}(z)^{-1}$) determine μ_+, and so $a_0^2 m(x+i0, J_0^+)$, and thus J_0^+ and a_0. Similarly, $m(z, J_0^-)^{-1}$ determine J_0^{-1}.

Remarks and Historical Notes. That reflectionless J's in the finite gap case are the isospectral torus (i.e., Theorem 7.5.1) goes back to Sodin–Yuditskii [413]; see also [163, 168, 436]. Our proof follows Remling [366].

7.6 THE DENISOV–RAKHMANOV–REMLING THEOREM

Given the last two sections, we immediately have the following beautiful result:

Theorem 7.6.1 (Denisov–Rakhmanov–Remling Theorem; Remling [366]). *Let $\mathfrak{e}$ be a finite gap set. Let J be a half-line Jacobi matrix with*

$$\sigma_{\text{ess}}(J) = \Sigma_{\text{ac}}(J) = \mathfrak{e} \tag{7.6.1}$$

Then, with $\mathcal{T}_\mathfrak{e}$ the isospectral torus,

$$\mathcal{R}(J) \subset \mathcal{T}_\mathfrak{e} \tag{7.6.2}$$

Remarks. 1. This is even interesting in case $\mathfrak{e} = [-2, 2]$ (due to Denisov [107]), in which case the conclusion is $a_n \to 1$, $b_n \to 0$ since $\mathcal{T}_\mathfrak{e}$ is a single point.

2. In colloquial language, J approaches $\mathcal{T}_\mathfrak{e}$ at infinity.

3. The periodic case shows that it can happen that $\mathcal{R}(J)$ is only a subset of $\mathcal{T}_\mathfrak{e}$.

Proof. Let $J_r \in \mathcal{R}(J)$. By Theorem 7.2.1, $\sigma_{\text{ess}}(J) = \mathfrak{e}$ implies

$$\sigma(J_r) \subset \mathfrak{e} \tag{7.6.3}$$

By Theorem 7.3.1 and $\Sigma_{\text{ac}}(J) = \mathfrak{e}$,

$$\Sigma_{\text{ac}}(J_r) \supset \mathfrak{e} \tag{7.6.4}$$

and, by Theorem 7.4.3, J_r is reflectionless. Thus,

$$\sigma(J_r) = \Sigma_{\text{ac}}(J_r) = \mathfrak{e} \tag{7.6.5}$$

and J_r is reflectionless. By Theorem 7.5.1,

$$J_r \in \mathcal{T}_\mathfrak{e} \tag{7.6.6}$$

□

Remarks and Historical Notes. Rakhmanov [358, 359] proved that, for OPUC, if $\Sigma_{\text{ac}}(\mathcal{C}) = \partial\mathbb{D}$, then $\alpha_n \to 0$. For alternate proofs and the involved history ([358] had an error!), see [400, Chapter 9]. Denisov [107] then proved that, for OPRL, if $\sigma_{\text{ess}}(J) = \Sigma_{\text{ac}}(J) = [-2, 2]$, then $a_n \to 1$, $b_n \to 0$.

Earlier, Bello–López [37], using ideas from López [286], had shown that if Σ_a is the σ_{ess} for the CMV matrix associated to $\alpha_n \equiv a > 0$ and if $\sigma(\mathcal{C}) = \Sigma_{\text{ac}}(\mathcal{C}) = \Sigma_a$, then $\{\alpha_n\}$ is in the *López class*

$$|\alpha_n| \to a \qquad \frac{\alpha_{n+1}}{\alpha_n} \to 1 \tag{7.6.7}$$

In [400], Simon realized that (7.6.7) was equivalent to saying that α_n approached the isospectral torus for Σ_a and he conjectured the result for the general periodic case. Damanik–Killip–Simon [97] proved this periodic conjecture using the magic formula machinery we discuss in Chapter 8, and they conjectured the result for general finite gap $\mathfrak{e}$. Their conjecture was then proven by Remling [366]. For more on the history of results on approach to the isospectral torus, see the Notes to Section 8.1.

Chapter Eight

Szegő and Killip–Simon Theorems for Periodic OPRL

In this chapter, we turn to a synthesis of the theory of periodic Jacobi matrices studied in Chapters 5 and 6 with the perturbation theory of Chapters 3 and 4.

8.1 OVERVIEW

We have looked at four results on perturbations of the Jacobi matrix, J_0, with $a_n \equiv 1, b_n \equiv 0$:

(i) Weyl-type results that $a_n \to 1, b_n \to 0 \Rightarrow \sigma_{\text{ess}}(J) = [-2, 2]$

(ii) Denisov–Rakhmanov-type results that $\sigma_{\text{ess}}(J) = \Sigma_{\text{ac}}(J) = [-2, 2]$ implies $a_n \to 1, b_n \to 0$

(iii) Szegő–Shohat–Nevai-type results relating a Szegő condition plus $\frac{1}{2}$ eigenvalue bounds to boundedness of $\sum_{n=1}^{N} \log(a_n)$

(iv) Killip–Simon-type results relating a pseudo-Szegő condition plus $\frac{3}{2}$ eigenvalue bounds to ℓ^2 conditions of the form

$$\sum_n (a_n - 1)^2 + b_n^2 < \infty \tag{8.1.1}$$

In this chapter, we want to focus on perturbation results of this type for general periodic J_0. A key initial question is what replaces the limit point $a_n \equiv 1, b_n \equiv 0$, for which it is not hard to see that J_0 alone is not enough. The answer, given what we have seen, especially since we addressed (i) and (ii) in Chapter 7, should be obvious: The single point J_0 for the case $a_n \equiv 1$, $b_n \equiv 0$ needs to be replaced instead by the isospectral torus. For a history that led to this realization, see the Notes.

We addressed (i) in Section 7.4 and (ii) in Section 7.6. Question (iii) will be addressed in Section 8.4 and question (iv) in Section 8.6. Remarkably, we will be able to do this by reducing things to an MOPRL involving perturbations of $A_n \equiv 1$, $B_n \equiv 0$. The key will be to form the discriminant, Δ_{J_0}, of J_0, and given a Jacobi matrix, J, to look at $\Delta_{J_0}(J)$. The key will be to show that $\Delta_{J_0}(J)$, which is a $p \times p$ block Jacobi matrix, has $A_n \to \mathbf{1}$, $B_n \to \mathbf{0}$ if and only if J approaches the isospectral torus. Indeed, we will prove in Section 8.2 that for a whole-line bounded Jacobi matrix (with $(Su)_n = u_{n-1}$ and $\mathfrak{e} = \sigma_{\text{ess}}(J_0)$) that

$$\Delta_{J_0}(J) = S^p + S^{-p} \Leftrightarrow J \in \mathcal{T}_{\mathfrak{e}} \tag{8.1.2}$$

something that has been dubbed the magic formula.

Section 8.3 discusses a technical issue relating the spectral measure for $\Delta_{J_0}(J)$ and for J. Section 8.5 relates a Hilbert–Schmidt condition on $\Delta_{J_0}(J) - (S^p + S^{-p})$

to ℓ^2 convergence of J to $\mathcal{T}_{\mathfrak{e}}$. Section 8.7 discusses the OPUC case, which turns out to also be related to MOPRL, not MOPUC.

Remarks and Historical Notes. The first perturbations of periodic problems where the proper set of limits was found was the OPUC case where $\alpha_n \equiv a$, a constant with $a \neq 0$. López, in a series of papers, some with collaborators [31, 37, 38, 285], with important followup by Khrushchev [219, 221], focused on what Khrushchev dubbed the López class, $\{\alpha_n\}_{n=0}^{\infty}$, so that for some $a > 0$,

$$|\alpha_n| \to a \qquad \frac{\alpha_{n+1}}{\alpha_n} \to 1 \tag{8.1.3}$$

Simon [400] realized that (8.1.3) is equivalent to α_n approaching the isospectral torus of $\alpha_n^{(0)} \equiv a$ (which is α_n's with $\alpha_n = ae^{i\theta}$ for a fixed θ) and conjectured OPUC analogs of all four results. These were then proven (for (i)) by Last–Simon [272] and by Damanik–Killip–Simon [97]. Most of this chapter will follow [97].

8.2 THE MAGIC FORMULA

We define $S\colon \ell^2(\mathbb{Z}) \to \ell^2(\mathbb{Z})$ by

$$(Su)_k = u_{k-1} \tag{8.2.1}$$

As we have discussed, the key to understanding perturbations of periodic Jacobi matrices is a characterization of the isospectral torus:

Theorem 8.2.1 (The Magic Formula [97]). *Let J be a bounded two-sided Jacobi matrix. Let Δ be the discriminant of a period p periodic Jacobi matrix J_0 with $\mathfrak{e} = \sigma_{\text{ess}}(J_0)$. Then*

$$J \in \mathcal{T}_{\mathfrak{e}} \Leftrightarrow \Delta(J) = S^p + S^{-p} \tag{8.2.2}$$

Remarks. 1. Δ is a polynomial so, by $\Delta(J)$, we mean the operator obtained by replacing the variable in $\Delta(z)$ by the operator J.

2. We emphasize that J is not assumed a priori to be periodic.

3. As we have discussed (see the end of Section 5.13), $\mathcal{T}_{\mathfrak{e}}$ can be viewed either as a class of one-sided matrices (using minimal Herglotz functions) or as two-sided reflectionless operators. Here, obviously, we have two-sided in mind.

We will first prove that

$$\Delta_{J_0}(J_0) = S^p + S^{-p} \tag{8.2.3}$$

which is a large part of the $\Rightarrow$ half of (8.2.2).

Proposition 8.2.2. (8.2.3) *holds.*

Proof. Both sides are periodic of period p, that is, commute with S^p so, as in the discussion in Section 5.3, we can use the Fourier transform, $\mathcal{F}$, of (5.3.18)/(5.3.19) to "diagonalize" them as matrices on $\mathbb{C}^p$. One sees directly that

$$(\mathcal{F}(S^p + S^{-p})\mathcal{F}^{-1} f)_n(\theta) = 2\cos(\theta) f_n(\theta) \tag{8.2.4}$$

On the other hand, by (5.3.24),

$$(\mathcal{F}\Delta_{J_0}(J_0)\mathcal{F}^{-1}f)_n(\theta) = [\Delta_{J_0}(J_0(\theta))f]_n(\theta) \tag{8.2.5}$$

which, by (5.4.9), is the right side of (8.2.4). □

Lemma 8.2.3. *Let J_0, J be two period p Jacobi matrices. The following are equivalent:*

$$\text{(i)} \qquad \sigma_{\text{ess}}(J) = \sigma_{\text{ess}}(J_0) \tag{8.2.6}$$

$$\text{(ii)} \qquad \Delta_J = \Delta_{J_0} \tag{8.2.7}$$

$$\text{(iii)} \qquad J \in \mathcal{T}_{\mathfrak{e}} \text{ where } \mathfrak{e} = \sigma_{\text{ess}}(J_0) \tag{8.2.8}$$

Remark. The essential spectra are the same for half- and whole-line periodic J's, so it does not matter in (8.2.6) which we mean!

Proof. (i) ⇒ (ii). Let $\mathfrak{e} = \sigma_{\text{ess}}(J_0)$. Then $\Delta_{J_0}(z) = \kappa z^p + \dots$ where $\kappa = C(\mathfrak{e})^{-1}$, the inverse of the logarithmic capacity of $\mathfrak{e}$. Moreover, the potential theorist's Green's function, $G_{\mathfrak{e}}(z)$, is related to Δ by (5.4.26) on account of Theorems 5.4.10 and 5.5.17. Put differently,

$$|\Delta_{J_0}(z)| = \exp(G_{\mathfrak{e}}(z)) + \exp(-G_{\mathfrak{e}}(z)) \tag{8.2.9}$$

Thus, $\mathfrak{e}$ determines $|\Delta_{J_0}(z)|$ and so, since Δ is a polynomial with leading positive coefficient, Δ_{J_0} and thus, (8.2.6) implies (8.2.7).

(ii) ⇒ (i). Immediate from (see Section 5.4)

$$\sigma_{\text{ess}}(J) = \Delta^{-1}([-2, 2]) \tag{8.2.10}$$

(i), (ii) ⇔ (iii). This depends on the definition of isospectral torus used, but all wind up being those periodic J's with $\sigma_{\text{ess}}(J) = \mathfrak{e}$, so (i) ⇔ (iii). □

Let Q be an operator on $\ell^2(\mathbb{Z})$. We say Q has *finite width* if there is a k so that

$$\text{supp}(u) \subset [n, m] \Rightarrow \text{supp}(Qu) \subset [n-k, m+k] \tag{8.2.11}$$

equivalently, if the matrix, Q_{mn}, of Q has

$$Q_{mn} = 0 \quad \text{if} \quad |m-n| \geq k+1 \tag{8.2.12}$$

A diagonal matrix has the form

$$D_{mn} = d_m \delta_{mn} \tag{8.2.13}$$

Q has finite width with k if and only if there are diagonal matrices $\{D^{(j)}\}_{j=-k}^{k}$ so that

$$Q = \sum_{j=-k}^{k} D^{(j)} S^j \tag{8.2.14}$$

Lemma 8.2.4 (Naĭman's Lemma). *Let Q be a bounded operator on ℓ^2 of finite width so that for some p,*

$$[Q, S^p + S^{-p}] = 0 \tag{8.2.15}$$

Then

$$[Q, S^p] = 0 \tag{8.2.16}$$

Remark. For Q of the form (8.2.14), (8.2.16) is equivalent to each $D^{(j)}$ having the form (8.2.13) with $d_m^{(j)}$ periodic, that is, $d_{m+p}^{(j)} = d_m^{(j)}$.

Proof. Write

$$Q = \sum_{j=-\ell}^{k} D^{(j)} S^j \tag{8.2.17}$$

with $D^{(k)} \neq 0 \neq D^{(-\ell)}$. (*Note*: $-\ell \leq k$, but k or ℓ is allowed to be negative.) Looking at matrix elements of $[Q, S^p + S^{-p}]_{mn}$ with $n = m + k + p$ shows that

$$D_{m\,m}^{(k)} = D_{m+p\,m+p}^{(k)} \tag{8.2.18}$$

so $[D^{(k)}, S^p] = 0$. Thus, $[D^{(k)}, S^{-p}] = 0$ also and

$$[Q - D^{(k)} S^k, S^p + S^{-p}] = 0 \tag{8.2.19}$$

By induction, one sees that each $D^{(j)}$ obeys

$$[D^{(j)}, S^p] = 0 \tag{8.2.20}$$

which implies (8.2.16). □

Lemma 8.2.5. *Let P be a polynomial and J a Jacobi matrix. Suppose $P(J) = 0$. Then P is the zero polynomial.*

Proof. If P is not the zero polynomial, then

$$P(z) = b_0 z^\ell + b_1 z^{\ell-1} + \cdots + b_\ell \tag{8.2.21}$$

for some $b_0 \neq 0$. But then

$$P(J)_{1\,\ell+1} = b_0 a_1 a_2 \ldots a_\ell \neq 0 \tag{8.2.22}$$

a contradiction to $P(J) = 0$. □

Proof of Theorem 8.2.1. If $J \in \mathcal{T}_{\mathfrak{e}}$, then by (8.2.7),

$$\Delta_{J_0}(J) = \Delta_J(J) = S^p + S^{-p}$$

by (8.2.3).

Conversely, suppose J is any two-sided Jacobi matrix and

$$\Delta_{J_0}(J) = S^p + S^{-p}$$

Since

$$[J, \Delta_{J_0}(J)] = 0 \tag{8.2.23}$$

we have

$$[J, S^p + S^{-p}] = 0 \tag{8.2.24}$$

By Naĭman's lemma,

$$[J, S^p] = 0 \tag{8.2.25}$$

that is, J is periodic with period p.

By (8.2.3),

$$\Delta_J(J) = S^p + S^{-p} \tag{8.2.26}$$

so if

$$P(z) = \Delta_J(z) - \Delta_{J_0}(z) \tag{8.2.27}$$

we have

$$P(J) = 0 \tag{8.2.28}$$

By Lemma 8.2.5, $P \equiv 0$, that is, $\Delta_J = \Delta_{J_0}$. By Lemma 8.2.3, $J \in \mathcal{T}_{J_0}$. □

Remarks and Historical Notes. Theorem 8.2.1 is due to Damanik–Killip–Simon [97]. Naĭman's lemma is from Naĭman [312], who had other ideas approaching the magic formula. The proof we give here using Naĭman's lemma follows a suggestion of L. Golinskii.

8.3 THE DETERMINANT OF THE MATRIX WEIGHT

The strategy is now clear. Take a half-line Jacobi matrix, J, which is "near" the isospectral torus at infinity. Then we expect $\Delta(J)$ to be near $S^p + S^{-p}$ at infinity. If we use sum rules for $\Delta(J)$, we can hope to get effective sum rules for J.

Specifically, let $d\rho_J$ be the spectral measure for J so

$$d\rho_J(x) = \omega(x)\,dx + d\rho_{J,\mathrm{s}}(x) \tag{8.3.1}$$

On the other hand, $\Delta(J)$ is a $p \times p$ block Jacobi matrix and the corresponding measure $d\rho_{\Delta(J)}$ is a $p \times p$ matrix-valued measure, which we can write

$$d\rho_{\Delta(J)}(E) = W(E)\,dE + d\rho_{\Delta(J),\mathrm{s}}(E) \tag{8.3.2}$$

The sum rules that we invoke from Chapter 4 involve $\det(W(E))$. So the question is how this is related to $\omega(x)$ for those p values of x solving $\Delta(x) = E$. This is what we want to compute in this section. Here is the main result:

Theorem 8.3.1. *Let J_0 be a period p Jacobi matrix with $\sigma_{\mathrm{ess}}(J_0) = \mathfrak{e}$ and discriminant Δ, so $\Delta^{-1}([-2,2]) = \mathfrak{e}$. Let $\{a_n^{(0)}, b_n^{(0)}\}_{n=1}^\infty$ be the Jacobi parameters of J_0. Let J be a half-line Jacobi matrix with*

$$\sigma_{\mathrm{ess}}(J) \subset \mathfrak{e} \tag{8.3.3}$$

and $d\rho_J(x)$ its spectral measure. Then $d\rho_J$ has the form (8.3.1) *with* $\mathrm{supp}(\omega) \subset \mathfrak{e}$. *Let $\{a_n, b_n\}_{n=1}^\infty$ be the Jacobi parameters for J. Then $\Delta(J)$ is a $p \times p$ matrix-valued Jacobi matrix with spectral measure $d\rho_{\Delta(J)}(E)$, where*

$$\sigma_{\mathrm{ess}}(\Delta(J)) \subset [-2, 2] \tag{8.3.4}$$

and $d\rho_{\Delta(J)}$ of the form (8.3.2). *Given $E \in (-2,2)$, let $x_1 < \cdots < x_p$ be the p solutions of*

$$\Delta(x) = E \tag{8.3.5}$$

Then

$$\det(W(E)) = \left(\prod_{j=1}^{p} [a_j^{2p-2j}]^{-1} [a_j^{(0)}]^p\right) \prod_{j=1}^{p} \omega(x_j) \tag{8.3.6}$$

Remark. $\mathfrak{e}$ and Δ only depend on the isospectral torus. The J_0-dependence of (8.3.6) is $(a_1^{(0)} \dots a_p^{(0)})^p = C(\mathfrak{e})^p$ is also only $\mathfrak{e}$-dependent.

Proof. We put $\Delta(J)$ into block form by placing $\delta_1, \dots, \delta_p$ into block 1, $\delta_{p+1}, \dots, \delta_{2p}$ into block 2, and so on. Thus, for $1 \le j, k \le p$,

$$\int F(E)(d\rho_\Delta(E))_{jk} = \langle \delta_j, F(\Delta(J))\delta_k \rangle \tag{8.3.7}$$

The orthogonality of $p_j(x)$ in $d\rho_J(x)$ implies orthogonality of $p_j(J)\delta_1$ in $\ell^2(\mathbb{Z})$, so

$$\delta_j = p_{j-1}(J)\delta_1 \tag{8.3.8}$$

Taking into account that Δ is a p to 1 map of $\mathfrak{e}^{\text{int}}$ to $(-2,2)$, we see that

$$W_{jk}(E) = \sum_{\ell=1}^{p} \omega(x_\ell)(|\Delta'(x_\ell)|)^{-1} p_{k-1}(x_\ell) p_{j-1}(x_\ell) \tag{8.3.9}$$

where we use

$$dx = dE\left(\frac{dE}{dx}\right)^{-1} = |\Delta'(x)|^{-1}\, dE \tag{8.3.10}$$

Thus, if

$$M_{k\ell} = p_{k-1}(x_\ell) \qquad k = 1, \dots, p;\ \ell = 1, \dots, p \tag{8.3.11}$$

and

$$A_{\ell m} = \delta_{\ell m}\omega(x_\ell)(|\Delta'(x_\ell)|)^{-1} \tag{8.3.12}$$

then

$$W = MAM^t \tag{8.3.13}$$

and

$$\det(W) = \det(M)^2 \det(A) \tag{8.3.14}$$

$$= \det(M)^2 \left(\prod_{k=1}^{p} \omega(x_k)\right)\left(\prod_{k=1}^{p} |\Delta'(x_k)|^{-1}\right) \tag{8.3.15}$$

To compute $\det(M)$, we note that

$$p_{k-1}(x_\ell) = \left(\prod_{j=1}^{k-1} a_j\right)^{-1} x_\ell^{k-1} + \text{lower order} \tag{8.3.16}$$

and the lower-order terms can be removed by subtracting rows. Thus,

$$\det(M) = \left[\prod_{k=1}^{p}\left(\prod_{j=1}^{k-1} a_j\right)^{-1}\right]\det(x_\ell^{k-1}) \tag{8.3.17}$$

$$= \left[\prod_{j=1}^{p} a_j^{p-j}\right]^{-1}\prod_{j>k}(x_j - x_k) \tag{8.3.18}$$

recognizing $\det(x_\ell^{k-1})$ as a Vandermonde determinant.

On the other hand, x_j solve $\Delta(x) - E = 0$, so

$$\Delta(x) - E = \left(\prod_{j=1}^{p} a_j^{(0)}\right)^{-1}\prod_{k=1}^{p}(x - x_k) \tag{8.3.19}$$

Thus,

$$\Delta'(x_j) = \left(\prod_{j=1}^{p} a_j^{(0)}\right)^{-1}\prod_{k\neq j}(x_j - x_k) \tag{8.3.20}$$

and

$$\left|\prod_{j>k}(x_j - x_k)\right|^2 = \prod_{j=1}^{p}[a_j^{(0)}]^p\prod_{j=1}^{p}|\Delta'(x_j)| \tag{8.3.21}$$

We conclude that

$$\det(M)^2 = \left(\prod_{j=1}^{p} a_j^{p-j}\right)^{-2}\prod_{j=1}^{p}[a_j^{(0)}]^p\prod_{j=1}^{p}|\Delta'(x_j')| \tag{8.3.22}$$

(8.3.14) and (8.3.22) imply (8.3.6). □

Corollary 8.3.2. *Under the hypotheses of Theorem 8.3.1,*

$$\int_{-2}^{2}(4 - E^2)^{-1/2}\log(\det(W(E)))\,dE > -\infty \tag{8.3.23}$$

if and only if

$$\int_{\sigma_{\text{ess}}(J)} \operatorname{dist}(x, \mathbb{R}\setminus\sigma_{\text{ess}}(J))^{-1/2}\log(\omega(x))\,dx > -\infty \tag{8.3.24}$$

Proof. By (8.3.6) and a change of variables, (8.3.23) is equivalent to

$$\int_{\sigma_{\text{ess}}(J)}(4 - \Delta(x)^2)^{-1/2}|\Delta'(x)|\log(\omega(x)) > -\infty \tag{8.3.25}$$

Near the edges of $\sigma_{\text{ess}}(J)$, including edges of open gaps, $(4 - \Delta(x)^2) \sim \operatorname{dist}(x, \mathbb{R}\setminus\sigma_{\text{ess}}(J))^{-1/2}$ and $\Delta'(x)$ is bounded above and away from zero.

At interior points other than closed gaps, $(4 - \Delta(x)^2)^{-1/2}$ and $|\Delta'(x)|$ are bounded above and away from zero.

At a closed gap, $4 - \Delta(x)^2$ has a double zero, so $|\Delta'(x)|$ has a simple zero cancelled by the first-order infinity in $(4 - \Delta(x)^2)^{-1/2}$. Thus, $\operatorname{dist}(x, \mathbb{R}\setminus\sigma_{\text{ess}}(J))^{1/2}$

$(4 - \Delta(x)^2)^{-1/2}|\Delta'(x)|$ is bounded above and away from zero globally on $\sigma_{\text{ess}}(J)$, and so (8.3.25) is equivalent to (8.3.24). □

If one looks at $(4 - \Delta(x)^2)^{\alpha}|\Delta'(x)|$, one only has the cancellation at closed gaps if $\alpha = -\frac{1}{2}$, but if all gaps are open, the above argument works on all of $\sigma_{\text{ess}}(J)$, and we obtain (we care mainly about $\alpha = \frac{1}{2}$ and $\alpha = -\frac{1}{2}$):

Corollary 8.3.3. *Let $\alpha > -1$ and let J_0 have all gaps open. Under the hypotheses of Theorem 8.3.1,*

$$\int_{-2}^{2} (4 - E^2)^{\alpha} \log(\det(W(E)))\, dE > -\infty \tag{8.3.26}$$

if and only if

$$\int_{\sigma_{\text{ess}}(J)} \text{dist}(x, \mathbb{R} \setminus \sigma_{\text{ess}}(J))^{\alpha} \log(\omega(x))\, dx > -\infty \tag{8.3.27}$$

Remarks and Historical Notes. These calculations are from Damanik–Killip–Simon [97].

8.4 A SHOHAT–NEVAI THEOREM FOR PERIODIC JACOBI MATRICES

Given the calculation of the last section, the magic formula and Theorem 4.5.1, it is easy to obtain a Szegő-type theorem, specifically an analog of the Shohat–Nevai theorem for perturbations of periodic Jacobi matrices. Our goal is to prove:

Theorem 8.4.1. *Let J_0 be a period p periodic Jacobi matrix with Jacobi parameters $\{a_n^{(0)}, b_n^{(0)}\}_{n=1}^{\infty}$. Let $\mathfrak{e} = \sigma_{\text{ess}}(J_0)$. Let J be a half-line Jacobi matrix with Jacobi parameters $\{a_n, b_n\}_{n=1}^{\infty}$ and*

$$\sigma_{\text{ess}}(J) = \mathfrak{e} \tag{8.4.1}$$

Suppose that

$$\sum_{E \in \sigma(J) \setminus \sigma_{\text{ess}}(J)} \text{dist}(E, \sigma_{\text{ess}}(J))^{1/2} < \infty \tag{8.4.2}$$

and the spectral measure $d\rho$ of J has the form

$$d\rho(x) = \omega(x)\, dx + d\rho_{\text{s}}(x) \tag{8.4.3}$$

Then

$$\int_{\sigma_{\text{ess}}(J)} \text{dist}(x, \mathbb{R} \setminus \sigma_{\text{ess}}(J))^{-1/2} \log(\omega(x)) > -\infty \tag{8.4.4}$$

if and only if

$$\limsup \prod_{j=1}^{m} \left(\frac{a_j}{a_j^{(0)}} \right) > 0 \tag{8.4.5}$$

Remarks. 1. Because $a_1^{(0)} \dots a_p^{(0)} = C(\mathfrak{e})^p$, (8.4.5) is equivalent to

$$\limsup \prod_{j=1}^{m} \left[\frac{a_j}{C(\mathfrak{e})} \right] > 0 \tag{8.4.6}$$

2. There is no assertion about a_j or b_j having a limit. Using very different methods, we will prove in Section 9.13 that there is $\{a_n^{(1)}, b_n^{(1)}\} \in \mathcal{T}_\mathfrak{e}$ so that $|a_n - a_n^{(1)}| + |b_n - b_n^{(1)}| \to 0$.

3. The hypotheses (8.4.1), (8.4.2), and (8.4.4) imply the hypotheses of Theorem 8.6.1, so if all gaps are open, the equivalent hypotheses (8.4.2)/(8.4.4) or (8.4.2)/(8.4.5) imply ℓ^2 approach to the isospectral torus as in that theorem.

Lemma 8.4.2. *Under the hypotheses of Theorem 8.4.1, suppose $\Delta_{J_0}(J)$ has Jacobi parameters $\{A_n, B_n\}_{n=1}^\infty$. Then*

$$\det(A_k) = \prod_{j=1}^{p} \prod_{\ell=1}^{p} \left(\frac{a_{kp+j-1+\ell}}{a_{kp+j-1+\ell}^{(0)}} \right) \tag{8.4.7}$$

Proof. Since A_k is lower triangular,

$$\det(A_k) = (A_k)_{11} (A_k)_{22} \dots (A_k)_{pp} \tag{8.4.8}$$

$$= (\Delta(J))_{(k-1)p+1\ (k-1)p+1+p} (\Delta(J))_{(k-1)p+2\ (k-1)p+2+p} \tag{8.4.9}$$

because of where A_k sits in $\Delta(J)$.

$$\Delta(J) = (a_1^{(0)} \dots a_p^{(0)})^{-1} J^p + \text{lower order} \tag{8.4.10}$$

so, for any m,

$$\Delta(J)_{m\, m+p} = (a_1^{(0)} \dots a_p^{(0)})^{-1}) J^p_{m\, m+1}$$

$$= (a_1^{(0)} \dots a_p^{(0)})^{-1} a_m a_{m+1} \dots a_{m+p-1} \tag{8.4.11}$$

Given that $a_m^{(0)}$ is periodic, (8.4.9) and (8.4.11) imply (8.4.7). □

Proof of Theorem 8.4.1. By the fact that $a_j/a_j^{(0)}$ is bounded above and away from 0, and (8.4.7), we see that

$$(8.4.5) \Leftrightarrow \limsup[\det(|A_1|) \dots \det(|A_n|)] > 0 \tag{8.4.12}$$

where we use the fact that since $\det(A_j) > 0$,

$$\det(|A_j|) = |\det(A_j)| = \det(A_j) \tag{8.4.13}$$

By Corollary 8.3.2, (8.3.23) $\Leftrightarrow$ (8.4.4). If we prove that

$$(8.4.2) \Rightarrow \sum_{\substack{E \notin [-2,2] \\ E \in \sigma(\Delta(J))}} (|E| - 2)^{1/2} < \infty \tag{8.4.14}$$

then Theorem 8.4.1 follows from Theorem 4.5.1.

By the spectral mapping theorem,

$$E \in \sigma(\Delta(J)) \Leftrightarrow E = \Delta(E') \text{ with } E' \in \sigma(J) \tag{8.4.15}$$

Moreover, since all gaps in which eigenvalues occur are open, there are $c > 0, d$ so that for all $E' \in \sigma(J) \setminus \sigma_{\text{ess}}(J)$,

$$c(|\Delta(E)| - 2) \leq \text{dist}(E, \sigma_{\text{ess}}(J)) \leq d(|\Delta(E)| - 2) \tag{8.4.16}$$

which verifies (8.4.14). □

Remarks and Historical Notes. The results in this section are from Damanik–Killip–Simon [97]. However, the periodic case is a special case of the finite gap case, where there is earlier work by Widom and Peherstorfer–Sodin–Yuditskii that overlaps Theorem 8.4.1. See the Notes to Section 9.13 for further discussion.

8.5 CONTROLLING THE ℓ^2 APPROACH TO THE ISOSPECTRAL TORUS

In this section, we will control the relation of Hilbert–Schmidt estimates on $\Delta_{J_0}(J) - (S^p + S^{-p})$ to ℓ^2 approach of the Jacobi parameters of J to the isospectral torus, $\mathcal{T}_\mathfrak{e}$, with $\mathfrak{e} = \sigma_{\text{ess}}(J)$. This is preliminary to proving a Killip–Simon-type theorem for periodic perturbations. We need to begin by considering the definition of the distance of the tail of J to $\mathcal{T}_\mathfrak{e}$.

Definition. Given two bounded sequences $\{a_n, b_n\}_{n=1}^\infty$ and $\{a'_n, b'_n\}_{n=1}^\infty$ of Jacobi parameters, we define

$$d_m((a,b),(a',b')) = \sum_{k=0}^{\infty} e^{-k}(|a_{m+k} - a'_{m+k}| + |b_{m+k} - b'_{m+k}|) \tag{8.5.1}$$

a metric defining the infinite product topology on $\{a_{n+m}, b_{n+m}\}_{n=1}^\infty$. We also define

$$\tilde{d}_m((a,b),(a',b')) = \sum_{k=0}^{p-1} (|a_{m+k} - a'_{m+k}| + |b_{m+k} - b'_{m+k}|) \tag{8.5.2}$$

Given a set, $\mathcal{T}$, of Jacobi parameters, we set

$$d_m((a,b),\mathcal{T}) = \inf\{d_m((a,b),(a',b')) \mid (a',b') \in \mathcal{T}\} \tag{8.5.3}$$

and similarly for $\tilde{d}_m$. Notice that because $\mathcal{T}_\mathfrak{e}$ is a translation invariant set and the translates of a bounded (a,b) lie in a compact set, we have that

$$\{\text{right limits of } (a,b)\} \subset \mathcal{T}_\mathfrak{e} \Leftrightarrow \lim_{m\to\infty} d_m((a,b),\mathcal{T}_\mathfrak{e}) = 0 \tag{8.5.4}$$

The main result of this section is the following:

Theorem 8.5.1. *Let J_0 be a period p periodic Jacobi matrix with all gaps open and let Δ_{J_0} be its discriminant. Let J be a bounded Jacobi matrix with Jacobi parameters $\{a_n, b_n\}_{n=1}^\infty$. Let A_n, B_n be the block Jacobi parameters of $\Delta_{J_0}(J)$. Then the following are equivalent:*

(i) *$\Delta_{J_0}(J) - S^p - S^{-p}$ is a Hilbert–Schmidt operator on $\ell^2(\{0,1,2,\dots\})$.*

(ii)

$$\sum_n \operatorname{Tr}(B_n^2 + (|A_n| - 1)^2) < \infty \tag{8.5.5}$$

(iii)

$$\sum_m d_m((a,b), \mathcal{T}_{J_0})^2 < \infty \tag{8.5.6}$$

(iv)

$$\sum_m \tilde{d}_m((a,b), \mathcal{T}_{J_0})^2 < \infty \tag{8.5.7}$$

We begin by proving equality of sums of d_m's and $\tilde{d}_m$ under great generality. This will require the following technical-looking result:

Lemma 8.5.2. *Fix $\varepsilon > 0$. Let $\{a_n^{(0)}, b_n^{(0)}\}_{n=1}^\infty$ be the Jacobi parameters of some period p periodic Jacobi matrix in an isospectral torus, $\mathcal{T}_\mathfrak{e}$. Let $(a_n, b_n)_{n=1}^\infty$ be a set of bounded Jacobi parameters with*

$$\varepsilon < a_n < \varepsilon^{-1} \tag{8.5.8}$$

There exists C depending only on ε and $\mathcal{T}_\mathfrak{e}$ so that for all m and all $n \ge m$,

$$|a_n - a_n^{(0)}| + |b_n - b_n^{(0)}| \le \tilde{d}_m((a,b), (a^{(0)}, b^{(0)})) + C \sum_{r=m}^{n-p+1} \tilde{d}_r((a,b), \mathcal{T}_\mathfrak{e}) \tag{8.5.9}$$

Proof. Decrease ε if necessary, so the a_n's of every element of $\mathcal{T}_\mathfrak{e}$ obeys (8.5.8). Define for $(b_1, \dots, b_p) \in \mathbb{R}^p$, $(a_1, \dots, a_p) \in (\varepsilon, \varepsilon^{-1})^p$,

$$f(a_1, \dots, a_p) = \sum_{j=1}^p [\log(a_j) - \log(a_j^{(0)})] \tag{8.5.10}$$

$$g(b_1, \dots, b_p) = \sum_{j=1}^p [b_j - b_j^{(0)}]$$

By (5.5.121), $\sum_{j=1}^p \log(a_j^{(1)}) = \sum_{j=1}^p \log(a_j^{(0)}) = \log(C(\mathfrak{e}))$ for any $\{a_j^{(1)}, b_j^{(1)}\}_{j=1}^p \in \mathcal{T}_\mathfrak{e}$. Since log is Lipschitz on $(\varepsilon, \varepsilon^{-1})$, we conclude that

$$f(a_m, \dots, a_{n+p-1}) \le C_1 \tilde{d}_m((a,b), \mathcal{T}_\mathfrak{e}) \tag{8.5.11}$$

By (5.4.13), $\Delta(x)$ determines $\sum_{j=1}^p b_j$ also, so $\sum_{j=1}^p b_j^{(1)} = \sum_{j=1}^p b_j^{(0)}$ for any $\{a_j^{(1)}, b_j^{(1)}\}_{j=1}^p \in \mathcal{T}_\mathfrak{e}$. As with (8.5.11), we obtain

$$g(b_m, \dots, b_{m+p-1}) \le \tilde{d}_m((a,b), \mathcal{T}_\mathfrak{e}) \tag{8.5.12}$$

Thus, by (8.5.12),

$$\begin{aligned} |b_n - b_{n-p}| &= |g(b_{n-p+1}, \dots, b_n) - g(b_{n-p}, \dots, b_{n-1})| \\ &\le \tilde{d}_{n-p+1}((a,b), \mathcal{T}_\mathfrak{e}) + \tilde{d}_{n-p}((a,b), \mathcal{T}_{J_0}) \end{aligned} \tag{8.5.13}$$

Similarly, using (8.5.11),

$$|\log(a_n) - \log(a_{n-p})| \le C_1[\tilde{d}_{n-p+1}((a,b),\mathcal{T}_\mathfrak{e}) + \tilde{d}_{n-p}((a,b),\mathcal{T}_\mathfrak{e})] \tag{8.5.14}$$

Since exp is Lipschitz on $(\log\varepsilon, \log\varepsilon^{-1})$,

$$|a_n - a_{n-p}| \le C_2[\tilde{d}_{n-p+1}((a,b),\mathcal{T}_\mathfrak{e}) + \tilde{d}_{n-p}((a,b),\mathcal{T}_\mathfrak{e})] \tag{8.5.15}$$

Thus, by periodicity of $a^{(0)}, b^{(0)}$, we see

$$\begin{aligned}|a_n - a_n^{(0)}| + |b_n - b_n^{(0)}| &\le |a_{n-p} - a_{n-p}^{(0)}| + |b_{n-p} - b_{n-p}^{(0)}| \\ &\quad + (1 + C_2)[\tilde{d}_{n-p+1}((a,b),\mathcal{T}_\mathfrak{e}) + \tilde{d}_{n-p}((a,b),\mathcal{T}_\mathfrak{e})]\end{aligned} \tag{8.5.16}$$

We can now prove (8.5.9) by induction. For $m \le n \le m+p-1$, the sum disappears on the right of (8.5.9) and the result is immediate from the definition of $\tilde{d}_m$. For $m+p \le n \le m+2p-1$, we obtain the result from the original case using (8.5.11). The general result follows by induction. □

Proposition 8.5.3. *Let J_0 be a period p periodic Jacobi matrix with isospectral torus $\mathcal{T}_\mathfrak{e}$. Then there is a constant C so that for all Jacobi parameters $(a_n, b_n)_{n=1}^\infty$ with* (8.5.8)*, we have*

$$e^{2(1-p)} \sum_m \tilde{d}_m((a,b),\mathcal{T}_\mathfrak{e})^2 \le \sum_m d_m((a,b),\mathcal{T}_\mathfrak{e})^2 \tag{8.5.17}$$

$$\le C \sum_m \tilde{d}_m((a,b),\mathcal{T}_\mathfrak{e})^2 \tag{8.5.18}$$

Proof. (8.5.17) is trivial since, except for a weight bounded below by $e^{-(p-1)}$, the sum in d_m includes all terms in $\tilde{d}_m$.

For the other direction, the lemma implies

$$d_m((a,b),\mathcal{T}_\mathfrak{e}) \le C_1 \sum_{j=0}^\infty e^{-j}\tilde{d}_{m+j}((a,b),\mathcal{T}_\mathfrak{e}) \tag{8.5.19}$$

Thus, by a Schwarz inequality and $\sum_{j=0}^\infty e^{-j} < \infty$,

$$d_m((a,b),\mathcal{T}_\mathfrak{e})^2 \le C_2 \sum_{j=0}^\infty e^{-j}\tilde{d}_{m+j}((a,b),\mathcal{T}_\mathfrak{e})^2 \tag{8.5.20}$$

so

$$\begin{aligned}\sum_m d_m((a,b),\mathcal{T}_\mathfrak{e})^2 &\le C_2 \sum_{m,j=0}^\infty e^{-j}\tilde{d}_{m+j}((a,b),\mathcal{T}_\mathfrak{e})^2 \\ &\le C_2 \sum_{n=0}^\infty \tilde{d}_n((a,b),\mathcal{T}_\mathfrak{e})^2 \Big(\sum_{j=0}^n e^{-j}\Big) \\ &\le C \sum_{n=0}^\infty \tilde{d}_n((a,b),\mathcal{T}_\mathfrak{e})^2\end{aligned}$$

□

Remark. While the technicalities may obscure this, the key fact that lets us use p-fold sums is that in $\mathcal{T}_{\mathfrak{e}}$, $(a_1^{(1)}, \dots, a_{p-1}^{(1)}; b_1^{(1)}, \dots, b_{p-1}^{(1)})$ determine $a_p^{(1)}$ and $b_p^{(1)}$ by the constancy of $\sum_{j=1}^p b_j$ and $\prod_{j=1}^p a_j$ over $\mathcal{T}_{\mathfrak{e}}$.

Having seen (iii) $\Leftrightarrow$ (iv) in Theorem 8.5.1, we turn to the other easy equivalence (i) $\Leftrightarrow$ (ii).

Proposition 8.5.4. *For any J_0, J we have* (i) $\Leftrightarrow$ (ii) *in Theorem 8.5.1.*

Remark. B_n and A_n as block Jacobi parameters for $\Delta_{J_0}(J)$ depend on J and J_0.

Proof. For two block Jacobi matrices, $\mathbb{J}$ and $\tilde{\mathbb{J}}$, since Hilbert–Schmidt norms are squares of matrix elements,

$$\|\mathbb{J} - \tilde{\mathbb{J}}\|_{\mathcal{J}_2}^2 = \sum_n \|B_n - \tilde{B}_n\|_{\mathcal{J}_2}^2 + 2\sum_n \|A_n - \tilde{A}_n\|_{\mathcal{J}2}^2 \tag{8.5.21}$$

Thus,

$$\|\Delta_{J_0}(J) - S^p - S^{-p}\|_{\mathcal{J}_2}^2 = \sum_n \operatorname{Tr}(B_n^2) + 2\operatorname{Tr}((A_n - \mathbf{1})^2) \tag{8.5.22}$$

This plus Theorem 4.6.7 implies (i) $\Leftrightarrow$ (ii). $\square$

We turn now to the most subtle part of Theorem 8.5.1, namely that (i) $\Leftrightarrow$ (iii), which will depend on the all-gaps-open hypothesis.

Lemma 8.5.5. *Let F be a C^∞ map of an open set $U \subset \mathbb{R}^n$ to $\mathbb{R}^\ell$ with $\ell < n$. Suppose for some $y_0 \in \mathbb{R}^\ell$,*

$$\mathcal{T} = F^{-1}(\{y_0\}) \tag{8.5.23}$$

is a smooth compact manifold of dimension $n - \ell$, and for all $x_0 \in \mathcal{T}$,

$$\operatorname{rank}((\nabla F)(x_0)) = \ell$$

Then for any compact neighborhood, K, of $\mathcal{T}$, there are constants c_K, d_K in $(0, \infty)$ so that for all $x \in K$,

$$c_K|F(x) - y_0| \le \operatorname{dist}(x, \mathcal{T}) \le d_K|F(x) - y_0| \tag{8.5.24}$$

Proof. That this holds locally near any $x_1 \in K$ follows from the implicit function theorem. Compactness then implies the global result on K. $\square$

For any $\{a_j, b_j\}_{j=1}^p$ in $((0, \infty) \times \mathbb{R})^p$, we can define $p + 1$ functions $c_0, \dots, c_p$ by

$$\Delta_{J_0(a,b)}(\lambda) = \sum_{k=0}^p c_k \lambda^k \tag{8.5.25}$$

where $J_0(a, b)$ is the periodic Jacobi matrix with parameters $\{a_j, b_j\}_{j=1}^p$. Let $F\colon (0, \infty) \times \mathbb{R}^p \to \mathbb{R}^{p+1}$ by $F_k(a, b) = c_k(a, b)$. Then

Proposition 8.5.6. *At any set of periodic Jacobi parameters for which J_0 has all gaps open,*

$$\operatorname{rank}[(\nabla F)(a, b)] = p + 1 \tag{8.5.26}$$

Proof. Since ∇F maps $\mathbb{R}^{2p}$ to $\mathbb{R}^{p+1}$, this is equivalent to saying $\{\nabla c_k\}_{k=0}^p$ as vectors in $\mathbb{R}^{2p}$ are linearly independent. That is the content of Theorem 6.9.1. □

Lemma 8.5.7. *Let χ_k be the projection in ℓ^2 onto $\{\delta_j\}_{j=1}^k$. For any compact subset, K, in $((0,\infty)\times\mathbb{R})^p$ of period p Jacobi matrices, there are constants c_K, d_K so for all $J \in K$ and $(y_0,\dots,y_p)\in\mathbb{C}^{p+1}$,*

$$c_K\left\| \sum_{j=1}^p y_j J^j \chi_{p+1}\right\|_{\mathcal{J}_2}^2 \le \sum_{j=0}^p |y_j|^2 \le d_K \left\| \sum_{j=1}^p y_j J^j \chi_{p+1}\right\|_{\mathcal{J}_2}^2 \tag{8.5.27}$$

Proof. $\{J^\ell \chi_{p+1}\}_{\ell=0}^p$ are linearly independent since $J^\ell\chi_{p+1}$ has nonzero elements in position 1 $\ell+1$ and zero elements in positions 1 $j+1$ for $j=\ell+1,\dots,p+1$. Thus, the matrix

$$M_{\ell k} = \operatorname{Tr}(\chi_{p+1} J^\ell J^k \chi_{p+1}) \qquad \ell, k = 0,\dots,p \tag{8.5.28}$$

is strictly positive definite. By continuity,

$$0 < \inf_{J\in K} \|M_{\ell k}\| \le \sup_{J\in K} \|M_{\ell k}\| < \infty \tag{8.5.29}$$

which leads directly to (8.5.27). □

Now view J, J_0, two period p periodic Jacobi matrices, as two-sided. Then $\Delta_{J_0}(J)$ is a two-sided block Jacobi matrix with constant A's and B's we will denote by $A_{J_0}(J)$, $B_{J_0}(J)$.

Proposition 8.5.8. *Let J_0 be a periodic Jacobi matrix with all gaps open and isospectral torus $\mathcal{T}_{\mathfrak{e}}$. Then for any compact neighborhood, K, of $\mathcal{T}_{\mathfrak{e}}$ in $((0,\infty)\times\mathbb{R})^p$, there are c_K, d_K in $(0,\infty)$ so that for all period p J with Jacobi parameters in K,*

$$\begin{aligned} c_K(\|A_{J_0}(J)-\mathbf{1}\|^2 + \|B_{J_0}(J)\|^2) &\le \operatorname{dist}(J,\mathcal{T}_{\mathfrak{e}})^2 \\ &\le d_K(\|A_{J_0}(J)-\mathbf{1}\|^2 + \|B_{J_0}(J)\|^2) \end{aligned} \tag{8.5.30}$$

Proof. Use $\sim$ to indicate two sides have a ratio bounded above and away from zero on compacts. Clearly, for any operator on ℓ^2,

$$\ell \to \|M\chi_\ell\|_{\mathcal{J}_2}$$

is monotone in ℓ (since it is the sum of the squares of rows $1,\dots,\ell$). Thus,

$$\begin{aligned} \|[\Delta_{J_0}(J) - (S^p+S^{-p})]\chi_p\|_{\mathcal{J}_2} &\le \|[\Delta_{J_0}(J) - (S^p+S^{-p})]\chi_{p+1}\|_{\mathcal{J}_2} \\ &\le \|[\Delta_{J_0}(J) - (S^p+S^{-p})]\chi_{2p}\|_{\mathcal{J}_2} \end{aligned} \tag{8.5.31}$$

while for $n=1,2$,

$$\|[\Delta_{J_0}(J) - (S^p+S^{-p})]\chi_{np}\|_{\mathcal{J}_2}^2 = n\|B_{J_0}(J)\|_{\mathcal{J}_2}^2 + 2n\|(A_{J_0}(J)-\mathbf{1})\|_{\mathcal{J}_2}^2 \tag{8.5.32}$$

so

$$\|A_{J_0}(J)-\mathbf{1}\|^2 + \|B_{J_0}(J)\|^2 \sim \|[\Delta_{J_0}(J) - S^p + S^{-p}]\chi_{p+1}\|^2 \tag{8.5.33}$$

By the magic formula,

$$[\Delta_{J_0}(J) - (S^p + S^{-p})]\chi_{p+1} = \sum_{\ell=0}^{p} c_\ell J^\ell \chi_{p+1} \tag{8.5.34}$$

where c_ℓ are the difference of the coefficients of the polynomials Δ_{J_0} and Δ_J. By Lemma 8.5.5,

$$\|[\Delta_{J_0}(J) - (S^p + S^{-p})]\chi_{p+1}\|^2 \sim \sum_{\ell=0}^{p} |c_\ell|^2 \tag{8.5.35}$$

By Lemma 8.5.5 and Proposition 8.5.6,

$$\sum_{\ell=0}^{p} |c_\ell|^2 \sim \text{dist}(J, \mathcal{T}_\mathfrak{e})^2 \tag{8.5.36}$$

(8.5.33), (8.5.35), and (8.5.36) imply (8.5.30). □

Proposition 8.5.9. *Let $k \leq \ell$. Then $\Delta_{J_0}(J)_{k\ell}$ for any bounded Jacobi matrix depends only on $\{b_j\}_{j=k-\alpha}^{\ell+\alpha}$ and $\{a_j\}_{j=k-\alpha}^{\ell+\alpha-1}$ where α is the greatest integer less than or equal to $\frac{1}{2}[p - (\ell - k)]$.*

Proof. Each J changes index by at most 1, so J^m, $m = 0, 1, \dots, p$, can change index by at most p steps of $0, \pm 1$, $\ell - k$ steps are needed to get from k to ℓ. The remaining steps have to go both up and back, so they cannot go higher than $\ell + \alpha$ or below $k - \alpha$. □

Corollary 8.5.10. *Fix J_0. Let $k \leq \ell$ and α given by Proposition 8.5.9. Then for any K and all $J, \tilde{J}$ whose Jacobi parameters obey*

$$\sup_j [|b_j| + |\tilde{b}_j| + |a_j| + |\tilde{a}_j|] \leq K \tag{8.5.37}$$

there is a C_K so that

$$|\Delta_{J_0}(J)_{k\ell} - \Delta_{J_0}(\tilde{J})_{k\ell}| \leq C_K \sup_{k-\alpha \leq j \leq \ell+\alpha} [|b_j - \tilde{b}_j| + |a_j - \tilde{a}_j|] \tag{8.5.38}$$

Proof. Immediate from Proposition 8.5.9, given that for J_0 fixed, $\Delta_{J_0}(J)_{k\ell}$ is a polynomial in a fixed number of variables with fixed coefficients. □

Lemma 8.5.11. (a) *For any Jacobi matrix, J, and $\ell = 1, 2, \dots$, $m = 1, 2, \dots$,*

$$(J^\ell)_{m\, m+\ell} = a_m a_{m+1} \dots a_{m+\ell-1} \tag{8.5.39}$$

and for $\ell = 2, 3, \dots$, $m = 1, 2, \dots$,

$$(J^\ell)_{m\, m+\ell-1} = a_m \dots a_{m+\ell-2} \left(\sum_{j=0}^{\ell-1} b_{m+j} \right) \tag{8.5.40}$$

(b) *For J_0 is periodic of period $p \geq 2$ and $m = 1, 2, \dots,$*

$$\Delta_{J_0}(J)_{m\,m+p} = \frac{(a_m \dots a_{m+p-1})}{[a_m^{(0)} \dots a_{m+p-1}^{(0)}]} \tag{8.5.41}$$

$$\Delta_{J_0}(J)_{m\,m+p-1} = [a_m^{(0)} \dots a_{m+p-1}^{(0)}]^{-1}(a_m \dots a_{m+p-2})\biggl(\sum_{j=0}^{p-1}(b_{m+j} - b_{m+j}^{(0)})\biggr) \tag{8.5.42}$$

Proof. (a) Since J can increase index by at most one,

$$(J^\ell)_{m\,m+\ell} = (J_{m\,m+1}) \dots (J_{m+\ell-1\,m+\ell}) \tag{8.5.43}$$

proving (8.5.39), while

$$(J^\ell)_{m\,m+\ell-1} = \sum_{j=0}^{\ell-1}(J^j)_{m\,m+j} J_{m+j\,m+j}(J^{\ell+j-1})_{m+j\,m+\ell-1} \tag{8.5.44}$$

which, given (8.5.39), proves (8.5.40).

(b) By (5.4.13),

$$\Delta_{J_0}(J) = (a_j^{(0)} \dots a_p^{(0)})^{-1}\biggl[J^p - \sum_{j=0}^{p-1} b_{j+1}^{(0)} J^{p-1} + O(J^{p-2})\biggr] \tag{8.5.45}$$

which, given (a), $(J^{p-k})_{m\,m+p} = (J^{p-k})_{m\,m+p-1} = 0$ if $k = 2, 3, \dots,$ and the periodicity of $a^{(0)}$ and $b^{(0)}$, implies (8.5.41) and (8.5.42). □

Lemma 8.5.12. *If $\Delta_{J_0}(J) - (S^p + S^{-p}) \in \mathcal{I}_2$, then*

(i) $$\sum_n (a_n a_{n+1} \dots a_{n+p-1} - a_n^{(0)} \dots a_{n+p-1}^{(0)})^2 < \infty \tag{8.5.46}$$

(ii) $$\sum_n \biggl(\sum_{j=0}^{p-1}(b_{n+j} - b_{n+j}^{(0)})\biggr)^2 < \infty \tag{8.5.47}$$

Proof. For a Hilbert–Schmidt operator, any subset of matrix elements lies in ℓ^2, so by (8.5.41),

$$\sum_n |a_n \dots a_{n+p-1}[a_n^{(0)} \dots a_{n+p-1}^{(0)}]^{-1} - 1|^2 < \infty \tag{8.5.48}$$

which, given that $a_n^{(0)} \dots a_{n+p-1}^{(0)}$ is n-independent, implies (8.5.46). Similarly, (8.5.42) implies (8.5.47) if we note that a_j bounded and $a_n \dots a_{n+p-1} \to a_1^{(0)} \dots a_p^{(0)} > 0$ implies $\inf(a_j) > 0$, so

$$\inf_m (a_m^{(0)} \dots a_{m+p-1}^{(0)})^{-1}(a_m \dots a_{m+p-2}) > 0 \tag{8.5.49}$$

□

Lemma 8.5.13. *If $\Delta_{J_0}(J) - (S^p + S^{-p}) \in \mathfrak{I}_2$, then*

$$\sum_n (a_{n+p} - a_n)^2 < \infty \tag{8.5.50}$$

$$\sum_n (b_{n+p} - b_n)^2 < \infty \tag{8.5.51}$$

Proof. Since a difference of ℓ^2 sequences is ℓ^2, (8.5.46) implies (since $a_n^{(0)}$ is periodic)

$$\sum_n (a_{n+p} - a_n)^2 (a_{n+1} \dots a_{n+p-1})^2 < \infty \tag{8.5.52}$$

which, given that $\inf(a_j) > 0$, implies (8.5.50).

Similarly, since

$$\sum_{j=0}^{p-1} (b_{n+1+j} - b_{n+j}) = b_{n+p} - b_n \tag{8.5.53}$$

(8.5.47) implies (8.5.51). □

Proof of Theorem 8.5.1. Given what we have proven already, we only need (iv) ⇒ (i) and (i) ⇒ (iii).

<u>(iv) ⇒ (i).</u> In Proposition 8.5.8, $\alpha \le \frac{1}{2}(p - (\ell - k))$, so

$$(\ell + \alpha) - (k - \alpha) = 2\alpha + \ell - k \le p \tag{8.5.54}$$

Since J is bounded and $\Delta_{J_0}(J)_{k\ell} - (S^p + S^{-p})_{k\ell}$ is a polynomial in at most p consecutive a, b pairs, which vanishes on $\mathcal{T}_{J_0}$, for some m (dependent on k, ℓ) and some C (independent of k, ℓ), we have that

$$|\Delta_{J_0}(J)_{k\ell} - (S^p + S^{-p})_{k\ell}| \le C\tilde{d}_m((a, b), \mathcal{T}_{J_0}) \tag{8.5.55}$$

Since a fixed m occurs for most p^2 $(k\ell)$ pairs ($m = k - \alpha$, so $m \le k \le m + p$ and $k \le \ell \le k + p$), with $k \le \ell$, we see (the 2 comes from $k \le \ell$ and $\ell \le k$ pairs)

$$\|\Delta_{J_0}(J) - (S^p + S^{-p})\|_{\mathcal{I}_2}^2 \le 2Cp^2 \sum_m \tilde{d}_m((a, b), \mathcal{T}_{J_0})^2 \tag{8.5.56}$$

<u>(i) ⇒ (iii).</u> Let $J^{(k)}$ be the period p Jacobi matrix that equals J on block k, that is, for $\ell = 1, \dots, p$,

$$b_\ell^{(k)} = b_{kp+\ell} \qquad a_\ell^{(k)} = a_{kp+\ell} \tag{8.5.57}$$

By Lemma 8.5.13 and the hypothesis (i),

$$\sum_n [\sup_{(k+1)p \le j \le (k+2)j-1} |b_j - b_j^{(k)}| + |a_j - a_j^{(k)}|]^2 < \infty \tag{8.5.58}$$

By Corollary 8.5.10, this implies that

$$\sum_n \|A_n(J) - A_{J_0}(J^{(n)})\|^2 + \|B_n(J) - B_{J_0}(J^{(n)})\|^2 < \infty \tag{8.5.59}$$

so that

$$\sum_n \|A_{J_0}(J^{(n)}) - \mathbf{1}\|^2 + \|B_{J_0}(J^{(n)})\|^2 < \infty \tag{8.5.60}$$

By Proposition 8.5.8, we see that

$$\sum_n \tilde{d}_{np}(J^{(n)}, \mathcal{T}_{\mathfrak{e}})^2 < \infty \tag{8.5.61}$$

On the other hand, (8.5.58) implies that for $k = 1, \dots, p$,

$$\sum_n \tilde{d}_{mp+k}(J, J^{(n)})^2 < \infty \tag{8.5.62}$$

By the triangle inequality,

$$\tilde{d}_{np+k}(J, \mathcal{T}_{\mathfrak{e}})^2 \le 2\tilde{d}_{np+k}(J, J^{(n)})^2 + 2\tilde{d}_{np+k}(J^{(n)}, \mathcal{T}_{\mathfrak{e}})^2 \tag{8.5.63}$$

so (8.5.61) and (8.5.62) imply (8.5.7). □

Remarks and Historical Notes. These results are from Damanik–Killip–Simon [97].

8.6 A KILLIP–SIMON THEOREM FOR PERIODIC JACOBI MATRICES

Here we will put together the magic formula, the Killip–Simon theorem for MOPRL, and the calculations in Sections 8.3 and 8.5 to prove (this is essentially what we earlier announced as Theorem 1.11.3):

Theorem 8.6.1 ([97]). *Let J_0 be a periodic Jacobi matrix with all gaps open. Let $\mathfrak{e} = \sigma_{\text{ess}}(J_0)$ and $\mathcal{T}_{\mathfrak{e}}$ the isospectral torus for J_0. Let J be a bounded Jacobi matrix with Jacobi parameters $\{a_n, b_n\}_{n=1}^\infty$ and spectral measure $d\rho$. Then*

$$\sum_{m=1}^\infty d_m((a,b), \mathcal{T}_{\mathfrak{e}})^2 < \infty \tag{8.6.1}$$

if and only if

(i)

$$\sigma_{\text{ess}}(J) = \mathfrak{e} \tag{8.6.2}$$

(ii)

$$\sum_{E \in \sigma(J) \setminus \mathfrak{e}} \text{dist}(E, \mathfrak{e})^{3/2} < \infty \tag{8.6.3}$$

(iii) *$d\rho$ has the form* (8.4.3) *with*

$$\int_{\mathfrak{e}} \text{dist}(x, \mathbb{R} \setminus \mathfrak{e})^{1/2} \log(\omega(x))\, dx > -\infty \tag{8.6.4}$$

Proof. By Theorem 8.5.1,

$$(8.6.1) \Leftrightarrow \Delta_{J_0}(J) - (S^p + S^{-p}) \in \mathcal{I}_2 \tag{8.6.5}$$

By the spectral mapping theorem,

$$(8.6.2) \Rightarrow \sigma_{\text{ess}}(\Delta_{J_0}(J)) = [-2, 2] \tag{8.6.6}$$

while

$$\sigma_{\text{ess}}(\Delta_{J_0}(J)) = [-2, 2] + (8.6.3) \Rightarrow (8.6.2) \tag{8.6.7}$$

By Corollary 8.3.3,

$$(8.6.4) \Leftrightarrow (8.3.26) \tag{8.6.8}$$

As in the proof in Section 8.4, all gaps open means

$$\text{dist}(E, \sigma_{\text{ess}}(J_0)) \approx \text{dist}(\Delta_{J_0}(E), [-2, 2]) \tag{8.6.9}$$

so

$$(8.6.3) \Leftrightarrow \sum_{E \in \sigma(\Delta_{J_0}(J)) \setminus [-2,2]} (|E| - 2)^{3/2} < \infty \tag{8.6.10}$$

We have thus proven equivalence of all conditions in the theorem and conditions on $\Delta_{J_0}(J)$, so Theorem 4.6.1 completes the proof of this theorem. □

Example 8.6.2. Let $J(t)$ be a curve in $\mathcal{T}_{\mathfrak{e}}$ (thought of as half-line Jacobi matrices) so $\|J'(t)\| = O(t^{-2/3})$. Thus,

$$\int \|J'(t)\|^2 \, dt < \infty \tag{8.6.11}$$

and $J(t)$ may not have a limit. For example, if we think of the torus $\mathcal{T}_{\mathfrak{e}}$ as $\mathbb{R}^\ell/\mathbb{Z}^\ell$, picking any unit vector $\eta \in \mathbb{R}^\ell$, we can take $J(t) = [t^{1/3}\eta]$, the equivalence class of $t^{1/3}\eta$, in which case $J(t)$ does not have a limit.

Let $\{a_n(t), b_n(t)\}_{n=1}^\infty$ be the Jacobi parameters of $J(t)$ and let J be the matrix with Jacobi parameters $a_n^J = a_n(n)$, $b_n^J = b_n(n)$. Then J is not asymptotic to any fixed $J_0 \in \mathcal{T}_{\mathfrak{e}}$, although all right limits lie in $\mathcal{T}_{\mathfrak{e}}$. Moreover, by (8.6.11), it is easy to see that

$$\sum_n \tilde{d}_m(J, J(n))^2 < \infty \tag{8.6.12}$$

so J obeys (8.6.1). Thus, in particular, $\Sigma_{\text{ac}}(J) = \mathfrak{e}$.

The point of this example is that one might have thought that while we can only prove in Theorem 7.6.1 that the right limits lie in $\mathcal{T}_{\mathfrak{e}}$, it might be that there is a single orbit as the limit points (as we will see (in Section 9.13) happens if a Szegő condition holds). This example shows that, in fact, the limit points can be the entire isospectral torus even though $\Sigma_{\text{ac}} = \sigma_{\text{ess}} = \mathfrak{e}$. □

Remarks and Historical Notes. This theorem and its proof are from Damanik–Killip–Simon [97].

8.7 SUM RULES FOR PERIODIC OPUC

We want to summarize here the main differences between the OPRL and OPUC results of the type discussed in Sections 8.4 and 8.6 and state, without proof, the OPUC results.

(1) For OPUC, the discriminant obeys

$$\overline{\Delta(z)} = \Delta(1/\bar{z}) \tag{8.7.1}$$

so $\Delta(z)$ is real on $\partial\mathbb{D}$. Thus, if $\mathcal{C}$ is a unitary CMV matrix, then

$$\Delta(\mathcal{C})^* = \Delta(\mathcal{C}) \tag{8.7.2}$$

Moreover, if $\mathcal{C}_0$ has period $p = 2k$, then $\Delta(z)$ has the form

$$\Delta_{\mathcal{C}_0}(z) = \sum_{j=-k}^{k} y_j z^j \tag{8.7.3}$$

for suitable y. Since $\mathcal{C}$ is five-diagonal, $\Delta_{\mathcal{C}_0}(\mathcal{C})$ has $2k = p$ diagonals above and below the main, so it is a block Jacobi matrix with $p \times p$ blocks.

There is still a magic formula. Namely, for any period $p = 2k$ CMV matrix, $\mathcal{C}_0$, and any two-sided CMV matrix, $\mathcal{C}$,

$$\Delta_{\mathcal{C}_0}(\mathcal{C}) = S^p + S^{-p} \tag{8.7.4}$$

if and only if $\mathcal{C} \in \mathcal{T}_\mathfrak{e}$ with $\mathfrak{e} = \sigma_{\text{ess}}(\mathcal{C}_0)$.

The only difference from OPRL is that A_n and B_n can have complex elements. However, B is still selfadjoint, that is, $B_n^\dagger = B_n$, and A_n is still lower triangular and positive on diagonal. The moral is that even for OPUC, it is MOPRL not MOPUC that is relevant!

(2) For OPRL, we have that if $\mathfrak{e}$ has all gaps open, the flows generated by the coefficients of Δ (other than the constants $(a_1 \dots a_p)^{-1}$ and $(a_1 \dots a_p)^{-1}(\sum_{j=1}^p b_j)$) are linearly independent. This was used critically in Section 8.5. For OPUC, it remains an open question to prove that the analog always holds. What is known (proven in Simon [400]) is that for a generic $\mathfrak{e}$, the normal bundle is spanned by the derivatives of the coefficients of Δ.

(3) In changing from d_m to $\tilde{d}_m$, we used the fact that in $\mathcal{T}_\mathfrak{e}$, $(a_j^{(0)}, b_j^{(0)})_{j=1}^{p-1}$ determine $a_p^{(0)}, b_p^{(0)}$. For OPUC, the analog is not true. But, of course, $\{\alpha_j^{(0)}\}_{j=1}^p$ determine $\alpha_{p+1}^{(0)}$, so if one defines $\tilde{d}_m$ as a sum over $p+1$, things work, but that change is needed.

Here are the two theorems:

Theorem 8.7.1. *Let $\mathcal{C}_0$ be a period p ($p = 2k$) periodic CMV matrix with Verblunsky coefficients $\{\alpha_n^{(0)}\}_{n=0}^\infty$ and let $\mathfrak{e} = \sigma_{\text{ess}}(\mathcal{C}_0)$. Let $\mathcal{C}_0$ be a CMV matrix with Verblunsky coefficients $\{\alpha_n\}_{n=0}^\infty$ and*

$$\sigma_{\text{ess}}(\mathcal{C}) = \mathfrak{e} \tag{8.7.5}$$

and spectral measure

$$d\mu = w(\theta)\,\frac{d\theta}{2\pi} + d\mu_s \tag{8.7.6}$$

Suppose that

$$\sum_{E\in\sigma(\mathcal{C})\setminus\mathfrak{e}} \operatorname{dist}(E,\sigma_{\text{ess}}(\mathcal{C}))^{1/2} < \infty \tag{8.7.7}$$

Then

$$\int_{\mathfrak{e}} \operatorname{dist}(e^{i\theta},\partial\mathbb{D}\setminus\mathfrak{e})^{-1/2}\log(w(\theta)) > -\infty \tag{8.7.8}$$

if and only if

$$\limsup \prod_{j=1}^{m}\biggl(\frac{\rho_j}{\rho_j^{(0)}}\biggr) > 0 \tag{8.7.9}$$

Theorem 8.7.2. *Fix $p = 2k$. There is a dense open set $U \subset \mathbb{D}^p$ so that if $\mathcal{C}_0$ is periodic with period p and with Verblunsky coefficients $\{\alpha_n\}_{n=0}^{p-1} \in U$, then with $\mathfrak{e} = \sigma_{\text{ess}}(\mathcal{C}_0)$ and $\mathcal{T}_{\mathfrak{e}}$ the isospectral torus, we have for any $\mathcal{C}$ with Verblunsky coefficients $\{\alpha_n\}_{n=1}^{\infty}$ that*

$$\sum_{m=1}^{\infty} d_m((\alpha),\mathcal{T}_{\mathfrak{e}})^2 < \infty \tag{8.7.10}$$

if and only if

(i)

$$\sigma_{\text{ess}}(\mathcal{C}) = \mathfrak{e} \tag{8.7.11}$$

(ii)

$$\sum_{E\in\sigma(\mathcal{C})\setminus\mathfrak{e}} \operatorname{dist}(E,\mathfrak{e})^{3/2} < \infty \tag{8.7.12}$$

(iii) *$d\mu$ has the form* (8.7.6) *with*

$$\int_{\mathfrak{e}} \operatorname{dist}(e^{i\theta},\partial\mathbb{D}\setminus\mathfrak{e})^{1/2}\log(w(x))\,dx > -\infty \tag{8.7.13}$$

Remarks and Historical Notes. The result on linear independence of flows is in Simon [400, Section 11.10]. The other results are from Damanik–Killip–Simon [97].

Chapter Nine

Szegő's Theorem for Finite Gap OPRL

9.1 OVERVIEW

In this chapter, we consider a general finite gap set, $\mathfrak{e}$, of the form

$$\mathfrak{e} = \bigcup_{j=1}^{\ell+1} [\alpha_j, \beta_j] \qquad \alpha_1 < \beta_1 < \alpha_2 < \cdots < \beta_{\ell+1} \tag{9.1.1}$$

and prove a Szegő–Shohat–Nevai theorem and Szegő asymptotics for suitable measures, μ, with $\sigma_{\text{ess}}(\mu) = \mathfrak{e}$. The key is to find an analog of the map $z \to z + z^{-1}$ of $\mathbb{D}$ to $\mathbb{C} \cup \{\infty\} \setminus [-2, 2]$, which was central to Chapter 3. Thus, we seek an analytic map

$$\mathbf{x}\colon \mathbb{D} \to \mathbb{C} \cup \{\infty\} \setminus \mathfrak{e} \tag{9.1.2}$$

Since the right side of (9.1.2) is not simply connected, we cannot hope that $\mathbf{x}$ is a bijection. Instead, we will want a many-to-one map. The inverse image, $\mathbf{x}^{-1}(w)$, of a single point will be a countable discrete set and we will deal with this set by finding a group, Γ, of analytic bijections of $\mathbb{D}$ so that

$$\mathbf{x}(z) = \mathbf{x}(w) \Leftrightarrow \exists \gamma \in \Gamma \quad \text{so that } w = \gamma(z) \tag{9.1.3}$$

Groups of analytic bijections of $\mathbb{D}$ with $\{\gamma(z)\}_{\gamma\in\Gamma}$ discrete are called Fuchsian groups. This approach to finite gap spectral theory was pioneered by Sodin–Yuditskii [413] and Peherstorfer–Yuditskii [343] and developed from a sum rule point of view by Christiansen–Simon–Zinchenko [86, 87, 88, 89].

In Sections 9.2–9.4, we discuss general analytic bijections on $\mathbb{D}$: individual maps in the first two sections and groups of such maps in the third. Section 9.5 constructs the map $\mathbf{x}$ and Section 9.6 studies the detailed structure for a finite gap set and its associated group and fundamental region. Section 9.7 finds functions vanishing at $\{\gamma(z_0)\}_{\gamma\in\Gamma}$ and relates these functions to a potential theory on $\mathfrak{e}$. Section 9.8 completes the general theory by proving a technical continuity result (that $\mathbf{x}$ and Γ are continuous in $\{\alpha_j, \beta_j\}_{j=1}^{\ell+1}$) that is important in Section 9.12.

An important role throughout is played by character automorphic functions, that is, analytic (or meromorphic or harmonic) functions, f, on $\mathbb{D}$ (or larger sets), which obey

$$f(\gamma(z)) = c(\gamma) f(z) \tag{9.1.4}$$

where $c\colon \Gamma \to \partial\mathbb{D}$ and obeys $c(\gamma\gamma') = c(\gamma)c(\gamma')$. c is a *character* of Γ and the set of all such characters, Γ^*, is isomorphic to the ℓ-dimensional torus. It is no coincidence that Γ^* is isomorphic to the isospectral torus and a natural map of $\mathbb{T}_{\mathfrak{e}}$ to Γ^* will play a central role in our proof of Szegő asymptotics.

In Section 9.9, we turn to applying the machinery developed earlier to spectral theory, proving a step-by-step sum rule, which, in Section 9.10, proves the following version of the Szegő–Shohat–Nevai theorem:

Theorem 9.1.1. *Let μ be a nontrivial probability measure on $\mathbb{R}$ with $\sigma_{\text{ess}}(\mu) = \mathfrak{e}$. Let $\{a_n, b_n\}_{n=1}^\infty$ be its Jacobi parameters and J its Jacobi matrix. Suppose μ has the form*

$$d\mu = w(x)\,dx + d\mu_s \tag{9.1.5}$$

and that

$$\sum_{E \in \sigma(J)\setminus\sigma_{\text{ess}}(J)} \text{dist}(E, \mathfrak{e})^{1/2} < \infty \tag{9.1.6}$$

Then

$$\int_{\mathfrak{e}} w(x)\text{dist}(x, \mathbb{R}\setminus\mathfrak{e})^{-1/2}\,dx > -\infty \tag{9.1.7}$$

if and only if

$$\limsup \frac{a_1 \dots a_n}{C(\mathfrak{e})^n} > 0 \tag{9.1.8}$$

If (9.1.6) and (9.1.7) (equivalently, (9.1.8)) hold, we say $\mu \in \text{Sz}(\mathfrak{e})$, the *Szegő class* for $\mathfrak{e}$. In Section 9.13, we will prove that if $\mu \in \text{Sz}(\mathfrak{e})$, there is $\{a_n^{(0)}, b_n^{(0)}\} \in \mathbb{T}_{\mathfrak{e}}$, an element of the isospectral torus, so that

$$\lim_{n\to\infty} |a_n - a_n^{(0)}| + |b_n - b_n^{(0)}| = 0 \tag{9.1.9}$$

a result that has not been proven using the methods of Chapter 8 even for the periodic case. To obtain this result (and an associated Szegő asymptotics on the polynomials), we rely on machinery developed in Sections 9.11 and 9.12. In Section 9.11, we define Θ-functions, natural character automorphic functions on $\mathbb{C}\cup\{\infty\}\setminus\Lambda(\Gamma)$. ($\Lambda(\Gamma)$ is the set of limit points of Γ, a closed nowhere dense subset of $\partial\mathbb{D}$ discussed in Section 9.4) with given zeros and poles. As a bonus of this theory, we will prove the case of Abel's theorem we need in Section 5.12. In Section 9.12, we associate a Jost function, yet another character automorphic function, to any $\mu \in \text{Sz}(\mathfrak{e})$. It will turn out that the $\{a_n^{(0)}, b_n^{(0)}\}_{n=1}^\infty$ of (9.1.9) is determined by the fact that the Jost functions for μ and for $\mu^{(0)}$ have the same character. Section 9.13 will also prove Szegő asymptotics for the polynomials and determine the asymptotics of the Jost function.

9.2 FRACTIONAL LINEAR TRANSFORMATIONS

Since we need more about fractional linear transformations (FLTs) than what is in the elementary books, our discussion begins with a rapid minicourse on the subject.

We will not describe the Riemann sphere in terms of stereographic projection but as $\mathbb{P}$, the complex projective line. In $\mathbb{C}^2 \setminus \{0\}$, we say u, v are equivalent, written $u \doteq v$, if and only if there is $\lambda \in \mathbb{C} \setminus \{0\}$ so $u = \lambda v$. It is easy to see this defines an equivalence relation. The equivalence classes, associated with (complex) lines in $\mathbb{C}^2$, are elements of $\mathbb{P}$.

$\mathbb{P}$ contains a distinguished element $\infty = [\binom{1}{0}]$, that is, the line with second coordinate 0. (Here $[\cdot]$ means equivalence class of $\cdot$.) $\mathbb{P} \setminus \{\infty\}$ can be put in one-one correspondence with $\mathbb{C}$ by associating z with $[\binom{z}{1}]$. Put equivalently, we can map $\mathbb{P} \setminus \{\infty\} \xrightarrow{\pi_\infty} \mathbb{C}$ by defining $\tilde{\pi}_\infty$ on $\mathbb{C}^2 \setminus \{u \in \mathbb{C}^2 \mid u_2 = 0\}$ by

$$\tilde{\pi}_\infty(u_1, u_2) = \frac{u_1}{u_2} \tag{9.2.1}$$

$\tilde{\pi}_\infty$ is constant on equivalence classes, and so induces π_∞ on $\mathbb{P} \setminus \{\infty\}$ by $\pi_\infty([u]) = \tilde{\pi}_\infty(u)$.

Similarly, if $0 \in \mathbb{P}$ is defined by $0 = \pi_\infty^{-1}(0)$, that is, $0 = [\binom{0}{1}]$, we can define $\tilde{\pi}_0$ on $\mathbb{C}^2 \setminus \{u \mid u_1 = 0\}$ by

$$\tilde{\pi}_0(u_1, u_2) = \frac{u_2}{u_1} \tag{9.2.2}$$

and induce π_0. The domains of π_0 and π_∞ overlap in $\mathbb{P} \setminus \{0, \infty\}$, which each maps to $\mathbb{C} \setminus \{0\}$, and

$$\pi_0 \pi_\infty^{-1} \colon \pi_\infty[\mathbb{P} \setminus \{0, \infty\}] \to \pi_0[\mathbb{P} \setminus \{0, \infty\}]$$

is, according to (9.2.1)/(9.2.2), given by

$$\pi_0 \pi_\infty^{-1}(z) = \frac{1}{z} \tag{9.2.3}$$

Thus, we have a local coordinate system with transition maps given by analytic functions, which defines a complex variables analog of manifolds called Riemann surfaces.

In practical terms, we associate $\mathbb{P}$ with $\mathbb{C} \cup \{\infty\}$, that is, we normally use $z = \pi_\infty(u)$ as our coordinates, shifting to $1/z$ as a coordinate near infinity.

If T is an invertible linear map on $\mathbb{C}^2$, clearly T maps $\mathbb{C}^2 \setminus \{\infty\}$ to itself and $u \doteq v$ implies $Tu \doteq Tv$, so T induces an invertible map f_T from $\mathbb{P}$ to itself. If $T = \left(\begin{smallmatrix} a & b \\ c & d \end{smallmatrix}\right)$, we have

$$f_T\left(\begin{bmatrix} z \\ 1 \end{bmatrix}\right) = \left[\begin{pmatrix} a & b \\ c & d \end{pmatrix}\begin{pmatrix} z \\ 1 \end{pmatrix}\right] = \left[\begin{pmatrix} \frac{az+b}{cz+d} \\ 1 \end{pmatrix}\right]$$

which in local coordinate z (from π_∞) is

$$f_T(z) = \begin{cases} \frac{az+b}{cz+d} & z \neq \infty, -\frac{d}{c} \\ \frac{a}{c} & \text{if } z = \infty \\ \infty & \text{if } z = -\frac{d}{c} \end{cases} \tag{9.2.4}$$

so FLTs are just the maps on the Riemann sphere induced by linear transformations. We already used this notion in Section 2.5.

Example 9.2.1. If $c = 0$, then $f_T(\infty) = \infty$ and f_T is the affine map

$$f_T(z) = \frac{a}{d}z + \frac{b}{d} \tag{9.2.5}$$

where $\det(T) \neq 0$ implies $d \neq 0 \neq a$. If $\det(T) = 1$, then $ad = 1$, so $d^{-1} = a$ and

$$f_T(z) = a^2 z + ba \tag{9.2.6}$$

We will summarize this below. As a second example,

$$f(z) = \frac{1}{z} \quad \text{or} \quad f(re^{i\theta}) = r^{-1}e^{-i\theta}$$

which inverses the radius and complex conjugates. To get pure inversion, one can define

$$r(z) = \frac{1}{\bar{z}} \tag{9.2.7}$$

which is the inversion in the unit circle, $\partial\mathbb{D}$. More generally, inversion in the circle $|z - z_0| = r$ has the form

$$T(z) = z_0 + \frac{r^2}{\bar{z} - \bar{z}_0} \tag{9.2.8}$$

Note that (9.2.7)/(9.2.8) are not analytic and not FLTs! □

We summarize the first example in

Proposition 9.2.2. *If* $f_T(\infty) = \infty$ *and* $\det(T) = 1$, *then*

$$f_T(z) = a^2 z + ab \tag{9.2.9}$$

A big point of defining FLTs as linear maps on $\mathbb{P}$ is that clearly,

$$f_T f_S = f_{TS} \tag{9.2.10}$$

The way to compose FLTs is matrix multiplication.

Proposition 9.2.3. *If* $T, S \in \mathbb{GL}(2, \mathbb{C})$ (2×2 *invertible matrices*), *then* $f_T = f_S$ *if and only if* $T = \lambda S$ *for some* $\lambda \in \mathbb{C} \setminus \{0\}$.

Proof. By using (9.2.10), it suffices to consider the case $S = \mathbf{1}$, since $f_T = f_S \Leftrightarrow f_{TS^{-1}} = f_{\mathbf{1}}$. But then $f_T(w) = w$ means $T\binom{w}{1} = \lambda_w \binom{w}{1}$ for $w = 0, z_0, 2z_0$. Since $\binom{z_0}{1} = \frac{1}{2}\binom{0}{1} + \frac{1}{2}\binom{2z_0}{1}$, we see (by looking at the two components) that

$$2\lambda_{2z_0} = 2\lambda_{z_0}$$
$$\lambda_0 + \lambda_{2z_0} = 2\lambda_{z_0}$$

which implies $\lambda_0 = \lambda_{z_0} = \lambda_{2z_0}$. Since z_0 is arbitrary, $T = \lambda_0 \mathbf{1}$. □

Given S, we can always pick λ so $\det(T) = 1$. We will henceforth do so unless noted explicitly. This determines T up to $\pm$ sign. Thus, if $\mathbb{SL}(2, \mathbb{C})$ is the set of 2×2 matrices of determinant 1, the map $T \to f_T$ is two-to-one with kernel $= \{\pm\mathbf{1}\}$, that is, if $\mathcal{F}$ is the group of all FLT, then

$$\mathcal{F} = \mathbb{SL}(2, \mathbb{C})/\{\pm 1\} \tag{9.2.11}$$

Thus, $\mathcal{F}$ is often called $\mathbb{PSL}(2, \mathbb{C})$.

One immediate advantage of the matrix connection is:

Lemma 9.2.4. $f_T([u]) = [u]$ *if and only if* u *is an eigenvector of* T.

Proof. $f_T([u]) = [u] \Leftrightarrow T(u) = \lambda u$. □

Proposition 9.2.5. *Let* $f \in \mathfrak{F}$ *and suppose* f *leaves three points fixed. Then* $f = \mathbf{1}$.

Proof. Any 2×2 matrix with three distinct eigenvectors (not counting multiples) is a multiple of the identity. □

Theorem 9.2.6. *Fix* w_0, w_1, w_2 *distinct. Then, for each distinct* z_0, z_1, z_2, *there is exactly one* $f \in \mathfrak{F}$ *with*

$$f(w_j) = z_j \tag{9.2.12}$$

for $j = 0, 1, 2$.

Proof. Uniqueness is immediate from Proposition 9.2.5. For if f_1 and f_2 solve (9.2.12), then $g = f_1 f_2^{-1}$ has z_0, z_1, z_2 as fixed points and so, by the proposition, $f_1 f_2^{-1} = \mathbf{1}$, proving uniqueness.

For existence, we note first that it suffices to handle the case $z_0 = 0, z_1 = 1, z_2 = \infty$. For if f takes (w_0, w_1, w_2) to $(0, 1, \infty)$ and g takes (z_0, z_1, z_2) to $(0, 1, \infty)$, then $g^{-1} f$ solves (9.2.12).

Given distinct w_0, w_1, w_2, the FLT

$$f(w) = \left(\frac{w_1 - w_2}{w_1 - w_0}\right)\left(\frac{w - w_0}{w - w_2}\right) \tag{9.2.13}$$

solves (9.2.12) with $(z_0, z_1, z_2) = (0, 1, \infty)$. □

As an immediate consequence, we see that $\mathfrak{F}$ is exactly the set of bianalytic homeomorphisms (aka conformal maps) of $\mathbb{P}$ to itself:

Corollary 9.2.7. *If* f *is a bijection of* $\mathbb{P}$ *to itself, which is analytic (in a local coordinate sense), then* $f \in \mathfrak{F}$.

Proof. Without loss, we can suppose f leaves $0, 1, \infty$ fixed since we can replace f by fg^{-1}, where $g \in \mathfrak{F}$ has $g(0) = f(0)$, $g(1) = f(1)$, $g(\infty) = f(\infty)$.

Since $f(\infty) = \infty$, $h(w) = \frac{1}{f(1/w)}$ is analytic near $w = 0$, has $h(0) = 0$ and $h'(0) \neq 0$ since it is single-valued near $w = 0$. It follows that $|h(w)| > C|w|$ near $w = 0$, so $|f(z)| \leq C^{-1}|z|$ near infinity. By Liouville's theorem, f is a degree one polynomial, hence $f(z) = z$ since there is a unique affine function with $f(0) = 0$, $f(1) = 1$. □

Remark. A useful way of thinking of this is that any analytic map of $\mathbb{P}$ to $\mathbb{P}$ is given by a rational function and it has to be a degree one polynomial if it is a bijection.

$\mathfrak{F}$ is a group and, as is with any group, its conjugacy classes are of interest.

Definition. $f, g \in \mathfrak{F}$ are called *conjugate* if and only if there is $h \in \mathfrak{F}$ so $hfh^{-1} = g$.

Theorem 9.2.8. *For $T, S \in \mathbb{SL}(2, \mathbb{C})$ neither equal to $\pm\mathbf{1}$, f_T is conjugate to f_S if and only if*

$$\mathrm{Tr}(S) = \pm \mathrm{Tr}(T) \tag{9.2.14}$$

and then we are in one of the following family of classes:

(i) Parabolic: $\mathrm{Tr}(S) = \pm 2$*; one conjugacy class; f has one fixed point. An element in the class is*

$$f(z) = z + 1 \tag{9.2.15}$$

(ii) Elliptic: $\mathrm{Tr}(S) \in (-2, 2)$, *so* $\mathrm{Tr}(S) = 2\cos\theta$. *Classes labeled by $\theta \in (0, \pi/2]$. f has two fixed points. An element in this class is*

$$f(z) = e^{2i\theta} z \qquad \theta \in (0, \pi/2] \tag{9.2.16}$$

(iii) Hyperbolic: $\mathrm{Tr}(S) \in \pm(2, \infty)$, *so* $\mathrm{Tr}(S) = \pm 2\cosh\varphi$, $\varphi \in (0, \infty)$. *f has two fixed points. An element in this class is*

$$f(z) = e^{-2\varphi} z \qquad \varphi \in (0, \infty) \tag{9.2.17}$$

(iv) Loxodromic: $\mathrm{Tr}(S) \in \{z \mid \mathrm{Im}(z) \neq 0\}$, *so* $\mathrm{Tr}(S) = \pm(e^{\alpha + i\theta} + e^{-\alpha - i\theta})$ *for some $\alpha \in (0, \infty)$ and $\theta \in (0, \pi)$. f has two fixed points. An element in this class is*

$$f(z) = e^{-2\alpha - 2i\theta} z \qquad \alpha \in (0, \infty),\ \theta \in (0, \pi) \tag{9.2.18}$$

Proof. If $f_T = f_W f_S f_W^{-1}$, then $T = \pm W S W^{-1}$, so of course, (9.2.14) holds. For the converse, we note that if we prove that one of models (9.2.15)–(9.2.18) is in each conjugacy class, we see there is only one conjugacy class with a given $\pm\mathrm{Tr}(T)$, proving that (9.2.14) implies conjugacy.

Every 2×2 matrix T has one or two (generically, two) eigenvectors, and so one or two fixed points, which we denote z_1, z_2 (if there is only one, we do not define z_2). If g maps z_1 to ∞ and z_2 to 0, then gTg^{-1} has $z_1 = \infty$, and if there is a z_2, it is 0. Thus, we can be sure there is a conjugate f to f_T with $f(\infty) = \infty$ and, unless $\mathrm{Tr}(T) = \pm 2$ (since $\det(T) = 1$ and two equal algebraic eigenvalues implies the eigenvalues are both $+1$ or both -1), $f(0) = 0$.

By Proposition 9.2.2, $f(\infty) = \infty$ implies f has the form (9.2.9). If $f(0) = 0$, $b = 0$ and thus

$$T = \begin{pmatrix} a & 0 \\ 0 & a^{-1} \end{pmatrix} \qquad f_T(z) = a^2 z \tag{9.2.19}$$

and $\mathrm{Tr}(T) = a + a^{-1}$.

Without loss, we can suppose $|a| \leq 1$, since interchanging 0 and ∞ interchanges a and a^{-1}. (ii), (iii), and (iv) correspond precisely to $|a| = 1$ with $a \neq \pm 1$, a real with $|a| < 1$, and $|a| < 1$, $\mathrm{Im}\, a \neq 0$, respectively. That leaves the case where T has a single eigenvector, which means there is a W with

$$\pm W T W^{-1} = \begin{pmatrix} 1 & 1 \\ 0 & 1 \end{pmatrix} \equiv S$$

which has $T_S(z) = z + 1$. $\square$

One advantage of these models is that they immediately make clear the asymptotics of $f^{(n)}(z) \equiv f \circ \cdots \circ f(z)$ repeated n times:

Theorem 9.2.9. *Let f be an element of $\mathfrak{F}$ with fixed points, z_1 and z_2 (if there is a second). Then*

(a) *If f is hyperbolic or loxodromic, for one of the fixed points, say z_1, we have $f^{(n)}(w) \to z_1$ as $n \to +\infty$ for any $w \neq z_2$, and for each fixed w, the approach is exponentially fast. As $n \to -\infty$, $f^{(n)}(w) \to z_2$ for any $w \neq z_1$.*

(b) *If f is parabolic, for any w, $f^{(n)}(w) \to z_1$ as $n \to \pm\infty$ and the approach is $O(1/n)$.*

(c) *If f is elliptic, either f is periodic, that is, $f^{(p)} = 1$ for some p, or else $f^{(n)}(w)$ is dense in an orbit, which is a closed curve, and $f^{(n)}(w)$ is almost periodic.*

Remarks. 1. Near any given point, we can measure distances in a local coordinate system, and ideas like "approach exponentially fast" are independent of coordinate system. There is a natural metric on $\mathbb{P}$, namely, $\rho([u],[v]) = \min(\|x - y\| \mid x \in [u],\, y \in [v],\, \|x\| = \|y\| = 1)$, and one could use that.

Alternatively, one can use chordal distance in stereographic projection, the so-called spherical metric $\sigma(z,w) = |z-w|/(1+|z|^2)^{1/2}(1+|w|^2)^{1/2}$. One can show $\sqrt{2}\,\sigma(z,w) \leq \rho([\begin{smallmatrix} z \\ 1 \end{smallmatrix}],[\begin{smallmatrix} w \\ 1 \end{smallmatrix}]) \leq 2\sigma(z,w)$.

2. The curves in (c) are circles. Our proof will show they are circles in a special case, and once one has Theorem 9.2.13 below, it will follow they are always circles.

Proof. For the last three models, where $z_2 = \infty$ and $z_1 = 0$ (in the terminology of this theorem), the claims are obvious, but the claims are preserved by conjugation. For the parabolic case, for $|w|$ near ∞, $\rho(w,\infty) \sim 1/w$ and $1/(w_0 + n)$ goes to ∞ as $O(n^{-1})$. □

Example 9.2.10. It pays to look at the parabolic case in more detail. Consider the case

$$f(z) = \frac{z}{z+1}$$

where $z = 0$ is the parabolic fixed point. $f = f_T$ with $T = \left(\begin{smallmatrix} 1 & 0 \\ 1 & 1 \end{smallmatrix}\right)$, so $T^n = \left(\begin{smallmatrix} 1 & 0 \\ n & 1 \end{smallmatrix}\right)$ and

$$f^{(n)}(i) = \frac{i}{ni+1} = \frac{n+i}{n^2+1}$$

The asymptotic approach is tangent to the real axis, but unlike the hyperbolic case where the approach to the asymptotic tangent is exponential, $\operatorname{Im} f^{(n)}(i) = O(1/n^2)$. It is easy to see that for any nonreal z_0, $f^{(n)}(z_0)$ has $\operatorname{Re} f^{(n)}(z_0) = O(1/n)$ and $|\operatorname{Im} f^{(n)}(z_0)| = O(1/n^2)$. The flow lines for hyperbolic and parabolic examples are shown in Figure 9.2.1. □

Note the name parabolic is *not* connected with the asymptotic parabolic relation of $\operatorname{Re} f^{(n)}$ and $\operatorname{Im} f^{(n)}$!

Next, we turn to the role of circles and lines under FLTs. $\mathbb{C}^2$ has its natural Euclidean inner product. Given a selfadjoint matrix B, $\langle u, Bu\rangle$ is not a function of

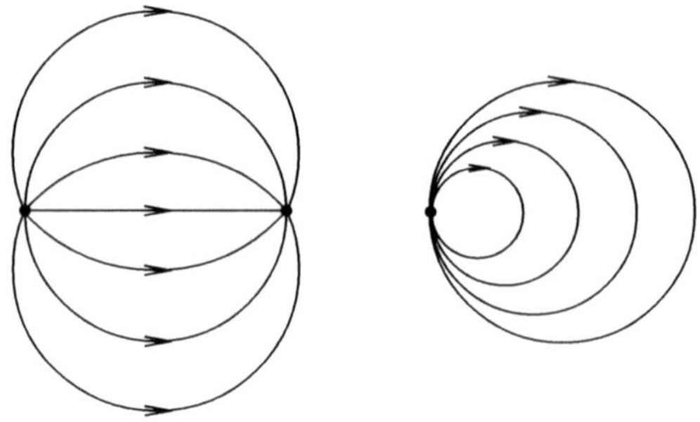

Figure 9.2.1. Flow lines for FLTs.

$[u]$, but since $\langle \lambda u, B\lambda u\rangle = |\lambda|^2 \langle u, Bu\rangle$ whether it is positive, zero, or negative, is constant on equivalence classes.

Theorem 9.2.11. *Let*

$$J = \begin{pmatrix} \alpha & \beta \\ \bar\beta & \gamma \end{pmatrix} \tag{9.2.20}$$

be a selfadjoint matrix (i.e., α, γ real). Let

$$C_J = \left\{ z \in \mathbb{C} \cup \{\infty\} \;\middle|\; \left\langle \begin{pmatrix} z \\ 1 \end{pmatrix}, J \begin{pmatrix} z \\ 1 \end{pmatrix} \right\rangle = 0 \right\} \tag{9.2.21}$$

$$C_J^{\pm} = \left\{ z \in \mathbb{C} \cup \{\infty\} \;\middle|\; \pm\left\langle \begin{pmatrix} z \\ 1 \end{pmatrix}, J \begin{pmatrix} z \\ 1 \end{pmatrix} \right\rangle > 0 \right\} \tag{9.2.22}$$

with the convention that, for $z = \infty$, we replace $\langle \binom{z}{1}, J\binom{z}{1}\rangle$ by $\langle \binom{1}{0}, J\binom{1}{0}\rangle$. Let $\det(J) < 0$. Then

(a) *If $\alpha = 0$, C_J is the straight line (real line, not complex line) ($z = x + iy$)*

$$(\operatorname{Re}\beta)x + (\operatorname{Im}\beta)y = -\frac{\gamma}{2} \tag{9.2.23}$$

(b) *If $\alpha \neq 0$, then C_J is the circle*

$$\left| z + \frac{\beta}{\alpha} \right|^2 = -\frac{\det(J)}{|\alpha|^2} \tag{9.2.24}$$

Every circle or line has this form. $C_J^{\pm}$ are the two connected components of $\mathbb{P} \setminus C_J$.

Remarks. 1. $\alpha = 0$ is equivalent to $\infty \in C_J$.

2. If $\alpha > 0$ (resp. $\alpha < 0$), C_J^+ is the outside (resp. inside) of C_J.

Proof.

$$\left\langle \begin{pmatrix} z \\ 1 \end{pmatrix}, J \begin{pmatrix} z \\ 1 \end{pmatrix} \right\rangle = \alpha\bar z z + 2\operatorname{Re}(\bar\beta z) + \gamma \tag{9.2.25}$$

If $\alpha = 0$, we get (9.2.23). If $\alpha \neq 0$,

$$\text{RHS of (9.2.25)} = \alpha \left| z + \frac{\beta}{\alpha} \right|^2 + \gamma - \frac{|\beta|^2}{\bar{\alpha}}$$

which leads to (9.2.24).

To get the line $ax + by = c$, take

$$J = \begin{pmatrix} 0 & a + ib \\ a - ib & -2c \end{pmatrix}$$

To get the circle $|z - z_0|^2 = r^2$, take

$$J = \begin{pmatrix} 1 & -z_0 \\ -\bar{z}_0 & -r^2 + |z_0|^2 \end{pmatrix}$$

□

Example 9.2.12. If

$$J = \begin{pmatrix} 1 & 0 \\ 0 & -1 \end{pmatrix} \tag{9.2.26}$$

then $\langle \binom{z}{1}, J\binom{z}{1} \rangle = |z|^2 - 1$. C_J is $\partial\mathbb{D}$; C_J^- is $\mathbb{D}$.

If

$$J = \begin{pmatrix} 0 & i \\ -i & 0 \end{pmatrix} \tag{9.2.27}$$

then $\langle \binom{z}{1}, J\binom{z}{1} \rangle = -2 \operatorname{Im} z$, so $C_J = \mathbb{R}$ and C_J^- is $\mathbb{C}_+$, the upper half-plane. □

Theorem 9.2.13. *$f \in \mathfrak{F}$ takes circles and lines into circles and lines. For any two circles (or circle and line, or two lines), there is an $f \in \mathfrak{F}$ taking the first to the second.*

Proof. Since

$$\left\langle T\begin{pmatrix} z \\ 1 \end{pmatrix}, JT\begin{pmatrix} z \\ 1 \end{pmatrix} \right\rangle = \left\langle \begin{pmatrix} z \\ 1 \end{pmatrix}, T^*JT\begin{pmatrix} z \\ 1 \end{pmatrix} \right\rangle \tag{9.2.28}$$

we have that

$$f_T[C_J] = C_{T^*JT} \tag{9.2.29}$$

proving circles/lines go to circles and lines.

Clearly, $\{z \mid |z - z_0| = r_0\}$ goes to $\{z \mid |z - z_1| = r_1\}$ under $f(z) = \frac{r_1}{r_0}(z - z_0 + z_1)$, so any circle can go to any other circle, and by a translation and rotation, any line goes to any line. Since $f(z) = 1/z$ takes $\{z \mid |z - 1| = 1\}$ to $\operatorname{Re} z = \frac{1}{2}$, we see that we can get from one particular circle to one particular line, so by the beginning of this paragraph, from any circle to any line. □

Example 9.2.14. Among the most famous FLTs are

$$f(z) = \frac{1}{i}\left[\frac{z - 1}{z + 1}\right] \tag{9.2.30}$$

and its inverse

$$f^{-1}(w) = \frac{1 - iw}{1 + iw} \tag{9.2.31}$$

f maps $\mathbb{D}$ (resp. $\partial\mathbb{D}$) to $\mathbb{C}_+$ (resp. $\mathbb{R}$) with

$$f(e^{i\theta}) = \tan\left(\frac{\theta}{2}\right)$$

and, of course, f^{-1} takes $\mathbb{C}_+$ to $\mathbb{D}$, and

$$f^{-1}(\tan(\psi)) = e^{2i\psi} \tag{9.2.32}$$

□

Next we want to look at setwise invariances of C_J and $C_J^{\pm}$ under some f_T. It will help to consider first the special case $\mathbb{C}_+$ and $\mathbb{R}$ considered in Example 9.2.12.

Proposition 9.2.15. *Let $A \in \mathbb{GL}(2, \mathbb{C})$. Then f_A maps $\mathbb{R} \cup \{\infty\}$ to itself if and only if there is a real matrix B and $e^{i\psi} \in \partial\mathbb{D}$ so that*

$$A = e^{i\psi} B \tag{9.2.33}$$

f_A maps $\mathbb{C}_+$ to $\mathbb{C}_+$ if $\det(B) > 0$ *and $\mathbb{C}_+$ to $\mathbb{C}_-$ if* $\det(B) < 0$*. In particular, $A \in \mathbb{SL}(2, \mathbb{C})$ maps $\mathbb{C}_+$ onto $\mathbb{C}_+$ if and only if $A \in \mathbb{SL}(2, \mathbb{R})$, that is, A has all real entries.*

Proof. Clearly, if B is real, f_B maps $\mathbb{R}$ to $\mathbb{R}$ and so, by projective equivalence, so does A of the form (9.2.33).

Conversely, if $f \in \mathfrak{F}$ maps $\mathbb{R} \cup \{\infty\}$ to itself, let $f^{-1}(0) = w_0$, $f^{-1}(1) = w_1$, $f^{-1}(\infty) = w_2$. Then f is given by (9.2.13) and so is f_C for $C \in \mathbb{GL}(2, \mathbb{R})$. Thus, $A = \lambda C = e^{i\psi} B$ with $B = |\lambda| C \in \mathbb{GL}(2, \mathbb{R})$.

If $\det(A) = 1$, the B in (9.2.33) has $\det(B)$ real, so $e^{i\psi}$ is ± 1 or $\pm i$. Thus, $\det(B) = 1$ or $\det(B) = -1$. If $B = \left(\begin{smallmatrix} a & b \\ c & d \end{smallmatrix}\right) \in \mathbb{GL}(2, \mathbb{R})$,

$$f_B(i) = \frac{ai + b}{ci + d}$$

so

$$\operatorname{Im} f_B(i) = \frac{ad - bc}{|ci + d|^2} = \frac{\det(B)}{|ci + d|^2} \tag{9.2.34}$$

so taking $\mathbb{C}_+$ to $\mathbb{C}_+$ (resp. $\mathbb{C}_-$) corresponds to $A \in \mathbb{SL}(2, \mathbb{R})$ (resp. $iA \in \mathbb{SL}(2, \mathbb{R})$). □

We are now ready for the main theorem on invariance of circles and disks:

Theorem 9.2.16. *Let C_J^+ be the disk or half-plane described by* (9.2.22). *Let $T \in \mathbb{SL}(2, \mathbb{C})$. Then f_T is a bijection of C_J^+ to itself if and only if*

$$T^* J T = J \tag{9.2.35}$$

If this happens for C_J and C_J^+, then T cannot be loxodromic.

If

$$T^*JT \geq J \tag{9.2.36}$$

then f_T maps C_J^+ into itself.

Remarks. 1. (9.2.35) can be rewritten

$$T^{-1} = J^{-1}T^*J \tag{9.2.37}$$

or

$$T^* = JT^{-1}J^{-1} \tag{9.2.38}$$

2. The parabolic model (9.2.15) and hyperbolic model (9.2.17) take $\mathbb{C}_+$ onto $\mathbb{C}_+$ and the elliptic model (9.2.16) takes $\mathbb{D}$ onto itself. By conjugacy, we see that any nonloxodromic $f \in \mathfrak{F}$ fixes some disk or half-plane.

Proof. By (9.2.29), we see (9.2.35) (resp. (9.2.36)) implies f_T maps C_J^+ bijectively to C_J^+ (resp. maps C_J^+ into C_J^+).

For the converse, let

$$J_r = \begin{pmatrix} 0 & i \\ -i & 0 \end{pmatrix} \tag{9.2.39}$$

so $\langle \binom{z}{1}, J_r \binom{z}{1} \rangle = 2\operatorname{Im} z$ and C_J^+ is $\mathbb{C}_+$. For any Hermitian J with $\det(J) < 0$, we can find $S \in \mathbb{SL}(2, \mathbb{C})$ so

$$J = S^*J_rS \tag{9.2.40}$$

For we can find $U \in \mathbb{SU}(2)$, so $U^{-1}JU = \left(\begin{smallmatrix} \alpha & 0 \\ 0 & \beta \end{smallmatrix}\right)$ for $\alpha > 0$, $\beta < 0$, and then if $V = U(\begin{smallmatrix} \alpha^{-1/2} & 0 \\ 0 & |\beta|^{-1/2} \end{smallmatrix})$, then $V^*JV = (\begin{smallmatrix} 1 & 0 \\ 0 & -1 \end{smallmatrix})$. Finally, there is a unitary, W, in $\mathbb{SU}(2)$, so $W^*\left(\begin{smallmatrix} 1 & 0 \\ 0 & -1 \end{smallmatrix}\right)W = J_r$. Thus, $S = W^{-1}V^{-1}$ yields (9.2.40).

Given S and T,

$$T^*JT = J \Leftrightarrow (STS^{-1})^*J_rSTS^{-1} = J_r \tag{9.2.41}$$

Thus, it suffices to show that for $T \in \mathbb{SL}(2, \mathbb{C})$,

$$f_T \text{ maps } \mathbb{C}_+ \text{ onto } \mathbb{C}_+ \Rightarrow T^*J_rT = J_r \tag{9.2.42}$$

A little calculation proves that for any $T \in \mathbb{SL}(2, \mathbb{C})$,

$$J_rT^{-1}J_r^{-1} = T^t \tag{9.2.43}$$

Thus,

$$\begin{aligned} T^*J_rT = J_r &\Leftrightarrow T^t = T^* \Leftrightarrow \bar{T} = T \\ &\Leftrightarrow T \in \mathbb{SL}(2, \mathbb{R}) \\ &\Leftrightarrow f_T \text{ maps } \mathbb{C}_+ \text{ to } \mathbb{C}_+ \end{aligned}$$

by Proposition 9.2.15.

Finally, if f_T has a fixed disk, there is a conjugate of f_T fixing $\mathbb{C}_+$, so a conjugate of T in $\mathbb{SL}(2, \mathbb{R})$. But the trace of $A \in \mathbb{SL}(2, \mathbb{R})$ is real, so T cannot be loxodromic. □

Remark. One reason we will discuss loxodromic maps so rarely is that we will be interested in FLTs, which map $\mathbb{D}$ to itself.

Example 9.2.12, continued. For J of the form (9.2.26), the T's that leave $\mathbb{D}$ invariant obey

$$T^* \begin{pmatrix} 1 & 0 \\ 0 & -1 \end{pmatrix} T = \begin{pmatrix} 1 & 0 \\ 0 & -1 \end{pmatrix} \tag{9.2.44}$$

This group is called $\mathbb{SU}(1, 1)$. □

Proposition 9.2.17. $T \in \mathbb{SU}(1, 1)$ *if and only if*

$$T = \begin{pmatrix} a & c \\ \bar{c} & \bar{a} \end{pmatrix} \tag{9.2.45}$$

with

$$|a|^2 - |c|^2 = 1 \tag{9.2.46}$$

Proof. That a T of the form (9.2.45)/(9.2.46) has $\det(T) = 1$ and obeys (9.2.38) are straightforward computation.

Conversely, if $T = \left(\begin{smallmatrix} a & c \\ b & d \end{smallmatrix}\right)$ has determinant 1, then $JT^{-1}J^{-1} = \left(\begin{smallmatrix} d & c \\ b & a \end{smallmatrix}\right)$, so by (9.2.38), we see (9.2.44) implies $d = \bar{a}$, $c = \bar{b}$. □

We have not discussed uniqueness of fixed circles because they are not; there is always an infinite family.

Theorem 9.2.18. (a) *If T is hyperbolic, T fixes all circles (or lines) through the two fixed points and no others.*

(b) *If T is elliptic, T fixes all circles orthogonal to all the circles (or lines) through its two fixed points and no others.*

(c) *If T is parabolic and its fixed point, z_0, is finite, T leaves fixed exactly one straight line, which goes through z_0. It leaves fixed all circles through z_0 tangent to this line and no other circles.*

Proof. Without loss, we can assume ∞ is a fixed point and if there is a second, it is zero; essentially, we can take the models in Theorem 9.2.8. For the statements to be proved are conjugacy invariant.

(a) The model is $T(z) = az$ with a real and less than 1. This clearly leaves each straight line through 0 invariant—precisely all "circles" through 0 and ∞. It is not hard to see that no other circle or line is invariant.

(b) The model is $T(z) = e^{2i\theta}z$. This leaves every circle centered at 0 invariant and no other circle or line. These circles are precisely the curves orthogonal to all lines through 0, which are the "circles" through 0 and ∞.

(c) The model is $z \to z + 1$. Its only invariant lines or circles are $\operatorname{Im} z = \frac{1}{2}a$ for a real. To see what happens when the fixed point is finite, move it to 0 and note $\operatorname{Im} \frac{1}{z} = \frac{1}{2}a \Leftrightarrow |z + \frac{i}{a}|^2 = \frac{1}{a^2} \Leftrightarrow |z + \frac{i}{a}| = \frac{1}{|a|}$, the circles through 0 tangent to $\mathbb{R}$. □

There is another way to understand invariance of circles and disks involving the cross-ratio:

Definition. If z_2, z_3, z_4 are distinct, one defines the *cross-ratio* by

$$(z_1, z_2, z_3, z_4) = \frac{z_1 - z_3}{z_1 - z_4} \frac{z_2 - z_4}{z_2 - z_3} \tag{9.2.47}$$

Remarks. 1. One takes the obvious limit if some z_j is ∞, for example, $(z_1, \infty, z_3, z_4) = (z_1 - z_3)/(z_1 - z_4)$.

2. There are some obvious covariances, for example,

$$(z_2, z_1, z_3, z_4) = (z_1, z_2, z_3, z_4)$$
$$(z_3, z_4, z_1, z_2) = (z_1, z_2, z_3, z_4)$$

In fact, there are four such covariances, so six values of (z_1, z_2, z_3, z_4) under the 24 permutations.

3. Notice that (9.2.13) can be rewritten

$$f(w) = (w, w_1, w_0, w_2)$$

allowing a reinterpretation of cross-ratios.

In $\mathbb{C}^2$, we can define the two-form $v \wedge w$ given $v, w \in \mathbb{C}^2$. Once one picks $\omega \neq 0$ in $\wedge^2(\mathbb{C}^2)$, we can define $v \times w$ to be the number

$$v \wedge w = (v \times w)\omega$$

For example, with the right choice of ω,

$$(v \times w) = v_1 w_2 - w_2 v_1 \tag{9.2.48}$$

Proposition 9.2.19. *Given v_1, v_2, v_3, v_4 in $\mathbb{C}^2$, the quantity*

$$\frac{(v_1 \times v_3)(v_2 \times v_4)}{(v_1 \times v_4)(v_2 \times v_3)} \tag{9.2.49}$$

is a function only of $[v_j]$. Moreover, if $v_j = \binom{z_j}{1}$, its value is the cross-ratio.

Proof. Since $v \times w$ is bilinear, (9.2.49) is invariant under $v_j \to \lambda_j v_j$ with $\lambda_j \in \mathbb{C}$. Since, for the choice (9.2.48), $\binom{z_1}{1} \times \binom{z_2}{1} = z_1 - z_2$, we have the cross-ratio formula. □

Theorem 9.2.20. *For any FLT, f, we have*

$$(f(z_1), f(z_2), f(z_3), f(z_4)) = (z_1, z_2, z_3, z_4) \tag{9.2.50}$$

Proof. Let $f = f_T$. Since $(Tv \wedge Tw) = \det(T)(v \wedge w)$, we see $Tv \times Tw = \det(T)(v \times w)$ and (9.2.46) is immediate from (9.2.49). □

Theorem 9.2.21. *Fix z_2, z_3, z_4 distinct. Then*

$$\{z \mid (z, z_2, z_3, z_4) \in \mathbb{R}\} \tag{9.2.51}$$

is the unique circle or line containing z_2, z_3, z_4.

Proof. $(z, 1, 0, \infty) \equiv z$, so the set in (9.2.51) is precisely the real axis, which is the unique circle or line containing $1, 0, \infty$. Now use Theorems 9.2.13 and 9.2.20. □

Remark. One can prove this theorem directly and use it for an alternate proof of Theorem 9.2.13.

The last topic in our presentation of FLTs is the study of reflections and the closely related issue of isometric circles and behavior of Euclidean lengths. One of our intermediate goals will be to generalize to the geometry associated with the group, $\mathfrak{F}$, the well-known fact about Euclidean geometry that any proper Euclidean motion is a product of two Euclidean reflections. (For two dimensions, the proof goes as follows: If $f(z) = z + z_0$, f is the product of reflection in the lines $\text{Re}(z\bar{z}_0) = 0$ and $\text{Re}(z\bar{z}_0) = \frac{1}{2}|z_0|^2$, while a rotation by angle θ is a product of reflections in two lines through the center of rotation with angle $\theta/2$.) We will see an element in $\mathfrak{F}$ is a product of two FLT reflections if and only if it is not loxodromic. But first we need to define FLT reflections.

Definition. An *antilinear* map on $\mathbb{C}^2$ is a map, T, that obeys $T(u + v) = u + v$, $T(\lambda u) = \bar{\lambda} u$ for $\lambda \in \mathbb{C}$.

The union of the sets of linear and antilinear invertible maps is a group. An antilinear map preserves lines and so also induces a map on $\mathbb{P}$. $\widetilde{\mathfrak{F}}$, the group of *extended FLTs*, is the set of all maps induced by linear and antilinear transformations.

c defined on $\mathbb{C}^2$ by $c\binom{u_1}{u_2} = \binom{\bar{u}_1}{\bar{u}_2}$ is antilinear; its induced map, which we will also call c, obeys

$$c(z) = \bar{z} \tag{9.2.52}$$

since $c\binom{z}{1} = \binom{\bar{z}}{1}$. If T is antilinear on $\mathbb{C}$, $A = Tc$ is linear so A has the form $\left(\begin{smallmatrix} a & b \\ c & d \end{smallmatrix}\right)$, so $T = Ac$ and

$$f_T(z) = \frac{a\bar{z} + b}{c\bar{z} + d} \tag{9.2.53}$$

where we can suppose $ad - bc = 1$. We will call such maps anti-FLTs.

Proposition 9.2.22. *For any three points in $\mathbb{C} \cup \{\infty\}$, there is a unique anti-FLT that fixes them. It pointwise fixes the circle or line they determine. Its square is the identity. Any circle or line has an anti-FLT that fixes it pointwise. If the circle is $|z - z_0| = r$, then the map is*

$$f(z) = z_0 + \frac{r^2}{\bar{z} - \bar{z}_0} \tag{9.2.54}$$

Remarks. 1. Especially since we talked about circles left setwise fixed by some $f \in \mathfrak{F}$, we emphasize in that earlier discussion we meant setwise fixed, that is, $z \in C \Rightarrow f(z) \in C$. Here we mean pointwise fixed, that is, $z \in C \Rightarrow z = f(z)$.

2. The map in (9.2.54) is called the *reflection* (or *inversion*) *in the circle* $|z - z_0| = r$.

Proof. By conjugacy, we can suppose the three points are $0, 1, \infty$. In that case, c leaves them fixed and $c^2 = 1$ and c leaves $\mathbb{R}$ pointwise fixed. If L, an anti-FLT, also leaves $0, 1, \infty$ fixed, Lc is an FLT leaving $0, 1, \infty$ fixed, hence, the identity, so

$$L = Lc^2 = (Lc)c = c$$

One could check (9.2.54) by a suitable mapping of the circle to $\mathbb{R}$, but it is easier to note that f is antilinear and

$$f(z_0 + re^{i\theta}) = z_0 + \frac{r^2}{r\,\overline{e^{i\theta}}} = z_0 + re^{i\theta} \qquad \square$$

There is a geometric connection between reflections and circles left setwise fixed by the reflections:

Theorem 9.2.23. *Let R be a reflection in the circle C_1 and suppose C_2 is another distinct circle. Then $R[C_2] = C_2$ (as sets) if and only if C_1 and C_2 intersect in two points and intersect orthogonally.*

Proof. By a conjugacy, we can suppose $C_1 = \mathbb{R}$ and $R = c$. If C_1 and C_2 intersect not at all or in a single point (including only at ∞), then C_2 lies on one side of C_1 (i.e., all in $\bar{\mathbb{C}}_+$ or $-\bar{\mathbb{C}}_+$) and so it cannot possibly be left invariant by c. Thus, C_1 and C_2 must intersect in two points, which, by conjugacy, we can take as 0 and ∞. Thus, C_2 is a straight line through 0. Such a line is invariant under c if and only if $C_2 = \mathbb{R}$ or $i\mathbb{R}$, so $C_2 \neq C_1 \Rightarrow C_2$ is orthogonal to C_1. $\square$

The following pieces of geometry will be very important in our analysis of the Fuchsian group associated to $\mathbb{C} \setminus \mathfrak{e}$:

Proposition 9.2.24. *Let f be the reflection in the circle $|z - z_0| = r$. Let z, w lie outside the disk $\{u \mid |u - z_0| \leq r\}$. Then*

$$|f(z) - f(w)| = \frac{r^2|z - w|}{|z - z_0|\,|w - z_0|} \tag{9.2.55}$$

Remark. (9.2.55) is always true! We state the result this way to emphasize the size contraction that takes place for distances outside the disk.

Proof. By scaling and translation, it suffices to consider the disk where $|u| = 1$, that is, $f(z) = 1/\bar{z}$. Then

$$|f(z) - f(w)| = \left|\frac{1}{\bar{z}} - \frac{1}{\bar{w}}\right| = \frac{|z - w|}{|z|\,|w|} \tag{9.2.56}$$

$\square$

We are heading toward proving a map is the product of two reflections if and only if it is nonloxodromic. Let us first look at products of reflections:

Theorem 9.2.25. *Let $f = R_1 R_2$ be a product of two reflections in circles or lines C_1, C_2. Then*

(a) *If C_1 and C_2 intersect in two points, f is elliptic.*
(b) *If C_1 and C_2 intersect in one point, f is parabolic.*
(c) *If C_1 and C_2 do not intersect, f is hyperbolic.*

Proof. Again, we have conjugacy conditions so we can move the intersection or other points where it is convenient.

(a) Move the intersection points to zero and infinity. C_1 and C_2 are now straight lines through 0. If they meet in angle θ, the product of reflections is rotation by angle 2θ, hence elliptic.

(b) Move the intersection point to infinity. Then C_1 and C_2 are parallel lines. The product of R_1 and R_2 is translation in the direction perpendicular to these lines by a distance twice the distance between them and is parabolic.

(c) By an FLT, we can move C_2 to $\mathbb{R}$ and C_1 to a circle about i of radius $r < 1$. Let $C^{(n)}$ be the image of $|z - i| \le r$ under f^n. Two points, z, w, in $C^{(0)}$ go to $\bar{z}, \bar{w} \in \mathbb{C}_-$, so $|\bar{z} - i| \ge 1$ and similarly for w. By (9.2.55), $|f(z) - f(w)| \le r^2|z - w|$. Thus, if $z, w \in C^{(0)}$, then $|f^{(n)}(z) - f^{(n)}(w)| \le r^{2n}|z - w|$. In particular, $|f^{(n+1)}(z) - f^{(n)}(z)| \le r^{2n}|f^{(1)}(z) - z| \le r^{2n+1}$. $f^{(n)}(z)$ converges to a point exponentially fast so $\mathfrak{F}$ is hyperbolic or loxodromic.

The ray $\{z \mid z = ia,\ a < 1\}$ is taken into itself under f, so the images of points in the intersection of $C^{(0)}$ and that ray approach the fixed point on the ray from a fixed direction. Hence, f is hyperbolic, not loxodromic. □

Remark. One can use the calculation in Proposition 9.2.31 below instead of these arguments, but we prefer the geometry.

Corollary 9.2.26. *A loxodromic map is not the product of two reflections.*

Theorem 9.2.27. *Any nonloxodromic map, f, is the product of two reflections. One of the reflections can be required to be in a circle or line containing any z_0 not a fixed point of f.*

First Proof. By conjugacy, we need only consider the basic models. As we have already seen, rotation about 0 (basic model for elliptic) is a product of a reflection in two lines through 0, one of which can be arbitrary (but it fixes the second). Similarly, $z_0 \to z_0 + 1$ is the product of reflections in the lines $\operatorname{Re} z = a$ and $\operatorname{Re} z = a + \frac{1}{2}$ with a arbitrary.

Finally, if $f_j(z) = r_j^2/\bar{z}$ for $j = 1, 2$, then $(f_1 \circ f_2)(z) = \frac{r_1^2}{r_2^2}z$, showing our hyperbolic model is a product of reflections, which can be arranged to contain any point different from 0 and ∞. □

This proof is simple; we will give a second proof not for the sake of a second proof, but because it introduces important notions. Euclidean distances, which are not invariant under most FLTs, will be critical here and will be important in Section 9.6.

While we have not mentioned it explicitly, we have used in passing that elements in $\widetilde{\mathfrak{F}}$ preserve orthogonality of curves; more generally, they all locally preserve angles (the $f \in \mathfrak{F}$ are conformal in that they also preserve orientation; the $f \in \widetilde{\mathfrak{F}} \backslash \mathfrak{F}$ are anticonformal in that they reverse orientation). Infinitesimal Euclidean lengths scale under f near z_0 by a factor $|f'(z_0)|$ where $f' = \partial f/\partial z$ if f is analytic and $\partial f/\partial \bar{z}$ is f anti-analytic.

Here is one important consequence of (anti)conformality:

Proposition 9.2.28. *Let $f \in \widetilde{\mathfrak{F}}$. Let $A = \{z_0 + re^{i\theta} \mid \theta_0 < \theta < \theta_1\}$ be an arc of the circle, C, of z with $|z - z_0| = r$. Suppose $f^{-1}(\infty) \notin C$ so $f[C]$ is also a circle. Then the angular fraction of (2π) subtended by $f[A]$ in $f[C]$ is*

$$\int_{\theta_0}^{\theta_1} |f'(z_0 + re^{i\theta})| \frac{d\theta}{2\pi} \Big/ \int_0^{2\pi} |f'(z_0 + re^{i\theta})| \frac{d\theta}{2\pi} \tag{9.2.57}$$

Proof. Immediate from the fact that angular fractions are ratios of arc length and the fact that f locally scales by f' in all directions. □

Here is a simple but basic calculation:

Proposition 9.2.29. *Let $f \in \widetilde{\mathfrak{F}}$ not have infinity as a fixed point. Let*

$$z_0 = f^{-1}(\infty) \tag{9.2.58}$$

Then for some r,

$$|f'(z)| = \frac{r^2}{|z - z_0|^2} \tag{9.2.59}$$

Proof. If f is given by (9.2.4) and $\det\left(\begin{smallmatrix} a & b \\ c & d \end{smallmatrix}\right) = 1$, a straightforward calculation shows

$$f'(z) = \frac{1}{(cz + d)^2} \tag{9.2.60}$$

so (9.2.59) holds with $r = c^{-1}$ and $z_0 = -\frac{d}{c}$, the point that goes into ∞ under f. The proof is identical in the anti-analytic case. □

The circle

$$\{z \mid |z - z_0| = r\} = \{z \mid |cz + d| = 1\} \tag{9.2.61}$$

is called the *isometric circle*. Distances inside this circle C expand under f, and outside C they compress. C is precisely the set of points where $|f'(z)| = 1$.

Theorem 9.2.30 (Ford's Theorem, Part 1). *Let $f \in \mathfrak{F}$ and let C be its isometric circle. Then $f[C]$ is a circle with the same radius but with center $f(\infty)$ and is the isometric circle for f^{-1}. Let R be the reflection in the circle C, and Q the reflection in the line, which is the perpendicular bisector of the line between $f^{-1}(\infty)$ and $f(\infty)$. For any θ, let A_θ be rotation by angle θ about $z_0 = f^{-1}(\infty)$, that is, $A_\theta(z) = z_0 + e^{i\theta}(z - z_0)$. Then for some θ,*

$$f = QRA_\theta \tag{9.2.62}$$

Remarks. 1. We will see shortly that if f is nonloxodromic, $\theta = 0$, that is, $f = QR$.

2. Since A_θ is a product of reflections, we see that any loxodromic f is a product of four reflections.

3. If $\tilde{R}$ is the reflection in $f[C]$, then $QRQ^{-1} = \tilde{R}$ and $f = \tilde{R}QA_\theta$.

4. If $f(\infty) = f^{-1}(\infty)$ (but ∞ is not fixed), we have a subtle situation since there is no line to bisect. Here is what is going on. If $f = f_T$ with $\det(T) = 1$

and $T = \left(\begin{smallmatrix} a & b \\ c & d \end{smallmatrix}\right)$, $f(\infty) = \frac{a}{c}$, and $f^{-1}(\infty) = -\frac{d}{c}$, so $f(\infty) = f^{-1}(\infty)$ means $a = -d$ or $\text{Tr}(T) = 0$. Thus, T has eigenvalues $\pm i$ and $f^2 = \mathbf{1}$. Let z_0, z_1 be the two fixed points of f. Let C be a circle of radius $\frac{1}{2}|z_1 - z_0|$ centered at $\frac{1}{2}(z_0 + z_1)$. Let R be the reflection through C and Q through the line through z_0 and z_1. Let $g = QR$, which also equals RQ in this case. g^2 is also the identity and g has the same fixed points as f since Q and R leave both points fixed. Thus, $f = g$ and C is the isometric circle.

5. Of course, $f(\infty) = \frac{a}{c}$ and the isometric circle for f^{-1} is $|cz - a| = 1$ as can also be seen by inverting the matrix.

Proof. Since f is isometric on C, $f[C]$, which is a circle, has the same size circumference, so same radius. Since f^{-1} maps $f[C]$ isometrically to C, $f[C]$ is the isometric circle for f^{-1}, so its center is $(f^{-1})^{-1}(\infty) = f(\infty)$. R maps C isometrically to itself and Q maps C isometrically to $f[C]$. Thus, QR and f are both isometries of C to $f[C]$, so $(QR)^{-1}f$ is an isometry of C to itself, hence, on C a rotation. Thus, for some A_θ, $A_\theta^{-1}(QR)^{-1}f$ leaves C pointwise fixed. Since C has more than two points, $A_\theta^{-1}(QR)^{-1}f = \mathbf{1}$, so (9.2.62) holds. □

Proposition 9.2.31 (Ford's Theorem, Part 2). *If* $z_1 \equiv f(\infty) \neq f^{-1}(\infty) \equiv z_2$ *and* f *is given by* (9.2.62), *then* $f = f_T$ *where* $T \in \mathbb{SL}(2, \mathbb{C})$ *and*

$$\text{Tr}(T) = 2\left[\frac{|z_1 - z_2|}{2r}\right]e^{-i\theta/2} \tag{9.2.63}$$

In particular, if f *is nonloxodromic, then* $\theta = 0$ *and*

$$f = QR \tag{9.2.64}$$

Remarks. 1. This provides another proof that any nonloxodromic map is a product of two reflections. Since a preliminary conjugation can take any point to infinity and Q leaves ∞ fixed, we can arrange for any nonfixed point to be on one of the reflection circles.

2. We see once again that if the two circles do not intersect, then f is hyperbolic (since then $\frac{1}{2}|z_1 - z_2| > r$ and $\text{Tr}(T) > 2$ by (9.2.63)), tangency means parabolic and intersection means elliptic.

Proof. Euclidean transformations preserve length and scalings do not change ratios, so without loss, we can make a preliminary conjugacy so that $f^{-1}(\infty) = i$, $f(\infty) = -i$, and thus Q is c, complex conjugation. Let r be the radius of the isometric circle, which is thus $|z - i| = r$. With these changes

$$A_\theta(z) = i + e^{i\theta}(z - i) \tag{9.2.65}$$

$$R(z) = i + \frac{r^2}{\bar{z} + i} \tag{9.2.66}$$

so

$$f(z) = QRA_\theta(z) = -i + \frac{r^2 e^{-i\theta}}{z - i} \tag{9.2.67}$$

$$= f_S(z) \tag{9.2.68}$$

where

$$S = \begin{pmatrix} -i & r^2e^{i\theta} - 1 \\ 1 & -i \end{pmatrix} \tag{9.2.69}$$

$\det(S) = -r^2e^{i\theta}$, so to get $T \in \mathbb{SL}(2, \mathbb{C})$, we take $T = S/(-ire^{i\theta/2})$, which has a trace given by (9.2.63) (since $|z_1 - z_2| = 2$). □

We emphasize that isometric circles are *not* preserved by FLTs that are not Euclidean motions, but their geometry can be very useful.

If C is the isometric circle of f, its (open) inside, D_i, will be called the *initial disk* and $C \equiv C_i$ the *initial circle*. $C_f \equiv f[C]$ will be called the *final circle* and its inside, D_f, the *final disk*. Here is the basic geometry:

Theorem 9.2.32. (a) *The initial circle, C_i, is mapped by f into the final circle, C_f. The exterior of the initial disk ($\mathbb{C} \setminus \bar{D}_i$) maps to the final disk, D_f, and the initial disk maps to the exterior, $\mathbb{C} \setminus \bar{D}_f$, of the final disk. For f^{-1}, just interchange "initial" and "final" in these statements.*

(b) *All fixed points of f lie in $\bar{D}_i \cup \bar{D}_f$.*

(c) *In the elliptic case, the two fixed points are the two points in which C_i and C_f intersect.*

(d) *In the parabolic case, the unique fixed point is the point in which C_i and C_f intersect.*

(e) *In the hyperbolic case, the fixed points are symmetric under Q, one lies in D_i and one in D_f, and they lie on the line segment strictly between the centers of the disks. The attracting fixed point lies in D_f and the other in D_i.*

Remark. We will actually show that fixed points lie in $D_f \cup \bar{D}_i$ and similarly in $D_i \cup \bar{D}_f$ (since $f(z) = z \Rightarrow f^{-1}(z) = z$). $(D_f \cup \bar{D}_i) \cap (D_f \cup D_i) = D_i \cup D_f \cup (\partial D_f \cap \partial D_i)$. We stated the simpler form in (b) since we analyze in more detail than in the other parts.

Proof. (a) is immediate from the QR representation (9.2.64) since R maps D_i to its exterior and Q maps D_i to D_f.

(b) If x lies in $\mathbb{C} \setminus \bar{D}_i$, Rx lies in D_i and QRx in D_f, so if $f(x) = x$, $x \in D_f$. Thus, $x \in D_f \cup \bar{D}_i$.

(c), (d) If C_f and C_i intersect (but are distinct), the intersection points are also on the line defining Q, so they are left invariant by both Q and R, and so by f. In the parabolic case, where the circles only touch at a single point, there is one fixed point; in the elliptic case, there are two of each. In both cases, the intersections account for all fixed points.

(e) In this case, the disks are disjoint. R maps points in D_f into D_i and then Q back into D_f. Thus, knowing all points in D_f—except possibly one gets mapped under iteration to the attracting fixed point—guarantees that this fixed point, call it z_0, lies in D_f. Since $RQ = f^{-1}$, $f(Qz_0) = QRQz_0 = Qf^{-1}(z_0) = Qz_0$, we see Qz_0 is also a fixed point, so the other fixed point lies in $Q[D_f] = D_i$.

Let w be the point on C_i on the segment from the center of C_i to the center of C_f. Let L be a half-line from w through C_f and off to ∞. R maps this to the segment from w to the center of C_i and Q maps that to the segment from Qw to the center of D_f. Thus, this segment is mapped into itself and so, as above, the attracting fixed point must lie in this segment. □

The argument behind the proof of (b), which says fixed points must lie in $\bar{D}_f \cup \bar{D}_i$, also shows that if $|f(z) - z|$ is small, z must be close to $\bar{D}_f \cup \bar{D}_i$:

Theorem 9.2.33. *Let $f \in \mathfrak{F}$ with $f(\infty) \neq \infty$. Let D_i and D_f be the initial and final disks. Then either $z \in D_i$ or*

$$\mathrm{dist}(z, D_f) \leq |z - f(z)| \tag{9.2.70}$$

Remarks. 1. Since we will talk about another metric in the next section, we emphasize that $\mathrm{dist}(\,\cdot\,, D_f)$ is here in the Euclidean metric.

2. This implies

$$\mathrm{dist}(z, D_f \cup D_i) \leq |z - f(z)| \tag{9.2.71}$$

Proof. If $z \notin D_i$, then $f(z) \in \bar{D}_f$, so (9.2.70) holds. □

By (9.2.60), we have

$$D_i = \{z \mid |f'(z)| > 1\} \tag{9.2.72}$$

and

$$\mathbb{C} \setminus \bar{D}_i = \{z \mid |f'(z)| < 1\} \tag{9.2.73}$$

Remarks and Historical Notes. Given how fundamental FLTs are to so many parts of mathematics, it is unfortunate how little they are discussed in basic texts (which, e.g., do not discuss the hyperbolic, parabolic, elliptic splitting), and that this discussion does not talk about projective space. The textbook description of the Riemann sphere is via stereographic projection—admittedly useful—but not as basic as the $\mathbb{P}$ point of view.

Most of the material in this section is classical (from the nineteenth century), although our discussion has some more modern elements. Key figures in these classical developments are Möbius, Schwarz, Klein, and especially Poincaré.

The use of isometric circles and the representation $f = QR$ for nonloxodromic transformations was emphasized especially by Ford; see, for example, [138].

If $\tilde{R} = QRQ$, the reflection in the isometric circle for f^{-1}, then $f^2 = QRQR = \tilde{R}R$, something that can easily be proven directly. It is simple in various ways to use geometric structures defined by f to get f^2 as a product of reflections. The neat thing about Ford's idea of using a perpendicular bisector is that it "takes the square root."

9.3 MÖBIUS TRANSFORMATIONS

In this section, we will discuss FLTs that take $\mathbb{D}$ onto $\mathbb{D}$ (equivalently, take $\mathbb{D}$ into $\mathbb{D}$ and $\partial\mathbb{D}$ to $\partial\mathbb{D}$). Of course, by Theorem 9.2.13, the FLTs, which are bijections of any disk or half-plane, are conjugate to bijections of the disk, so this section could also describe analytic bijections of, say, $\mathbb{C}_+$. That said, there are often good reasons to study $\mathbb{C}_+$ (as we will explain in the Notes). But we will need $\mathbb{D}$ later, so we study these maps in this guise.

An FLT, which is a bijection of $\mathbb{D}$, we will call a *Möbius transformation*. We use $\mathcal{M}$ for the family of Möbius transformations. This is nonstandard terminology since "Möbius transformation" is typically used as a synonym for FLT, but it is useful to have a standard term.

It will be very useful to have Möbius transformations that map any point in $\mathbb{D}$ to any other point. As usual, if we do it for a fixed endpoint, we can do it for any other, for if f_{z_0} takes z_0 to 0, then $f_{w_0}^{-1} f_{z_0}$ maps z_0 to w_0.

Proposition 9.3.1. *Let $z_0 \in \mathbb{D}$. Then*

$$f_{z_0}(z) = \frac{z - z_0}{1 - \bar{z}_0 z} \tag{9.3.1}$$

maps $\mathbb{D}$ onto $\mathbb{D}$ and has $f_{z_0}(z_0) = 0$.

Proof. f is analytic in $\{z \mid |z| < |z_0|^{-1}\}$ and so in a neighborhood of $\overline{\mathbb{D}}$. Moreover, $|f_{z_0}(e^{i\theta})| = |e^{i\theta} - z_0|/|e^{-i\theta} - \bar{z}_0| = 1$, so by the maximum principle, f maps $\mathbb{D}$ into $\mathbb{D}$. But by calculating, $f_{-z_0} \cdot f_{z_0} = 1$ since $\left(\begin{smallmatrix} 1 & -z_0 \\ -\bar{z}_0 & 1 \end{smallmatrix}\right)\left(\begin{smallmatrix} 1 & z_0 \\ \bar{z}_0 & 1 \end{smallmatrix}\right) = 1 - |z_0|^2 \left(\begin{smallmatrix} 1 & 0 \\ 0 & 1 \end{smallmatrix}\right)$, so f is an analytic bijection of $\mathbb{D}$. Clearly, $f_{z_0}(z_0) = 0$. □

The second main result that we will need to analyze all Möbius transformations is a general one about analytic bijections, which we do not know a priori are FLTs restricted to $\mathbb{D}$:

Theorem 9.3.2. *If $f\colon \mathbb{D} \to \mathbb{D}$ is an analytic bijection and $f(0) = 0$, then for some $\theta \in [0, 2\pi)$,*

$$f(z) = e^{i\theta} z \tag{9.3.2}$$

Proof. We begin with the Schwarz lemma (Proposition 2.3.4), which implies that $|f(z)| \le |z|$. But since f^{-1} also maps $\mathbb{D}$ to $\mathbb{D}$ and $f^{-1}(0) = 0$, we have that $|f^{-1}(z)| \le |z|$. Setting $w = f^{-1}(z)$, we see $|w| \le |f(w)|$, so $|f(z)/z| = 1$ on $\mathbb{D}$. By the maximum principle, $f(z)/z$ is constant. □

Theorem 9.3.3. *If $f\colon \mathbb{D} \to \mathbb{D}$ is an analytic bijection, then f is a Möbius transformation. In fact, if $f(z_0) = 0$, then for some $\theta \in [0, 2\pi)$,*

$$f(z) = e^{i\theta} f_{z_0}(z) \tag{9.3.3}$$

where f_{z_0} is given by (9.3.1).

Proof. $f f_{z_0}^{-1}$ maps $\mathbb{D}$ onto $\mathbb{D}$ and takes 0 to 0, so this follows from Proposition 9.3.1 and Theorem 9.3.2. □

The remarkable fact about this is that analytic bijections of $\mathbb{D}$ automatically have meromorphic continuations to all of $\mathbb{P}$. This is not quite as surprising as it might seem at first. If $|z_n| \to 1$, $f(z_n)$ cannot converge to a point, w_0, in $\mathbb{D}$ because $f(z)$ near w_0 means z must be near $f^{-1}(w_0)$, and so must have $|z|$ near $|f^{-1}(w_0)|$. Thus,

$|f(z)| \to 1$ as $|z| \to 1$. If we knew f had a continuous extension of $\overline{\mathbb{D}}$ to $\overline{\mathbb{D}}$, then we could extend f to $\mathbb{C} \cup \{\infty\}$ by

$$f(z) = \overline{f(1/\bar{z})}^{-1} \tag{9.3.4}$$

which is trivially meromorphic in $\mathbb{D} \cup \mathbb{C} \setminus \overline{\mathbb{D}}$ and analytic across $\partial\mathbb{D}$ by the reflection principle and the fact that $|f(e^{i\theta})| = 1$. There is a version of the Schwarz reflection principle that only requires that $\operatorname{Im} g$ vanishes. That can be applied to $i \log|f|$. In any event, we have (9.3.4) for any Möbius transformation.

In the last section, we saw that FLTs could be labeled by three complex variables, $f(0)$, $f(1)$, $f(\infty)$, so $\mathfrak{F}$ has real dimension 6. Here we saw that Möbius transformations are parametrized by one complex variable $z_0 = f^{-1}(0)$ and one real variable, so $\mathcal{M}$ is three-dimensional. Moreover, we see $\mathcal{M}$ topologically is $\mathbb{D} \times \partial\mathbb{D}$.

By Theorem 9.2.16, any $f \in \mathcal{M}$ is nonloxodromic. $f(z) = e^{i\theta} z$ is elliptic (it is f_T for $T = \begin{pmatrix} e^{i\theta/2} & 0 \\ 0 & e^{-i\theta/2} \end{pmatrix}$ has $\det(T) = 1$ and $\operatorname{Tr}(T) \in [-2, 2]$). f_{z_0} is hyperbolic since it is f_T for $T = (1 - |z_0|^2)^{-1/2} \begin{pmatrix} 1 & -z_0 \\ -\bar{z}_0 & 1 \end{pmatrix}$ has $\det(T) = 1$ and $\operatorname{Tr}(T) = 2/(1 - |z|^2)^{1/2} > 2$. The parabolic example is

$$f(z) = \frac{(1+i)z - i}{iz + 1 - i}$$

($T = \begin{pmatrix} 1+i & -i \\ i & 1-i \end{pmatrix}$ has determinant 1 and trace 2, and a little calculation shows $|f(e^{i\theta})| = 1$.) Thus, all nonloxodromic possibilities occur. Here is what one can say about fixed points:

Theorem 9.3.4. *Let $f \in \mathcal{M}$ not be the identity. Then*

(a) *If f is elliptic, it has one fixed point at z_0 in $\mathbb{D}$ and one fixed point in $\mathbb{C} \setminus \overline{\mathbb{D}}$ at $1/\bar{z}_0$.*

(b) *If f is hyperbolic or parabolic, all the fixed points of f lie in $\partial\mathbb{D}$.*

Proof. By (9.3.4), if $f \in \mathcal{M}$ has a fixed point z_0, then $1/\bar{z}_0$ is also a fixed point, so if there is a fixed point not in $\partial\mathbb{D}$, there is one, call it z_0 in $\mathbb{D}$.

If $f(z_0) = z_0$, then $h \equiv g_{-z_0}^{-1} f g_{-z_0}$ maps zero to zero, and so is $h(z) = e^{i\theta} z$, which is elliptic, and thus f is elliptic. This proves (b).

All that remains is the proof that elliptic elements of $\mathcal{M}$ cannot have their fixed points on $\partial\mathbb{D}$. As we have seen, if f has a fixed point off $\partial\mathbb{D}$, it has a second at the reflected point. Thus, if f has a fixed point on $\partial\mathbb{D}$, it must have two. Let g be a map in $\mathfrak{F}$ that takes these two fixed points to zero and infinity and some other point, z_2, on $\partial\mathbb{D}$ to ± 1. g thus maps $\partial\mathbb{D}$ to $\mathbb{R}$ and so if we pick the ± 1 for $g(z_2)$ properly, $\mathbb{D}$ maps to $\mathbb{C}_+$. Since $h \equiv g f g^{-1}$ fixes zero and infinity and is elliptic, it has the form $h(z) = e^{i\theta} z$. No such map takes $\mathbb{C}_+$ to $\mathbb{C}_+$, which proves (a). □

Remark. We will see later (see the discussion after Proposition 9.3.8) a geometric way to understand why parabolic and hyperbolic maps have their fixed points on $\partial\mathbb{D}$.

Obviously, if $f, g \in \mathcal{M}$ are conjugate in $\mathcal{M}$, they are conjugate in $\mathfrak{F}$ but, in principle (and in practice!), they could be conjugate in $\mathfrak{F}$ but not in $\mathcal{M}$. Put differently, if $C \subset \mathfrak{F}$ is a class in $\mathfrak{F}$ and $C \cap \mathcal{M} \neq \emptyset$, $C \cap \mathcal{M}$ is one or more classes in $\mathcal{M}$. Here is the breakdown:

Theorem 9.3.5. (a) *Each hyperbolic conjugacy class in $\mathfrak{F}$ intersects $\mathcal{M}$. Two hyperbolic elements in $\mathcal{M}$ are conjugate in $\mathcal{M}$ if and only if they are conjugate in $\mathfrak{F}$. Hyperbolic conjugacy classes in $\mathcal{M}$ are labeled by $a \in (0, 1)$ with*

$$f_a(z) = \frac{z-a}{1-az} \tag{9.3.5}$$

The associated T in $\mathbb{SU}(1,1)$ *has* $\mathrm{Tr}(T) = 2/(1-|a|^2)^{1/2}$.

(b) *Each elliptic conjugacy class in $\mathfrak{F}$ intersects $\mathcal{M}$, and for $\theta \in (0, \pi/2)$, its intersection is two classes in $\mathcal{M}$ labeled by $\pm\theta$. The $\mathfrak{F}$-class with $\theta = \pi/2$* ($\mathrm{Tr}(T) = 0$) *intersects $\mathcal{M}$ in a single class of $\mathcal{M}$. All elliptic classes are labeled by $\theta \in \pm(0, \pi/2)$. An element in the class is*

$$f_\theta(z) = e^{2i\theta} z \tag{9.3.6}$$

The associated trace is $2\cos\theta$.

(c) *The single parabolic class in $\mathfrak{F}$ intersects $\mathcal{M}$ and the intersection is two classes of $\mathcal{M}$ of which representative elements are*

$$f_\pm(z) = \frac{(1 \pm i)z \mp i}{iz + 1 \mp i} \tag{9.3.7}$$

These have $\mathrm{Tr}(T) = 2$.

Remark. The $f_\pm$ in (9.3.7) has

$$f_\pm^{(n)}(0) = \frac{n^2}{1+n^2} \pm \frac{in}{1+n^2}$$

and iterates approach 1 asymptotically tangent to $\partial\mathbb{D}$ but from the top (resp. bottom) for f_+ (resp. f_-). In $\mathfrak{F}$, they are conjugate via $g(z) = z^{-1}$, but that maps $\mathbb{D}$ to $\mathbb{C} \setminus \overline{\mathbb{D}}$ and is not in $\mathcal{M}$.

Proof. (a), (c) It is easier to look at the conjugate of $\mathcal{M}$ that maps $\mathbb{C}_+$ to $\mathbb{C}_+$, that is, $\mathbb{SL}(2,\mathbb{R})$. In the hyperbolic case, we can find a conjugate in $\mathbb{SL}(2,\mathbb{R})$ that takes any hyperbolic map to one whose fixed points are 0 and ∞ and with 0 the attracting fixed point. The classes in $\mathbb{SL}(2,\mathbb{R})$ are thus $z \mapsto az$ with $a \in (0,1)$, as they are in $\mathbb{SL}(2,\mathbb{C})$. In the parabolic case, we can take the fixed point to infinity. The map is then $T_b(z) = z + b$ with $b \in \mathbb{R} \setminus 0$. By a scaling map in $\mathbb{SL}(2,\mathbb{R})$, we can conjugate that to $T_{\pm 1}$ but T_{+1} and T_{-1} are not conjugate in $\mathbb{SL}(2,\mathbb{R})$. The conjugacy in $\mathbb{SL}(2,\mathbb{C})$ is by $z \to -z$, which maps $\mathbb{C}_+$ to $\mathbb{C}_-$.

(b) By conjugating with f_{z_0}, we can suppose the elliptic map has zero as a fixed point, so of the form (9.3.6). For distinct θ's, these are not conjugate in $\mathcal{M}$, although conjugation with $z \to 1/z$ takes f_θ to $f_{-\theta}$. □

Next, we want to discuss the Ford representation when $f \in \mathcal{M}$. Note that $f \in \mathcal{M}$ has $f(\infty) = \infty$ if and only if $f(0) = 0$, so the condition that f not leave ∞ fixed is $f(z) \not\equiv e^{i\theta} z$.

Theorem 9.3.6. *Let $f \in \mathcal{M}$ not be a rotation about* 0. *Then the isometric circle of f has a center outside $\overline{\mathbb{D}}$ and is orthogonal to $\partial\mathbb{D}$. $z = 0$ lies outside both the initial and final disks for f and on the (Euclidean) perpendicular bisection of the line between the center of D_i and D_f. $f(0)$ lies in D_f.*

Proof. We know f maps $\mathbb{C}\setminus\overline{\mathbb{D}}$ to itself, so $f^{-1}(\infty) \in \mathbb{C}\setminus\overline{\mathbb{D}}$, which says the center of the circle lies outside $\overline{\mathbb{D}}$. We know $f = f_T$ for $T = \left(\begin{smallmatrix} a & c \\ \bar{c} & \bar{a} \end{smallmatrix}\right)$. Then $f^{-1}(\infty) = -\frac{\bar{a}}{\bar{c}}$ and $f(\infty) = \frac{a}{\bar{c}}$. Since $|f^{-1}(\infty)| = |f(\infty)|$, they are equidistant from 0, which means that 0 lies on the perpendicular bisector of the line between $f^{-1}(\infty)$ and $f(\infty)$. Thus, in the Ford factorization of $f = QR$, Q maps $\mathbb{D}$ to $\mathbb{D}$, so $R = QF$ maps $\mathbb{D}$ to $\mathbb{D}$. By Theorem 9.2.23, the isometric circle is orthogonal to $\partial\mathbb{D}$.

With $f = f_T$ and $T = \left(\begin{smallmatrix} a & c \\ \bar{c} & \bar{a} \end{smallmatrix}\right)$ and $|a|^2 - |c|^2 = 1$, we have that $|\bar{c}z + \bar{a}| = 1$ is the isometric circle. Since $|\bar{c} \cdot 0 + \bar{a}| = |a| > 1$, (if f is not a rotation), 0 is outside $\bar{D}_i$. Since D_f is the initial circle for f^{-1}, 0 is also outside $\bar{D}_f$. $f(0) \in D_f$ since $\mathbb{C}\setminus \bar{D}_i$ is mapped to D_f by f (see Theorem 9.2.32). □

Remarks. 1. There is a quantitative way of seeing that $f(0)$ lies inside D_f, namely, since $f(0) = \frac{c}{\bar{a}}$, $f(\infty) = \frac{a}{\bar{c}}$, so $|f(0) - f(\infty)| = \frac{1}{|ca|}$ since $|a|^2 - |c|^2 = 1$. On the other hand, r_f is $|c|^{-1}$, so $|a| > 1$ implies $|f(0) - f(\infty)| < r_f$.

2. This theorem illustrates Theorem 9.2.32. If f is parabolic, C_i and C_f intersect on $\partial\mathbb{D}$ (since C_i and C_f are orthocircles). If T is elliptic, C_i and C_f intersect in points inside and outside. If T is hyperbolic, the line from center to center intersects $\partial\mathbb{D}$, giving the fixed point on that line segment.

Definition. An *orthocircle* is a circle or line in $\mathbb{C}$ that intersects $\partial\mathbb{D}$ in two points with orthogonal intersections.

The *extended Möbius transformations* are those *extended FLTs* that map $\mathbb{D}$ onto $\mathbb{D}$. The set of such maps we denote by $\widetilde{\mathcal{M}}$. Since c is such a map, one easily sees:

Proposition 9.3.7. *Every $f \in \widetilde{\mathcal{M}}$ is of the form g or gc for some $g \in \mathcal{M}$. A reflection is an extended Möbius transformation if and only if the line or circle in which one reflects is an orthocircle.*

Proof. The first statement is immediate and the second follows from Theorem 9.2.23. □

One big difference between $\mathcal{M}$ and $\mathfrak{F}$ is that there is a Riemannian metric (on $\mathbb{D}$) that is left fixed by all elements of $\mathcal{M}$, while there cannot be such a metric on $\mathbb{P}$ left invariant by all elements of $\mathfrak{F}$ since:

Proposition 9.3.8. *If (X, ρ) is a metric space, $f: X \to X$ an isometry (i.e., $\rho(f(x), f(y)) = \rho(x, y)$ for all x, y), then there cannot be an x_0 and $x_\infty \neq x_0$ so that $f^{(n)}(x_0) \to x_\infty$.*

Proof. Since f is continuous, $f(f^{(n)}(x_0)) \to f(x_\infty)$ but $f(f^{(n)}(x_0)) = f^{(n+1)}(x_0)$, so x_∞ is a fixed point. But then $\rho(f^{(n+1)}(x_0), x_\infty) = \rho(f^{(n+1)}(x_0), f(x_\infty)) = \rho(f^{(n)}(x_0), x_\infty) = \cdots = \rho(x_0, x_\infty) \neq 0$. Thus, $f^{(n)}(x_0)$ does not converge to x_∞. This contradiction proves the result. □

Thus, isometries cannot have attracting fixed points, so there is no metric (let alone Riemann metric) on $\mathbb{P}$ in which hyperbolic or parabolic maps are isometries. The reason we can define a metric on $\mathbb{D}$ in which hyperbolic or parabolic maps are isometries is that the attracting fixed points are not in $\mathbb{D}$ (but in $\partial\mathbb{D}$). This will not be a problem because the metric will diverge as we approach $\partial\mathbb{D}$.

The following calculation is the key to the invariant metric:

Theorem 9.3.9. *Let f be an extended Möbius transformation. Then*

$$|f'(z)| = \frac{1 - |f(z)|^2}{1 - |z|^2} \tag{9.3.8}$$

Proof. If g is an antilinear extended Möbius transformation, then $f = cg$ is in $\mathcal{M}$ and $|f'(z)| = |g'(z)|$ and $|f(z)| = |g(z)|$, so (9.3.8) for f implies it for g, that is, we can suppose $f \in \mathcal{M}$, that is, $f = f_T$ with

$$T = \begin{pmatrix} a & c \\ \bar{c} & \bar{a} \end{pmatrix} \tag{9.3.9}$$

where $\det(T) = |a|^2 - |c|^2$.

As we computed in (9.2.60),

$$|f'(z)| = \frac{1}{|\bar{c}z + \bar{a}|^2} \tag{9.3.10}$$

On the other hand, since (the cross-terms cancel)

$$|az + c|^2 - |\bar{c}z + \bar{a}|^2 = (|a|^2 - |c|^2)(|z|^2 - 1)$$

we see that

$$|f(z)|^2 - 1 = \frac{|z|^2 - 1}{|\bar{c}z + \bar{a}|^2} \tag{9.3.11}$$

(9.3.11) and (9.3.10) imply (9.3.8). □

The standard Euclidean Riemannian structure will be called $(dz)^2$. The *Poincaré metric* on $\mathbb{D}$ is defined to be the one associated to the Riemann structure

$$(1 - |z|^2)^{-2}\,(dz)^2 \tag{9.3.12}$$

Put differently, the length of a smooth curve $\gamma\colon [0, 1] \to \mathbb{D}$ is

$$L(\gamma) = \int_0^1 |\gamma'(s)|(1 - |\gamma(s)|^2)^{-1}\, ds \tag{9.3.13}$$

and

$$\rho(x, y) = \inf\{L(\gamma) \mid \gamma(0) = x,\ \gamma(1) = y\} \tag{9.3.14}$$

Theorem 9.3.10. *Let $g \in \widetilde{\mathcal{M}}$. Then g preserves the Poincaré–Riemann structure* (9.3.12)*, the length* (9.3.13)*, and the metric* (9.3.14).

Proof. It suffices to prove preservation of the Riemann structure. Since g is conformal or anticonformal, it preserves angles, so we need only show infinitesimal lengths get mapped properly. The mapping is, of course, by $|f'(z)|$. (9.3.8) is precisely this statement, that is,

$$\frac{|df|}{1-|f|^2} = \frac{|dz|}{1-|z|^2} \tag{9.3.15}$$

□

The metric has a $\frac{1}{2}(1-|z|)^{-1}$ divergence as $|z| \to 1$ whose integral diverges logarithmically, so we expect $\rho(0,z)$ to look like $\frac{1}{2}\log(1-|z|)^{-1}$ as $|z| \uparrow 1$. That is part of the following:

The set $\mathbb{D}$ with the Poincaré metric is called the $\mathbb{D}$-*model of the hyperbolic plane.*

Theorem 9.3.11. (i) *The geodesic from* 0 *to* $z \in \mathbb{D}$ *is the straight line segment between them.*

(ii) *We have that* $\rho(z,0)$ *is given by*

$$\tanh(\rho(z,0)) = |z| \tag{9.3.16}$$

so that as $|z| \uparrow 1$,

$$\rho(z,0) = \tfrac{1}{2}\log((1-|z|)^{-1}) + \tfrac{1}{2}\log 2 + O(1-|z|) \tag{9.3.17}$$

(iii) *For any* $z, w \in \mathbb{D}$,

$$\tanh(\rho(z,w)) = \frac{|z-w|}{|1-\bar{z}w|} \tag{9.3.18}$$

(iv) *The geodesics in the* $\mathbb{D}$-*model of the hyperbolic plane are precisely segments of the orthocircles.*

Proof. (i) Because the Poincaré metric is conformal, for any curve from 0 to z, if $\hat{z} = z/|z|$, then

$$\begin{aligned}|\gamma'(s)|^2 &= [\mathrm{Re}(\gamma'(s)\hat{z})]^2 + [\mathrm{Im}(\gamma'(s)\hat{z})]^2 \\ &\geq \mathrm{Re}(\gamma'(s)\hat{z})^2\end{aligned} \tag{9.3.19}$$

that is, the infinitesimal length is larger than its radial component. Since the metric is invariant under rotations,

$$|\gamma'(s)| \geq \frac{1}{1-|\gamma(s)|^2}\left|\frac{d|\gamma(s)|}{ds}\right| \tag{9.3.20}$$

with equality only if $\arg(\gamma(s))$ is constant. This shows the minimal length path has $\arg(\gamma(s))$ constant, and so it is the straight line.

(ii) By (i), $\gamma(s) = sz$, so

$$|\gamma'(s)| = \frac{|z|}{1-|\gamma(s)|^2}$$

and thus

$$\begin{aligned}\rho(0,z) &= \int_0^1 \frac{|z|\,ds}{1-|zs|^2} = \int_0^{|z|} \frac{dy}{1-y^2} \\ &= \mathrm{arctanh}(|z|)\end{aligned}$$

since $\frac{d}{dy}\mathrm{arctanh}(y) = (1-y^2)^{-1}$. This proves (9.3.16).

To get (9.3.17), we note (9.3.16) with $|z| = r$, we have

$$\frac{1 - e^{-2\rho}}{1 + e^{-2\rho}} - r \tag{9.3.21}$$

so

$$(1-r)^{-1} = \left[\frac{2e^{-2\rho}}{1 + e^{-2\rho}}\right]^{-1} = \tfrac{1}{2}\, e^{2\rho} + \tfrac{1}{2} \tag{9.3.22}$$

which implies (9.3.17).

(iii) By the invariance of ρ under $f \in \mathcal{M}$,

$$\begin{aligned}\rho(z, w) &= \rho(f_z(z), f_z(w)) \\ &= \rho\biggl(0, \frac{w - z}{1 - \bar{z}w}\biggr)\end{aligned}$$

so (9.3.16) implies (9.3.18).

(iv) The geodesic from z to w is taken into the geodesic from 0 to $g_z(w)$ by g_z. Thus, this geodesic is the image under g_z^{-1} of a diameter, so a segment of an orthocircle. □

Remark. A convenient way of rewriting (9.3.21) is

$$e^{-2\rho(0,z)} = \frac{1 - |z|}{1 + |z|} \tag{9.3.23}$$

Notice that given an orthocircle and a point not on that circle, we can find multiple orthocircles that contain the point but do not intersect the original circle, for by a Möbius transformation, we can suppose the point is 0 and it is obvious that multiple diameters avoid a given orthocircle. That is, if parallel lines mean infinite geodesics, which are nonintersecting, Euclid's fifth postulate fails. This is a homogeneous geometry that is a realization of Lobachevsky's plane.

Analogous to the fact that $\mathcal{M}$ is the set of holomorphic bijections of $\mathbb{D}$, we can describe all isometries.

Theorem 9.3.12. *Let $f : \mathbb{D} \to \mathbb{D}$ be any continuous function, which is an isometry in the Poincaré metric. Then $f \in \widetilde{\mathcal{M}}$.*

Remark. Since we have seen all $f \in \widetilde{\mathcal{M}}$ are isometries, we see $\widetilde{\mathcal{M}}$ is the set of all isometries.

Proof. Let $f(0) = z_0$, $f(\frac{1}{2}) = w_0$. Then $(g_{z_0} \circ f)(0) = 0$. Since $g_{z_0} \circ f$ is an isometry, $\rho((g_{z_0} \circ f)(\frac{1}{2}), 0) = \rho((g_{z_0} \circ f)(\frac{1}{2}), (g_{z_0} \circ f)(0)) = \rho(\frac{1}{2}, 0)$. Since $\rho(w, 0)$ is a monotone function of $|w|$, $|(g_{z_0} \circ f)(\frac{1}{2})| = \frac{1}{2}$. Thus, by following g_{z_0} by a rotation about zero, we find $h \in \mathcal{M}$, so $h \circ f$ takes 0 to 0 and $\frac{1}{2}$ to $\frac{1}{2}$.

It thus takes the geodesic from 0 to $\frac{1}{2}$ and its continuation setwise to itself, that is, $h \circ f$ maps $(-1, 1)$ to itself. Since $h \circ f$ is one-one and continuous, either $h \circ f[\mathbb{C}_+ \cap \mathbb{D}] \subset \mathbb{C}_+ \cap \mathbb{D}$ or in $\mathbb{C}_- \cap \mathbb{D}$. By replacing h by ch, we can be sure the image is in $\mathbb{C}_+ \cap \mathbb{D}$, that is, we can find $h \in \widetilde{\mathcal{M}}$ so that

$$(h \circ f)(0) = 0 \qquad (h \circ f)(\tfrac{1}{2}) = \tfrac{1}{2} \qquad (h \circ f)(\mathbb{C}_+ \cap \mathbb{D}) \subset \mathbb{C}_+ \cap \mathbb{D}$$

If we prove $h \circ f$ is the identity, then $f = h^{-1} \in \widetilde{\mathcal{M}}$.

Let w lie in $\mathbb{C}_+ \cap \mathbb{D}$. The two sets $S_0 = \{w_1 \mid \rho(w_1, 0) = \rho(w, 0)\}$ and $S_1 = \{w_1 \mid \rho(w_1, \frac{1}{2}) = \rho(w, \frac{1}{2})\}$ are circles (S_0 is a circle by (9.3.16) and S_1 is an image under a Möbius transformation of a circle about 0, and so a circle).

These circles are distinct (look at their real points) and contain w and $\bar{w}$. Since circles can intersect in at most two points, $S_1 \cap S_0 = \{w, \bar{w}\}$. But $(h \circ f)(w) \in S_1 \cap S_0$ and is in $\mathbb{C}_+$ so $(h \circ f)(w) = w$. Thus, $h \circ f = \mathbf{1}$ on $\mathbb{C}_+ \cap \mathbb{D}$ and similarly on $\mathbb{C}_- \cap \mathbb{D}$ and so, by continuity, on $\mathbb{D}$. □

Next, we want to look at which points in $\mathbb{D}$ are closer to z than w. For Euclidean geometry, this is answered by the perpendicular bisector. The same is true here but the bisector is an orthocircle:

Theorem 9.3.13. *Fix $z_0 \neq z_1$ both in $\mathbb{D}$. Then*

$$\{w \mid \rho(w, z_0) = \rho(w, z_1)\}$$

is an orthocircle. Removing this orthocircle from $\mathbb{D}$ yields two open connected components with z_0 and z_1 in the two components. In the component with z_0, we have $\rho(w, z_0) < \rho(w, z_1)$, and vice versa within the other.

Proof. Suppose first $z_0 = ia$, $z_1 = -ia$ with $0 < a < 1$ and $\operatorname{Im} w > 0$, $w \in \mathbb{D}$. We claim $\rho(w, z_0) < \rho(w, z_1)$. By (9.3.18), this is equivalent to

$$|(w - ia)(1 + ia\bar{w})| < |(w + ia)(1 - ia\bar{w})| \tag{9.3.24}$$

$$\text{LHS} = A + B \qquad \text{RHS} = A - B$$

where

$$A = -ia + ia|w|^2 \qquad B = w + a^2\bar{w}$$

A is pure imaginary, so

$$|\operatorname{Re}(A + B)| = |\operatorname{Re} B| = |\operatorname{Re}(A - B)|$$

Since $|w| < 1$ and $|a| < 1$, $\operatorname{Im} A < 0$, and since $\operatorname{Im} w > 0$, $\operatorname{Im} B > 0$. Thus, $|\operatorname{Im}(A + B)| < |\operatorname{Im}(A - B)|$, proving (9.3.24).

This proves the result in the special case $z_0 = ia$, $z_1 = -ia$. In general, let w be the geodesic midpoint of the geodesic from z_0 to z_1. Let $g \in \mathcal{M}$ take w to 0. Since it preserves geodesics and hyperbolic lengths, it must map z_0 and z_1 to equidistant points from 0 on the same line through zero. By a further rotation, we see any pair is equivalent to the special case under a hyperbolic isometry. □

Corollary 9.3.14. *Let r be a reflection in an orthocircle, C. Let w, z be on the same side of C (and not on C). Then*

$$\rho(w, z) < \rho(w, r(z)) \tag{9.3.25}$$

Proof. Since ρ is preserved by $\gamma \in \mathcal{M}$, we can suppose the orthocircle is $(-1, 1)$. Then C is the perpendicular bisector of points equidistant from z, $r(z) = \bar{z}$, and (9.3.25) is the final assertion of the theorem. □

Theorem 9.3.15. *For any $f \in \mathcal{M}$, the hyperbolic perpendicular bisection of the hyperbolic line from* 0 *to* $f(0)$ *is the part of the boundary,* ∂D_f, *of the final circle,* D_f, *inside* $\mathbb{D}$.

Proof. f^{-1} is the reflection in ∂D_f followed by reflection in the line, L, which is the Euclidean bisector of the line between the centers of D_f and D_i. By Theorem 9.3.6, $0 \in L$, so for $w \in \mathbb{D} \cap \partial D_f$,

$$|f^{-1}(w)| = |w| \tag{9.3.26}$$

Since $\rho(0, z)$ is a function of $|z|$ only, we have

$$\rho(0, f^{-1}(w)) = \rho(0, w) \tag{9.3.27}$$

But since f is a ρ-isometry,

$$\rho(f(0), w) = \rho(0, f^{-1}(w)) \tag{9.3.28}$$

Thus, w lies on the hyperbolic perpendicular bisector. □

Remarks and Historical Notes. The fact that $\mathbb{D}$ has a metric in which all fractional linear automorphisms are isometries is a discovery of Poincaré. This metric has constant curvature -1. It is a remarkable fact that the other two simply connected Riemann surfaces (namely, $\mathbb{C}$ and $\mathbb{C} \cup \{\infty\}$) have natural constant curvature metrics—the flat metric on $\mathbb{C}$ and the spherical metric on $\mathbb{C} \cup \{\infty\}$. However, in these other cases, there are automorphisms that are not isometries.

For further discussion of the group $\mathbb{SU}(1, 1)$, see Sections 10.4 and 10.5 of [400].

The study of subgroups of $\mathbb{SL}(2, \mathbb{R}) \cong \mathbb{SU}(1, 1)$ has arithmetic significance because it contains matrices with integral coefficients. Indeed, $\mathbb{SL}(2, \mathbb{Z})$, the 2×2 matrices of determinant 1 with integral coefficients, is a subgroup. For this reason, the upper half-plane model is often more popular than the disk model.

Katok [216] proves Theorem 9.3.13 in the UHP model where the calculation is less messy.

9.4 FUCHSIAN GROUPS

In this section, we will say something about general Fuchsian groups as a preliminary to the study in the next two sections of the ones of interest for finite gap Jacobi matrices. This will hardly be a comprehensive look at the subject—our example, as we will explain in the next two sections, will be infinitely nicer than more typical cases, so we can avoid discussions of all sorts of subtleties. Our main theme here will be equivalences of various measures of discreteness and of a critical number called the Poincaré index.

Given $f \in \mathcal{M}$, there are various measures of how "large" f is, that is, how far it is from the identity. We can write $f = f_T$ with $\det(T) = 1$ and use $\|T\|$; we can look at $(1 - |f(0)|)^{-1}$, $e^{2\rho(f(0),0)}$, or $|f'(0)|^{-1}$, or replace $f(0)$ by $f(z)$ for some other $z \in \mathbb{D}$. Our initial goal will be to prove an equivalence in the quantitative sense of upper and lower bounds on ratios. We begin with what happens at a fixed

z for a single f:

Theorem 9.4.1. *Let $f = f_T$ lie in $\mathcal{M}$. Then:*

$$\text{(a)} \qquad 1 - |f(z)| \le |f'(z)| \le \frac{2}{1-|z|^2}\,(1 - |f(z)|) \tag{9.4.1}$$

$$\text{(b)} \qquad \tfrac{1}{2}(1 - |f(z)|) \le e^{-2\rho(f(z),0)} \le (1 - |f(z)|) \tag{9.4.2}$$

$$\text{(c)} \qquad (\|T\|_2^2 + 2)^{-1} = \tfrac{1}{4}(1 - |f(0)|^2) \tag{9.4.3}$$

where $\det(T) = 1$, $\|\cdot\|_2$ *is Hilbert–Schmidt norm, and* ρ *is the Poincaré metric.*

Remark. All norms on 2×2 matrices are equivalent, so for any norm, (9.4.3) says $1 - |f(0)| \sim \|T\|^{-2}$ in the sense that the ratio in either direction is bounded by some constant.

Proof. (a) By (9.3.8),

$$1 - |f(z)| \le (1 - |f(z)|)(1 + |f(z)|) = (1 - |z|^2)|f'(z)| \le |f'(z)|$$

and

$$|f'(z)| = \frac{1 + |f(z)|}{1 - |z|^2}\,(1 - |f(z)|) \le \frac{2}{1-|z|^2}\,(1 - |f(z)|)$$

(b) (9.3.23) implies (9.4.2) if we note that

$$\tfrac{1}{2} \le \frac{1}{1 + |f(z)|} \le 1$$

(c) If $T = \left(\begin{smallmatrix} \alpha & \gamma \\ \bar\gamma & \bar\alpha \end{smallmatrix}\right)$, then $|f(0)| = |\frac{\gamma}{\alpha}|$, so

$$1 - |f(0)|^2 = 1 - \left|\frac{\gamma}{\alpha}\right|^2 = \frac{1}{|\alpha|^2}$$

while

$$\|T\|_2^2 = 2|\alpha|^2 + 2|\gamma|^2 = 4|\alpha|^2 - 2$$ □

Corollary 9.4.2. *Fix $z_0 \in \mathbb{D}$ and $\varepsilon > 0$. Then $\{f_T \mid |f_T(z_0)| \le 1 - \varepsilon\}$ is compact in $\mathcal{M}$.*

Proof. The set is clearly closed. By (9.4.2) and $|\rho(f(z_0), 0) - \rho(f(0), 0)| \le \rho(f(z_0), f(0)) = \rho(z_0, 0)$, we see $1 - |f(0)|$ is bounded away from 0 on the set in question. So, by (9.4.3), $\|T\|$ is bounded above, implying compactness. □

The following shows all quantities are comparable as z, w run through fixed compact subsets of $\mathbb{D}$:

Theorem 9.4.3. *For any $f \in \mathcal{M}$ and $z, w \in \mathbb{D}$,*

$$e^{-2\rho(z,w)} \le \frac{e^{-2\rho(0,f(z))}}{e^{-2\rho(0,f(w))}} \le e^{2\rho(z,w)} \tag{9.4.4}$$

Proof. By the triangle inequality and the fact that f is a ρ-isometry,

$$|\rho(0, f(z)) - \rho(0, f(w))| \le \rho(f(z), f(w)) = \rho(z, w)$$ □

We will soon use Γ for certain countable subgroups of $\mathcal{M}$. But for a while, we will use Γ to denote a countable family in $\mathcal{M}$ that need not (yet) be a group.

Theorem 9.4.4. *Let Γ be a countable subset of $\mathcal{M}$. Then the following are equivalent:*

(i) *For every $z_0 \in \mathbb{D}$ and every $r < 1$, $\{f \in \Gamma \mid |f(z_0)| < r\}$ is finite.*
(ii) *For one $z_0 \in \mathbb{D}$ and every $r < 1$, $\{f \in \Gamma \mid |f(z_0)| < r\}$ is finite.*
(iii) *For every compact subset $K \subset \mathbb{D}$ and every $r < 1$, $\{f \in \Gamma \mid \inf_{z_0 \in K} |f(z_0)| < r\}$ is finite.*
(iv) *For every $z_0 \in \mathbb{D}$ and every $\eta > 0$, $\{f \in \Gamma \mid |f'(z_0)| > \eta\}$ is finite.*
(v) *For one $z_0 \in \mathbb{D}$ and every $\eta > 0$, $\{f \in \Gamma \mid |f'(z_0)| > \eta\}$ is finite.*
(vi) *For every compact $K \subset \mathbb{D}$ and every $\eta > 0$, $\{f \in \Gamma \mid \sup_{z_0 \in K} |f'(z_0)| > \eta\}$ is finite.*
(vii) *For every C, $\{f \in \Gamma \mid f = f_T,\ T \in \mathbb{SL}(2,\mathbb{C}),\ \|T\| < C\}$ is finite.*

Remarks. 1. We do not include $e^{-2\rho(f(z),0)}$ results since they are trivially equivalent to (i), (ii).

2. For families of FLTs, discreteness of orbits (a condition like (i)) implies an analog of (vii) but not vice versa.

3. If these conditions hold, we say Γ is a *discrete family*.

Proof. Immediate from Theorems 9.4.1 and 9.4.3. $\square$

Definition. A *Fuchsian group*, Γ, is a discrete subgroup of $\mathcal{M}$.

Theorem 9.4.4 does not use the quantitative equivalence of Theorems 9.4.1 and 9.4.3. The following does:

Theorem 9.4.5. *Let Γ be a discrete family in $\mathcal{M}$. Fix $s > 0$. Then the following are equivalent:*

$$\text{(i)} \quad \sum_{f\in\Gamma} (1 - |f(z)|)^s < \infty \ \textit{for one } z \in \mathbb{D} \tag{9.4.5}$$

$$\text{(ii)} \quad \sum_{f\in\Gamma} (1 - |f(z)|)^s < \infty \ \textit{for all } z \in \mathbb{D}$$

$$\text{(iii)} \quad \sum_{f\in\Gamma} |f'(z)|^s < \infty \ \textit{for one } z \in \mathbb{D} \tag{9.4.6}$$

$$\text{(iv)} \quad \sum_{f\in\Gamma} |f'(z)|^s < \infty \ \textit{for all } z \in \mathbb{D}$$

$$\text{(v)} \quad \sum_{f\in\Gamma} e^{-2s\rho(0,f(z))} < \infty \ \textit{for some } z \in \mathbb{D} \tag{9.4.7}$$

$$\text{(vi)} \quad \sum_{f\in\Gamma} e^{-2s\rho(0,f(z))} < \infty \ \textit{for all } z \in \mathbb{D}$$

$$\text{(vii)} \quad \sum_{T \mid f_T \in \Gamma} \|T\|^{-2s} < \infty$$

Proof. Again immediate from Theorems 9.4.1 and 9.4.3. □

For Fuchsian groups, the series in (9.4.5)–(9.4.7) for $z = 0$ are, depending on the author, called *Poincaré series*. The inf over s for which these sums converge is called the *critical exponent*. If it converges for some s, we will say that s is a Poincaré exponent. Convergence for $s = 1$ implies Blaschke products

$$B_\Gamma(z, w) \equiv \prod_{f \in \Gamma} b_{f(w)}(z) \tag{9.4.8}$$

converge where $b_w(z)$ is given by (2.3.67). We will also see later that it is important for the groups we consider that the critical exponent is strictly less than 1. Poincaré [347] used his series to construct automorphic functions; see the Notes.

Example 9.4.6. Let Γ be a Fuchsian group with a single generator, f. If f is elliptic, it must be periodic to assure discreteness, and all series are finite. If f is hyperbolic, $f^{(n)}(z)$ approaches a limit in $\partial\mathbb{D}$ exponentially fast as $n \to \pm\infty$ (different limits for $+\infty$ and $-\infty$), so $1 - |f^{(n)}(z)| \le e^{-C|n|}$ and the critical index is 0. If f is parabolic, $1 - |f^{(n)}(z)|$ is $O(n^{-2})$ and the critical index is $\frac{1}{2}$. Since this example is a little subtle, let us give the details.

Parabolic elements of $\mathbb{SU}(1, 1)$ have the form

$$T = \pm \begin{pmatrix} 1 + ia & ae^{i\psi} \\ ae^{-i\psi} & 1 - ia \end{pmatrix}$$

for some $a \in \mathbb{R}$ and $\psi \in [0, 2\pi)$. The unique eigenvector in this case is $(1 \; -ie^{-i\psi})^t$. For this T we have (with $\pm$ taken to be $+$)

$$T^n = \begin{pmatrix} 1 + ina & nae^{i\psi} \\ nae^{-i\psi} & 1 - ina \end{pmatrix}$$

Picking $a = 1$, $e^{-i\psi} = i$, we have

$$f^{(n)}(z) = \frac{(1 - in)z - in}{inz + (1 - in)} \tag{9.4.9}$$

Thus,

$$f^{(n)}(0) = \frac{n^2 - in}{1 + n^2} \tag{9.4.10}$$

The fixed point is $1 = w_\infty$ and $f^{(n)}(0) \to w_\infty$. We have

$$|w_\infty - f^{(n)}(0)| = \frac{1}{(1 + n^2)^{1/2}} \tag{9.4.11}$$

$$1 - |f^{(n)}(0)|^2 = \frac{1}{1 + n^2} \tag{9.4.12}$$

As claimed, $1 - |f^{(n)}(0)|^2 = O(1/n^2)$ even though the distance to the fixed point is $O(1/n)$. The asymptotic direction is $\partial\mathbb{D}$. This is the phenomenon explained in Example 9.2.10. □

We are heading toward a proof that if f_n is a sequence in $\mathcal{M}$ with $|f_n(0)| \to 1$, then $|f_n(0) - f_n(z)| \to 0$ for any fixed z (so orbits in Fuchsian groups will have the same limit points). The idea will be that near $\partial\mathbb{D}$, the Euclidean distance is much smaller than the hyperbolic distance, at least if the hyperbolic distance is not too large. The following expresses this idea quantitatively:

Proposition 9.4.7. *Let $z, w \in \mathbb{D}$. Then*

$$\rho(z, w) \le \frac{|z - w|}{1 - \max(|z|, |w|)^2} \tag{9.4.13}$$

while

$$|z - w| \le (1 - \max(|z|, |w|)^2)\rho(z, w)\, e^{4\rho(z,w)} \tag{9.4.14}$$

Proof. The (Euclidean) straight line from z to w is a possible trial geodesic for the hyperbolic metric, so its hyperbolic length bounds $\rho(z, w)$, that is, with $\gamma(t) = tz + (1 - t)w$,

$$\rho(z, w) \le \int_0^1 \frac{|d\gamma(t)|}{1 - |\gamma(t)|^2} \le \big(\sup_t [1 - |\gamma(t)|^2]^{-1}\big)|z - w|$$

which is (9.4.13).

Similarly, suppose $|z| \ge |w|$. With f_{z_0} given by (9.3.1), the hyperbolic geodesic from z to w is

$$\begin{aligned} \gamma(t) &= f_{-z}(tf_z(w)) \\ &= \frac{z + t\zeta}{1 + \bar{z}t\zeta} \end{aligned} \tag{9.4.15}$$

where $\zeta = f_z(w)$. Using this as a trial for the Euclidean distance,

$$\begin{aligned} |z - w| &\le \int_0^1 |d\gamma(t)| \\ &\le \max(1 - |\gamma(t)|^2)\rho(z, w) \end{aligned} \tag{9.4.16}$$

since $\int_0^1 |d\gamma(t)|/(1 - |\gamma(t)|^2 = \rho(z, w)$ since it is the length of the geodesic.

By (9.4.15),

$$1 - |\gamma(z)|^2 = \frac{(1 - |z|^2)(1 - |t\zeta|^2)}{|1 + \bar{z}t\zeta|^2} \le \frac{1 - |z|^2}{(1 - |\zeta|)^2} \tag{9.4.17}$$

But, by (9.3.23),

$$1 - |\zeta| \ge e^{-2\rho(\zeta,0)} = e^{-2\rho(z,w)} \tag{9.4.18}$$

(9.4.16)–(9.4.18) imply (9.4.14). □

Remark. The occurrence of $e^{4\rho}$ might be surprising, but it is needed! For if $w = 0$, then $|z - w| = |z|$, $\rho(z, 0) \sim \frac{1}{2}(1 - |z|)$ (by (9.3.17)). Thus, for $\rho(z, w)$ large, we must cancel the $1 - |z|^2$, which requires $e^{4\rho}$! For the application we want, ρ is bounded and $e^{4\rho}$ is harmless.

Theorem 9.4.8. *Let* $\{f_n\}_{n=1}^\infty$ *be a family in* $\mathcal{M}$ *with* $\lim f_n(z_0) = w_0 \in \partial\mathbb{D}$ *for some* $z_0 \in \mathbb{D}$. *Then*

$$f_n(z) \to w_0$$

as $n \to \infty$ *for each* $z \in \mathbb{D}$, *uniformly in compact subsets of* $\mathbb{D}$.

First Proof. Since f_n is a hyperbolic isometry, $\rho(f_n(z), f_n(\zeta)) = \rho(z, \zeta)$, so by (9.4.14),

$$|f_n(z) - f_n(\zeta)| \le (1 - |f_n(z)|^2)\rho(z, \zeta)e^{4\rho(z,\zeta)} \tag{9.4.19}$$

so $|f_n(z)| \to 1 \Rightarrow |f_n(z) - f_n(\zeta)| \to 0$ uniformly on compact subsets. □

Second Proof. By Theorems 9.4.1 and 9.4.3, $|f_n(z)| \to 1$ implies for any compact $K \subset \mathbb{D}$,

$$\sup_{\zeta \in K} |f_n'(\zeta)| \to 0 \tag{9.4.20}$$

Thus, if $z, \zeta \in \{\eta \mid |\eta| \le r < 1\}$,

$$|f_n(z) - f_n(\zeta)| \le |z - \zeta| \sup_{|\eta| \le r} |f_n'(\eta)| \to 0$$ □

Henceforth, we will use Γ to denote only Fuchsian groups and we will generally use the symbol γ for a generic element in Γ.

Definition. A point, $w_0 \in \partial\mathbb{D}$, is called a *limit point* for Γ if and only if there exists $\{\gamma_n\}_{n=1}^\infty \subset \Gamma$ so $\gamma_n(0) \to w_0$. The set of all limit points is denoted by $\Lambda(\Gamma)$. A point in $\partial\mathbb{D} \setminus \Lambda(\Gamma)$ is called an *ordinary point*.

By Theorem 9.4.8, the limit points are the same if we take limit points of any $\gamma_n(z)$ with $z \in \mathbb{D}$. By compactness, there are always limit points so long as Γ is not finite—we will discuss this below.

Notice that

$$\Lambda(\Gamma) = \overline{\{\gamma(0) \mid \gamma \in \Gamma\}} \cap \partial\mathbb{D} \tag{9.4.21}$$

and so $\Lambda(\Gamma)$ is always closed in $\mathbb{C}$ and in $\partial\mathbb{D}$. Also, notice that, as we have seen, $\gamma(0)$ lies in the final disk for γ and so in the initial disk for γ^{-1}. Moreover, the radii of these disks go to zero as $\gamma(0) \to \partial\mathbb{D}$. Thus, we have

Proposition 9.4.9. *For any Fuchsian group,*

$$\Lambda(\Gamma) = \overline{\{\text{center of isometric circles for } \gamma \in \Gamma\}} \cap \partial\mathbb{D} \tag{9.4.22}$$

$$= \overline{\{\text{center of final disks for } \gamma \in \Gamma\}} \cap \partial\mathbb{D} \tag{9.4.23}$$

One other immediate result about Γ is

Proposition 9.4.10. *For any Fuchsian group, the fixed points of all hyperbolic and parabolic elements are limit points.*

Proof. For parabolic γ, $\gamma^n(0)$ converges to the unique fixed point as $|n| \to \infty$. For hyperbolic γ, $\gamma^{\pm n}(0)$ converges to the two fixed points as $n \to +\infty$. □

Limit points help us understand when we can extend comparison results for $\gamma'(z)$ and $\gamma'(0)$ to z's in $\partial\mathbb{D}$.

Theorem 9.4.11. (a) *For all $z \in \partial\mathbb{D}$ and $f \in \mathcal{M}$, we have that*

$$|f'(0)| \leq 4|f'(z)| \tag{9.4.24}$$

(b) *Let K be a compact subset of the ordinary points for Γ. Then there is a constant $C < \infty$ so that for all $\gamma \in \Gamma$,*

$$\sup\{|\gamma'(z)| \mid z \in \overline{\mathbb{D}},\ \arg z \in K\} \leq C|\gamma'(0)| \tag{9.4.25}$$

Proof. By (9.2.59), for any $f \in \mathcal{M}$ with $w_f \equiv f^{-1}(\infty)$ the center of the isometric circle,

$$\frac{|f'(0)|}{|f'(z)|} = \frac{|z - w_f|^2}{|w_f|^2} \tag{9.4.26}$$

To get (9.4.23), we note that since $|z| = 1$ and $|w_f| > 1$ (see Theorem 9.3.6),

$$\frac{|z - w_f|}{|w_f|} \leq 1 + \frac{|z|}{|w_f|} \leq 2$$

To get (9.4.25), let

$$S = \{z \in \overline{\mathbb{D}} \mid \arg z \in K\} \cup \{0\}$$

By Proposition 9.4.9 and Theorem 9.3.6, $\overline{\{w_\gamma \mid \gamma \in \Gamma,\ \gamma \neq 1\}} \cap S = \emptyset$, so since both sets are closed,

$$d \equiv \min(|z - w_\gamma| \mid z \in S,\ \gamma \in \Gamma \setminus \{1\}) > 0$$

Since $|w_\gamma| = |\gamma(0)|^{-1}$ (see (9.3.4)), we get, by (9.4.26),

$$\frac{|\gamma'(z)|}{|\gamma'(0)|} = \frac{1}{|z - w_\gamma|^2 |\gamma(0)|^2} \tag{9.4.27}$$

so (9.4.25) holds with

$$C = d^{-2}[\inf(|\gamma(0)|^2 \mid \gamma \in \Gamma \setminus \{1\})] \tag{9.4.28}$$

□

Remark. If there are $\gamma \in \Gamma$ with $\gamma(0) = 0$, γ' is constant and we can drop them from consideration in (9.4.28) and earlier in the proof.

If $|\cdot|$ is Lebesgue measure of a set in $\partial\mathbb{D}$, we have

$$\frac{|\gamma[K]|}{|K|} = \frac{1}{|K|} \int_K |\gamma'(e^{i\theta})|\, d\theta \tag{9.4.29}$$

so, by (9.4.25), if K is disjoint from the limit points,

$$\tfrac{1}{4}|\gamma'(0)| \leq \frac{|\gamma[K]|}{|K|} \leq C|\gamma'(0)| \tag{9.4.30}$$

Thus:

Theorem 9.4.12. *Let $K \subset \partial\mathbb{D}$ be a compact subset of the regular points for a Fuchsian group Γ. Then for each $s > 0$,*

$$\sum_{\gamma\in\Gamma} |\gamma'(0)|^s < \infty \Leftrightarrow \sum_{\gamma\in\Gamma} |\gamma[K]|^s < \infty \tag{9.4.31}$$

Next, we want to study possible sets that can be $\Lambda(\Gamma)$ for some Γ. For this, it will be useful to note that since Γ is a set of maps each analytic in some neighborhood of $\overline{\mathbb{D}}$, they define maps of $\partial\mathbb{D}$ to $\partial\mathbb{D}$. Clearly, if $\gamma_n(0) \to w_0$, then $(\gamma \circ \gamma_n)(0) \to \gamma(w_0)$, so this action maps $\Lambda(\Gamma)$ onto itself, and since γ is invertible on $\partial\mathbb{D}$, of $\partial\mathbb{D} \setminus \Lambda(\Gamma)$ to itself. Here is a key fact:

Lemma 9.4.13 (Three-Point Lemma)**.** *Let $w_0 \in \Lambda(\Gamma)$. Let w_1, w_2 be points in $\partial\mathbb{D}$ so w_0, w_1, w_2 are distinct. Then there exists $\gamma_n \in \Gamma$ so that either $\gamma_n(w_1) \to w_0$ or $\gamma_n(w_2) \to w_0$.*

Remark. If γ_0 is a hyperbolic map with fixed points w_0 and w_1 and $\Gamma = \{\gamma_0^n \mid n \in \mathbb{Z}\}$, then there is no $\gamma_n \in \Gamma$ so $\gamma_n(w_1) \to w_0$. This shows we need two extra points in general.

Proof. By passing to a subsequence, we can find $\gamma_n \in \Gamma$ so $\gamma_n(0) \to w_0$ and $\gamma_n^{-1}(\infty) \to w_3$ for some $w_3 \in \partial\mathbb{D}$. Clearly, since $w_1 \neq w_2$, one is distinct from w_3, say, w_1 is. Since $|\gamma_n(0)| \to 1$, the radius, r_n, of the isometric circle of γ_n goes to zero. Thus, for n large, w_1 is not in the initial disk. By Theorem 9.2.32, for such n, both 0 and w_1 map into the final disk, so $|\gamma_n(0) - \gamma_n(w_1)| \leq 2r_n \to 0$. $\square$

We need a final technical result before we can get to a more thorough analysis of $\Lambda(\Gamma)$:

Lemma 9.4.14. (a) *If $f \in \mathcal{M}$ is elliptic and $g \in \mathcal{M}$ does not leave fixed the fixed points of f, then $gfg^{-1}f^{-1}$ is hyperbolic.*

(b) *If f and g are FLTs with g parabolic and f not fixing the point left invariant by g, then fg^n is hyperbolic for large n.*

Proof. (a) By conjugation, we can suppose $f(0) = 0$ so $f = f_T$ with $T = \left(\begin{smallmatrix} e^{i\theta} & 0 \\ 0 & e^{-i\theta} \end{smallmatrix}\right)$. Let $g = f_S$ where $S = \left(\begin{smallmatrix} a & c \\ \bar{c} & \bar{a} \end{smallmatrix}\right)$. Then $TS^{-1}T^{-1} = \left(\begin{smallmatrix} \bar{a} & -e^{2i\theta}c \\ -e^{-2i\theta}\bar{c} & a \end{smallmatrix}\right)$ and

$$\mathrm{Tr}(STS^{-1}T^{-1}) = 2[|a|^2 - \sin 2\theta |c|^2] > 2$$

since $\theta \neq 0, \pi$ (since $f \neq 1$), $|a|^2 - |c|^2 = 1$, and $c \neq 0$ since $g(0) = \frac{c}{\bar{a}} \neq 0$. Thus, $gfg^{-1}f^{-1}$ is hyperbolic.

(b) By a conjugation, g leaves ∞ fixed, so $g = f_S$ with $S = \left(\begin{smallmatrix} 1 & 1 \\ 0 & 1 \end{smallmatrix}\right)$ and $f = f_T$ with $T = \left(\begin{smallmatrix} a & b \\ c & d \end{smallmatrix}\right)$ where $c \neq 0$ since $f(\infty) \neq \infty$. We have

$$|\mathrm{Tr}(TS^n)| = |a + d + cn| > 2$$

for n large. $\square$

Theorem 9.4.15. *Let Γ be a Fuchsian group.*

(a) $\Lambda(\Gamma)$ *is empty if and only if* Γ *is a finite cyclic group with an elliptic generator.*

(b) $\Lambda(\Gamma)$ *is a single point if and only if* Γ *is an infinite cyclic group with a parabolic generator.*

(c) *If* $\Lambda(\Gamma)$ *has at least two points,* Γ *has hyperbolic elements and the fixed points of these elements are dense in* $\Lambda(\Gamma)$.

Proof. (a) By Proposition 9.4.10, if $\Lambda(\Gamma)$ is empty, Γ can only contain elliptic elements. By (a) of the last lemma, those have to have common fixed points, so Γ is a subgroup of a group of two-dimensional rotations. The only such groups that are discrete are the finite cyclic groups.

(b) The group cannot contain any hyperbolic elements since they have two fixed points in $\Lambda(\Gamma)$. It cannot contain only elliptic elements since it if does, they either have common fixed points, in which case $\Lambda(\Gamma)$ is empty, or some distinct fixed points, in which case, there are hyperbolic elements by (a) of the last lemma.

Thus, Γ has parabolic elements. Those elements must fix the unique point of $\Lambda(\Gamma)$.

By (b) of the last lemma, it cannot also have elliptic elements because elliptic elements in $\mathcal{M}$ do not have fixed points in $\partial\mathbb{D}$, so Γ would have hyperbolic elements.

Thus, by the analysis in Example 9.4.6, Γ is a subgroup of $\{f_T \mid T = \left(\begin{smallmatrix} 1+ia & ae^{i\psi} \\ ae^{-i\psi} & 1-ia \end{smallmatrix}\right)$ with ψ fixed. This group is isomorphic to $\mathbb{R}$ under the variable a. All discrete subgroups of $\mathbb{R}$ are cyclic.

(c) By the analysis above, if Γ has a parabolic element and $\Lambda(\Gamma)$ has more than one point, it will have hyperbolic elements guaranteed by (b) of the last lemma. If it has an elliptic element and $\Lambda(\Gamma)$ is nonempty, it will have hyperbolic elements generated by (a) of the last lemma. Thus, Γ has hyperbolic elements as claimed.

If $\Lambda(\Gamma)$ has exactly two points, they must be the fixed points of this hyperbolic element—proving the second assertion in this part.

If $\Lambda(\Gamma)$ has a point, w_0, which is not a hyperbolic fixed point, there must be two hyperbolic fixed points, w_1, w_2, associated with a hyperbolic element, γ_0. By the three-point lemma, there is γ_n so $\gamma_n(w_1)$ or $\gamma_n(w_2)$ converges to w_0. But $\gamma_n(w_j)$ are the fixed points of the hyperbolic element, $\gamma_n\gamma_0\gamma_n^{-1}$. □

Theorem 9.4.16. *The set* $\Lambda(\Gamma)$ *is one of the following possibilities:*

(a) *The empty set, in which case* Γ *is a finite cyclic group with an elliptic generator.*

(b) *A single point, in which case* Γ *is an infinite cyclic case with a parabolic generator.*

(c) *Two points.*

(d) *A nowhere dense, perfect set (aka a Cantor set).*

(e) *The whole circle.*

Remarks. 1. In case (e), Γ is called a *type 1 Fuchsian group*; in cases (a)–(d), a *type 2 Fuchsian group*.

2. We will analyze the specific cases in (c) after the theorem.

3. Recall that a perfect set, S, is a closed set where any $x \in S$ is a limit point of points in $S \setminus \{x\}$. Such sets are always uncountable.

Proof. It suffices to show that if $\Lambda(\Gamma)$ has three or more points but is not all of $\partial\mathbb{D}$, then it is nowhere dense and perfect. Since it is not all of $\partial\mathbb{D}$, its complement, which is open, has two distinct points, w_1 and w_2.

If $w_0 \in \Lambda(\Gamma)$, by the three-point lemma, there exist γ_n so $\gamma_n(w_1) \to w_0$ or $\gamma_n(w_2) \to \gamma_0$. Since $\gamma_n(\partial\mathbb{D} \setminus \Lambda(\Gamma)) = \partial\mathbb{D} \setminus \Lambda(\Gamma)$, each $\gamma_n(w_j) \in \partial\mathbb{D} \setminus \Lambda(\Gamma)$ so $\partial\mathbb{D} \setminus \Lambda(\Gamma)$ is dense, and thus, $\Lambda(\Gamma)$ is nowhere dense.

To see that $\Lambda(\Gamma)$ is perfect, let $w_0 \in \Lambda(\Gamma)$. If w_0 is not a hyperbolic fixed point, it is a limit of such points by Theorem 9.4.15 and so a limit of other points in $\Lambda(\Gamma)$.

If w_0 is a hyperbolic fixed point for $\gamma \in \Gamma$, since $\Lambda(\Gamma)$ has at least three points, there is a point, w_1, in $\Lambda(\Gamma)$ neither w_0 nor the other fixed point of γ, so either $\gamma_0^n(w_1) \to w_0$ or $\gamma_0^{-n}(w_1) \to w_0$. In either case, the points are not w_0 (since it is fixed by γ_0). Thus, w_0 is a limit of other points of $\Lambda(\Gamma)$, and so the set is perfect. □

Example 9.4.17. We will analyze the case of two limit points. Before beginning, it pays to note that if we classify up to conjugacy, the classes with $\Lambda(\Gamma)$ empty are one for each $n \in \mathbb{Z}$, the order of the group, since all elliptic elements in $\mathcal{M}$ of exact order n are conjugate. For $\#(\Lambda(\Gamma)) = 1$, there is a single class since all parabolic elements are conjugate.

For calculations, it is easier to use conjugacy from $\mathbb{SU}(1,1)$ to $\mathbb{SL}(2,\mathbb{R})$, in which case we can assume the fixed points are 0 and ∞. The only possible T's where f_T map the set of two fixed points to themselves are

$$T(a) = \begin{pmatrix} a & 0 \\ 0 & a^{-1} \end{pmatrix} \qquad S(b) = \begin{pmatrix} 0 & b \\ b^{-1} & 0 \end{pmatrix}$$

with $a, b \in (0,\infty)$.

One class of discrete examples are infinite cyclic groups, which have a single hyperbolic generator. Since $\{a, a^{-1}\}$ is a conjugacy invariant, these are classified by $a \in (1,\infty)$ with the group being $\gamma_n(z) = a^{2n}z$, $n \in \mathbb{Z}$.

Since $T(c)S(b)T(c)^{-1} = S(c^2 b)$ while $T(c)T(a)T(c)^{-1} = T(a)$, up to conjugacy, we can suppose, if the group is not infinite cyclic, that it contains $S(1)$. We then get a class of groups isomorphic to $\mathbb{Z} \circledS \mathbb{Z}_2$ with

$$\gamma_n^+(z) = a^{2n} z \qquad \gamma_n^-(z) = -\frac{a^{2n}}{z}$$

Again, $a \in (1,\infty)$ is a conjugacy invariant. □

Next, we prove two general results about Poincaré indices:

Theorem 9.4.18 (Poincaré [347]). *For any Fuchsian group, the Poincaré series*

$$\sum_{\gamma\in\Gamma} |\gamma'(0)|^s \tag{9.4.32}$$

converges for $s = 2$.

Proof. By Theorem 9.4.4(iii), for each r, $\{\gamma \in \Gamma \mid \gamma(z_0) = z_0 \text{ for some } |z_0| < r\}$ is finite, so the set of points in $\mathbb{D}$ left fixed by some nonidentity $\gamma \in \Gamma$ is discrete. Pick z_0 so $\gamma(z_0) \neq z_0$ for all $\gamma \in \Gamma$, $\gamma \neq 1$.

Since $\{\gamma(z_0) \mid \gamma \in \Gamma\}$ is discrete, we have

$$\delta = \min_{\gamma \neq 1} \rho(z_0, \gamma(z_0)) > 0 \tag{9.4.33}$$

Let

$$Q = \left\{ w \in \mathbb{D} \;\middle|\; \rho(z_0, w) < \frac{\delta}{2} \right\} \tag{9.4.34}$$

We claim

$$\gamma \neq \gamma' \Rightarrow \gamma[Q] \cap \gamma'[Q] = \emptyset \tag{9.4.35}$$

For if $w \in \gamma[Q] \cap \gamma'[Q]$, then $\rho(w, \gamma(z_0)) < \delta/2$ and $\rho(w, \gamma'(z_0)) < \delta/2$, so $\rho(\gamma(z_0), \gamma'(z_0)) = \rho(z_0, \gamma^{-1}\gamma'(z_0)) < \delta$ violating the definition (9.4.33). Thus, (9.4.35) holds.

Since the $\{\gamma[Q]\}$ are disjoint and lie in $\mathbb{D}$, with vol$(\cdot)$ the Euclidean volume,

$$\sum_{\gamma \in \Gamma} \operatorname{vol}(\gamma[Q]) < \infty \tag{9.4.36}$$

Since γ is conformal and this is two-dimensional volume,

$$\begin{aligned} \operatorname{vol}(\gamma[Q]) &= \int_Q |\gamma'(z)|^2 \, d^2z \\ &\geq \big[\min_{z \in Q} |\gamma'(z)|^2\big] \operatorname{vol}(Q) \\ &\geq C|\gamma'(0)|^2 \end{aligned} \tag{9.4.37}$$

where we use

$$\begin{aligned} \min_{z \in Q} |\gamma'(z)|^2 &\geq \min_{z \in Q} (1 - |\gamma(z)|)^2 && \text{(by (9.4.1))} \\ &\geq \min_{z \in Q} e^{-4\rho(\gamma(z),0)} && \text{(by (9.4.2))} \\ &\geq e^{-4\rho(\gamma(0),0)} A && \text{(by the triangle inequality)} \end{aligned}$$

where

$$A = \exp\big(-4 \max_{z \in Q} \rho(z, 0)\big)$$

Thus,

$$\begin{aligned} \min_{z \in Q} |\gamma'(z_0)|^2 &\geq Ae^{-4\rho(\gamma(0),0)} \\ &\geq \frac{A}{4} (1 - |\gamma(0)|)^2 && \text{(by (9.4.2))} \\ &\geq \frac{A}{16} |\gamma'(0)|^2 && \text{(by (9.4.3))} \end{aligned}$$

verifying (9.4.37).

Clearly, (9.4.37) plus (9.4.36) imply the Poincaré series converges for $s = 2$. □

Theorem 9.4.19 (Burnside [67, 68]). *For any type 2 Fuchsian group, the Poincaré series* (9.4.32) *converges for* $s = 1$.

This will depend on

Lemma 9.4.20. *Let* $z_0 \in \partial\mathbb{D}$ *be an ordinary point for a Fuchsian group,* Γ. *Then there exists* $\delta > 0$ *so that if*

$$I = \{z_0 e^{i\theta} \mid |\theta| \le \delta\} \tag{9.4.38}$$

then

$$\gamma[I] \cap I = \emptyset \tag{9.4.39}$$

for all $\gamma \neq 1$ *in* Γ.

Proof. If not, then there exist $\gamma_n \in \Gamma$ different from 1 and $w_n \in \partial\mathbb{D}$ so that $|w_n - z_0| < 1/n$ and $|\gamma_n(w_n) - z_0| < 1/n$. We first claim that

$$1 - |\gamma_n(0)| \to 0 \tag{9.4.40}$$

for if not, since Γ is discrete, there is a subsequence $\gamma_{n(j)} = \gamma_0$ for some $\gamma_0 \in \Gamma$ and then taking $n \to \infty$, $\gamma_0(z_0) = z_0$, implying z_0 is a limit point (by Proposition 9.4.10). But z_0 is, by hypothesis, not a limit point. Thus, (9.4.40) holds.

By Theorem 9.2.30 and $|\gamma_n(w_n) - w_n| < 2/n$, we see either $w_n \in D_i(\gamma_n)$, the initial circle of γ_n, or w_n is within $2/n$ of the final circle, $D_f(\gamma_n)$, of γ_n.

Thus,

$$\operatorname{dist}(z_0, D_i(\gamma_n) \cup D_f(\gamma_n)) \le \frac{3}{n} \tag{9.4.41}$$

By Theorem 9.2.32, $\gamma_n(0)$ lies in $D_f(\gamma_n)$ and $\gamma_n^{-1}(0)$ lies in $D_i(\gamma_n)$, so if r_n is the radius of $D_i(\gamma_n)$,

$$\operatorname{dist}(z_0, \gamma_n(0)) \le 2r_n + \frac{3}{n} \quad \text{or} \quad \operatorname{dist}(z_0, \gamma_n^{-1}(0)) \le 2r_n + \frac{3}{n}$$

By (9.4.40), $r_n \to 0$ so z_0 is a limit point of Γ. This contradiction proves that (9.4.39) holds for some δ. $\square$

Proof of Theorem 9.4.19. Find I of the form (9.4.14) so that (9.4.39) holds and so that $\tilde{I} = \{z_0 e^{i\theta} \mid |\theta| \le \delta/2\}$ is in the ordinary points. As in the proof of Theorem 9.4.19, (9.4.39) implies $\gamma[\tilde{I}] \cap \gamma'[\tilde{I}] = \emptyset$ for all $\gamma \neq \gamma^{-1}$ in Γ, so if $|\cdot|$ is now a one-dimensional Lebesgue measure on $\partial\mathbb{D}$,

$$\sum_{\gamma \in \Gamma} |\gamma(\tilde{I})| \le 2\pi \tag{9.4.42}$$

so (9.4.31) implies (9.4.32) converges for $s = 1$. $\square$

As a final topic in this section, we want to discuss fundamental domains for Γ.

Definition. Let Γ be a Fuchsian group. A *fundamental domain for* Γ is a closed set $\Omega \subset \mathbb{D}$ so that

$$\text{(i)} \qquad \Omega = \overline{\Omega^{\text{int}}} \tag{9.4.43}$$

$$\text{(ii)} \qquad \gamma[\Omega^{\text{int}}] \cap \Omega^{\text{int}} = \emptyset \text{ for all } \gamma \in \Gamma,\ \gamma \neq 1 \tag{9.4.44}$$

$$\text{(iii)} \qquad \bigcup_{\gamma} \gamma[\Omega] = \mathbb{D} \tag{9.4.45}$$

Remarks. 1. The term "closed" here means in the relative topology on $\mathbb{D}$, not necessarily closed in $\mathbb{C}$.

2. Thus, Ω contains one point from "most" orbits but can contain multiple points from orbits that intersect $\partial\Omega$.

3. In Section 9.6, we will consider "fundamental domains," which are not closed but rather picked so each orbit contains exactly one point in the domain.

Definition. Given a point, $z_0 \in \mathbb{D}$, and Fuchsian group, Γ, the *Dirichlet domain* $D_{z_0}(\Gamma)$ is defined by

$$D_{z_0}(\Gamma) = \{w \in \mathbb{D} \mid \rho(w, z_0) = \inf_{\gamma \in \Gamma} \rho(w, \gamma(z_0))\} \tag{9.4.46}$$

We will also let

$$\mathring{D}_{z_0}(\Gamma) = \{w \in \mathbb{D} \mid \rho(w, z_0) < \rho(w, \gamma(z_0)) \text{ for all } \gamma \in \Gamma,\ \gamma \neq 1\} \tag{9.4.47}$$

We define $\mathcal{I} = \{z_0 \in \mathbb{D} \mid \exists \gamma \neq 1,\ \gamma \in \Gamma,\ \gamma(z_0) = z_0\}$, which we have proven earlier, is always a discrete set.

Most Dirichlet domains are fundamental:

Theorem 9.4.21. *For any $z_0 \in \mathbb{D} \setminus \mathcal{I}$ and $r < 1$, there is a finite set, S, of $\gamma \in \Gamma$ so that*

$$D_{z_0}(\Gamma) \cap \{z \mid |z| \leq r\} = \bigcap_{\gamma \in S} \{w \mid \rho(w, z_0) \leq \rho(w, \gamma(z_0))\} \tag{9.4.48}$$

$\mathring{D}_{z_0}$ is the interior of $D_{z_0}(\Gamma)$ and is dense in $D_{z_0}(\Gamma)$. $D_{z_0}(\Gamma)$ is a fundamental domain.

Proof. Fix z_0 and r. By discreteness, the set S with

$$\bigl\{\gamma \bigm| \min_{|w| \leq r} \rho(\gamma(z_0), w) \leq \max_{|w| \leq r} \rho(z_0, w)\bigr\}$$

is finite, so if $\gamma \notin S$,

$$\{z \mid |z| < r\} \subset \{w \mid \rho(w, z_0) \leq \rho(w, \gamma(z_0))\}$$

and therefore (9.4.48) holds.

The right-hand side of (9.4.48) is a subset of $\mathbb{D}$ bounded by a finite number of arcs from $\{z \mid |z| = r\}$ and arcs from the orthocircles, which, by Theorem 9.3.13, are the set of points equidistant from z_0 and $\gamma(z_0)$ for some $\gamma \in S$, $\gamma \neq 1$, and the interior of $D_{z_0}(\Gamma) \cap \{z \mid |z| \leq r\}$ is the "inside" of this boundary curve. It follows that $\mathring{D}_{z_0}(\Gamma)$ given by (9.4.47) is the interior of $D_{z_0}(\Gamma)$ and is dense in it. Clearly, (9.4.44) holds for $\mathring{D}_{z_0}(\Gamma)$ and (9.4.45) for $D_{z_0}(\Gamma)$. $\square$

Finally, we can describe $D_0(\Gamma)$ in terms of isometric circles:

Theorem 9.4.22. *For $\gamma \in \Gamma$, a Fuchsian group so $0 \notin \mathcal{I}$, let $D_i(\gamma)$ be the (open) initial disk and $D_f(\gamma)$ be the (open) final disk. Then*

$$D_0(\Gamma) = \bigcap_{\substack{\gamma\in\Gamma \\ \gamma\neq 1}} (\mathbb{D} \setminus [D_f(\gamma)]) \tag{9.4.49}$$

$$= \bigcap_{\substack{\gamma\in\Gamma \\ \gamma\neq 1}} (\mathbb{D} \setminus [D_i(\gamma)]) \tag{9.4.50}$$

Remark. Because of (9.4.50), $D_0(\Gamma)$ is sometimes called the *Ford fundamental domain.*

Proof. Since $D_i(\gamma) = D_f(\gamma^{-1})$, (9.4.49) is equivalent to (9.4.50). By Theorem 9.3.15,

$$\mathbb{D} \setminus D_f(\gamma) = \{z \mid \rho(z,0) \leq \rho(z_0, \gamma(0))\}$$ □

Thus, by (9.2.73),

$$D_0(\Gamma) = \bigcap_{\gamma\neq 1} \{z \in \mathbb{D} \mid |\gamma'(z)| < 1\} \tag{9.4.51}$$

Remarks and Historical Notes. For further discussion of Fuchsian groups, see Beardon [36], Ford [138], Katok [216], and Tsuji [446].

Poincaré series first came up in his attempts to construct automorphic functions. If f is any analytic function, $\sum_{\gamma\in\Gamma} f(\gamma(z))$ will be automorphic.

9.5 COVERING MAPS FOR MULTICONNECTED REGIONS

Our main goal in this section is to discuss the following result:

Theorem 9.5.1. *Let $\mathfrak{e}$ be a closed subset of the Riemann sphere $\mathbb{C} \cup \{\infty\}$ so that*

$$\mathcal{S}_+ \equiv (\mathbb{C} \cup \{\infty\}) \setminus \mathfrak{e} \tag{9.5.1}$$

is connected. Suppose that $\mathfrak{e}$ contains at least three points. Then there exists a Fuchsian group, Γ, and function

$$\mathbf{x}\colon \mathbb{D} \to \mathcal{S}_+ \tag{9.5.2}$$

which is locally one-one, that is, $\mathbf{x}'$ is everywhere nonvanishing, and so that

$$\mathbf{x}(z) = \mathbf{x}(w) \Leftrightarrow \exists\, \gamma \in \Gamma \text{ with } \gamma(z) = w \tag{9.5.3}$$

Remark. $\mathbf{x}'$ everywhere nonvanishing at (9.5.3) implies each $\gamma \in \Gamma$ is parabolic or hyperbolic.

We will provide a proof of the special case when

$$\mathfrak{e} \text{ has a component with more than one point} \tag{9.5.4}$$

Of course, since components are connected, such a component is uncountable if not a single point. In our applications, $\mathfrak{e}$ is a finite union of nontrivial closed intervals in $\mathbb{R}$, so (9.5.4) holds. In the Notes, we will discuss the general case. As we also explain there, the conclusion of the theorem fails if $\mathfrak{e}$ has only one or two points.

The proof and interpretation of the theorem depend on the theory of covering spaces, which in turn relies on the theory of fundamental groups. We assume familiarity with the necessary homotopy theory; see the Notes. We will provide a synopsis of the main parts of the covering space theory that we need.

Definition. Let X be an arcwise connected space. A *covering space* is an arcwise connected space, Y, and a map, f (the covering map), $f\colon Y \to X$ so that $\text{Ran}(f) = X$, and for every $x \in X$, there is an open arcwise connected neighborhood, U, of x so that $f^{-1}[U]$ is a union of disjoint arcwise connected sets $\{U_\alpha\}_{\alpha\in A}$ with f a homeomorphism of U_α and U.

(i) It is not hard to see that any continuous curve in X, $\gamma\colon [0,1] \to X$ can be lifted to Y, that is, for any $y_0 \in f^{-1}(\gamma(0))$, there is $\tilde{\gamma}\colon [0,1] \to Y$ so $f \circ \tilde{\gamma} = \gamma$ and $\tilde{\gamma}(0) = y_0$. This lift is unique. Similarly, any homotopy in X can be lifted to Y.

(ii) Pick base points, x_0 and y_0, in X and Y with $f(y_0) = x_0$. Any closed loop, γ, in Y with $\gamma(0) = \gamma(1) = y_0$ is mapped into one, $f \circ \gamma$ in X. The fact that homotopies lift shows that on the level of equivalence classes, this map is injective, that is, $f_*\colon \pi_1(Y, y_0) \to \pi_1(X, x_0)$ maps one-one to an image subgroup $G_Y \equiv f_*[\pi_1(Y, y_0)]$. Let $y_1 \in f^{-1}[\{x_0\}] \equiv F_Y$, the fiber over x_0, and let γ be a curve with $\gamma(0) = y_0$, $\gamma(1) = y_1$. Then $f \circ \gamma$ is a closed loop in X and the lifting of homotopies shows it is nontrivial in $\pi_1(X, x_0)$ if $y_1 \neq y_0$. Indeed, one shows this association of points in F_Y into classes of elements of $\pi_1(X, x_0)$ is a bijection of F_Y and left cosets of G_Y, that is, to $\pi_1(X, x_0)/G_Y$.

(iii) Two covers $Y_1 \overset{f_1}{\longrightarrow} X$, $Y_2 \overset{f_2}{\longrightarrow} X$ are called isomorphic if there is a homeomorphism, $Q\colon Y_1 \to Y_2$ so $f_2 \circ Q = f_1$. The analysis in (ii) shows this happens if and only if $G_{Y_1} = G_{Y_2}$ and that, then, Q is uniquely determined (to make this precise, one needs to speak of spaces with distinguished points, so $y_1 \in Y_1$, $y_2 \in Y_2$, $f_j(y_j) = x_0$, and $Q(y_1) = y_2$). Moreover, if any $x \in X$ has a simply connected neighborhood, then every subgroup, G, of $\pi_1(X, x_0)$ enters as some G_Y. Thus, in that case, there is a one-one correspondence between subgroups of $\pi_1(X, x_0)$ and equivalence classes of covering maps.

(iv) In particular, if we demand $G_Y = \{1\}$, so $\pi_1(Y, y_0) = \{1\}$, we get a distinguished cover called the *universal covering space*, which is equivalent to any simply connected cover, that is, cover with $\pi_1(Y, y_0) = \{1\}$.

(v) Each element $[\gamma]$ of $\pi_1(X, x_0)$ induces a map $\tau_{[\gamma]}$ on Y, called the *deck transformation*, that obeys

$$f \circ \tau_{[\gamma]} = f \tag{9.5.5}$$

determined by also requiring if γ is a loop in X, with $\gamma(0) = \gamma(1) = x_0$, its lift $\tilde{\gamma}$ with $\tilde{\gamma}(0) = y_0$ has $\gamma(1) = \tau_\gamma(y_0)$. τ_γ is the identity map if and only if $[\gamma] \in G_Y$

and any other $\tau_{[\gamma]}$ leaves no points fixed. Thus, $\pi_1(X, x_0)/G_Y$ acts simply on Y and orbits $\{\tau_{[\gamma]}(y) \mid \gamma \in \pi_1(X, x_0)\}$ are all of $f^{-1}[\{f(y)\}]$. In particular, if Y is the universal cover, $\pi_1(X, x_0)$ acts transitively on each $f^{-1}[\{f(y)\}]$ and

$$f(y_1) = f(y_0) \Leftrightarrow \exists\, [\gamma] \in \pi_1(X, x_0) \text{ s.t. } \tau_{[\gamma]}(y_1) = y_0 \tag{9.5.6}$$

(vi) If X is a connected Riemann surface, that is, a one-dimensional complex manifold with a distinguished set of charts whose transition functions are all analytic, then the fact that covering maps are local homeomorphisms allows one to make any cover $Y \xrightarrow{f} X$ into a Riemann surface in such a way that f is analytic. It is then easy to see that the deck transformations are bianalytic homeomorphisms of Y.

(vii) By combining the uniqueness of universal covering spaces with the local analytic structures, one sees that if $f, g\colon \mathbb{D} \to X$ are both analytic covering maps, then there is a Möbius transformation, h, with

$$f \circ h = g \tag{9.5.7}$$

for uniqueness implies there is a homeomorphism h, and the fact that f locally has an analytic inverse implies h is analytic locally, and so analytic globally.

The relevance of this to Theorem 9.5.1 is now clear. The fact that Γ is a Fuchsian group, where each $\gamma \in \Gamma$ ($\gamma \neq e$) has no fixed points, lets one find for any $z_0 \in \mathbb{D}$, a disk, D, about z_0 so $\{\gamma(D)\}_{\gamma \in \Gamma}$ are disjoint, which implies that x is a covering map. Since $\mathbb{D}$ is simply connected, it is the universal cover. On the other hand, if the universal cover is $\mathbb{D}$, the covering map can be taken as x and the family of deck transformations as Γ. Thus, Theorem 9.5.1 is equivalent to the statement that as Riemann surfaces, the cover of $\mathcal{S}_+$ is $\mathbb{D}$. One proof of the theorem (discussed in the Notes) relies on the fact (due to Poincaré) that the only simply connected Riemann surfaces are $\mathbb{D}$, $\mathbb{C}$, and the Riemann sphere.

Instead, we use a proof going back to Radó [357], which is based on the usual proof of the Riemann mapping theorem. We begin by describing that proof not merely because we will use the Riemann mapping theorem in our proof but because most of the steps in the proof of the Riemann mapping theorem are identical to steps in the proof we will give of Theorem 9.5.1. One downside is that, because it relies on a compactness argument, our proof is not constructive.

Consider three properties of a connected Riemann surface:

(i) The surface is *topologically simply connected* in the sense that any closed curve (with base point) is homotopic to the trivial curve.
(ii) Contour integrals around closed contours of functions analytic on the surface are zero; we call such surfaces *holomorphically simply connected.*
(iii) When the surface is a subset of $\mathbb{C} \cup \{\infty\}$, its complement is connected.

It is fairly easy to see that (ii) $\Leftrightarrow$ (iii) (see the reference in the Notes) and that (i) $\Rightarrow$ (ii).

Theorem 9.5.2 (Riemann Mapping Theorem)**.** *Let $\Omega \subset \mathbb{C} \cup \{\infty\}$ be a connected open region so that $\mathbb{C} \cup \{\infty\} \setminus \Omega$ has at least two points and so that Ω is analytically simply connected. Then there is an analytic bijection $h\colon \mathbb{D} \to \Omega$.*

This theorem then implies (ii) $\Rightarrow$ (i). We can suppose without loss, by a preliminary fractional linear transformation, that $\infty \notin \Omega$ and thus, that $\Omega \subsetneq \mathbb{C}$. Instead of constructing h, we construct its inverse, so pick $z_0 \in \Omega$ finite and define

$$\mathcal{R} = \{f\colon \Omega \to \mathbb{D} \mid f(z_0) = 0,\ f'(z_0) > 0,\ f(z) = f(w) \Rightarrow z = w\} \tag{9.5.8}$$

We will prove:

(a) $\mathcal{R}$ is nonempty.

(b) $\mathcal{R} \cup \{f \equiv 0\}$ is compact.

(c) Given $f \in \mathcal{R}$, if $\operatorname{Ran}(f) \neq \mathbb{D}$, then there exist $g \in \mathcal{R}$ and $\varphi\colon \mathbb{D} \to \mathbb{D}$, $\varphi(z) \not\equiv z$ so that

$$f = \varphi \circ g \tag{9.5.9}$$

(d) These imply Theorem 9.5.2.

Simple connectedness comes in via:

Lemma 9.5.3. *Let f be an analytic function on a holomorphically simply connected region or simply connected and connected Riemann surface, Ω, which is everywhere nonvanishing. Then there is an analytic function, g, on Ω with $g^2 = f$.*

Remark. There are exactly two such g's. We will write them as $\pm\sqrt{f}$.

Proof. Pick $z_0 \in \Omega$ and α_0 so $\alpha_0^2 = f(z_0)$. Define

$$g(z) = \alpha_0 \exp\biggl(\frac{1}{2} \oint_{z_0}^{z} \frac{f'(w)}{f(w)}\, dw\biggr) \tag{9.5.10}$$

Since f is nonvanishing, f'/f is analytic on Ω and, by the holomorphic simple connectivity, the integral is a single-valued analytic function equal to (a branch of) $\log[f(z)/f(z_0)]$. □

We can now do step (a).

Lemma 9.5.4. *$\mathcal{R}$ is nonempty if $\Omega \subsetneq \mathbb{C}$.*

Proof. Pick $w_0 \notin \Omega$. Thus, $f(z) = z - w_0$ is nonvanishing and, by Lemma 9.5.3, we can find an analytic function, $g(z)$, so $g(z)^2 = f(z)$. If $g(z_1) = \pm g(z_2)$, then $z_1 - w_0 = z_2 - w_0$, so $z_1 = z_2$. Thus, if $w_1 = g(z_1)$, then $-w_1 \notin \operatorname{Ran}(g)$. Since g is analytic and nonconstant, $\operatorname{Ran}(g)$ is open, so for some δ, $\{w \mid |w - w_1| < \delta\} \subset \operatorname{Ran}(g)$. By the above, $\{-w \mid |w - w_1| < \delta\} \cap \operatorname{Ran}(g) = \emptyset$, that is, on Ω,

$$\left|\frac{1}{g(z) + w_1}\right| \le \frac{1}{\delta}$$

so $h(z) \equiv \delta/(g(z) + w_1)$ maps Ω to $\mathbb{D}$. By composing h with a suitable Möbius transformation, we get F taking Ω to $\mathbb{D}$ with $F(z_0) = 0$ and $F'(z_0) > 0$. Since h is one-one, so is F. □

Next we do step (c).

Lemma 9.5.5. *Let $f \in \mathcal{R}$ and suppose there exists $w_0 \in \mathbb{D} \setminus \mathrm{Ran}(f)$. Then there exists $g \in \mathcal{R}$ and $\varphi\colon \mathbb{D} \to \mathbb{D}$ so $\varphi(z) \not\equiv z$, and so* (9.5.9) *holds. In particular (with strict inequality),*

$$g'(z_0) > f'(z_0) \tag{9.5.11}$$

Proof. Let

$$T_1(z) = \frac{z - w_0}{1 - \bar{w}_0 z} \tag{9.5.12}$$

so $T_1 \circ f$ is nonvanishing, and thus we can pick a branch of

$$H(z) \equiv \sqrt{T_1(f(z))} \tag{9.5.13}$$

which maps to $\mathbb{D}$ since $w \in \mathbb{D} \Rightarrow \pm\sqrt{w} \in \mathbb{D}$. Let

$$T_2(z) = \frac{|H'(z_0)|}{H'(z_0)} \frac{z - H(z_0)}{1 - \overline{H(z_0)}z} \tag{9.5.14}$$

(note that H is one-one, so $H'(z_0) \neq 0$) and let

$$g(z) = (T_2 \circ H)(z) \tag{9.5.15}$$

Define

$$\varphi(z) = T_1^{-1}((T_2^{-1}(z))^2) \tag{9.5.16}$$

so $\varphi\colon \mathbb{D} \to \mathbb{D}$ and $\varphi'(T_2(0)) = 0$ so $\varphi(z) \not\equiv z$.

By construction, (9.5.9) holds, so by the chain rule,

$$f'(z_0) = \varphi'(0)g'(z_0) \tag{9.5.17}$$

By (9.5.15),

$$g'(z_0) = T_2'(H(z_0))H'(z_0) = \frac{|H'(z_0)|}{1 - |H(z_0)|^2} \tag{9.5.18}$$

so

$$\varphi'(0) > 0 \tag{9.5.19}$$

Since φ is analytic in a neighborhood of $\mathbb{D}$ and maps $\overline{\mathbb{D}}$ to $\overline{\mathbb{D}}$,

$$\varphi'(0) = \int_0^{2\pi} e^{-i\theta} \varphi(e^{i\theta}) \frac{d\theta}{2\pi} \tag{9.5.20}$$

has $|\varphi'(0)| \leq 1$ with equality only if $\varphi(z_0) = cz$ for $|c| = 1$, inconsistent with $\varphi'(T_2(0)) = 0$. Thus,

$$\varphi'(0) < 1 \tag{9.5.21}$$

so, by (9.5.17), we have (9.5.11). □

Remarks. 1. The square root is used in constructing g to be sure that the inverse is two-to-one so $\varphi'(0) < 1$.

2. Rather than rely on a theoretical proof that (9.5.11) holds, one can do a direct calculation to find that

$$g'(z_0) = \frac{1 + |w_0|}{2\sqrt{|w_0|}} f'(z_0) \tag{9.5.22}$$

We prefer the indirect argument rather than rely on a calculation that "happens to work."

Proof of Theorem 9.5.2. Let f_n be a sequence of functions in $\mathcal{R}$ that converge to some function, f, uniformly on compact subsets of Ω. Then for each fixed $w \in \Omega$, $f_n(z) - f_n(w)$ has a single zero at $z = w$, so by Hurwitz's theorem, either f has the same property or else $f(z) \equiv f(w)$, so $f \equiv 0$ (since $f(z_0) = 0$). Thus, either $f \equiv 0$ or $f \in \mathcal{R}$. It follows that $\mathcal{R} \cup \{0\}$ is closed in this topology. By Montel's theorem, $\mathcal{R}$ is compact in this topology of uniform convergence on compact subsets since

$$\mathcal{R} \subset \{f \mid \|f\|_\infty \leq 1\}$$

By compactness and by continuity of $f \mapsto f'(z_0)$, we can find $f_0 \in \mathcal{R}$ with

$$f_0'(z_0) = \sup_{f \in \mathcal{R} \cup \{0\}} \{f'(z_0)\} \tag{9.5.23}$$

Since $\mathcal{R}$ is not empty, $f_0'(z_0) > 0$ and $f_0 \in \mathcal{R}$. If f_0 is not onto $\mathbb{D}$, we can, by Lemma 9.5.5, find $g \in \mathcal{R}$ so $g'(z_0) > f_0(z_0)$, violating (9.5.23). Thus, f_0 is a bijection and $h = f_0^{-1}$ provides the required map of $\mathbb{D}$ to Ω. □

Proof of Theorem 9.5.1 when (9.5.4) *holds.* Let $\pi : \mathcal{U} \to \mathcal{S}_+$ be the universal covering space of $\mathcal{S}_+$. As before, suppose $\infty \notin \mathcal{S}_+$ and pick some $z_0 \in \mathcal{U}$. We will define

$$\mathcal{R} = \{f : \mathcal{U} \to \mathbb{D} \mid f(z_0) = 0,\ f'(z_0) > 0,\ f(z) = f(w) \Rightarrow \pi(z) = \pi(w)\}$$

Step (c) in the earlier strategy holds without any change—the argument that proves Lemma 9.5.5 only needed Ω simply connected and $\mathcal{U}$ is simply connected. Step (b) is also essentially unchanged: Montel's and Hurwitz's theorems remain true on $\mathcal{U}$, and if $f_n \in \mathcal{R}$ and w is fixed, the zeros of $f_n(z) - f_n(w)$ are contained in $\pi^{-1}[\{\pi(w)\}]$.

That leaves step (a). In the general case, this requires an argument exploiting the elliptic modular function; see the Notes. Given our assumption (9.5.4), it is easy. Let $\mathfrak{e}_1$ be the assumed component with more than one point. Then $\mathbb{C} \cup \{\infty\} \setminus \mathfrak{e}_1 \supset \mathcal{S}_+$ and is simply connected. So, by the Riemann mapping theorem (indeed, by part (a) of its proof!), there is a one-one $f_0 : \mathbb{C} \cup \{\infty\} \setminus \mathfrak{e}_1 \to \mathbb{D}$ with $f_0(z_0) = 0$, $f_0'(z_0) > 0$. Let $f = f_0 \circ \pi$, so $f \in \mathcal{R}$.

Following the proof of Theorem 9.5.2, we see that there exists $f : \mathcal{U} \to \mathbb{D}$, which is onto and in $\mathcal{R}$. For all $z \in \mathbb{D}$, π is constant on $f^{-1}[\{z_0\}]$, so we can define $\mathbf{x}$ to be this common value.

By construction, f is locally one-one and π is locally one-one, so $\mathbf{x}$ is locally one-one. For given $w \in \mathbb{D}$, pick z in $\mathcal{U}$ with $g(z) = w$ and a neighborhood U of z and which g and π are one-one. So on $g[U]$, $f = g^{-1} \circ \pi$ is one-one.

Given z_0 in $\mathcal{S}_+$, let U be a connected open neighborhood of z_0 so $\pi^{-1}(U)$ is a collection of connected open sets, $\{U_\alpha\}_{\alpha\in A}$, which are disjoint in $\mathcal{U}$ so that π is a homeomorphism on each U_α to U. Thus, if $\alpha, \beta \in A$, there is a unique homeomorphism $\pi_{\alpha\beta}\colon U_\alpha \to U_\beta$ to $\pi \circ \pi_{\alpha\beta} = \pi$. We claim $\{z \in U_\alpha \mid f(\pi_{\alpha\beta}(z)) = f(z)\}$ is both open and closed. By continuity of f and π, it is obviously closed. On the other hand, suppose that $f(\pi_{\alpha\beta}(z)) = f(z)$ and that $z_n \to z$ has $f(\pi_{\alpha\beta}(z_n)) \neq f(z_n)$. Then since x is locally one-one, $x(f(\pi_{\alpha\beta}(z_n))) \neq x(f(z_n))$ for n large. But $x(f(z_n)) = \pi(z_n)$ and $\pi(\pi_{\alpha\beta}(z_n)) = \pi(z_n)$, so we have a contraction.

Thus, for each α, β, either $f \circ \pi_{\alpha\beta} = f$ on U_β or $f[U_\alpha]$ and $f[U_\beta]$ are disjoint. This shows that x is a covering map. Thus, the fundamental group of $\mathcal{S}_+$ acts as a Fuchsian group and (9.5.3) holds. □

Remarks and Historical Notes. The Riemann mapping theorem appeared in 1851 in Riemann's inaugural dissertation [368], but its proof depended on ideas (which he called Dirichlet's principle) that at the time were not rigorous and even now rely on regularity of the boundary. The first general proof was found by Osgood [330] in 1900 (see Walsh [456] for Osgood's proof in modern language). Osgood was isolated in the U.S. and his proof not widely noted—the now standard proof, which we give here, is based in part on ideas of Carathéodory [72] and Koebe [237, 238] in 1912–1915 and first appeared in full in the paper of Radó quoted later.

The uniformization theorem, sometimes called the Poincaré or Klein–Poincaré theorem, states that every simply connected Riemann surface is analytically equivalent to one of three standard models: the Riemann sphere, $\mathbb{C}$, or $\mathbb{D}$. It is due to Poincaré [348] based in part on results of Klein [230], with important clarifications by Koebe [233, 234, 235, 236].

As we have discussed, the fundamental group acts on the universal cover of a Riemann surface as a group of analytic isomorphisms with no fixed points. The Riemann sphere has no analytic isomorphisms with no fixed points, so it is not the universal cover of any surface but itself. The only analytic isomorphisms of $\mathbb{C}$ with no fixed points are of the form $z \to z + a$ for some a in $\mathbb{C}$. The only discrete subgroups are isomorphic to $\mathbb{Z}$ or to $\mathbb{Z}^2$. The quotient by $\mathbb{Z}^2$ is a torus, and by $\mathbb{Z}$ a cylinder, which is the same as the once-punctured plane. All other Riemann surfaces have $\mathbb{D}$ as universal cover, providing one proof of Theorem 9.5.1. This also shows that Theorem 9.5.1 fails if $\mathfrak{e}$ has only one or two points.

The idea of using the standard proof of the Riemann mapping theorem that we use to prove Theorem 9.5.1 is from Radó [357] (see also [179]) who says he used in part ideas of Fejér and F. Riesz.

To get the full version of Theorem 9.5.1, one needs to use the elliptic modular function, $\lambda(\tau)$, defined on the upper half-plane, $\mathbb{C}_+$ (see, e.g., Ahlfors [7] for the definition and proof of properties). Let Γ be the group of fractional linear transformations induced by the elements of $\mathbb{SL}(2, \mathbb{C})$, $\left(\begin{smallmatrix} a & b \\ c & d \end{smallmatrix}\right)$ where a, d are odd integers and b, c are even integers. Then

$$\lambda(\tau) = \lambda(\tau') \Leftrightarrow \exists \gamma \in \Gamma \text{ s.t. } \gamma(\tau) = \tau' \tag{9.5.24}$$

and $\mathrm{Ran}(\lambda) = \mathbb{C} \setminus \{0, 1\}$. Since $\mathbb{C}_+$ is simply connected (and analytically isomorphic to $\mathbb{D}$), this provides an explicit model where the set $\mathfrak{e}$ in (9.5.1) is

$\{0, 1, \infty\} = \mathfrak{e}_0$. One can also construct λ by using the Riemann mapping theorem on the region $\Omega = \{z \mid |z| > 1, -1 < \text{Re}\, z < 1, \text{Im}\, z > 0\}$ and reflection symmetry; see [448].

For general $\mathfrak{e}$ with at least three points, by a fractional linear transformation, we can suppose $\mathfrak{e}_0 \subset \mathfrak{e}$. For any $z_0 \in \mathbb{C} \setminus \mathfrak{e}$, find $w_0 \in \mathbb{C}_+$ with $\lambda(w_0) = z_0$ and let f be a local inverse of λ, defined originally near z_0 (by (9.5.24) and the fact that nonidentity elements in Γ have no fixed points, λ' is everywhere nonvanishing). Using (9.5.24), it is not hard to see that f can be continued along any curve in $\mathcal{S}_+$, although it will be a multivalued function on $\mathcal{S}_+$. On $\mathcal{U}$, it defines a single-valued function by the monodromy theorem (see Ahlfors [7]).

By construction, $\lambda \circ f(z) = \pi(z)$ so f on $\mathcal{U}$ obeys

$$f(w) = f(z) \Rightarrow \pi(z) = \pi(w) \tag{9.5.25}$$

By composing f with a suitable fractional linear transformation mapping $\mathbb{C}_+$ to $\mathbb{D}$, we get an element in $\mathcal{R}$. The rest of the proof is then unchanged.

For books on basic topology—fundamental group and covering spaces—see [24, 143, 311, 451]. For background in complex analysis (such as Montel's and Hurwitz's theorems), see Ahlfors [7], Stein–Shakarchi [419], and Lang [266].

It is interesting to see how starting with $f\colon \mathcal{S}_+ \to \mathbb{D}$ (but not onto) constant on the fibers $\pi^{-1}[\{z_0\}]$, we get an f, which is a bijection of $\mathcal{U}$ and $\mathbb{D}$. In step (c), when we take the square roots, we essentially halve the set of points where f has a given value.

9.6 THE FUCHSIAN GROUP OF A FINITE GAP SET

We specialize to $\mathfrak{e}$, a finite gap set of the form (5.12.1). We normalize the covering map $\mathbf{x}\colon \mathbb{D} \to \mathbb{C} \cup \{\infty\} \setminus \mathfrak{e} \equiv \mathcal{S}_+$ by requiring

$$\mathbf{x}(0) = \infty \qquad \lim_{\substack{z \to 0 \\ z \neq 0}} z\mathbf{x}(z) > 0 \tag{9.6.1}$$

By Theorem 9.5.1, there is a unique such map and an associated Fuchsian group, Γ, which is isomorphic to $\pi_1(\mathcal{S}_+)$, and so a free nonabelian group on ℓ generators. Since any $\gamma \in \Gamma$ acts freely on $\mathbb{D}$, $0 \notin \mathcal{I} = \emptyset$, so there is an associated Ford fundamental domain, $D_0(\Gamma)$. Our goal in this section is to study the group, Γ, and the fundamental domain, $D_0(\Gamma)$. In particular, we will prove a theorem critical to step-by-step sum rules that the Poincaré critical exponent is strictly smaller than 1.

We will begin by analyzing a fundamental domain, $\mathcal{F}$, which will turn out to be essentially $D_0(\Gamma)$ (more precisely, $\mathcal{F}^{\text{int}}$ will be $\overset{\circ}{D}_0(\Gamma)$). Consider in $\mathcal{S}_+$, $\mathcal{P} \equiv \mathbb{C} \cup \{\infty\} \setminus [\alpha_1, \beta_{\ell+1}]$, that is, we remove the ℓ gaps, $\cup_{j=1}^{\ell} (\beta_j, \alpha_{j+1})$, from $\mathcal{S}_+$. $\mathcal{P}$ is connected and simply connected. For any $z_0 \in \mathcal{P}$, all curves $\gamma\colon [0, 1] \to \mathcal{P}$ with $\gamma(0) = \infty$ and $\gamma(1) = z_0$ are homotopic, so the lift to the universal cover, $\tilde{\gamma}$, with $\tilde{\gamma}(0) = 0 \in \mathbb{D}$ has $\tilde{\gamma}(1)$, the same for all such γ's. This allows us to define a unique branch of $\mathbf{x}^{-1}$ on $\mathcal{P}$ whose range is connected and contains 0. The image of this branch we will call $\mathcal{F}^{\text{int}}$ (for now, ^{int} is a symbol; later it will be the interior of $\mathcal{F}$). Thus, $\mathcal{F}^{\text{int}}$ is a connected open subset of $\mathbb{D}$ for which $\mathbf{x}$ is a bijection of $\mathcal{F}^{\text{int}}$ and $\mathcal{P}$.

Consider first what $\mathbf{x}$ does to $(-1, 1)$. Since $\mathcal{S}_+$ is invariant under complex conjugation, $\overline{\mathbf{x}(\bar{z})}$ is also a locally bijective map of $\mathbb{D}$ to $\mathcal{S}_+$, which clearly obeys (9.6.1). Thus, by uniqueness, $\mathbf{x}$ must obey

$$\overline{\mathbf{x}(\bar{z})} = \mathbf{x}(z) \tag{9.6.2}$$

Thus, $\mathbf{x}$, and so $\mathbf{x}'$, is real on $(-1, 1) \setminus \{\infty\}$. By (9.6.1), $\mathbf{x}'(w) < 0$ for w real and near zero, so since $\mathbf{x}'$ is never zero or ∞, we see that

$$\mathbf{x}'(w) < 0 \qquad \text{if } w \in (0, 1) \cup (-1, 0) \tag{9.6.3}$$

Thus, $\mathbf{x}$ maps $(0, 1)$ to a part of $(\beta_{\ell+1}, \infty)$ in a monotone decreasing way, so

$$\lim_{w \downarrow 0} \mathbf{x}(w) = \infty \tag{9.6.4}$$

We claim that

$$\lim_{w \uparrow 1} \mathbf{x}(w) = \beta_{\ell+1} \tag{9.6.5}$$

for if the limit (which exists by monotonicity) were some $y > \beta_{\ell+1}$, we would be unable to lift the curve in $\mathcal{P}$ that runs from ∞ to y.

Thus, one inverse image of the curve in $\mathcal{S}_+$ that runs from $\beta_{\ell+1}$ up to ∞ and then from $-\infty$ up to α_1 is exactly $(-1, 1)$ (run from 1 to -1). By the action of the Fuchsian group, the other inverse images are images of $(-1, 1)$ under Möbius transformations, and so a set of orthocircles.

Pick some point z_0 in the gap (β_1, α_2). There is a covering map $\tilde{\mathbf{x}}\colon \mathbb{D} \to \mathcal{S}_+$ with $\tilde{\mathbf{x}}(0) = z_0$ and $\tilde{\mathbf{x}}'(0) < 0$. As above, this map must take $(-1, 1)$ onto (β_1, α_2) and all inverse images of (β_1, α_2) are orthocircles. But $\mathbf{x}$ and $\tilde{\mathbf{x}}$ are related by a Möbius transformation by Remark (vii) in the last section. Thus, under $\mathbf{x}^{-1}$ also, all images of (β_1, α_2) are orthocircles. We have thus proven:

Proposition 9.6.1. *The inverse images under* $\mathbf{x}$ *of any gap* (β_j, α_{j+1}), $j = 1, \dots, \ell$, *or of* $(\beta_{\ell+1}, \infty) \cup \{\infty\} \cup (\infty, \alpha_1)$ *are a family of orthocircles.*

Note that since $\mathbf{x}'(z) < 0$ for $w \in (-1, 1)$, near $(-1, 1)$, $\mathbf{x}(z)$ reverses the sign of $\operatorname{Im} z$. By continuity, one sees that $\mathbf{x}^{-1}$ maps $\mathcal{P} \cap \mathbb{C}_+$ onto $\mathbb{D} \cap \mathbb{C}_-$ and $\mathcal{P} \cap \mathbb{C}_-$ onto $\mathbb{D} \cap \mathbb{C}_+$.

Consider now what happens as $z \in \mathcal{P} \cap \mathbb{C}_-$ approaches a gap. Since $\mathbf{x}$ is a covering map, $\mathbf{x}^{-1}$ has a limit, which lies in an inverse image of a gap—thus, in an orthocircle that lies entirely in $\mathbb{D} \cap \mathbb{C}_+$. By (9.6.2), as we approach the gap from the other side, $\mathbf{x}^{-1}$ goes to the conjugate orthocircle. The boundary of $\mathcal{F}^{\text{int}}$ is thus 2ℓ orthocircles. Since there are bands between gaps, these orthocircles are a finite distance apart. We thus have shown that

Proposition 9.6.2. *In* $\mathbb{D}$, *the topological boundary of* $\mathcal{F}^{\text{int}}$ *consists of* ℓ *orthocircles in* $\mathbb{C}_+$ *and their complex conjugates. There is a finite distance in* $\overline{\mathbb{D}}$ *between the ends of distinct orthocircles.*

We will use $C_1^+, \dots, C_\ell^+$ to denote the orthocircles in $\mathbb{C}_+ \cap \mathbb{D}$, labeled going clockwise. We let $C_j^- = \overline{C_j^+}$ be their conjugates. $C_j^\pm$ are arcs of full circles, which

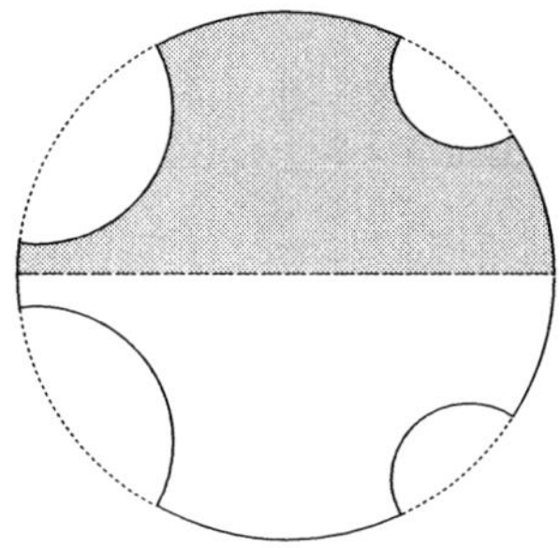

Figure 9.6.1. The fundamental region.

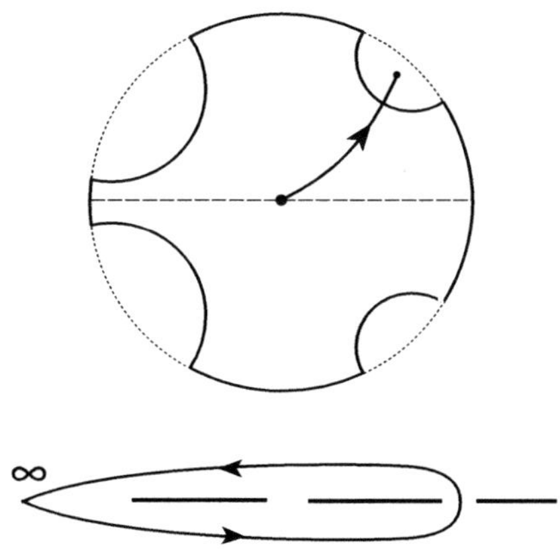

Figure 9.6.2. Fuchsian group generators.

we denote by $\widetilde{C}_j^\pm$ and call *complete orthocircles*. Thus, $C_j^\pm = \widetilde{C}_j^\pm \cap \mathbb{D}$. Notice also that $\gamma(\widetilde{C}_j^\pm) = \widetilde{\gamma(C_j^\pm)}$.

Thus, there are 2ℓ orthocircles and their interiors removed to get $\mathcal{F}^{\text{int}}$. Figure 9.6.1 shows the way this looks for a case with $\ell = 2$. The shaded region is the inverse image of $\mathcal{P} \cap \mathbb{C}_-$.

Consider now a curve in $\mathcal{S}_+$ as shown in the lower half of Figure 9.6.2 starting at ∞, going in $\mathbb{C}_-$ to a gap, crossing the gap, and returning to ∞ in $\mathbb{C}_+$. The lift leaves $\mathcal{P}$ when the gap is crossed. The lift is thus shown in the upper half of Figure 9.6.2.

If we had used $\tilde{\mathbf{x}}$, a cover map taking $(-1, 1)$ to the crossed gap, the two halves would be complex conjugate, so in Figure 9.6.2, the two pieces of lift curve are images under inversion in the orthocircle corresponding to the gap. In particular, the other endpoint is just the image of 0 under this inversion.

The same argument shows that for any point on $(-1, 1)$, the image under the deck transformation associated to this curve is inversion in the circle. Let γ be the deck transformation and r^+ reflection in the circle. Then $\gamma^{-1}r^+$ is a conjugate linear extended FLT, which leaves $(-1, 1)$ fixed. It must be complex conjugation

$$c(z) = \bar{z} \tag{9.6.6}$$

Thus, $\gamma^{-1}r^+ = c$, so $r^+\gamma = c$ or $\gamma = r^+c$ (we used here $(r^+)^2 = (c)^2 = 1$). We have thus proven:

Theorem 9.6.3. *Let r_j^+ be the inversions in C_j^+ for $j = 1, \dots, \ell$. Let c be given by* (9.6.6). *Let*

$$\gamma_j = r_j^+ c \tag{9.6.7}$$

Then Γ is the free nonabelian group generated by $\{\gamma_1, \dots, \gamma_\ell\}$.

If r_j^- is reflection in C_j^-, then $cr_j^+c = r_j^-$, so by $(r_j^+)^2 = c^2 = 1$, we see

$$\gamma_j^{-1} = r_j^- c \tag{9.6.8}$$

We can now define $\mathcal{F}$ by

$$\mathcal{F} = \mathcal{F}^{\text{int}} \cup \bigcup_{j=1}^{\ell} C_j^+ \tag{9.6.9}$$

Thus, $\mathcal{F}$ is a strict fundamental domain in the sense that it contains one point from each orbit $\{\gamma(z)\}_{\gamma\in\Gamma}$. Its interior is indeed $\mathcal{F}^{\text{int}}$. We will use $\overline{\mathcal{F}}$ in two different ways: sometimes the closure in $\mathbb{D}$, that is,

$$\overline{\mathcal{F}} = \mathcal{F}^{\text{int}} \cup \bigcup_{j=1}^{\ell} (C_j^+ \cup C_j^-) \tag{9.6.10}$$

and sometimes the closure in $\overline{\mathbb{D}}$, including some boundary points in $\partial\mathbb{D}$.

We will return to $\mathcal{F}$ and Γ shortly, but first we want to use $\mathcal{F}$ to extend $\mathbf{x}$ beyond $\mathbb{D}$. Let z_n lie in $\mathcal{F}$ with $|z_n| \to 1$. $\mathbf{x}(z_n)$ lies in the Riemann sphere, which is compact, so without loss, we can pass to a subsequence so that $\mathbf{x}(z_n)$ has a limit, x_∞. Suppose $x_\infty \in \mathcal{S}_+$. There is then $z_\infty \in \mathcal{F}$ so that $\mathbf{x}(z_\infty) = x_\infty$. But $\mathbf{x}$ is one-one on $\mathcal{F}$ so all nearby points for $\mathbf{x}(z)$ have z near z_∞, that is, $|z_n| \to |z_\infty| < 1$. It follows that $x_\infty \in \mathfrak{e}$ and, in particular, is real. In particular, since all limit points are real, we see that $\operatorname{Im} \mathbf{x}(z) \to 0$ as $|z| \to 1$ with $z \in \mathcal{F}$.

It follows from the strong form of the reflection principle (see Ahlfors [7, Theorem 4.24]) that if we define $\mathbf{x}$ on $\mathbb{C} \setminus \overline{\mathbb{D}}$ with values in $\mathbb{C} \cup \{\infty\}$ by

$$\mathbf{x}(z) = \overline{\mathbf{x}(1/\bar{z})} \tag{9.6.11}$$

then $\mathbf{x}$ can be continued across $\partial\overline{\mathcal{F}} \cap \partial\mathbb{D}$ (where here $\overline{\mathcal{F}}$ means in $\overline{\mathbb{D}}$). Combining (9.6.2) and (9.6.11), we get

$$\mathbf{x}(1/z) = \mathbf{x}(z) \tag{9.6.12}$$

Given that we have continued outside $\mathbb{D}$, it will be useful to define extended versions of $\mathcal{F}^{\text{int}}$ and $\mathcal{F}$. By $\widetilde{\mathcal{F}}^{\text{int}}$, we mean the union of $\mathcal{F}^{\text{int}}$, $\{z \mid \bar{z}^{-1} \in \mathcal{F}^{\text{int}}\}$, and the interior in $\partial\mathbb{D}$ of $\partial\mathbb{D} \cap \overline{\mathcal{F}}$. By $\widetilde{\mathcal{F}}$, we mean, following (9.6.9),

$$\widetilde{\mathcal{F}} = \widetilde{\mathcal{F}}^{\text{int}} \cup \bigcup_{j=1}^{\ell} \widetilde{C}_j^+ \tag{9.6.13}$$

$\widetilde{\mathcal{F}}^{\rm int}$ is, indeed, the interior of $\widetilde{\mathcal{F}}$. Moreover, for any distinct $\gamma, \gamma' \in \Gamma$, $\gamma[\mathcal{F}] \cap \gamma'[\mathcal{F}] = \emptyset$ and

$$\bigcup_{\gamma\in\Gamma} \gamma[\widetilde{\mathcal{F}}] = \mathbb{C} \cup \{\infty\} \setminus \Lambda(\Gamma) \tag{9.6.14}$$

We claim that

Theorem 9.6.4. *Let $\Lambda(\Gamma)$ be the set of limit points for Γ. Then* $\mathbf{x}$*, defined by* (9.6.11)*, can be defined on $\partial\mathbb{D} \setminus \Lambda(\Gamma)$ so that* $\mathbf{x}$ *is analytic from $\mathbb{C} \cup \{\infty\} \setminus \Lambda(\Gamma)$ to $\mathbb{C} \cup \{\infty\}$. Moreover,*

(i) *If $\gamma \in \Gamma$, defined from $\mathbb{D}$ to $\mathbb{D}$, is extended (analytically) to a map of $\mathbb{C} \cup \{\infty\}$ to $\mathbb{C} \cup \{\infty\}$, then for all $z \in \mathbb{C} \cup \{\infty\} \setminus \Lambda$ and all $\gamma \in \Gamma$,*

$$\mathbf{x}(\gamma(z)) = \mathbf{x}(z) \tag{9.6.15}$$

(ii) $\mathbf{x}'(z) \neq 0$ *so long as* $\mathbf{x}(z) \notin \{\alpha_j, \beta_j\}_{j=1}^{\ell+1}$.

(iii) *At points with $\mathbf{x}(z) \in \{\alpha_j, \beta_j\}_{j=1}^{\ell+1}$ (necessarily $z \in \partial\mathbb{D}$), we have*

$$\mathbf{x}'(z) = 0 \qquad \mathbf{x}''(z) \neq 0 \tag{9.6.16}$$

Proof. As we explained, analyticity across $\partial\mathcal{F} \cap \partial\mathbb{D}$ follows from the reflection principle. (i) follows from the fact that it holds for $z \in \mathbb{D}$ by analytic continuation. (9.6.15) then implies analyticity across $\cup_\gamma \gamma[\partial\overline{\mathcal{F}} \cap \partial\mathbb{D}] = \partial\mathbb{D} \setminus \Lambda$.

Now let $C_j^{\pm}$ denote the full orthocircle, not just the part in $\mathbb{D}$. Then x is real exactly on

$$\bigcup_{\gamma\in\Gamma} \gamma \left[\mathbb{R} \cup \bigcup_{j=1}^{\ell} (C_j^+ \cup C_j^-) \right] \cup \partial\mathbb{D} \setminus \Lambda(\Gamma) \tag{9.6.17}$$

The first union is over disjoint sets and the last set intersects all the others orthogonally. The set in (9.6.17) is displayed in Figure 9.6.3.

$\mathbf{x}$ is locally one-one on $\mathbb{D}$, so $\mathbf{x}'(z) \neq 0$ for $z \in \mathbb{D}$ and then, by (9.6.11), on $\mathbb{C}\setminus\overline{\mathbb{D}}$. As an analytic function, if $\mathbf{x}(z) - \mathbf{x}(z_0)$ has a kth order zero at z_0, there are $2k$ asymptotic rays at relative angle $2\pi/2k$ near z_0 on which $\mathbf{x}$ is real. Thus, $\mathbf{x}'(z) \neq 0$ on all points in (9.6.17), except the points in

$$\left(\bigcup_{\gamma\in\Gamma} \gamma \left[\mathbb{R} \cup \bigcup_{j=1}^{\ell} (C_j^+ \cup C_j^-) \right] \right) \cap (\partial\mathbb{D} \setminus \Lambda(\Gamma)) \tag{9.6.18}$$

where four real rays come in at 90° angles. At these points, the zero of $\mathbf{x}(z) - \mathbf{x}(z_0)$ is double, so $\mathbf{x}''(z) \neq 0$.

If we note that the set in (9.6.18) is exactly $\mathbf{x}^{-1}(\{\alpha_j, \beta_j\}_{j=1}^{\ell+1})$, we have (ii) and (iii). □

Remark. This says $\mathbf{x}^{-1}$ has square root behavior at points in $\{\alpha_j, \beta_j\}_{j=1}^{\ell+1}$.

Thus, $\mathbf{x}$ is locally one-one on the complement of the set in (9.6.18) and it is locally two-one at those points. But the image points, $\{\alpha_j, \beta_j\}_{j=1}^{\ell+1}$, are precisely the

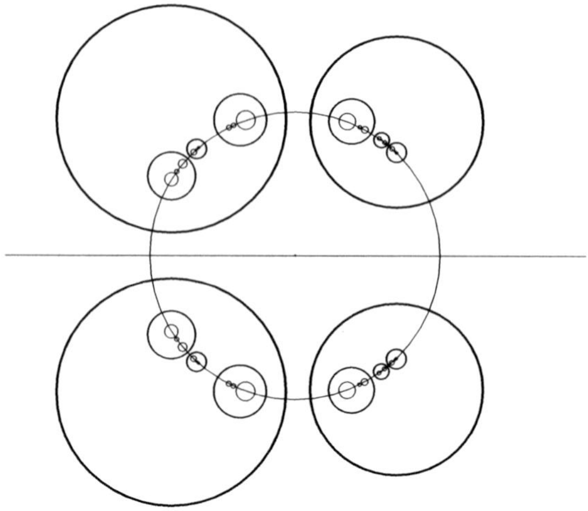

Figure 9.6.3. Three generations of $\gamma[C_j^\pm]$.

branch points of $\mathcal{S}$, so we introduce a modified map, $\mathbf{x}^\sharp$, to be a map from $\mathbb{C} \setminus \Lambda(\Gamma)$ to $\mathcal{S}$ and define it so that (9.6.11) is replaced by

$$\mathbf{x}^\sharp(1/\bar{z}) = \overline{\tau(\mathbf{x}(z))} \tag{9.6.19}$$

where $\tau(z_+) = z_-$ is the reflection on $\mathcal{S}$ discussed in Section 5.12. (9.6.19) is for $z \in \mathbb{D}$. For $z \in \mathbb{D}$, we also have

$$\mathbf{x}^\sharp(z) = \mathbf{x}(z) \tag{9.6.20}$$

interpreting $\mathbb{C} \cup \{\infty\} \setminus \mathfrak{e}$ as $\mathcal{S}_+$. Then we have proven that

Theorem 9.6.5. $\mathbf{x}^\sharp : \mathbb{C} \cup \{\infty\} \setminus \Lambda(\Gamma) \to \mathcal{S}$ *is a covering map.*

Of course, $\mathbb{C} \cup \{\infty\} \setminus \Lambda(\Gamma)$ is not simply connected, so this is not the universal cover.

Example 9.6.6 (One gap set)**.** Let $\ell = 1$. Then $\pi_1(\mathcal{S}^+)$ is $\mathbb{Z}$ while $\mathcal{S}$ is a torus so $\pi_1(\mathcal{S})$ is $\mathbb{Z}^2$. Γ has a single hyperbolic generator, γ_1, and $\Gamma = \{(\gamma_1)^n \mid n \in \mathbb{Z}\} \cong \mathbb{Z}$. Unlike the case $\ell \geq 2$ where $\Lambda(\Gamma)$ is infinite, in this case there are only two limit points: the two fixed points of γ_1. Notice that $\mathbb{C} \cup \{\infty\}$ with two points removed is homeomorphic to the punctured plane, $\mathbb{C} \setminus \{0\}$, so its π_1 is $\mathbb{Z}$. As a covering map, $\mathbf{x}^\sharp$ induces a map of $\pi_1(\mathbb{C} \cup \{\infty\} \setminus \Lambda(\Gamma)) = \mathbb{Z}$ to $\pi_1(\mathcal{S}) = \mathbb{Z}^2$. This image is the group generated by a loop around one band. The loops around the gap on both sheets generate the quotient and label $\gamma \in \Gamma$. □

We return to the group, Γ, and its action on $\mathcal{F}$ and on $\mathbb{D}$.

Proposition 9.6.7. *Let* $\{\gamma_j\}_{j=1}^\ell$ *be the generators given by* (9.6.7)*. Then every element* $\gamma \in \Gamma$ *can be written uniquely as*

$$\gamma = \alpha_{w(\gamma)} \dots \alpha_2 \alpha_1 \tag{9.6.21}$$

where each α_k is a γ_j or γ_j^{-1} with the convention that for no $j = 1, \dots, w(\gamma) - 1$ is $\alpha_{j+1}\alpha_j = 1$. If, for $k = 0, 1, 2, \dots,$

$$\Gamma_k = \{\gamma \mid w(\gamma) = k\} \tag{9.6.22}$$

then for $k \geq 1$,

$$\#\Gamma_k = (2\ell)(2\ell - 1)^{k-1} \tag{9.6.23}$$

In addition, any $\gamma \in \Gamma_{2m}$ has a unique representation,

$$\gamma = s_1 \dots s_{2m} \tag{9.6.24}$$

where each s_k is an $r_j^{\pm}$, and for all $j = 1, \dots, 2m - 1$, $s_{j+1} \neq s_j$. Similarly, any $\gamma \in \Gamma_{2m+1}$ has the form

$$\gamma = s_1 \dots s_{2m+1}c \tag{9.6.25}$$

Remarks. 1. $w(\gamma)$ is called the *word length* or *length* of γ.

2. Among all representations of γ as a product of γ_j's and γ_j^{-1}'s, (9.6.21) is the one of minimal length.

Proof. Γ is the free nonabelian group generated by $\{\gamma_j\}_{j=1}^{\ell}$, so any γ has a product representation of the form (9.6.21). If some $\alpha_{j+1}\alpha_j = 1$, remove them and so end up with a shorter representation of that form.

Since Γ is free, all such products in Γ_k are distinct (with the no $\alpha_{j+1}\alpha_j = 1$ condition). α_1 can be chosen in 2ℓ ways. Since $\alpha_2 \neq \alpha_1^{-1}$, it can only be chosen in $(2\ell - 1)$ ways. This leads to (9.6.23).

Given (9.6.8) and $(r_j^+)^2 = c^2 = 1$, we get

$$\gamma_j = cr_j^- \qquad \gamma_j^{-1} = cr_j^+ \tag{9.6.26}$$

In addition,

$$cr_j^{\pm}c = r_j^{\mp} \tag{9.6.27}$$

Thus, any representation of the form (9.6.21) leads to one of the form (9.6.24)/(9.6.25). □

Later, we will need the fact that $w(\gamma^n)$ grows linearly in n. We are heading toward a proof that

$$w(\gamma^n) \geq |n| - 1 + w(\gamma) \tag{9.6.28}$$

Call γ *solid* if the representation (9.6.21) has $\alpha_1\alpha_{w(\gamma)} \neq 1$ or if $w(\gamma) = 1$.

Lemma 9.6.8. *Any γ has the form*

$$\gamma = \gamma_0\gamma_1\gamma_0^{-1} \tag{9.6.29}$$

where $\gamma_1 \neq 1$ and is solid.

Proof. There is a first k_0 with

$$\alpha_{k_0} \neq \alpha_{w(\gamma)+1-k_0}^{-1} \tag{9.6.30}$$

for if $w(\gamma)$ is odd, $k_0 = \frac{1}{2}(w(\gamma) - 1)$ works. If $w(\gamma)$ is even, then (9.6.30) holds for $k_0 = w(\gamma)/2$ since $\alpha_{j+1}\alpha_j \neq 1$. Let

$$\gamma_0 = (\alpha_{k_0-1} \dots \alpha_1)^{-1} \tag{9.6.31}$$

if $k_0 \neq 1$ and $\gamma_0 = 1$ if k_0 is 1. Let

$$\gamma_1 = \alpha_{w(\gamma)+1-k_0} \dots \alpha_{k_0} \tag{9.6.32}$$

By (9.6.30), γ_1 is solid and not 1. By construction, (9.6.29) holds. □

Proposition 9.6.9. *Given γ, find a representation* (9.6.29) *with γ_1 solid and let*

$$s(\gamma) = w(\gamma_1) \tag{9.6.33}$$

Then

$$\gamma^n = \gamma_0 \gamma_1^n \gamma_0^{-1} \tag{9.6.34}$$

is the (9.6.21) *representation of γ^n so*

$$\begin{aligned} w(\gamma^n) &= 2w(\gamma_0) + |n| w(\gamma_1) \\ &= w(\gamma) + (|n| - 1)s(\gamma) \end{aligned} \tag{9.6.35}$$

In particular, (9.6.28) *holds.*

Proof. Since γ_1 is solid, the (9.6.21) representation of γ_1^n is just n times that of γ_1 repeated. □

We next want to define some subsets of $\mathbb{D}$ that keep track of how many γ_j's or γ_j^{-1}'s we need to get to these sets, starting in $\mathcal{F}$. Since $\mathcal{F}$ is a fundamental domain,

$$\mathbb{D} = \bigcup_{\gamma \in \Gamma} \gamma[\mathcal{F}] \tag{9.6.36}$$

and the union is over disjoint sets. We define

$$\mathbb{D}_k = \bigcup_{\gamma:\, w(\gamma) \leq k} \gamma[\overline{\mathcal{F}}] \tag{9.6.37}$$

and

$$\mathcal{R}_k = \mathbb{D} \setminus \mathbb{D}_k \tag{9.6.38}$$

Returning to Figure 9.6.3, $\mathbb{D}_0 = \overline{\mathcal{F}}$ is the intersection of $\mathbb{D}$ and the exterior of the four big circles and $\mathcal{R}_0$ is the part of $\mathbb{D}$ inside those circles. $\mathbb{D}_1$ is the exterior of the $12 = 4 \times 3$ next biggest circles and $\mathcal{R}_1$ the interior of the 12 circles. $\mathbb{D}_1 \setminus \mathbb{D}_0$ are the four images of $\mathcal{F}$ under $\gamma_1, \gamma_2, \gamma_1^{-1}, \gamma_2^{-1}$ (up to some edges). The interior of the $36 = 4 \times 3^2$ smallest circles is $\mathcal{R}_2$ and their complement is $\mathbb{D}_2$.

Finally, let $\overline{\mathcal{R}_k}$ be the closure of $\mathcal{R}_k$ in $\overline{\mathbb{D}}$ and

$$\partial\mathcal{R}_k = \overline{\mathcal{R}_k} \cap \partial\mathbb{D} \tag{9.6.39}$$

We are heading toward a proof of a major geometric theorem, which will be critical in our proof of step-by-step sum rules.

Theorem 9.6.10 (Beardon's Theorem). *For some positive constants C_0, C_1, we have*

$$|\partial\mathcal{R}_k| \le C_0 e^{-C_1 k} \tag{9.6.40}$$

Remark. $|\cdot|$ means $\frac{d\theta}{2\pi}$ measure.

As we will see below (see Theorem 9.6.13 and the Notes), this is equivalent to the fact that there is a Poincaré index, s, for Γ with $s < 1$—and it is in this form that Beardon stated his theorem (for more general Fuchsian groups).

As noted, $\partial\mathcal{R}_k$ contains $2\ell(2\ell-1)^{k-1}$ arcs. It is not hard to see that the maximum radius of disks in $\mathcal{R}_k$ decays exponentially (see Lemma 9.6.16 below), while the number of arcs grows exponentially. (9.6.40) says the size decrease wins out by a bit.

We note that

$$\Lambda(\Gamma) = \bigcap_k \overline{\mathcal{R}}_k \tag{9.6.41}$$

$$= \bigcap_k \partial\mathcal{R}_k \tag{9.6.42}$$

Before turning to a proof of Theorem 9.6.10, we want to note a number of consequences. The first can be proven without using anything as powerful as (9.6.40), but since we have it, we will use it.

Corollary 9.6.11. *Every $\gamma \in \Gamma$, $\gamma \neq 1$, is hyperbolic.*

Remarks. 1. Indeed, our proof shows that if $T_\gamma \in \mathbb{SU}(1,1)$ is defined by $\gamma = f_{T_\gamma}$, then $\inf_{\gamma\neq 1}|\text{Tr}(T_\gamma)| > 2$.

2. In the lead-up to the proof of Theorem 9.6.10, we will prove (9.6.43).

Proof. The length of the arcs in $\partial\mathcal{R}_k$ is comparable to the radii of the circles in $\mathcal{R}_k$. Thus, (9.6.40) implies

$$\sup\{|w-z| \mid w, z \text{ inside the same circle of } \overline{\mathcal{R}}_k\} \le \widetilde{C}_0 e^{-C_1 k} \tag{9.6.43}$$

By construction, $\gamma^n(0)$ lies in one of these circles for $k = w(\gamma^n)$ and, as we will show below (see (9.6.48)), $\gamma^n(0)/|\gamma^n(0)|$ lies in the same circle. Thus,

$$1 - |\gamma^n(0)| \le \widetilde{C}_0 e^{-C_1 w(\gamma^n)} \tag{9.6.44}$$

By (9.6.28),

$$1 - |\gamma^n(0)| \le \widetilde{C}_0 e^{-C_1(n-1)} \tag{9.6.45}$$

which implies that approach to the limit is exponential, so γ is hyperbolic. □

The main use that we will have for Theorem 9.6.10 is

Theorem 9.6.12. *Let f be an analytic function so that for some $C > 0$,*

$$\{z \mid |\text{Im}\, f(z)| > Cn\} \subset \mathcal{R}_n \tag{9.6.46}$$

Then

$$f \in \bigcap_{p<\infty} H^p(\mathbb{D}) \tag{9.6.47}$$

Proof. Let D_0 be a disk, which is a connected component of $\mathcal{R}_n$. Since its boundary is orthogonal to $\partial\mathbb{D}$, the radii from 0 to the two intersections with $\partial\mathbb{D}$ are tangent, so

$$z \in \mathcal{R}_n \Rightarrow \frac{z}{|z|} \in \partial\mathcal{R}_n \tag{9.6.48}$$

From (9.6.46) and (9.6.47), we conclude that for any r,

$$\{e^{i\theta} \mid |\text{Im}\, f(re^{i\theta})| > Cn\} \subset \partial\mathcal{R}_n \tag{9.6.49}$$

so, by (9.6.40),

$$|\{e^{i\theta} \mid |\text{Im}\, f(re^{i\theta})| > Cn\}| \le C_0 e^{-C_1 n} \tag{9.6.50}$$

Thus, for any $p < \infty$,

$$\sup_r \int |\text{Im}\, f(re^{i\theta})|^p \,\frac{d\theta}{2\pi} < \infty \tag{9.6.51}$$

By M. Riesz's theorem (Proposition 2.3.8), $f \in \bigcap_{p<\infty} H^p(\mathbb{D})$. □

As a final consequence of Theorem 9.6.10:

Theorem 9.6.13. *There exists an $s < 1$, so*

$$\sum_{\gamma\in\Gamma} (1 - |\gamma(0)|)^s < \infty \tag{9.6.52}$$

Remark. As we will sketch in the Notes, (9.6.52) for some $s < 1$ implies (9.6.40).

Proof. Each $\gamma \in \Gamma$ maps $\mathcal{F}$ into a region bounded by 2ℓ orthocircles, one which we will call C_γ outside, the other $2\ell - 1$. For example, for γ_j, C_γ is C_j^+, and for γ_j^{-1}, C_γ is C_j^-.

Let r_γ be the radius of C_γ. Since r_γ is comparable in size to the intersection of the inside of C_γ and $\partial\mathbb{D}$, we have, by (9.6.40), that

$$\sum_{\gamma:\, w(\gamma)\ge k} r_\gamma \le \widetilde{C}_0 e^{-C_1 k} \tag{9.6.53}$$

which also implies

$$r_\gamma \le \widetilde{C}_0 e^{-C_1 w(\gamma)} \tag{9.6.54}$$

On the other hand, since $0 \in \mathcal{F}$, $\gamma(0)$ is inside C_γ, and since C_γ is an orthocircle, $\gamma(0)/|\gamma(0)|$ is also inside C_γ. Thus,

$$1 - |\gamma(0)| \le 2r_\gamma \tag{9.6.55}$$

Fix $s < 1$ and let $p = 1/s$ and $q = 1/(1-s)$, the dual L^p index. By Hölder's inequality,

$$\begin{aligned}\sum_{\gamma:\, w(\gamma)=k} (2r_\gamma)^s &\le \biggl(\sum_{\gamma:\, w(\gamma)=k} (2r_\gamma)^{sp}\biggr)^{1/p} \biggl(\sum_{\gamma:\, w(\gamma)=k} 1\biggr)^{1/q} \\ &= (2\widetilde{C}_0 e^{-C_1 k})^s (2\ell)^{k(1-s)}\end{aligned} \tag{9.6.56}$$

We used here $\#(\gamma \mid w(\gamma) = k) \le (2\ell)^k$. Pick $s < 1$ so that

$$C_1 s - (\log(2\ell))(1 - s) \equiv d > 0 \tag{9.6.57}$$

which can be done since it is positive at $s = 1$ and continuous in s. Then

$$\sum_{\gamma:\, w(\gamma)=k} (2r_\gamma)^s \le (2\widetilde{C}_0)^s e^{-dk} \tag{9.6.58}$$

Thus, by (9.6.55),

$$\sum_{\gamma} (1 - |\gamma(0)|)^s < \infty \tag{9.6.59}$$

□

We are now ready to turn to the proof of Theorem 9.6.10. The basic idea is the following. The boundary of $\mathcal{R}_k$ is $(2\ell)(2\ell-1)^{k-1}$ orthocircles $\{C_\gamma\}$, which we have seen can be labeled by $\gamma \in \Gamma$ with $w(\gamma) = k$. We will let A_γ be the arc of $\partial\mathbb{D}$ cut off by the interior of C_γ. (9.6.40) is equivalent to

$$\sum_{\gamma:\, w(\gamma)=k} |A_\gamma| \le C_0 e^{-C_1 k} \tag{9.6.60}$$

$\gamma[\mathcal{F}]$ has a boundary, which is 2ℓ orthocircles: C_γ and $2\ell - 1$ orthocircles from $\partial\mathcal{R}_{k+1}$. Thus, A_γ is broken into $2\ell - 1$ $A_{\gamma'}$'s and 2ℓ intervals, which are "lost" in going from $\mathcal{R}_k$ to $\mathcal{R}_{k+1}$. Call the union of these 2ℓ lost arcs (with apology to Indiana Jones), Q_γ. Suppose we prove that there is a γ-independent constant, $q \in (0, 1)$, so that

$$\frac{|Q_\gamma|}{|A_\gamma|} \ge q \tag{9.6.61}$$

Then using

$$\mathcal{R}_k \setminus \mathcal{R}_{k+1} = \bigcup_{\gamma:\, w(\gamma)=k} Q_\gamma \tag{9.6.62}$$

it is immediate from (9.6.61) that

$$|\mathcal{R}_{k+1}| \le (1 - q)|\mathcal{R}_k| \tag{9.6.63}$$

which leads to (9.6.40).

To understand (9.6.61), we note that for any γ, there is a $\tilde\gamma$ with $w(\tilde\gamma) = w(\gamma) - 1$ so that C_γ is obtained from $C_{\tilde\gamma}$ by applying an inversion about some circle $C_j^\pm$. Because we will prove $|A_\gamma|$ goes to zero exponentially, lengths of sets in C_γ will change by a uniform factor up to errors, which are exponentially small, which will show for all k large and some C_2,

$$\frac{|Q_\gamma|}{|A_\gamma|} \ge (1 - e^{-C_2 k})\, \frac{|Q_{\tilde\gamma}|}{|A_{\tilde\gamma}|} \tag{9.6.64}$$

which will lead to (9.6.61) since $\prod_{k=1}^\infty (1 - e^{-C_2 k}) > 0$. This completes the sketch of what follows.

We will need the following, which is essentially a restatement of Proposition 9.2.28:

Proposition 9.6.14. *Let η be a conjugate analytic function of z in a neighborhood of $\partial\mathbb{D}$, $\eta' = \partial\eta/\partial\bar{z}$ its derivative. Suppose η maps $\partial\mathbb{D}$ to itself and let $Q \subset \partial\mathbb{D}$. Then*

$$|\eta[Q]| = \int_Q |\eta'(e^{i\theta})| \, \frac{d\theta}{2\pi} \tag{9.6.65}$$

Remark. $|\cdot|$ is in $\frac{d\theta}{2\pi}$ measure.

Proof. η is anticonformal, so it infinitesimally stretches or contracts distances by $|\eta'|$. Since $d\theta$ is arclength in Euclidean metric, (9.6.65) is immediate. □

Corollary 9.6.15. *Under the hypotheses of Proposition 9.6.14, if Q_1, Q_2 are any two subsets of $\partial\mathbb{D}$, then*

$$\frac{|\eta[Q_1]|}{|\eta[Q_2]|} \geq \frac{\inf_{Q_1} |\eta'(e^{i\theta})|}{\sup_{Q_2} |\eta'(e^{i\theta})|} \frac{|Q_1|}{|Q_2|} \tag{9.6.66}$$

Proof. Immediate from (9.6.65), which implies

$$\sup_Q |\eta'(e^{i\theta})| \, |Q| \geq |\eta[Q]| \geq \inf_Q |\eta'(e^{i\theta})| \, |Q| \tag{9.6.67}$$

□

Let C_γ be the outer circle of $\gamma[\mathcal{F}]$, as discussed in the proof of Theorem 9.6.13 and let A_γ be the arc of $\partial\mathbb{D}$ inside C_γ. We need to prove $|A_\gamma|$ decreases exponentially in $w(\gamma)$. Let $r_j^\pm$ be the reflections in $C_j^\pm$. $(r_j^\pm)'$ has $C_j^\pm$ as isometric circle. Outside $C_j^\pm$, $|(r_j^\pm)'| < 1$. Let

$$b = \max_{j,\pm} \left[\max_{e^{i\theta} \text{ inside some other } C_k^\pm} |(r_j^\pm)'(e^{i\theta})| \right] \tag{9.6.68}$$

Since the $C_j^\pm$ are a strictly positive distance from each other,

$$b < 1 \tag{9.6.69}$$

Lemma 9.6.16. *For any $\gamma \neq 1$ in Γ,*

$$|A_\gamma| \leq b^{w(\gamma)-1} \tag{9.6.70}$$

Proof. This is trivial for $w(\gamma) = 1$, so we need only prove it by induction. Suppose we know it for all γ' with $w(\gamma') = w(\gamma) - 1$. Since $c[\mathcal{F}] = \mathcal{F}$, by (9.6.24)/(9.6.25), there are $s_1, \dots, s_{w(\gamma)} \in \{r_j^\pm\}_{j=1}^\ell$, so

$$\gamma[\mathcal{F}] = s_1 \dots s_{w(\gamma)}[\mathcal{F}] \tag{9.6.71}$$

Let C_s be the circle for which s is the reflection. We claim $\gamma[\mathcal{F}]$ is inside the circle C_{s_1}. For $s_{w(\gamma)}[\mathcal{F}]$ is inside $C_{s_{w(\gamma)}}$. Since $s_{w(\gamma)-1} \neq s_{w(\gamma)}$, it is outside

$C_{s_{w(\gamma)-1}}$, so $s_{w(\gamma)-1}s_{w(\gamma)}[\mathcal{F}]$ is inside $C_{s_{w(\gamma)-1}}$. In this way, one sees inductively that $s_j \dots s_{w(\gamma)}[\mathcal{F}]$ is inside C_{s_j}. In particular,

$$A_\gamma = s_1[A_{\tilde{\gamma}}] \tag{9.6.72}$$

with $\tilde{\gamma} = s_2 \dots s_{w(\gamma)}$ (if $w(\gamma)$ is odd) or $s_2 \dots s_{w(\gamma)}c$ (if $w(\gamma)$ is even). $A_{\tilde{\gamma}}$ is inside a different circle from C_{s_1} as we have proven.

By Proposition 9.6.14 and the definition of b, this implies

$$|A_\gamma| \le b|A_{\tilde{\gamma}}| \tag{9.6.73}$$

By induction,

$$|A_{\tilde{\gamma}}| \le b^{w(\gamma)-2} \tag{9.6.74}$$

so (9.6.70) holds. □

Remark. Proposition 9.2.24 is another expression of the fact that $r_j^\pm$ contract distances outside and a finite distance away from $C_j^\pm$.

Proof of Theorem 9.6.10. Each $r_j^\pm$ is anti-analytic in a neighborhood of $\partial\mathbb{D}$ and so C^2. Moreover, $|(r_j^\pm)'|$ is strictly positive on $\partial\mathbb{D}$. Thus, for suitable C, we have for all $\eta \in \{r_j^\pm\}$ that

$$\inf_{|e^{i\theta}-e^{i\varphi}|\le\delta} \frac{|\eta'(e^{i\theta})|}{|\eta'(e^{i\varphi})|} \ge 1 - C\delta \tag{9.6.75}$$

It follows from this and Corollary 9.6.15 that if $Q \subset A_\gamma$, then for all $s = r_j^\pm$,

$$\frac{|s[Q]|}{|s(A_\gamma)|} \ge 1 - Cb^{w(\gamma)-1} \tag{9.6.76}$$

where we used Lemma 9.6.16.

Pick n_0 so $Cb^{n_0} < \frac{1}{2}$ and let

$$f = \prod_{n=n_0}^{\infty} (1 - Cb^n) > 0 \tag{9.6.77}$$

by $b < 1$.

Let

$$Q_\gamma = \partial\gamma[\mathcal{F}] \cap \partial\mathbb{D} \tag{9.6.78}$$

which is 2ℓ arcs between the $2\ell - 1$ orthocircles inside C_γ. For each γ, $|Q_\gamma| > 0$, so

$$m = \min_{w(\gamma)\le n_0} \frac{|Q_\gamma|}{|A_\gamma|} > 0 \tag{9.6.79}$$

as a finite min of positive numbers. By (9.6.77) and (9.6.21), for any γ, we have

$$\frac{|Q_\gamma|}{|A_\gamma|} \ge mf \tag{9.6.80}$$

Write $\gamma' \rhd \gamma$ if $\gamma' = \tilde\gamma\gamma$ with $w(\gamma') = w(\tilde\gamma) + w(\gamma)$, that is, $\gamma' = \alpha_{w(\gamma')}\dots\alpha_1$ and $\gamma = \alpha_{w(\gamma)}\dots\alpha_1$ (same α's). Given γ, the γ''s with $\gamma' \rhd \gamma$ and $w(\gamma') = w(\gamma)+1$ number $2\ell - 1$ and the corresponding A'_γ's are the arcs between the area making up Q_γ, that is,

$$\sum_{\substack{\gamma' \rhd \gamma \\ w(\gamma')=w(\gamma)+1}} |A'_\gamma| \le (1-mf)|A_\gamma| \tag{9.6.81}$$

This implies, by summing (9.6.81) over all words of length k,

$$\sum_{\gamma:\, w(\gamma)=k+1} |A_\gamma| \le (1-mf) \sum_{\gamma:\, w(\gamma)=k} |A_\gamma| \tag{9.6.82}$$

so, by induction,

$$\sum_{\gamma:\, w(\gamma)=k} |A_\gamma| \le (1-mf)^k \tag{9.6.83}$$

But

$$\partial\mathcal{R}_k = \bigcup_{w(\gamma)=k} A_\gamma \tag{9.6.84}$$

so (9.6.83) implies

$$|\partial\mathcal{R}_k| \le (1-mf)^k \tag{9.6.85}$$

which proves (9.6.40). □

This completes what we want to prove about the Γ and $\mathcal{F}$ associated to a finite gap set. The reader may have noticed that we did not use any isometric circles. We end this section with some alternate proofs that use that technology.

Proposition 9.6.17. *Let γ_1 be solid. Then $D_f(\gamma_1)$ and $D_i(\gamma_1)$ lie inside distinct $C_j^\pm$ and so are disjoint.*

Remark. By Theorem 9.2.32, this implies γ_1 is hyperbolic. Thus, by Lemma 9.6.8, any $\gamma \ne 1$ in Γ is hyperbolic, that is, we have a second proof of Corollary 9.6.11.

Proof. Suppose first $w(\gamma_1)$ is even. Then $\gamma_1 = s_1 \dots s_{2m}$ with each s_k one of $r_j^\pm$ and no $s_{j+1} = s_j$. Thus, as above, if $s_1 = r_j^\pm$, then $\gamma_1(0)$ lies inside $C_{j_1}^\pm$. But, by Theorem 9.4.22, all $D_f(\gamma_1)$'s lie inside some $C_k^\pm$ and, by Theorem 9.3.6, $\gamma_1(0)$ lies in $D_f(\gamma_1)$. We conclude $D_f(\gamma_1)$ lies inside $C_{j_1}^\pm$. Similarly, since $D_i(\gamma_1) = D_f(\gamma_f^{-1})$, we see $D_i(\gamma_1)$ is inside $C_{j_{2m}}^\pm$ since $\gamma_1^{-1} = s_{2m}\dots s_1$. Since $s_{2m} \ne s_1$, the initial and final circles lie inside distinct $C_j^\pm$ as claimed.

The analysis in the odd case is similar. □

Finally, we want to provide a different proof of the key Theorem 9.6.10:

Lemma 9.6.18. *Let $\gamma \in \mathcal{M}$ have $\gamma(0) \ne 0$ and let $\theta(z)$ be the angle (in $(-\pi, \pi]$) between $z \in \partial\mathbb{D}$ and the ray from 0 through the center of $D_i(\gamma)$. Then $|\gamma'(z)|$ is a function of $|\theta(z)|$ only and monotone decreasing as $|\theta(z)|$ increases.*

Proof. By covariance, we can suppose the center of $D_i(\gamma)$ is on $(1,\infty)$ at β. Then, by (9.2.61),

$$\begin{aligned}|\gamma'(z)| &= c^{-2}|e^{i\theta(z)}-\beta|^{-1}\\ &= c^{-2}(1+\beta^2-2\beta\cos(\theta(z)))^{-1}\end{aligned} \tag{9.6.86}$$

is clearly monotone decreasing in $\theta(z)$. □

Sketch of Proof of Theorem 9.6.10. For $j=1,\dots,\ell$, let C_j^+ be the jth orthocircle in $\mathbb{C}_+$ and let A_j^+ be the arc in $\partial\mathbb{D}$ it cuts off. Let $Q_j^{\ell,r}$ be the two arcs in $\partial\mathfrak{F}\cap\partial\mathbb{D}$ adjacent to A_j^+ on the left and right (so Q_j^ℓ is between C_{j-1}^+ and C_j^+ with $C_0^+\equiv C_1^-$ and Q_j^r between C_j^+ and C_{j+1}^+ with $C_{\ell+1}^+\equiv C_\ell^-$). Let

$$q_j=\frac{|Q_j^+\cup Q_j^-|}{|\partial\mathbb{D}\setminus A_j|} \tag{9.6.87}$$

be the fraction of the remainder of $\partial\mathbb{D}$ taken by $Q_j^+\cup Q_j^-$.

Let

$$q=\min_j q_j>0 \tag{9.6.88}$$

We will prove that for any γ,

$$\frac{|Q_\gamma|}{|A_\gamma|}\geq q \tag{9.6.89}$$

from which

$$|\partial\mathcal{R}_k|\leq(1-q)^k \tag{9.6.90}$$

as in the other proof.

As noted above, $D_i(\gamma)$ lies inside some $C_j^\pm$, say C_j^+ for simplicity of notation. Thus, ∂C_j^+ goes under γ into C_γ, $\partial\mathbb{D}\setminus A_j$ into all of A_γ, and $Q_j^\pm$ into parts of Q_γ. Since $|\gamma'|$ is decreasing by the lemma and $A_j^\pm$ is closest to $D_i(\gamma)$, we have that

$$\frac{|\gamma(Q_j^\pm)|}{|\gamma(\partial\mathbb{D}\setminus A_j)|}\geq\frac{|Q_j^\pm|}{|A_j|} \tag{9.6.91}$$

which implies

$$\frac{|Q_\gamma|}{|A_\gamma|}\geq\frac{|\gamma(Q_j^+)\cup\gamma(Q_j^-)|}{|\gamma(A_j)|}\geq\frac{|Q_j^+\cup Q_j^-|}{|A_j|}\geq q_j\geq q$$

proving (9.6.88). □

Remarks and Historical Notes. The use of explicit covering maps in spectral theory and the structure of Fuchsian groups goes back to Sodin–Yuditskii [413] and has been developed by Peherstorfer–Yuditskii [343, 344] and Christiansen–Simon–Zinchenko [86, 87, 88, 89].

The basic picture with orthocircles in complex symmetric positions, one in $\mathbb{C}_+$ for each gap, is from [413]. The importance of Beardon's theorem in the finite gap case is due to Christiansen–Simon–Zinchenko [87, 88].

What we call Beardon's theorem is a special case of a much more general theorem of Beardon [35]: he proved that any finitely generated Fuchsian group, Γ, for which $\Lambda(\Gamma)$ is not dense in $\partial\mathbb{D}$, has a Poincaré index of convergence $s < 1$. He also proved that this implies the set of limit points has Hausdorff dimension less than one. For more on Hausdorff dimensions of limit sets of Fuchsian groups, see Patterson [332] and Sullivan [424]. Beardon's general result is much more difficult to prove because of the need to accommodate parabolic and elliptic elements.

Our proof of Theorem 9.6.10 here is new and was arrived at in discussion with Jacob Christiansen and Maxim Zinchenko.

We proved that (9.6.40) implies (9.6.52) for some $s < 1$. One can go backwards and show (9.6.52) for $s < 1$ implies (9.6.40). For $|A_\gamma|$ is comparable to $1 - |\gamma(0)|$ so (9.6.40) is equivalent to

$$\sum_{\gamma:\, w(\gamma)=k} 1 - |\gamma(0)| \le \widetilde{C}_0 e^{-kC_1} \tag{9.6.92}$$

On the other hand, one proves

$$1 - |\gamma(0)| \le D_0 e^{-w(\gamma)D_1} \tag{9.6.93}$$

so

$$\begin{aligned}\sum_{\gamma:\, w(\gamma)=k} 1 - |\gamma(0)| &\le \Bigl[\sum_{\gamma:\, w(\gamma)=k} (1 - |\gamma(0)|)^s\Bigr](D_0 e^{-kD_1})^{1-s} \\ &\le D_0^{1-s} e^{-k(1-s)D_1} \sum_\gamma (1 - |\gamma(0)|)^- s \end{aligned} \tag{9.6.94}$$

so (9.6.52) implies (9.6.92).

9.7 BLASCHKE PRODUCTS AND GREEN'S FUNCTIONS

The analog of what we did for a single interval is that, given a measure, $d\mu$, with $\sigma_{\text{ess}}(d\mu) = \mathfrak{e}$, we form its m-function, $m(z)$, on $\mathbb{C} \setminus \sigma(d\mu)$, meromorphic on $\mathbb{C} \setminus \mathfrak{e}$ and define on $\mathbb{D}$,

$$M(z) = -m(\mathbf{x}(z)) \tag{9.7.1}$$

This function is *automorphic* in that for all γ,

$$M(\gamma(z)) = M(z) \tag{9.7.2}$$

That is, automorphic functions, f, are defined on $\mathbb{D}$ and obey

$$f(\gamma(z)) = f(z) \tag{9.7.3}$$

for all $\gamma \in \Gamma$ and $z \in \mathbb{D}$. We will mainly want to consider meromorphic functions obeying (9.7.3), but occasionally we will also want to allow f to be a real harmonic or subharmonic function.

One of the first things we want to do is remove zeros and poles. For example, even if there were no bound states, we needed to consider $M(z)/z$ in case

$\mathfrak{e} = [-2, 2]$. As in that case, m has a zero at ∞, so M has a zero at $z = 0$. But then, by (9.7.2), it has zeros at all points in $\{\gamma(0)\}_{\gamma\in\Gamma}$. So we have to divide out by an infinity of zeros even in the simplest cases. That will lead us to Blaschke products and, as a bonus, we will find a remarkably simple connection to the logarithmic potential for $\mathfrak{e}$.

Recall that $s = 1$ is a Poincaré index if for one, and hence all, $z_0 \in \mathbb{D}$, we have

$$\sum_{\gamma\in\Gamma} (1 - |\gamma(z_0)|) < \infty \tag{9.7.4}$$

and, in particular, if Γ is of the second kind, (9.7.4) holds (see Theorem 9.4.19). This is, of course, exactly a Blaschke condition, (2.3.69). Thus, by Proposition 2.3.16,

Theorem 9.7.1. *If Γ is a Fuchsian group for which* (9.7.4) *holds for one, and hence all, $z_0 \in \mathbb{D}$, the function* (*b defined by* (2.3.67))

$$B(z, z_0) = \prod_{\gamma\in\Gamma} b(z, \gamma(z_0)) \tag{9.7.5}$$

is an absolutely convergent product, which defines a function of z on $\mathbb{D}$ analytic there, vanishing exactly at the points $\{\gamma(z_0)\}_{\gamma\in\Gamma}$ with simple zeros there. Moreover, if $\Lambda(\Gamma) \neq \partial\mathbb{D}$, then B has an analytic continuation to a neighborhood of $\partial\mathbb{D}\setminus\Lambda(\Gamma)$. On $\partial\mathbb{D} \setminus \Lambda(\Gamma)$,

$$|B(e^{i\theta}, z_0)| = 1 \tag{9.7.6}$$

$B(\,\cdot\,, z_0)$ then also has a meromorphic continuation to $(\mathbb{C}\cup\{\infty\}) \setminus \Lambda(\Gamma)$ with poles exactly at $\{1/\overline{\gamma(z_0)}\}_{\gamma\in\Gamma}$ where all poles are simple.

Remark. $B(\,\cdot\,, z_0)$ is called a *Fuchsian Blaschke product*, or sometimes just a Blaschke product. The case $z_0 = 0$ is special, so we will write

$$B(z) \equiv B(z, z_0 = 0) \tag{9.7.7}$$

Proof. By (3.3.3) and (3.3.4), one has for any $z_0 \in \mathbb{D}$ and $z \in \mathbb{C} \setminus \{\bar{z}_0^{-1}\}$ that

$$|b_{z_0}(z) - 1| \leq \frac{1 + |z|}{|1 - z\bar{z}_0|}(1 - |z_0|) \tag{9.7.8}$$

from which one concludes that for $z \notin \overline{\{\overline{\gamma(z_0)}^{-1}\}_{\gamma\in\Gamma}} = \Lambda(\Gamma) \cup \{\overline{\gamma(z_0)}^{-1}\}_{\gamma\in\Gamma} \equiv \mathcal{P}(z_0)$, we have

$$\sum_{\gamma\in\Gamma} |b_{\gamma(z_0)} - 1| < \infty \tag{9.7.9}$$

with a bound uniform on compact subsets of $\mathbb{C} \setminus \mathcal{P}$.

It follows that the product converges uniformly on compacts of the open set $\mathbb{C}\setminus\mathcal{P}$, which includes $\partial\mathbb{D} \setminus \Lambda(\Gamma)$. Since $|b_{z_0}(e^{i\theta})| = 1$ and $|b_{\gamma(z_0)}(e^{i\theta})| = 1$, the uniform convergence implies (9.7.6).

By Hurwitz's theorem, the only zeros in $\mathbb{C} \setminus \mathcal{P}$ are at $\{\gamma(z_0)\}_{\gamma\in\Gamma}$.

From (9.7.6) and the fact that $\partial\mathbb{D} \setminus \Lambda(\Gamma)$ is open in $\partial\mathbb{D}$ and nonempty, we get, by the reflection principle,

$$B(z, z_0) = \overline{B(1/\bar{z}, z_0)}^{-1} \tag{9.7.10}$$

initially for $z \in (\mathbb{C} \setminus \mathcal{P}) \cup \mathbb{C} \setminus \overline{\mathbb{D}}$. This then implies the claim about poles. $\square$

Since the set of zeros of $B(\,\cdot\,, z_0)$ is invariant under all $\gamma \in \Gamma$, one might guess that this is true of B itself. We will see this is true for $|B(\,\cdot\,, z_0)|$ but not for the phase.

Definition. A *character* of a Fuchsian group, Γ, is group homomorphism of Γ to $\partial\mathbb{D}$ viewed as a multiplicative group. Γ^* is the group of all characters of Γ under pointwise multiplication. Given $\omega \in \Gamma^*$, a function f on $\mathbb{D}$ is called *character automorphic* with character ω if

$$f(\gamma(z)) = \omega(\gamma) f(z) \tag{9.7.11}$$

for all $\gamma \in \Gamma$, $z \in \mathbb{D}$. f is called character automorphic if and only if it is character automorphic for some $\omega \in \Gamma^*$.

For a finite gap set, Γ is generated by $\{\gamma_j\}_{j=1}^{\ell}$. So, since $\partial\mathbb{D}$ is abelian, $\{\omega(\gamma_j)\}_{j=1}^{\ell}$ determine ω. Since Γ is free, any values in $\partial\mathbb{D}$ are allowed, that is, if $(\alpha_1, \dots, \alpha_\ell) \in \partial\mathbb{D}^\ell$, then there is a unique character with

$$\omega_\alpha(\gamma_j) = \alpha_j \tag{9.7.12}$$

and this describes all characters. Thus, $\Gamma^* \cong (\partial\mathbb{D})^\ell$, a torus of the same dimension as the the isospectral torus. We will eventually see that this is no coincidence!

Theorem 9.7.2. *For any $z_0 \in \mathbb{D}$, there is a character $\omega_{z_0} \in \Gamma^*$ so*

$$B(\gamma(z), z_0) = \omega_{z_0}(\gamma) B(z, z_0) \tag{9.7.13}$$

$z_0 \to \omega_{z_0}$ is continuous in z_0 and obeys

$$\omega_{\gamma(z_0)} = \omega_{z_0} \tag{9.7.14}$$

Proof. We claim first that for any $z_1 \in \mathbb{D}$ and γ, there is $\alpha_{\gamma,z_1} \in \partial\mathbb{D}$ with

$$b(\gamma(z), z_1) = \alpha_{\gamma,z_1} b(z, \gamma^{-1}(z_1)) \tag{9.7.15}$$

For

$$g(z) = \frac{b(\gamma(z), z_1)}{b(z, \gamma^{-1}(z_1))} \tag{9.7.16}$$

is a ratio of functions analytic in a neighborhood of $\overline{\mathbb{D}}$, each with a single simple zero at $\gamma(z) = z_1$, that is, $z = \gamma^{-1}(z_1)$. Thus, g is analytic and nonvanishing on $\overline{\mathbb{D}}$. Since $|g(z)| = 1$ on $\partial\mathbb{D}$, g has a meromorphic continuation to $\mathbb{C} \cup \{\infty\}$ given by

$$g(z) = (\overline{g(1/\bar{z})})^{-1} \tag{9.7.17}$$

outside $\overline{\mathbb{D}}$. But g is nonvanishing on $\overline{\mathbb{D}}$, so g is entire and bounded, hence a constant α_{γ,z_1}. But $|g(z)| = 1$ on $\partial\mathbb{D}$, so $\alpha_{\gamma,z_1} \in \partial\mathbb{D}$.

Now fix $\gamma_0 \in \Gamma$. Then, by (9.7.15),

$$b(\gamma_0(z), \gamma(z_0)) = \alpha_{\gamma_0, \gamma(z_0)} b(z, \gamma_0^{-1}\gamma(z_0)) \tag{9.7.18}$$

As γ runs through all of Γ, $\gamma_0^{-1}\gamma$ does also. So, by uniform convergence of the product,

$$B(\gamma_0(z), z_0) = \omega_{z_0}(\gamma_0) B(z, z_0) \tag{9.7.19}$$

where for now $\omega_{z_0}(\gamma_0)$ is just some number in $\partial\mathbb{D}$.

But

$$\begin{aligned} B(\gamma_0\gamma_1(z), z_0) &= \omega_{z_0}(\gamma_0) B(\gamma_1(z), z_0) \\ &= \omega_{z_0}(\gamma_0)\omega_{z_0}(\gamma_1) B(z, z_0) \end{aligned} \tag{9.7.20}$$

proving that $\omega \in \Gamma^*$.

Since $z_0 \to B(z, z_0)$ is continuous for any $z \in \mathbb{D}$ and $B(z, \gamma(z_0)) = B(z, z_0)$, we see that $z_0 \to \omega_{z_0}$ is continuous and that (9.7.14) holds. □

We want to note a corollary of (9.7.15):

Proposition 9.7.3. *For any type 2 Fuchsian group, one has*

$$|B(z)| = \prod_{\gamma \in \Gamma} |\gamma(z)| \tag{9.7.21}$$

for all $z \in (\mathbb{C} \cup \{\infty\}) \setminus \Lambda(\Gamma)$.

Proof. By convergence of the product defining B and analyticity, it suffices to prove this for $z \in \mathbb{D}$. By (9.7.15),

$$\begin{aligned} |b(z, \gamma(0))| &= |b(\gamma^{-1}(z), 0)| \\ &= |\gamma^{-1}(z)| \end{aligned} \tag{9.7.22}$$

Since γ^{-1} runs through Γ as γ does, (9.7.21) follows (using $1 - |w| \le |1 - w|$ for $|w| < 1$). □

One might worry that B is really fully automorphic and it is just our proof that is lacking. After some notation, we will show that is an unfounded worry. Henceforth, we suppose Γ is the Fuchsian group of a finite gap covering map. Define $Q_1, \dots, Q_{\ell+1}$ arcs on $\partial\mathbb{D} \cap \mathbb{C}_+$ as follows: Q_1 runs from 1 to the right endpoint of C_ℓ^+, Q_2 from the left endpoint of C_ℓ^+ to the right of $C_{\ell-1}^+, \dots, Q_\ell$ from C_2^+ to C_1^+ and $Q_{\ell+1}$ from the left endpoint of C_1^+ to -1.

Proposition 9.7.4. *Fix* $z_0 \in (-1, 1)$. *Let* $\Delta_1, \Delta_2, \dots, \Delta_{\ell+1}$ *be the change of* arg $B(e^{i\theta}, z_0)$ *as* $e^{i\theta}$ *runs counterclockwise along* $Q_1, Q_2, \dots, Q_{\ell+1}$. *Then*

$$\text{(i)} \qquad 0 < \Delta_j < \pi \tag{9.7.23}$$

$$\text{(ii)} \qquad \sum_{j=1}^{\ell+1} \Delta_j = \pi \tag{9.7.24}$$

$$\text{(iii)} \qquad \omega_{z_0}(\gamma_j) = \exp\left(2i \sum_{k=1}^{\ell+1-j} \Delta_k\right) \tag{9.7.25}$$

Remark. In particular, by (9.7.23)/(9.7.24), $\omega_{z_0}(\gamma_j) \neq 1$, so $B(z, z_0)$ is not automorphic.

Proof. We first claim that for any $z_1 \in \mathbb{D}$,

$$b(z, \bar{z}_1) = \overline{b(\bar{z}, z_1)} \tag{9.7.26}$$

as follows from the definition or by noting, as in the proof of (9.7.15), that the two are equal up to phase but both are positive at $z = 0$.

Second, since $c\gamma_j c = \gamma_j^{-1}$, we see $\{c\gamma c\}_{\gamma\in\Gamma}$ runs through Γ as γ runs through Γ. Thus, if $z_0 \in (-1, 1)$, $\{\gamma(z_0)\}_{\gamma\in\Gamma}$ and $\{\gamma(\bar{z}_0)\}_{\gamma\in\Gamma}$ are the same. In particular, for such z_0,

$$B(z, z_0) = \overline{B(\bar{z}, z_0)} \tag{9.7.27}$$

Thus, for $z \in (-1, 1)$, $B(z, z_0)$ is real.

By (9.7.13),

$$B(\gamma_j(z), z_0) = \omega_{z_0}(\gamma_j) B(z, z_0) \tag{9.7.28}$$

By (9.6.7), this implies that if x is real, then

$$B(r_j^+(x), z_0) = \omega_{z_0}(\gamma_j)\, \overline{B(x, z_0)} \tag{9.7.29}$$

(since $\overline{B(x, z_0)} = B(x, z_0)$). This in turn implies that for all $z \in \mathbb{D}$,

$$B(r_j^+(z), z_0) = \omega_{z_0}(\gamma_j)\, \overline{B(z, z_0)} \tag{9.7.30}$$

for both sides are anti-analytic in z and agree if $z \in (-1, 1)$.

Suppose for $z \notin \{\gamma(z_0)\}_{\gamma\in\Gamma}$ we write

$$B(z, z_0) = |B(z, z_0)| A(z, z_0) \tag{9.7.31}$$

Then, by (9.7.30), if $r_j(z) = z$, that is, $z \in C_j^+$, $A(z, z_0)$ is constant, and for such z,

$$A(z, z_0)^2 = \omega_{z_0}(\gamma_j) \tag{9.7.32}$$

Consider tracking $\arg B(z, z_0)$, as z follows a path from 1 to -1, going successively through $Q_1, C_\ell^+, Q_2, C_{\ell-1}^+, \dots, C_1^+, Q_{\ell+1}$ on a curve we call η. On each Q_j, $\arg B$ is increasing, for $|B(e^{i\theta}, z_0)| = 1$ and $|B(re^{i\theta}, z_0)| < 1$ for $r < 1$ implies $\frac{\partial}{\partial r}|B(re^{i\theta}, z_0)| < 0$, which, by the Cauchy–Riemann equations, imply

$$\frac{\partial}{\partial\theta} \arg B(e^{i\theta}, z_0) > 0 \tag{9.7.33}$$

Thus, $\Delta_j > 0$, and since $\arg B$ is constant on each C_j^+, the change of $\arg B$ along the curve η is $\Delta_1 + \cdots + \Delta_{\ell+1}$. If we follow η by $\bar{\eta}$ run backwards, the change is the same by (9.7.27), so the closed curve running from 1 to 1 along $\partial\mathcal{F}$ is $2(\Delta_1 + \cdots + \Delta_{\ell+1})$.

By the argument principle, the change is also $2\pi \times$ number of zeros in $\mathcal{F}^{\text{int}}$ which is 2π since the only zero is at z_0. This proves (9.7.24), which in turn implies $\Delta_j < \pi$ since $\Delta_j > 0$.

By construction, the constant argument on C_j^+ is $\sum_{k=1}^{\ell+1-j} \Delta_k$, so by (9.7.32), we obtain (9.7.25). □

Our next topic concerns the connection of $B(z)$ to the potential theorist's Green's function, $G_{\mathfrak{e}}(z)$, discussed in Section 5.5 (see (5.5.110))—recall for $\mathfrak{e}$, a finite gap set, it is the unique positive harmonic function of $\mathbb{C} \setminus \mathfrak{e}$ so that $\lim_{z\to\mathfrak{e}} G_{\mathfrak{e}}(z) = 0$ and $G_{\mathfrak{e}}(z) = \log(|z|) + O(1)$ as $|z| \to \infty$; indeed (see (5.5.110)), with $C(\mathfrak{e})$ the capacity of $\mathfrak{e}$,

$$G_{\mathfrak{e}}(z) = \log|z| - \log(C(\mathfrak{e})) + O\left(\frac{1}{z}\right) \tag{9.7.34}$$

We will also need a symbol for $\lim_{z\to 0,\, z\neq 0} z\mathbf{x}(z)$, so we define x_∞ by requiring

$$\mathbf{x}(z) = \frac{x_\infty}{z} + O(1) \tag{9.7.35}$$

near $z = 0$.

Theorem 9.7.5. *Let $\mathfrak{e}$ be a finite gap set and $B(z)$ the associated Blaschke product for $z_0 = 0$. Then*

$$|B(z)| = e^{-G_{\mathfrak{e}}(\mathbf{x}(z))} \tag{9.7.36}$$

In particular,

(i)

$$B(z) = \frac{C(\mathfrak{e})}{x_\infty}\, z + O(z^2) \tag{9.7.37}$$

(ii) *For $z_0 = 0$, the numbers Δ_j of Proposition 9.7.4 are given by*

$$\Delta_j = \pi\rho_{\mathfrak{e}}(\mathfrak{e}_{\ell+1-j}) \tag{9.7.38}$$

where $\rho_{\mathfrak{e}}$ is the equilibrium measure, and $\mathfrak{e}_j = [\alpha_j, \beta_j]$ is the jth interval in $\mathfrak{e}$.

Proof. By (9.7.19), $|B(z)|$ is automorphic, so there exists a real-valued function β on $(\mathbb{C} \cup \{\infty\}) \setminus \mathfrak{e}$ with values in $[0, 1)$

$$\beta(\mathbf{x}(z)) = |B(z)| \tag{9.7.39}$$

For $z_1 \neq 0$, $b(0, z_1) = |z_1|$ and $b(z, 0) = z$, so

$$B(z) = \left(\prod_{\gamma\neq 1} |\gamma(0)|\right) z + O(z^2) \tag{9.7.40}$$

which implies that near $x = \infty$ in $\mathbb{C}$,

$$-\log(\beta(x)) = \log|x| - \log\left(x_\infty \prod_{\gamma\neq 1} |\gamma(0)|\right) + o\left(\frac{1}{x}\right) \tag{9.7.41}$$

Away from $z \in \{\gamma(0)\}_{\gamma\in\Gamma}$, $|B(z)|$ is nonvanishing, so $-\log(\beta(x))$ is a positive harmonic function on $\mathbb{C} \setminus \mathfrak{e}$. Since $|B(re^{i\theta})| \to 1$ as $r \uparrow 1$ with $e^{i\theta} \in \partial\mathcal{F} \cap \partial\mathbb{D}$, as $x \to \mathfrak{e}$,

$$-\log(\beta(x)) \to 0 \tag{9.7.42}$$

Thus, by the unique specification of $G_{\mathfrak{e}}$, we have

$$-\log(\beta(x)) = G_{\mathfrak{e}}(x) \tag{9.7.43}$$

which is (9.7.36).

(9.7.34), (9.7.40), and (9.7.41) then imply (9.7.37) as well as

$$\prod_{\gamma \neq 1} |\gamma(0)| = \frac{C(\mathfrak{e})}{x_\infty} \tag{9.7.44}$$

Finally, by looking at the curve in Figure 9.6.2 and (9.7.25), we see that

$$2 \sum_{k=1}^{\ell+1-j} \Delta_k \tag{9.7.45}$$

is the change of the argument of the multivalued analytic function whose magnitude is $e^{-G_{\mathfrak{e}}(x)}$ under the curve in the lower half of Figure 9.6.2. This implies, using a Cauchy–Riemann equation, that

$$\Delta_j = \int_{\alpha_{\ell+1-j}}^{\beta_{\ell+1-j}} \frac{\partial G_{\mathfrak{e}}}{\partial n}(x)\, dx \tag{9.7.46}$$

(the 2 in (9.7.45) and the two sides of the contour cancel to give a single integral over the top of the cut).

By (5.6.7) for $x \in \mathfrak{e}^{\text{int}}$,

$$\frac{\partial G_{\mathfrak{e}}}{\partial n}(x) = \pi \rho_{\mathfrak{e}}(x) \tag{9.7.47}$$

with $\rho_{\mathfrak{e}}$ the density of $d\rho_{\mathfrak{e}}$ and thus, (9.7.46) is (9.7.38). □

This will let us compute integrals of automorphic functions over $\partial\mathbb{D}$!

Theorem 9.7.6. *Let $\mathfrak{e}$ be a finite gap set and $d\rho_{\mathfrak{e}}$ its equilibrium measure. Then*

$$\int_{\partial\mathbb{D}} f(\mathbf{x}(e^{i\theta}))\, \frac{d\theta}{2\pi} = \int_{\mathfrak{e}} f(x)\, d\rho_{\mathfrak{e}}(x) \tag{9.7.48}$$

where this holds for any continuous function, f, on $\mathfrak{e}$ and also for any positive measurable function (with integrals allowed to be infinite). This implies $f(\mathbf{x}(e^{i\theta})) \in L^p(\partial\mathbb{D}, \frac{d\theta}{2\pi})$ if and only if $f(x) \in L^p(\mathfrak{e}, d\rho_{\mathfrak{e}})$.

Remark. The explicit formula (5.4.96) for $\frac{d\rho_{\mathfrak{e}}}{dx}$ (which only depends on the fact that $\int \frac{d\rho_{\mathfrak{e}}(x)}{x-z}$ has pure imaginary boundary values on $\mathfrak{e}$ and so works in all finite gap situations) and (9.7.48) implies

$$\int_{\partial\mathbb{D}} |f(\mathbf{x}(e^{i\theta}))|\, \frac{d\theta}{2\pi} < \infty \Leftrightarrow \int_{\mathfrak{e}} |f(x)| \text{dist}(x, \mathbb{R} \setminus \mathfrak{e})^{-1/2}\, dx < \infty \tag{9.7.49}$$

Proof. If we prove it for continuous f's, we get it for characteristic functions of open sets by taking decreasing monotone limits, and then for general positive functions by taking increasing monotone limits.

Let $\mathcal{A} = \partial\mathcal{F} \cap \partial\mathbb{D}$ and $\mathcal{A}_\gamma = \gamma[\mathcal{A}]$ so $\partial\mathbb{D} \setminus \Lambda(\mathcal{F})$ is the disjoint union of $\mathcal{A}_\gamma$ over $\gamma \in \Gamma$, that is,

$$\int_{\partial\mathbb{D}} f(\mathbf{x}(e^{i\theta}))\,\frac{d\theta}{2\pi} = \sum_\gamma \int_{\mathcal{A}_\gamma} f(\mathbf{x}(e^{i\theta}))\,\frac{d\theta}{2\pi} \tag{9.7.50}$$

Since γ is a smooth function from $\mathcal{A}$ to $\mathcal{A}_\gamma$ and $f(x(\gamma(e^{i\theta}))) = f(x(e^{i\theta}))$, we see

$$\int_{\mathcal{A}_\gamma} f(\mathbf{x}(e^{i\theta}))\,\frac{d\theta}{2\pi} = \int_{\mathcal{A}} f(\mathbf{x}(e^{i\theta}))|\gamma'(e^{i\theta})|\,\frac{d\theta}{2\pi} \tag{9.7.51}$$

Since $|\gamma(e^{i\theta})| = 1$, we see

$$|\gamma'(e^{i\theta})| = \frac{\partial \arg\gamma(e^{i\theta})}{\partial\theta} \tag{9.7.52}$$

where we use

$$\frac{\partial \arg\gamma(e^{i\theta})}{\partial\theta} \geq 0 \tag{9.7.53}$$

because

$$\frac{\partial}{\partial r}\,|\gamma(re^{i\theta})|\bigg|_{r=1} \geq 0 \tag{9.7.54}$$

(since $|\gamma(re^{i\theta})| < 1 = |\gamma(e^{i\theta})|$ if $r < 1$).

By (9.7.21),

$$\frac{\partial}{\partial r}\,\log|B(re^{i\theta})| = \sum_{\gamma\in\Gamma} \frac{\partial}{\partial r}\,\log|\gamma(re^{i\theta})| \tag{9.7.55}$$

which leads, via a Cauchy–Riemann equation, to

$$\frac{\partial}{\partial\theta}\,\arg B(e^{i\theta}) = \sum_{\gamma\in\Gamma} |\gamma'(e^{i\theta})| \tag{9.7.56}$$

From (9.7.50), (9.7.51), and (9.7.56), we deduce

$$\begin{aligned}\int_{\partial\mathbb{D}} f(\mathbf{x}(e^{i\theta}))\,\frac{d\theta}{2\pi} &= \int_{\mathcal{A}} f(\mathbf{x}(e^{i\theta}))\,\frac{d\arg B(e^{i\theta})}{d\theta}\,\frac{d\theta}{2\pi} \\ &= \int_{\mathfrak{e}} f(x)\,\frac{d\arg B(x^{-1}(u))}{du}\,\frac{du}{\pi}\end{aligned} \tag{9.7.57}$$

$(2\pi)^{-1}$ becomes $(\pi)^{-1}$ because $\mathbf{x}^{-1}$ maps the $u + i0$ to $\mathcal{A} \cap \mathbb{C}_+$ and $u - i0$ to $\mathcal{A} \cap \mathbb{C}_-$, so the single integral over $\mathfrak{e}$ gets counted twice when we integrate over $\mathcal{A}$.

By a Cauchy–Riemann equation,

$$\begin{aligned}\frac{d\arg B(\mathbf{x}^{-1}(u))}{du} &= -\frac{\partial}{\partial n}\,\log|B(\mathbf{x}^{-1}(u))| \\ &= \frac{\partial}{\partial n}\,G_{\mathfrak{e}}(u)\end{aligned} \tag{9.7.58}$$

by (9.7.36). By (9.7.47),

$$\frac{\partial}{\partial n} G_{\mathfrak{e}}(u) \frac{du}{\pi} = \rho_{\mathfrak{e}}(u)\, du \tag{9.7.59}$$

so

$$\text{RHS of (9.7.57)} = \int f(u)\, d\rho_{\mathfrak{e}}(u) \tag{9.7.60}$$

proving (9.7.48). □

There is a version of (9.7.48) that holds for noninvariant functions. Namely, given any function $g \in L^1(\partial\mathbb{D}, \frac{d\theta}{2\pi})$, we define

$$\tilde{g}(e^{i\theta}) = \frac{\sum_{\gamma\in\Gamma} g(\gamma(e^{i\theta}))|\gamma'(e^{i\theta})|}{\sum_{\gamma\in\Gamma}|\gamma'(e^{i\theta})|} \tag{9.7.61}$$

which is invariant under γ, so there is h on $\mathfrak{e}$ with

$$h(\mathbf{x}(e^{i\theta})) = \tfrac{1}{2}[\tilde{g}(e^{i\theta}) + \tilde{g}(e^{-i\theta})] \tag{9.7.62}$$

and then

$$\int g(e^{i\theta}) \frac{d\theta}{2\pi} = \int_{\mathfrak{e}} h(x)\, d\rho_{\mathfrak{e}}(x) \tag{9.7.63}$$

Note that if $g \in C(\partial\mathbb{D})$, then $h \in C(\mathfrak{e})$.

As a final topic, we want to consider when infinite products and alternating products of $B(z, z_k)$ converge. Since $B(z, \gamma(0)) = B(z, 0)$ and $\sum_{\gamma\in\Gamma}(1-|\gamma(0)|) < \infty$, we cannot hope that $\sum(1 - |z_k|) < \infty$ is enough with no restrictions on z_k. But if we restrict to $z_k \in \mathcal{F}$, it is sufficient. Here is a pair of relevant theorems:

Theorem 9.7.7. *Let* $\{z_k\}_{k=1}^{\infty}$ *all lie in* $\mathcal{F}$. *If*

$$\sum_k (1 - |z_k|) < \infty \tag{9.7.64}$$

then $\prod_{k=1}^{K} B(z, z_k)$ *is absolutely convergent as* $K \to \infty$ *for all* $z \in \mathbb{D}$, *that is,*

$$\sum_k (1 - |B(z, z_k)|) < \infty \tag{9.7.65}$$

uniformly on compact subsets of $\mathbb{D}$. *If* (9.7.64) *fails, then uniformly on compact subsets of* $\mathbb{D}$, $\prod_{k=1}^{K} B(z, z_k) \to 0$.

Proof. Since $|b(z, \gamma(z_k))| \le 1$ for $z \in \mathbb{D}$, we have

$$|B(z, z_k)| \le |b(z, z_k)|$$

Thus, by Proposition 2.3.16(i), if (9.7.64) fails, $\prod_k |B(z, z_k)| \to 0$.

Conversely, by Proposition 2.3.16(iv), we need only prove that

$$\sum_{\substack{z_k \\ \gamma\in\Gamma}} (1 - |\gamma(z_k)|) < \infty \tag{9.7.66}$$

to imply the absolute convergence of the product. Since $\sum_{\gamma\in\Gamma}(1-|\gamma(0)|)<\infty$, we can drop any $z_k=0$ terms and so suppose the sum is over those z_k with $z_k\neq 0$. Then $\inf_{k,\gamma}|\gamma(z_k)|>0$, so (9.7.66) is equivalent to

$$\prod_{\substack{z_k\neq 0\\ \gamma\in\Gamma}}|\gamma(z_k)|>0 \tag{9.7.67}$$

or equivalently, by (9.7.21), to

$$\prod_{z_k\neq 0}|B(z_k)|>0 \tag{9.7.68}$$

or equivalently to

$$\sum_k (1-|B(z_k)|)<\infty \tag{9.7.69}$$

B is analytic in a closed neighborhood, N, of $\overline{\mathcal{F}}$ (closure in $\overline{\mathbb{D}}$). We can suppose this neighborhood has the property that for some $\varepsilon>0$, $\omega\in\mathbb{D}\cap N$ and $|\omega|>1-\varepsilon$ implies $\omega/|\omega|\in N$. Since B is analytic on N, $\sup_{\omega\in N}|B'(\omega)|<\infty$, so for some C and all ω with $|\omega|>1-\varepsilon$,

$$\begin{aligned} 1-|B(\omega)| &\le |B(\omega/|\omega|)-B(\omega)| && (9.7.70)\\ &\le C(1-|\omega|) && (9.7.71)\end{aligned}$$

In proving (9.7.70), we used $|B(\omega/|\omega|)|=1$.

Since only finitely many z_k have $1-|z_k|>\varepsilon$, we have, by the hypothesis $z_k\in\mathcal{F}$,

$$\begin{aligned}\sum_{z_k}(1-|B(z_k)|) &\le \text{const}+\sum_{|z_k|>1-\varepsilon}(1-|B(z_k)|)\\ &\le \text{const}+C\sum_{z_k}(1-|z_k|) && (9.7.72)\\ &<\infty && (9.7.73)\end{aligned}$$

proving (9.7.69). □

Theorem 9.7.8. *Let $\{x_k\}_{k=1}^\infty$ be a sequence of points in $\mathbb{R}\setminus\mathfrak{e}$. Let $z_k\in\mathcal{F}$ be picked so $\mathbf{x}(z_k)=x_k$. Then*

$$\sum_k(1-|z_k|)<\infty \Leftrightarrow \sum_k \operatorname{dist}(x_k,\mathfrak{e})^{1/2}<\infty \tag{9.7.74}$$

Proof. We note that $|z_k|\to 1$ implies $x_k\to\mathfrak{e}$, so we can consider separately x_k's converging to a fixed α_j or β_j. By Theorem 9.6.4 near any such α_j or β_j, say α_j,

$$\mathbf{x}(z)=x(\alpha_j)+c(z-\alpha_j)^2+O((z-\alpha_j)^3) \tag{9.7.75}$$

So, since the $C_j^\pm$ and $(-1,1)$ are orthocircles,

$$1-|z_k|=c^{-1/2}\operatorname{dist}(x_k,\mathfrak{e})^{1/2}+O(\operatorname{dist}(x_k,\mathfrak{e})^{3/2}) \tag{9.7.76}$$

(c depends on the branch point). (9.7.74) thus holds. □

From (9.7.76), (9.7.36), and the fact that B' is nonvanishing on $\partial\mathbb{D} \setminus \mathcal{L}$, we see, near any $x_0 \in \{\alpha_j, \beta_j\}_{j=1}^{\ell+1}$ for $x \notin \mathfrak{e}$, that

$$G_{\mathfrak{e}}(x) = c(x_0)|x - x_0|^{1/2} + O(|x - x_0|^{3/2}) \tag{9.7.77}$$

This also follows from the explicit form of the Stieltjes transform of $d\nu_{\mathfrak{e}}$ (see (5.4.88)) and the relation of $G_{\mathfrak{e}}$ and this Stieltjes transform.

As a final topic, we turn to alternating Blaschke products like those treated in Theorem 3.3.2.

Theorem 9.7.9. *Let $\eta \in \{\alpha_j, \beta_j\}_{j=1}^{\ell+1}$. Let $\{\zeta_j\}_{j=1}^{\infty}$ and $\{\rho_j\}_{j=1}^{\infty}$ be a sequence of reals with $\zeta_j \to \eta$, $\rho_j \to \eta$. If η is an α_j (if $j = 1$, we mean $-\infty$ for β_{j-1}),*

$$\beta_{j-1} < \zeta_1 < \rho_1 < \zeta_2 < \cdots < \alpha_j \tag{9.7.78}$$

or

$$\beta_{j-1} < \rho_1 < \zeta_1 < \rho_2 < \cdots < \alpha_j \tag{9.7.79}$$

If ζ is a β_j (if $j = \ell - 1$, we mean ∞ for α_{j+1}),

$$\alpha_{j+1} > \zeta_1 > \rho_1 > \zeta_2 > \cdots > \beta_j \tag{9.7.80}$$

or

$$\alpha_{j+1} > \rho_1 > \zeta_2 > \rho_2 > \cdots > \beta_j \tag{9.7.81}$$

Let n, $\{z_j\}_{j=1}^{\infty}$, $\{p_j\}_{j=1}^{\infty}$ be the unique points in $\mathcal{F}$ with

$$\mathbf{x}(n) = \eta \qquad \mathbf{x}(z_j) = \zeta_j \qquad \mathbf{x}(p_j) = \rho_j \tag{9.7.82}$$

Then, as $N \to \infty$,

$$\prod_{j=1}^{N} \frac{B(z, z_j)}{B(z, p_j)} \to B_\infty(z) \tag{9.7.83}$$

uniformly on compact subsets of

$$\mathbb{C} \cup \{\infty\} \Big\backslash \Big[\mathcal{L} \cup \bigcup_{\gamma \in \Gamma} (\{\gamma(p_j)\}_{j=1}^{\infty} \cup \{\gamma(z_j^{-1})\}_{j=1}^{\infty} \cup \gamma(n))\Big] \tag{9.7.84}$$

to an analytic function with only simple poles at

$$\bigcup_{\gamma \in \Gamma} [\{\gamma(p_j)\}_{j=1}^{\infty} \cup \{\gamma(z_j^{-1})\}_{j=1}^{\infty}] \tag{9.7.85}$$

B_∞ is nonvanishing in (9.7.84) *except at*

$$\bigcup_{\gamma \in \Gamma} [\{\gamma(z_j)\}_{j=1}^{\infty} \cup \{\gamma(p_j^{-1})\}_{j=1}^{\infty}] \tag{9.7.86}$$

Moreover, on $\partial\mathbb{D} \setminus [\mathcal{L} \cup \cup_{\gamma \in \Gamma} \{\gamma(n)\}]$,

$$|z| = 1 \Rightarrow |B_\infty(z)| = 1 \tag{9.7.87}$$

Finally, we have that if $\arg B_\infty$ *is defined on* $\mathcal{F}$ *by requiring* $\arg B_\infty(0) = 0$*, then there is a* Γ*-dependent constant,* C_Γ*, so that*

$$z \in \mathcal{F} \Rightarrow |\arg B_\infty(z)| \leq C_\Gamma \tag{9.7.88}$$

If we place a cut along the orthocircle, which contains the $\{z_j\}_{j=1}^\infty$ *from* z_1 *(in case* (9.7.78) *or* (9.7.80)*) or* p_1 *(in case* (9.7.79) *or* (9.7.81)*) to* n *and all its images under* $\gamma \in \Gamma$*, to get a region* $\mathcal{B}$*, which is simply connected and on which* B_∞ *is analytic and nonvanishing, then*

$$z \in \mathcal{B} \setminus \mathcal{R}_{n+1} \Rightarrow |\arg B_\infty(z)| \leq (2n+1)C_\Gamma \tag{9.7.89}$$

Remarks. 1. In the case of Theorem 3.3.2, we have z_j real, so we could replace $\bar{z}_j^{-1}$ by z_j. Here z_j may lie on some $C_k^\pm$ so not be real. However, by (9.6.9),

$$(\gamma_k^\pm)^{-1}(z_j) = \bar{z}_j \tag{9.7.90}$$

so $\{\gamma(z_j^{-1})\}_{\gamma\in\Gamma} = \{\gamma(\bar{z}_j^{-1})\}_{\gamma\in\Gamma}$.

2. For simplicity of notation, we henceforth restrict to the case (9.7.78) or (9.7.80).

To begin the proof, we need an analog of the functions $\tilde{b}(z, x)$.

Proposition 9.7.10. (i) *Let* $\{a_j\}_{j=1}^\infty$, $\{b_j\}_{j=1}^\infty$ *be sets in* $\mathbb{C}$ *with no* a_j *equal to a* b_k *and*

$$\sum_{j=1}^\infty |a_j - b_j| < \infty \tag{9.7.91}$$

Then uniformly on compact subsets of $\mathbb{C} \cup \{\infty\} \setminus \overline{\{b_j\}_{j=1}^\infty}$*, we have that*

$$\prod_{j=1}^N \frac{z - a_j}{z - b_j} \tag{9.7.92}$$

converges uniformly and absolutely. The only zeros are at $\{a_j\}_{j=1}^\infty$.

(ii) *For* $\zeta, \omega \in \mathbb{C} \setminus \mathcal{L}$ *distinct so* $\infty \notin \{\gamma(\zeta)\}_{\gamma\in\Gamma} \cup \{\gamma(\omega)\}_{\gamma\in\Gamma}$ *and all* $z \in \mathbb{C} \cup \{\infty\} \setminus [\mathcal{L} \cup \{\gamma(\omega)\}_{\gamma\in\Gamma}]$,

$$\prod_{\substack{\gamma\in\Gamma \\ \omega(\gamma)\leq n}} \frac{z - \gamma(\zeta)}{z - \gamma(\omega)} \tag{9.7.93}$$

converges uniformly and absolutely as $n \to \infty$*. We write* $\Xi(z; \zeta, \omega)$ *for the limit*

$$\Xi(z; \zeta, \omega) = \prod_{\gamma\in\Gamma} \frac{z - \gamma(\zeta)}{z - \gamma(\omega)} \tag{9.7.94}$$

(iii) *For any* $z_0 \in \mathbb{D}$, $z_0 \neq 0$,

$$B(z, z_0) = \left[\prod_{\gamma\in\Gamma} |\gamma(z_0)|\right]^{-1} \Xi(z; z_0, \bar{z}_0^{-1}) \tag{9.7.95}$$

(iv) *For* $\zeta, \omega \in \mathbb{C} \cup [\mathcal{L} \cup \{\gamma(0)\}_{\gamma\in\Gamma} \cup \{\gamma(\infty)\}_{\gamma\in\Gamma}]$, $\Xi(z; \zeta, \omega)$ *is jointly meromorphic in* z, ζ, ω.

Remark. By (9.7.21), the product in (9.7.95) is $|B(z_0)|$.

Proof. (i) Since

$$\left|1 - \frac{(z-a_j)}{(z-b_j)}\right| = \frac{|a_j - b_j|}{|z-b_j|} \le \frac{|a_j - b_j|}{\min_k |z-b_k|} \tag{9.7.96}$$

we get the absolute convergence by (9.7.91).

(ii) Since $\zeta, \omega \notin \overline{\{\gamma^{-1}(\infty)\}_{\gamma\in\Gamma}}$, we can find a smooth curve $c(t)$ with $c(0) = \zeta$, $c(1) = \omega$ so

$$\inf_{\substack{\gamma\neq 1;\, \gamma\in\Gamma \\ t\in[0,1]}} |c(t) - \gamma^{-1}(\infty)| = Q > 0 \tag{9.7.97}$$

By (9.4.26), for $\gamma \neq 1$,

$$\frac{|\gamma'(c(t))|}{|\gamma'(0)|} = \frac{|\gamma^{-1}(\infty)|^2}{|c(t) - \gamma^{-1}(\infty)|^2} \tag{9.7.98}$$

$$\le \frac{\sup_{\gamma\neq 1} |\gamma^{-1}(\infty)|^2}{Q^2} \equiv Q_1 \tag{9.7.99}$$

Thus,

$$\left|\frac{d}{dt}\gamma(c(t))\right| \le |c'(t)| Q_1 |\gamma'(0)| \tag{9.7.100}$$

so with $Q_2 = Q_1 \int_0^1 |c'(t)|\, dt$,

$$|\gamma(\zeta) - \gamma(\omega)| \le Q_2 |\gamma'(0)| \tag{9.7.101}$$

Since $\sum_{\gamma\in\Gamma} |\gamma'(0)| < \infty$, (9.7.101) implies

$$\sum_{\gamma\in\Gamma} |\gamma(\zeta) - \gamma(\omega)| < \infty \tag{9.7.102}$$

that is, (9.7.91), so (i) $\Rightarrow$ (ii).

(iii) We have

$$-\frac{z_0}{|z_0|}(1 - \bar{z}_0 z) = |z_0|(z - \bar{z}_0^{-1}) \tag{9.7.103}$$

so

$$b_{z_0}(z) = \frac{1}{|z_0|}\left(\frac{z - z_0}{z - \bar{z}_0^{-1}}\right) \tag{9.7.104}$$

which leads to (9.7.95).

(iv) This is clearly true for finite products, and so for the limit. □

Lemma 9.7.11. *Fix* Q *a compact subset of a single* $\overline{C_\ell^\pm}$ *(closure in* $\partial\mathbb{D}$*) or* $(0, 1]$ *or* $[-1, 0)$ *and* K *a compact subset of* $\mathbb{C}$ *with*

$$K \cap \bigcup_{\gamma\in\Gamma} \gamma[Q] \cup \gamma[Q^{-1}] \cup \mathcal{L} = \emptyset \tag{9.7.105}$$

Then there is a C so that for all $\zeta, \omega \in Q \cap \mathbb{D}$ and $z \in K$,

$$\left| 1 - \frac{B(z,\zeta)}{B(z,\omega)} \right| \leq C|\zeta - \omega| \tag{9.7.106}$$

Proof. By (9.7.105),

$$\inf_{\substack{z \in K \\ \omega \in Q}} |B(z,\omega)| > 0$$

so it suffices to prove

$$|B(z,\zeta) - B(z,\omega)| \leq C_1|\zeta - \omega| \tag{9.7.107}$$

which, by (9.7.104) and

$$\inf_{\omega \in Q} |B(\omega)| > 0 \tag{9.7.108}$$

is implied by

$$||B(\zeta)| - |B(\omega)|| \leq C_2|\zeta - \omega| \tag{9.7.109}$$

$$|\Xi(z;\zeta,\bar{\zeta}^{-1}) - \Xi(z;\omega,\bar{\omega}^{-1})| \leq C_3|\zeta - \omega| \tag{9.7.110}$$

To prove (9.7.109), we use the fact that

$$||B(\zeta)| - |B(\omega)|| \leq |B(\zeta) - B(\omega)| \tag{9.7.111}$$

and that B is analytic in a neighborhood of $\overline{C_\ell^\pm}$.

For (9.7.110), we use the fact that when (9.7.90) holds, $\Xi(z;\zeta,\eta)$ is jointly analytic in all variables in a neighborhood of $z \in K$, $\zeta \in Q$, $\eta \in \bar{Q}^{-1}$, so (9.7.110) holds. □

Lemma 9.7.12. *Let C be a circle, $\{z = z_0 + re^{i\theta}\}$, in $\mathbb{C}$ and f a smooth function on C. Define*

$$\mathrm{Var}_C(f) = \int_0^{2\pi} \left| \frac{d}{d\theta} f(z) \right| d\theta \tag{9.7.112}$$

be the total variation of f over C. If $w \notin$ closed disk surrounded by C and r is the radius of C and

$$f_w(z) = \arg(w - z) \tag{9.7.113}$$

then

$$\text{(i)} \qquad \mathrm{Var}_C(f_w) \leq 2\pi \tag{9.7.114}$$

$$\text{(ii)} \qquad \mathrm{Var}_C(f_w) \leq \frac{4r}{\mathrm{dist}(w, C)} \tag{9.7.115}$$

Proof. Let z_0, z_1 be the two points on C where the lines from w through z_j are tangent to C. Order them so the clockwise arc from z_0 to z_1 goes through the point, z_2, on C closest to w (see Figure 9.7.1). Let θ_0 be the angle between the lines from w to the center of C and the line from w to z_1. Let θ_1 be $\arg(w - z_2)$.

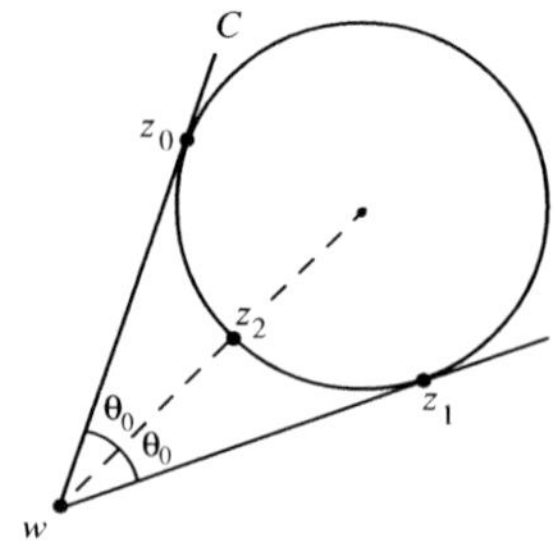

Figure 9.7.1. Point on a circle.

Then $\arg(w - z)$ goes from $\theta_1 - \theta_0$ to $\theta_1 + \theta_0$, monotonically increasing as z runs from z_0 to z_1 and monotonically decreasing from $\theta_1 + \theta_0$ to $\theta_1 - \theta_0$ as z completes the circuit, that is,

$$\mathrm{Var}_C(f_w) = 4\theta_0 \tag{9.7.116}$$

Since $\theta_0 \le \pi/2$, (9.7.114) is immediate.

Let $\tilde{z} = \frac{1}{2}(z_0 + z_1)$. Then

$$\tan(\theta_0) = \frac{|z_1 - \tilde{z}|}{|w - \tilde{z}|} \tag{9.7.117}$$

$$\le \frac{r}{\mathrm{dist}(w, C)} \tag{9.7.118}$$

(9.7.115) follows from this, (9.7.116), and (for $y > 0$)

$$\tan^{-1}(y) = \int_0^y \frac{dx}{1 + x^2} \le y \tag{9.7.119}$$

□

It will be useful to discuss total variations over arcs of C also. Recall in Theorem 9.6.13, we used r_γ for the radius of the orthocircle C_γ.

Lemma 9.7.13. *For any $z \in \mathcal{F}$ and ζ in some $C_j^\pm$, let*

$$f_z(\zeta) = \arg(B(z, \zeta)) \tag{9.7.120}$$

Then

$$\mathrm{Var}_{C_j^\pm}(f_z) \le 4\ell\pi + \sum_{\gamma:\, w(\gamma) \ge 2} \frac{4r_\gamma}{d} \tag{9.7.121}$$

where

$$d = \min\biggl(|z - w| \biggm| z \in \mathcal{F},\ w \in \bigcup_{w(\gamma)=2} C_\gamma\biggr) \tag{9.7.122}$$

If $A^\pm = \pm(0, 1)$, $\operatorname{Im} z > 0$, *$z \in \mathcal{F}$, and $f_z(\zeta)$ is given by* (9.7.120) *for $\zeta \in A^\pm$, then*

$$\mathrm{Var}_{A^\pm}(f_z) \le \pi + \textit{RHS of } (9.7.121) \tag{9.7.123}$$

Proof. As ζ runs through the part of some orthocircle, C, inside $\mathbb{D}$, $\bar{\zeta}^{-1}$ runs through the part of the same orthocircle outside $\mathbb{D}$. Thus, for z fixed in $\mathbb{D}$ outside C, if

$$g_z(\zeta) = \arg\left(\frac{z-\zeta}{z-\bar{\zeta}^{-1}}\right) \tag{9.7.124}$$

and $h_z(\zeta) = \arg(z-\zeta)$, then

$$\mathrm{Var}_{C\cap\mathbb{D}}(g_z) \le \mathrm{Var}_C(h_z) \tag{9.7.125}$$

Since $\prod_{\gamma\in\Gamma}|\gamma(z_0)|^{-1}$ is positive, by (9.7.95) and (9.7.125),

$$\mathrm{Var}_{C_j^{\pm}}(B(z,\,\cdot\,)) \le \sum_{\gamma} \mathrm{Var}_{\widetilde{(C_j^{\pm})}}(\arg(z-\,\cdot\,)) \tag{9.7.126}$$

where $\widetilde{\gamma(C_j^{\pm})}$ is the complete orthocircle containing $\gamma(C_j^{\pm})$. Since $\gamma(C_j^{\pm})$ is inside C_γ, its radius and its distance from $z\in\mathcal{F}$ is bounded by the same for C_γ. Thus, $4\ell\pi$ bounds the 2ℓ terms in (9.7.126) with $w(\gamma)=1$ and, by (9.7.115), the sum over $w(\gamma)\ge 2$ is bounded by the sum in (9.7.121).

For $A^{\pm}$, we have $\gamma(A^{\pm})$ is inside C_γ for $\gamma\ne 1$, so the sum over γ's with $\gamma\ne 1$ is bounded by the right side of (9.7.121). The $A^{\pm}$ term is bounded by π as in Theorem 3.3.6. □

Proof of Theorem 9.7.9. By Lemma 9.7.11, for any compact K in the set (9.7.84),

$$\sum_j \left|1-\frac{B(z,z_j)}{B(z,p_j)}\right| \le \sum_j C(K)|z_j-p_j| \tag{9.7.127}$$

since $|z_j-p_j|\le$ arclength on that $C_k^{\pm}$, which contains all the z_j and p_j. But, by the interlacing property, these arcs are disjoint, so their sum is bounded by the total arclength of $C_k^{\pm}$. Thus, the sum converges uniformly on K and so, all the analyticity properties and also (9.7.87) hold. Thus, we need only prove the statements about $\arg B_\infty$.

(9.7.88) follows from Lemma 9.7.13 since the arg of a finite product is bounded by a sum of args of single ratios—which is precisely what a bounded variation condition bounds. Thus, (9.7.88) holds with C_Γ given by the right-hand side of (9.7.123).

Because each $B(\,\cdot\,,z_j)$ is character automorphic, so is $B_\infty(z)$ as a uniform limit. Thus,

$$\max_{z,w\in\gamma[\mathcal{F}]} |\arg B_\infty(z)-\arg B_\infty(w)|$$

is γ-independent, and so bounded by $2C_\Gamma$. If $z\in\mathcal{B}\cap(\mathcal{R}_n\setminus\mathcal{R}_{n+1})$, there is a path from 0 to z that goes through part of $\mathcal{F}, \gamma^{(1)}(\mathcal{F}), \gamma^{(2)}(\mathcal{F}),\dots,\gamma^{(n)}(\mathcal{F})$ where $w(\gamma^{(j)})=j$ successively. The change of arg B_∞ is at most C_Γ in $\mathcal{F}$ and $2C_\Gamma$ in $\gamma^{(j)}(\mathcal{F})$, so at most $(2n+1)C_\Gamma$. □

Remarks and Historical Notes. The connection between $s=1$ Poincaré convergence and convergence of Blaschke products is classical. Indeed, Poincaré used his series to construct automorphic functions. The connection of $B(z)$ to the potential theorist's Green's function is also part of standard lore; see, for example, Tsuji [446].

Theorem 9.7.7 is from the work of Sodin–Yuditskii [413] and Peherstorfer–Yuditskii [343, 344] who also have calculations similar to (9.7.48). The present proofs we give of Theorem 9.7.6 and Theorem 9.7.9 are from the work of Christiansen–Simon–Zinchenko [86, 87, 88, 89].

One can use (9.7.48) to define a natural map that is an analog of the Szegő mapping of Section 1.9. The idea is that, under this map, Sz: $d\rho_{\mathfrak{e}}$ goes to $\frac{d\theta}{2\pi}$ and $g(x)\,d\rho_{\mathfrak{e}}(x)$ goes to $g(\mathbf{x}(e^{i\theta}))\frac{d\theta}{2\pi}$. This plus continuity determines this mapping. Put differently, there is a map, $\mathbf{x}_\sharp^*\colon \mathcal{M}_{+,1}(\partial\mathbb{D}) \to \mathcal{M}_{+,1}(\mathfrak{e})$ by $\rho = \mathbf{x}_\sharp^*(\mu)$ given by

$$\int h(x)\,d\rho = \int h(\mathbf{x}(e^{i\theta}))\,d\mu(\theta) \tag{9.7.128}$$

This map is many-to-one. But it is one-one if we restrict to quasi-invariant measures, that is, measures with

$$\mu(-\theta) = \mu(\theta) \tag{9.7.129}$$

and for all $\gamma \in \Gamma$,

$$d\mu(\arg(\gamma(e^{i\theta}))) = |\gamma'(e^{i\theta})|\,d\mu(\theta) \tag{9.7.130}$$

9.8 CONTINUITY OF THE COVERING MAP

Fix ℓ and let $Q_\ell \subset \mathbb{R}^{2\ell+2}$ be all $(2\ell+2)$-tuples $(\alpha_1, \dots, \beta_{\ell+1})$ obeying (5.12.2). In this section, we want to consider the dependence of the basic objects of this chapter, the covering map, $\mathbf{x}$, the Fuchsian group generators, $\{\gamma_j\}_{j=1}^{\ell}$, and the Blaschke factors, $B(z,w)$, on $q \in Q_\ell$. So we will often write $\mathbf{x}_q(z)$, $\gamma_j^{(q)}(z)$, $B_q(z)$, $B_q(z,w)$. Our main goal in this section is to prove that

Theorem 9.8.1.

$$\text{(i)} \qquad q \mapsto \mathbf{x}_q(\cdot) \tag{9.8.1}$$

$$\text{(ii)} \qquad q \mapsto \gamma_j^{(q)}(\cdot) \tag{9.8.2}$$

$$\text{(iii)} \qquad q \mapsto B_q(\,\cdot\,, w) \tag{9.8.3}$$

are continuous as maps in $q \in Q_\ell$ to analytic functions in the topology of uniform convergence on compact subsets of $\mathbb{D}$.

Remark. γ_j, $B(\,\cdot\,, w)$ have values in $\mathbb{D}$ but $\mathbf{x}_q$ has values in $\mathbb{C}\cup\{\infty\}$, so we mean uniform in the proper local coordinates on $\mathbb{C}\cup\{\infty\}$ (to handle poles).

This is the kind of result that one is tempted to prove via Goldberger's method (see the Notes): "The argument is via the method of *reductio ad absurdum*—suppose the result is false. Why, that's absurd!" The proof, while not difficult, is not so short. Two keys will be that if f_n is a sequence of analytic functions from

$\mathbb{D} \to \mathbb{D}$, then there is a subsequence, $n(j)$, so $f_{n(j)}$ converges uniformly on compact subsets of $\mathbb{D}$ either to another analytic function of $\mathbb{D}$ to $\mathbb{D}$ or to a constant function with value in $\partial\mathbb{D}$ (Montel's theorem). The second is that if $f_n \to f$ uniformly on compact subsets of a region Ω and if $z_n \to z$ in Ω, then $f_n(z_n) \to f(z)$ (because Cauchy estimates imply equicontinuity).

We will let $q_n \to q_\infty$ in Q_ℓ and use $\mathbf{x}_n$ for $\mathbf{x}_{q_n}$, $\mathbf{x}_\infty$ for $\mathbf{x}_{q_\infty}$, $\gamma_j^{(n)}$ for $\gamma_j^{(q_n)}$, $\mathfrak{e}_n$, $\mathfrak{e}_\infty$ for the associated subsets of $\mathbb{R}$, and so on. The idea of the proof will involve showing that any limit point, $\tilde{\mathbf{x}}_\infty$, of the $\mathbf{x}_n$ is $\mathbf{x}_\infty$. To do this, we will need a way of identifying covering maps. Here is the result we will use:

Proposition 9.8.2. *Let* $\mathbf{x}\colon \mathbb{D} \to \mathbb{C} \cup \{\infty\}$ *have the following properties:*

(a) $\mathbf{x}'(z) \neq 0$ *for all* z *with* $\mathbf{x}(z) \neq \infty$, *and at any point with* $\mathbf{x}(z_0) = \infty$, *the pole is simple.*

(b) $\mathbf{x}(0) = \infty$; *the residue at* 0 *is in* $(0, \infty)$.

(c) *There is a Fuchsian group,* Γ, *with*

$$\mathbf{x}(z) = \mathbf{x}(w) \Leftrightarrow \exists \gamma \in \Gamma \quad \text{so that} \quad w = \gamma(z) \tag{9.8.4}$$

Then $\mathbf{x}$ *is the covering map of* $\mathbb{D}$ *onto* $\operatorname{Ran}(\mathbf{x})$ *and* Γ *the associated Fuchsian group.*

Proof. By (a) and (b), $\mathbf{x}$ is locally one-one and $\mathbf{x}$ has the normalization we have demanded for covering maps, so we need only confirm for any $z_0 \in \mathbb{D}$, $\mathbf{x}(z_0)$ has an open neighborhood, N, so $\mathbf{x}^{-1}[N]$ is a disjoint union of open sets on which $\mathbf{x}$ is one-one.

Since $\mathbf{x}$ is locally one-one, for no $\gamma \in \Gamma$, $\gamma \neq 1$, can we have $\gamma(z_0) = z_0$. Thus,

$$r = \min_{\gamma \in \Gamma} \rho(\gamma(z_0), z_0) > 0 \tag{9.8.5}$$

where ρ is the hyperbolic metric.

For $\gamma \in \Gamma$, let

$$M_\gamma = \left\{ w \;\middle|\; \rho(w, \gamma(z_0)) < \frac{r}{2} \right\} \tag{9.8.6}$$

Since

$$\rho(\gamma^{-1}(w), z_0) = \rho(w, \gamma(z_0)) \tag{9.8.7}$$

$$\gamma^{-1}[M_\gamma] = M_1 \tag{9.8.8}$$

so, by (9.8.4),

$$\mathbf{x}[M_\gamma] = \mathbf{x}[M_1] = N \tag{9.8.9}$$

Also, by (9.8.4)

$$\mathbf{x}^{-1}[N] = \cup_\gamma M_\gamma \tag{9.8.10}$$

Next, if $\gamma \neq 1$,

$$M_1 \cap M_\gamma = \emptyset \tag{9.8.11}$$

since $w \in M_1 \cap M_\gamma$ implies

$$\rho(z_0, \gamma(z_0)) \leq \rho(z_0, w) + \rho(w, \gamma(z_0)) < r$$

violating (9.8.5).

Since

$$M_\gamma \cap M_{\gamma'} = \gamma[M_1 \cap M_{\gamma^{-1}\gamma'}]$$

we see the M_γ are disjoint. Thus, N is the required neighborhood.

Finally, (9.8.8) and (9.8.11) imply that if $w \in M_1$ and $\gamma \neq \mathrm{id}$, then $\gamma(w) \notin M_1$, so $w, w_1 \in M_1$, $w \neq w_1$ implies $\gamma(w) \neq w_1$, and thus, by (9.8.4), $\mathbf{x}(w) \neq \mathbf{x}(w')$, that is, $\mathbf{x}$ is one-one on M_1, and so on each M_γ. □

Next, we want to construct limits of $\mathbf{x}_n$. Fix an interval $[c, d] \in \mathfrak{e}_\infty^{\text{int}}$. For n large, $[c, d] \subset \mathfrak{e}_n$ also. Let $G\colon \mathbb{D} \to \mathbb{C} \cup \{\infty\} \setminus [c, d]$ be the standard conformal bijection with $G(0) = \infty$ and the residue at ∞ positive (i.e., $G(z) = C(z + z^{-1}) + D$ for suitable C, D) and G^{-1} is its inverse. Let

$$g_n(z) = G^{-1}(\mathbf{x}_n(z)) \tag{9.8.12}$$

which maps $\mathbb{D}$ to $\mathbb{D}$ and 0 to 0.

By compactness, $\{g_n\}$ have a limit point, g_∞, in the topology of uniform convergence on compact subsets of $\mathbb{D}$, and since $g_n(0) = 0$, we have that $g_\infty(0) = 0$. Thus, g_∞ maps to $\mathbb{D}$. We therefore define

$$\tilde{\mathbf{x}}_\infty(z) = G(g_\infty(z)) \tag{9.8.13}$$

If we prove $\tilde{\mathbf{x}}_\infty = \mathbf{x}_\infty$, then by compactness again, we have convergence of the original sequence. We will abuse notation by still using $\mathbf{x}_n$ for the subsequence picked to converge.

Proposition 9.8.3. (i) $\tilde{\mathbf{x}}_\infty(0) = \infty$.

(ii) *Either $\tilde{\mathbf{x}}_\infty(z) \equiv \infty$ or else $\tilde{\mathbf{x}}_\infty$ is locally one-one (in the sense that $\tilde{\mathbf{x}}'_\infty \neq 0$ at nonpoles and all poles are simple).*

(iii) *If $\tilde{\mathbf{x}}_\infty \not\equiv \infty$, then the residue at $z = 0$ is strictly positive.*

(iv)

$$\operatorname{Ran}(\tilde{\mathbf{x}}_\infty) \subset \mathbb{C} \cup \{\infty\} \setminus \mathfrak{e}_\infty \tag{9.8.14}$$

Proof. (i)–(iii) It is immediate from $g_\infty(0) = 0$, that either g_∞ is locally one-one or identically 0 by applying Hurwitz's theorem to $g'_n \to g'_\infty$ and that $\tilde{\mathbf{x}}_\infty$ has a positive residue at 0 if and only if $g'_\infty(0) > 0$.

(iv) Since $\operatorname{Ran}(g_\infty) \subset \mathbb{D}$, $\operatorname{Ran}(\tilde{\mathbf{x}}_\infty) \subset \mathbb{C} \cup \{\infty\} \setminus [c, d]$. Let $[\tilde{c}, \tilde{d}]$ be another interval in $\mathfrak{e}_\infty^{\text{int}}$ and $\tilde{G}$ the associated conformal map from $\mathbb{D}$ to $\mathbb{C} \cup \{\infty\} \setminus [\tilde{c}, \tilde{d}]$. Then $\tilde{G}^{-1}(\mathbf{x}_n) \to \tilde{G}^{-1}(\tilde{\mathbf{x}}_\infty)$ near $z = 0$, and so on all of $\mathbb{D}$. Thus, $\operatorname{Ran}(\tilde{G}^{-1}(\tilde{\mathbf{x}}_\infty)) \subset \mathbb{D}$, so $[\tilde{c}, \tilde{d}] \cap \operatorname{Ran}(\tilde{\mathbf{x}}_n) = \emptyset$.

It follows that

$$\operatorname{Ran}(\tilde{\mathbf{x}}_\infty) \subset \mathbb{C} \cup \{\infty\} \setminus \mathfrak{e}_\infty^{\text{int}} \tag{9.8.15}$$

But if $\tilde{\mathbf{x}}_\infty \not\equiv \infty$, its range is open, so either way, (9.8.14) holds. □

Proposition 9.8.4. *We have*

$$\operatorname{Ran}(\tilde{\mathbf{x}}_\infty) \supset \mathbb{C} \cup \{\infty\} \setminus [\alpha_1^{(\infty)}, \beta_{\ell+1}^{(\infty)}] \tag{9.8.16}$$

In particular, $\tilde{\mathbf{x}}_\infty \not\equiv \infty$, so $\tilde{\mathbf{x}}_\infty$ is locally one-one with positive residue at ∞.

Proof. As constructed in Section 9.6, $\mathbf{x}_n$ has a unique inverse, $\mathbf{y}_n$, from $X_n \equiv \mathbb{C} \cup \{\infty\} \setminus [\alpha_1^{(n)}, \beta_{\ell+1}^{(n)}]$ into $\mathbb{D}$ with

$$\mathbf{y}_n(\infty) = 0 \tag{9.8.17}$$

It is inverse in the sense that

$$\mathbf{x}_n(\mathbf{y}_n(z)) = z \tag{9.8.18}$$

for all $z \in X_n$. Of course, $\text{Ran}(\mathbf{y}_n) = \mathcal{F}_n^{\text{int}}$.

Since $\{\mathbf{y}_n\}$ are uniformly bounded, and for any $z \in X_\infty \equiv \mathbb{C}\cup\{\infty\}\setminus[\alpha_1^{(\infty)}, \beta_{\ell+1}^{(\infty)}]$, eventually $z \in X_n$, by passing to a subsequence, we can suppose $\mathbf{y}_n$ has a limit, $\mathbf{y}_\infty$, uniformly on compact subsets of X_∞. By (9.8.17), $\mathbf{y}_\infty$ maps to $\mathbb{D}$ and, by the uniform convergence within $\mathbb{D}$,

$$\tilde{\mathbf{x}}_\infty(\mathbf{y}_\infty(z)) = z \tag{9.8.19}$$

which proves (9.8.16).

Clearly, $\tilde{\mathbf{x}}_\infty \not\equiv \infty$ and so, by Proposition 9.8.3, completes the proof of the last statement. □

Remark. While it is not essential (since passing to subsequences finitely many times is harmless), we note that once we see that $\tilde{\mathbf{x}}_\infty$ is locally one-one near ∞, we see all solutions of (9.8.19) with $\mathbf{y}_\infty(\infty) = 0$ are equal near ∞, and so equal. Thus, $\mathbf{y}_n \to \mathbf{y}_\infty$ without the need to pass to a subsequence. The same is true of the $\gamma_j^{(\infty)}$ discussed below.

Proposition 9.8.5. (i)

$$\text{Ran}(\tilde{\mathbf{x}}_\infty) = \mathbb{C} \cup \{\infty\} \setminus \mathfrak{e} \tag{9.8.20}$$

(ii) *We have that*

$$\sup_{j,n} |\gamma_j^{(n)}(0)| < 1 \tag{9.8.21}$$

Proof. If $[c, d] \subset G_j^{(\infty)}$, for some j, for n large, $[c, d] \cap \mathfrak{e}^{(n)} = \emptyset$. Let $\tilde{X}_n$ be X_n with $[\alpha_1^{(n)}, \beta_1^{(n)}]$ replaced by $([\alpha_1^{(n)}, \beta_1^{(n)}] \setminus [c, d]) \cup \{w \mid |w - \frac{1}{2}[c + d]| = \frac{1}{2}|d - c|;\ \text{Im}\, w \geq 0\}$, that is, the interval pushed into a semicircle in the upper half-plane. Because $\tilde{X}_n$ is simply connected, there is a unique map, $\tilde{\mathbf{y}}_n : \tilde{X}_n \to \mathbb{D}$, so $\tilde{\mathbf{y}}_n$ obeys (9.8.17) and (9.8.18). Near infinity, $\tilde{\mathbf{y}}_n$ converges to $\tilde{\mathbf{y}}_\infty$ so $\tilde{\mathbf{y}}_n$ converges to $\tilde{\mathbf{y}}_\infty$ on $\tilde{X}_\infty$, which agrees with $\mathbf{y}_\infty$ in $\tilde{X}_\infty \setminus \{w \mid |w - \frac{1}{2}(c+d)| \leq \frac{1}{2}|d - c|;\ \text{Im}\, w \geq 0\}$. Since $\tilde{\mathbf{x}}_\infty \circ \tilde{\mathbf{y}}_\infty(z) = z$, we see $[c, d] \subset \text{Ran}(\mathbf{x}_\infty)$. Since $[c, d]$ is an arbitrary interval in any gap and we have (9.8.14) and (9.8.16), we conclude (9.8.20).

Since $C_j^{(n)+}$ is the hyperbolic perpendicular bisector of $0, \gamma_j^{(n)}(0)$, we have

$$w \in C_j^{(n)+} \Rightarrow \rho(0, \gamma(0)) \leq 2\rho(0, w) \tag{9.8.22}$$

By construction, if $[c, d] \subset G_j^{(\infty)}$, $\tilde{y}_n(\frac{1}{2}(c + d)) \in C_j^{(n)+}$, so

$$\lim_{n\to\infty} \rho(0, \gamma_j^{(n)}(0)) \leq 2\rho(0, \tilde{y}_\infty(\tfrac{1}{2}\,(c + d))) \tag{9.8.23}$$

This holds for each j and proves (9.8.20). □

Let $\Gamma^{(n)}$ be the Fuchsian group associated to $\mathbb{C}\cup\{\infty\}\setminus\mathfrak{e}^{(n)}$. We will need to look at limits of $\Gamma^{(n)}$ as $n\to\infty$. For this, the following will be useful:

Proposition 9.8.6. (i) *As $n\to\infty$, $B_n(z)$ has a limit $\tilde{B}_\infty(z)$ (uniformly on compact subsets of $\mathbb{D}$), which is not identically* 0.

(ii)

$$\sup_n \sum_{\gamma\in\Gamma^{(n)}} (1-|\gamma(0)|)<\infty \tag{9.8.24}$$

Proof. (i) We can find $\varepsilon>0$, so for large n, $\text{Ran}(\mathbf{y}_n)\supset\{z\mid |z|<2\varepsilon\}$. Since $\mathbf{x}_n$ is one-one on $\text{Ran}(\mathbf{y}_n)$, we see B_n is nonvanishing on $\{z\mid 0\le|z|\le\varepsilon\}$. Since $B_n'(0)>0$, this implies for $|z|<1$ that

$$B_n(z\varepsilon)=z\exp\biggl(\int\frac{e^{i\theta}+z}{e^{i\theta}-z}\log|B_n(\varepsilon e^{i\theta})|\,\frac{d\theta}{2\pi}\biggr) \tag{9.8.25}$$

By (9.7.36) and Proposition 5.6.2,

$$|B_n(\varepsilon e^{i\theta})|=\exp(-G_{\mathfrak{e}_n}(\mathbf{x}_n(\varepsilon e^{i\theta})))\to\exp(-G_{\mathfrak{e}}(\tilde{\mathbf{x}}_\infty(\varepsilon e^{i\theta})) \tag{9.8.26}$$

so, by (9.8.25), $B_n(z)$ converges for $|z|<\varepsilon$. By boundedness of $B_n(z)$ uniformly in n and $|z|<1$, we get convergence on all $\mathbb{D}$. Since (9.8.26) implies the limit $\tilde{B}_\infty$ is nonvanishing on $\{z\mid 0<|z|<\varepsilon\}$, we see $\tilde{B}_\infty$ is not identically zero.

(ii) By Hurwitz's theorem, $B_\infty(z)/z$ has a nonzero value at $z=0$, so by (9.7.40),

$$\inf_n \prod_{\substack{\gamma\in\Gamma^{(n)}\\ \gamma\neq 1}} |\gamma(0)|>0 \tag{9.8.27}$$

For all real y, $e^y\ge 1+y$ (by convexity), so $e^{(w-1)}\ge w$. So for $0<w<1$, $e^{(1-w)}\le w^{-1}$ and

$$\exp\biggl(\sum_{\substack{\gamma\in\Gamma\\ \gamma\neq 1}}(1-|\gamma(0)|)\biggr)\le\prod_{\substack{\gamma\in\Gamma\\ \gamma\neq 1}}|\gamma(0)|^{-1} \tag{9.8.28}$$

and (9.8.27) implies (9.8.24). □

By Corollary 9.4.2 and (9.8.20), by passing to a subsequence, we can suppose for each $j=1,\dots,\ell$ that there is $\tilde{\gamma}_j^{(\infty)}\in\mathcal{M}$ so

$$\gamma_j^{(n)}\to\gamma_j^{(\infty)} \tag{9.8.29}$$

Let $\tilde{\Gamma}^{(\infty)}$ be the free group generated by $\{\gamma_j^{(\infty)}\}_{j=1}^{\ell}$. By (9.8.24),

$$\sum_{\gamma\in\tilde{\Gamma}^{(\infty)}}(1-|\gamma(0)|)<\infty \tag{9.8.30}$$

so $\tilde{\Gamma}^{(\infty)}$ is Fuchsian.

Proposition 9.8.7. *If $\gamma \in \tilde{\Gamma}^{(\infty)}$, there exist $\gamma_n \in \Gamma^{(n)}$ so $\gamma_n \to \gamma$. Conversely, if γ_n is a sequence in $\Gamma^{(n)}$ and $\gamma_n^{(\infty)}$ has a limit in $\mathbb{D}$, then $\gamma_{n(j)} \to \gamma$ for some $\gamma \in \tilde{\Gamma}^{(\infty)}$.*

Proof. If $\gamma \in \tilde{\Gamma}^{(\infty)}$ is a finite word in $\{\gamma_j^{(\infty)\pm 1}\}_{j=1}^{\ell}$, it is a limit of the same word in $\{\gamma_j^{(n)\pm 1}\}_{j=1}^{\ell}$.

For the converse, we note that, by Corollary 9.3.14, if $w(\tilde{\gamma})$ is the word length of $\tilde{\gamma} \in \Gamma^{(n)}$, we can find $1, \gamma_{(1)}, \dots, \gamma_{(w-1)} \in \Gamma^{(n)}$ with $|\gamma_{(j)}(0)| \leq |\tilde{\gamma}(0)|$ and $w(\gamma_{(j)}) = j$ for we can write $\tilde{\gamma}(0) = r_1 \dots r_{w(\gamma)}(0)$ where r_k is a reflection in a C_j^+ and the $r_{k+1} \dots r_{w(\gamma)}(0)$ is outside the circle in which r_k is a reflection. Thus,

$$\sum_{\gamma \in \Gamma^{(n)}} |1 - \gamma(0)| \geq w(\tilde{\gamma})(1 - |\tilde{\gamma}(0)|) \tag{9.8.31}$$

Now suppose $\gamma_n \to \gamma$ so $\gamma_n(0) \to z_\infty \in \mathbb{D}$. By (9.8.24) and (9.8.31), $\sup_n w(\gamma_n) \equiv W < \infty$. There are only finitely many word patterns of length W or less, so one must get repeated infinitely often, and that provides the subsequence. □

Proposition 9.8.8. *$\tilde{\mathbf{x}}_\infty$ has the property that*

$$\tilde{\mathbf{x}}_\infty(z) = \tilde{\mathbf{x}}_\infty(w) \Leftrightarrow \exists\, \gamma \in \tilde{\Gamma}^{(\infty)} \quad \textit{so that} \quad w = \gamma(z) \tag{9.8.32}$$

Proof. If $\gamma \in \tilde{\Gamma}^{(\infty)}$, there exist $\gamma_n \in \Gamma^{(n)}$ so $\gamma_n \to \gamma$. Thus, $\gamma_n(z) \to \gamma(z) \in \mathbb{D}$ so $\mathbf{x}_n(\gamma_n(z)) \to \tilde{\mathbf{x}}_\infty(\gamma(z))$ and $\mathbf{x}_n(\gamma_n(z)) = \mathbf{x}_n(z)$ implies

$$\tilde{\mathbf{x}}_\infty(\gamma(z)) = \tilde{\mathbf{x}}_\infty(z) \tag{9.8.33}$$

Conversely, let z, w be such that the left-hand side of (9.8.32) holds. Since $\mathbf{x}_n(w) - \mathbf{x}_n(z) \to 0$ and $\mathbf{x}_n(w), \mathbf{x}_n(z)$ have locally one-one limits, so $\mathbf{x}_n$ is uniformly locally invertible, there exists $w_n \to w$ so $\mathbf{x}_n(w_n) = \mathbf{x}_n(z)$. Thus, there is $\gamma_n \in \Gamma^{(n)}$ with $w_n = \gamma(z) \to w$. By Proposition 9.8.7, there is $\gamma \in \tilde{\Gamma}^{(n)}$ with $\gamma(z) = w$. □

Proof of Theorem 9.8.1. Let $\tilde{\mathbf{x}}_\infty$ be a limit point of the $\mathbf{x}_n$'s. As discussed above, $\tilde{\mathbf{x}}_\infty$ obeys all the hypotheses of Proposition 9.8.2 with $\text{Ran}(\tilde{\mathbf{x}}_\infty) = \mathbb{C} \cup \{\infty\} \setminus \mathfrak{e}_\infty$ and $\Gamma = \Gamma^{(\infty)}$. Thus, $\tilde{\mathbf{x}}_\infty = \mathbf{x}_\infty$ and $\tilde{\Gamma}^{(\infty)} = \Gamma^{(\infty)}$. By compactness, we conclude $\mathbf{x}_n \to \mathbf{x}_\infty$ and $\gamma_j^{(n)} \to \gamma_j^{(\infty)}$ uniformly on compacts. This implies convergence of finite Blaschke products associated to a set of words in $\mathbb{F}_n$. By (9.8.24), these finite Blaschke products converge to $B(z, w)$ uniformly in n. Thus, $B_n(z, w) \to B_\infty(z, w)$. □

Remarks and Historical Notes. Theorem 9.8.1 is a special case of a result of Hejhal [194] who noted that one could also base a proof on ideas of Ahlfors–Bers [8]. Hejhal's method is different from the one in this section that describes joint work with Jacob Christiansen and Maxim Zinchenko. M. Goldberger is a distinguished theoretical physicist with a running gag about "Goldberger's method" as we have quoted it—it expressed the notion of many theoretical physicists that mathematical statements that are "obviously" true do not require proof!

9.9 STEP-BY-STEP SUM RULES FOR FINITE GAP JACOBI MATRICES

With the covering map in hand, we can follow the by now standard path to get nonlocal step-by-step sum rules and from that the step-by-step sum rule that will yield a Szegő–Shohat–Nevai-type theorem in the next section. The disappointment is that we do not know how to get a Killip–Simon-type sum rule.

Theorem 9.9.1 (Nonlocal finite gap step-by-step sum rule). *Let $\mathfrak{e}$ be a finite gap set and J a Jacobi matrix with $\sigma_{\text{ess}}(J) = \mathfrak{e}$. Let $\mathbf{x}$ be the covering map for $\mathbb{C} \cup \{\infty\} \setminus \mathfrak{e}$ and let $\{p_j\}_{j=1}^{N_1}, \{z_j\}_{j=1}^{N_2}$ be a counting of the points in $\mathcal{F}$, which go, under $\mathbf{x}$, into the eigenvalues of J (for p_j) and J_1 (for z_j). Let $M(z)$ be given by (9.7.1) and let B_∞ be the alternating Blaschke product for the z's and p's given by Theorem 9.7.9. Then up to sets of $\frac{d\theta}{2\pi}$ measure zero,*

$$\{\theta \mid \operatorname{Im} M(e^{i\theta}) \neq 0\} = \{\theta \mid \operatorname{Im} M_1(e^{i\theta}) \neq 0\} \tag{9.9.1}$$

and

$$\log\biggl(\frac{\operatorname{Im} M(e^{i\theta})}{\operatorname{Im} M_1(e^{i\theta})}\biggr) \in \bigcap_{p<\infty} L^p\biggl(\partial\mathbb{D}, \frac{d\theta}{2\pi}\biggr) \tag{9.9.2}$$

Moreover,

$$a_1 M(z) = B(z) B_\infty(z) \exp\biggl(\int \frac{e^{i\theta}+z}{e^{i\theta}-z} \log\biggl(\frac{\operatorname{Im} M(e^{i\theta})}{\operatorname{Im} M_1(e^{i\theta})}\biggr) \frac{d\theta}{4\pi}\biggr) \tag{9.9.3}$$

Remarks. 1. Theorem 9.7.9 deals with a single gap and endpoint of that gap. In the above, we may need $2\ell + 2$ such "B_∞'s" to get the B_∞ of this theorem.

2. As usual, the left side of (9.9.2) is really a function, which is given as shown only on the set of (9.9.1).

Proof. Let

$$g(z) = \frac{a_1 M(z)}{B(z) B_\infty(z)} \tag{9.9.4}$$

which is a character automorphic function on $\mathbb{D}$ with no poles or zeros (after fixing the removable singularities at $\{\gamma(p_j)\}_{j=1,\dots,\gamma\in\Gamma}$). Clearly, on $\mathcal{F}^{\text{int}}$,

$$|\arg(M(z))| \leq \pi \tag{9.9.5}$$

and on $\mathcal{F}$, by (9.7.30),

$$|\arg(B(z) B_\infty(z))| \leq \widetilde{C}_\Gamma \tag{9.9.6}$$

Thus, on $\mathcal{F}$,

$$|\arg(g(z))| \leq (\widetilde{C}_\Gamma + \pi) \tag{9.9.7}$$

Since g is character automorphic, the variation of $\arg(g(z))$ over any $\overline{\gamma(\mathcal{F})}$ is at most $2(\widetilde{C}_\Gamma + \pi)$, so

$$z \in \mathbb{D} \setminus \mathcal{R}_{n+1} \Rightarrow |\arg(g(z))| \leq (2n+1)(\widetilde{C}_\Gamma + \pi) \tag{9.9.8}$$

or

$$\{z \mid |\arg(g(z))| \geq (2n+1)(\widetilde{C}_\Gamma + \pi)\} \subset \mathcal{R}_{n+1} \tag{9.9.9}$$

By Theorem 9.6.12,

$$\log(g) \in \bigcap_{p<\infty} H^p(\mathbb{D}) \tag{9.9.10}$$

Since

$$(a_1|M(e^{i\theta})|)^2 = \frac{\operatorname{Im} M(e^{i\theta})}{\operatorname{Im} M_1(e^{i\theta})} \tag{9.9.11}$$

we obtain (9.9.1), (9.9.2), (9.9.3), just as in the proof of Theorem 3.4.1. □

We emphasize that unlike the previous case considered, where $\arg(g)$ is bounded, here $\arg(g)$ is not bounded, but it is still in h_p for all $p < \infty$.

Lemma 9.9.2. *$a_1 M(z)/B(z)$ has a removable singularity at $z = 0$ and*

$$\left.\frac{a_1 M(z)}{B(z)}\right|_{z=0} = \frac{a_1}{C(\mathfrak{e})} \tag{9.9.12}$$

Proof. By (9.7.37),

$$B(z) = \frac{C(\mathfrak{e})}{x_\infty} z + O(z^2) \tag{9.9.13}$$

Moreover, near $x = \infty$,

$$m(x) = -\frac{1}{x} + O(x^{-2}) \tag{9.9.14}$$

and

$$\mathbf{x}(z) = \frac{x_\infty}{z} + O(1) \tag{9.9.15}$$

so

$$M(z) = \frac{z}{x_\infty} + O(z^2) \tag{9.9.16}$$

We obtain (9.9.12) immediately from (9.9.13) and (9.9.16). □

Theorem 9.9.3 (C_0 finite gap step-by-step sum rule)**.** *Under the hypotheses of Theorem 9.9.1, we have that*

$$-\log\left(\frac{a_1}{C(\mathfrak{e})}\right) = Z_{\mathbf{x}}(J_1 \mid J) + \sum G_{\mathfrak{e}}(\mathbf{x}(z_j)) - G_{\mathfrak{e}}(\mathbf{x}(p_j)) \tag{9.9.17}$$

where $Z_{\mathbf{x}}$ is given by (3.4.27) (*but with a different meaning of M, M_1*).

Remarks. 1. In (3.4.27), M means $-m(z+z^{-1})$, while here it means $-m(\mathbf{x}(z))$, which is why we use $Z_{\mathbf{x}}$.

2. The sum in (9.9.17) is a finite sum (looking at sequences outside $\mathfrak{e}$ but approaching each point of $\partial\mathfrak{e}$) of alternating sums since $G_{\mathfrak{e}}$ is monotone in a neighborhood (in $\mathbb{R} \setminus \mathfrak{e}$) of each point of $\partial\mathfrak{e}$.

Proof. Given the lemma and (9.7.36) plus $B(0, z) = |B(z)|$ (see (9.7.21)), this is precisely $\log|\cdot|$ of (9.9.3) evaluated at $z = 0$. □

Remarks and Historical Notes. This result is from Christiansen–Simon–Zinchenko [87, 88].

9.10 THE SZEGŐ–SHOHAT–NEVAI THEOREM FOR FINITE GAP JACOBI MATRICES

Our goal in this section is to prove Theorem 9.1.1.

Theorem 9.10.1 (Szegő–Shohat–Nevai Theorem for Finite Gap Sets). *Let* $\mathfrak{e}$ *be a finite gap set and*

$$d\mu = w\,dx + d\mu_{\rm s} \tag{9.10.1}$$

a measure with associated Jacobi matrix J *with Jacobi parameters* $\{a_n, b_n\}_{n=1}^\infty$, *and so that*

$$\sigma_{\rm ess}(J) \subseteq \mathfrak{e} \tag{9.10.2}$$

Suppose that

$$\sum_{E\in\sigma(J)\setminus\sigma_{\rm ess}(J)} \operatorname{dist}(E, \mathfrak{e})^{1/2} < \infty \tag{9.10.3}$$

Then

$$\int \operatorname{dist}(x, \mathbb{R}\setminus\mathfrak{e})^{-1/2} \log(w(x))\,dx > -\infty \Leftrightarrow \limsup_{n\to\infty} \frac{a_1\dots a_n}{C(\mathfrak{e})^n} > 0 \tag{9.10.4}$$

Moreover, if one and hence both sides of (9.10.4) *are valid, then*

$$0 < \liminf \frac{a_1\dots a_n}{C(\mathfrak{e})^n} \le \limsup \frac{a_1\dots a_n}{C(\mathfrak{e})^n} < \infty \tag{9.10.5}$$

Remarks. 1. Notice that (9.10.4) has lim sup, not lim inf, that is, (9.10.4) is equivalent to

$$\int \operatorname{dist}(x, \mathbb{R}\setminus\mathfrak{e})^{-1/2} \log(w(x))\,dx = -\infty \Leftrightarrow \lim \frac{a_1\dots a_n}{C(\mathfrak{e})^n} = 0 \tag{9.10.6}$$

2. By Remark 1 and (9.10.5), we see that if (9.10.2)/(9.10.3) hold, then either the lim sup $< \infty$ or the limit is 0, and so lim sup $< \infty$! Thus,

$$(9.10.4) + (9.10.5) \Rightarrow \limsup \frac{a_1\dots a_n}{C(\mathfrak{e})^n} < \infty \tag{9.10.7}$$

3. We will eventually prove $\frac{a_1\dots a_n}{C(\mathfrak{e})^n}$ is asymptotically almost periodic (see Section 9.13).

4. The class of measures, μ, with (9.10.2) so that (9.10.3) and both (or either) condition in (9.10.4) we call the *Szegő class for* $\mathfrak{e}$, denoted $\operatorname{Sz}(\mathfrak{e})$. If $\mu \in \operatorname{Sz}(\mathfrak{e})$, we say the Szegő condition holds.

One step in the proof requires us to go from (9.10.3) to uniform control on such sums for some approximations and/or restrictions of J, so we begin with this issue.

Proposition 9.10.2. *Let A be a bounded selfadjoint operator with a gap (a,b) in its essential spectrum. Define for any such operator, C,*

$$\Sigma_{(a,b)}(C) = \sum_{E\in(a,b)\cap\sigma(C)} \operatorname{dist}(E,\mathbb{R}\setminus(a,b))^{1/2} \tag{9.10.8}$$

(counting eigenvalues up to multiplicity). Then

(i) *If B is another bounded selfadjoint operator so*

$$\operatorname{rank}(B-A) \le r < \infty \tag{9.10.9}$$

then

$$\Sigma_{(a,b)}(B) \le \Sigma_{(a,b)}(A) + r\left(\frac{|b-a|}{2}\right)^{1/2} \tag{9.10.10}$$

(ii) *If P is an orthogonal projection so that*

$$\operatorname{rank}(PA(1-P)) = r < \infty \tag{9.10.11}$$

then

$$\Sigma_{(a,b)}(PAP) \le \Sigma_{(a,b)}(A) + r\left(\frac{|b-a|}{2}\right)^{1/2} \tag{9.10.12}$$

Proof. (i) By induction, we can suppose $r=1$ and then that $B-A\ge 0$ (by replacing A, B by $\frac{1}{2}(a+b)-A$, $\frac{1}{2}(a+b)-B$). Label the eigenvalues of A by

$$\cdots \le E_{-1}(A) \le \tfrac{1}{2}(a+b) \le E_0(A) \le E_1(A) \le \cdots \tag{9.10.13}$$

which can be infinite in number on one or both sides, or finite, but by the essential spectrum assumption, a countable ordered set exhausts the spectrum in (a,b).

Since $B-A$ is rank one and positive, we can label all eigenvalues of B so that

$$E_j(A) \le E_j(B) \le E_{j+1}(A) \tag{9.10.14}$$

Clearly, for $j\ge 0$,

$$\operatorname{dist}(E_j(B),\mathbb{R}\setminus(a,b)) \le \operatorname{dist}(E_j(A),\mathbb{R}\setminus(a,b)) \tag{9.10.15}$$

and for $j\le -1$,

$$\operatorname{dist}(E_{j-1}(B),\mathbb{R}\setminus(a,b)) \le \operatorname{dist}(E_j(A),\mathbb{R}\setminus(a,b)) \tag{9.10.16}$$

Thus,

$$\Sigma_{(a,b)}(B) \le \operatorname{dist}(E_{-1}(B),\mathbb{R}\setminus(a,b))^{1/2} + \Sigma_{(a,b)}(A) \tag{9.10.17}$$

which implies (9.10.10) for $r=1$ if we note that no single eigenvalue contributes more than $(|b-a|/2)^{1/2}$ to the sum.

(ii) By adding constants and scaling, we can suppose

$$b=-a=1 \tag{9.10.18}$$

Define for any nonnegative bounded selfadjoint operator C with $\sigma_{\text{ess}}(C) \cap [0,1) = \emptyset$,

$$\tilde{\Sigma}(C) = \sum_{\substack{E \in \sigma(C) \\ 0 \le E \le 1}} \left(1 - \sqrt{E}\right)^{1/2} \tag{9.10.19}$$

Then, clearly,

$$\Sigma_{(-1,1)}(A) = \tilde{\Sigma}(A^2) \tag{9.10.20}$$

Moreover, the same argument as in (i) shows that if

$$\operatorname{rank}(C - D) = r \tag{9.10.21}$$

then

$$\tilde{\Sigma}(C) \le r + \tilde{\Sigma}(D) \tag{9.10.22}$$

By the min-max principle, the eigenvalues of PCP are bounded below by those of C for any projection P, so

$$\tilde{\Sigma}(PCP) \le \tilde{\Sigma}(C) \tag{9.10.23}$$

On the other hand,

$$PA^2P - (PAP)^2 = PA(1-P)AP \tag{9.10.24}$$

has rank r. Thus,

$$\begin{aligned} \Sigma_{(-1,1)}(PAP) &= \tilde{\Sigma}((PAP)^2) && \text{(by (9.10.20))} \\ &\le r + \tilde{\Sigma}(PA^2P) && \text{(by (9.10.22))} \\ &\le r + \tilde{\Sigma}(A^2) && \text{(by (9.10.23))} \\ &= r + \Sigma_{(-1,1)}(A) && \text{(by (9.10.20))} \end{aligned}$$

proving (9.10.12) in this case. □

Here is a consequence of this theorem:

Corollary 9.10.3. *Let A be a bounded selfadjoint operator with a gap, (a,b), in its essential spectrum. Suppose $\Sigma_{(a,b)}(A)$ given by* (9.10.8) *is finite. For any ε and r, there is a δ so that for all projections, P, obeying* (9.10.11)*, we have*

$$\sum_{E \in [(a,a+\delta] \cup [b-\delta,b)] \cup \sigma(PAP)} \operatorname{dist}(E, \mathbb{R} \setminus (a,b))^{1/2} < \varepsilon \tag{9.10.25}$$

Proof. First pick δ_1 so (9.10.25) holds for ε replaced by $\varepsilon/2$, $\delta = \delta_1$, and $P \equiv 1$, which is possible since $\Sigma_{(a,b)}(A) < \infty$. By Proposition 9.10.2,

$$\Sigma_{(a,a+\delta)}(PAP) \le \Sigma_{(a,a+\delta)}(A) + \left(\frac{r\delta}{2}\right)^{1/2}$$

so we pick $\delta \le \delta_1$ so that $(r\delta/2)^{1/2} < \varepsilon/4$. Then (9.10.25) holds. □

We begin by relating $Z(J \mid J_1)$ to relative entropy and relative entropy to the Szegő condition on the left side of (9.10.4). Our reference measure will be $d\rho_{\mathfrak{e}}$, the equilibrium measure for $\mathfrak{e}$. Of course, by (2.2.1),

$$S(\rho_{\mathfrak{e}} \mid \mu) = \int_{\mathfrak{e}} \log\left(\frac{w(x)}{\rho_{\mathfrak{e}}(x)}\right) d\rho_{\mathfrak{e}} \tag{9.10.26}$$

Proposition 9.10.4. *We have that*

(i) $$\int \log(w(x)) \operatorname{dist}(x, \mathbb{R} \setminus \mathfrak{e})^{-1/2} > -\infty \Leftrightarrow S(\rho_{\mathfrak{e}} \mid \mu) > -\infty \tag{9.10.27}$$

(ii) $$Z(J \mid J_1) = \tfrac{1}{2} S(\rho_{\mathfrak{e}} \mid \mu_1) - \tfrac{1}{2} S(\rho_{\mathfrak{e}} \mid \mu) \tag{9.10.28}$$

where μ_1 is the spectral measure for J_1, the once-stripped Jacobi matrix.

Remarks. 1. (9.10.28) is only true if $S(\rho_{\mathfrak{e}} \mid \mu) > -\infty$. More properly, it should say $S(\rho_{\mathfrak{e}} \mid \mu) > -\infty \Leftrightarrow S(\rho_{\mathfrak{e}} \mid \mu_1) > -\infty$, and then (9.10.28) holds.

2. In (3.6.8), we had an extra $-\frac{1}{2} \log 2$ because we really had two reference measures in mind, $d\rho_{\mathfrak{e}}$ and the free Jacobi measure, $d\mu_0$.

Proof. (i) By Theorem 5.5.22 and (5.5.138), we have

$$C \operatorname{dist}(x, \mathbb{R} \setminus \mathfrak{e})^{-1/2} \le \rho_{\mathfrak{e}}(x) \le D \operatorname{dist}(x, \mathbb{R} \setminus \mathfrak{e})^{-1/2} \tag{9.10.29}$$

so $\int |\log(\rho_{\mathfrak{e}}(x))| \, d\rho_{\mathfrak{e}}(x) < \infty$ and (9.10.26) implies (9.10.27).

(ii) By (2.3.56),

$$Z(J \mid J_1) = \frac{1}{2} \int_{\mathfrak{e}} \log\left(\frac{\operatorname{Im} m_1(x+i0)}{\operatorname{Im} m(x+i0)}\right) d\rho_{\mathfrak{e}}(x) \tag{9.10.30}$$

which, by

$$\operatorname{Im} m_\mu(x+i0) = \pi w_\mu(x) \tag{9.10.31}$$

and (9.10.26), implies (9.10.28). □

We can now rewrite (9.9.17).

Proposition 9.10.5. *If $S(\rho_{\mathfrak{e}} \mid \mu) > -\infty$, then*

$$\frac{a_1 \dots a_n}{C(\mathfrak{e})^n} = K \exp\left(\tfrac{1}{2} S(\rho_{\mathfrak{e}} \mid \mu) - \tfrac{1}{2} S(\rho_{\mathfrak{e}} \mid \mu_n)\right) \tag{9.10.32}$$

where μ_n is the spectral measure of J_n, the n-times stripped J, and

$$K = \exp\left(\sum_j [G_{\mathfrak{e}}(E_j(J)) - G_{\mathfrak{e}}(E_j(J_n))]\right) \tag{9.10.33}$$

Remarks. 1. Included is

$$S(\rho_{\mathfrak{e}} \mid \mu) > -\infty \Leftrightarrow S(\rho_{\mathfrak{e}} \mid \mu_n) > -\infty \tag{9.10.34}$$

2. $E_j(J)$ are the eigenvalues of J outside $\mathfrak{e}$. The sum in (9.10.33) may only be conditionally convergent.

Proof. For $n = 1$, immediate from (9.9.17) and (9.10.28). For general n, we iterate and take products. (9.10.34) follows from $|Z(J \mid J_1)| < \infty$ always. $\square$

Proposition 9.10.6. *If* (9.10.5) *holds, there is a constant,* C_1, *depending only on* $\mathfrak{e}$ *and the sum in* (9.10.5) *so that*

$$\limsup \frac{a_1 \dots a_n}{C(\mathfrak{e})^n} \le C_1 \exp(\tfrac{1}{2} S(\rho_\mathfrak{e} \mid \mu)) \tag{9.10.35}$$

In particular, $\Rightarrow$ *holds in* (9.10.6).

Proof. Let $J_\mathfrak{e}$ be the Jacobi matrix whose spectral measure is the equilibrium measure $d\rho_\mathfrak{e}$ and let $\{a_n^{(e)}, b_n^{(e)}\}_{n=1}^\infty$ be its Jacobi parameters. Define $J^{(n)}$ to be the Jacobi matrix with parameters

$$a_j^{(n)} = \begin{cases} a_j & j = 1, \dots, n \\ a_{j-n}^{(e)} & j = n+1, \dots \end{cases} \tag{9.10.36}$$

$$b_j^{(n)} = \begin{cases} b_j & j = 1, \dots, n \\ b_{j-n}^{(e)} & j = n+1, \dots \end{cases} \tag{9.10.37}$$

We claim that

$$C_1 = \sup_n \exp\biggl(\sum_j G_\mathfrak{e}(E_j(J^{(n)}))\biggr) < \infty \tag{9.10.38}$$

Accepting this, let us prove (9.10.35).

Notice that $(J^{(n)})_n$, the n-times stripped $J^{(n)}$, is $J_\mathfrak{e}$. Since $J_\mathfrak{e}$ has no eigenvalues outside $\mathfrak{e}$, (9.10.32) implies

$$\frac{a_1 \dots a_n}{C(\mathfrak{e})^n} = K_n \exp(\tfrac{1}{2} S(\rho_\mathfrak{e} \mid \mu^{(n)})) \tag{9.10.39}$$

where

$$K_n = \exp\biggl(\sum_j G_\mathfrak{e}(E_j(J^{(n)}))\biggr) \tag{9.10.40}$$

Notice next that $J^{(n)} \to J$ strongly, so $d\mu^{(n)} \xrightarrow{w} d\mu$. By the upper semicontinuity of S,

$$\limsup_{n\to\infty} S(\rho_\mathfrak{e} \mid \mu^{(n)}) \le S(\rho_\mathfrak{e} \mid \mu) \tag{9.10.41}$$

(9.10.39), (9.10.38), and (9.10.41) imply (9.10.35). Thus, we are reduced to proving (9.10.38).

The sum over E_j is a sum over ℓ gaps plus sums over two fixed intervals above $\beta_{\ell+1}$ and below α_1 (since it is easy to see $\sup_n \|J^{(n)}\| < \infty$). On each interval, we can use that on each $[-R, R]$,

$$C_R \operatorname{dist}(x, \mathfrak{e})^{1/2} \le G_\mathfrak{e}(x) \le D_R \operatorname{dist}(x, \mathfrak{e})^{1/2} \tag{9.10.42}$$

(see (9.7.77)) to control sums of $\operatorname{dist}(x, \mathfrak{e})^{1/2}$ in place of sums of $G_\mathfrak{e}$.

Let $\tilde{J}^{(n)}$ be $J^{(n)}$ with $a_n^{(n)}$ be replaced by 0. By (9.10.10),

$$\Sigma_{(a,b)}(J^{(n)}) \le \Sigma_{(a,b)}(\tilde{J}^{(n)}) + 2\biggl(\frac{|b-a|}{2}\biggr)^{1/2} \tag{9.10.43}$$

since $\operatorname{rank}(J^{(n)} - \tilde{J}^{(n)}) = 2$.

$\tilde{J}_n$ is a direct sum of $J^{(e)}$, which has no eigenvalues in (a, b) and $P_n J P_n$ where P_n is the projection onto $\{\delta_j\}_{j=1}^n$. Thus,

$$\Sigma_{(a,b)}(\tilde{J}^{(n)}) = \Sigma_{(a,b)}(P_n J P_n) \tag{9.10.44}$$

Since $P_n J(1 - P_n)$ is rank 1, (9.10.12) implies that

$$\Sigma_{(a,b)}(P_n J P_n) \le \Sigma_{(a,b)}(J) + \biggl(\frac{|b-a|}{2}\biggr)^{1/2} \tag{9.10.45}$$

Thus, by (9.10.43), (9.10.44), and (9.10.45),

$$\sup_n \Sigma_{(a,b)}(J^{(n)}) \le \Sigma_{(a,b)}(J) + 3\biggl(\frac{|b-a|}{2}\biggr)^{1/2} \tag{9.10.46}$$

which leads to (9.10.38). □

Proposition 9.10.7. *If* (9.10.5) *holds, there is a nonzero constant,* C_2*, depending only on* $\mathfrak{e}$ *and the sum in* (9.10.5) *so that*

$$\liminf \frac{a_1 \dots a_n}{C(\mathfrak{e})^n} \ge C_2 \exp(\tfrac{1}{2} S(\rho_{\mathfrak{e}} \mid \mu)) \tag{9.10.47}$$

In particular, $\Leftarrow$ *holds in* (9.10.6).

Proof. If $S(\rho_{\mathfrak{e}} \mid \mu) = -\infty$, (9.10.47) is trivial. So we suppose $S(\rho_{\mathfrak{e}} \mid \mu) > -\infty$. Since $S(\rho_{\mathfrak{e}} \mid \mu_n) \le 0$, $\exp(-\frac{1}{2} S(\rho_{\mathfrak{e}} \mid \mu_n)) \ge 1$ so, by (9.10.32)/ (9.10.35),

$$\frac{a_1 \dots a_n}{C(\mathfrak{e})^n} \ge C_2 \exp(\tfrac{1}{2} S(\rho_{\mathfrak{e}} \mid \mu))$$

where (since $G_{\mathfrak{e}} \ge 0$)

$$C_2 = \exp\biggl(-\sup_n \biggl(\sum_j G_{\mathfrak{e}}(E_j(J_n))\biggr)\biggr) \tag{9.10.48}$$

so (9.10.47) is equivalent to

$$\sup_n \biggl(\sum_j G_{\mathfrak{e}}(E_j(J_n))\biggr) < \infty \tag{9.10.49}$$

By (9.10.42), this follows if we show that for each of $\ell + 2$ intervals (ℓ gaps plus intervals adjacent to α_1 and $\beta_{\ell+1}$) that

$$\sup_n \Sigma_{(a,b)}(J_n) < \infty \tag{9.10.50}$$

But if P_n is the projection onto $\{\delta_j\}_{j=n+1}^\infty$, then

$$J_n = P_n J P_n$$

Since $P_n J(1 - P_n)$ is rank 1, (9.10.12) implies

$$\Sigma_{(a,b)}(J_n) \le \Sigma_{(a,b)}(J) + \left(\frac{|b-a|}{2}\right)^{1/2} \tag{9.10.51}$$

proving (9.10.50). □

Proof of Theorem 9.10.1. We have proven (9.10.6), which is equivalent to (9.10.4).

As noted in (9.10.27), the left side of (9.10.6) is equivalent to $S(\rho_{\mathfrak{e}} \mid \mu) > -\infty$. By (9.10.35)/(9.10.47), this implies (9.10.5). □

Remarks and Historical Notes. The approach in this section is from Christiansen–Simon–Zinchenko [88], but large parts predate their work. With no bound states and with no singular part, that the left-hand side of (9.10.4) implies (9.10.5) is due to Widom [459]. Widom considered general sets of finitely many arcs and Szegő proved asymptotics of polynomials. Aptekarev [21] specified the impact on Jacobi parameters in the OPRL case. Peherstorfer–Yuditskii [343], using the framework of Sodin–Yuditskii [413], recovered Widom's results and extended them to certain infinite gaps sets. In [344], they considered the general condition (9.10.3) on bound states.

While [88] were the first to state $S(\rho_{\mathfrak{e}} \mid \mu) = -\infty \Rightarrow \frac{a_1 \dots a_n}{C(\mathfrak{e})^n} \to 0$, Peherstorfer noted that one can obtain it also from the results of [343, 344] (see [88] for details).

[88] also prove that $S(\rho_{\mathfrak{e}} \mid \mu) > -\infty$ and $\limsup_{n\to\infty} \frac{a_1 \dots a_n}{C(\mathfrak{e})^n} < \infty$ implies (9.10.3). The proof uses the same sum rule ideas we discuss in this section.

9.11 THETA FUNCTIONS AND ABEL'S THEOREM

Blaschke products allow us to specify arbitrary points $z_0 \in \mathcal{F}$ and find f analytic in $\mathbb{D}$ with zeros only at $\{\gamma(z_0)\}_{\gamma\in\Gamma}$. The resulting functions can be meromorphically continued to $\mathbb{C} \cup \{\infty\} \setminus \Lambda(\Gamma)$ and they still have zeros only at $\{\gamma(z_0)\}_{\gamma\in\Gamma}$. But the poles lie at $\{\gamma(\bar{z}_0)^{-1}\}_{\gamma\in\Gamma}$.

In this section, one of our main goals will be to break this rigid connection between zeros and poles and allow poles instead at $\{\gamma(z_1)\}_{\gamma\in\Gamma}$ where z_1 may not be $\bar{z}_0^{-1}$. We will only accomplish this when both z_0 and z_1 lie in the same complete orthocircles $\widetilde{C}_j^{\pm}$. This will suffice for our applications.

The corresponding "theta functions" will be character automorphic. If a product of these theta functions has trivial character, that is, is character automorphic, then it defines a meromorphic function on $\mathcal{S}$ and this will give us a handle on the existence part of Abel's theorem. We will only get Abel's theorem when the zeros and poles lie in the G_j, but that suffices for the application in Theorem 5.12.10.

How might one get rid of the poles at $\{\gamma(\bar{z}_0)^{-1}\}$? If $z_0 \in \widetilde{C}_j^{\pm}$ for some j, then $\mathbf{x}(z_0) = \mathbf{x}(\overline{z_0^{-1}})$, so $B(z, z_0)(\mathbf{x}(z) - \mathbf{x}(z_0))$ has no pole at $\bar{z}_0^{-1}$ (it has a pole at $z = 0$ and its images—we will worry about that soon), but it has a double zero at each $\{\gamma(z_0)\}_{\gamma\in\Gamma}$. Thus, we need to be able to take square roots of functions with only double zeros and poles.

Lemma 9.11.1. *Let f be a character automorphic meromorphic function on $\mathbb{C} \cup \{\infty\} \setminus \Lambda(\Gamma)$ that obeys*

(i) *The only zeros and poles lie on $\cup_{j=1}^{\ell} \widetilde{C}_j^{\pm}$ and its images under Γ.*
(ii) *Every zero and pole of f has even order.*
(iii) *If $D_j^{\pm}$ is a counterclockwise contour just outside $\widetilde{C}_j^{\pm}$ (e.g., a circle with the same center and a slightly longer radius), then*

$$\oint_{D_j^{\pm}} \frac{f'(z)}{f(z)}\, dz = 0 \tag{9.11.1}$$

(iv)

$$f(0) > 0 \tag{9.11.2}$$

Then there is a unique character automorphic function g (denoted by $\sqrt{f}$) so

$$g(z)^2 = f(z) \qquad g(0) > 0 \tag{9.11.3}$$

Proof. By (9.11.1),

$$h(z) = \log(f(0)) + \oint_0^z \frac{f'(w)}{f(w)}\, dw \tag{9.11.4}$$

defines a single-valued function on $\widetilde{\mathcal{F}}^{\text{int}}$ where any contour staying in $\widetilde{\mathcal{F}}^{\text{int}}$ can be used in (9.11.4).

On $\widetilde{\mathcal{F}}^{\text{int}}$, define

$$g(z) = \exp(\tfrac{1}{2} h(z)) \tag{9.11.5}$$

which obeys (9.11.3). Since all zeros and poles on $\widetilde{C}_j^{\pm}$ are of even order, g can be meromorphically continued to a neighborhood, N, of $\overline{\widetilde{\mathcal{F}}}$.

For each j,

$$S_j \equiv \{z \in \widetilde{\mathcal{F}}^{\text{int}} \mid \gamma_j(z) \in N\} \tag{9.11.6}$$

is nonempty and open, and by decreasing N, we can suppose it is connected. By hypothesis, $g(\gamma_j(z))^2 = c_f(\gamma_j) g(z)^2$ for all $z \in S_j$. By continuity and compactness of S_j, we can find a square root, $c_g(\gamma_j)$, of $c_f(\gamma_j)$ so that for $z \in S_j$, $g(\gamma_j(z)) = c_g(\gamma_j) g(z)$. This allows a unique character automorphic extension of g to $\mathbb{C} \cup \{\infty\} \setminus \Lambda(\Gamma)$. □

We need a base point on each $\widetilde{C}_j^{+}$ where we will place the pole. Once we have a function with an arbitrary zero and such a pole, we can take a ratio of two such to move the pole. For each $j = 1, \dots, \ell$, ζ_j will be the unique point in $\widetilde{C}_j^{+}$ with

$$\mathbf{x}(\zeta_j) = \beta_j \tag{9.11.7}$$

ζ_j lies in $\partial\mathbb{D}$.

Theorem 9.11.2. *Let $y \in G_j$, some gap in $\mathcal{S}$. Let ζ be the unique point in C_j^{+} with*

$$\mathbf{x}(\zeta) = y \tag{9.11.8}$$

Then there exists a unique function $\Theta_0(\,\cdot\,; y)$ *meromorphic on* $\mathbb{C} \cup \{\infty\} \setminus \Lambda(\Gamma)$ *so that*

(i) Θ_0 *has simple zeros at* $\{\gamma(\zeta)\}_{\gamma\in\Gamma} = \{z \in \mathbb{C} \mid \mathbf{x}(z) = y\}$ *and simple poles at* $\{\gamma(\zeta_j)\}_{\gamma\in\Gamma} = \{z \in \mathbb{C} \mid \mathbf{x}(z) = \beta_j\}$ *and no other zeros and poles.*

(ii) Θ_0 *is character automorphic.*

(iii)

$$\Theta_0(0; y) = 1 \tag{9.11.9}$$

Moreover, Θ_0 *is continuous in* y *as a function from* $\mathbb{C} \cup \{\infty\} \setminus \Lambda(\Gamma)$ *to* $\mathbb{C} \cup \{\infty\}$ *in the topology of uniform convergence on compacts.*

Remarks. 1. Of course, if $y = \beta_j$, the conditions on zeros and poles conflict. We set $\Theta_0(z; \beta_j) \equiv 1$.

2. We use Θ_0 since we will define a slightly different Θ below (see (9.12.24)).

Proof. We will prove existence and continuity now and defer the proof of uniqueness.

Define $\eta(z)$ by

$$\eta(z) = \begin{cases} B(z, \zeta) & \text{if } \zeta \in \mathbb{D} \\ 1 & \text{if } \zeta \in \partial\mathbb{D} \\ B(z, \bar{\zeta}^{-1})^{-1} & \text{if } \zeta \in \mathbb{C} \setminus \overline{\mathbb{D}} \end{cases} \tag{9.11.10}$$

and f by

$$f(z) = \frac{\mathbf{x}(z) - \mathbf{x}(\zeta)}{\mathbf{x}(z) - \beta_j}\, \eta(z)\eta(0)^{-1} \tag{9.11.11}$$

with $f \equiv 1$ if $\zeta = \zeta_j$.

It is easy to see that f is continuous in y. Moreover, we claim that f obeys all the hypotheses of Lemma 9.11.1. Indeed, it has double poles at $\{\gamma(\zeta_j)\}_{\gamma\in\Gamma}$ and nowhere else since the pole of η at $\bar{\zeta}^{-1}$ is cancelled by the zeros of $\mathbf{x}(\cdot) - \mathbf{x}(\zeta)$ there and it has double zeros at $\{\gamma(\zeta)\}_{\gamma\in\Gamma}$. Thus, conditions (i) and (ii) hold. (9.11.2) holds since

$$f(0) = 1 \tag{9.11.12}$$

on account of $\mathbf{x}(0) = \infty$.

To check (9.11.1), let $\tilde{f}(z) = (\mathbf{x}(z) - \mathbf{x}(\zeta))/(\mathbf{x}(z) - \mathbf{x}(\zeta_j))$. Then

$$\frac{f'}{f} = \frac{\tilde{f}'}{\tilde{f}} + \frac{\tilde{\eta}'}{\eta} \tag{9.11.13}$$

So we need only prove (9.11.2) for f replaced by $\tilde{f}$ and by η.

$\tilde{f}$ is real on $\partial\mathbb{D}$ and $D_j^{\pm}$ are conjugate symmetric, so the $\tilde{f}$ integral is zero. For η, it suffices to prove it for B replaced by a finite product $\prod_{\gamma\in G} b(\,\cdot\,, \gamma(\zeta))$, and then (9.11.1) follows by noting the number of zeros and poles inside D_j^{+} cancel.

Finally, f is clearly character automorphic since B is character automorphic and $\mathbf{x}$ is automorphic. Thus, we can apply Lemma 9.11.1 and define

$$\Theta_0(z; y) = \sqrt{f(z)} \tag{9.11.14}$$

It has the required properties and is continuous in y since f is. □

Let $\mathfrak{A}_j(y_j) \in \Gamma^*$ be the character of $\Theta_0(\,\cdot\,; y)$, that is,

$$\mathfrak{A}_j(y_j)(\gamma) = \frac{\Theta_0(\gamma(0); y)}{\Theta_0(0; y)} = \Theta_0(\gamma(0); y) \tag{9.11.15}$$

Recall that in Section 5.12, we defined $\mathbb{T}_\mathfrak{e} = G_1 \times \cdots \times G_\ell$. Define

$$\widetilde{\mathfrak{A}}\colon \mathbb{T}_\mathfrak{e} \to \Gamma^* \tag{9.11.16}$$

by

$$\widetilde{\mathfrak{A}}(\vec{y}) = \mathfrak{A}_1(y_1) \dots \mathfrak{A}_\ell(y_\ell) \tag{9.11.17}$$

and

$$\widetilde{\Theta}_0(z; \vec{y}) = \prod_{j=1}^{\ell} \Theta_0(z; y_j) \tag{9.11.18}$$

Note that in Section 5.12, we used $\mathfrak{A}$ and $\tilde{a}$ for different maps with the same significance.

Theorem 9.11.3. *$\widetilde{\mathfrak{A}}$ is a real analytic homeomorphism of the ℓ-dimensional tori $\mathbb{T}_\mathfrak{e}$ and Γ^*.*

Remark. By real analytic, we mean given locally by convergent Taylor series in real coordinates describing the tori.

Proof. $\Theta_0(z; y)$ is real analytic in y, so by (9.11.15), $\mathfrak{A}_j(\cdot)(\gamma_k)$ is real analytic, and so therefore is $\widetilde{\mathfrak{A}}$. By degree theory as explained in Section 5.12, if we prove that $\widetilde{\mathfrak{A}}$ is one-one, then it is onto, and the theorem is proven.

Suppose $\vec{y}$ and $\vec{w}$ are in $\mathbb{T}_\mathfrak{e}$ so

$$\widetilde{\mathfrak{A}}(\vec{y}) = \widetilde{\mathfrak{A}}(\vec{w}) \tag{9.11.19}$$

and that $k = \#\{j \mid y_j \neq w_j\}$. Consider

$$g(z) = \frac{\widetilde{\Theta}_0(z; \vec{y})}{\widetilde{\Theta}_0(z; \vec{w})} \tag{9.11.20}$$

By (9.11.19), g is automorphic, so there is a meromorphic function G on $\mathcal{S}$, so

$$g(z) = G(\mathbf{x}^\sharp(z)) \tag{9.11.21}$$

The poles at ζ_j cancel in g, so g has exactly k zeros and k poles on $\cup_{j=1}^{\ell} C_j^+$ and thus, G has exactly $k \leq \ell$ zeros and poles. By the theory in Section 5.12, G is root free and so it must have an even number of zeros and poles on each G_j. Since g has exactly zero or one zero or pole on each C_j^+, we see that G has no zeros and poles, that is, $k = 0$, so $\vec{y} = \vec{w}$. □

As an immediate consequence, we get that

Theorem 9.11.4. *Let f be analytic and nonvanishing on $\mathbb{C} \cup \{\infty\} \setminus \Lambda(\Gamma)$ and suppose that f is character automorphic. Then f is constant.*

Proof. Since $\widetilde{\mathfrak{A}}$ is onto, we can find $\vec{y} \in \mathbb{T}_{\mathfrak{e}}$ so that $\mathfrak{A}(\vec{y})$ is the character of f. Then $g = f\Theta_0^{-1}(\,\cdot\,, \vec{y})$ is automorphic and the function G of (9.11.21) has at most ℓ zeros and poles, so it is square root free—and that is impossible by the same argument as above, unless $\vec{y} = (\beta_1, \dots, \beta_\ell)$ and $g = f$ is constant. □

Corollary 9.11.5 (Uniqueness part of Theorem 9.11.2). *The function Θ_0 of Theorem 9.11.2 is the unique function obeying* (i)–(iii) *of that theorem. Moreover, Θ_0 is real, that is,*

$$\overline{\Theta_0(z;\, y_j)} = \Theta_0(\bar{z};\, y_j) \tag{9.11.22}$$

Proof. Let h obey (i), (ii), (iii) and let $f = h/\Theta_0$. Then f has no zeros and poles, is character automorphic, and thus constant by the above theorem. Since $f(0) = 1$, we see $h = \Theta_0$.

If $\zeta \in C_j^+$, $\gamma_j^-(\zeta) = \bar{\zeta}$, and $\{\bar{\gamma}(\zeta)\}_{\gamma\in\Gamma} = \{\gamma(\zeta)\}_{\gamma\in\Gamma}$. It follows that $\overline{\Theta_0(\bar{z};\, y_j)}$ has the same zeros and poles as $\Theta_0(z;\, y_j)$. So, by the first part of the corollary, it must equal $\Theta_0(z;\, y)$. □

We are now ready to prove the special case of Abel's theorem as used in Section 5.12.

Definition. By a divisor, we mean a finite subset $\Delta \subset \cup_{j=1}^{\ell} G_j$ and an assignment of a nonzero, n_x, to each $x \in \Delta$ plus an assignment of integers, $n_{\infty_\pm}$ to $\infty_\pm$. We require

$$n_{\infty_-} = -n_{\infty_+} \tag{9.11.23}$$

and, for $j = 1, \dots, \ell$,

$$\sum_{x\in G_j} n_x = 0 \tag{9.11.24}$$

We write

$$n_{\infty_+}\delta_{\infty_+} + n_{\infty_-}\delta_{\infty_-} + \sum_{x\in\Delta} n_x\delta_x \tag{9.11.25}$$

as the formal divisor.

Definition. By a *special meromorphic function*, we mean a meromorphic function, f, on $\mathcal{S}$, all of whose zeros and poles lie on $\cup_{j=1}^{\ell} G_j \cup \{\infty_+\} \cup \{\infty_-\}$, and if n_x is the order of the zero at x ($n_x < 0$ means a pole), then n_x obeys (9.11.23) and (9.11.24).

We define $\mathfrak{A}(\infty_\pm)$ by letting $\omega_0 \in \Gamma^*$ be the character of $B(\cdot)$ and $\omega \in \Gamma^*$ a solution of $\omega^2 = \omega_0$ (there are 2^ℓ such solutions) and setting

$$\mathfrak{A}(\infty_\pm) = \omega^{\pm 1}$$

Theorem 9.11.6 (Abel's Theorem for Special Meromorphic Functions). *If f is a special meromorphic function and n_x is the order of its poles and zeros, then*

$$\prod_{x\in\Delta\cup\{\infty_\pm\}} \mathfrak{A}(x)^{n_x} = \mathbf{1} \tag{9.11.26}$$

Conversely, if n_x, $x \in \Delta \cup \{\infty_\pm\}$, where $\Delta \subset \cup_{j=1}^{\ell} G_j$ is finite, obeys (9.11.26), *then there is a unique (up to a multiplicative constant) special meromorphic function, f, whose divisor is* (9.11.25).

Proof. Given n_x obeying (9.11.23)/(9.11.24), let g be the meromorphic function on $\mathbb{C} \cup \{\infty\} \setminus \Lambda(\Gamma)$

$$g(z) = B(z)^{n_{\infty_+}} \prod_{x \in \Delta} \Theta_0(z; \zeta(x))^{n_x} \tag{9.11.27}$$

where $\zeta(x)$ is the unique $\zeta \in \cup_{j=1}^{\ell} C_j^+$ with $\mathbf{x}(\zeta) = x$. Then g is character isomorphic with character

$$\mathfrak{A}_g \equiv \prod_{x \in \Delta \cup \{\infty_\pm\}} \mathfrak{A}(x)^{n_x} \tag{9.11.28}$$

(since $n_{\infty_+} = -n_{\infty_-}$ and $\omega^2 = \omega_0$, we get the character of B to the n_{∞_+} power).

If (9.11.26) holds, then g is automorphic and there is a special meromorphic function, f, with

$$g(z) = f(\mathbf{x}^\sharp(z)) \tag{9.11.29}$$

proving existence. Uniqueness is obvious, since the ratio of two functions with the same n_x is an analytic function on $\mathcal{S}$ with no zeros and poles, hence constant.

If f is a special meromorphic function with divisor (9.11.25) and g is given by (9.11.27), then

$$\frac{g(z)}{f(\mathbf{x}^\sharp(z))} \equiv h(z)$$

is character automorphic with no zeros and poles, hence constant by Theorem 9.11.4. Thus, g is automorphic, that is, (9.11.26) holds. □

Remark. The existence proof is constructive, that is, we have shown

$$f(\mathbf{x}^\sharp(z)) = \text{LHS of (9.11.27)} \tag{9.11.30}$$

We can thus write down the m-function for elements of the isospectral torus in terms of theta functions.

Theorem 9.11.7. *Let $(p_1, \dots, p_\ell) \in \mathbb{T}_\mathfrak{e}$. Let $(z_1, \dots, z_\ell) \in \mathbb{T}_\mathfrak{e}$ be determined by*

$$\widetilde{\mathfrak{A}}(\vec{z}) = \widetilde{\mathfrak{A}}(\vec{p})\mathfrak{A}(\infty_-)\mathfrak{A}(\infty_+)^{-1} \tag{9.11.31}$$

Then, $m_p(z)$, the minimal Herglotz function with poles at $(p_1, \dots, p_\ell)$, is given by

$$m_p(\mathbf{x}(\zeta)) = -C(\mathfrak{e})^{-1} \frac{B(\zeta)\widetilde{\Theta}_0(\zeta; \vec{z})}{\widetilde{\Theta}_0(\zeta; \vec{p})} \tag{9.11.32}$$

Proof. The two sides of (9.11.32) have the same zeros and poles, so by Theorem 9.11.6, they agree up to a constant. Near $\zeta = 0$,

$$m_p(\mathbf{x}^\sharp(\zeta)) = -\frac{\zeta}{x_\infty} + O(\zeta^2) \tag{9.11.33}$$

while, by (9.7.37) and (9.11.9),

$$\frac{B(\zeta)\widetilde{\Theta}_0(\zeta;\vec{z})}{\widetilde{\Theta}_0(\zeta;\vec{p})} = \frac{C(\mathfrak{e})}{x_\infty}\zeta + O(\zeta^2)$$

so (9.11.32) holds. □

In the next section, we will use $U(\vec{p}) = \vec{z}$ for the map $\vec{p} \to \vec{z}$ given by (9.11.31). Note that under $\widetilde{\mathfrak{A}}$, U is just multiplication by the character of B.

We also want to note the following:

Theorem 9.11.8. *Parametrize $y_j \in G_j$ by a point in $\partial\mathbb{D}$ by letting p_j be the point in $\widetilde{C}_j^+$ farthest from 0, c_j the center of $\widetilde{C}_j^+$, and $\zeta_j = c_j + e^{i\theta_j}(p_j - c_j)$. We can use these as uniform parameters as we vary $\alpha_1, \beta_1, \dots, \alpha_{\ell+1}, \beta_{\ell+1}$, the edges of the bands. Then as functions of $q = (\alpha_1, \dots, \beta_{\ell+1})$ in the set Q_ℓ of Section 9.8 and $\theta_1, \dots, \theta_\ell \in \mathbb{D}^\ell$, $\Theta_0(z; \vec{y}; q)$ is continuous in θ and q.*

Proof. Direct from the construction and the continuity results of Section 9.8. □

Remarks and Historical Notes. The history of elliptic functions involved the study of doubly periodic entire functions on $\mathbb{C}$, that is, functions on the torus, which is our $\mathcal{S}$ with $\ell = 1$. Jacobi constructed elliptic functions as products and quotients of building blocks we now call Jacobi theta functions. These were only doubly periodic up to a phase (not actually a character)—the condition on the zeros and poles we call Abel's theorem came from a requirement that these phases cancel. Thus, building blocks like those in this section have come to be called theta functions. Poincaré also had products related to our Blaschke products that he called theta functions. So the name is quite natural.

For their Riemann surface, which could be infinite genus, Sodin–Yuditskii [413] defined theta functions for zeros in the gap. Our construction here of Θ_0, following Christiansen–Simon–Zinchenko [87], is motivated by, but distinct from, the Sodin–Yuditskii construction. The $\mathfrak{A}(\infty)$ of [87] is our $\mathfrak{A}(\infty_+)\mathfrak{A}(\infty_-)^{-1}$.

Widom [460] and Aptekarev [21] use Riemann theta functions, which are distinct from what we use here and which do not lead to character automorphic functions when lifted. They allow poles and zeros off the real axis on $\mathcal{S}$ but are not as natural from the covering space point of view we use.

See the Notes to Section 5.12 for a discussion of references for Abel's theorem.

9.12 JOST FUNCTIONS AND THE JOST ISOMORPHISM

In this section, we will associate to each measure, $\mu \in \text{Sz}(\mathfrak{e})$, which obeys the Szegő condition, a character automorphic function, $u(z; J_\mu)$, on $\mathbb{D}$ which we will call the *Jost function*. In the Notes, we discuss the connection to the Jost function of Section 3.7. This function will allow us to define a modified Weyl solution called the Jost solution whose asymptotics, as in Section 3.7, will be important in establishing Szegő asymptotics.

In the case of μ in the isospectral torus, we will see that the Jost function can be expressed in terms of theta functions and that it continues to a meromorphic function on $\mathbb{C} \cup \{\infty\} \setminus \Lambda(\Gamma)$. Moreover, we will prove the important fact that the map from the isospectral torus to Γ^*, obtained by mapping $\vec{y} \in \mathbb{T}_{\mathfrak{e}}$ into the character of the Jost function of the associated almost periodic Jacobi matrix, is an isomorphism, which we will call the *Jost isomorphism.* As a bonus, we will prove that, for $\{a_n, b_n\}_{n=1}^{\infty}$ in the isospectral torus, $n \to a_1 \dots a_n / C(\mathfrak{e})^n$ is almost periodic.

Jost functions require the choice of a base point whose eigenvalues determine poles of the Jost function. It will be natural to make these poles as far from $z = 0$ as possible and to pick the base point in the isospectral torus. Thus, for each $j = 1, \dots, \ell$, we let $\rho_j \in \widetilde{C}_j^+$ be the point farthest from 0, that is, $|\rho_j| = \sup_{\zeta \in \widetilde{C}_j^+} |\zeta|$. We let $r_j \in G_j$ be defined by

$$r_j = \mathbf{x}^\sharp(\rho_j) \tag{9.12.1}$$

r_j lies in $\mathcal{S}_-$. Let $\vec{\rho} = (\rho_1, \dots, \rho_\ell)$ and $\vec{r} = (r_1, \dots, r_\ell) \in \mathbb{T}_{\mathfrak{e}}$. For $\vec{y} \in \mathbb{T}_{\mathfrak{e}}$, we let $m_{\vec{y}}$ be the m-function for the associated Jacobi matrix, $J_{\vec{y}}$.

Definition. Let $\mu \in \mathrm{Sz}(\mathfrak{e})$. For each eigenvalue (= point mass), $p_j \in \mathbb{R} \setminus \mathfrak{e}$, let $\pi_j \in (-1, 1) \cup \bigcup_{j=1}^{\ell} C_j^+$ be the point with

$$\mathbf{x}(\pi_j) = p_j \tag{9.12.2}$$

The *Jost function* for μ is defined, for $z \in \mathbb{D}$, by

$$u(z; \mu) = \biggl[\prod_j B(z, \pi_j)\biggr] \exp\biggl(\frac{1}{4\pi} \int \frac{e^{i\theta} + z}{e^{i\theta} - z} \log\biggl[\frac{w_{\vec{r}}(\mathbf{x}(e^{i\theta}))}{w(\mathbf{x}(e^{i\theta}))}\biggr] d\theta\biggr) \tag{9.12.3}$$

where w is the weight for μ and $w_{\vec{r}}$ for $m_{\vec{r}}$.

Notice that since $\mu \in \mathrm{Sz}(\mathfrak{e})$, the π_j obey a Blaschke condition and the product in (9.12.3) converges by Theorem 9.7.7. Since $\mu \in \mathrm{Sz}(\mathfrak{e})$ and the measures in the isospectral torus obey a Szegő condition (by Theorem 5.13.6),

$$\left|\log\biggl[\frac{w_{\vec{r}}(\mathbf{x}(e^{i\theta}))}{w(\mathbf{x}(e^{i\theta}))}\biggr]\right| \in L^1\biggl(\frac{d\theta}{2\pi}\biggr) \tag{9.12.4}$$

by Theorem 9.7.6.

Theorem 9.12.1. *For any $\mu \in \mathrm{Sz}(\mathfrak{e})$, the Jost function, $u(z; \mu)$, is analytic in $\mathbb{D}$, character automorphic, and its zeros lie exactly at $\{\gamma(\pi_j)\}_{j;\gamma \in \Gamma}$. Moreover, u is real on $(-1, 1)$ so*

$$u(\bar{z}; \mu) = \overline{u(z; \mu)} \tag{9.12.5}$$

Proof. Analyticity and the zero position is immediate from the L^1 condition (9.11.4) and the Blaschke condition. Reality on $(-1, 1)$ is immediate from symmetry of the set of zeros and $w_{\cdot}(\mathbf{x}(e^{i\theta})) = \overline{w_{\cdot}(\mathbf{x}(e^{i\theta}))}$ for $\cdot = \vec{r}$ or nothing. The Blaschke factors are character automorphic and the product is convergent, so that u is character automorphic follows from the lemma below. $\square$

Lemma 9.12.2. *Let f be a real-valued L^1 function on $\partial\mathbb{D}$ so that*

$$f(\gamma(e^{i\theta})) = f(e^{i\theta}) \tag{9.12.6}$$

for all $\gamma \in \Gamma$. Define for $z \in \mathbb{D}$,

$$S_f(z) = \exp\biggl(\int \frac{e^{i\theta}+z}{e^{i\theta}-z}\, f(e^{i\theta})\, \frac{d\theta}{2\pi}\biggr) \tag{9.12.7}$$

Then S_f is character automorphic.

Proof. Suppose first $f \in L^\infty$, so

$$\exp(-\|f\|_\infty) \le |S_f(z)| \le \exp(\|f\|_\infty) \tag{9.12.8}$$

since $\text{Re}(\frac{e^{i\theta}+z}{e^{i\theta}-z})$ is a Poisson kernel, which is positive and whose integral is 1. Fix $\gamma \in \Gamma$ and let

$$h(z) = \frac{S_f(\gamma(z))}{S_f(z)} \tag{9.12.9}$$

so that, by (9.12.8), $\exp(-2\|f\|_\infty) \le |h(z)| \le \exp(2\|f\|_\infty)$.

By (9.12.7) and Proposition 2.3.11 for a.e. θ,

$$\lim_{r\uparrow 1} |S_f(re^{i\theta})| = \exp(f(e^{i\theta})) \tag{9.12.10}$$

so, by (9.12.6), for a.e. θ,

$$\lim_{r\uparrow 1} |h(re^{i\theta})| = 1 \tag{9.12.11}$$

Thus, $\log|h(z)|$ is a bounded harmonic function whose boundary values vanish on $\partial\mathbb{D}$, so $|h(z)| = 1$, that is, $h(z) = e^{i\psi_\gamma}$ for some ψ_γ, and thus,

$$S_z(\gamma(z)) = e^{i\psi_\gamma} S_f(z) \tag{9.12.12}$$

which shows S_f is character isomorphic.

For general $f \in L^1$, we can find $f_n \to f$ in L^1 with $f_n \in L^\infty$. Then, $S_{f_n} \to S_f$ uniformly on compact subsets of $\mathbb{D}$. By the compactness of Γ^*, this implies S_f is also character automorphic. □

The following is a rewriting of Theorem 9.9.1:

Theorem 9.12.3. *Let $\mu \in \text{Sz}(\mathfrak{e})$ and let $M(z;\mu)$ be its M-function, $\{a_j, b_j\}_{j=1}^\infty$ its Jacobi parameters, and $u(z;\mu)$ its Jost function. Let μ_1 be the first stripped measure, that is, the measure with Jacobi parameters $\{a_{j+1}, b_{j+1}\}_{j=1}^\infty$. Then*

$$a_1 M(z;\mu) = \frac{B(z)u(z;\mu_1)}{u(z;\mu)} \tag{9.12.13}$$

In particular, if $\mathcal{J}(\mu)$ is the character of $u(\,\cdot\,;\mu)$, then

$$\mathcal{J}(\mu_1) = \mathcal{J}(\mu)\mathfrak{A}(\infty)^{-2} \tag{9.12.14}$$

Note that the reference measure drops out of the ratio (9.12.13). Recall (see Proposition 3.2.5) that the Weyl solution is given by ($n = 0, 1, \dots$)

$$g_{n-1}(x; \mu) = \langle \delta_n, (J_\mu - x)^{-1} \delta_1 \rangle \tag{9.12.15}$$

We define the Jost solution by lifting to $\mathbb{D}$ and multiplying by $u(z; \mu)$ (to cancel the poles of g by zeros of u!).

Definition. Let $\mu \in \mathrm{Sz}(\mathfrak{e})$. The *Jost solution*, $u_n(z; \mu)$, for $n = 0, \dots$ is defined for $z \in \mathbb{D}$ by

$$u_n(z; \mu) = -u(z; \mu) g_{n-1}(\mathbf{x}(z); \mu) \tag{9.12.16}$$

where

$$g_{n=-1}(z) = -a_0^{-1} \tag{9.12.17}$$

Remarks. 1. g_n has poles at eigenvalues of J_μ and u has zeros; in (9.12.16) we cancel the poles by the zeros.

2. Of course, a_0 is not defined by μ and is only relevant to the extent we want (9.12.19) below to hold at $n = 1$. One often takes $a_0 \equiv 1$, but for elements of the isospectral torus, there is another natural definition.

Theorem 9.12.4. *Let $\mu \in \mathrm{Sz}(\mathfrak{e})$. Let μ_n be the n-times stripped measures (i.e., with Jacobi parameters $\{a_{j+n}, b_{j+n}\}_{j=1}^\infty$). Then*

$$u_n(z; \mu) = a_n^{-1} B(z)^n u(z; \mu_n) \tag{9.12.18}$$

For all $z \in \mathbb{D}$, $\{u_n(z; \mu)\}_{n=1}^\infty$ is a nonzero solution of

$$(J - \mathbf{x}(z))u = 0 \tag{9.12.19}$$

which is ℓ^2 at infinity.

Remark. (9.12.19) holds for $n = 1, 2, \dots$ but at $n = 1$, we include an $a_0 u_0$ term.

Proof. By Theorem 3.7.6 (or (5.4.57)),

$$m(x; J_{\mu_n}) = -\frac{g_n(z)}{a_n g_{n-1}(x)} \tag{9.12.20}$$

which, taking into account the minus sign in $M(z) = -m(\mathbf{x}(z))$, leads to (for $n \geq 1$)

$$\begin{aligned} g_n(\mathbf{x}(z)) &= a_n M(z; \mu_n) g_{n-1}(\mathbf{x}(z)) \\ &= -a_n \dots a_1 M(z; \mu_n) \dots M(z; \mu_1) M(z; \mu) \end{aligned} \tag{9.12.21}$$

where the minus sign comes from $M(z; \mu) = -m(\mathbf{x}(z)) = -g_0(\mathbf{x}(z))$.

Thus, by the definition (9.12.16),

$$\begin{aligned} a_n u_n(z; \mu) &= a_n a_{n-1} \dots a_1 M(z; \mu_{n-1}) \dots M(z; \mu) u(z; \mu) \\ &= B(z)^n \left[\left(\frac{u(z; \mu_n)}{u(z; \mu_{n-1})} \right) \left(\frac{u(z; \mu_{n-1})}{u(z; \mu_{n-2})} \right) \cdots \right] u(z; \mu) \\ &= B(z)^n u(z; \mu_n) \end{aligned} \tag{9.12.22}$$

proving (9.12.18).

Since u_n is a multiple of g_{n-1}, it solves the difference equation. If $u_0(z) = 0$, then $M(z)$ has a pole, so $u_1(z) \neq 0$, showing that we have a nonzero solution of the difference equation. □

Remark. (9.12.21) is essentially an extension of Corollary 3.7.7.

Having presented the Jost function and solution for general $\mu \in \mathrm{Sz}(\mathfrak{e})$ (something we will return to in the next section), we turn to μ associated to an element of the isospectral torus. We use $u(z; \vec{y})$ for $u(z; \mu_{\vec{y}})$ when $\vec{y} \in \mathbb{T}_\mathfrak{e}$.

Initially, via (9.12.16), $u_n(z; \mu)$ is only defined for $n \geq 0$. But by (9.12.18), which says, for $\vec{y} \in \mathbb{T}_\mathfrak{e}$, that

$$u_n(z; \vec{y}) = a_n^{-1} B(z)^n u(z; U^n(\vec{y})) \tag{9.12.23}$$

allows one to define u_n for all $n \in \mathbb{Z}$ and it obeys $(J(\vec{y}) - \mathbf{x}(z))u = 0$ as a difference equation on all of $\mathbb{Z}$.

Here the theta functions enter. Having moved the base points for u, we want to move them for Θ, so we define for $y_j \in G_j$,

$$\Theta(z; y_j) = \frac{\Theta_0(z; y_j)}{\Theta_0(z; p_j)} \tag{9.12.24}$$

which has its zeros still at ζ_j (with $\mathbf{x}(\zeta_j) = y_j$) but its pole now at π_j. Of course, we also define for $\vec{y} \in \mathbb{T}_\mathfrak{e}$,

$$\widetilde{\Theta}(z; \vec{y}) = \prod_{j=1}^{\ell} \Theta(z; y_j) \tag{9.12.25}$$

Theorem 9.12.5. *For $\vec{y} \in \mathbb{T}_\mathfrak{e}$, define $U(\vec{y})$ as the point of $\mathbb{T}_\mathfrak{e}$ corresponding to the once-stripped Jacobi matrix of $J_{\vec{y}}$ (i.e., if $\{a_j(\vec{y}), b_j(\vec{y})\}_{j=1}^\infty$ are the Jacobi parameters of $J_{\vec{y}}$, $a_j(U(\vec{y})) = a_{j+1}(\vec{y})$, $b_j(U(\vec{y})) = b_{j+1}(\vec{y})$). Then there is a continuous strictly positive function, φ, on $\mathbb{T}_\mathfrak{e}$ so that*

$$u(z; \vec{y}) = \varphi(\vec{y})\widetilde{\Theta}(z; \vec{y}) \tag{9.12.26}$$

Moreover, φ obeys

$$\varphi(\vec{p}) = 1 \tag{9.12.27}$$

$$\frac{\varphi(U(\vec{y}))}{\varphi(\vec{y})} = \frac{a_1(\vec{y})}{C(\mathfrak{e})} \tag{9.12.28}$$

Before turning to the proof of this theorem, we note:

Corollary 9.12.6. *For any fixed $\vec{y} \in \mathbb{T}_\mathfrak{e}$,*

$$n \to \frac{a_1(\vec{y}) \dots a_n(\vec{y})}{C(\mathfrak{e})^n} \tag{9.12.29}$$

is almost periodic.

Proof. By (9.12.28),

$$\text{RHS of (9.12.29)} = \frac{\varphi(U^n(\vec{y}))}{\varphi(\vec{y})} \tag{9.12.30}$$

is almost periodic since, under the Abel map, U is translation on the torus by a fixed amount. □

We also note that (9.12.26) says that for points in the isospectral torus, $u(z)$ has an analytic continuation from $\mathbb{D}$ to a neighborhood of $\overline{\mathbb{D}} \setminus \Lambda(\Gamma)$; indeed, a meromorphic continuation to all of $\mathbb{C} \cup \{\infty\} \setminus \Lambda(\Gamma)$ with poles only at images of $\{\pi_j\}_{j=1}^{\ell}$ under Γ.

Proof of Theorem 9.12.5. We will first note that (9.12.26) holds at $\vec{y} = \vec{p}$, then on the orbit $\{U^n(\vec{p})\}_{n\in\mathbb{Z}}$, and then, everywhere, by a double density theorem.

Since $J_{\vec{y}=\vec{p}}$ has no eigenvalues in gaps and $\vec{p}$ is the base point, $u(z; \vec{y} = \vec{p}) = 1$. By construction, $\widetilde{\Theta}(z; \vec{p}) = 1$, so (9.12.26) with (9.12.27) holds. Next, we claim that if (9.12.26) holds at some point $\vec{y}$, it holds at $U(\vec{y})$ with $\varphi(U(\vec{y}))$ given by (9.12.28). For, by Theorem 9.11.7,

$$\frac{\widetilde{\Theta}(z; U(\vec{y}))}{\widetilde{\Theta}(z; \vec{y})} = M_{\vec{y}}(z) B(z)^{-1} C(\mathfrak{e}) \tag{9.12.31}$$

On the other hand, by Theorem 9.12.3,

$$\frac{u(z; U_{\vec{y}})}{u(z; \vec{y})} = a_1 M_{\vec{y}}(z) B(z)^{-1} \tag{9.12.32}$$

Thus,

$$\frac{u(z; U(\vec{y}))}{\widetilde{\Theta}(z; U(\vec{y}))} = \frac{a_1(\vec{y})}{C(\mathfrak{e})} \frac{u(z; \vec{y})}{\widetilde{\Theta}(z; \vec{y})} \tag{9.12.33}$$

This proves our claim that if (9.12.26) holds at $\vec{y}$ and $\varphi(U(\vec{y}))$ is *defined* by (9.12.28), then (9.12.26) holds at $U(\vec{y})$. It also proves that if (9.12.26) holds at $U(\vec{y})$, then it holds at $\vec{y}$ with $\varphi(\vec{y})$ defined by (9.12.28).

Now suppose $q \in Q_\ell$ so that all the harmonic measures of $\{[\alpha_j, \beta_j]\}_{j=1}^{\ell}$ are rationally independent. Then $\{U^n(\vec{p})\}_{n\in\mathbb{Z}}$ are distinct and dense in $\mathbb{T}_\mathfrak{e}$. By the above, it follows that (9.12.26) holds at all these points. Define φ on $\mathbb{T}_\mathfrak{e}$ by

$$\varphi(\vec{y}) \equiv u(z = 0; \vec{y}) \tag{9.12.34}$$

Since u is continuous in $\vec{y}$, φ is continuous on $\mathbb{T}_\mathfrak{e}$ and, by (9.12.28), it agrees with the previous definition on $\{U^n(\vec{p})\}_{n\in\mathbb{Z}}$. By continuity of φ, u, and Θ in $\vec{y}$, and the density of $\{U^n(\vec{p})\}_{n\in\mathbb{Z}}$, we get (9.12.26) on all of $\mathbb{T}_\mathfrak{e}$.

Next, note that, by our proof of Theorem 5.6.1, $Q_\ell^{(0)} \equiv \{q \in Q_\ell \mid \{[\alpha_j, \beta_j]\}_{j=1}^{\ell}$ are rationally independent$\}$ is dense in Q_ℓ. Since φ is given by (9.12.34), u and $\widetilde{\Theta}$ are all continuous in q (if we parametrize y_j by θ_j as in Theorem 9.11.8), we get (9.12.26) for all of Q_ℓ from the formula for $Q_\ell^{(0)}$. □

Theorem 9.12.7. *The map $\mathcal{J}$ from $\mathbb{T}_\mathfrak{e}$ to Γ^*, given by taking $\vec{y} \in \mathbb{T}_\mathfrak{e}$ to the character of $u(\,\cdot\,; \vec{y})$, is an isomorphism called the Jost isomorphism.*

Proof. By (9.12.26) and (9.12.24), $\mathcal{J}$ differs from $\widetilde{\mathfrak{A}}$ multiplication by a fixed group element, that is,

$$\mathcal{J}(\vec{y}) = \widetilde{\mathfrak{A}}(\vec{y})\widetilde{\mathfrak{A}}(\vec{p})^{-1} \tag{9.12.35}$$

Thus, this theorem is a restatement of Theorem 9.11.3. □

We can use this to specify the asymptotics of $p_n(z; \vec{y})$, the OPs for μ, when $\vec{y} \in \mathbb{T}_\mathfrak{e}$ and $z \notin \sigma(J_{\vec{y}})$.

Theorem 9.12.8. *Let $\vec{y} \in \mathbb{T}_{\mathfrak{e}}$. Then for any z with $\mathbf{x}(z) \notin \sigma(J_{\vec{y}})$, $p_n(\mathbf{x}(z); \vec{y})B(z)^n$ is asymptotically almost periodic. In addition, for any compact $K \subset \mathbb{C}\setminus[\alpha_1, \beta_{\ell+1}]$, there is a constant $C > 0$ so that for any n and all $z \in \mathbb{D}$ with $x(z) \in K$,*

$$C|B(z)|^n \leq |p_n(\mathbf{x}(z); \vec{y})| \leq C^{-1}|B(z)|^n \tag{9.12.36}$$

Remark. This replaces Szegő asymptotics, which says in the case $\mathfrak{e} = [-2, 2]$ that $p_n(\mathbf{x}(z))/B(z)^n$ has a nonzero limit as $n \to \infty$.

Proof. Let $\vec{\tilde{y}}$ be the element in $\mathbb{T}_{\mathfrak{e}}$ (see the remark after the theorem) with

$$b_n(\vec{\tilde{y}}) = b_{-n}(\vec{y}) \qquad a_n(\vec{\tilde{y}}) = a_{1-n}(\vec{y}) \tag{9.12.37}$$

and let

$$\tilde{u}_n(z; \vec{y}) = u_{-n}(z; \vec{\tilde{y}}) \tag{9.12.38}$$

Then $\tilde{u}_n$ is also a solution of $(J_{\vec{y}} - \mathbf{x}(z))u = 0$—it is also the Weyl solution at $-\infty$. Since p_{n-1} solves the half-line case and $u_n, \tilde{u}_n$ are independent (since one is ℓ^2 at $+\infty$ and one at $-\infty$ and $\mathbf{x}(z) \notin \sigma(J_{\vec{y}})$), we see

$$p_{n-1}(\mathbf{x}(z)) = \alpha(z)u_n(z) + \beta(z)\tilde{u}_n(z) \tag{9.12.39}$$

where α, β are given by Wronskians and are analytic in $z \in \mathbb{D}$. Since only u_n is ℓ^2 at $+\infty$, the zeros of β are exactly z's with $\mathbf{x}(z) \in \sigma(J_{\vec{y}})$. As $n \to \infty$, $u_n B^n \to 0$, while $\tilde{u}_n B^{-n}$ has an almost periodic limit. So if $\beta(z) \neq 0$ asymptotically,

$$p_{n-1}(\mathbf{x}(z))B(z)^{n-1} \sim B(z)^{-1}\beta(z)a_n^{-1}u(z; U^{-n}(\vec{\tilde{y}})) \tag{9.12.40}$$

is asymptotically almost periodic and (9.12.36) holds. Since $u(z; \vec{y})$ has zeros only in $\mathbf{x}^{-1}(\cup G_j)$, we get (9.12.36). □

Remarks. 1. Given that $u(z; \vec{y})$ can be written in terms of theta functions, (9.12.40) gives "explicit" asymptotics for $p_n(x)$ to the extent that theta functions are explicit.

2. If $p_n(\mathbf{x}(z))$ is replaced by $|p_n(\mathbf{x}(z))|^2 + |p_{n-1}(\mathbf{x}(z))|^2$, we can replace $K \subset \mathbb{C} \setminus [\alpha_1, \beta_{\ell+1}]$ by $K \subset \mathbb{C} \setminus \sigma(J_{\vec{y}})$.

3. Since $p_{-1} = 0$, $p_0 = 1$, the Wronskians in (9.12.39) are easy to compute:

$$p_{n-1}(\mathbf{x}(z)) = \frac{a_0[\tilde{u}_0(z)u_n(z) - u_0(z)\tilde{u}_n(z)]}{W(\tilde{u}_n, u_n)} \tag{9.12.41}$$

The fact that u can be analytically continued from $\mathbb{D}$ to a neighborhood of $\overline{\mathbb{D}} \setminus \Lambda(\Gamma)$ allows one to control the limits of the Jost solutions as one approaches the spectrum, and thereby the spectral theorist's Green's function. The following are proven in [87]:

Theorem 9.12.9 ([87]). *For $x \in \mathfrak{e}$ and $\vec{y} \in \mathbb{T}_{\mathfrak{e}}$, define*

$$u_n^+(x; \vec{y}) = u_n(\mathbf{z}(x); \vec{y}) \tag{9.12.42}$$

where $\mathbf{z}(x) \in \overline{\mathcal{F}}$ is determined by $\mathbf{x}(\mathbf{z}(x)) = x$. Let

$$u_n^-(x; \vec{y}) = \overline{u_n^+(x; \vec{y})}$$

Then there are constant C_1 and C_2 so that

(i) *Uniformly in $\vec{y} \in \mathbb{T}_{\mathfrak{e}}$, $x \in \mathfrak{e}^{\text{int}}$, and n,*

$$|u_n^{\pm}(x; \vec{y})| \leq C_1 \text{dist}(x; \mathbb{R} \setminus \mathfrak{e})^{-1/2} \tag{9.12.43}$$

(ii) *Uniformly in $\vec{y} \in \mathbb{T}_{\mathfrak{e}}$, $x \in \mathfrak{e}$, and n,*

$$|u_n^{\pm}(x; \vec{y})| \leq C_2(|n+1|) \tag{9.12.44}$$

Remark. Similar bounds hold for $p_n(x)$.

For $\vec{y} \in \mathbb{T}_{\mathfrak{e}}$, we use $J_{\vec{y}}$ for the half-line Jacobi matrix and $\tilde{J}_{\vec{y}}$ for the whole-line Jacobi matrix (using the almost periodic construction of the Jacobi parameters). The spectral theorist's Green's functions are given by

$$G_{nm}(z) = \langle \delta_n (J_{\vec{y}} - z)^{-1} \delta_m \rangle \qquad n, m = 1, 2, \dots$$

$$\widetilde{G}_{nm}(z) = \langle \delta_n, (\tilde{J}_{\vec{y}} - z)^{-1} \delta_m \rangle \qquad n, m \in \mathbb{Z}$$

Theorem 9.12.10 ([87]). *There is a constant C_3 so that uniformly in $\vec{y} \in \mathbb{T}_{\mathfrak{e}}$ and $x \in \mathbb{R} \setminus \mathfrak{e}$,*

$$|\widetilde{G}_{nm}(x)| \leq C_3 \exp(-G_{\mathfrak{e}}(x)|n-m|) \text{dist}(x, \mathfrak{e})^{-1/2} \tag{9.12.45}$$

where $G_{\mathfrak{e}}$ is the potential theorist's Green's function.

We say a point, $x_0 \in \{\alpha_j, \beta_j\}_{j=1}^{\ell+1}$, is a *resonance* for $J_{\vec{y}}$ if $m_{\vec{y}}(z)$ has a pole there in the sense of the Riemann surface (i.e., $|m(z)| \sim C|x_0 - z|^{-1/2}$ for z near x_0). If x_0 is not a resonance, we say it is nonresonant. If x_0 is nonresonant for x_0 at an edge of a gap, we let I be the open interval that has one end at x_0 and the other at half the distance to any pole of $m_y(z)$ in the gap, to the other end of the gap if there is no pole, and one unit from x_0 if x_0 is α_1 or $\beta_{\ell+1}$.

Theorem 9.12.11 ([87]). *Let $\vec{y} \in \mathbb{T}_{\mathfrak{e}}$ be fixed and $x_0 \in \{\alpha_j, \beta_j\}_{j=1}^{\ell+1}$ nonresonant. There are constants C_4, C_5 so that for $x \in I$, the interval above, and for all $n, m \in \{1, 2, \dots\}$,*

$$|G_{nm}(x)| \leq C_4 \min(n, m) \tag{9.12.46}$$

$$|G_{nm}(x)| \leq C_5 |x - x_0|^{-1/2} \tag{9.12.47}$$

Remarks and Historical Notes. This section describes material from [87, 88], mainly [87]. If $\mathfrak{e} = [-2, 2]$, the isospectral torus has a single point with Jacobi parameters given by $a_n \equiv 1$, $b_n \equiv 0$. (3.7.42) shows that our Jost function specialized to this case agrees with the Jost function of Section 3.7.

9.13 SZEGŐ ASYMPTOTICS

One of our main goals in this section is to prove the following:

Theorem 9.13.1 ([88]). *Let μ be a measure in* Sz($\mathfrak{e}$) *for a finite gap set, $\mathfrak{e}$, with Jacobi parameters $\{a_n, b_n\}_{n=1}^{\infty}$. Let $\{\tilde{a}_n, \tilde{b}_n\}_{n=1}^{\infty}$ be the Jacobi parameters of the unique* *element of $\mathbb{T}_{\mathfrak{e}}$ whose Jost function has the same character as the Jost function of μ. Then, as $n \to \infty$,*

$$|a_n - \tilde{a}_n| + |b_n - \tilde{b}_n| \to 0 \tag{9.13.1}$$

Remark. In this form, the theorem is due to Christiansen–Simon–Zinchenko [88]; but, as we will explain in the Notes, the essence is an earlier result of Widom [460] and Peherstorfer–Yuditskii [343, 344].

It is the proof here that is different from the earlier work.

The key to the proof of this is

Theorem 9.13.2. *Let J be a Jacobi matrix whose spectral measure, μ, is in* $\mathrm{Sz}(\mathfrak{e})$ *and let $u(z;\mu)$ be its Jost function and μ_n the measure of the n-times stripped Jacobi matrix. Suppose for some $\{a_n^\sharp, b_n^\sharp\}_{n=1}^\infty$ in the isospectral torus, $u^\sharp(z)$ is its Jost function, and for some $n_k \to \infty$, we have that*

$$a_{n+n_k} \to a_n^\sharp \qquad b_{n+n_k} \to b_n^\sharp \tag{9.13.2}$$

as $k \to \infty$. Then, uniformly on compact subsets of $\mathbb{D}$,

$$u(z;\mu_{n_k}) \to u^\sharp(z) \tag{9.13.3}$$

We will prove this later in the section—we want to show it implies Theorem 9.13.1 and explore the consequences of these two theorems for further asymptotic results, including Szegő asymptotics for the OPs.

Proof of Theorem 9.13.1 given Theorem 9.13.2. Suppose (9.13.1) fails. By compactness of bounded sequences in the product topology, there is a sequence $\{a_n^\sharp, b_n^\sharp\}_{n=1}^\infty$ and some n_k so that (9.13.2) holds, and for some n_0,

$$|a_{n_0}^\sharp - \tilde{a}_{n_0+n_k}| - |b_{n_0}^\sharp - \tilde{b}_{n_0+n_k}| \to d \neq 0 \tag{9.13.4}$$

By the Denisov–Rakhmanov–Remling theorem, $\{a_n^\sharp, b_n^\sharp\}_{n=1}^\infty$ lies in the isospectral torus.

Since Γ^* is compact, without loss, we can pass to a further subsequence and suppose $\mathfrak{A}(\infty)^{-2n_k}$ has a limit $c \in \Gamma^*$. By the Abel theorem analysis of the isospectral torus, that means $\{\tilde{a}_{n+n_k}, \tilde{b}_{n+n_k}\}_{n=1}^\infty$ also has a limit, call it $\{a_n^\natural, b_n^\natural\}_{n=1}^\infty$.

By Theorems 9.13.2 and 9.12.3, if $\mathcal{J}(\{a_n, b_n\}_{n=1}^\infty)$ is the character of the associated Jost function, then

$$\mathcal{J}(\{a_n, b_n\}_{n=1}^\infty)c = \mathcal{J}(\{a_n^\sharp, b_n^\sharp\}_{n=1}^\infty) \tag{9.13.5}$$

$$\mathcal{J}(\{\tilde{a}_n, \tilde{b}_n\}_{n=1}^\infty)c = \mathcal{J}(\{a_n^\natural, b_n^\natural\}_{n=1}^\infty) \tag{9.13.6}$$

Thus,

$$\mathcal{J}(\{a_n^\sharp, b_n^\sharp\}) = \mathcal{J}(\{a_n^\natural, b_n^\natural\}_{n=1}^\infty) \tag{9.13.7}$$

Thus, by Theorem 9.12.7, $\{a_n^\sharp, b_n^\sharp\}_{n=1}^\infty = \{a_n^\natural, b_n^\natural\}_{n=1}^\infty$, violating (9.13.4). This contradiction shows that (9.13.1) holds. □

Remark. We can summarize the proof by saying the Denisov–Rakhmanov–Remling theorem implies the Jacobi parameters must approach the isospectral torus and the character of the Jost function specifies which orbit.

We can use these theorems to deduce Szegő asymptotics for any $p_n(x;\mu)$ with $\mu \in \mathrm{Sz}(\mathfrak{e})$:

Theorem 9.13.3 (Szegő Asymptotics). *Let $\mu \in \mathrm{Sz}(\mathfrak{e})$. Let $\vec{y} \in \mathbb{T}_\mathfrak{e}$ be the point in the isospectral torus with*

$$\lim_{n\to\infty} |a_n(\mu) - a_n(\vec{y})| + |b_n(\mu) - b_n(\vec{y})| = 0 \tag{9.13.8}$$

Then for all $x \in \mathbb{C} \setminus [\alpha_1, \beta_{\ell+1}]$, we have

$$\frac{p_n(x;\mu)}{p_n(x;\vec{y})} \to \frac{u(\mathbf{z}(x);\mu)}{u(\mathbf{z}(x);\vec{y})} \tag{9.13.9}$$

as $n \to \infty$, where $\mathbf{z}(x)$ is the unique point in $\mathcal{F}^{\mathrm{int}}$ with $\mathbf{x}(\mathbf{z}(x)) = x$. In particular, $p_n(x)B(\mathbf{z}(x))$ is asymptotically almost periodic as $n \to \infty$.

Remark. Given (9.12.40), (9.13.9) gives explicit asymptotics for $p_n(x;\mu)$.

Proof [88]. Since $x \notin [\alpha_1, \beta_{\ell+1}]$, the Green's functions $\langle \delta_j, (J-x)^{-1}\delta_j\rangle$ and $\langle \delta_j, (\tilde{J}_y - x)^{-1}\delta_j\rangle$ are nonvanishing (if $\mathrm{Im}\, x \neq 0$, $\mathrm{Im}(J-x)^{-1} = \mathrm{Im}\, x[(J-\bar{x})^{-1}(J-x)^{-1}]$ and $\|(J-x)\varphi\| \ge \mathrm{Im}\, x\|\varphi\|$, so $|\mathrm{Im}\langle\delta_j, (J-x)^{-1}\delta_j\rangle| \ge |\mathrm{Im}\, x|^{-1}$, and if $x \le \alpha_1$ or $x \ge \beta_{\ell+1}$, $\|(J-x)^{-1}\varphi\| \ge \mathrm{dist}(x, [\alpha_j, \beta_j])^{-1}$). It follows that

$$\lim_{n\to\infty} \frac{\langle \delta_n, (J-x)^{-1}\delta_n\rangle}{\langle \delta_n, (J_{\vec{y}}-x)^{-1}\delta_n\rangle} = 1 \tag{9.13.10}$$

But, since $a_n(p_n u_n - p_{n-1}u_{n+1})|_{n=0} = a_0 u(z(x);\mu)$, we see

$$\langle \delta_n, (J-x)^{-1}\delta_n\rangle = \frac{p_{n-1}(x;\mu)u_n(z(x);\mu)}{a_0 u(z(x);\mu)} \tag{9.13.11}$$

so (9.13.10) and (9.13.3) imply (9.13.9), □

Corollary 9.13.4. *Let $\mu \in \mathrm{Sz}(\mathfrak{e})$ and $\vec{y} \in \mathbb{T}_\mathfrak{e}$ with (9.13.8). Then, as $n \to \infty$,*

$$\frac{a_1(\vec{y})\dots a_n(\vec{y})}{a_1(\mu)\dots a_n(\mu)} \to \frac{u(0;\mu)}{u(0;\vec{y})} \tag{9.13.12}$$

In particular (by Corollary 9.12.6), $a_1(\mu)\dots a_n(\mu)/C(\mathfrak{e})^n$ is asymptotically almost periodic.

Proof. For $z \in \mathbb{D}$, let

$$f_n(z) = \frac{p_n(\mathbf{x}(z);\mu)}{p_n(\mathbf{x}(z);\vec{y})} \tag{9.13.13}$$

The last theorem implies that for some $\varepsilon > 0$ and $0 < |z| < \varepsilon$, we have

$$f_n(z) \to \frac{u(z;\mu)}{u(z;\vec{y})} \tag{9.13.14}$$

and the proof shows the convergence is uniform on $\{z \mid \frac{\varepsilon}{2} < |z| < \varepsilon\}$. By the maximum principle, (9.13.14) holds for $z = 0$. But

$$f_n(0) = \text{LHS of (9.13.12)}$$

so (9.13.14) for $z = 0$ is (9.13.12). □

We now turn to the proof of Theorem 9.13.2. We write the Jost function (9.12.3) as the product $\beta(z;\mu)$ of the Blaschke factors and $\varepsilon(z;\mu)$, the exponential, which we call the Blaschke part and the entropy part of the Jost function, respectively. We will prove (9.13.2) by proving convergence of the two parts separately. By Corollary 9.10.3, one has uniform control over the tail of the contributions to the Blaschke part, so it suffices to prove convergence of individual eigenvalues, that is, the first part of the following implies (9.13.15).

Proposition 9.13.5. *Let* (9.13.2) *hold. Let* (β_j, α_{j+1}) *be a gap in* $\mathfrak{e}$ *and let* $\varepsilon > 0$ *be such that neither* $\beta_j + \varepsilon$ *nor* $\alpha_{j+1} - \varepsilon$ *is an eigenvalue of* $J^\sharp$. *Then, as* $k \to \infty$, *the number of eigenvalues of* J_{n_k} *in* $I_\varepsilon \equiv (\beta_j + \varepsilon, \alpha_{j+1} - \varepsilon)$ *is the number of eigenvalues of* $J^\sharp$ *in* I_ε *and the eigenvalues converge. In particular, uniformly for compact subsets of* $\mathbb{D}$,

$$\beta(z;\mu_{n_k}) \to \beta^\sharp(z) \tag{9.13.15}$$

Remark. Of course, since $J^\sharp$ is in the isospectral torus, it has either 0 or 1 eigenvalue in I_ε.

Proof. Let $\lambda^\sharp$ be an eigenvalue of $J^\sharp$ in I_ε and $u^\sharp$ a unit vector so that

$$J^\sharp u^\sharp = \lambda^\sharp u^\sharp \tag{9.13.16}$$

Then

$$\lim_{k\to\infty} \|(J_{n_k} - \lambda^\sharp)u^\sharp\| = 0 \tag{9.13.17}$$

so if ε_k is the norm on the left, J_{n_k} has spectrum in $(\lambda^\sharp - \varepsilon_k, \lambda^\sharp + \varepsilon_k)$. Since J_{n_k} only has eigenvalues in I_ε, for k large, J_{n_k} has an eigenvalue near $\lambda^\sharp$.

Let λ_{n_k} be eigenvalues of J_{n_k} in I_ε so that $\lambda_{n_k} \to \lambda^\sharp$. We will prove that $\lambda^\sharp$ is an eigenvalue of $J^\sharp$ and that the corresponding one-dimensional spectral projections converge. It is easy to see that this and the initial part of this proof shows, for large k, that J_{n_k} has exactly one eigenvalue in I_ε for each eigenvalue of $J^\sharp$, and we have the claimed convergence.

Pick the unique ℓ^2 element, $u^{(k)}$, with

$$J_{n_k} u^{(k)} = \lambda_{n_k} u^{(k)} \tag{9.13.18}$$

and

$$u^{(k)}_{j=1} > 0 \qquad \|u^{(k)}\| = 1 \tag{9.13.19}$$

Let $u^{(\sharp)}$ be a weak limit point of $u^{(k)}$; we will continue to use $u^{(k)}$ for the subsequence that converges. Then

$$J^\sharp u^\sharp = \lambda^\sharp u^\sharp \tag{9.13.20}$$

Suppose we prove

$$\|u^\sharp\| = 1 \tag{9.13.21}$$

Then $u^\sharp \neq 0$, so $\lambda^\sharp$ is an eigenvalue of $J^\sharp$, and since (9.13.21) holds,

$$\|u^{(k)} - u^\sharp\|^2 = 2 - 2\lim\langle u^{(k)}, u^\sharp\rangle = 0 \tag{9.13.22}$$

so we have the claimed convergence of spectral projections. Thus, we need only verify that (9.13.21) is true. If it is false, then for $\varepsilon > 0$,

$$\|u^\sharp\| = 1 - \varepsilon \tag{9.13.23}$$

and for each m, with χ_m the characteristic function of $\{1, \dots, m\}$,

$$\lim_{k\to\infty} \|\chi_m u^{(k)}\| \le 1 - \varepsilon \tag{9.13.24}$$

so we can find $m_k \to \infty$ and $\delta > 0$ so that for all k,

$$\|(1 - \chi_{m_k})u^{(k)}\| \ge \delta \tag{9.13.25}$$

Picking ℓ_k with $m_k/2 \le \ell_k \le m_k$ with minimum value of $|u_\ell^{(k)}| + |u_{\ell+1}^{(k)}|$ in the range, we see that

$$|u_{\ell_k}^{(k)}| + |u_{\ell_k+1}^{(k)}| \to 0 \tag{9.13.26}$$

Let

$$v_j^{(k)} = \begin{cases} u_j^{(k)} & j \ge \ell_k + 1 \\ 0 & j < \ell_k \end{cases} \tag{9.13.27}$$

By (9.13.26),

$$\|(J_{n_k} - \lambda_{n_k})v_j^{(k)}\| \to 0 \tag{9.13.28}$$

and if

$$w_j^{(k)} = \begin{cases} v_{j-n_k}^{(k)}/\|v^{(k)}\| & j \ge n_k \\ 0 & j \le n_k \end{cases} \tag{9.13.29}$$

then, by (9.13.25) and (9.13.28),

$$\|(J - \lambda_{n_k})w^{(k)}\| \to 0 \tag{9.13.30}$$

Since $w^{(k)} \to 0$ weakly and $\|w^{(k)}\| = 1$, we conclude that $\lambda^\sharp \in \sigma_{\text{ess}}(J)$, contrary to $\sigma_{\text{ess}}(J) \subset \mathfrak{e}$. This contradiction proves (9.13.21) and completes the proof of the theorem. □

To control the entropy part, we need the following:

Proposition 9.13.6 (Simon–Zlatoš [410]). *Let X be a compact Hausdorff space and S given by* (2.2.1). *Let μ be a fixed probability measure, $\nu_n \to \nu_\infty$ weakly for probability measures, and let*

$$d\nu_n = f_n\, d\mu + d\nu_{n;\text{s}}$$

for $\nu_{n;\text{s}}$ singular with respect to $d\mu$. Suppose

$$S(\mu \mid \nu_n) \to S(\mu \mid \nu) \tag{9.13.31}$$

with all S's finite. Then

$$\log(f_n)\, d\mu \xrightarrow{w} \log(f)\, d\mu \tag{9.13.32}$$

Proof. Suppose first that w is continuous and strictly positive on X. Then by upper semicontinuity (Theorem 2.2.3),

$$\limsup S(w\mu \mid \nu_n) \le S(w\mu \mid \nu) \tag{9.13.33}$$

or

$$\limsup \int \log(f_n) w \, d\mu \le \int \log(f) w \, d\mu \tag{9.13.34}$$

For arbitrary real-valued continuous g, let $w = 2\|g\|_\infty \pm g$ in (9.13.34) to conclude the claimed weak convergence. □

Proposition 9.13.7. *To prove convergence of the entropy part, $\varepsilon(z; \mu_{n_k})$, to $\varepsilon^\sharp(z)$ uniformly on compact subsets of $\mathbb{D}$, it suffices to prove that*

$$\lim_{k\to\infty} \int \log(|\mathrm{Im}\, M_{\mu_k}(e^{i\theta})|) \, \frac{d\theta}{2\pi} = \int \log(|\mathrm{Im}\, M_{\mu}(e^{i\theta})|) \, \frac{d\theta}{2\pi} \tag{9.13.35}$$

Proof. By (9.10.28), (9.13.35) is equivalent to

$$\lim_{k\to\infty} S(\rho_{\mathfrak{e}} \mid \mu_k) = S(\rho_{\mathfrak{e}} \mid \mu^\sharp) \tag{9.13.36}$$

By Proposition 9.13.6, this implies for any continuous function, h, on $\mathfrak{e}$ that with w_k and $w^\sharp$ the weights of μ_k and $\mu^\sharp$,

$$\int h(x) \log\biggl(\frac{w_{\bar{r}}(x)}{w_k(x)}\biggr) \, d\rho_{\mathfrak{e}} \to \int h(x) \log\biggl(\frac{w_{\bar{r}}(x)}{w^\sharp(x)}\biggr) \, d\rho_{\mathfrak{e}}(x) \tag{9.13.37}$$

By (9.7.63), this implies that, uniformly for compact subsets of $z \in \mathbb{D}$, $\varepsilon(z; \mu_{n_k})$ converges to $\varepsilon^\sharp(z)$. □

As noted, (9.13.35) is equivalent to (9.13.36). By upper semicontinuity of the entropy, we have

$$\limsup_{k\to\infty} S(\rho_{\mathfrak{e}} \mid \mu_k) \le S(\rho_{\mathfrak{e}} \mid \mu^\sharp) \tag{9.13.38}$$

so it suffices to prove that

$$\liminf_{k\to\infty} S(\rho_{\mathfrak{e}} \mid \mu_k) \ge S(\rho_{\mathfrak{e}} \mid \mu^\sharp) \tag{9.13.39}$$

By passing to a subsequence where $S(\rho_{\mathfrak{e}} \mid \mu_k)$ converges to the lim inf, we can also suppose

$$\tau_k \equiv \frac{a_1 \dots a_{n_k}}{C(E)^{n_k}} \tag{9.13.40}$$

has a limit τ_∞ (since $\{\tau_k\}_{k=1}^\infty$ are bounded by Theorem 9.10.1).

Define, for $k < \ell$, a Jacobi matrix $J^{(k,\ell)}$ by

$$a_m^{(k,\ell)} = \begin{cases} a_{n_k+m} & 1 \le m \le n_\ell - n_k \\ a_m^\sharp & m > n_\ell - n_k \end{cases} \tag{9.13.41}$$

$$b_m^{(k,\ell)} = \begin{cases} b_{n_k+m} & 1 \le m \le n_\ell - n_k \\ b_m^\sharp & m > n_\ell - n_k \end{cases} \tag{9.13.42}$$

that is, first replace (a, b) by $(a^\sharp, b^\sharp)$ for indices $n_\ell + 1$ or more, and then strip off n_k rows and columns. By iterating the step-by-step sum rule to go from $J^{(n_j, n_k)}$ to $J^\sharp$ (by stripping off $n_\ell - n_k$ rows and columns), one gets, via (9.9.17) and (9.10.28),

$$\frac{\tau_\ell}{\tau_k} = \frac{\beta(0; \mu^\sharp)}{\beta(0; \mu_{k,\ell})} \exp[\tfrac{1}{2}S(\rho_\mathfrak{e} \mid \mu_{k,\ell}) - \tfrac{1}{2}S(\rho_\mathfrak{e} \mid \mu^\sharp)] \tag{9.13.43}$$

Lemma 9.13.8.

$$\lim_{\ell\to\infty} \beta(0; \mu_{k,\ell}) = \beta(0; \mu_k) \tag{9.13.44}$$

Proof. By Corollary 9.10.3, we need only prove convergence of eigenvalues. The proof follows that of Proposition 9.13.5. The elimination of (9.13.24) is even easier than in that proposition. By (9.13.2), any eigenvector that lives far from $m = 0$ gives essential spectral values for J_{n_k}. □

The following completes the proof of Theorem 9.13.2:

Proposition 9.13.9. (9.13.39) *holds.*

Proof. By upper semicontinuity of the entropy,

$$\limsup_{\ell\to\infty} S(\rho_\mathfrak{e} \mid \mu_{k,\ell}) \le S(\rho_\mathfrak{e} \mid \mu_k) \tag{9.13.45}$$

Thus, taking $\ell \to \infty$ in (9.13.43) and using (9.13.44) gives

$$\exp[\tfrac{1}{2}S(\rho_\mathfrak{e} \mid \mu_k) - \tfrac{1}{2}S(\rho_\mathfrak{e} \mid \mu^\sharp)] \ge \frac{\tau_\infty}{\tau_k} \frac{\beta(0; \mu_k)}{\beta(0; \mu^\sharp)} \tag{9.13.46}$$

Picking a subsequence so $S(\rho_\mathfrak{e} \mid \mu_k)$ goes to the lim inf and using $\tau_k \to \tau_\infty$ and (9.13.14) at $z = 0$, we get

$$\exp[\tfrac{1}{2} \liminf S(\rho_\mathfrak{e} \mid \mu_k) - \tfrac{1}{2}S(\rho_\mathfrak{e} \mid \mu^\sharp)] \ge 1$$

which is (9.13.39). □

Remarks and Historical Notes. This section follows Christiansen–Simon–Zinchenko [88]. In 1967, Widom [460] proved Szegő asymptotics for orthogonal polynomials for measures in the complex plane supported on a finite union of analytic curves obeying a Szegő condition. This was made precise for finite gap sets on $\mathbb{D}$ by Aptekarev [21], who noted the consequences for the Jacobi parameters. Peherstorfer–Yuditskii [343, 344] extended this to certain infinite gap sets and to allow eigenvalues outside $\mathfrak{e}$ obeying a Blaschke condition. Widom, Aptekarev, and Peherstorfer–Yuditskii all use a variational approach with a character-dependent variational principle.

For problems with varying but smooth weights, Riemann–Hilbert problem asymptotics have been used by Deift et al. [103] and Aptekarev–Lysov [22] to get

Szegő asymptotics. The variable weights are technically harder to control and Riemann–Hilbert methods can also be used for fixed analytic weights, as is made explicit in [22].

[88] also has an L^2 asymptotic result analogous to (3.7.39), namely,

$$\int_{\mathfrak{e}} \left| p_n(x) - \frac{2\,\mathrm{Im}[\bar{u}(\mathbf{z}(x))u_{n+1}(\mathbf{z}(x),\vec{y})]}{W(u_n^-(\mathbf{z}(x),\vec{y}),u_n^+(\mathbf{z}(x),\vec{y}))} \right|^2 w(x)\,dx \to 0 \tag{9.13.47}$$

and

$$\int |p_n(x)|^2\,d\mu_{\mathrm{s}}(x) \to 0 \tag{9.13.48}$$

Chapter Ten

A.C. Spectrum for Bethe–Cayley Trees

In this final chapter, we discuss some work of Denisov [108, 110] about the use of sum rules in the study of perturbed Laplacians on what physicists call Bethe lattices and mathematicians call Cayley trees—so we will call them Bethe–Cayley trees. These have a kind of coefficient stripping, so a one-dimensional aspect, but they are not fully one-dimensional in part because a single spectral measure does not describe them. For this reason, the results will be one-sided: from coefficient information to spectral data and not vice versa.

10.1 OVERVIEW

We begin by describing the rooted Bethe–Cayley tree of degree $d = 1, 2, \dots,$ which is a graph that is homogeneous and then cut in half. We have a root, which we will denote by ϕ. It has d neighbors, which we denote $(0), \dots, (d-1)$. Each of these d neighbors, (j), has $d+1$ neighbors, ϕ, and d neighbors $(j0), \dots, (jd-1)$. At level k, there are d^k elements (we have described levels $k = 0, 1, 2$ above) described by $(j_1 \dots j_k)$ with $j_\ell \in \{0, \dots, d-1\}$. Two elements are neighbors if one is obtained from the other by deleting the last j or by adding a j at the end. The root has d neighbors and all other vertices have $d+1$ neighbors. For a vertex α, we will use $\#(\alpha)$ for the level of α, that is, its distance from ϕ.

If β is obtained from α by adding extra j's at the end, we write $\alpha \lhd \beta$ or $\beta \rhd \alpha$ (with the convention that $\alpha \lhd \alpha$). If neither holds, we write $\alpha \nsim \beta$. We use $|\alpha - \beta|$ for the length of the shortest path from α to β. Thus, $\#(\alpha) = |\alpha - \phi|$.

$d = 1$ is, of course, the half-line. Figure 10.1.1 shows the case $d = 2$ displaying levels 0–3.

Given such a tree, $\mathbb{B}$, we let $\ell^2(\mathbb{B})$ be the sequences labeled by vertices, α, in $\mathbb{B}$. The free Hamiltonian is defined by

$$(H_0 u)(\alpha) = \sum_{|\beta - \alpha| = 1} u(\beta) \tag{10.1.1}$$

Given a function V on $\mathbb{B}$, we let

$$(Hu)(\alpha) = (H_0 u)(\alpha) + V(\alpha) u(\alpha) \tag{10.1.2}$$

Thus, we have an analog of a Jacobi matrix where $a_n \equiv 1$, and we use $V(\alpha)$ instead of b_n. We will only consider the case where V is a bounded function so

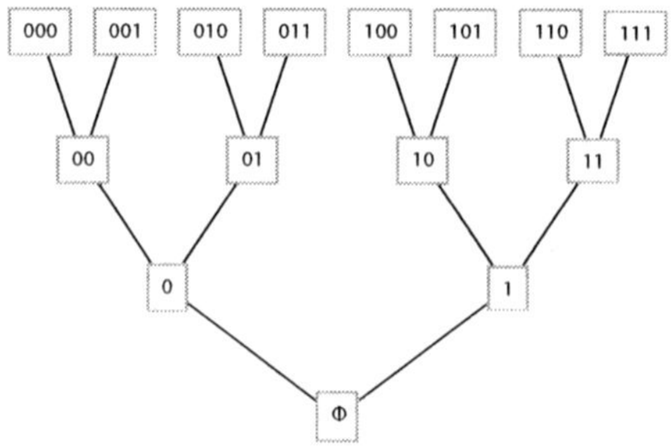

Figure 10.1.1. A Bethe–Cayley tree of degree two.

H is a bounded operator on $\ell^2(\mathbb{B})$. By the spectral measure, $d\mu$, we mean the one given by

$$\langle \delta_\phi, (H-z)^{-1}\delta_\phi \rangle = \int \frac{d\mu(x)}{x-z} \tag{10.1.3}$$

We will see (in Section 10.2) that for H_0,

$$\sigma(H_0) = \left[-2\sqrt{d}, 2\sqrt{d}\,\right] \tag{10.1.4}$$

with spectrum that is purely absolutely continuous of infinite multiplicity. Moreover,

$$d\mu_0(x) = \frac{1}{2d\pi}(4d - x^2)^{1/2}\,dx \tag{10.1.5}$$

which is just a scaling of the free Jacobi matrix.

The main theorem we will focus on in this chapter is:

Theorem 10.1.1 (Denisov [108]). *Let V be bounded and obey*

$$\sum_{n=0}^\infty \frac{1}{d^n} \sum_{\alpha|\#(\alpha)=n} |V(\alpha)|^2 < \infty \tag{10.1.6}$$

Then H has a.c. spectrum of infinite multiplicity on $[-2\sqrt{d}, 2\sqrt{d}\,]$.

Remarks. 1. We are not claiming that H only has spectrum in this interval.

2. We will see in Section 10.2 that this is optimal in the sense that if $|V(\alpha)|^2$ is replaced by $|V(\alpha)|^p$ with any $p > 2$, then it can happen that H has no a.c. spectrum.

We have simplified the presentation by taking the analog of the case $a_n \equiv 1$. We will also simplify things by henceforth taking $d = 2$ in the text, leaving general d to the Notes.

We begin the sketch of the strategy of the proof of Theorem 10.1.1 by noting the following: Suppose we prove that H has a.c. spectrum of multiplicity at least k under (10.1.6). In H, drop the links of ϕ to (0) and (1). This is a rank four perturbation, so by Corollary 7.3.4, the a.c. spectrum and its multiplicity is unchanged.

But if $\mathbb{B}_0$ (resp. $\mathbb{B}_1$) is the subset of $\mathbb{B}$ with $\#(\alpha) \geq 1$ and $\alpha_1 = 0$ (resp. $\alpha_1 = 1$), $\ell^2(\mathbb{B}) = \mathbb{C} \oplus \ell^2(\mathbb{B}_0) \oplus \ell^2(\mathbb{B}_1)$ and the modified H, call it $\widetilde{H}$, is a direct sum of $V(\phi) \oplus H^{(0)} \oplus H^{(1)}$ where $H^{(j)}$ is an operator like H but on $\ell^2(\mathbb{B}_j)$. However, the potentials for $H^{(j)}$ also satisfy (10.1.6). Thus, $\widetilde{H}$ has a.c. multiplicity at least $2k$. Therefore, it suffices to show the a.c. multiplicity is at least 1, for then, iterating this argument, we get multiplicities at least 2, 4, 8, ..., and so infinite.

In terms of the relative entropy, (2.2.1), we will prove that

$$S(\mu_0 \mid \mu) \geq -\frac{1}{4} \sum_{n=0}^{\infty} \frac{1}{d^n} \sum_{\alpha \mid \#(\alpha)=n} |V(\alpha)|^2 \tag{10.1.7}$$

so (10.1.6) implies this entropy is finite, so if

$$d\mu = W\, d\mu_0 + d\mu_s$$

then $W(x) > 0$ for a.e. $x \in [-2\sqrt{2}, 2\sqrt{2}]$, which implies a.c. spectrum of multiplicity at least 1.

In fact, we will eventually prove a much stronger result than Theorem 10.1.1. Consider all paths out to infinity, that is, sequences $\omega \equiv (\omega_1, \omega_2, \dots) \in \{0,1\}^{\mathbb{N}} \equiv \Omega$. Associate such an α with the walk through nearest neighbors

$$\alpha_\omega(j) = \begin{cases} \phi & j = 0 \\ (\omega_1 \dots \omega_j) & j \geq 1 \end{cases} \tag{10.1.8}$$

Put the infinite product measure, ν, on Ω with $\nu(\omega_j = 0) = \nu(\omega_j = 1) = \frac{1}{2}$. Moreover, define

$$\|V_\omega\|^2 = \sum_{j=0}^{\infty} |V(\alpha_\omega(j))|^2 \tag{10.1.9}$$

Then it is not hard to see that

$$\sum_{n=0}^{\infty} \frac{1}{d^n} \sum_{\alpha \mid \#(\alpha)=n} |V(n)|^2 = \int \|V_\omega\|^2\, d\nu(\omega) \tag{10.1.10}$$

The following improves Theorem 10.1.1:

Theorem 10.1.2 (Denisov–Kiselev [110]). *Let V be bounded and suppose $\nu(\{\omega \mid \|V_\omega\| < \infty\}) > 0$. Then H has a.c. spectrum of infinite multiplicity on $[-2\sqrt{2}, 2\sqrt{2}\,]$.*

Again, the key will be an improved sum rule. Instead of (10.1.7), we will prove

$$\exp(S(\mu_0 \mid \mu)) \geq \int \exp(-\tfrac{1}{4} \|V_\omega\|^2)\, d\nu(\omega) \tag{10.1.11}$$

By Jensen's inequality ($\log \int e^f\, d\eta \geq \int f\, d\eta$ for real-valued f and $d\eta$ a probability measure), (10.1.11) implies (10.1.7).

In Section 10.2, we will discuss the free Laplacian, while in Section 10.3, we will find the coefficient stripping formula, which will lead to a step-by-step sum rule in

Section 10.4. We will use the step-by-step sum rule to prove Theorem 10.1.1 in Section 10.5 and Theorem 10.1.2 in Section 10.6.

Remarks and Historical Notes. As noted, Theorem 10.1.1 is from Denisov [108] and Theorem 10.1.2 from Denisov–Kiselev [110]. Higher-order sum rules for trees have been found by Kupin [261].

Bethe lattices were introduced by Bethe [45]. The name Cayley tree comes from the fact that the Cayley graph of a free nonabelian group on p-generators is the unrooted Cayley tree with degree $2p$. Other than [108, 110], most of the spectral theory literature on Bethe–Cayley trees concerns Anderson localization; see [9, 10, 17, 142, 151, 228, 229]. For other papers on spectral theory on trees, see [58, 59, 414].

10.2 THE FREE HAMILTONIAN AND RADIALLY SYMMETRIC POTENTIALS

The lattice $\mathbb{B}$ has a huge symmetry group. The map T_ϕ interchanges the two branches coming out of ϕ, that is, $T_\phi(\phi) = \phi$, and for any $\alpha \neq \phi$, $T_\phi(\alpha) = (t(\alpha_1), \alpha_0, \dots, \alpha_{\#(\alpha)})$ where $t(0) = 1$, $t(1) = 0$. Similarly, one can define T_α for any $\alpha \in \mathbb{B}$, which leaves invariant α and any point not comparable or smaller than α and interchanges the two branches coming out of α. Let T_α generate a map U_α on $\ell^2(\mathbb{B})$ (by $(U_\alpha\varphi)(\beta) = \varphi(T_\alpha(\beta))$). All the U_α's commute with H_0, the free Hamiltonian.

In this section, we will use this symmetry to do a complete spectral analysis of H_0 and, as a bonus, reduce any $H_0 + V$, where

$$V(\alpha) = v(\#(\alpha)) \tag{10.2.1}$$

is radially symmetric, to a direct sum of Jacobi matrices.

Let $\mathcal{H}_s$ be the totally symmetric functions in $\ell^2(\mathbb{B})$, that is, there is $f\colon \{0, 1, \dots\} \to \mathbb{C}$ so

$$\varphi(\alpha) = f(\#(\alpha)) \tag{10.2.2}$$

Clearly, for $\varphi \in \mathcal{H}_s$,

$$\|\varphi\|^2 = \sum_{n=0}^{\infty} 2^n |f(n)|^2 \tag{10.2.3}$$

Fix $\alpha_0 \in \mathbb{B}$. Let $\mathcal{H}_{\alpha_0}$ be the odd functions based on α_0, that is, there is $f\colon \{0, 1, \dots\} \to \mathbb{C}$ so

$$\varphi(\beta) = \begin{cases} 0 & \text{if } \beta \not\sim \alpha_0 \text{ or } \beta \lhd \alpha_0 \\ f(\#(\beta) - \#(\alpha_0) - 1) & \text{if } \beta \rhd (\alpha_0, 0) \\ -f(\#(\beta) - \#(\alpha_0) - 1) & \text{if } \beta \rhd (\alpha_0, 1) \end{cases} \tag{10.2.4}$$

that is, f lives on the two trees coming out of the node α_0 and is antisymmetric under the switching of these two trees. Of course, if $\varphi \in \mathcal{H}_\alpha$, then

$$\|\varphi\|^2 = 2 \sum_{n=0}^{\infty} 2^n |f(n)|^2 \tag{10.2.5}$$

Define

$$U_{\mathrm{s}} \colon \mathcal{H}_{\mathrm{s}} \to \ell^2(\mathbb{N}) \tag{10.2.6}$$

(where $\mathbb{N} = \{0, 1, 2, \dots\}$) by the inverse of the map $g \mapsto \varphi$ with

$$\varphi(\alpha) = 2^{-n/2} g(\#(\alpha)) \tag{10.2.7}$$

and similarly,

$$U_\alpha \colon \mathcal{H}_\alpha \to \ell^2(\mathbb{N}) \tag{10.2.8}$$

as the inverse map to φ given by (10.2.4) with

$$g(n) = 2^{(n+1)/2} f(n) \tag{10.2.9}$$

Theorem 10.2.1. *We have*

$$\ell^2(\mathbb{B}) = \mathcal{H}_{\mathrm{s}} \oplus \bigoplus_{\alpha \in \mathbb{B}} \mathcal{H}_\alpha \tag{10.2.10}$$

that is, $\mathcal{H}_{\mathrm{s}} \perp \mathcal{H}_\alpha \perp \mathcal{H}_\beta$ *for all* α *and all* $\beta \neq \alpha$, *and these spaces span* $\ell^2(\mathbb{B})$.

Proof. The orthogonality to $\mathcal{H}_{\mathrm{s}}$ is a simple calculation (given the sign flip in (10.2.4)). When $\mathcal{H}_\alpha$ and $\mathcal{H}_\beta$ have disjoint subspaces if $\alpha \sim \beta$ and when $\alpha \lhd \beta$, the same sign flip yields orthogonality.

To get the spanning fact, note first that we can write

$$\ell^2(\mathbb{B}) = \bigoplus_n \ell^2(\mathbb{B}^{(n)}) \tag{10.2.11}$$

where

$$\mathbb{B}^{(n)} = \{\alpha \mid \#(\alpha) = n\} \tag{10.2.12}$$

so

$$\dim(\ell^2(\mathbb{B}^{(n)})) = 2^n \tag{10.2.13}$$

On the other hand,

$$\dim(\mathcal{H}_\alpha \cap \ell^2(\mathbb{B}^{(n)})) = \begin{cases} 0 & \text{if } \#(\alpha) \geq n \\ 1 & \text{if } \#(\alpha) < n \end{cases} \tag{10.2.14}$$

and that (the 1 comes from $\dim(\mathcal{H}_{\mathrm{s}} \cap \ell^2(\mathbb{B}^{(n)})) = 1$) $1 + \#(\alpha \mid \#(\alpha) < n) = 1 + 1 + \cdots + 2^{n-1} = 2^n$, accounting for (10.2.13). □

Theorem 10.2.2. *Let* V *obey* (10.2.1). *Then* H_0 *and* V *leave* $\mathcal{H}_{\mathrm{s}}$ *and each* $\mathcal{H}_\alpha$ *invariant. Moreover, on* $\ell^2(\mathbb{N})$,

$$U_\alpha H_0 U_\alpha^{-1} = \sqrt{2}\, J_0 \tag{10.2.15}$$

where J_0 is the matrix

$$J_0 = \begin{pmatrix} 0 & 1 & 0 & \dots \\ 1 & 0 & 1 & \dots \\ 0 & 1 & 0 & \dots \\ \vdots & \vdots & \vdots & \ddots \end{pmatrix} \tag{10.2.16}$$

and

$$U_\alpha V U_\alpha^{-1} = \begin{pmatrix} v(\#(\alpha)+1) & & \\ & v(\#(\alpha)+2) & \\ & & \ddots \end{pmatrix} \tag{10.2.17}$$

Moreover, $U_{\mathrm{s}} H_0 U_{\mathrm{s}}^{-1} = \sqrt{2} J_0$ *and* $U_{\mathrm{s}} V U_{\mathrm{s}}$ *has the form* (10.2.17) *where* $\#(\alpha)$ *is replaced by* -1.

Proof. Because V acts pointwise and, by (10.2.1), preserves the symmetries, the results for V are immediate.

As for H_0, using the formula for φ in terms of f ((10.2.2) or (10.2.4)), we see $H_0\varphi$ has the same form for some $\tilde{f}$ (by the minus symmetry, the bottom point α has $(H_0\varphi)(\alpha) = 0$ for $\varphi \in \mathcal{H}_\alpha$) and

$$\tilde{f}(n) = 2f(n+1) + f(n-1) \tag{10.2.18}$$

(with $f(-1) \equiv 0$).

Since g and f are related by (10.2.9) (or for $\mathcal{H}_2$, $g(n) = 2^{n/2} f(n)$), we see

$$(U_\alpha H_0 U_\alpha^{-1} g)(n) = \sqrt{2}\,(g(n+1) + g(n-1)) \tag{10.2.19}$$

which is (10.2.15). □

Corollary 10.2.3.

$$\sigma(H_0) = [-2\sqrt{2}, 2\sqrt{2}\,] \tag{10.2.20}$$

and is purely a.c. spectrum of infinite multiplicity.

Proof. As we have seen in (1.10.4), $\sigma(J_0) = [-2, 2]$ and is purely a.c. Thus, since each $H_0 \restriction \mathcal{H}_\alpha$ and $H_0 \restriction H_{\mathrm{s}}$ is unitarily equivalent to $\sqrt{2} J_0$, we get the result. □

Corollary 10.2.4. *There exist $V(\alpha)$'s obeying*

$$|V(\alpha)| \le C(1 + \#(\alpha))^{-1/2} \tag{10.2.21}$$

and, in particular, for all $p > 2$,

$$\sum_{n=0}^{\infty} 2^{-n} \sum_{\alpha \mid \#(\alpha) = n} |V(\alpha)|^p < \infty \tag{10.2.22}$$

so that $H_0 + V$ has only pure point spectrum in $[-2\sqrt{2}, 2\sqrt{2}\,]$.

Proof. Immediate from the last theorem and Theorem 3.5.6. □

Finally, we can compute $d\mu_0$ and its Stieltjes transform directly:

Theorem 10.2.5. *The spectral measure, $d\mu_0$, for H_0 is given by* (10.1.5) (*with* $d = 2$). *In particular,*

$$\langle \delta_\phi, (H_0 - z)^{-1}\delta_\phi \rangle = \frac{-z + \sqrt{z^2 - 8}}{4} \tag{10.2.23}$$

Remark. We will give another proof of this in Corollary 10.3.3.

Proof. Since $\delta_\phi \in \mathcal{H}_s$ and H_0 leaves $\mathcal{H}_s$ invariant, we have $U_s H_0 U_s^{-1} = \sqrt{2} J_0$ and $U_s \delta_\phi = \delta_1$, this follows by scaling (10.1.3) and the fact that the m-function for J_0, by (3.2.28), solves

$$m(z) = \frac{1}{-z - m(z)}$$

with $m(z) = -z^{-1} + O(z^{-2})$ at infinity. □

Remarks and Historical Notes. The form of $\sigma(H_0)$ and m (Theorem 10.2.5) goes back at least to Acosta–Klein [5]. Corollary 10.2.4 is noted by Denisov [108]. For an interesting application of the reduction to one-dimensional problems, see [59].

10.3 COEFFICIENT STRIPPING FOR TREES

As we have seen in Chapters 2 and 3, a key first step is coefficient stripping—and we consider that in this section. For Jacobi matrices, removing the root and the link to it yields another Jacobi matrix; but for our Bethe–Cayley tree with $d = 2$, removing ϕ and links to it yields two operators, $H^{(0)}$ on $\ell^2(\mathbb{B}_0)$ and $H^{(1)}$ on $\ell^2(\mathbb{B}_1)$. Let $m(z)$ be the function in (10.1.3) and $m^{(0)}(z)$, $m^{(1)}(z)$ the analogs for $H^{(0)}$ and $H^{(1)}$. Then we will prove

Theorem 10.3.1. *We have for any V that*

$$m(z) = \frac{1}{-z + V(\phi) - m^{(0)}(z) - m^{(1)}(z)} \tag{10.3.1}$$

To prove this, it is useful to consider extended lattices $\widetilde{\mathbb{B}}_0 = \phi \cup \mathbb{B}_0$, $\widetilde{\mathbb{B}}_1 = \phi \cup \mathbb{B}_1$, and $\widetilde{\mathbb{B}} = \psi \cup \mathbb{B}$, which is $\mathbb{B}$ with a node, ψ, added that connects only to ϕ. Define $\widetilde{H}$ ($\widetilde{H}^{(0)}$, $\widetilde{H}^{(1)}$) like H but with a coupling to the extra node added (i.e., $(\widetilde{H}g)(\phi) = g(\psi) + (H\dot{g})(\phi)$ and $(\widetilde{H}g)(\psi) = g(\phi)$). We have that

Lemma 10.3.2. *Define u on $\widetilde{\mathbb{B}}$ by*

$$u(\alpha) = \langle \delta_\alpha, (H - z)^{-1}\delta_\phi \rangle \tag{10.3.2}$$

$$u(\psi) = -1 \tag{10.3.3}$$

Then u obeys

$$\widetilde{H}u(\alpha) = zu(\alpha) \tag{10.3.4}$$

for $\alpha \in \mathbb{B}$ and is the unique such function obeying (10.3.3) *and $u \in \ell^2(\widetilde{\mathbb{B}})$.*

Proof. $\check{u} \equiv u \restriction \mathbb{B}$ obeys

$$H\check{u}(\alpha) = z\check{u}(\alpha) + \delta_{\alpha\phi} \tag{10.3.5}$$

which is (10.3.4), given (10.3.3). Moreover, given (10.3.3), (10.3.4) is equivalent to (10.3.5), so the uniqueness follows from the invertibility of $(H - z)$, which implies there is a unique ℓ^2 solution of

$$(H - z)\check{u} = \delta_\phi \tag{10.3.6}$$

□

First Proof of Theorem 10.3.1. Let u be given by (10.3.2). Then $u^{(0)} \equiv u \restriction \widetilde{\mathbb{B}}_0$ solves (for $\alpha \in \mathbb{B}_0$)

$$\widetilde{H}^{(0)}u^{(0)}(\alpha) = zu^{(0)}(\alpha) \tag{10.3.7}$$

and is in $\ell^2(\widetilde{\mathbb{B}}_0)$, but it does not obey the analog of (10.3.3). However, $-u(\phi)^{-1}u^{(0)}$ does and still solves (10.3.7) and is in $\ell^2(\widetilde{\mathbb{B}}_0)$. It follows that

$$m^{(0)}(z) = -\frac{u((0))}{u(\phi)} \tag{10.3.8}$$

and similarly,

$$m^{(1)}(z) = -\frac{u((1))}{u(\phi)} \tag{10.3.9}$$

Moreover,

$$u(\phi) = m(z) \tag{10.3.10}$$

Thus,

$$(V(\phi) - z)u(\phi) + u((0)) + u((1)) - 1 = 0 \tag{10.3.11}$$

becomes (dividing by $u(\phi)$)

$$V(\phi) - z - m^{(0)}(z) - m^{(1)}(z) = m(z)^{-1} \tag{10.3.12}$$

which is (10.3.1). □

Second Proof of Theorem 10.3.1. Here is a more operator theoretic argument. Let Γ be the part of H_0 that links ϕ to (0) and (1), that is,

$$\Gamma\delta_\phi = \delta_{(0)} + \delta_{(1)} \qquad \Gamma\delta_{(0)} = \delta_\phi \qquad \Gamma\delta_{(1)} = \delta_\phi \tag{10.3.13}$$

and otherwise $\Gamma = 0$. Thus,

$$\widetilde{H} = H - \Gamma \tag{10.3.14}$$

$$= V(0) \oplus H^{(0)} \oplus H^{(1)} \tag{10.3.15}$$

acting on $\mathbb{C} + \ell^2(\mathbb{B}_0) \oplus \ell^2(\mathbb{B}_1)$.

By the resolvent identity,

$$(H - z)^{-1} = (\widetilde{H} - z)^{-1} - (H - z)^{-1}\Gamma(\widetilde{H} - z)^{-1} \tag{10.3.16}$$

and, of course, by (10.3.15),

$$(\widetilde{H} - z)^{-1} = (V(0) - z)^{-1} \oplus (H^{(0)} - z)^{-1} \oplus (H^{(1)} - z)^{-1} \tag{10.3.17}$$

Apply (10.3.16) to δ_ϕ and to $\delta_{(0)}$ to get

$$(H - z)^{-1}\delta_\phi = (V(0) - z)^{-1}[\delta_\phi - (H - z)^{-1}[\delta_{(0)} + \delta_{(1)}]] \tag{10.3.18}$$

$$(H - z)^{-1}\delta_{(0)} = -m^{(0)}(z)(H - z)^{-1}\delta_\phi \tag{10.3.19}$$

and similarly for $\delta_{(1)}$.

Taking inner products with δ_ϕ, we get from the second equation for $j = 0, 1$,

$$\langle \delta_\phi, (H - z)^{-1}\delta_{(j)}\rangle = -m^{(j)}(z)m(z) \tag{10.3.20}$$

and then from the first equation that

$$m(z) = (V(0) - z)^{-1}[1 + m(z)(m^{(0)}(z) + m^{(1)}(z))] \tag{10.3.21}$$

Multiplying by $m(z)^{-1}(V(0) - z)$, we get

$$V(0) - z = m(z)^{-1} + m^{(0)}(z) + m^{(1)}(z) \tag{10.3.22}$$

which is equivalent to (10.3.1). □

Remark. Both these methods can be used to prove Theorem 3.2.4.

As an application of Theorem 10.3.1, we get a second proof of (10.2.21):

Corollary 10.3.3. *The free m-function is given by*

$$m_0(z) = \frac{-z + \sqrt{z^2 - 8}}{4} \tag{10.3.23}$$

Proof. In this case, $V(\phi) = 0$ and $m(z) = m^{(1)}(z) = m^{(2)}(z)$, so solves

$$m(z) = \frac{1}{-z - 2m(z)} \tag{10.3.24}$$

which gives a quadratic equation solved by (10.3.23). □

From (10.3.1), using $m^{(j)}(z) = -z^{-1} + O(z^{-2})$,

$$\begin{aligned} m(z) &= (-z)^{-1}(1 - V(\phi)z^{-1} + 2z^{-2} + O(z^{-3})) \\ &= -z^{-1} - V(\phi)z^{-2} - (2 + V(\phi)^2)z^{-3} + O(z^{-4}) \end{aligned} \tag{10.3.25}$$

Remarks and Historical Notes. Theorem 10.3.1 goes back at least to Klein [229]; see also [142]. Klein gave what we call the second proof.

For general degrees, there are d subtrees with m-function $\{m^{(j)}(z)\}_{j=0}^{d-1}$ and the coefficient stripping formula is

$$m(z)^{-1} = -z + V(\phi) - \sum_{j=0}^{d-1} m^{(j)}(z) \tag{10.3.26}$$

10.4 A STEP-BY-STEP SUM RULE FOR TREES

Our goal in this section is to prove a step-by-step inequality:

Theorem 10.4.1 (Denisov [108]). *Let V be a bounded function on $\mathbb{B}$ with $|V(\alpha)| \to 0$ as $\#(\alpha) \to \infty$. Let μ be the spectral measure for $H_0 + V$ and $\mu^{(j)}$, $j = 1, 2$, for $H^{(j)}$ and $\ell^2(\mathbb{B}_j)$. Then*

$$S(\mu_0 \mid \mu) \geq S(\mu_0 \mid \tfrac{1}{2}(\mu^{(0)} + \mu^{(1)})) - \tfrac{1}{4}V(\phi)^2 \tag{10.4.1}$$

This will come from a sum rule:

Theorem 10.4.2. *Under the hypotheses of Theorem 10.4.1,*

$$S(\mu_0 \mid \mu) = S(\mu_0 \mid \tfrac{1}{2}\mu^{(0)} + \tfrac{1}{2}\mu^{(1)}) - \tfrac{1}{4}V(\phi)^2 - 2\sum_k Y(\widetilde{E}_k) + 2\sum_k Y(E_k) \tag{10.4.2}$$

where $\{E_k\}$ are the eigenvalues of H outside $(-2\sqrt{2}, 2\sqrt{2})$ and $\{\widetilde{E}_k\}$ are the eigenvalues of $H^{(0)} \oplus H^{(1)}$ outside $(-2\sqrt{2}, 2\sqrt{2})$ and where

$$Y(E) = F\big(E/\sqrt{2}\big) \tag{10.4.3}$$

and F is given by (1.10.9).

Remarks. 1. Since $|V(\alpha)| \to 0$ as $\#(\alpha) \to \infty$, $\sigma_{\text{ess}}(H) = \sigma_{\text{ess}}(H_0)$, so the spectrum outside $[-2\sqrt{2}, 2\sqrt{2}\,]$ is discrete.

2. As usual, we should label $E_j^{\pm}$, $\widetilde{E}_j^{\pm}$ with $E^+ > 2\sqrt{2}$ and $E^- < -2\sqrt{2}$. As we will see, $E_j^+ > \widetilde{E}_j^+$, $E_j^- < \widetilde{E}_j^-$, and so $Y(\widetilde{E}_j^{\pm}) \leq Y(E_j^{\pm})$. By $\sum_k Y(E_k) - \sum_k Y(\widetilde{E}_k)$, we mean

$$\sum_{j,\pm} (Y(E_j^{\pm}) - Y(\widetilde{E}_j^{\pm}))$$

which is a sum of positive terms which could be ∞. Actually, we only apply the step-by-step sum rule directly to cases where V has finite support and the spectrum outside $[-2\sqrt{2}, 2\sqrt{2}\,]$ is finite.

Theorem 10.4.2 $\Rightarrow$ Theorem 10.4.1. $H^{(0)} \oplus H^{(1)} = PHP$ where P is the projection onto $\ell^2(\mathbb{B}^{(0)}) \oplus \ell^2(\mathbb{B}^{(1)})$. So by the min-max principle,

$$\pm\widetilde{E}_j^{\pm} \leq \pm E_j^{\pm} \tag{10.4.4}$$

F, and so Y, is monotone (see (3.5.10)), so

$$Y(\widetilde{E}_j^{\pm}) \leq Y(E_j^{\pm}) \tag{10.4.5}$$

and so

$$\sum_k Y(E_k) - \sum_k Y(\widetilde{E}_k) \geq 0 \tag{10.4.6}$$

so (10.4.2) implies (10.4.1). □

Remark. It may appear that we have thrown away something useful but, as we will explain in the Notes to Section 10.5, we have not.

Proof of Theorem 10.4.2. This closely follows the proof of Corollary 3.4.7. Define

$$M(z) = -m(\sqrt{2}\,(z + z^{-1})) \tag{10.4.7}$$

In comparison with (3.4.26), we have, by (10.3.25),

$$\log\left(\frac{M(z)}{z}\right) = -\frac{1}{2}\log(2) + \frac{V(\phi)}{\sqrt{2}}\,z + \frac{V(\phi)^2}{4}\,z^2 + O(z^3) \tag{10.4.8}$$

On the other hand, by (10.3.1),

$$\frac{\operatorname{Im}(M(e^{i\theta}))}{|M(e^{i\theta})|^2} = \operatorname{Im}(M^{(0)}(e^{i\theta}) + M^{(1)}(e^{i\theta})) \tag{10.4.9}$$

By directly following the proof of the P_2 sum rule, that is, combining the zeroth- and second-order Poisson–Jensen formulae, one gets

$$\begin{aligned} -\frac{1}{4\pi}\int_0^{2\pi} & \log\left(\frac{\operatorname{Im}(M^{(0)}) + \operatorname{Im}(M^{(1)})}{\operatorname{Im}(M)}\right)(e^{i\theta})(1-\cos 2\theta)\,d\theta \\ &= -\frac{1}{2}\log(2) - \frac{V(\phi)^2}{8} + \sum_k Y(\widetilde{E}_k) - \sum_k Y(E_k) \end{aligned} \tag{10.4.10}$$

Since

$$\frac{1}{4\pi}\int_0^{2\pi} \log(\tfrac{1}{2})(1 - 2\cos\theta)\,d\theta = \tfrac{1}{2}\log(2) \tag{10.4.11}$$

we can drop the $-\frac{1}{2}\log(2)$ from the right side of (10.4.10) if we replace $\operatorname{Im}(M^{(0)} + M^{(1)})$ by $\operatorname{Im}((M^{(0)} + M^{(1)})/2)$.

Following the identification of energies in Section 3.4, we get (10.4.2). □

Remarks and Historical Notes. This section follows Denisov [108].

10.5 THE GLOBAL ℓ^2 THEOREM

In this section, we will prove Theorem 10.1.1 by proving (10.1.7). Recall (Theorem 2.2.3) that $S(\mu \mid \nu)$ is jointly concave, so

$$S(\mu_0 \mid \tfrac{1}{2}(\mu^{(0)} + \mu^{(1)})) \geq \tfrac{1}{2}S(\mu_0 \mid \mu^{(0)}) + \tfrac{1}{2}S(\mu_0 \mid \mu^{(1)}) \tag{10.5.1}$$

Proposition 10.5.1. *Let V have compact support. Then*

$$S(\mu_0 \mid \mu) \geq -\frac{1}{4}\sum_{n=0}^{\infty}\frac{1}{2^n}\sum_{\#(\alpha)=n} V(\alpha)^2 \tag{10.5.2}$$

Remark. Since V has compact support, the sum in (10.5.2) is finite.

Proof. By (10.4.1) and (10.5.1),

$$S(\mu_0 \mid \mu) \geq -\tfrac{1}{4}V(\phi)^2 - \tfrac{1}{2}S(\mu_0 \mid \mu^{(0)}) - \tfrac{1}{2}S(\mu_0 \mid \mu^{(1)}) \tag{10.5.3}$$

For $\alpha \in \mathbb{B}$, let

$$\mathbb{B}_\alpha = \{\beta \mid \alpha \lhd \beta\} \tag{10.5.4}$$

and $H^{(\alpha)}$, the operator on $\ell^2(\mathbb{B}_\alpha)$ obtained by letting P_α be the projection of $\ell^2(\mathbb{B}_\alpha)$ to $\ell^2(\mathbb{B})$ and setting

$$H^{(\alpha)} = P_\alpha H P_\alpha \restriction \ell^2(\mathbb{B}_\alpha) \tag{10.5.5}$$

that is, H with all links from $\mathbb{B}_\alpha$ to $\mathbb{R}\setminus\mathbb{B}_\alpha$ dropped. Let $\mu^{(\alpha)}$ be the spectral measure for $H^{(\alpha)}$ and $\delta_{[\alpha]}$.

By induction, from (10.5.3) we get for any k,

$$S(\mu_0 \mid \mu) \geq -\frac{1}{2^k} \sum_{\#(\alpha)=k} S(\mu_0 \mid \mu^{(\alpha)}) - \frac{1}{4}\sum_{n=0}^{k-1} \frac{1}{2^n} \sum_{\#(\alpha)=n} V(\alpha)^2 \tag{10.5.6}$$

Since V has compact support, we can find k so $V(\beta) = 0$ if $\#(\beta) \geq k$. For such k, $\mu^{(\alpha)} = \mu_0$ if $\#(\alpha) = k$, so (10.5.6) is (10.5.2). $\square$

Theorem 10.5.2 (implies Theorem 10.1.1)**.** (10.5.2) *holds for any bounded V. In particular, if*

$$\sum_{n=0}^{\infty} \frac{1}{2^n} \sum_{\#(\alpha)=n} V(\alpha)^2 < \infty \tag{10.5.7}$$

then H has a.c. spectrum of infinite multiplicity on $[-2\sqrt{2}, 2\sqrt{2}\,]$.

Proof. For any k, define $V^{[k]}$ by

$$V^{[k]}(\alpha) = \begin{cases} V(\alpha) & \text{if } \#(\alpha) \leq k \\ 0 & \text{if } \#(\alpha) > k \end{cases} \tag{10.5.8}$$

and let $\mu^{[k]}$ be the measure for $H^{[k]} = H_0 + V^{[k]}$.

For any $\eta \in \ell^2(\mathbb{B})$, we have $\|(V - V^{[k]})\eta\| \to 0$, so

$$H^{[k]} \to H \quad \text{strongly} \tag{10.5.9}$$

Since the H's are uniformly bounded, $(H^{[k]} - z)^{-1} \to (H - z)^{-1}$ strongly, so in the weak (vague) topology on measures,

$$\mu^{[k]} \xrightarrow{w} \mu \tag{10.5.10}$$

By the weak upper semicontinuity of S (see Theorem 2.2.3), we have

$$S(\mu_0 \mid \mu) \geq \limsup_{k\to\infty} S(\mu_0 \mid \mu^{[k]}) \tag{10.5.11}$$

so (10.5.2) for $V^{[k]}$ implies it for V.

Given this, if (10.5.7) holds, then $S(\mu_0 \mid \mu) > -\infty$, so $\Sigma_{\text{ac}}(\mu) \supset [-2, \sqrt{2}, 2\sqrt{2}\,]$. By the argument in Section 10.1, we get infinite multiplicity. $\square$

Remarks and Historical Notes. The results of this section are from Denisov [108]. One can see why this method does not yield much information about eigenvalues. For after one step, one finds

$$\begin{aligned} 2\sum_{j,\pm} Y(E_j^{\pm}) - S(\mu_0 \mid \mu) &\leq \tfrac{1}{4}V(\phi)^2 \\ &+\tfrac{1}{2}\Bigl(2\sum_{j,\pm} Y(\widetilde{E}_j^{\pm}) - S(\mu_0 \mid \mu^{(1)}) - S(\mu_0 \mid \mu^{(2)})\Bigr) + \sum_{j,\pm} Y(\widetilde{E}_j^{\pm}) \end{aligned} \tag{10.5.12}$$

One can iterate this but $\frac{1}{2}$ of the $Y(\widetilde{E}_j^{\pm})$, $\frac{1}{4}$ of the Y's at the next level, and so on will be left over. So one does not get a bound on $\sum_{j,\pm} Y(E_j^{\pm})$ in terms of $\sum_{n=0}^{\infty} 2^{-n} \sum_{\alpha \mid \#(\alpha)=n} |V(\alpha)|^2$.

However, since $S \leq 0$, we have

$$S(\mu_0 \mid \mu) \geq S(\mu_0 \mid \mu^{(0)}) + S(\mu_0 \mid \mu^{(1)}) \tag{10.5.13}$$

and this can be iterated to get

$$\sum_{j,\pm} Y(E_j^{\pm}) \leq \tfrac{1}{4} \sum_{\alpha} |V(\alpha)|^2 \tag{10.5.14}$$

It remains to be seen if this can be improved.

10.6 THE LOCAL ℓ^2 THEOREM

In this section, we will prove Theorem 10.1.2 by proving (10.1.11). The key is the following, whose proof we defer:

Theorem 10.6.1. *Consider $S(\mu \mid \nu)$ where μ and ν are probability measures. For μ fixed, the map $\eta \to e^{S(\mu|\eta)}$ is concave, that is, for η_0, η_1 probability measures and $0 < \theta < 1$, we have*

$$e^{S(\mu|\theta\eta_1+(1-\theta)\eta_0)} \geq \theta\, e^{S(\mu|\eta_1)} + (1-\theta)e^{S(\mu|\eta_0)} \tag{10.6.1}$$

Theorem 10.6.2 (implies Theorem 10.1.2)**.** (10.1.11) *holds for any bounded V. In particular, if*

$$\nu(\{\omega \mid \|V_\omega\| < \infty\}) > 0 \tag{10.6.2}$$

then H has a.c. spectrum of infinite multiplicity on $[-2\sqrt{2}, 2\sqrt{2}\,]$.

Proof. By (10.6.1) and (10.4.1), we have

$$e^{S(\mu_0|\mu)} \geq \tfrac{1}{2}\, e^{-\frac{1}{4}V(\phi)^2}[e^{S(\mu_0|\mu^{(0)})} + e^{S(\mu_0|\mu^{(1)})}] \tag{10.6.3}$$

So, by iterating,

$$\exp((S(\mu_0 \mid \mu)) \geq \frac{1}{2^k} \sum_{\#(\alpha)=k} \exp\left[S(\mu_0 \mid \mu^{(\alpha)}) - \frac{1}{4} \sum_{j=0}^{k} V((\alpha_1, \dots, \alpha_j))^2 \right] \tag{10.6.4}$$

where the $j = 0$ term in the sum is $V(\phi)^2$.

If V has compact support, by taking large k, we get (10.1.11) and then we get (10.1.11) in general by taking limits using upper semicontinuity of S.

If (10.6.2) holds, then the right side of (10.1.11) is strictly positive, so $S(\mu_0 \mid \mu) > 0$, which implies that $[-2\sqrt{2}, 2\sqrt{2}\,] \subset \Sigma_{\text{ac}}(\mu)$, showing that the a.c. spectrum of H contains that interval with multiplicity 1.

The proof of infinite multiplicity has to be modified slightly compared to the argument in Section 10.1. Removing the connections to ϕ breaks the lattice in two. However, (10.6.2) does not necessarily imply that for both $j = 0$ and 1 that

$$2\nu(\{\omega \mid \omega \rhd (j) \text{ and } \|V_\omega\| < \infty\}) > 0 \tag{10.6.5}$$

so one cannot conclude each of $H^{(0)}$ and $H^{(1)}$ has a.c. spectrum.

But if

$$\nu(\{\omega \mid \|V_\omega\| < \infty\}) > 2^{-k} \tag{10.6.6}$$

removing connections to nodes, α, with $\#(\alpha) \geq k - 1$ breaks $\mathbb{B}$ into 2^k lattices, and at least two of them must have paths with $\|V_\omega\| < \infty$ and positive ν measure (since *all* the paths in a single sublattice only have measure 2^{-k}). Thus, we see that if (10.6.2) implies a.c. spectrum of multiplicity at least ℓ, it implies multiplicity at least 2ℓ. So, as in Section 10.1, we get infinite multiplicity. □

We turn to the proof of Theorem 10.6.1:

First Proof of Theorem 10.6.1. If

$$\nu_j^{(\varepsilon)} = (1 + \varepsilon)^{-1}(\nu_j + \varepsilon\mu) \tag{10.6.7}$$

then, by using the monotone convergence theorem,

$$S(\mu \mid \theta\nu_1 + (1 - \theta)\nu_0) = \lim_{\varepsilon \downarrow 0} S(\mu \mid \theta\nu_1^{(\varepsilon)} + (1 - \theta)\nu_0^{(\varepsilon)}) \tag{10.6.8}$$

so, without loss, we can suppose

$$d\nu_j = w_j \, d\mu + d\mu_s^{(j)} \tag{10.6.9}$$

with

$$\inf_x w_j(x) > 0$$

In that case,

$$S(\theta) \equiv S(\mu \mid \theta\nu_1 + (1 - \theta)\nu_2) = \int \log(\theta w_1 + (1 - \theta)w_2) \, d\mu \tag{10.6.10}$$

is a C^∞ function of θ and

$$S'(\theta) = \int \frac{(w_1 - w_2)}{\theta w_1 + (1-\theta) w_2} \, d\mu \tag{10.6.11}$$

$$S''(\theta) = -\int \frac{(w_1 - w_2)^2}{(\theta w_1 + (1-\theta)(w_2))^2} \, d\mu \tag{10.6.12}$$

By the Schwarz inequality,

$$-S''(\theta) - S'(\theta)^2 \geq 0 \tag{10.6.13}$$

Let

$$g(\theta) = e^{S(\theta)} \tag{10.6.14}$$

Then, by (10.6.13),

$$g''(\theta) = [S''(\theta) + S'(\theta)^2] e^{S(\theta)} \leq 0 \tag{10.6.15}$$

so g is concave in θ, as claimed. □

Our second proof depends on a variational principle for S that complements (2.2.5)/(2.2.6):

Proposition 10.6.3. *Let μ, ν be probability measures. Then the relative entropy obeys (see* (2.2.21))

$$S(\mu \mid \nu) = \inf_{g \in C(X)} \left[\log \int e^g \, d\nu - g \, d\mu \right] \tag{10.6.16}$$

Proof. Let $\mathcal{S}(f; \mu, \nu)$ be given by (2.2.6) and let

$$\mathcal{G}(g) = \log \int e^g \, d\nu - \int g \, d\mu \tag{10.6.17}$$

We claim that

$$\mathcal{S}(e^g; \mu, \nu) \geq \mathcal{G}(g) \geq S(\mu \mid \nu) \tag{10.6.18}$$

from which (10.6.16) is immediate from (2.2.5).

By concavity of $\log(y)$, we have that

$$\log(y) \leq y - 1 \tag{10.6.19}$$

so

$$\int e^g \, d\nu - 1 \geq \log\left(\int e^g \, d\nu \right) \tag{10.6.20}$$

which implies the first inequality in (10.6.18).

For the second inequality, note it is trivial if $S(\mu \mid \nu) = -\infty$. If $S(\mu \mid \nu) > -\infty$, then μ is ν a.c. So write

$$d\nu = \left(\frac{d\mu}{d\nu} \right)^{-1} d\mu + d\mu_s \tag{10.6.21}$$

Thus, by Jensen's inequality (i.e., $\log \int e^h \, d\eta \geq \int h \, d\eta$), we have that

$$\begin{aligned}\log\left(\int e^g \, d\nu\right) &= \log\left(\int e^g \left(\frac{d\mu}{d\nu}\right)^{-1} d\mu + \int e^g \, d\mu_s\right) \\ &\geq \log\left(\int \exp\left(g - \log\left(\frac{d\mu}{d\nu}\right)\right) d\mu\right) \\ &\geq \int g \, d\mu + S(\mu \mid \nu) \qquad (10.6.22)\end{aligned}$$

completing the proof of (10.6.18). □

Second Proof of Theorem 10.6.1. By (10.6.16),

$$e^{S(\mu|\nu)} = \inf_{g \in C(X)} \left[\int e^g \, d\nu \Big/ \exp\left(\int g \, d\mu\right)\right] \qquad (10.6.23)$$

is an inf of linear functionals of ν, and so concave. □

Remarks and Historical Notes. The results of this section are from Denisov–Kiselev [110]. They have a different proof of Theorem 10.6.1 using Young's inequality that for $x, y > 0$ and $p^{-1} + q^{-1} = 1$, $p, q > 1$, we have

$$xy \leq \frac{x^p}{p} + \frac{y^q}{q}$$

They optimize over p.

Theorem 10.6.1 has the smell of something that must be well known in the entropy literature. I consulted several different experts in various aspects of entropy and got the same reply: "I agree this must be well known but I haven't seen it. I suggest you ask so-and-so." When I asked so-and-so, I got the same answer!

Bibliography

[1] M. J. Ablowitz and J. F. Ladik, *Nonlinear differential-difference equations*, J. Math. Phys. **16** (1975), 598–603. (Cited on 416.)

[2] M. J. Ablowitz and J. F. Ladik, *Nonlinear differential-difference equations and Fourier analysis*, J. Math. Phys. **17** (1976), 1011–1018. (Cited on 416.)

[3] M. J. Ablowitz and J. F. Ladik, *A nonlinear difference scheme and inverse scattering*, Studies in Appl. Math. **55** (1976), 213–229. (Cited on 416.)

[4] R. Abraham and J. E. Marsden, *Foundations of Mechanics*, 2nd edition, revised and enlarged, Benjamin/Cummings, Reading, MA, 1978. (Cited on 382, 387.)

[5] V. Acosta and A. Klein, *Analyticity of the density of states in the Anderson model on the Bethe lattice*, J. Statist. Phys. **69** (1992), 277–305. (Cited on 597.)

[6] S. Agmon, *Lectures on Exponential Decay of Solutions of Second-Order Elliptic Equations: Bounds on Eigenfunctions of N-body Schrödinger Operators*, Mathematical Notes, **29**, Princeton University Press, Princeton, NJ; University of Tokyo Press, Tokyo, 1982. (Cited on 425.)

[7] L. V. Ahlfors, *Complex Analysis. An Introduction to the Theory of Analytic Functions of One Complex Variable*, McGraw–Hill, New York, 1978. (Cited on 320, 351, 444, 524, 525, 528.)

[8] L. Ahlfors and L. Bers, *Riemann's mapping theorem for variable metrics*, Annals of Math. (2) **72** (1960), 385–404. (Cited on 561.)

[9] M. Aizenman, R. Sims, and S. Warzel, *Stability of the absolutely continuous spectrum of random Schrödinger operators on tree graphs*, Probab. Theory Related Fields **136** (2006), 363–394. (Cited on 594.)

[10] M. Aizenman and S. Warzel, *The canopy graph and level statistics for random operators on trees*, Math. Phys. Anal. Geom. **9** (2006), 291–333. (Cited on 594.)

[11] N. I. Akhiezer, *On a proposition of A. N. Kolmogorov and a proposition of M. G. Kreĭn*, Doklady Akad. Nauk SSSR (N.S.) **50** (1945), 35–39 [Russian]. (Cited on 207.)

[12] N. I. Akhiezer, *Continuous analogues of orthogonal polynomials on a system of intervals*, Dokl. Akad. Nauk SSSR **141** (1961), 263–266 [Russian]. (Cited on 359.)

[13] N. I. Akhiezer, *The Classical Moment Problem and Some Related Questions in Analysis*, Hafner, New York, 1965; Russian original, 1961. (Cited on 73, 202, 207.)

[14] N. I. Akhiezer and I. M. Glazman, *Theory of Linear Operators in Hilbert Space, Vols. 1 and 2*, Ungar, New York, 1961. (Cited on 2, 65.)

[15] N. I. Akhiezer and M. G. Krein, *Über Fouriersche Reihen beschränkter summierbarer Funktionen und ein neues Extremumproblem. II*, Communications Kharkoff (4) **10** (1934), 3–32; Supplement, ibid. (4) **12** (1935), 37–40. (Cited on 283.)

[16] N. I. Akhiezer and M. Krein, *Some Questions in the Theory of Moments*, Transl. Math. Monographs, **2**, American Mathematical Society, Providence, RI, 1962; Russian original, 1938. (Cited on 425.)

[17] C. Allard and R. Froese, *A Mourre estimate for a Schrödinger operator on a binary tree*, Rev. Math. Phys. **12** (2000), 1655–1667. (Cited on 594.)

[18] G. S. Ammar and W. B. Gragg, *Schur flows for orthogonal Hessenberg matrices*, in "Hamiltonian and Gradient Flows, Algorithms and Control," pp. 27–34, Fields Inst. Commun. **3** (1994), American Mathematical Society, Providence, RI. (Cited on 417.)

[19] V. V. Andrievskii and H.-P. Blatt, *Discrepancy of Signed Measures and Polynomial Approximation*, Springer Monographs in Mathematics, Springer-Verlag, New York, 2002. (Cited on 305, 319.)

[20] P. M. Anselone, *Collectively Compact Operator Approximation Theory and Applications to Integral Equations*, Prentice–Hall, Englewood Cliffs, NJ, 1971. (Cited on 425.)

[21] A. I. Aptekarev, *Asymptotic properties of polynomials orthogonal on a system of contours, and periodic motions of Toda chains*, Math. USSR Sb. **53** (1986), 233–260; Russian original in Mat. Sb. (N.S.) **125(167)** (1984), 231–258. (Cited on 304, 305, 570, 576, 589.)

[22] A. I. Aptekarev and V. G. Lysov, *Systems of Markov functions generated by graphs and the asymptotics of their Hermite–Padé approximants*, Math. Sbornik. **201** (2010), 183–234. (Cited on 589, 590.)

[23] A. Aptekarev and E. Nikishin, *The scattering problem for a discrete Sturm–Liouville operator*, Mat. Sb. **121(163)** (1983), 327–358. (Cited on 229, 237, 239.)

[24] M. A. Armstrong, *Basic Topology*, corrected reprint of the 1979 original, Undergraduate Texts in Mathematics, Springer-Verlag, New York-Berlin, 1983. (Cited on 525.)

[25] V. I. Arnold, *Mathematical Methods of Classical Mechanics*, corrected reprint of the 2nd (1989) edition, Graduate Texts in Mathematics, **60**, Springer-Verlag, New York, 1997. (Cited on 387.)

[26] V. I. Arnold, V. V. Kozlov, and A. Neishtadt, *Mathematical Aspects of Classical and Celestial Mechanics*, [Dynamical systems. III], 3rd edition, Encyclopaedia of Mathematical Sciences, **3**, Springer-Verlag, Berlin, 2006. (Cited on 382, 387.)

[27] N. Aronszajn and W. F. Donoghue, *On exponential representations of analytic functions in the upper half-plane with positive imaginary part*, J. Anal. Math. **5** (1956/57), 321–388. (Cited on 283, 431.)

[28] A. Avila, J. Bochi, and D. Damanik, *Cantor spectrum for Schrödinger operators with potentials arising from generalized skew-shifts*, Duke Math. J. **146** (2009), 253–280. (Cited on 371.)

[29] A. Avila, J. Bochi, and D. Damanik, *Opening gaps in the spectrum of strictly ergodic Schrödinger operators*, preprint (Cited on 371.)

[30] A. Avila, Y. Last, and B. Simon, *Bulk universality and clock spacing of zeros for ergodic Jacobi matrices with a.c. spectrum*, to appear in Analysis & PDE. (Cited on 222, 223, 225, 226, 227.)

[31] D. Barrios Rolanía and G. López Lagomasino, *Ratio asymptotics for polynomials orthogonal on arcs of the unit circle*, Constr. Approx. **15** (1999), 1–31. (Cited on 457.)

[32] G. Baxter, *A convergence equivalence related to polynomials orthogonal on the unit circle*, Trans. Amer. Math. Soc. **99** (1961), 471–487. (Cited on 42.)

[33] D. Bättig, A. M. Bloch, J.-C. Guillot, and T. Kappeler, *On the symplectic structure of the phase space for periodic KdV, Toda, and defocusing NLS*, Duke Math. J. **79** (1995), 549–604. (Cited on 387.)

[34] D. Bättig, B. Grébert, J.-C. Guillot, and T. Kappeler, *Fibration of the phase space of the periodic Toda lattice*, J. Math. Pures Appl. (9) **72** (1993), 553–565. (Cited on 387.)

[35] A. F. Beardon, *Inequalities for certain Fuchsian groups*, Acta Math. **127** (1971), 221–258. (Cited on 540.)

[36] A. F. Beardon, *The Geometry of Discrete Groups*, corrected reprint of the 1983 original, Graduate Texts in Mathematics, **91**, Springer-Verlag, New York, 1995. (Cited on 518.)

[37] M. Bello Hernández and G. López Lagomasino, *Ratio and relative asymptotics of polynomials orthogonal on an arc of the unit circle*, J. Approx. Theory **92** (1998), 216–244. (Cited on 454, 457.)

[38] M. Bello Hernández and E. Miña Díaz, *Strong asymptotic behavior and weak convergence of polynomials orthogonal on an arc of the unit circle*, J. Approx. Theory **111** (2001), 233–255. (Cited on 457.)

[39] C. Bennewitz and W. N. Everitt, *Some remarks on the Titchmarsh–Weyl m-coefficient*, in "Tribute to Åke Pleijel: Proceedings of the Pleijel Conference," pp. 49–108, Department of Mathematics, University of Uppsala, Sweden, 1979. (Cited on 73.)

[40] P. Bérard, *Variétés riemanniennes isospectrales non isométriques*, Séminaire Bourbaki, Vol. 1988/89, Astérisque No. 177–178 (1989), Exp. No. 705, 127–154. (Cited on 2.)

[41] Ju. M. Berezans'ki, *Expansions in Eigenfunctions of Selfadjoint Operators*, Transl. Math. Monographs, **17**, American Mathematical Society, RI, 1968. (Cited on 228.)

[42] C. Berg, *Indeterminate moment problems and the theory of entire functions*, Proc. Internat. Conference on Orthogonality, Moment Problems and Continued Fractions (Delft, 1994), J. Comput. Appl. Math. **65** (1995), 27–55. (Cited on 202, 207.)

[43] S. Bernstein, *Sur les polynomes orthogonaux relatifs à un segment fini*, Journ. de Math. elem. (9) **9** (1930), 127–177. (Cited on 74.)

[44] A. S. Besicovitch, *Almost Periodic Functions*, Cambridge University Press, Cambridge, 1932. (Cited on 375.)

[45] H. A. Bethe, *Statistical theory of superlattices*, Proc. Roy. Soc. London Ser. A **150** (1935), 552–575. (Cited on 594.)

[46] O. Blumenthal, *Ueber die Entwicklung einer willkürlichen Funktion nach den Nennern des Kettenbruches für* $\int_{-\infty}^{0} \frac{\varphi(\xi)\,d\xi}{z-\xi}$, Ph.D. dissertation, Göttingen, 1898. (Cited on 19.)

[47] S. Bochner, *Beiträge zur Theorie der Fastperiodischen Funktionen, I*, Math. Ann. **96** (1927), 119–147. (Cited on 375.)

[48] S. Bochner, *Über Sturm–Liouvillesche polynomsysteme*, Math. Z. **29** (1929), 730–736. (Cited on 10.)

[49] S. Bochner, *Abstrakte Fastperiodische Funktionen*, Acta Math. **61** (1933), 149–184. (Cited on 375.)

[50] A. B. Bogatyrëv, *On the efficient computation of Chebyshev polynomials for several intervals*, Sb. Math. **190** (1999), 1571–1605; Russian original in Mat. Sb. **190** (1999), no. 11, 15–50. (Cited on 306, 312.)

[51] H. Bohr, *Zur Theorie der Fastperiodischen Funktionen. II. Zusammenhang der fastperiodischen Funktionen mit Funktionen von unendlich vielen Variabeln; gleichmässige Approximation durch trigonometrische Summen*, Acta Math. **46** (1925), 101–214. (Cited on 375.)

[52] H. Bohr, *Fastperiodischen Funktionen*, Springer-Verlag, Berlin, 1934. (Cited on 375.)

[53] H. Bohr, *Almost Periodic Functions*, Chelsea, New York, 1951. (Cited on 375.)

[54] W. M. Boothby, *An Introduction to Differentiable Manifolds and Riemannian Geometry*, 2nd edition, Pure and Applied Mathematics, **120**, Academic Press, Orlando, FL, 1986. (Cited on 382.)

[55] G. Borg, *Eine Umkehrung der Sturm–Liouvilleschen Eigenwertaufgabe. Bestimmung der Differentialgleichung durch die Eigenwerte*, Acta Math. **78** (1946), 1–96. (Cited on 283.)

[56] S. V. Breimesser and D. B. Pearson, *Asymptotic value distribution for solutions of the Schrödinger equation*, Math. Phys. Anal. Geom. **3** (2000), 385–403. (Cited on 451.)

[57] S. V. Breimesser and D. B. Pearson, *Geometrical aspects of spectral theory and value distribution for Herglotz functions*, Math. Phys. Anal. Geom. **6** (2003), 29–57. (Cited on 451.)

[58] J. Breuer, *Localization for the Anderson model on trees with finite dimensions*, Ann. Henri Poincaré **8** (2007), 1507–1520. (Cited on 594.)

[59] J. Breuer, *Singular continuous spectrum for the Laplacian on certain sparse trees*, Comm. Math. Phys. **269** (2007), 851–857. (Cited on 594, 597.)

[60] J. Breuer, Y. Last, and B. Simon, *The Nevai condition*, to appear in Const. Approx. (Cited on 213.)

[61] J. Breuer, E. Ryckman, and B. Simon, *Equality of the spectral and dynamical definitions of reflection*, Comm. Math. Phys. **295** (2010), 531–550. (Cited on 282.)

[62] J. Breuer, E. Ryckman, and M. Zinchenko, *Right limits and reflectionless measures for CMV matrices*, Comm. Math. Phys. **292** (2009), 1–28. (Cited on 451.)

[63] J. Breuer and B. Simon, *Natural boundaries and spectral theory*, preprint. (Cited on 418, 434.)

[64] R. Brooks, *Constructing isospectral manifolds*, Amer. Math. Monthly **95** (1988), 823–839. (Cited on 2.)

[65] R. Brooks, R. Gornet, and W. H. Gustafson, *Mutually isospectral Riemann surfaces*, Adv. in Math. **138** (1998), 306–322. (Cited on 2.)

[66] R. K. Bullough, N. M. Bogoliubov, G. D. Pang, and J. Timonen, *Quantum repulsive nonlinear Schrödinger models and their "superconductivity,"* Chaos Solitons Fractals **5** (1995), 2639–2656. (Cited on 417.)

[67] W. Burnside, *On a class of automorphic functions*, Proc. London Math. Soc. **23** (1891), 49–88. (Cited on 516.)

[68] W. Burnside, *Further note on automorphic functions*, Proc. London Math. Soc. **23** (1891), 281–295. (Cited on 516.)

[69] M. J. Cantero, L. Moral, and L. Velázquez, *Measures and para-orthogonal polynomials on the unit circle*, East J. Approx. **8** (2002), 447–464. (Cited on 118.)

[70] M. J. Cantero, L. Moral, and L. Velázquez, *Five-diagonal matrices and zeros of orthogonal polynomials on the unit circle*, Linear Algebra Appl. **362** (2003), 29–56. (Cited on 30, 103.)

[71] M. J. Cantero and B. Simon, *Poisson brackets of orthogonal polynomials*, J. Approx. Theory **158** (2009), 3–48. (Cited on 397, 398, 413, 417.)

[72] C. Carathéodory, *Untersuchungen über die konformen Abbildungen von festen und veränderlichen Gebieten*, Math. Ann. **72** (1912), 107–144. (Cited on 524.)

[73] B. Carl and I. Stephani, *Entropy, Compactness and the Approximation of Operators*, Cambridge Tracts in Mathematics, **98**, Cambridge University Press, Cambridge, 1990. (Cited on 51.)

[74] T. Carleman, *Zur theorie der linearen integralgleichungen*, Math. Z. **9** (1921), 196–217. (Cited on 22.)

[75] T. Carleman, *Les fonctions quasi analytiques*, Collection Borel, Gauthier–Villars, Paris, 1926. (Cited on 202.)

[76] K. M. Case, *Orthogonal polynomials from the viewpoint of scattering theory*, J. Math. Phys. **15** (1974), 2166–2174. (Cited on 150, 163, 182.)

[77] K. M. Case, *Orthogonal polynomials, II*, J. Math. Phys. **16** (1975), 1435–1440. (Cited on 163, 182.)

[78] S. N. Chandler-Wilde and M. Lindner, *Sufficiency of Favard's condition for a class of band-dominated operators on the axis*, J. Funct. Anal. **254** (2008), 1146–1159. (Cited on 425.)

[79] S. N. Chandler-Wilde and M. Lindner, *Limit Operators, Collective Compactness, and the Spectral Theory of Infinite Matrices*, submitted to Memoirs of the AMS. (Cited on 425.)

[80] P. L. Chebyshev, *Théorie des mécanismes connus sous le nom de parallélogrammes*, Mémoires présentés à l'Academie Impériale des Sciences de St-Pétersbourg **VII** (1854), 539–568. (Cited on 319.)

[81] P. L. Chebyshev, *Sur les questions de minima qui se rattachent à la représentation approximative des fonctions*, Mémoires de l'Academie Impériale des Sciences de St-Pétersbourg, Sixiéme série. Sciences mathématiques et physiques **VII** (1859), 199–291. (Cited on 319.)

[82] T. S. Chihara, *An Introduction to Orthogonal Polynomials*, Mathematics and Its Applications, **13**, Gordon and Breach, New York-London-Paris, 1978. (Cited on 10.)

[83] T. S. Chihara, *Indeterminate symmetric moment problems*, J. Math. Anal. Appl. **85** (1982), 331–346. (Cited on 202.)

[84] J. S. Christiansen, *The moment problem associated with the q-Laguerre polynomials*, Constr. Approx. **19** (2003), 1–22. (Cited on 207.)

[85] J. S. Christiansen, *The moment problem associated with the Stieltjes–Wigert polynomials*, J. Math. Anal. Appl. **277** (2003), 218–245. (Cited on 206, 207.)

[86] J. S. Christiansen, B. Simon, and M. Zinchenko, *Finite gap Jacobi matrices: An announcement*, J. Comput. Appl. Math. **233** (2009), 652–662. (Cited on ix, 477, 539, 556.)

[87] J. S. Christiansen, B. Simon, and M. Zinchenko, *Finite gap Jacobi matrices, I. The isospectral torus*, to appear in Constr. Approx. (Cited on ix, 477, 539, 556, 564, 576, 582, 583.)

[88] J. S. Christiansen, B. Simon, and M. Zinchenko, *Finite gap Jacobi matrices, II. The Szegő class*, to appear in Constr. Approx. (Cited on ix, 182, 477, 539, 556, 564, 570, 583, 584, 585, 589, 590.)

[89] J. S. Christiansen, B. Simon, and M. Zinchenko, *Finite gap Jacobi matrices, III. Beyond the Szegő class*, in preparation. (Cited on ix, 477, 539, 556.)

[90] S. Clark, F. Gesztesy, H. Holden, and B. M. Levitan, *Borg-type theorems for matrix-valued Schrödinger operators*, J. Differential Equations **167** (2000), 181–210. (Cited on 283.)

[91] C. V. Coffman, *Asymptotic behavior of solutions of ordinary difference equations*, Trans. Amer. Math. Soc. **110** (1964), 22–51. (Cited on 182.)

[92] J. M. Combes and L. Thomas, *Asymptotic behaviour of eigenfunctions for multiparticle Schrödinger operators*, Comm. Math. Phys. **34** (1973), 251–270. (Cited on 149, 425.)

[93] L. Conlon, *Differentiable Manifolds*, 2nd edition, Birkhäuser Advanced Texts: Basel Textbooks, Birkhäuser, Boston, 2001. (Cited on 382.)

[94] C. Corduneanu, *Almost Periodic Functions*, Interscience, New York, 1968. (Cited on 375.)

[95] W. Craig, *The trace formula for Schrödinger operators on the line*, Comm. Math. Phys. **126** (1989), 379–407. (Cited on 279, 283.)

[96] D. Damanik and R. Killip, *Half-line Schrödinger operators with no bound states*, Acta Math. **193** (2004), 31–72. (Cited on 36.)

[97] D. Damanik, R. Killip, and B. Simon, *Perturbations of orthogonal polynomials with periodic recursion coefficients*, Annals of Math. **171** (2010), 1931–2010. (Cited on ix, 40, 41, 150, 157, 229, 244, 246, 249, 455, 457, 460, 463, 465, 473, 474, 476.)

[98] D. Damanik, A. Pushnitski, and B. Simon, *The analytic theory of matrix orthogonal polynomials*, Surveys in Approximation Theory **4** (2008), 1–85. (Cited on 150, 228, 234, 239.)

[99] D. Damanik and B. Simon, *Jost functions and Jost solutions for Jacobi matrices, I. A necessary and sufficient condition for Szegő asymptotics*, Invent. Math. **165** (2006), 1–50. (Cited on 98, 147, 173, 174, 175, 182.)

[100] D. Damanik and B. Simon, *Jost functions and Jost solutions for Jacobi matrices, II. Decay and analyticity*, Int. Math. Res. Not. **2006**, Article ID 19396, 32 pages, 2006. (Cited on 42.)

[101] E. B. Davies and B. Simon, *Scattering theory for systems with different spatial asymptotics on the left and right*, Comm. Math. Phys. **63** (1978), 277–301. (Cited on 282.)

[102] P. Deift, T. Kriecherbauer, K. T-R McLaughlin, S. Venakides, and X. Zhou, *Strong asymptotics of orthogonal polynomials with respect to exponential weights*, Comm. Pure Appl. Math. **52** (1999), 1491–1552. (Cited on 132.)

[103] P. Deift, T. Kriecherbauer, K. T-R McLaughlin, S. Venakides, and X. Zhou, *Uniform asymptotics for polynomials orthogonal with respect to varying exponential weights and applications to universality questions in random matrix theory*, Comm. Pure Appl. Math. **52** (1999), 1335–1425. (Cited on 589.)

[104] P. Deift, L. C. Li, and C. Tomei, *Toda flows with infinitely many variables*, J. Funct. Anal. **64** (1985), 358–402. (Cited on 402, 403, 405, 407.)

[105] P. Deift, T. Nanda, and C. Tomei, *Ordinary differential equations and the symmetric eigenvalue problem*, SIAM J. Numer. Anal. **20** (1983), 1–22. (Cited on 402, 403.)

[106] P. Delsarte, Y. V. Genin, and Y. G. Kamp, *Orthogonal polynomial matrices on the unit circle*, IEEE Trans. Circuits and Systems **CAS-25** (1978), 149–160. (Cited on 113, 229.)

[107] S. A. Denisov, *On Rakhmanov's theorem for Jacobi matrices*, Proc. Amer. Math. Soc. **132** (2004), 847–852. (Cited on 19, 454.)

[108] S. A. Denisov, *On the preservation of absolutely continuous spectrum for Schrödinger operators*, J. Funct. Anal. **231** (2006), 143–156. (Cited on 591, 592, 594, 597, 600, 601, 603.)

[109] S. A. Denisov, *On a conjecture by Y. Last*, J. Approx. Theory **158** (2009), 194–213. (Cited on 91.)

[110] S. A. Denisov and A. Kiselev, *Spectral properties of Schrödinger operators with decaying potentials*, in "Spectral Theory and Mathematical Physics: A Festschrift in Honor of Barry Simon's 60th birthday," pp. 565–589, Proc. Sympos. Pure Math., **76.2**, American Mathematical Society, Providence, RI, 2007. (Cited on 591, 593, 594, 606.)

[111] S. A. Denisov and S. Kupin, *Asymptotics of the orthogonal polynomials for the Szegő class with a polynomial weight*, J. Approx. Theory **139** (2006), 8–28. (Cited on 173.)

[112] H. Dette and W. J. Studden, *The Theory of Canonical Moments With Applications in Statistics, Probability, and Analysis*, John Wiley, New York, 1997. (Cited on 29.)

[113] W. F. Donoghue, *On the perturbation of spectra*, Comm. Pure Appl. Math. **18** (1965), 559–579. (Cited on 431.)

[114] J. L. Doob, *Measure Theory*, Graduate Texts in Mathematics, **143**, Springer-Verlag, New York, 1994. (Cited on 130.)

[115] B. A. Dubrovin, V. B. Matveev, and S. P. Novikov, *Nonlinear equations of Korteweg–de Vries type, finite-band linear operators and Abelian varieties*, Uspekhi Mat. Nauk **31** (1976), no. 1(187), 55–136 [Russian]. (Cited on 359.)

[116] I. Dumitriu and A. Edelman, *Matrix models for beta ensembles*, J. Math. Phys. **43** (2002), 5830–5847. (Cited on 397.)

[117] A. J. Durán and P. López-Rodríguez, *Orthogonal matrix polynomials: Zeros and Blumenthal's theorem*, J. Approx. Theory **84** (1996), 96–118. (Cited on 239.)

[118] P. L. Duren, *Theory of H^p Spaces*, Pure and Applied Mathematics, **38**, Academic Press, New York-London, 1970. (Cited on 58, 65.)

[119] H. Dym and V. Katsnelson, *Contributions of Issai Schur to analysis*, in "Studies in memory of Issai Schur" (Chevaleret/Rehovot, 2000), pp. xci–clxxxviii, Progr. Math., **210**, Birkhäuser, Boston, 2003. (Cited on 65.)

[120] M. S. P. Eastham, *The Spectral Theory of Periodic Differential Equations*, Scottish Academic Press, Edinburgh, 1973. (Cited on 282.)

[121] L. H. Eliasson, *Floquet solutions for the 1-dimensional quasi-periodic Schrödinger equation*, Comm. Math. Phys. **146** (1992), 447–482. (Cited on 371.)

[122] R. S. Ellis, *Entropy, Large Deviations, and Statistical Mechanics*, (reprint of the 1985 original), Classics in Mathematics, Springer-Verlag, Berlin, 2006. (Cited on 51.)

[123] P. Erdős and P. Turán, *On interpolation. III. Interpolatory theory of polynomials*, Annals of Math. (2) **41** (1940), 510–553. (Cited on 305, 327.)

[124] W. N. Everitt, *A personal history of the m-coefficient*, J. Comput. Appl. Math. **171** (2004), 185–197. (Cited on 73.)

[125] G. Faber, *Über Tschebyscheffsche Polynome*, J. Reine Angew. Math. **150** (1919), 79–106. (Cited on 305, 319.)

[126] H. M. Farkas and I. Kra, *Riemann Surfaces*, Graduate Texts in Mathematics, **71**, Springer, New York-Berlin, 1980. (Cited on 348, 359.)

[127] J. Favard, *Sur les équations différentielles linéaires à coefficients presque-périodiques*, Acta Math. **51** (1927), 31–81. (Cited on 425.)

[128] J. Favard, *Sur les polynômes de Tchebycheff*, C. R. Acad. Sci. Paris **200** (1935), 2052–2055. (Cited on 18.)

[129] L. Faybusovich and M. Gekhtman, *On Schur flows*, J. Phys. A **32** (1999), 4671–4680. (Cited on 36, 417.)

[130] L. Faybusovich and M. Gekhtman, *Elementary Toda orbits and integrable lattices*, J. Math. Phys. **41** (2000), 2905–2921. (Cited on 408.)

[131] L. Faybusovich and M. Gekhtman, *Poisson brackets on rational functions and multi-Hamiltonian structure for integrable lattices*, Phys. Lett. A **272** (2000), 236–244. (Cited on 397.)

[132] M. Fekete, *Über die Verteilung der Wurzeln bei gewissen algebraischen Gleichungen mit ganzzahligen Koeffizienten*, Math. Z. **17** (1923), 228–249. (Cited on 319.)

[133] E. Findley, *Universality for local Szegő measures*, J. Approx. Theory **155** (2008), 136–154. (Cited on 134, 142, 221.)

[134] H. Flaschka, *The Toda lattice, II. Existence of integrals*, Phys. Rev. B (3) **9** (1974), 1924–1925. (Cited on 380, 397, 404, 410, 413, 416.)

[135] H. Flaschka, *Discrete and periodic illustrations of some aspects of the inverse method*, in "Dynamical Systems, Theory and Applications," pp. 441–466, Lecture Notes In Physics, **38**, Springer, Berlin, 1975. (Cited on 283, 409, 413.)

[136] H. Flaschka and D. W. McLaughlin, *Canonically conjugate variables for the Korteweg–de Vries equation and the Toda lattice with periodic boundary conditions*, Progr. Theoret. Phys. **55** (1976), 438–456. (Cited on 359.)

[137] I. Fonseca and W. Gangbo, *Degree Theory in Analysis and Applications*, Oxford Lecture Series in Mathematics and Its Applications, **2**, The Clarendon Press, Oxford University Press, New York, 1995. (Cited on 360.)

[138] L. R. Ford, *Automorphic Functions*, 2nd edition, Chelsea, New York, 1951. (Cited on 496, 518.)

[139] P. J. Forrester and E. M. Rains, *Jacobians and rank 1 perturbations relating to unitary Hessenberg matrices*, Int. Math. Res. Not. **2006**, Art. ID 48306, 36 pp. (Cited on 397.)

[140] J. G. F. Francis, *The QR transformation: a unitary analogue to the LR transformation. I*, Comput. J. **4** (1961/1962), 265–271. (Cited on 390.)

[141] G. Freud, *Orthogonal Polynomials*, Pergamon Press, Oxford-New York, 1971. (Cited on 10, 97, 132, 227.)

[142] R. Froese, D. Hasler, and W. Spitzer, *Transfer matrices, hyperbolic geometry and absolutely continuous spectrum for some discrete Schrödinger operators on graphs*, J. Funct. Anal. **230** (2006), 184–221. (Cited on 594, 599.)

[143] W. Fulton, *Algebraic Topology. A First Course*, Graduate Texts in Mathematics, **153**, Springer-Verlag, New York, 1995. (Cited on 525.)

[144] J.-P. Gabardo, *A maximum entropy approach to the classical moment problem*, J. Funct. Anal. **106** (1992) 80–94. (Cited on 207.)

[145] C. S. Gardner, J. M. Greene, M. D. Kruskal, and R. M. Miura, *Method for solving the Korteweg–deVries equation*, Phys. Rev. Lett. **19** (1967), 1095–1097. (Cited on 387, 404.)

[146] J. B. Garnett, *Bounded Analytic Functions*, Pure and Applied Mathematics, **96**, Academic Press, New York-London, 1981. (Cited on 106.)

[147] M. Gekhtman and I. Nenciu, *Multi-Hamiltonian structure for the finite defocusing Ablowitz–Ladik equation*, Comm. Pure Appl. Math. **62** (2009), 147–182. (Cited on 397, 398, 417.)

[148] M. I. Gekhtman and M. Z. Shapiro, *Noncommutative and commutative integrability of generic Toda flows in simple Lie algebras*, Comm. Pure Appl. Math. **52** (1999), 53–84. (Cited on 408.)

[149] I. M. Gel'fand, *Expansion in series of eigenfunctions of an equation with periodic coefficients*, Dokl. Akad. Nauk SSSR **73** (1950), 1117–1120. (Cited on 263.)

[150] I. M. Gel'fand, D. Raikov, and G. Shilov, *Commutative Normed Rings*, Chelsea, New York, 1964; Russian orignal, 1960. (Cited on 377.)

[151] V. Georgescu and S. Golénia, *Isometries, Fock spaces, and spectral analysis of Schrödinger operators on trees*, J. Funct. Anal. **227** (2005), 389–429. (Cited on 594.)

[152] V. Georgescu and A. Iftimovici, *Crossed products of C^*-algebras and spectral analysis of quantum Hamiltonians*, Comm. Math. Phys. **228** (2002), 519–560. (Cited on 425.)

[153] J. S. Geronimo, *Polynomials orthogonal on the unit circle with random recurrence coefficients*, in "Methods of Approximation Theory in Complex Analysis and Mathematical Physics" (Leningrad, 1991), pp. 43–61, Lecture Notes in Mathematics, **1550**, Springer, Berlin, 1993. (Cited on 73, 74.)

[154] J. S. Geronimo and K. M. Case, *Scattering theory and polynomials orthogonal on the real line*, Trans. Amer. Math. Soc. **258** (1980), 467–494. (Cited on 182.)

[155] J. S. Geronimo and W. Van Assche, *Orthogonal polynomials on several intervals via a polynomial mapping*, Trans. Amer. Math. Soc. **308** (1988), 559–581. (Cited on 323.)

[156] Ya. L. Geronimus, *Generalized orthogonal polynomials and the Christoffel–Darboux formula*, C. R. (Doklady) Acad. Sci. URSS (N.S.) **26** (1940), 847–849. (Cited on 29.)

[157] Ya. L. Geronimus, *Sur quelques propriétés des polynômes orthogonaux généralisés*, C. R. (Doklady) Acad. Sci. URSS (N.S.) **29** (1940), 5–8. (Cited on 29.)

[158] Ya. L. Geronimus, *On polynomials orthogonal on the circle, on trigonometric moment problem, and on allied Carathéodory and Schur functions*, Mat. Sb. **15** (1944), 99–130 [Russian]. (Cited on 73, 74, 79, 80.)

[159] Ya. L. Geronimus, *On the trigonometric moment problem*, Annals of Math. (2) **47** (1946), 742–761. (Cited on 36.)

[160] Ya. L. Geronimus, *Polynomials Orthogonal on a Circle and Their Applications*, Amer. Math. Soc. Translation **1954** (1954), no. 104, 79 pp. (Cited on 36, 73, 74.)

[161] Ya. L. Geronimus, *Orthogonal Polynomials: Estimates, Asymptotic Formulas, and Series of Polynomials Orthogonal on the Unit Circle and on an Interval*, Consultants Bureau, New York, 1961. (Cited on 73, 74.)

[162] F. Gesztesy, N. J. Kalton, K. A. Makarov, and E. Tsekanovskii, *Some applications of operator-valued Herglotz functions*, Operator Theory, System Theory and Related Topics (Beer-Sheva/Rehovot, 1997), pp. 271–321, Oper. Theory Adv. Appl. **123**, Birkhäuser, Basel, 2001. (Cited on 239.)

[163] F. Gesztesy, M. Krishna, and G. Teschl, *On isospectral sets of Jacobi operators*, Comm. Math. Phys. **181** (1996), 631–645. (Cited on 278, 454.)

[164] F. Gesztesy and B. Simon, *Rank one perturbations at infinite coupling*, J. Funct. Anal. **128** (1995), 245–252. (Cited on 275, 431.)

[165] F. Gesztesy and B. Simon, *The xi function*, Acta Math. **176** (1996), 49–71. (Cited on 3.)

[166] F. Gesztesy, B. Simon, and G. Teschl, *Spectral deformations of one-dimensional Schrödinger operators*, J. Anal. Math. **70** (1996), 267–324. (Cited on 3.)

[167] F. Gesztesy and E. Tsekanovskii, *On matrix-valued Herglotz functions*, Math. Nachr. **218** (2000), 61–138. (Cited on 239.)

[168] F. Gesztesy and P. Yuditskii, *Spectral properties of a class of reflectionless Schrödinger operators*, J. Funct. Anal. **241** (2006), 486–527. (Cited on 454.)

[169] F. Gesztesy and M. Zinchenko, *Local spectral properties of reflectionless Jacobi, CMV, and Schrödinger operators*, J. Differential Equations **246** (2009), 78–107. (Cited on 452.)

[170] I. C. Gohberg and M. G. Krein, *Introduction to the Theory of Linear Nonselfadjoint Operators*, Transl. Math. Monographs, **18**, American Mathematical Society, Providence, RI, 1969. (Cited on 22, 37, 103.)

[171] I. Gohberg and L. A. Sakhnovich (editors), *Matrix and operator valued functions*, The Vladimir Petrovich Potapov memorial volume, Operator Theory: Advances and Applications, **72**, Birkhäuser Verlag, Basel, 1994. (Cited on 229.)

[172] M. Goldstein and W. Schlag, *On resonances and the formation of gaps in the spectrum of quasi-periodic Schrödinger equations*, to appear in Annals of Math. (Cited on 371.)

[173] L. Golinskii, *Schur functions, Schur parameters and orthogonal polynomials on the unit circle*, Z. Anal. Anwendungen **12** (1993), 457–469. (Cited on 74.)

[174] L. Golinskii, *Quadrature formula and zeros of para-orthogonal polynomials on the unit circle*, Acta Math. Hungar. **96** (2002), 169–186. (Cited on 118.)

[175] L. Golinskii, *Absolutely continuous measures on the unit circle with sparse Verblunsky coefficients*, Mat. Fiz. Anal. Geom. **11** (2004), 408–420. (Cited on 48.)

[176] L. Golinskii, *Schur flows and orthogonal polynomials on the unit circle*, Mat. Sbornik **197** (2006), 41–62. (Cited on 403, 417.)

[177] L. Golinskii and P. Nevai, *Szegő difference equations, transfer matrices and orthogonal polynomials on the unit circle*, Comm. Math. Phys. **223** (2001), 223–259. (Cited on 48, 73, 74.)

[178] L. Golinskii and A. Zlatoš, *Coefficients of orthogonal polynomials on the unit circle and higher order Szegő theorems*, Constr. Approx. **26** (2007), 361–382. (Cited on 91.)

[179] G. M. Goluzin, *Geometric Theory of Functions of a Complex Variable*, Transl. Math. Monographs, **26**, American Mathematical Society, Providence, RI, 1969. (Cited on 319, 524.)

[180] C. Gordon and Z. Szabó, *Isospectral deformations of negatively curved Riemannian manifolds with boundary which are not locally isometric*, Duke Math. J. **113** (2002), 355–383. (Cited on 2.)

[181] C. Gordon, D. Webb, and S. Wolpert, *Isospectral plane domains and surfaces via Riemannian orbifolds*, Invent. Math. **110** (1992), 1–22. (Cited on 2.)

[182] R. M. Gray, *Entropy and Information Theory*, Springer-Verlag, New York, 1990. (Cited on 51.)

[183] R. E. Greene and S. G. Krantz, *Function Theory of One Complex Variable*, 3rd edition, Graduate Studies in Mathematics, **40**, American Mathematical Society, Providence, RI, 2006. (Cited on 10.)

[184] U. Grenander and G. Szegő, *Toeplitz Forms and Their Applications*, 2nd edition, Chelsea, New York, 1984; 1st edition, University of California Press, Berkeley-Los Angeles, 1958. (Cited on 106.)

[185] P. Griffiths and J. Harris, *Principles of Algebraic Geometry*, John Wiley & Sons, New York, 1978. (Cited on 359.)

[186] A. Grothendieck, *Produits tensoriels topologiques et espaces nucléaires*, Mem. Amer. Math. Soc. **1955** (1955), 140 pp. (Cited on 22.)

[187] A. Grothendieck, *La théorie de Fredholm*, Bull. Soc. Math. France **84** (1956), 319–384. (Cited on 22.)

[188] F. A. Grünbaum and L. Haine, *A theorem of Bochner, revisited*, in "Algebraic Aspects of Integrable Systems," pp. 143–172, Progr. Nonlinear Differential Equations Appl., **26**, Birkhäuser, Boston, 1997. (Cited on 10.)

[189] V. Guillemin and A. Pollack, *Differential Topology*, Prentice–Hall, Englewood Cliffs, NJ, 1974. (Cited on 360.)

[190] H. Hamburger, *Über eine Erweiterung des Stieltjesschen Momentproblems*, Math. Ann. **81** (1920), 235–319; **82** (1921), 120–164, 168–187. (Cited on 206.)

[191] G. Hamel, *Über die lineare Differentialgleichung zweiter ordnung mit periodischen Koeffizienten*, Math. Ann. **73** (1913), 371–412. (Cited on 282.)

[192] P. Hartman and A. Wintner, *Asymptotic integrations of linear differential equations*, Amer. J. Math. **77** (1955), 45–86; errata, 404. (Cited on 182.)

[193] O. Haupt, *Über lineare homogene Differentialgleichungen 2. Ordnung mit periodischen Koeffizienten*, Math. Ann. **79** (1919), 278–285. (Cited on 282.)

[194] D. A. Hejhal, *Universal covering maps for variable regions*, Math. Z. **137** (1974), 7–20. (Cited on 561.)

[195] S. Helgason, *Analysis on Lie groups and homogeneous spaces*, Conference Board of the Mathematical Sciences Regional Conference Series in Mathematics, **14**, American Mathematical Society, Providence, RI, 1972. (Cited on 390.)

[196] L. L. Helms, *Introduction to Potential Theory*, Pure and Applied Mathematics, **22**, Wiley–Interscience, New York, 1969. (Cited on 305, 323.)

[197] H. Helson and D. Lowdenslager, *Prediction theory and Fourier series in several variables*, Acta Math. **99** (1958), 165–202. (Cited on 103, 106, 108.)

[198] A. Henrici and T. Kappeler, *Global action-angle variables for the periodic Toda lattice*, Int. Math. Res. Not. IMRN **2008**, Art. ID rnn031, 52 pp. (Cited on 387.)

[199] D. Herbert and R. Jones, *Localized states in disordered systems*, J. Phys. C: Solid State Phys. **4** (1971), 1145–1161. (Cited on 305.)

[200] H. Hochstadt, *On the theory of Hill's matrices and related inverse spectral problems*, Linear Algebra and Appl. **11** (1975), 41–52. (Cited on 282, 283.)

[201] D. Hundertmark and B. Simon, *Lieb–Thirring inequalities for Jacobi matrices*, J. Approx. Theory **118** (2002), 106–130. (Cited on 172, 173.)

[202] N. E. Hurt, *Geometric Quantization in Action*, Mathematics and Its Applications (East European Series), **8**, Reidel Publishing, Dordrecht-Boston, 1983. (Cited on 387.)

[203] I. A. Ibragimov, *A theorem of Gabor Szegő*, Mat. Zametki **3** (1968), 693–702 [Russian]. (Cited on 24, 42.)

[204] M. E. H. Ismail, *Classical and Quantum Orthogonal Polynomials in One Variable*, Encyclopedia of Mathematics and its Application, **98**, Cambridge University Press, Cambridge, 2009. (Cited on 10.)

[205] R. B. Israel, *Convexity in the Theory of Lattice Gases*, Princeton Series in Physics, Princeton University Press, Princeton, NJ, 1979. (Cited on 51.)

[206] C. G. J. Jacobi, *Über die Reduction der quadratischen Formen auf die kleinste Anzahl Glieder*, J. Reine Angew. Math. **39** (1848), 290–292. (Cited on 150.)

[207] K. Jacobs, *Measure and Integral*, Probability and Mathematical Statistics, Academic Press, New York-London, 1978. (Cited on 130.)

[208] V. Jakšić and Y. Last, *A new proof of Poltoratskii's theorem*, J. Funct. Anal. **215** (2004), 103–110. (Cited on 452.)

[209] S. Jitomirskaya, *Ergodic Schrödinger operators (on one foot)*, in "Spectral Theory and Mathematical Physics: A Festschrift in Honor of Barry Simon's 60th Birthday," pp. 613–647, Proc. Symp. Pure Math., **76.2**, American Mathematical Society, Providence, RI, 2007. (Cited on 227.)

[210] W. B. Jones, O. Njåstad, and W. J. Thron, *Moment theory, orthogonal polynomials, quadrature, and continued fractions associated with the unit circle*, Bull. London Math. Soc. **21** (1989), 113–152. (Cited on 118.)

[211] R. Jost, *Über die falschen Nullstellen der Eigenwerte der S-Matrix*, Helvetica Phys. Acta **20** (1947), 256–266. (Cited on 182.)

[212] M. Kac, *Can one hear the shape of a drum?*, Amer. Math. Monthly **73** (1966), no. 4, part II, 1–23. (Cited on 2.)

[213] S. Karlin and W. J. Studden, *Tchebycheff Systems: With Applications in Analysis and Statistics*, Pure and Applied Mathematics, **15**, John Wiley, New York-London-Sydney, 1966. (Cited on 29.)

[214] T. Kato, *On finite-dimensional perturbations of self-adjoint operators*, J. Math. Soc. Japan **9** (1957), 239–249. (Cited on 431.)

[215] T. Kato, *Perturbation Theory for Linear Operators*, 2nd edition, Grundlehren der Mathematischen Wissenschaften, **132**, Springer, Berlin-New York, 1976. (Cited on 239, 243, 259, 409, 431.)

[216] S. Katok, *Fuchsian Groups*, University of Chicago Press, Chicago, 1992. (Cited on 505, 518.)

[217] Y. Katznelson, *An Introduction to Harmonic Analysis*, 2nd corrected edition, Dover Publications, New York, 1976. (Cited on 142.)

[218] A. Ya. Khinchin, *Continued Fractions*, reprint of the 1964 translation, Dover Publications, Mineola, NY, 1997. (Cited on 80.)

[219] S. Khrushchev, *Schur's algorithm, orthogonal polynomials, and convergence of Wall's continued fractions in $L^2(\mathbb{T})$*, J. Approx. Theory **108** (2001), 161–248. (Cited on 80, 118, 457.)

[220] S. Khrushchev, *A singular Riesz product in the Nevai class and inner functions with the Schur parameters in $\bigcap_{p>2}\ell^p$*, J. Approx. Theory **108** (2001), 249–255. (Cited on 47.)

[221] S. Khrushchev, *Classification theorems for general orthogonal polynomials on the unit circle*, J. Approx. Theory **116** (2002), 268–342. (Cited on 96, 97, 457.)

[222] R. Killip, *Spectral theory via sum rules*, in "Spectral Theory and Mathematical Physics: A Festschrift in Honor of Barry Simon's 60th birthday," pp. 907–930, Proc. Sympos. Pure Math., **76.2**, American Mathematical Society, Providence, RI, 2007. (Cited on 22.)

[223] R. Killip and I. Nenciu, *Matrix models for circular ensembles*, Int. Math. Res. Not. **50** (2004), 2665–2701. (Cited on 36, 398, 416.)

[224] R. Killip and I. Nenciu, *CMV: The unitary analogue of Jacobi matrices*, Comm. Pure Appl. Math. **60** (2007), 1148–1188. (Cited on 398, 408, 417.)

[225] R. Killip and B. Simon, *Sum rules for Jacobi matrices and their applications to spectral theory*, Annals of Math. (2) **158** (2003), 253–321. (Cited on ix, 39, 47, 52, 84, 86, 91, 144, 163, 173, 182.)

[226] A. A. Kirillov, *Lectures on the Orbit Method*, Graduate Studies in Mathematics, **64**, American Mathematical Society, Providence, RI, 2004. (Cited on 387.)

[227] A. Kiselev, Y. Last, and B. Simon, *Modified Prüfer and EFGP transforms and the spectral analysis of one-dimensional Schrödinger operators*, Comm. Math. Phys. **194** (1998), 1–45. (Cited on 48, 167.)

[228] A. Klein, *Spreading of wave packets in the Anderson model on the Bethe lattice*, Comm. Math. Phys. **177** (1996), 755–773. (Cited on 594.)

[229] A. Klein, *Extended states in the Anderson model on the Bethe lattice*, Adv. in Math. **133** (1998), 163–184. (Cited on 594, 599.)

[230] F. Klein, *Neue Beiträge zur Riemann'schen Functionentheorie*, Math. Ann. **21** (1883), 141–218. (Cited on 524.)

[231] K. Knopp, *Mengentheoretische Behandlung einiger Probleme der diophantischen Approximation und der transfiniten Wahrscheinlichkeiten*, Math. Ann. **95** (1925), 409–426. (Cited on 79.)

[232] H. Koch, *Number Theory. Algebraic Numbers and Functions*, Graduate Studies in Mathematics, **24**, American Mathematical Society, Providence, RI, 2000. (Cited on 256.)

[233] P. Koebe, *Über die Uniformisierung beliebiger analytischer Kurven*, Nachr. K. Ges. Wissenschaft. Göttinger Math. Phys. Kl. (1907), 191–210. (Cited on 524.)

[234] P. Koebe, *Über die Uniformisierung beliebiger analytischer Kurven. Zweite Mitteilung*, Nachr. K. Ges. Wissenschaft. Göttinger Math. Phys. Kl. (1907), 633–669. (Cited on 524.)

[235] P. Koebe, *Über die Uniformisierung beliebiger analytischer Kurven. Dritte Mitteilung*, Nachr. K. Ges. Wissenschaft. Göttinger Math. Phys. Kl. (1908), 337–358. (Cited on 524.)

[236] P. Koebe, *Über die Uniformisierung beliebiger analytischer Kurven. Vierte Mitteilung*, Nachr. K. Ges. Wissenschaft. Göttinger Math. Phys. Kl. (1909), 324–361. (Cited on 524.)

[237] P. Koebe, *Abhandlungen zur Theorie der konformen Abbildung, I. Die Kreisabbildung des allgemeinsten einfach und zweifach zusammenhängenden schlichten Bereichs und die Ränderzuordnung bei konformer Abbildung*, J. Math. **145** (1915), 177–225. (Cited on 524.)

[238] P. Koebe, *Abhandlungen zur Theorie der konformen Abbildung, II. Die Fundamentalabbildung beliebiger mehrfach zusammenhängender schlichter Bereiehe nebst einer Anwendung auf die Bestimmung algebraischer Funktionen zu gegebener Riemannscher Fläche*, Acta Math. **40** (1916), 251–290. (Cited on 524.)

[239] A. N. Kolmogorov, *Stationary sequences in Hilbert space*, Bull. Univ. Moscow **2** (1941), 40 pp. [Russian]. (Cited on 103, 206.)

[240] P. Koosis, *Introduction to H^p Spaces*, London Mathematical Society Lecture Note Series, **40**, Cambridge University Press, Cambridge, 1980. (Cited on 65.)

[241] B. Kostant, *The solution to a generalized Toda lattice and representation theory*, Adv. in Math. **34** (1979), 195–338. (Cited on 407.)

[242] B. Kostant, *Flag manifold quantum cohomology, the Toda lattice, and the representation with highest weight ρ*, Selecta Math. (N.S.) **2** (1996), 43–91. (Cited on 408.)

[243] S. Kotani, *Ljapunov indices determine absolutely continuous spectra of stationary random one-dimensional Schrödinger operators*, in "Stochastic Analysis" (Katata/Kyoto, 1982), pp. 225–247, North–Holland Mathematical Library, **32**, North–Holland, Amsterdam, 1984. (Cited on 433, 451.)

[244] S. Kotani, *Jacobi matrices with random potentials taking finitely many values*, Rev. Math. Phys. **1** (1989), 129–133. (Cited on 433, 451.)

[245] R. Kozhan, *Szegő asymptotics for matrix-valued measures with countably many bound states*, to appear in J. Approx. Theory. (Cited on 229.)

[246] H. A. Kramers, *Das Eigenwertproblem im eindimensionalen periodischen Kraftfelde*, Physica **2** (1935), 483–490. (Cited on 282.)

[247] W. Krawcewicz and J. Wu, *Theory of Degrees With Applications to Bifurcations and Differential Equations*, Canadian Mathematical Society Series of Monographs and Advanced Texts, John Wiley & Sons, New York, 1997. (Cited on 360.)

[248] Y. Kreimer, Y. Last, and B. Simon, *Monotone Jacobi parameters and non-Szegő weights*, J. Approx. Theory **157** (2009), 144–171. (Cited on 87.)

[249] M. G. Krein, *On a generalization of some investigations of G. Szegő, V. Smirnoff and A. Kolmogoroff*, C. R. (Doklady) Acad. Sci. URSS (N.S.) **46** (1945), 91–94. (Cited on 103, 206.)

[250] M. G. Krein, *On a problem of extrapolation of A. N. Kolmogorov*, Dokl. Akad. Nauk SSSR **46** (1945), 306–309. (Cited on 203, 206.)

[251] M. G. Krein, *Infinite J-matrices and a matrix moment problem*, Dokl. Akad. Nauk SSSR **69** (1949), 125–128 [Russian]. (Cited on 228.)

[252] M. G. Krein, *The ideas of P. L. Čebyšev and A. A. Markov in the theory of limiting values of integrals and their further development*, Amer. Math. Soc. Transl. (2) **12** (1959), 1–121; Russian original in Uspekhi Matem. Nauk (N.S.) **6** (1951), 3–120. (Cited on 29.)

[253] M. G. Krein and A. A. Nudel'man, *The Markov moment problem and extremal problems. Ideas and problems of P. L. Čebyšev and A. A. Markov and their further development*, Transl. Math. Monographs, **50**, American Mathematical Society, Providence, RI, 1977; Russian original in Izdat. Nauka, Moscow, 1973, 551 pp. (Cited on 29, 227.)

[254] I. M. Krichever, *Algebraic curves and nonlinear difference equations*, Uspekhi Mat. Nauk **33** (1978), no. 4(202), 215–216 [Russian]. (Cited on 359.)

[255] I. M. Krichever, *Appendix to "Theta-functions and nonlinear equations" by B. A. Dubrovin*, Russian Math. Surveys **36** (1981), 11–92 (1982); Russian original in Uspekhi Mat. Nauk **36** (1981), no. 2(218), 11–80. (Cited on 359.)

[256] A. B. Kuijlaars and M. Vanlessen, *Universality for eigenvalue correlations from the modified Jacobi unitary ensemble*, Int. Math. Res. Not. **30** (2002), 1575–1600. (Cited on 132.)

[257] P. P. Kulish, *Quantum difference nonlinear Schrödinger equation*, Lett. Math. Phys. **5** (1981), 191–197. (Cited on 417.)

[258] S. Kupin, *On sum rules of special form for Jacobi matrices*, C. R. Math. Acad. Sci. Paris **336** (2003), 611–614. (Cited on 173.)

[259] S. Kupin, *On a spectral property of Jacobi matrices*, Proc. Amer. Math. Soc. **132** (2004), 1377–1383. (Cited on 173.)

[260] S. Kupin, *Spectral properties of Jacobi matrices and sum rules of special form*, J. Funct. Anal. **227** (2005), 1–29. (Cited on 173.)

[261] S. Kupin, *Absolutely continuous spectrum of a Schrödinger operator on a tree*, J. Math. Phys. **49** (2008), 113506-1–113506-10. (Cited on 594.)

[262] V. G. Kurbatov, *On the invertibility of almost periodic operators*, Math. USSR Sb. **67** (1990), 367–377. (Cited on 425.)

[263] H. J. Landau, *Maximum entropy and the moment problem*, Bull. Amer. Math. Soc. **16** (1987), 47–77. (Cited on 29, 305.)

[264] N. S. Landkof, *Foundations of Modern Potential Theory*, Springer-Verlag, Berlin-New York, 1972. (Cited on 295, 305, 323.)

[265] S. Lang, *Introduction to Diophantine Approximations*, Addison–Wesley, Reading, MA, 1966. (Cited on 256.)

[266] S. Lang, *Complex Analysis*, 4th edition, Graduate Texts in Mathematics, **103**, Springer-Verlag, New York, 1999. (Cited on 525.)

[267] S. Lang, *Introduction to Differentiable Manifolds*, 2nd edition, Universitext, Springer-Verlag, New York, 2002. (Cited on 382.)

[268] A. Laptev, S. Naboko, and O. Safronov, *On new relations between spectral properties of Jacobi matrices and their coefficients*, Comm. Math. Phys. **241** (2003), 91–110. (Cited on 173.)

[269] Y. Last, *On the measure of gaps and spectra for discrete 1D Schrödinger operators*, Comm. Math. Phys. **149** (1992), 347–360. (Cited on 282.)

[270] Y. Last, *Destruction of absolutely continuous spectrum by perturbation potentials of bounded variation*, Comm. Math. Phys. **274** (2007), 243–252. (Cited on 91.)

[271] Y. Last and B. Simon, *Eigenfunctions, transfer matrices, and absolutely continuous spectrum of one-dimensional Schrödinger operators*, Invent. Math. **135** (1999), 329–367. (Cited on 418, 425, 426, 429, 431.)

[272] Y. Last and B. Simon, *The essential spectrum of Schrödinger, Jacobi, and CMV operators*, J. Anal. Math. **98** (2006), 183–220. (Cited on 418, 425, 457.)

[273] Y. Last and B. Simon, *Fine structure of the zeros of orthogonal polynomials, IV. A priori bounds and clock behavior*, Comm. Pure Appl. Math. **61** (2008), 486–538. (Cited on 132.)

[274] P. D. Lax, *Integrals of nonlinear equations of evolution and solitary waves*, Comm. Pure Appl. Math. **21** (1968) 467–490. (Cited on 387, 403, 404.)

[275] E. Levin and D. S. Lubinsky, *Universality limits involving orthogonal polynomials on the unit circle*, Comput. Methods Funct. Theory **7** (2007), 543–561. (Cited on 123, 132.)

[276] E. Levin and D. S. Lubinsky, *Applications of universality limits to zeros and reproducing kernels of orthogonal polynomials*, J. Approx. Theory **150** (2008), 69–95. (Cited on 132.)

[277] N. Levinson, *The Wiener RMS (root-mean square) error criterion in filter design and prediction*, J. Math. Phys. Mass. Inst. Tech. **25** (1947), 261–278. (Cited on 29.)

[278] B. M. Levitan, *Inverse Sturm–Liouville Problems*, VNU Science Press, Utrecht, 1987. (Cited on 359.)

[279] B. M. Levitan and V. V. Zhikov, *Almost Periodic Functions and Differential Equations*, Cambridge University Press, Cambridge, 1982. (Cited on 375.)

[280] L.-C. Li, *Some remarks on CMV matrices and dressing orbits*, Int. Math. Res. Not. **2005**, no. 40, 2437–2446. (Cited on 408.)

[281] V. B. Lidskii, *Non-selfadjoint operators with a trace*, Dokl. Akad. Nauk SSSR **125** (1959), 485–487 [Russian]. (Cited on 22.)

[282] E. H. Lieb and M. Loss, *Analysis*, 2nd edition, Graduate Studies in Mathematics, **14**, American Mathematical Society, Providence, RI, 2001. (Cited on 135.)

[283] N. G. Lloyd, *Degree Theory*, Cambridge Tracts in Mathematics, **73**, Cambridge University Press, Cambridge-New York-Melbourne, 1978. (Cited on 360.)

[284] L. H. Loomis and S. Sternberg, *Advanced Calculus*, revised edition, Jones and Bartlett Publishers, Boston, 1990. (Cited on 382, 387.)

[285] G. López Lagomasino, *Strong convergence for sequence of polynomials, orthonormal with respect to varying measures on an interval*, Cienc. Mat. (Havana) **7** (1986), 3–16. (Cited on 457.)

[286] G. López Lagomasino, *Asymptotics of polynomials orthogonal with respect to varying measures*, Constr. Approx. **5** (1989), 199–219. (Cited on 454.)

[287] D. S. Lubinsky, *Universality limits in the bulk for arbitrary measures on compact sets*, J. Anal. Math. **106** (2008), 373–394. (Cited on x, 132, 222, 224, 227.)

[288] D. S. Lubinsky, *A new approach to universality limits involving orthogonal polynomials*, Annals of Math. **170** (2009), 915–939. (Cited on x, 132, 218.)

[289] D. S. Lubinsky and P. Nevai, *Sub-exponential growth of solutions of difference equations*, J. London Math. Soc. (2) **46** (1992), 149–160. (Cited on 213.)

[290] G. L. Luke (editor), *Representation Theory of Lie Groups*, Proc. SRC/LMS Research Symposium (Oxford, 1977), London Mathematical Society Lecture Note Series, **34**, Cambridge University Press, Cambridge-New York, 1979. (Cited on 408.)

[291] A. Lyapunov, *Problème général de la stabilité du movement*, Ann. Fac. Sci. Univ. Toulouse (2) **9** (1907), 203–474. (Cited on 282.)

[292] W. Magnus and S. Winkler, *Hill's Equation*, Interscience Tracts in Pure and Applied Mathematics, **20**, Interscience Publishers, New York, 1966. (Cited on 282.)

[293] S. V. Manakov, *Complete integrability and stochastization of discrete dynamical systems*, Soviet Phys. JETP **40** (1974), 269–274. (Cited on 380.)

[294] M. Măntoiu, *C^*-algebras, dynamical systems at infinity and the essential spectrum of generalized Schrödinger operators*, J. Reine Angew. Math. **550** (2002), 211–229. (Cited on 425.)

[295] F. Marcellán and R. Álvarez-Nodarse, *On the "Favard theorem" and its extensions*, J. Comput. Appl. Math. **127** (2001), 231–254. (Cited on 18.)

[296] A. A. Markov, *Démonstration de certaines inégalités de M. Tchébychef*, Math. Ann. **24** (1884), 172–180. (Cited on 227.)

[297] J. Marsden and T. Ratiu, *Introduction to Mechanics and Symmetry. A Basic Exposition of Classical Mechanical Systems*, 2nd edition, Texts in Applied Mathematics, **17**, Springer-Verlag, New York, 1999. (Cited on 382.)

[298] A. Martínez-Finkelshtein, *Equilibrium problems of potential theory in the complex plane*, in "Orthogonal Polynomials and Special Functions," pp. 79–117, Lecture Notes in Mathematics, **1883**, Springer, Berlin, 2006. (Cited on 305.)

[299] J. C. Mason and D. C. Handscomb, *Chebyshev Polynomials*, Chapman & Hall/CRC, Boca Raton, FL, 2003. (Cited on 312, 319.)

[300] A. Máté and P. Nevai, *Bernstein's inequality in L^p for $0 < p < 1$ and $(C, 1)$ bounds for orthogonal polynomials*, Annals of Math. (2) **111** (1980), 145–154. (Cited on 134, 142.)

[301] A. Máté and P. Nevai, *Remarks on E. A. Rakhmanov's paper: "The asymptotic behavior of the ratio of orthogonal polynomials"* [Mat. Sb. (N.S.) 103(145) (1977), no. 2, 237–252; MR 56 #3556], J. Approx. Theory **36** (1982), 64–72. (Cited on 97.)

[302] A. Máté, P. Nevai, and V. Totik, *Szegő's extremum problem on the unit circle*, Annals of Math. (2) **134** (1991), 433–453. (Cited on 36, 132, 133, 142, 220.)

[303] J. L. McCauley, *Classical Mechanics. Transformations, Flows, Integrable and Chaotic Dynamics*, Cambridge University Press, Cambridge, 1997. (Cited on 382.)

[304] H. P. McKean and P. van Moerbeke, *The spectrum of Hill's equation*, Invent. Math. **30** (1975), 217–274. (Cited on 359, 411, 416.)

[305] J. W. Milnor, *Topology From the Differentiable Viewpoint*, revised reprint of the 1965 original, Princeton Landmarks in Mathematics, Princeton University Press, Princeton, NJ, 1997. (Cited on 360.)

[306] R. Miranda, *Algebraic Curves and Riemann Surfaces*, Graduate Studies in Mathematics, **5**, American Mathematical Society, Providence, RI, 1995. (Cited on 346, 359.)

[307] J. Moser, *Finitely many mass points on the line under the influence of an exponential potential—an integrable system*, in "Dynamical Systems, Theory and Applications," pp. 467–497, Lecture Notes in Physics, **38**, Springer, Berlin, 1975. (Cited on 387, 397, 402.)

[308] J. Moser, *Three integrable Hamiltonian systems connected with isospectral deformations*, Adv. in Math. **16** (1975), 197–220. (Cited on 404.)

[309] E. M. Muhamadiev, *On invertibility of differential operators in the space of continuous functions bounded on the real axis*, Soviet Math. Dokl. **12** (1971), 49–52. (Cited on 425.)

[310] E. M. Muhamadiev, *On the invertibility of elliptic partial differential operators*, Soviet Math. Dokl. **13** (1972), 1122–1126. (Cited on 425.)

[311] J. R. Munkres, *Topology*, 2nd edition, Prentice–Hall, Upper Saddle River, NJ, 2000. (Cited on 525.)

[312] P. B. Naĭman, *On the theory of periodic and limit-periodic Jacobian matrices*, Soviet Math. Dokl. **3** (1962), 383–385; Russian original in Dokl. Akad. Nauk SSSR **143** (1962), 277–279. (Cited on 460.)

[313] I. P. Natanson, *Constructive Function Theory, Vol. II: Approximation in Mean*, Ungar, New York, 1965. (Cited on 18.)

[314] F. Nazarov, F. Peherstorfer, A. Volberg, and P. Yuditskii, *On generalized sum rules for Jacobi matrices*, Int. Math. Res. Not. **2005**, 155–186. (Cited on 173.)

[315] F. Nazarov, A. Volberg, and P. Yuditskii, *Reflectionless measures with a point mass and singular continuous component*, preprint. (Cited on 452.)

[316] I. Nenciu, *Lax pairs for the Ablowitz–Ladik system via orthogonal polynomials on the unit circle*, Ph.D. dissertation, California Institute of Technology, 2005. (Cited on 416.)

[317] I. Nenciu, *Lax pairs for the Ablowitz–Ladik system via orthogonal polynomials on the unit circle*, Int. Math. Res. Not. **2005**, no. 11, 647–686. (Cited on 416.)

[318] I. Nenciu, *Poisson brackets for orthogonal polynomials on the unit circle*, preprint. (Cited on 417.)

[319] I. Nenciu and B. Simon, unpublished. (Cited on 416, 417.)

[320] P. Nevai, *Orthogonal polynomials*, Mem. Amer. Math. Soc. **18** (1979), no. 213, 185 pp. (Cited on 10, 36, 132, 182, 207, 213.)

[321] P. Nevai, *Géza Freud, orthogonal polynomials and Christoffel functions. A case study*, J. Approx. Theory **48** (1986), 167 pp. (Cited on 10, 132.)

[322] P. Nevai, *Orthogonal polynomials, recurrences, Jacobi matrices, and measures*, in "Progress in Approximation Theory" (Tampa, FL, 1990), pp. 79–104, Springer Series in Computational Mathematics, **19**, Springer, New York, 1992. (Cited on 173.)

[323] P. Nevai and V. Totik, *Orthogonal polynomials and their zeros*, Acta Sci. Math. (Szeged) **53** (1989), 99–104. (Cited on 42, 96.)

[324] P. Nevai, V. Totik, and J. Zhang, *Orthogonal polynomials: their growth relative to their sums*, J. Approx. Theory **67** (1991), 215–234. (Cited on 210, 213.)

[325] R. Nevanlinna, *Asymptotische Entwickelungen beschränkter Functionen und das Stieltjessche Momentenproblem*, Ann. Acad. Sci. Fenn. A **18** (1922), Nr. 5. (Cited on 73, 202.)

[326] J. M. Nunes da Costa and P. A. Damianou, *Toda systems and exponents of simple Lie groups*, Bull. Sci. Math. **125** (2001), 49–69. (Cited on 408.)

[327] M. Ohya and D. Petz, *Quantum Entropy and Its Use*, Texts and Monographs in Physics, Springer-Verlag, Berlin, 1993. (Cited on 51.)

[328] M. A. Olshanetsky and A. M. Perelomov, *Explicit solutions of classical generalized Toda models*, Invent. Math. **54** (1979), 261–269. (Cited on 408.)

[329] P. J. Olver, *Orthogonal bases and the QR algorithm*, unpublished notes; `http://www.math.umn.edu/~olver/aims_/qr.pdf`. (Cited on 390.)

[330] W. F. Osgood, *On the existence of Green's function for the most general simply connected plane region*, Trans. Amer. Math. Soc. **1** (1900), 310–314. (Cited on 524.)

[331] W. Parry, *Entropy and Generators in Ergodic Theory*, W. A. Benjamin, New York-Amsterdam, 1969. (Cited on 51.)

[332] S. J. Patterson, *The limit set of a Fuchsian group*, Acta Math. **136** (1976), 241–273. (Cited on 540.)

[333] D. B. Pearson, *Singular continuous measures in scattering theory*, Comm. Math. Phys. **60** (1978), 13–36. (Cited on 451.)

[334] D. B. Pearson, *Value distribution and spectral analysis of differential operators*, J. Phys. A **26** (1993), 4067–4080. (Cited on 451.)

[335] D. B. Pearson, *Value distribution and spectral theory*, Proc. London Math. Soc. (3) **68** (1994), 127–144. (Cited on 451.)

[336] F. Peherstorfer, *On the asymptotic behaviour of functions of the second kind and Stieltjes polynomials and on the Gauss–Kronrod quadrature formulas*, J. Approx. Theory **70** (1992), 156–190. (Cited on 73, 98, 306, 319.)

[337] F. Peherstorfer, *Orthogonal and extremal polynomials on several intervals*, in "Proc. Seventh Spanish Symposium on Orthogonal Polynomials and Applications (VII SPOA)" (Granada, 1991), J. Comput. Appl. Math. **48** (1993), 187–205. (Cited on 312.)

[338] F. Peherstorfer, *A special class of polynomials orthogonal on the unit circle including the associated polynomials*, Constr. Approx. **12** (1996), 161–185. (Cited on 79, 368.)

[339] F. Peherstorfer, *Deformation of minimal polynomials and approximation of several intervals by an inverse polynomial mapping*, J. Approx. Theory **111** (2001), 180–195. (Cited on 312.)

[340] F. Peherstorfer, *Inverse images of polynomial mappings and polynomials orthogonal on them*, in "Proc. Sixth International Symposium on Orthogonal Polynomials, Special Functions and their Applications" (Rome, 2001), J. Comput. Appl. Math. **153** (2003), 371–385. (Cited on 312.)

[341] F. Peherstorfer and R. Steinbauer, *Orthogonal polynomials on the circumference and arcs of the circumference*, J. Approx. Theory **102** (2000), 96–119. (Cited on 91.)

[342] F. Peherstorfer and P. Yuditskii, *Asymptotics of orthonormal polynomials in the presence of a denumerable set of mass points*, Proc. Amer. Math. Soc. **129** (2001), 3213–3220. (Cited on 182.)

[343] F. Peherstorfer and P. Yuditskii, *Asymptotic behavior of polynomials orthonormal on a homogeneous set*, J. Anal. Math. **89** (2003), 113–154. (Cited on ix, 172, 174, 182, 477, 539, 556, 570, 584, 589.)

[344] F. Peherstorfer and P. Yuditskii, *Remark on the paper "Asymptotic behavior of polynomials orthonormal on a homogeneous set,"* arXiv math.SP/0611856. (Cited on 539, 556, 570, 584, 589.)

[345] O. Perron, *Die Lehre von den Kettenbrüchen*, 2nd edition, Teubner, Leipzig, 1929. (Cited on 18.)

[346] F. Pintér and P. Nevai, *Schur functions and orthogonal polynomials on the unit circle*, in "Approximation Theory and Function Series," Bolyai Soc. Math. Stud., **5**, pp. 293–306, János Bolyai Mathematical Society, Budapest, 1996. (Cited on 80.)

[347] H. Poincaré, *Mémoire sur les fonctions fuchsiennes*, Acta Math. **1** (1882), 193–294. (Cited on 508, 514.)

[348] H. Poincaré, *Sur l'uniformisation des fonctions analytiques*, Acta Math. **31** (1908), 1–63. (Cited on 524.)

[349] A. G. Poltoratski, *The boundary behavior of pseudocontinuable functions*, St. Petersburg Math. J. **5** (1994) 389–406. (Cited on 436, 452.)

[350] A. G. Poltoratski and C. Remling, *Reflectionless Herglotz functions and Jacobi matrices*, Comm. Math. Phys. **288** (2009), 1007–1021. (Cited on 435, 452.)

[351] A. Poltoratski, B. Simon, and M. Zinchenko, *The Hilbert transform of a measure*, to appear in J. Anal. Math. (Cited on 452.)

[352] G. Pólya, L'Intermédiaire des Mathématiciens **21** (1914), S. 27 (Question 4340). (Cited on 23.)

[353] G. Pólya and G. Szegő, *Problems and Theorems in Analysis. I. Series, Integral Calculus, Theory of Functions*, reprint of the 1978 English translation, Classics in Mathematics, Springer-Verlag, Berlin, 1998. *Problems and Theorems in Analysis. II. Theory of Functions, Zeros, Polynomials, Determinants, Number Theory, Geometry*, reprint of the 1976 English translation, Classics in Mathematics, Springer-Verlag, Berlin, 1998. (Cited on 22.)

[354] V. P. Potapov, *The multiplicative structure of J-contractive matrix functions*, Amer. Math. Soc. Transl. (2) **15** (1960), 131–243; Russian original in Trudy Moskov. Mat. Obšč. **4** (1955), 125–236. (Cited on 229.)

[355] L. Pukánszky, *Characters of Connected Lie Groups*, Mathematical Surveys and Monographs, **71**, American Mathematical Society, Providence, RI, 1999. (Cited on 387.)

[356] V. S. Rabinovich, *Essential spectrum of perturbed pseudodifferential operators. Applications to the Schrödinger, Klein–Gordon, and Dirac operators*, Russian J. Math. Phys. **12** (2005), 62–80. (Cited on 425.)

[357] T. Radó, *Über die Fundamentalabbildungen schlichter Gebiete*, Acta Litt. ac. Scient. Univ. Hung. **1** (1923), 240–251. (Cited on 520, 524.)

[358] E. A. Rakhmanov, *On the asymptotics of the ratio of orthogonal polynomials*, Math. USSR Sb. **32** (1977), 199–213. (Cited on 19, 96, 454.)

[359] E. A. Rakhmanov, *On the asymptotics of the ratio of orthogonal polynomials, II*, Math. USSR Sb. **46** (1983), 105–117. (Cited on 19, 96, 454.)

[360] T. Ransford, *Potential Theory in the Complex Plane*, Press Syndicate of the University of Cambridge, New York, 1995. (Cited on 305, 323.)

[361] M. Reed and B. Simon, *Methods of Modern Mathematical Physics, I: Functional Analysis*, Academic Press, New York, 1972. (Cited on 2.)

[362] M. Reed and B. Simon, *Methods of Modern Mathematical Physics, II. Fourier Analysis, Self-Adjointness*, Academic Press, New York, 1975. (Cited on 227.)

[363] M. Reed and B. Simon, *Methods of Modern Mathematical Physics, III: Scattering Theory*, Academic Press, New York, 1978. (Cited on 431.)

[364] M. Reed and B. Simon, *Methods of Modern Mathematical Physics, IV: Analysis of Operators*, Academic Press, New York, 1978. (Cited on 18, 19, 239, 243, 244, 252, 259, 263, 334, 409.)

[365] C. Remling, *The absolutely continuous spectrum of one-dimensional Schrödinger operators*, Math. Phys. Anal. Geom. **10** (2007), 359–373. (Cited on 451.)

[366] C. Remling, *The absolutely continuous spectrum of Jacobi matrices*, preprint. (Cited on ix, 418, 433, 434, 435, 451, 452, 454, 455.)

[367] N. Reshetikhin, *Integrability of characteristic Hamiltonian systems on simple Lie groups with standard Poisson Lie structure*, Comm. Math. Phys. **242** (2003), 1–29. (Cited on 408.)

[368] B. Riemann, *Grundlagen für eine allgemeine Theorie der Funktionen einer veränderlichen complexen Grösse*, Inaugural dissertation, Göttingen, 1851. (Cited on 524.)

[369] F. Riesz and B. Sz.-Nagy, *Functional Analysis*, Ungar, New York, 1955. (Cited on 2.)

[370] T. J. Rivlin, *Chebyshev Polynomials. From Approximation Theory to Algebra and Number Theory*, 2nd edition, Wiley, New York, 1990. (Cited on 319.)

[371] E. Routh, *On some properties of certain solutions of a differential equation of the second order*, Proc. London Math. Soc. **16** (1884), 245–261. (Cited on 10.)

[372] W. Rudin, *Real and Complex Analysis*, 3rd edition, McGraw–Hill, New York, 1987. (Cited on 58, 65, 142, 225.)

[373] D. Ruelle, *Statistical Mechanics: Rigorous Results*, W. A. Benjamin, New York-Amsterdam, 1969. (Cited on 51.)

[374] D. Ruelle, *Thermodynamic Formalism. The Mathematical Structures of Equilibrium Statistical Mechanics*, 2nd edition, Cambridge Mathematical Library, Cambridge University Press, Cambridge, 2004. (Cited on 51.)

[375] E. Ryckman, *A spectral equivalence for Jacobi matrices*, J. Approx. Theory **146** (2007), 252–266. (Cited on 42.)

[376] E. Ryckman, *A strong Szegő theorem for Jacobi matrices*, Comm. Math. Phys. **271** (2007), 791–820. (Cited on 42.)

[377] C. Ryll-Nardzewski, *On the ergodic theorems. II. Ergodic theory of continued fractions*, Studia Math. **12** (1951), 74–79. (Cited on 79.)

[378] E. B. Saff and V. Totik, *Logarithmic Potentials With External Fields*, Grundlehren der Mathematischen Wissenschaften, **316**, Springer-Verlag, Berlin, 1997. (Cited on 305, 319.)

[379] K. Schiefermayr, *A lower bound for the minimum deviation of the Chebyshev polynomial on a compact real set*, East J. Approx. **14** (2008), 223–233. (Cited on 319.)

[380] I. Schur, *Über Potenzreihen, die im Innern des Einheitskreises beschränkt sind, I*, J. Reine Angew. Math. **147** (1917), 205–232. English translation in "I. Schur Methods in Operator Theory and Signal Processing" (edited by I. Gohberg), pp. 31–59, Operator Theory: Advances and Applications, **18**, Birkhäuser, Basel, 1986. (Cited on 65, 80, 239.)

[381] I. Schur, *Über die rationalen Darstellungen der allgemeinen linearen Gruppe*, S'ber. Akad. Wiss. Berlin (1927), 58–75; Ges. Abh. III, 68–85. (Cited on 37.)

[382] J. Sherman, *On the numerators of the convergents of the Stieltjes continued fractions*, Trans. Amer. Math. Soc. **35** (1933) 64–87. (Cited on 18.)

[383] B. A. Shipman, *The geometry of the full Kostant–Toda lattice of* sl(4, **C**), J. Geom. Phys. **33** (2000), 295–325 (Cited on 408.)

[384] J. A. Shohat, *Théorie Générale des Polinomes Orthogonaux de Tchebichef*, Mémorial des Sciences Mathématiques, **66**, pp. 1–69, Paris, 1934. (Cited on 36.)

[385] J. A. Shohat and J. D. Tamarkin, *The Problem of Moments*, American Mathematical Society Mathematical Surveys, vol. II, American Mathematical Society, New York, 1943. (Cited on 202.)

[386] M. A. Shubin, *The Favard–Muhamadiev theory and pseudodifferential operators*, Soviet Math. Dokl. **16** (1975), 1646–1649. (Cited on 425.)

[387] M. A. Shubin, *Almost periodic functions and partial differential operators*, Russian Math. Surveys **33** (1978), 1–52. (Cited on 425.)

[388] D. J. Simms and N. M. J. Woodhouse, *Lectures in Geometric Quantization*, Lecture Notes in Physics, **53**, Springer-Verlag, Berlin-New York, 1976. (Cited on 387.)

[389] B. Simon, *Notes on infinite determinants of Hilbert space operators*, Adv. in Math. **24** (1977), 244–273. (Cited on 22.)

[390] B. Simon, *Trace Ideals and Their Applications*, London Mathematical Society Lecture Note Series, **35**, Cambridge University Press, Cambridge-New York, 1979; 2nd edition, Mathematical Surveys and Monographs, **120**, American Mathematical Society, Providence, RI, 2005. (Cited on 22, 103, 431.)

[391] B. Simon, *Kotani theory for one-dimensional stochastic Jacobi matrices*, Comm. Math. Phys. **89** (1983), 227–234. (Cited on 451.)

[392] B. Simon, *The Statistical Mechanics of Lattice Gases, Vol. I*, Princeton University Press, Princeton, NJ, 1993. (Cited on 51.)

[393] B. Simon, *Operators with singular continuous spectrum, I. General operators* Annals of Math. (2) **141** (1995), 131–145. (Cited on 48.)

[394] B. Simon, *Representations of Finite and Compact Groups*, Graduate Studies in Mathematics, **10**, American Mathematical Society, Providence, RI, 1996. (Cited on 375.)

[395] B. Simon, *The classical moment problem as a self-adjoint finite difference operator*, Adv. in Math. **137** (1998), 82–203. (Cited on 73, 202, 207.)

[396] B. Simon, *A canonical factorization for meromorphic Herglotz functions on the unit disk and sum rules for Jacobi matrices*, J. Funct. Anal. **214** (2004), 396–409. (Cited on 47, 144, 157, 163.)

[397] B. Simon, *Ratio asymptotics and weak asymptotic measures for orthogonal polynomials on the real line*, J. Approx. Theory. **126** (2004), 198–217 (Cited on 97.)

[398] B. Simon, *OPUC on one foot*, Bull. Amer. Math. Soc. **42** (2005), 431–460. (Cited on 30, 79.)

[399] B. Simon, *Orthogonal Polynomials on the Unit Circle, Part 1: Classical Theory*, AMS Colloquium Series, **54.1**, American Mathematical Society, Providence, RI, 2005. (Cited on x, 14, 23, 24, 29, 30, 42, 47, 49, 74, 79, 84, 86, 91, 96, 97, 103, 106, 108, 118, 165, 167, 169, 182, 268, 378.)

[400] B. Simon, *Orthogonal Polynomials on the Unit Circle, Part 2: Spectral Theory*, AMS Colloquium Series, **54.2**, American Mathematical Society, Providence, RI, 2005. (Cited on ix, x, 19, 24, 30, 31, 33, 36, 47, 48, 65, 84, 91, 96, 97, 147, 157, 173, 175, 182, 281, 304, 359, 368, 377, 378, 416, 417, 425, 451, 454, 455, 457, 475, 476, 505.)

[401] B. Simon, *The sharp form of the strong Szegő theorem*, in "Geometry, Spectral Theory, Groups, and Dynamics," pp. 253–275, Contemp. Math. **387**, American Mathematical Society, Providence, RI, 2005. (Cited on 24.)

[402] B. Simon, *Fine structure of the zeros of orthogonal polynomials, I. A tale of two pictures*, Electron. Trans. Numer. Anal. **25** (2006), 328–268. (Cited on 132.)

[403] B. Simon, *CMV matrices: Five years after*, J. Comput. Appl. Math. **208** (2007), 120–154. (Cited on 30, 103.)

[404] B. Simon, *Equilibrium measures and capacities in spectral theory*, Inverse Problems and Imaging **1** (2007), 713–772. (Cited on 150, 305, 327, 344.)

[405] B. Simon, *Rank one perturbations and the zeros of paraorthogonal polynomials on the unit circle*, J. Math. Anal. Appl. **329** (2007), 376–382. (Cited on 118.)

[406] B. Simon, *Zeros of OPUC and long time asymptotics of Schur and related flows*, Inverse Problems and Imaging **1** (2007), 189–215. (Cited on 402, 403, 407, 417.)

[407] B. Simon, *The Christoffel–Darboux kernel*, in "Perspectives in PDE, Harmonic Analysis and Applications," pp. 295–335, Proc. Sympos. Pure Math., **79**, American Mathematical Society, Providence, RI, 2008. (Cited on 108, 113, 213, 227.)

[408] B. Simon, *Two extensions of Lubinsky's universality theorem*, J. Anal. Math. **105** (2008), 345–362. (Cited on 333.)

[409] B. Simon, *Weak convergence of CD kernels and applications*, Duke Math. J. **146** (2009), 305–330. (Cited on 123, 134, 142, 220.)

[410] B. Simon and A. Zlatoš, *Sum rules and the Szegő condition for orthogonal polynomials on the real line*, Comm. Math. Phys. **242** (2003), 393–423. (Cited on 47, 91, 144, 163, 173, 587.)

[411] B. Simon and A. Zlatoš, *Higher-order Szegő theorems with two singular points*, J. Approx. Theory **134** (2005), 114–129. (Cited on 91.)

[412] A. Sinap, *Gaussian quadrature for matrix valued functions on the real line*, J. Comput. Appl. Math. **65** (1995), 369–385. (Cited on 239.)

[413] M. Sodin and P. Yuditskii, *Almost periodic Jacobi matrices with homogeneous spectrum, infinite-dimensional Jacobi inversion, and Hardy spaces of character-automorphic functions*, J. Geom. Anal. **7** (1997), 387–435. (Cited on ix, 278, 452, 453, 454, 477, 539, 556, 570, 576.)

[414] M. Solomyak, *On the spectrum of the Laplacian on regular metric trees*, Waves Random Media **14** (2004), S155–S171. (Cited on 594.)

[415] M. Spivak, *Calculus on Manifolds. A Modern Approach to Classical Theorems of Advanced Calculus*, W. A. Benjamin, New York-Amsterdam, 1965. (Cited on 382.)

[416] M. Spivak, *A Comprehensive Introduction to Differential Geometry, Volume I*, 2nd edition, Publish or Perish, Wilmington, DE, 1979. (Cited on 360, 382.)

[417] H. Stahl and V. Totik, *General Orthogonal Polynomials*, in "Encyclopedia of Mathematics and its Applications," **43**, Cambridge University Press, Cambridge, 1992. (Cited on 150, 305, 327, 344.)

[418] R. P. Stanley, *Enumerative Combinatorics. Vol. 1*, corrected reprint of the 1986 original, Cambridge Studies in Advanced Mathematics, **49**, Cambridge University Press, Cambridge, 1997; *Vol. 2*, Cambridge Studies in Advanced Mathematics, **62**, Cambridge University Press, Cambridge, 1999. (Cited on 339, 413.)

[419] E. M. Stein and R. Shakarchi, *Complex Analysis*, Princeton University Press, Princeton, NJ, 2003. (Cited on 525.)

[420] F. Stenger, *Numerical Methods Based on Sinc and Analytic Functions*, Springer, New York, 1993. (Cited on 227.)

[421] T. Stieltjes, *Quelques recherches sur la théorie des quadratures dites mécaniques*, Ann. Sci. École Norm. Sup. (3) **1** (1884), 409–426. (Cited on 227.)

[422] T. Stieltjes, *Recherches sur les fractions continues*, Ann. Fac. Sci. Univ. Toulouse **8** (1894–1895), J76–J122; ibid. **9**, A5–A47. (Cited on 18, 150, 201, 202, 206.)

[423] M. H. Stone, *Linear Transformations in Hilbert Spaces and Their Applications to Analysis*, Amer. Math. Soc. Colloq. Publ., **15**, American Mathematical Society, New York, 1932. (Cited on 18, 202.)

[424] D. Sullivan, *The density at infinity of a discrete group of hyperbolic motions*, Inst. Hautes Études Sci. Publ. Math. **50** (1979), 171–202. (Cited on 540.)

[425] W. W. Symes, *Systems of Toda type, inverse spectral problems, and representation theory*, Invent. Math. **59** (1980), 13–51. (Cited on 405, 407.)

[426] W. W. Symes, *The QR algorithm and scattering for the finite nonperiodic Toda lattice*, Phys. D **4** (1981/82), 275–280. (Cited on 405, 407.)

[427] Z. I. Szabó, *A cornucopia of isospectral pairs of metrics on spheres with different local geometries*, Annals of Math. (2) **161** (2005), 343–395. (Cited on 2.)

[428] G. Szegő, *Ein Grenzwertsatz über die Toeplitzschen Determinanten einer reellen positiven Funktion*, Math. Ann. **76** (1915), 490–503. (Cited on 23.)

[429] G. Szegő, *Über Orthogonalsysteme von Polynomen*, Math. Z. **4** (1919), 139–151. (Cited on 26.)

[430] G. Szegő, *Beiträge zur Theorie der Toeplitzschen Formen I, II*, Math. Z. **6** (1920), 167–202; **9** (1921), 167–190. (Cited on 24, 26, 29, 91, 97, 103, 108, 173.)

[431] G. Szegő, *Über den asymptotischen Ausdruck von Polynomen, die durch eine Orthogonalitätseigenschaft definiert sind*, Math. Ann. **86** (1922), 114–139. (Cited on 30, 31, 35, 91, 182.)

[432] G. Szegő, *Bemerkungen zu einer Arbeit von Herrn M. Fekete: Über die Verteilung der Wurzeln bei gewissen algebraischen Gleichungen mit ganzzahligen Koeffizienten*, Math. Z. **21** (1924), 203–208. (Cited on 305, 319.)

[433] G. Szegő, *On certain Hermitian forms associated with the Fourier series of a positive function*, Comm. Sém. Math. Univ. Lund **1952** (1952), Tome Supplementaire, 228–238. (Cited on 24.)

[434] G. Szegő, *Orthogonal Polynomials*, Amer. Math. Soc. Colloq. Publ., **23**, American Mathematical Society, Providence, RI, 1939; 3rd edition, 1967. (Cited on 10, 29, 30, 31, 35, 36, 74, 97, 108, 113, 319.)

[435] R. Szwarc, *Uniform subexponential growth of orthogonal polynomials*, J. Approx. Theory **81** (1995), 296–302. (Cited on 213.)

[436] G. Teschl, *Jacobi Operators and Completely Integrable Nonlinear Lattices*, Mathematical Surveys and Monographs, **72**, American Mathematical Society, Providence, RI, 2000. (Cited on 282, 454.)

[437] W. Thirring, *A Course in Mathematical Physics. Vol. I. Classical Dynamical Systems*, Springer-Verlag, New York-Vienna, 1978. (Cited on 382.)

[438] D. J. Thouless, *Electrons in disordered systems and the theory of localization*, Phys. Rep. **13** (1974), 93. (Cited on 305.)

[439] E. C. Titchmarsh, *The Theory of Functions*, 2nd edition, Oxford University Press, Oxford, 1932. (Cited on 227.)

[440] E. C. Titchmarsh, *On expansions in eigenfunctions. IV*, Quart. J. Math., Oxford Ser. **12** (1941), 33–50. (Cited on 73.)

[441] M. Toda, *Theory of Nonlinear Lattices*, 2nd edition, Springer Series in Solid-State Sciences, **20**, Springer, Berlin, 1989. (Cited on 282.)

[442] V. Totik, *Orthogonal polynomials with ratio asymptotics*, Proc. Amer. Math. Soc. **114** (1992), 491–495. (Cited on 47.)

[443] V. Totik, *Asymptotics for Christoffel functions for general measures on the real line*, J. Anal. Math. **81** (2000), 283–303. (Cited on 323, 343.)

[444] V. Totik, *Polynomial inverse images and polynomial inequalities*, Acta Math. **187** (2001), 139–160. (Cited on 305, 306, 312, 323.)

[445] V. Totik, *Chebyshev constants and the inheritance problem*, to appear in J. Approx. Theory. (Cited on 306, 312.)

[446] M. Tsuji, *Potential Theory in Modern Function Theory*, reprint of the 1959 original, Chelsea, New York, 1975. (Cited on 305, 319, 518, 556.)

[447] J. L. Ullman, *On the regular behaviour of orthogonal polynomials*, Proc. London Math. Soc. (3) **24** (1972), 119–148. (Cited on 305, 327.)

[448] D. C. Ullrich, *Complex Made Simple*, Graduate Studies in Mathematics, **97**, American Mathematical Society, Providence, RI, 2008. (Cited on 525.)

[449] W. Van Assche, *Invariant zero behaviour for orthogonal polynomials on compact sets of the real line*, Bull. Soc. Math. Belg. Ser. B **38** (1986), 1–13. (Cited on 327.)

[450] P. van Moerbeke, *The spectrum of Jacobi matrices*, Invent. Math. **37** (1976), 45–81. (Cited on 282, 359, 410, 411, 413, 414, 416.)

[451] V. A. Vassiliev, *Introduction to Topology*, Student Mathematical Library, **14**, American Mathematical Society, Providence, RI, 2001. (Cited on 525.)

[452] S. Verblunsky, *On positive harmonic functions: A contribution to the algebra of Fourier series*, Proc. London Math. Soc. (2) **38** (1935), 125–157. (Cited on 29, 30, 79.)

[453] S. Verblunsky, *On positive harmonic functions* (*second paper*), Proc. London Math. Soc. (2) **40** (1936), 290–320. (Cited on 29, 30, 47, 51, 74, 86, 106.)

[454] J. von Neumann and E. Wigner, *Über das Verhalten von Eigenwerten bei adiabatischen Prozesses*, Phys. Z. **30** (1929), 467–470. (Cited on 253.)

[455] H. S. Wall, *Analytic Theory of Continued Fractions*, Van Nostrand, New York, 1948; AMS Chelsea, Providence, RI, 2000. (Cited on 80, 150.)

[456] J. L. Walsh, *History of the Riemann mapping theorem*, Amer. Math. Monthly **80** (1973), 270–276. (Cited on 524.)

[457] H. Weyl, *Über beschraänkte quadratische Formen, deren Differenz vollstetig ist*, Rend. Circ. Mat. Palermo **27** (1909), 373–392. (Cited on 19.)

[458] H. Weyl, *Über gewöhnliche Differentialgleichungen mit Singularitäten und die zugehörigen Entwicklungen willkürlicher Funktionen*, Math. Ann. **68** (1910), 220–269. (Cited on 73.)

[459] H. Widom, *Polynomials associated with measures in the complex plane*, J. Math. Mech. **16** (1967), 997–1013. (Cited on 305, 327, 570.)

[460] H. Widom, *Extremal polynomials associated with a system of curves in the complex plane*, Adv. in Math. **3** (1969), 127–232. (Cited on 305, 312, 576, 584, 589.)

[461] A. Wintner, *Spektraltheorie der unendlichen Matrizen. Einführung in den analytischen Apparat der Quantenmechanik*, Hirzel, Leipzig, 1929. (Cited on 18.)

[462] M.-W. L. Wong, *First and second kind paraorthogonal polynomials and their zeros*, J. Approx. Theory **146** (2007), 282–293. (Cited on 113, 118.)

[463] N. M. J. Woodhouse, *Geometric quantization*, 2nd edition, Oxford Mathematical Monographs, Oxford Science Publications, The Clarendon Press, Oxford University Press, New York, 1992. (Cited on 387.)

[464] J. Zhang, *Relative growth of linear iterations and orthogonal polynomials on several intervals*, Linear Algebra Appl. **186** (1993), 97–115. (Cited on 213.)

[465] A. Zlatoš, *Sum rules for Jacobi matrices and divergent Lieb–Thirring sums*, J. Funct. Anal. **225** (2005), 371–382. (Cited on 173.)

Author Index

Subject Index